AQUAPONICS
HOW TO DO EVERYTHING
From Backyard to Profitable Business

David H. Dudley, PE

HOWARD PUBLISHING

Aquaponics How to Do Everything from Backyard to Profitable Business

Copyright © 2017 by David H. Dudley

All rights reserved. No part of this book shall be reproduced, stored in a retrieval system, or transmitted by any means—electronic, mechanical, photocopying, recording, or otherwise—without written permission from the publisher. No patent liability is assumed with respect to the use of the information contained herein. Although every precaution has been taken in the preparation of this book, the publisher and author assume no responsibility for errors or omissions. Neither is any liability assumed for damages resulting from the use of the information contained herein.

ISBN: 978-0-9985377-0-2 (paperback)
ISBN: 978-0-9969090-7-5 (hardback, color version)
ISBN: 978-0-9969090-9-9 (ePUB, color version).
Also avilable for Kindle.

Published by:
Howard Publishing

CONTENTS

Dedication i
Acknowledgments iii
Note to Reader v
Introduction 1

PART I **Aquaponic Fundamentals** **5**
    CHAPTER 1 Aquaponics Defined 7
    CHAPTER 2 Benefits of Aquaponics 9
    CHAPTER 3 History of Aquaponics 13
    CHAPTER 4 Aquaponics Globally and the Big Picture 15
    CHAPTER 5 Toxic Food vs. Healthy Organic Aquaponic Food 33
    CHAPTER 6 The Different Types of Aquaponic Systems ... 55
    CHAPTER 7 Aquaponics for You 59
    CHAPTER 8 Location, Location, Location 63
    CHAPTER 9 Nitrogen Cycle 67

PART II **Plants and Fish** **75**
    CHAPTER 10 Plants/Crops—Keys to Success 77
    CHAPTER 11 Fish—Everything You Need to Know 95
    CHAPTER 12 Fish Feed 113

PART III **Components of Aquaponics Used in Aquaponics** **129**
    CHAPTER 13 Equipment & Component Overview 131
    CHAPTER 14 Plumbing 139
    CHAPTER 15 Fish Tanks 153
    CHAPTER 16 Liner Material 165
    CHAPTER 17 Making a Water Tight Container 169
    CHAPTER 18 Growing Media for Plants 173
    CHAPTER 19 Pumps & Choosing the Right Pump 179
    CHAPTER 20 Filtration (Mechanical, Biofiltration, Natural) ... 183
    CHAPTER 21 Greenhouses 193
    CHAPTER 22 Alternative Energy Options & Operating Off-The-Grid ... 215

Part IV	**N.F.T. & D.W.C. Design and Layout**	225
	Chapter 23 Nutrient Film Technique	227
	Chapter 24 Deep Water Culture / Raft System	245
Part V	**Flood-and-Drain System Design and Layout**	257
	Chapter 25 Flood-and-Drain System Instructions & Procedures	259
	Chapter 26 Flood-and-Drain System Drain Options, Plans, and Instructions	283
	Chapter 27 How to Make a Auto Siphon (Bell Siphon)	291
	Chapter 28 Design Plans and Construction Details DIY Flood-and-Drain System	299
	Chapter 29 Design Plans and Construction Details Using IBC Containers Flood-and-Drain System	313
Part VI	**Operation and Maintenance**	335
	Chapter 30 Starting and Managing Your Aquaponics System	337
	Chapter 31 Water Quality	353
	Chapter 32 Fish Breeding, Fish Reproduction, and Raising Your Own Crop of Fish	373
	Chapter 33 Greenhouse Energy Management	383
	Chapter 34 Canning and Saving Produce from Your Harvest	395
Part VII	**Making Money and Earning a Profit from Aquaponics**	403
	Chapter 35 Cost-Benefit Analysis (Aquaponic Economics)	405
	Chapter 36 Creating a Profitable Aquaponics Business	409
	Chapter 37 Selling Your Aquaponic Products	437
	Chapter 38 Bartering Your Aquaponic Products	463
	Chapter 39 Marketing and Selling Fish	467
	Chapter 40 Sales & Selling: Being Successful	477
	Chapter 41 Customer Service Recommendations and Rewards	491
	Chapter 42 Time Management	499
	Chapter 43 Aquaponics Business Plan (A Real-World 'Go-By' Example)	511
Part VIII	**Appendix**	549
	Appendix 1 Aquaponic Resources	551
	Appendix 2 References	555
	Appendix 3 Vendors	559
	Appendix 4 Best Places for You to Live in America	563
	Appendix 5 Best Places to Live for Autism Cognitive & Physical Disabilities	565
	Appendix 6 Aquaponic Design Plans	567
	Appendix 7 Aquaponics: Plans and Instructions	569
	Appendix 8 Earning Money from Aquaponics	571
	Appendix 9 Recommended Resources	575
	Appendix 10 Conversion Units	579
	Appendix 11 Encouragement & Keys to Success	581

DEDICATION

This book is dedicated to every person, young and old, employed and unemployed, educated and those with little schooling, who desire to eat truly healthy organic food, dream of becoming better off financially, who genuinely care about the environment, and aspire to build a happy, successful, and rewarding life. It doesn't matter whether you live in the city, suburbs, or country; this book was written with the hope and intention of enabling you to bring your dreams to fruition, directly where you are in life and location.

ACKNOWLEDGMENTS

I owe a tremendous amount of gratitude to my dear wife, awesome kids, and gracious parents for their love, encouragement, patience, and sacrifices during the creation of this book and in life. They have enriched my journey though life beyond measure.

I am also indebted to the University of Oklahoma, the University of Tennessee, and Sacramento State University, as well as the faculty of these fine institutions, for their valuable contributions in the way of information, knowledge, and guidance. Thanks also to all the graduate students who assisted me along the way by conducting an extensive amount of research and providing critical data for this book.

I would like to thank my previous and current engineering employers, peers, and co-workers with whom I have learned much. Their mentorship and trust in allowing me the opportunity to work on and manage all types of engineering projects over the past two decades has provided me with real world experience and equipped me with the skill set needed to understand and address the scientific aspects of aquaponics and share this information with others in a practical way.

I would also like to thank all of those who have taught me so much about nutrition, farming, vertical gardening, aquaculture, hydroponics, aquaponics, sustainable agriculture methods, marketing, and business over the past five decades. I beg forgiveness for not mentioning you individually as the list is too long and, quite frankly, I have simply forgotten some names. Nevertheless, your time, wisdom, instruction, and influence in my life is appreciated more than words can ever say.

In addition, I must thank many of the associations, organizations, and alternative media sources that are dedicated to providing truth. Because of their efforts, we are able to get the real story without a government political agenda and mainstream media bias. Further acknowledgement, and other resource information, about these indispensable groups is provided in the Appendix.

I greatly appreciate Michelle Wagner, Nicole Avers, and Cheryl Goldman for their assistance with the illustrations presented within this book; as well RM Drafting & Design, LLC for their assistance with the CAD work. I must also thank Mitzi Cunningham and Robert Nash for their professional assistance in proofreading this book for grammatical errors.

I am deeply appreciative to Stephanie Anderson for all of her hard work, patience, attention to detail, and professionalism during the formatting process of this book. Stephanie singlehandedly over a thousand complex manuscript pages, tables, and illustrations in to a premium quality valuable resource which will benefit many people. Stephanie also served as an extremely helpful publishing consultant during the formation of this book; and I am now pleased to consider her a true friend.

Lastly, my deepest gratitude and greatest acknowledgement is reserved to my Lord and Savior, Jesus Christ. Only through His grace, provisions, forgiveness, and work in my life is anything I do possible. I fully recognize and appreciate that God is in control. I am at awe, and undeserving, of the unconditional love, mercy, and the blessings He has bestowed upon me. All praise for this book must go to Him. Thank you.

NOTE TO READER

I appreciate the fact that for some the main objective for acquiring this book was to access the user-friendly, reliable aquaponic system design plans included within and immediately get started on building an aquaponic system. Although I applaud your enthusiasm, I would like to encourage all readers to obtain a good understanding of aquaponics before taking on the construction process. Such will ensure you achieve optimum success in aquaponics by preventing aggravation and saving you a lot of money and time in the long run.

For example, obtaining design plans to build a sail boat and constructing the sail boat according to those plans would result in much wasted time and frustration once out on the lake without realistic expectations and the knowledge of how to operate a sailing vessel. The 'would be sailor' would undoubtedly experience set-backs and potentially costly failures.

Great advice on any project is to "start with the end in mind." Staying focused on that goal, and understanding what is involved in achieving that goal, enables us to have greater success. Immediately jumping into a "build" project without a solid foundation of the fundamentals and knowing what is involved is not a prudent strategy, regardless of the undertaking.

With that said, I hope you will invest some time better educating yourself on the material preceding the chapters with design details. Although this book addresses everything one could possibly need to be successful in aquaponics, I want to assure you it is not necessary to memorize the material in this book. Aquaponics really is not that difficult. With a basic understanding of the fundamentals, anyone can be successful.

But keep in mind that success means different things to different people. For some, success in aquaponics is having a cool hobby and a fun conversation topic. For others, it means lowering food costs and supplementing food supply with year-round organically grown healthy produce and fish. For others, it means growing enough extra food to sell, barter, and give to friends/family. For some, a profitable aquaponic business is the meaning of success. It really does not matter which one of the above best fits your definition of success in aquaponics. They are all possible, and I am confident this book will provide you with everything you need to know in order to achieve and enjoy your aspirations for success in aquaponics.

INTRODUCTION

While aquaponics has been around for thousands of years, only during the past century, and especially the past decade, has it really taken off. There are good reasons as to why aquaponics is on the rise worldwide. The noble benefits associated with it being a sustainable farming practice, requiring very little water, and having the multitude of other environmentally friendly advantages makes it an extremely appealing agricultural method. Although these are persuasive forces, when it comes right down to it, most people get into aquaponics for more personal reasons, such as the attraction of having a constantly available supply of healthy, organic food; to lower their food costs; the ability to produce much higher yields compared to those of traditional farming; having a known food supply with a much higher nutrient density content than mass-produced crops; and/or the benefits of being able to convert even just a small part of their land into a viable revenue-generating space.

During the past century, and even more so over the last two decades, the appeal of aquaponics has become increasingly more popular because of the readily available assimilation of information from the internet; increasing media attention; the advancement of new technologies, i.e., plastics, fiberglass, electric pumps; and the relatively low cost of materials.

Moreover, as people become more concerned about the high cost of food and more educated about our toxic food system and the detrimental impact it has upon their family's health, they seek better ways to meet their dietary requirements. Aquaponics offers the perfect solution to all of these concerns.

Indeed, raising your own healthy food is an extremely rewarding process. Saving big money while consuming your own homegrown, organic healthy food at much higher nutrient density content than conventional grocery store food is priceless; and aquaponics in of itself is really an interesting process.

As a matter of fact, you will find that aquaponics is a stimulating conversation topic for people from all walks of life, and most everyone is eager to learn more about it. You will also get many requests from people wanting to see your system in operation. This social element adds even more joy beyond the satisfaction derived from growing your own healthy food.

I have also thoroughly enjoyed bringing my family into aquaponics and teaching my children many of the valuable lessons associated with the operation, i.e., plants/fish, water dynamics, harvest, work ethic, responsibilities, business lessons. It is a fun family affair and a wonderful way to connect with your spouse and children. Aquaponics truly is a win-win method all the way around for many reasons.

I fell in love with aquaponics for all of the above reasons. As a professionally licensed civil engineer, with a strong background in nutrition/dietetics, decades of experience in aquaculture, a passion for vegetable gardening, hydroponics and vertical gardening, I spent years studying aquaponics from all these perspectives.

I discovered many ways to set up and operate an aquaponics system. I researched every aquaponic method and system out there while on my own journey to discover the best way to raise healthy, organic, nutrient-rich food at maximum yields. After successfully achieving those goals I began searching for the best way to transfer those methods into a commercial operation that would produce a respectable profit margin. During the process, and while pursuing a parallel career path serving as the Construction Manager on the Oklahoma Aquarium project, designing a $5M aquaculture farm for white sturgeon, serving as the Engineering Manager overseeing four large aquaculture farms for the largest caviar producer in the United States, and designing a large fishing clinic facility for the U.S. Dept. of Wildlife, I have been blessed to become not only an expert in aquaponics, but a consultant in the industry. I have helped many people not only get into aquaponics but turn their operations into a profitable business. I trust this book will help you achieve success in aquaponics, as well. The more wholesome, organic healthy food we can make available, the better for all.

An aquaponics setup can be as small as a 20-gallon aquarium or as big as several acres. The bulk of this book was written for those desiring to raise food fish (although pet fish, such as Koi is also a viable option) and editable produce (although other income earning crops are also addressed in this book). It was also written to help people set up a fairly decent sized aquaponic operation on their property. Apartment dwellers and those desiring a small system can benefit from the principles and instruction provided in this book; however, the book speaks very little to those extremely small systems, i.e., 20-gallon aquariums directly, and focuses more on small/mid-size systems for those with property, i.e. homeowners, and larger aquaponic systems.

Over the years it has become quite apparent that most people desiring to get into aquaponics are of the Do-It-Yourself camp. Yes, some do purchase package units, but the vast majority of aquaponic enthusiasts build their own systems, either because they cannot afford an expensive packaged system or because they see the simplicity of it. The problem—and why there are so many different types of aquaponic setups out there—is that the vast majority of information on how to develop and maintain an aquaponics operation is either too expensive for most people to acquire, requires a substantial sacrifice of time to learn (university courses or weeks at an onsite training course), or is found to be based upon information that is not reliable, thorough, or user-friendly.

Seeing the need for a one-stop source for "everything you need to know," laid out in one efficient, easy-to-understand resource is what inspired me to prepare this book (a "how-to" instructional guide). I have made every attempt to cover all aspects of aquaponics and present the information in a way anyone can understand. The magnitude of all that is involved can seem somewhat mind boggling at first, but you should not get discouraged or feel overwhelmed by it all. Quite frankly, it really is a simple process once you learn the fundamentals. This book not only addresses the fundamentals but shows you how to do aquaponics using the most efficient and effective means possible, whether you are doing it just to supplement your dietary needs, feed your family, or to make it a profitable business.

In addition to being a licensed professional engineer who loves farming, as mentioned above I was also working in the aquaculture/aquaponics industry which provided me the opportunity to gain much

INTRODUCTION

experience about critical life support systems and very unique aquatic construction methods. Overseeing four large aquaculture farms that were providing 80 percent of the caviar consumed in the U.S., as well as shipping fish meat and caviar all over the world, I have been directly responsible for overseeing all infrastructure and equipment needed to provide optimal conditions for rearing fish.

I obtained a bachelor's degree in nutrition/dietetics in the 1980s after which I worked as a professional nutritionist for several years before returning to school to study engineering. However, my interest in nutrition has remained high over the years. I have always enjoyed researching and reading about nutrition, and over the past two decades I have invested much time acquiring additional nutrition and health-related certificates. As a result, as you read this book you will find I could not resist the temptation to provide additional information regarding the health benefits of aquaponics and to call attention to some of the problems associated with our current food industry.

My hope and intention is for this book to provide you with the knowledge and confidence needed to achieve success in the most efficient cost effective way possible. I believe that you will find that I have covered all issues associated with aquaponics in an easy to understand user-friendly approach so as to meet that objective.

On one final note, I am in the process of creating a website at *www.FarmYourSpace.com* which will have additional valuable information posted such as photos and videos related to aquaponics, vertical gardening techniques, and other beneficial ideas to help folks achieve optimal success growing food within their own space, and maximize that space, regardless of size. I would encourage you to visit the website.

Thank you for checking out this book. I welcome your feedback and comments. My desire as an aquaponics operator and consultant to the industry is to continue improving so I can better serve others. Aquaponics is, by all accounts, most certainly a worthwhile endeavor. It is also made up of a special community of people who typically are better able to see the 'big picture' in regards to the problems associated with our current food supply, appreciate good health, care about the environmental health of our planet, and desire to improve their economic situation. I truly hope this book is the catalyst that brings your dreams to fruition. I wish you the very best of success.

PART I

Aquaponic Fundamentals

CHAPTER 1

Aquaponics Defined

Overview

Aquaponics is essentially the combination of Aquaculture and Hydroponics. Aquaculture, also known as aquafarming, is the farming of marine or freshwater food fish or shellfish—such as oysters, clams, salmon, and trout—under controlled conditions. Hydroponics is a method of growing plants using mineral nutrient solutions, in water, without soil. Terrestrial plants may be grown with their roots either in the mineral nutrient solution only or in an inert medium, such as perlite, gravel, mineral wool, expanded clay pebbles or coconut husk.

Both aquaculture and hydroponics have some operational downsides. Hydroponics requires expensive nutrients to feed the plants and also requires periodic flushing of the systems—a process that can lead to waste disposal issues. Re-circulating aquaculture needs to have excess nutrients removed from the system; normally, this means that a percentage of the water is removed, generally on a daily basis. This nutrient rich water then needs to be disposed of and replaced with clean fresh water.

While re-circulating aquaculture and hydroponics are both very efficient methods of producing fish and vegetables, when we look at combining the two, their individual negative aspects are turned into positives. The positive aspects of both aquaculture and hydroponics are retained and the negative aspects no longer exist. Aquaponics can be either a simple or a very complex operation. The plants extract the water and nutrients they need to grow, thus cleaning the water for the fish. There are bacteria that live on the surface of the growbed media. These bacteria convert ammonia wastes from the fish into nitrates that can be used by the plants. The conversion of ammonia into nitrates is often termed "the nitrogen cycle."

Growbeds, filled with a media such as gravel or expanded clay pebbles, are a common method of growing plants in an aquaponic system, but there are many different methods that can be used. In fact any method of hydroponic growing can be adapted to aquaponics. Plants can be grown in floating foam rafts that sit on the water surface. Vegetables can also be grown using N.F.T. (Nutrient Film Technique), or through various other methods using a "run to waste" style of growing. This is done by circulating the fish tank water through the media where vegetables are grown.

Many different species of fish can be grown in an aquaponic system, and your species selection will depend on a number of factors that include your local government regulations. Quite high stocking densities of fish can be grown in an aquaponic system, and, because of the recirculating nature of the systems, very little water is used. Research has shown that an aquaponic system uses about 1/10 of the water used to grow vegetables in the ground. An aquaponic system can be incredibly productive. For instance, 125 lbs. (approx. 57kg) of fish and hundreds of kilograms of vegetables can be produced within 6 months in an area the size of an average carport, 13-ft x 26-ft (approx. 4 m x 8 m). Best of all, a typical small aquaponics system as was just described requires no bending, weeding or fertilizers, and uses only about the same power it takes to operate two household light bulbs.

Summary

Aquaponics consists of two main parts, with the aquaculture part for raising aquatic animals and the hydroponics part for growing plants. Aquatic effluents, resulting from uneaten feed and waste from raising aquatic animals such as fish, accumulate in most aquaculture systems. In high concentrations, without proper filtration, the effluent-rich water becomes toxic to the aquatic animals, but these effluents, in controlled aquaponic operations, provide nutrients essential for plant growth.

CHAPTER 2

Benefits of Aquaponics

Sustainability Benefits of Aquaponics Farming

There are many benefits to aquaponics. As a matter of fact, the advantages of aquaponic farming have been clearly demonstrated in regards to environmental, social, economic, and health aspects. The section below, along with the remaining parts of this chapter, addresses those benefits.

- Conservation through constant water reuse and recycling. Aquaponics preserves our fresh water resources and can produce food in water scarce environments. Aquaponics uses up to 90 percent less water than does traditional soil-based farming.
- Organic fertilization of plants accomplished with natural fish emulsion.
- The elimination of solid waste disposal efficiently accomplished, as compared to most other types of farming.
- The reduction of needed cropland to produce crops. Aquaponic farming methods yield larger harvest per square foot of grow space compared to soil based farming yield. In soil, plants compete for limited nutrients. In Aquaponics, nutrients are constantly produced throughout the system. These nutrients are delivered to the plant roots via the water column. Since there is no competition for nutrients, plants can be grown very close together with no nutrient loss.
- The overall reduction of the environmental footprint of crop production. Aquaponics uses less land and can even be applied indoors in urban settings to produce fresh fish and organic produce.
- Aquaponically grown produce is fresh, guaranteed organic, and has minimal impact on the environment.
- Aquaponics systems are pure and healthy. If we used chemical pesticides, herbicides, or fertilizers, our fish would die. Not only do you know for certain your produce is chemical free but you know that your fish are fresh and clean as well! No mercury or heavy metals are present in your food supply.
- Greenhouses ensure optimal dependable food production, food security against possible contamination, and vast reduction of travel time, distribution and fuel costs.
- Building small efficient commercial installations near markets reduces food transport miles.
- Reduction of pathogens that often plague aquaculture production systems.
- Reduction of erosion by eliminating the need to plough the soil.

Feed Conversion Ratios

Animals must be fed. Raising animals and growing the feed to sustain them takes land and tremendous amounts of energy resources. Obviously, minimizing these requirements is best for our planet as a whole. With such in mind, it is encouraging to see how raising fish has the lowest negative footprint over that of other animal products. Following is a list of the average pounds of feed needed to produce 1 lb. of product for various animal groups:

- Fish = 1.7 lbs
- Chicken = 2.4 lbs
- Turkey = 5.2 lbs
- Pork = 4.9 lbs
- Lamb = 8.0 lbs
- Beef = 9.0 lbs

Summary of Benefits

- Healthy food (100 percent organic food)
- Saves money
- Helps the environment
- Grow a tremendous amount of food in a relatively small area.
- Uses very little energy and can even be solar- or wind-powered should the "grid" become unusable or undesirable.
- Grow your own food year-round for your own needs, to use for bartering, or sell to others. Very important should the there be a disruption in commerce or should you just desire to become more self-reliant.

Benefits of Aquaponics— Itemized

Aquaponics is a high efficiency food-production method for growing food year round, in any climate. Following is an itemized list of aquaponic benefits:

- Aquaponics is the most energy efficient farming method in the world
- Most aquaponic systems can be modularized for ease of construction and operation
- It is relatively easy to replace mechanical components
- Automated water circulation—plants never need watering
- Waist high work area for efficiency and reduced injury potential
- No weeding necessary
- No need to clean medium/gravel
- No need to clean fish tanks
- No need for pesticides, herbicides, or fungicides.
- No hormones, harsh chemicals or antibiotics used.
- Ability to grow a wide variety of plants based on preference or demand.
- Entire system can be remotely monitored to notify you if any needs arise
- Waste produced is nontoxic and is a very efficient fertilizer for plants growing in ground
- Solar or alternative-energy power can be utilized.
- Uses 90 to 95 percent less water than does traditional in-ground growing methods.
- Precious topsoil is not needed.
- Utilizes locally available materials for construction.
- Can be installed in almost any location.
- Food can be harvested quickly and often.
- Does not require a higher education or skilled labor to operate.
- Time demands and labor needs are small once operational.
- Can be installed indoors in warehouses, basements, etc., with the use of grow lights
- Excellent Return on Investment—three to five years depending on markets and varieties of vegetables grown.
- Systems are expandable and scalable.
- Can be used as an educational tool. The educational component, teaching others about aquaponics, can also be an additional source of income.
- It is a fast growing food system that provides the freshest produce.
- Transportation costs are negligible.

CHAPTER 2: BENEFITS OF AQUAPONICS

- Produce plantings can be seasonally adjusted to maximize their value in the marketplace
- Additional enhanced methods for aquaponics, including revolutionary greenhouse cover and specialized induction lighting, speeds production by its ideal growing conditions.
- Aquaponics growing produces healthy food at the maximum possible growth rate.
- Healthy organic food is in high demand. This food will save your family money and enhance your family's health. Extra food can be sold, or used to barter a good value on a variety of goods and services provided by others.

AQUAPONICS VS TRADITIONAL AGRICULTURE

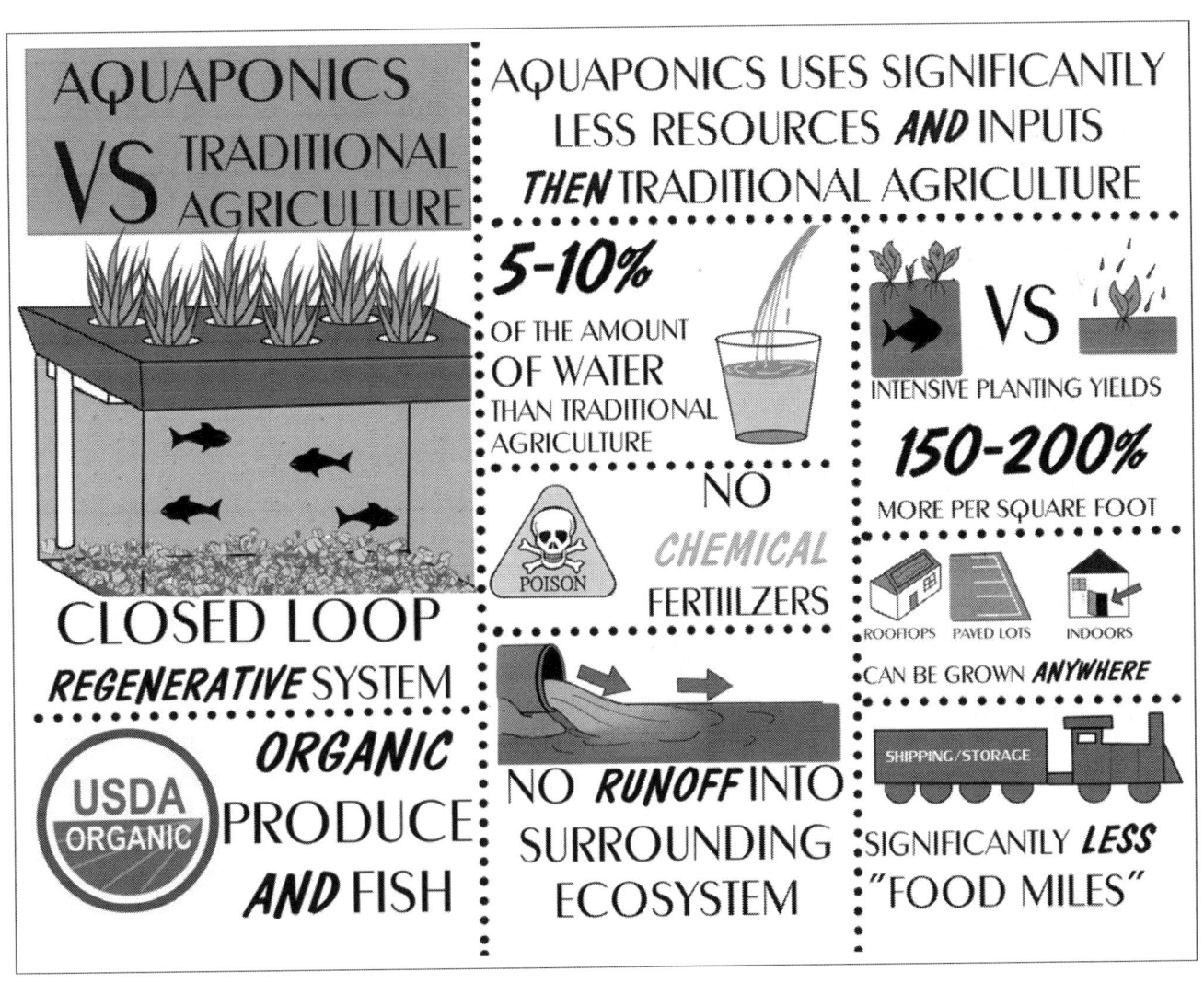

AQUAPONICS USES SIGNIFICANTLY LESS RESOURCES *AND* INPUTS *THEN* TRADITIONAL AGRICULTURE

5-10% OF THE AMOUNT OF WATER THAN TRADITIONAL AGRICULTURE

NO CHEMICAL FERTIILZERS

VS INTENSIVE PLANTING YIELDS 150-200% MORE PER SQUARE FOOT

ROOFTOPS   PAVED LOTS   INDOORS
CAN BE GROWN **ANYWHERE**

CLOSED LOOP REGENERATIVE SYSTEM

USDA ORGANIC PRODUCE AND FISH

NO **RUNOFF** INTO SURROUNDING ECOSYSTEM

SHIPPING/STORAGE
SIGNIFICANTLY **LESS** "FOOD MILES"

FIGURE 1.

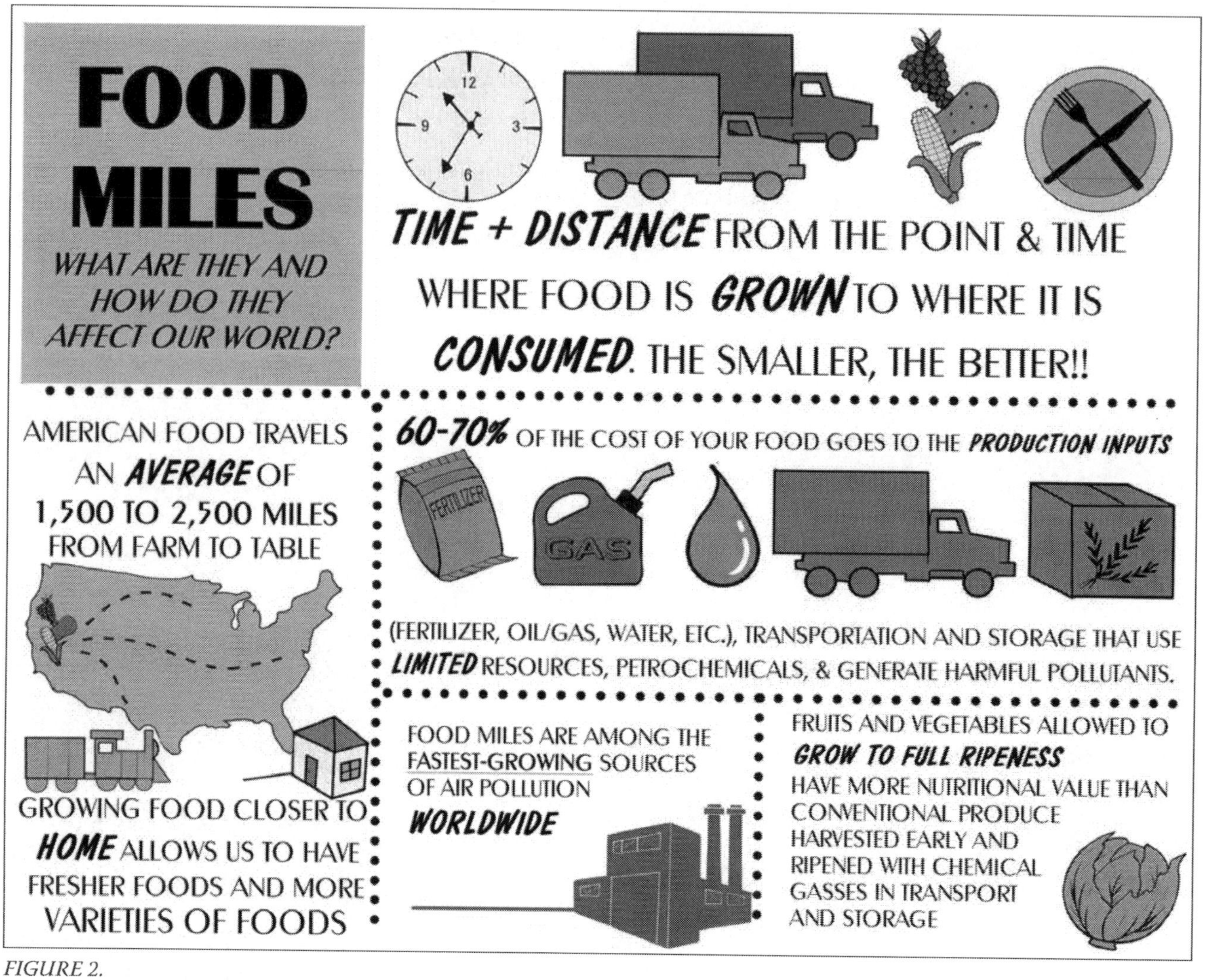

PART I: AQUAPONIC FUNDAMENTALS

FOOD MILES
WHAT ARE THEY AND HOW DO THEY AFFECT OUR WORLD?

TIME + DISTANCE FROM THE POINT & TIME WHERE FOOD IS **GROWN** TO WHERE IT IS **CONSUMED**. THE SMALLER, THE BETTER!!

AMERICAN FOOD TRAVELS AN **AVERAGE** OF 1,500 TO 2,500 MILES FROM FARM TO TABLE

GROWING FOOD CLOSER TO **HOME** ALLOWS US TO HAVE FRESHER FOODS AND MORE VARIETIES OF FOODS

60-70% OF THE COST OF YOUR FOOD GOES TO THE **PRODUCTION INPUTS** (FERTILIZER, OIL/GAS, WATER, ETC.), TRANSPORTATION AND STORAGE THAT USE **LIMITED** RESOURCES, PETROCHEMICALS, & GENERATE HARMFUL POLLUTANTS.

FOOD MILES ARE AMONG THE FASTEST-GROWING SOURCES OF AIR POLLUTION **WORLDWIDE**

FRUITS AND VEGETABLES ALLOWED TO **GROW TO FULL RIPENESS** HAVE MORE NUTRITIONAL VALUE THAN CONVENTIONAL PRODUCE HARVESTED EARLY AND RIPENED WITH CHEMICAL GASSES IN TRANSPORT AND STORAGE

FIGURE 2.

CHAPTER 3

History of Aquaponics

Beginnings of Aquaponics

Aquaponics has ancient roots, although there is some debate concerning its first use in history. However, the confirmed history of aquaponics goes back over 1,500 years.

Most scholars and anthropologists believe that the notion of using fish waste to fertilize plants has its roots in early Asian and South American civilizations. For instance, in early Asia, the existence of cultivated rice in fields, in combination with fish, is an example of early aquaponics. These farming systems employed a system of integrated aquaculture in which finfish, catfish, ducks and plants co-existed.

The ducks were housed in cages over the finfish ponds, and then the finfish processed the uneaten food and waste from the ducks. In a lower pond, the catfish lived on the wastes that had flowed from the finfish pond. The water from the catfish ponds was used to irrigate rice and vegetable crops.

The Inca's of Peru also practiced a form of aquaponics. They dug oval ponds near their mountain dwellings, leaving an island in the center. After the ponds filled, they added fish. Migratory geese flew in and ate from the water while roosting on the island. Fish were readily at hand, natural fertilizer was left behind by the geese and their system supplied food

FIGURE 3. The Tomb of Nakht, 1500 BC, contains a tilapia hieroglyph just above the head of the central figure.

to more people in the Peruvian highlands than had any other method of farming.

From what has been learned so far, it appears that the early Aztec cultivated agricultural islands known as *chinampa* for agricultural use. In this system, plants were raised on stationary islands in lake shallows formed by waste materials dredged from the chinampa canals and surrounding cities. The canals and lake shallows irrigated the plants.

Aquaponics Today

In recent years, aquaponics has gained in popularity due to ecological benefits and easy methods for establishing a sustainable farming system that provides higher yields than do traditional farming methods. The purpose is to preserve water and to reduce pollution in waterways that receive chemicals and fertilizers from traditional methods. Current methods help reduce water pollution by recycling water and wastes, thus making recycled water available for other uses.

Aquaponics is one of the fastest growing segments of agriculture and is becoming more popular at an astonishing rate. Aquaponic growers range from those with small garage and backyard operations to commercial endeavors employing large tank systems for culturing fish combined with greenhouse and open-environment systems for plant production. Aquaponics' current popularity stems from people's interest in participating in sustainable and urban applications of agriculture, and from the concept of the advantages provided when you "grow your own healthy more affordable food."

CHAPTER 4

Aquaponics Globally and the Big Picture

Worldwide Aquaponics

U.S. sales of organic food and beverages grew from $1 billion in 1990 to $26.7 billion in 2010. Sales in 2010 represented 7.7 percent growth over 2009 sales. Experiencing the highest growth in sales during 2010 were organic fruits and vegetables, up 11.8 percent over 2009 sales.

Consumers are demanding more from organic food products; they are increasingly looking at ethical sourcing, traceability, the carbon footprint, sustainability and corporate social responsibility when making their food buying decisions.

According to Organic Monitor estimates, global organic sales reached $54.9 billion in 2009 up from, $50.9 billion in 2008. The countries with the largest organic food markets are the United States, Germany and France. The highest per capita consumption of organic food is in Denmark, Switzerland and Austria.

Mass market retailers (mainstream supermarkets, club/warehouse stores, and mass merchandisers) in 2010 sold 54 percent of all organic food. Natural retailers were next, selling 39 percent of total organic food sold. Other sales occurred via export, the internet, farmers' markets, Community Supported Agriculture (CSA), mail order, boutique and specialty stores.

Aquaponics, a seemingly unfamiliar word, has become more popular over the past few years, and appears to be something that people have at least heard of these days. With food and resources growing scarce in many parts of the world, aquaponics offers an ideal solution. In a nut shell, aquaponics is a sustainable healthy food production alternative that aims to conserve and reduce the amount of unnecessary resources. It allows virtually anyone to produce his/her own organic food while maintaining a sustainable way of life, at a low cost.

This method of food production combines aquaculture and agriculture without the use of soil. Plants, fish, herbs, fruits, and vegetables to thrive in this system. Aquaponics uses only 10 percent of the water typically used for plant production and just a fraction of the water used for fish in most traditional aquaculture farming operations. Unlike traditional food production systems, aquaponics uses no unhealthy chemicals, which have also proven to be harmful to the environment.

One of the most beneficial aspects of aquaponics is that virtually anyone can purchase and set up an aquaponics system from home. Aquaponics mimics the productive systems found in nature, relying on man-made nutrients to create a 'perfect' balance within the system. Combining fish and plant production, this system allows for optimum growth by providing each species the nutrients that it needs. The plants provide the fish with clean, purified water while the fish's effluent gives the plants much needed nutrients. Although aquaponics may seem quite complicated at first, it is basically a simple cycle between the plants and the fish.

This method of sustainable food production does not stop at a reduction in water usage. Recycled material such as plastic drums, containers, and pipes are often used as equipment for these systems. This strategy reduces water usage, uses recyclable material, and provides a more sustainable way to produce healthy food at its finest.

Seriously disturbing weather and rain patterns happen globally. Massive crop failures, due to drought, are quite common. With crop failures increasingly prevalent, hunger edges closer to us all.

Even the so-called 'drought resistant' genetically modified crops are not immune to the major droughts that ravage our planet. Conventional agriculture, whether it be organic, GMO or something in between, depends on reservoirs with increasing use demands and rapidly depleting underground water supplies. The vast majority of fertilizer used in agriculture today is made from natural gas and oil dependent feedstock. All of these ingredients are finite; meaning they are limited and will keep going up in price until they are eventually depleted. The cost is always passed on to the consumer.

In many parts of the world, the cost for basic staple foods already exceeds personal incomes. As a result hunger has become a normal daily experience for many people because food grown conventionally is becoming too expensive, and conventional agriculture is vulnerable to unpredictable weather. Food insecurity and hunger is becoming a reality for everyone.

Furthermore, many billion-dollar corporations are buying, or have already purchased food companies, farms (large and small) huge tracts of land, food distribution companies, and water resources. This practice has been, and is, taking place all over the world—in both developing countries and in rich countries such as the United States.

How, then, are we going to eat? While debates rage and conventional agriculture goes on with business as usual, crops being bombarded

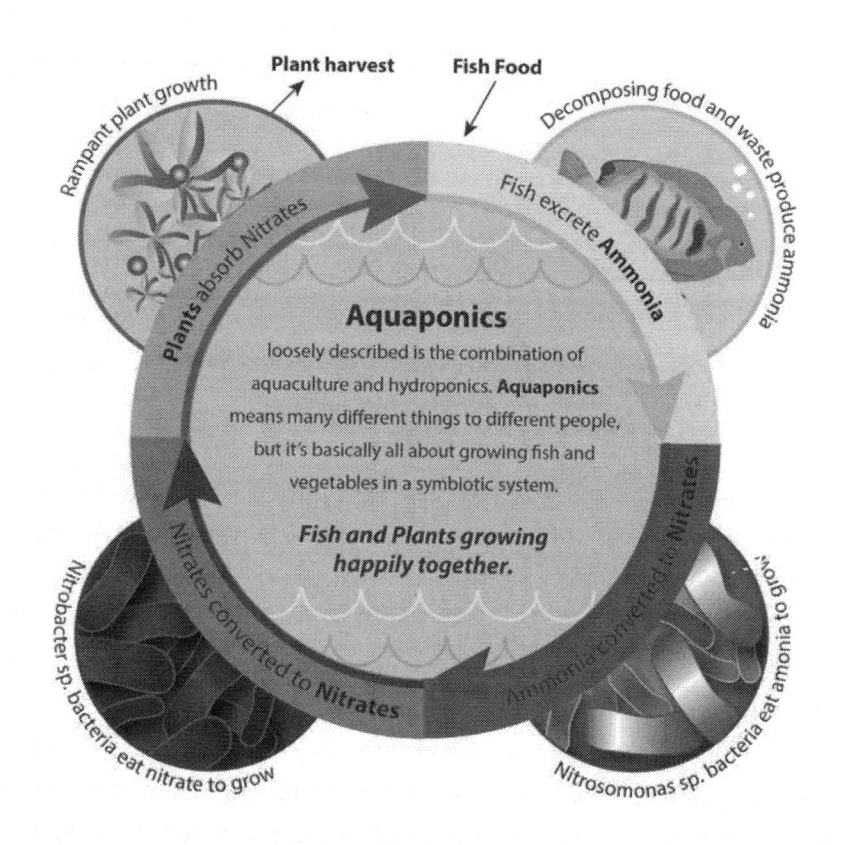

FIGURE 4. Aquaponics Life Cycle

16

with droughts, storms, toxic pesticides, and crop-devastating diseases immune to modern controls, hunger is increasing its grip on growing human populations across the globe. Conventional agriculture is proving totally unsustainable, if not downright unworkable in current climate conditions. Hunger is even making its presence known among many within the so-called rich Western countries. Global food prices are rising at a steady 140 percent annually. That includes the food prices in your local corner market.

The answer has already been invented and is catching on. It is totally sustainable. It does not use soil, toxic pesticides, herbicides, artificial fertilizers, harmful chemicals, genetically modified crops or fish, or antibiotics. It wastes 90 percent less water than conventional agriculture methods. It does not need expensive, unhealthy genetically modified seeds or plants. It is far more generally adopted by farmers worldwide, and it promises to end food insecurity for millions of people.

It is a way to raise fish and vegetables intensively in the same recirculating water. It marries fish farming and hydroponics and by doing so, gets rid of the endemic problems of both technologies. The vegetables clean the water for the fish, so there are no uncontrolled effluent discharges to the environment. The fish fertilize the water for the vegetables, so no artificial fertilizers need be bought. Plants thrive and grow up to twice as fast as they do in traditional methods. Therefore, this new revolution is taking the world by storm.

Aquaponics farms both the fish and the vegetables sustainably. Aquaponic systems growing fish and vegetables together can be certified organic. Everything grows in tanks in aquaponics, and water evaporation is greatly reduced by shading the fish tanks and enclosing the water almost completely in hydroponic type raceways, so at least 90 percent less water wastage occurs than in conventional agriculture. Water is constantly recycled between the fish and the vegetables.

Aquaponics is a proven technology that has been used commercially since the 1970s. It uses around 17 percent of the energy used by conventional farming, since no trucks, tractors, or other machinery is necessary. As a modest user of energy, it is very suitable to using alternative energy sources such as wind power or solar panels. Working with the plants and harvesting is quick and easy, since everything is done at a comfortable ergonomically correct waist-high level.

In summary, aquaponics is an efficient way to grow highly productive, sustainable, organic and healthy food. It is industrial agriculture gone green. It is renewable food. It fits snugly into a climate controlled greenhouse, and reduces water usage by at least 90 percent. It is also an environmentally friendly operation that benefits people, the community, and the planet.

The Dire Circumstances

Many scientist and government officials predict that events over the next forty years, largely of humanity's doing, will converge to put into question our ability to feed ourselves.

Foreign Owned/Controlled Farmland and Food Supply

Another disturbing practice, occurring at an exponential rate over the last decade, is the purchase of real estate by foreign countries and companies. A slew of foreigners—primarily Chinese state corporations and Gulf sheiks— are buying up farmland throughout the world at an accelerated pace to acquire as much precious soil, farmland, and water as possible. This phenomenon is known as "land grabbing," This practice displaces family farms and drives up food costs. Large companies and foreign countries are rapidly obtaining the ability to control food supply and distribution. The economic outlook of forthcoming higher food prices in the near future is alarming.

Data shows that this troubling trend, of foreign governments with trillion dollar budgets and large foreign corporate companies with million dollar budgets, purchasing enormous amounts of precious limited farmland. Although this has been occurring since the 70's it has been increasing at an exponential rate over the past decade. For instance, in 2011 foreign ownership of agricultural land increased by 1,490,781 acres in the United States alone. Unfortunately, most American's are ill-informed about this issue, and so many other critical matters, due to mainstream media's continued failure to truly and accurate inform the public on important issues.

The USDA released a report in 2013 (most recent report as of this writing) detailing foreign holdings of U.S. agricultural land as of December 2011. "Foreign persons" are defined as individuals who are not citizens of the U.S., foreign businesses and governments that have their principal place of business in a foreign country and U.S. entities in which there is a significant foreign interest. The report showed that foreign investors held an interest in 25.7 million acres of U.S. agricultural land (forest land and farmland) as of December 31, 2011. This is an increase of approximately 1.5 million acres from December 31, 2010.

The report also stated that foreign persons have reported acreage holdings in all 50 States and Puerto Rico. Forest land accounted for 54 percent of all foreign held agricultural acreage, cropland for 19 percent, and pasture and other agricultural land for 27 percent. Together, 9,511,437 acres or 36 percent of foreign-held acres are owned by individuals or entities from these countries. Foreign entities are also buying up critical farmland and water source acreage in enormous quantities in Mexico, Central America, South America, Caribbean, Asia, and Africa. Again, the pace and amount of land being grabbed is astonishing and truly disturbing.

As of 2013 "water grabbing" by corporations amounted to 454 billion cubic meters per year globally. Cooperate and foreign investors from seven countries—the United States, United Arab Emirates, India, United Kingdom, Egypt, China and Israel—accounted for 60 percent of the water acquired under these deals.

Between 2000 and 2012 nearly two-thirds of the land being purchased was in Eastern Africa and Southeast Asia. During this period over 205 million acres of land were been purchased by foreigners and large corporations. About 62 percent of these deals were in Africa, totaling about 138 million acres. In September 2010 the World Bank, showed that over 460,000 square kilometers (180,000 sq mi) or 46,000,000 hectares (110,000,000 acres) in large-scale farmland acquisitions or negotiations were announced between October 2008 and August 2009 alone. More than one economist has stated that China is buying Africa to feed its rapidly growing population.

These large land grabs push out small farmers and destabilizes the local economy. In Sudan, for instance, the local population is becoming increasingly dependent on food aid and international food subsidies because the land grabbers are pushing out small farmers, and the produce being harvested is shipped to markets in other parts of the world. Evidence also shows that these large land grabs lead to lost natural ecosystems, as a result of farming at such a large commercial scale. Another problem resulting from these land grabs is the large-scale displacement of local peoples without adequate compensation, in either land or money. These displacements often result in resettlement in marginal lands, loss of livelihoods especially in the case of pastoralists, and the erosion of social networks. Lastly, the reduction of available land drives up land prices and is going to make it all the more difficult for the average person to afford real estate.

Some examples for foreign corporate land purchases include the company Cargill purchased 775,000 acres of Brazil's valuable soybean farmland. Nile Trading and Development purchase of 1,482,632 acres of east Africa's rich farmland. BHP Billiton, a large mining company, purchase of 877,000 acres in Indonesia. Ted Turner of AOL and CNN fame, purchase of 111,000 acres in Argentina. The South Korean corporation Daewoo purchase of 1.3 million hectares, half of all Madagascar's agricultural land, to produce corn and palm oil. This is just a small fraction of some of the land grab transactions.

Economics: Food Demand Increases

"The two root causes of our environmental crisis—exploding population growth and wasteful consumption of irreplaceable resources. Over-consumption and overpopulation underlie every environmental problem we face today"
–Jacques Cousteau

Economics and how human behavior could be synthesized down to lines on a graph is fascinating. Take the famous graph of the law of supply and demand. It says that if demand rises and supply stays constant, prices will go up. There are more dollars "chasing" the same number of goods.

There are several main drivers of the projected increase in global demand for food in the next forty years: global population growth, increasing standards of living for developing nations, and depletion of resources.

Population Growth

First, demographers' project the worldwide population will grow from the current seven billion to over nine billion by 2050. This means that we will average adding an additional 75 million people to the world, or the equivalent of the population of Turkey, every year for the next forty (see Figure 5).

According to the Population Reference Bureaus' 2013 World Population Data Sheet, the majority of global growth over the next 40 years will happen in Africa, Latin American, Asia, and the Caribbean. China, India and Nigeria are projected to account for 37 per cent of the increase of nearly 2.5 billion

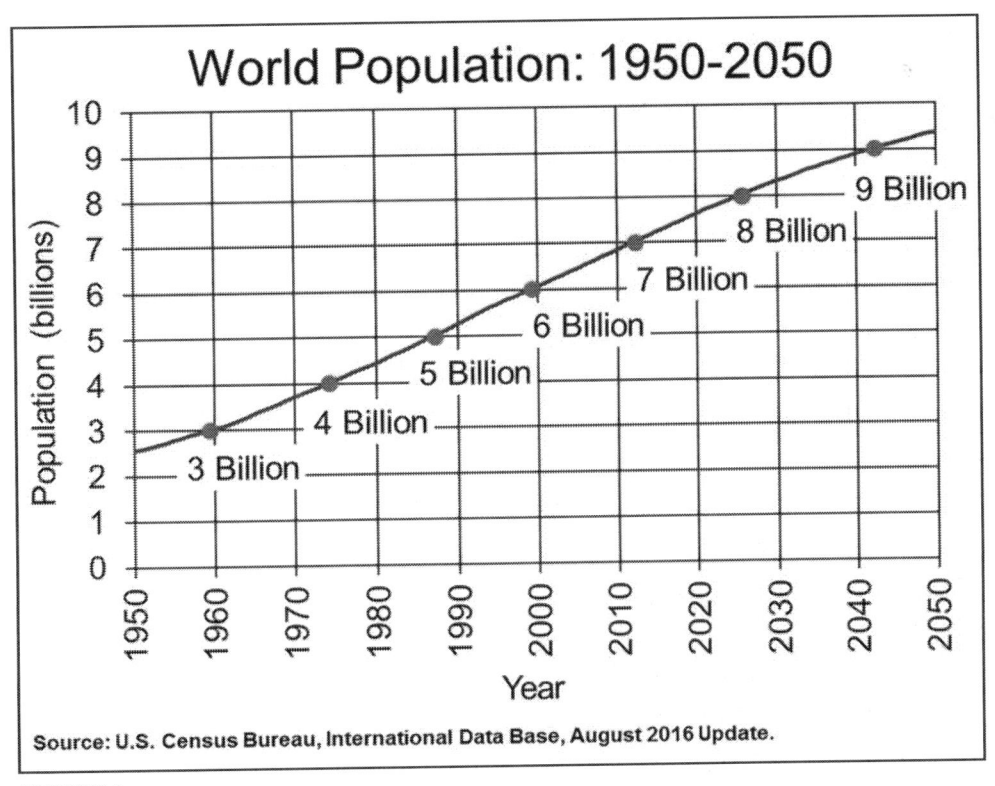

FIGURE 5.

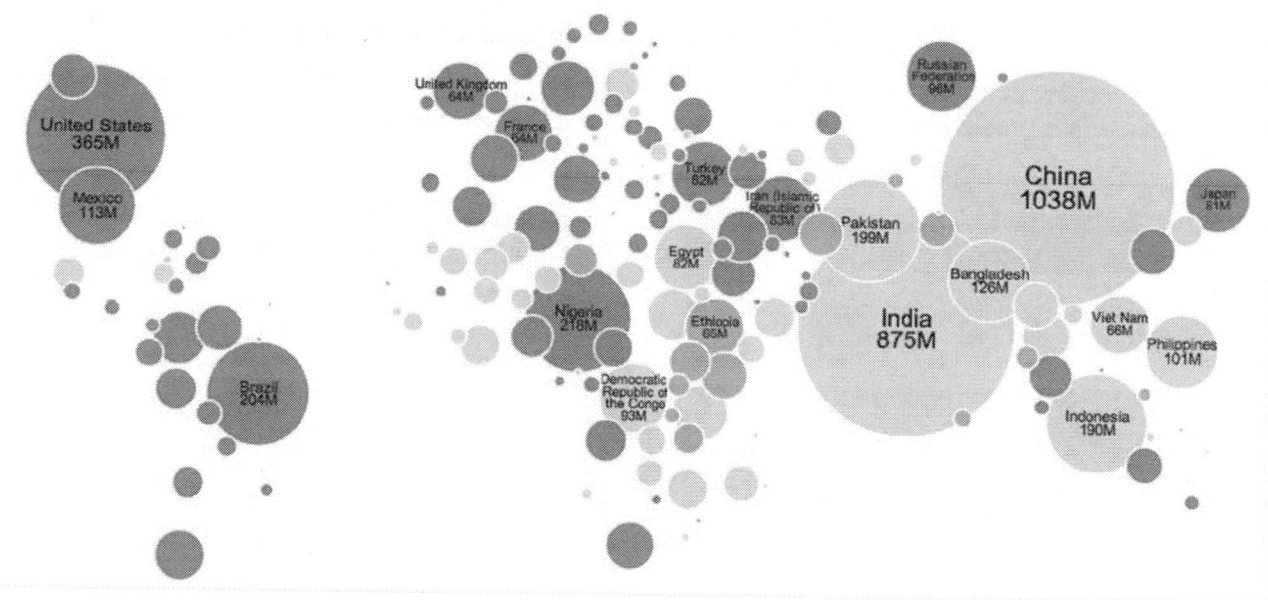

FIGURE 6. World estimated population in 2040

people in the urban population by 2050. Between 2014 and 2050, the urban areas are expected to grow by 404 million people in India, 292 million in China and 212 million in Nigeria. Seven other countries, the Democratic Republic of Congo, Ethiopia, the United Republic of Tanzania, Bangladesh, Indonesia and Pakistan, and the United States of America, are projected to contribute more than 50 million each to the urban increment and will constitute together another 20 per cent of the total increase in urban population.

Growth in the developed world is expected to be much smaller, but take place primarily in the United States and Canada. To put this in perspective, Canada's population is expected to increase from about 35 million to 37 million by 2050, but the population of Uganda (in Africa) is expected to rise from about 37 million to almost 130 million by 2050, more than tripling.

The dilemma lies in the fact that we are, perhaps, already beyond our planet's ability to support even the life we have now. According to the World Wildlife Fund's Living Planet Report 2012, "we are using 50 percent more resources than the Earth can provide. What this means is that we are now consuming planetary resources faster than they are being regenerated,

FIGURE 7.

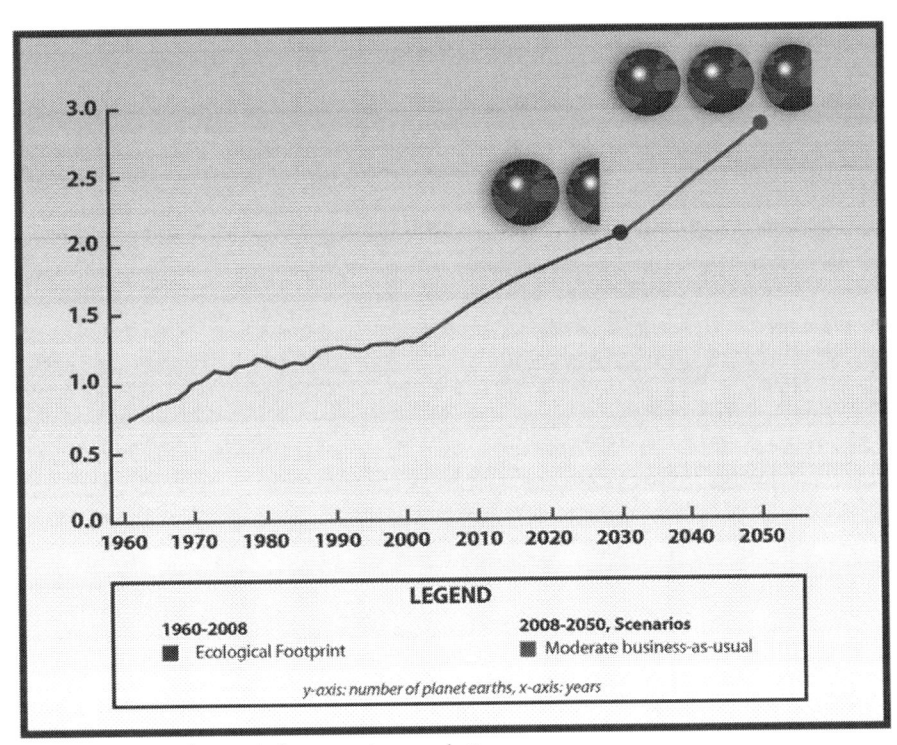

FIGURE 8. *Earths needed to sustain population*

including the planet's ability to process waste. Now consider that most of the upcoming population growth will take place in areas of the world that are least able to support that growth. Again, from the WWF report: "Ever-growing human demand for resources...not only further threatens biodiversity but also our own species; future security, health and well-being."

Increasing Standards of Living for Developing Nations

The second big impact on world food demand is globalization. Rising standards of living, especially in China and India where the populations are the highest, will increase demand for a more American lifestyle, especially regarding meat consumption. The technological revolution that is connecting us all is leveling the playing field and opening up a world of possibilities to everyone, no matter what the development status of the country. This opens the eyes of millions across the globe to what a better life looks like, but it brings with it significant change.

This eye opening is also driving ever-increasing demand. Consumption in the developed world is the primary source of strain on the earth's resources. This problem multiplies when we try to equally elevate the standards of the developing world. What if other countries were able to increase their standard of living to match that of North America? A 2008 Associated Press article, revealed that each American consumed as much as 13 Chinese or 31 Indians. A Statistical Review of World Energy 2012 found that 90 percent of the increased world consumption is occurring in China and India. If the Chinese consumed the way we do, we would roughly double world consumption rates. If India as well as China were to catch up, world consumption rates would triple. If the whole developing world were suddenly to catch up, world rates would increase at least eleven fold.

It is certainly fair that the citizens of the developing world have the right to improve their circumstances, but can the world really afford to have the entire population at the same consumption rate? Again, some say we are already beyond the planet's biological capacity to support itself. Imagine if the entire planet suddenly achieved a much higher standard of living.

Economics: Food Supply Decreases

As discussed above there are two reasons global food demand will increase over the next forty years: population growth and increasing standards of living.

FIGURE 9. Train in India

FIGURE 10. Beach in China

Based on supply and demand, let's now examine why supply—our ability to grow food at the same rate we have been growing it—is destined to decline.

Our current model of industrial agriculture depends on three factors that are no longer in place: cheap fossil fuels, unlimited water and a stable climate. Following, we will look at how each of these inputs affects our food supply.

Petroleum Use in Agriculture

As farmers know well, every step in the agricultural process utilizes fossil fuel. From planting (tractors, fertilizers, weed and pest control) to harvesting, factory processing and delivering, all the products and all the steps are dependent upon petroleum. Currently, we use ten calories of energy for every one calorie of food we produce worldwide. This is only sustainable if there is an unlimited supply of cheap, renewable energy. Petroleum, the energy engine of agriculture, is not a renewable resource. While scientists, researchers and activists are debating the exact date of "peak oil," no one disagrees that our supplies are limited. If the supply of something is limited and the demand increases, then the price will rise. Because oil is so intricately woven into every aspect of our current food-production system, increasing oil prices have a direct impact upon the price of food.

Around seven calories of fossil fuels are currently required to produce just one calorie of food. This formula works well as long as fossil fuels are abundant and affordable, but that situation is not going to last much longer. Out of sheer desperation to keep production levels high, energy companies are turning to highly disruptive exploration techniques like fracking which use a cocktail of toxic chemicals, and mining oil sand beds which causes major environmental damage to large geographical areas.

All the really easy (and therefore cheap) oil has already been pumped out of the ground. Now the energy industry is spending more and more money and resources for every barrel of oil they can recover. When cheap oil collapses, cheap food will no longer be available. The planting, harvesting, storage and transportation of food requires enormous quantities of fossil fuels under our current agricultural system.

Other alternative power sources such as wind and wave energy are simply not capable of replacing our current fossil fuel economy—not by a long shot. Solar power is often talked about as a replacement for fossil fuels, but such delusions are mere pipe dreams. For starters, solar panels require rare earth minerals from China in order to be manufactured. Energy derived from solar panels cannot be easily stored or transported, and there are no electric farm tractors because the energy expenditure of tractors is very high and requires high-density fuels.

People who do not own tractors usually have no clue about this. They think tractors can be outfitted with electric motors. Technology is a long way from developing an electric vehicle motor and battery

FIGURE 11.

FIGURE 12.

bank that can replace the petroleum-powered tractor engines that run our agricultural system today. Many of today's tractors have engines which produce over 200 horsepower.

Figures 11 and 12 show a few of the typical high petroleum machines used to produce our food.

Petroleum Use in Aquaponics vs. Traditional Agriculture

In aquaponics, energy consumption is needed to power the pump, sometimes for temperature control, and possibly a few additional hours of supplemental lighting during the winter months; however, those energy needs are small and can generally come from renewable sources. For instance, sustainable sources of heat like geothermal, solar and rocket heaters are already being used to heat aquaponic fish tanks. Aquaponic operators have also discovered techniques including insulating, burying and covering the fish tanks that significantly helps prevent heat loss.

Additionally, there is no soil to till, so there is no longer a need to use tractors and gas-powered farm equipment. Most large-scale commercial aquaponics operations typically employ either a raft method, where the plants float in water until they are harvested, or a nutrient flow technique, where nutrient rich water flows along plant grow channels. Most small and mid-size aquaponic systems are a flood-and-drain operation, where plants are grown in a media filled grow bed. These systems will be discussed extensively in the chapters ahead, but for now it is enough to know that none of the aquaponic systems used requires the labor commitment found in soil-based farming. Since there are no weeds in aquaponics, there is no need to mechanically remove them or spray herbicides. The plant nutrients and water are both integral to an aquaponics system, so there is no need for petroleum-based fertilizers or truck-mounted irrigators. Additionally, aquaponically grown plants are either growing in waist-high grow beds or in rafts floating in water, which makes them easier to harvest than soil-grown plants.

Aquaponic farms also play an important role in reducing world petroleum consumption as they are completely site agnostic. This means that aquaponic systems can be set up anywhere with an appropriate climate for the plants or here one can be artificially established (i.e. greenhouse). Aquaponics is very well adapted to providing food to local communities that may not have fertile soil available for growing. It is especially important that food-growing facilities be established closer to the people rather than trucking food in from distant locations, since over half of the population now lives in cities. Currently, most of our produce is shipped hundreds, if not thousands of miles. Growing food within population centers would save a tremendous amount of fuel.

Water Use in Agriculture

Over 70 percent of our Earth's surface is covered by water. Of all that water on earth 97.5 percent is salt water. Nearly 70 percent of the fresh water is frozen in the ice of Greenland and Antarctica and most of the remainder is either captured in our soil as soil moisture or is in parts of groundwater aquifers too deep for realistic access. Only 2.5 percent of the earth's water is fresh water. Of that, less than one percent of the earth's water is actually available for human use.

Unfortunately, people are using that one percent at an increasing rate. In a speech on February 5, 2009, United Nations Deputy Secretary-General Asha-Rose Migiro warned that two-thirds of the world's population will face a lack of water in less than twenty years if current trends in climate change, population growth, rural-to-urban migration and consumption continue. Also interesting to note is that agriculture consumes roughly three-quarters of the world's fresh water. The proportion is higher in Africa and the United States, with as much as 90 percent of our water use being for agricultural purposes (UN News Center, 2009; USDA, 2004).

When the earth had six billion inhabitants, we used nearly 30 percent of the world's accessible, renewable water supply. Projections for 2025 state we will be using 70 percent, but like other natural resources, water use will not be evenly distributed. (Friedman, 2009) Fresh water is made available to us in three ways: rain, surface water (lakes, reservoirs, streams, etc.) and near-surface groundwater aquifers. All three sources are currently being threatened by climate change, overuse and pollution. If the groundwater is not recharged at the same rate that it is being withdrawn, it becomes depleted and eventually disappears.

About one third of the world is completely dependent upon groundwater, the second largest reserve of fresh water after the frozen polar icecaps and glaciers. As the global population increases, and our need for water rises accordingly, our groundwater supplies will decrease just like our petroleum supplies. Already evident in China, groundwater levels in some areas have dropped at the rate of 5-feet (1.5 meters) per year over the past ten years. (Worm, 2004) According to the Ministry of Water Resources, as of 2009, 90 percent of that fresh water was polluted. (*Nature* 466, 2010)

In the United States, 40 percent of our fresh water comes from groundwater supplies. The most famous example is the Ogallala Aquifer, the nation's largest aquifer underlying some 250,000 square miles stretching from Texas to South Dakota. More than 90 percent of the groundwater pumped from the Ogallala is used for agricultural irrigation.. Since it was first tapped in 1911, six percent of the aquifer has become unusable because of depletion. At the current rate of draw, scientists estimate that another six percent will become unusable every 25 years moving forward.

Aquifer depletion has drastic consequences that go well beyond the obvious lack of water. The land above a depleted aquifer can turn into a sinkhole and become dangerous and unusable. If an aquifer is close to an ocean, lowering the water level can destabilize the barriers between the aquifer and the salt water. This results in seepage of ocean water into the aquifer, and the remaining water in the aquifer becomes unusable as a fresh water source.

Worse yet, the water we can use is being polluted through the very agriculture that it nurtures. The 2004 National Water Quality Inventory reported that agricultural pollution is the leading source of water quality issues on surveyed rivers and lakes, the second largest source of impairments to wetlands and a major

FIGURE 13. Irrigating crop

FIGURE 14. Crop irrigation

FIGURE 15. *Lake water recedes from heavy water use*

FIGURE 16. *Typical crop irrigation method*

contributor to contamination of surveyed estuaries and ground water. What is this type of pollution? It is pollution that comes from a wide array of sources instead of a single "point" like a factory or a sewage treatment plant. Agricultural activities that cause this problem include poorly located or managed animal feeding operations; overgrazing; plowing too often or at the wrong time; and improper, excessive or poorly timed application of pesticides, irrigation water and fertilizer. Pollutants that result from farming and ranching include sediment, nutrients, pathogens, pesticides, metals and salts. The consequence is widespread water pollution and degradation of our lakes, streams and groundwater. (US Environmental Protection Agency, 2005)

Water Use in Aquaponics vs. Traditional Agriculture

Modern agricultural methods waste an incredible amount of water. Water is either sprayed or flooded through fields where a huge amount either evaporates into the air or seeps past the plant roots onto the surrounding ground or into the water table, pulling chemical fertilizers, herbicides and pesticides down with it.

Aquaponics, on the other hand, is a closed, recirculating system. The only water that leaves the system is the small amounts taken up by the plants, transpired through the plant leaves, or that evaporated from the top of the tank. This allows aquaponics to use less than one tenth the amount of water a comparable soil-based garden would use.

Aquaponics is even more water thrifty than it's horticultural cousin, hydroponics. With hydroponics being a completely human-managed, chemical-based system, the nutrients regularly become unbalanced. At a particular stage in the process these nutrient chemicals build up to toxic levels. Every two to four weeks, the entire nutrient solution reservoir needs to be pumped out and replaced with fresh chemicals. This nutrient waste is full of chemical mineral salts that need to be carefully disposed of and prevented from running into our sanitary sewer system, streams, or seeping into our groundwater.

Conversely, aquaponics is an organic ecosystem in which the nutrients are balanced naturally, so there is never any toxic nutrient levels. In fact, since the water in an aquaponics system is so full of healthy biological constitutes, it is possible to never discharge the water from your fish tank. Even if a rare problem event occurred requiring a water exchange, the discharge from an aquaponics system is completely organic and would not have any negative pollution impacts.

Climate, Pollution, and Agriculture

"If we want things to stay as they are, things will have to change."
—Giuseppe di Lampedusa,
  The Leopard, 1958

A joint statement by 21 national science academies to the 2007 G8 summit declared, "It is unequivocal that the climate is changing....." (National Science Academy, 2007). The earth is becoming warmer. Studies show that the average temperature of the earth is on the increase. Since thermometer records became available in 1860, the eleven hottest years have all occurred between 1995 and 2011, with 2010 being the hottest.

This change in the earth's climate threatens our ability to produce food throughout the world. The Food and Agriculture Organization (FAO) warns that a global temperature increase of even two degrees Celsius over preindustrial levels could reduce crop yields as much as 35 percent in some countries, especially Africa, Asia and the Middle East. (Smith, 2010). Increasing temperatures will lead to changes in many aspects of weather, including wind patterns, the amount and type of precipitation and the types and frequency of severe weather events. Such climate destabilization could have dramatic unpredictable environmental, social and economic consequences.

Current conventional agriculture methods are recognized as one of the biggest contributors to air pollution — producing tremendous amounts of carbon dioxide (CO_2), nitrous oxide (N_2O) and methane (CH_4). Carbon dioxide is released by all the fossil fuels. Nitrous oxide comes from chemical-based fertilizers, and has 296 times the negative atmospheric impact of CO_2. Methane comes from livestock operations and has up to 25 times the negative atmospheric impact potential of CO_2. (Smith, 2010)

Deforestation

In 2000, nearly 40 percent of the earth's land has already been cleared for agriculture. By 2005, the estimate was closer to 50 percent (National Geographic News, 2005) Agriculture currently uses 60 times more land than urban and suburban areas combined, and covers an area about the size of Africa. Yet the demand for farmland is growing. In our search for more fertile soil, we are turning to the soils of the tropical rain forests.

According to Rainforest Facts, more than an acre and a half of tropical rain forest is being cleared every second of every day. Rainforests once covered 14 percent of the earth's land surface; now they cover a mere 6 percent and experts estimate that the last remaining rainforests could be consumed in less than 40 years. (Rainforest Facts, 2012)

Unfortunately, by clearing tropical rain forests to create more farmland to feed ourselves, we are

FIGURE 17. Deforestation of native habitat for agriculture. Loss of plants and wildlife habitat. Erosion of topsoil over time.

tearing down our best chance to solve the climate problem and filter our air. The rain forests are our greatest source of the air we breathe because of their tremendous efficiency in converting carbon dioxide into oxygen. With less rainforests, our condition only worsens.

Overfishing the Oceans

Before moving on to aquaponics as part of the solution, it is important to look at one more critical food supply issue: the overfishing of our oceans. Overfishing occurs when more fish are caught than the population can replace through natural reproduction. (WWF, 2013)

Our oceans could arguably be one of the last wild sources of food on our planet, and we are quickly emptying them of fish. Today, 85 percent of the world's fisheries are either fully exploited, overexploited or have collapsed. The global fishing fleet is operating at two and a half times the sustainable level—there are simply too many boats chasing a dwindling number of fish. (Monterey Bay Aquarium, 2013) To date, we have lost at least 2,048 fish species that we know of due to overfishing. In a study at the National Center of Ecological Analysis and Synthesis at the University of California, scientists projected that, barring significant changes, the oceans would become barren of fish by 2048.

Billions of fish worldwide are killed for food every year. Unless current fishing rates are drastically reduced, scientists predict that every species of wild-caught seafood will collapse by the year 2050. Excessive fish depletion can widely be attributed to the dual culprits of advances in fishing technology and techniques on the one hand, coupled with minimal regulations and restrictions to allow fish to repopulate.

Sadly, some reports state we only eat about 10 percent of all the marine life that is killed in order to feed us. And although other reports are more conservative, saying only about 25 percent of marine life is discarded after being caught, that is still nearly 27 million tons of fish.

Large computerized ships trawl the deep seas with miles of netting that can obliterate 130 tons of fish in a single sweep. Bottom trawlers cause massive destruction, scraping the sea bottom and destroying miles and miles of coral, sponges, and non-target bottom dwelling fish which are simply discarded as collateral damage. The fact is that the depletion we are seeing is happening because we have become incredibly efficient at harvesting the entire depth of the ocean populated by fish we eat. Even species once left mostly untouched are now in danger of depletion. We now have the technology and fishing fleet capacity to catch four times the current supply of fish.

In fact, fishing is one of the world's most wasteful and destructive industries. Every year, more than seven million tons of so-called "by-catch", (perhaps more accurately described as "by-kill") is inadvertently caught and wantonly destroyed; including over 300,000 sea animals such as non-target fish species, sea turtles, dolphins, whales, sharks, albatrosses and other sea birds. Every year 7.3 million tons of marine life is caught unintentionally by the fishing industry just to be callously thrown back dead, considered an acceptable loss in the industrial pursuit for profit.

Endangered species are also vulnerable to industry nets and other gear. Just as the dolphins still suffocate and die in the tuna industry nets, sea turtles are killed by the millions in the nets of the shrimping industry. In fact, for every pound of shrimp netted in the Gulf of Mexico, four pounds of "by-kill" is wasted and thrown back overboard dead.

Not only are we depleting the oceans of fish at an alarming rate for human consumption, but half the world's fish catch is fed to livestock! In fact, more fish are consumed by U.S. livestock than by the entire human population of all the countries of Western Europe combined.

To solve this crisis, various governmental agencies have attempted to set annual quotas by species to prevent a complete collapse. For example, the International Commission for the Conservation of Atlantic Tuna (ICCAT) sets the limits on bluefin tuna. However, these limits appear to be based more on politics than science. According to many biologist and other scientist these limits are set far above recommended recovery levels. Furthermore, these limits are largely ignored by commercial fishing companies. It is estimated that at least 20 percent of the fish caught worldwide, and as much as 50 percent in some areas, are caught illegally (WWF, 2013).

Therefore, we need to either stop eating so much fish or turn to aquaculture and farm more of our fish on land so the ocean species can recover. Since the health benefits of fish are so compelling, it is unlikely we are going to eat less fish anytime soon. Fish provide an excellent source of omega-3 fatty acids, vitamins and minerals that benefit a person's overall health. The American Heart Association recommends at least two servings of fish per week to help prevent heart disease, lower blood pressure and reduce the risk of heart attacks and strokes. Compared with other sources of animal-based proteins, all of which are full of saturated fats, fish is the healthy alternative.

Fish are also vastly more efficient sources of protein than other forms of animal protein. Currently 37 percent of the world grain harvest is being used to produce animal protein, (Brown, 2003) and in the United States, that number reaches nearly 70 percent (Earthsave, 2013). Unfortunately, most of that gain is going to feed cattle, the most inefficient source of animal protein. Imagine what would happen if the 700 million tons of grain used annually to feed livestock were fed to fish and chicken.

Fish covert the majority of their caloric intake into ample flesh that is edible, usually tasty, rich in protein and flush with heart-healthy oil. This streamlined translation, if responsibly managed and harnessed by humans, has the potential to improve the environment while generating more protein with fewer resources.

The global demand for fish has increased dramatically. In its 2012 report on "The State of the World's Fisheries and Aquaculture" the FAO revealed that global average consumption of fish has hit a record of over 41 pounds (18 kg) per person per year. Over 40 percent of that fish is now being supplied through aquaculture. Fish farming is the fastest growing area of animal food production, having increased at a 7.5 percent annual rate since 2009.

Unfortunately, some types of aquaculture only relieves little or no strain on the supply of ocean fish, since the feed used in aquaculture operations uses ocean-harvested fish to create the fish meal which is its protein base. Furthermore, some large scale ocean aquaculture operations generate more ammonia and feces waste than many human population centers.

In addition, fish farming, whether done in cages and pens floating off the coastline or in re-circulating tanks inland, produces substantial amounts of waste. These farms have even been described by some as floating confined animal feeding operations, comparing their waste production and pollution with that of the cattle and poultry industries.

Finally, while you would think that the ability to grow fish in ponds and tanks would be great for local fish production, in the United States only 10 percent of the farmed fish we eat is produced domestically. As of 2010, China produced over 60 percent of the farm-raised fish in the world. (FAO, 2012). The transport of this product from halfway around the world results in more unnecessary atmospheric pollution and consumption of precious limited resources.

Differing Views on the Global Warming Debate

There are several different views regarding 'Global Warming'. Although now the most current political correct catch phrase, is 'Climate Change'. Regardless of what the most outspoken pundits call it, they are referring to the same thing. The various views are summarized on the following page.

Point of View #1: Many believe that global warming started long before the "Industrial Revolution" and the invention of the internal combustion engine. They argue that global warming began long ago, as the earth started warming its way out of the so called 'Ice Age'— a time when much of North America, Europe, and Asia is said to lay buried beneath great sheets of glacial ice- and continues today as part of a natural process.

Point of View #2: Same view as above, adding that the industrial revolution and human activities are now accelerating global warming and climate change.

Point of View #3: The "Industrial Revolution", CO2 producing human activities, and destruction of our natural resources is the cause of global warming and the world will continue to warm at a devastating exponential rate.

Point of View #4: Global warming is all hype. Planet earth has always experienced variations in climate; and a few decades (or centuries) of warming (or cooling) is just a natural process.

Point of View #5: There is no global warming, it is just bad science and media hype.

Point of View #6: There is no global warming, but the new world order globalist, mainstream media, and government leaders are propagandizing it to push their agenda.

Our Responsibility

Regardless of which view you and I have in regards to the climate change issue, we all share one thing in common. Each one of us has an important moral and ethical responsibility to take good care of this planet during our lifetime. We need to be concerned about the planet, do our part to help out, and reduce our negative impact upon wildlife, oceans, forests, grasslands, and the environment as a whole. We should make every effort to minimize pollution, destruction of natural habitat, consumption of resources, and waste. We should always be striving to do positive things that directly benefit our planet, and the natural environment, regardless of the global warming issue.

A good rule of thumb is to **always leave a place (and a person) better off than when you arrived.** This perspective, when applied, is a win-win for all. Build others up with an encouraging word, a helping hand, good advice, constructive instruction, and/or an empathetic ear and you will make a positive difference in their lives. This positive impact in the life of another in turn spreads into families and communities like the ripples of a rock thrown into a pond. Being a good steward of this planet — constantly trying to reduce consumption of natural resources, trying not to pollute the planet, making a consistent effort to help the environment—likewise benefits earth's inhabitants now and in the future. Such should not be considered as a good deed, but instead viewed as simply honoring our responsibilities.

It has been said that a winner is someone who picks up a piece of trash that is not his or hers. As an individual, we cannot dictate government policy, force companies to operate differently, control others, or convince everyone of our climate change viewpoint, but we can make a big difference in regards to our individual impact upon this planet — for better or worse. Our lifestyle will result in either negative consequences or positive benefits for the wildlife, natural habitat, preservation of species, and quality of life for future generations of this planet. The responsibility is ours — an individual choice that truly makes a difference.

The Aquaponic Solution

Food prices have risen, climate has changed; and the supply and demand structures have changed. So is there hope for the future? While aquaponics can not address the dual demand pressures of population growth and increasing global standards of living, it does offer solutions to many problems.

Climate, Pollution, and Aquaponics vs. Traditional Agriculture

Aquaponics is a food-growing system that can have negligible impact on our environment, especially if the pumps and heaters are powered through renewable energy sources. Except for purely wild food-growing systems, such as the ocean, and most permaculture techniques, no other food system can likely make that claim.

On the other hand, as previously addressed, traditional agriculture is the single largest contributor of CO2 emissions, while simultaneously consuming and requiring more natural habitats for growing crops and raising cattle. The main pollutant sources are CO2 emissions from all the petroleum being used in farm production and food transportation, methane from cattle production, and nitrous oxide from over-fertilizing. Aquaponics requires none of these inputs. These pollutants are non-existent or negligible in aquaponics. The need for petroleum is next to nothing in Aquaponics (maybe a backup generator on rare occasions or electricity from a petroleum power plant). Furthermore, fish do not produce methane as cattle do, and there is no chance of over-fertilizing an aquaponics system.

Perhaps, most importantly, aquaponic systems can be started anywhere. This alleviates the need to clear jungles and forests and allows the focus to be on urban centers, like old factories or warehouse buildings as the farms of the future. While not necessarily suited to growing vast fields of grain, aquaponics allows growth of any vegetable and many types of fruit crops, and does so in a way that is even more productive on a square foot basis, even in an urban setting. Aquaponics can produce 50,000 pounds of tilapia and 100,000 pounds of vegetables per year in a single acre of space. By contrast, one grass-fed cow requires eight acres of grassland. Another way of looking at it is that over the course of a year, aquaponics will generate about 35,000 pounds of edible flesh per acre, while the grass-fed beef will generate about 75 pounds in the same space.

It is not an unrealistic notion to think at least some portion of our food can be produced in our urban centers. In fact, it is not unusual for city dwellers to grow a meaningful portion of the food they eat. Hong Kong and Singapore already both produce more than 20 percent of their meat and vegetables within the city limits. With aquaponics, production yields are much greater, and done correctly have the potential of allowing a city to basically feed itself.

Aquaponics vs. Aquaculture

With the projected dramatic decrease of the ocean's fish supply, we must turn to other options to continue enjoying the healthful benefits of eating fish. As with any intensive animal-raising operation though, the central problem is how dispose of the waste without harming the environment. Fortunately, aquaponics solves this problem. Aquaponics takes the potentially toxic waste water from an aquaculture system and creates an organic nutrient for a hydroponic system. This acts as a biofilter for the aquaculture system and purifies the water that goes back to the fish. By seeking a solution through bio-mimicry techniques and observing nature, scientists found the solution in polyculture. Aquaponics is a beautiful intertwining of fish and plant production.

Making a Positive Difference

"When the wind changes direction, there are those who build walls and those who build windmills."
—Chinese proverb

Yes, there is hope. Mayor Barrett of Milwaukee recently gave the final approval for a zoning variance to enable Growing Power to build the world's first vertical farm. The five-story structure will be built to grow plants and fish together, with the water flowing from one story to another. It is a building entirely dedicated to the growing of fresh food in an area of Milwaukee surrounded by low-income housing, liquor stores and mini-marts.

By learning more about aquaponics, you are choosing to become a part of the solution.

FIGURE 18. *Modern building with plans on balconies and roof. Solar panels on the rgith side of the building also serve as carports.*

CHAPTER 5

Toxic Food vs. Healthy Organic Aquaponic Food

Crop Considerations

Organic food has become a fairly common preference for most people. Despite the growing popularity of organic products there is still a large segment of society, including those who purchase organics, who cannot explain why organic food may be better than non-organic food and do not understand many of the other dangers added to our food: antibiotics, growth hormones, preservatives, arsenic, mercury, and the other 30+ heavy metals commonly found in our foods.

U.S. sales of organic food and beverages grew from $1 billion in 1990 to $26.7 billion in 2010, according to the Organic Trade Association, an industry group. In addition to attracting newcomers like Sprouts, the growth in the organic food segment has prompted some traditional grocery stores to rethink their product lineups and are increasingly offering more organic choices. SPINS, a reporting firm which tracks the natural products industry, stated that the organic sector has shown "resilience to the recession,"

FIGURE 19. Pesticide being sprayed on food crop.

because consumers are becoming more educated regarding the dangers of pesticide use and genetically modified crops. In other words, the demand for organic crops is increasing, and the prices for such continue to rise.

One important thing to keep in mind when desiring to grow edible plants is whether or not the plant is readily available in your market area. Ideally, it makes sense to focus on growing organic plants that are not readily available in your area, and/or those organic crops that cost the most in your location.

There is a list of plants referred to as the 'dirty dozen' (which has actually been expanded to 16) that should only be eaten if organic. The dirty list of plants are typically laced with unhealthy levels of heavy metals, pesticides, fungicides, and/or herbicides; poisons that are very harmful to the human body. The most polluted plants, and recommendations for the best plants to grow, are discussed below and in 'Chapter 10: Plants/Crops—Keys to Success'.

Studies link exposure to these toxins to cancer, birth defects, stillbirth, infertility, and damage to the brain and nervous system (including Parkinson's disease). These substances are designed to kill, and because their mode of action is not specific to one species, they often kill or harm organisms other than pests. The application of these toxic chemicals makes its way into our food chain. Even exposure to low doses can cause a range of neurological health effects such as memory loss, loss of coordination, reduced speed of response to stimuli, reduced visual ability, altered or uncontrollable mood and general behavior, and reduced motor skills. These symptoms are often very subtle and may not be recognized by the medical community as a clinical effect. Other possible health effects include asthma, allergies, hypersensitivity, and hormone disruption.

What Are GMOs and Why Are GMOs Bad?

According to the Food and Drug Administration, genetically modified organisms (GMO)—such as plants or animals—have been genetically engineered to create new characteristics. In other words, a specific gene is added to an organism to produce a new trait.

Most of these new unnatural gene alterations allow the plant to be sprayed with pesticides, herbicides, and/or fungicides without being killed. Meanwhile, everything else around the plant dies. Unfortunately, the plant absorbs the toxic chemicals and associated heavy metals, which are then passed on to you and your family.

Where I live in the California valley—also known as the Salad Bowl of America since most of the nation's produce is grown here—it is not uncommon to see a food crop being sprayed with various toxic chemical applications three or four times a season. In addition, with a mild climate farmers are able to grow up to three crops a season, depending on the types of crops being grown. As a result, the same

FIGURE 20. Crop duster spraying pesticide on food crop.

farm land often receives a chemical application many times during the year. These chemicals get more and more concentrated in the soil over time. So, not only are these poisons and associated heavy metals entering the plant through direct application, but they are absorbed by the plant roots from the soil.

When we first relocated to California from Oklahoma, we were absolutely thrilled to discover such an abundance of fresh fruits and vegetables. Just about every variety of vegetable and fruit can be found at local stores throughout the year. However, it did not take long for our bubble of joy to pop. On just about every drive out of town we would see crops being sprayed or chemicals being added to the water used for irrigation. The amount of chemicals we see being regularly applied to crops within our area of California alone is truly astonishing. This was very influential as to why I got into aquaponics so heavily.

The California counties known for growing agricultural crops average over 250 tons of pesticides, herbicides, and fungicides applied to crops annually. That is not a typo—over 250 tons of chemicals are applied annually on farms in each of California's agriculture counties!

Yet Florida applies more than eight times the amount of pesticide and herbicides on tomatoes as does California. In order to get a successful crop of tomatoes, the official Florida handbook for tomato growers lists 110 different fungicides, pesticides and herbicides which can be applied to a tomato field over the course of the growing season, and many of those are what the Pesticide Action Network calls 'bad actors' —the worst of the worst in the agricultural chemical arsenal.

FIGURE 21. Pesticides being sprayed on an orchard.

Washing produce does minimal good, as the plants actually absorb these chemicals into their cells. The most recent developments in GMO technology actually programs the plant itself to produce toxic substances that repel or kill insects. Genetically modified organisms are found in a huge variety of food products from baby food to fruit juice, and many believe they are dangerous to your health.

GMOs Harmful Health Impacts (Bottom Line)

Studies on GMO are almost infinite. Many studies are performed by companies and industries that profit from GMOs, whereas others are conducted by those that are opposed to GMOs. Therefore, getting a true unbiased and untainted representation of the impacts GMOs have on human health can be a challenge. However, if one will honestly examine the evidence produced from 'true' scientific studies, which are not influenced by money backers and tainted with biased data collection or have subjective conclusions, it is obvious that GMOs can result in negative health impacts.

Regardless, basic common sense should be enough to discourage us from the consumption of GMO foods. It does not take a brain scientist, a university study, or a rocket engineer telling us that eating foods sprayed with harmful chemicals is not healthy. Why would any reasonable person want to eat food sprayed with pesticides, herbicides, and/or fungicides?

An excellent resource for accurate scientific evidence on the negative impact GMOs have on human health is the Organic Consumers Association — a non-profit 501(c)3 public interest organization campaigning for health, justice, and sustainability (http://www.organicconsumers.org).

The website "Natural News" (http://www.naturalnews.com) also has many articles based upon unbiased science which provide the real truth about GMO foods. Those that educate themselves about the harmful effects of GMO foods by not relying on mainstream media or government propaganda, and who truly look into how pro-GMO studies are funded easily become non-GMO advocates.

GMOs Harmful Environmental Impact

In addition to the potential health risks, there is solid evidence that GMOs are bad for the environment. The Union of Concerned Scientists (UCSUSA) mentions six ways GMOs might have a negative environmental impact.

1. GMO crops could become weeds.
2. New genes could move to wild plants, causing those plants to become weeds.
3. Crops that produce viruses may lead to new, stronger viruses.
4. GMO plants created to release toxins threaten wildlife.
5. GMO crops could disturb the eco-system in an unpredictable way.
6. GMO crops threaten crop diversity.

The UCSUSA also states there is no consistent monitoring program in the United States, and there may be further negative impacts yet to be detected.

Harmful GMOs Everywhere

Eighty percent of the food supply in North America contains GMOs, and there is currently no national labelling requirement for GMO identification. While you may associate GMOs with only corn or soy products, they are found in many others, especially processed foods, which includes, but is not limited to, those in the chart below.

- Vegetable oils
- Lecithin
- Artificial sweeteners
- Flavor enhancers
- Cereals
- Sugar
- Dairy products
- Beets
- Papaya
- Squash

Some GMO crops have been designed to produce a protective pesticide-like by-product in their cells which is also detrimental to human health. Aquaponics is about being organic and growing nutrient dense environmentally friendly foods, whereas GMO plants are completely contrary to logical reasoning, in regards to good health, concern for the environment, nutrient density, and optimal plant production. Educated and responsible citizens make an effort to not do anything that further promotes the GMO industry or proliferates GMO plants.

Being GMO-Free

There are several ways to avoid genetically altered foods:

- Purchase only organic foods.
- Reduce your consumption of processed foods, including store bakery goods wherever possible.
- Avoid artificial sweeteners, especially aspartame.
- Keep in mind that vitamin supplements may also contain GMO soy or corn ingredients, so you should contact the manufacturer when in doubt.

- While genetically modified animals are not approved for humans to eat, many GMO crops are used to feed animals that are later consumed by humans. The concentration of heavy metals and the toxins increase as they move up the food chain; with humans being at the end of the food chain. Unfortunately, many of these chemicals and heavy metals are not easily removed from the body after they are consumed—becoming significantly concentrated in us—leading to a higher propensity to health problems.
- To avoid eating animal products containing GMOs, purchase only organic, wild, or 100 percent grass-fed animals or those that were only fed organic feed.

There is sufficient reason to be concerned about GMO effects on health and the environment. Buying organic and arming yourself with the knowledge necessary to make informed choices, will help you avoid potential health risks to you and those you care about. In addition, every time you buy an organic product, your dollars serve as a voice to further support and strengthen the organic market. Being knowledgeable about GMOs and their negative impacts will also help you better market your produce.

Most Polluted Fruits and Vegetables

The Environmental Working Group (EWG) studied 100,000 produce pesticide reports from the U.S. Department of Agriculture and the U.S. Food and Drug Administration to create a list of dirtiest produce. The EWG recommends purchasing organic or locally grown varieties of the following products; doing so can lower pesticide intake by as much as 80 percent.

Celery

Research showed that a single celery stalk had 13 pesticides, while a large sample of celery contained as many 67 pesticides. Chemicals cling directly to this vegetable as it has no protective skin, and its stems cup inward, making it difficult to wash the entire surface of the stalk. If you like this crunchy veggie, go organic.

Peaches

Peaches are laced with 67 different chemicals, placing them second on the list of most contaminated fruits and vegetables. They have soft, fuzzy skin; a delicate structure;, and high susceptibility to most pests, causing them to be sprayed much more frequently.

Strawberries

This red, juicy fruit has a soft, seedy skin, allowing easier absorption of pesticides. Research showed that strawberries contained 53 pesticides. Buy organic strawberries or shop for a more naturally grown crop at a local farmer's market.

Apples

Apples are high-maintenance fruit, which means they need many pesticides to stave off mold, pests, and diseases. The EWG found 47 different kinds of pesticides on apples, and while produce washes can help remove some of the residue, they are not 100 percent effective.

Blueberries (domestic)

These antioxidant-rich berries have a thin layer of skin that allows chemicals to easily contaminate the fruit. Domestic blueberries were loaded with 13 pesticides on a single sample, according to the EWG. Imported blueberries had similar contaminants.

Bell Pepper

This vegetable is highly susceptible to pesticides. According to the EWG, sweet bell peppers showed traces of 63 types of pesticides. As with all fruits and vegetables, even though some pesticides can be washed away, many still remain.

Spinach, Kale/Collard Greens

Spinach has been found to be loaded with 45 different kinds of pesticides. Kale/collard greens have been found to be loaded with 57 different kinds of pesticides.

Grapes

These tiny fruit have extremely thin skins, allowing for easy absorption of pesticides. Imported and domestic varieties typically have significant pesticide residue.

Potatoes

The potato is highly laced with an average of 36 different types of pesticides to prevent pests and diseases. Although potatoes will not grow in an aquaponics system, one should avoid consuming potatoes or potato products that are not organic.

Cherries

Cherries, like blueberries, strawberries, and peaches, have a thin coating of skin, often not enough to protect the fruit from harmful pesticides. Research has shown that cherries grown in the U.S. had three times the amount of pesticides as imported cherries. Because cherries contain ellagic acid, an antioxidant that neutralizes carcinogens, they should not be abandoned. However, one would be prudent to buy organic or imported cherries over our domestic varieties.

Dirty Produce— Organic Substitute

The following list of produce has been shown to have the greatest concentration of harmful pesticides, and should only be consumed if grown organically. Since the pesticide, herbicide, and/or fungicide is often absorbed by the plant, simply washing the produce will not remove the al of the toxic chemicals.

1. Peaches
2. Apples
3. Bell Peppers
4. Celery
5. Nectarines
6. Strawberries
7. Cherries
8. Kale / Lettuce / Spinach
9. Carrots
10. Pears
11. Cucumbers
12. Spinach
13. Blueberries
14. Grapes
15. Potatoes
16. Collard Greens

GMO Labeling

Chemical manufacturers, Big Agriculture and junk food companies spent more than $70 million spreading misinformation in 2012 to narrowly defeat GMO food labeling ballot initiatives in California and Washington. They continue to spend hundreds of millions of dollars in campaigning to defeat state GMO labeling laws in other states.

Currently, 64 countries have policies requiring the labeling of Genetically Engineered (GE) foods, including all our trading partners in Asia, the European Union and even countries where GE crops are a large part of the economy, such as Brazil. While there may be universal agreement on labeling abroad, here in the U.S. multinational agribusinesses and food corporations spend millions of dollars in campaign contributions, lobbying efforts, and false advertising to block state labeling initiatives.

As a result of this outpouring of cash to defeat labeling, Americans have been inundated with a number of myths intended to defeat state labeling efforts. These myths need to be corrected.

> **MYTH #1: American consumers will see a huge spike in food prices if companies are required to disclose the GE ingredients that are already in their products.** The fact is companies change their labels all the time without causing a spike in the price of food. In fact, most companies do not print labels more than one year in advance for regulatory or marketing purposes. The establishment of a mandatory labeling standard for GE foods

could easily fall within a company's regular label refresh cycle. The cost of modifying food labels is negligible.

MYTH #2: Farmers are opposed to mandatory labeling. While some may oppose labeling, major farm groups like the National Farmers Union, the National Family Farm Coalition and the National Black Farmers Association support mandatory labeling and have opposed legislation intended to block state laws in the absence of a national standard. With mandatory labeling there will be no new financial cost to farmers, and given that food companies already produce foods for a variety of markets, there should not be additional segregation costs. Rather the information that is already being captured within the food supply chain will simply be provided to consumers on the end product.

MYTH #3: We can rely on voluntary measures alone. In the 14 years that the FDA has allowed companies to voluntarily label foods produced using genetic engineering, not one single company has done so. Similarly, we cannot merely rely on the use of voluntary absence claims like "GMO-Free." While such marketing claims allow companies to distinguish themselves in the marketplace, they are not a substitution for mandatory disclosure, because consumers are not given the full universe of information. Creating a federal standard for voluntary marketing claims will do nothing to address the overwhelming demand for labeling and likewise will do nothing to address consumer confusion that has festered in the absence of mandatory labeling.

MYTH #4: w But in the U.S. we do not label dangerous food; we take it off the shelf. Rather, foods produced using genetic engineering are fundamentally different at the molecular and genetic level than those produced using conventional breeding methods. Mandatory labeling of GE foods is essential for preventing consumer deception and will allow consumers to make informed choices about the products they are buying and feeding their families.

FACT: What is lost in these industry-driven myths (false advertisements) is the fundamental issue of equality. Food issues can all too often turn into issues of class. Each and every person should have the right and the information available to know what they are buying and feeding to their families, regardless of where they shop and where they live.

It is time for Congress and all governments to provide leadership on an issue that impacts each and every person. Any costs would be negligible; and the benefits of labeling are numerous. Establishing a responsible national labeling standard that informs consumers of what is in the food they are buying and feeding their families is only right and fair.

Other Food Label Problems

Food labels are supposed to be there to help you make healthy, informed decisions about what you eat, in terms of not only the calories, sodium and fat content, but also the ingredients. However, a recent report released by the Government Accountability Office (GAO) gave the FDA a failing grade when it comes to preventing false and misleading labeling. The GAO also reported that the FDA does not track the correction of labeling violations, which means even if a food manufacturer is known to be using

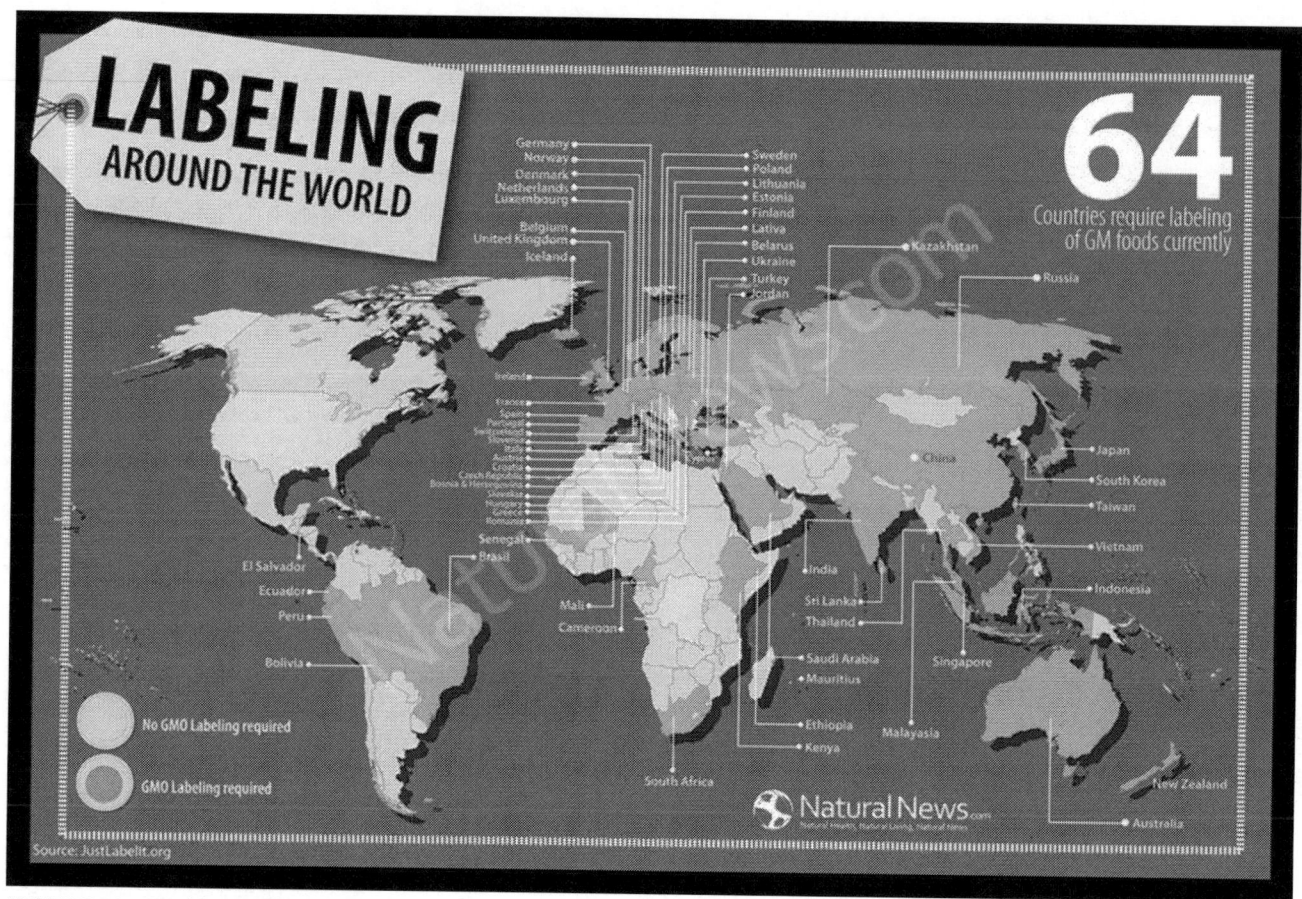

FIGURE 22. *Worldwide GMO Labeling, GMOs are not labeled in the USA.*

inaccurate labels, no one is checking up to make sure the problem is fixed.

Recently, "Good Morning America" hired a lab to test a dozen packaged food products to see if the nutrients matched the labels. All 12 products had label inaccuracies of some sort and three were actually off by more than 20 percent on items like sodium and total fat.

The US Government and GMOs

If you're wondering *why* the United States leads the world in GMO crop acreage, it's because the United States Department of Agriculture (USDA) and the FDA are heavily influenced by Monsanto, which spends millions of dollars lobbying the U.S. government for favorable legislation that supports the spread of their toxic products every year. Monsanto spends $1.4 to $2.5 million annually lobbying the federal government. Furthermore, they invest enormous sums of money on campaign contributions to politicians running for office that show favoritism towards their agenda.

In addition, the U.S. Food and Drug Administration (FDA), the USDA, and the U.S. Trade Representative all have a special set of revolving doors leading straight to Monsanto, which has allowed this transnational giant to gain phenomenal authority and influence. Former Monsanto employees currently hold positions in US government agencies such as the Food and Drug Administration (FDA), United States Environmental Protection Agency (EPA) and the Supreme Court. Of few of these include:

- Clarence Thomas, Associate Justice of the Supreme Court.
- Michael R. Taylor, Deputy Commissioner for Foods at the USDA.

CHAPTER 5: TOXIC FOOD VS. HEALTHY ORGANIC AQUAPONIC FOOD

- Ann Veneman, former Executive Director of UNICEF and US Secretary of Agriculture.
- Linda Fisher, Deputy Administrator of EPA.
- Michael Friedman, Deputy Director of the EPA.
- William D. Ruckelshaus, Administrator of EPA.
- Mickey Kantor, US Secretary of Commerce.
- Linda Fisher, Deputy Administrator of EPA. (Ms. Fisher has been back and forth between positions at Monsanto and the EPA)
- Former Secretary of Defense Donald Rumsfeld was chairman and chief executive officer of G. D. Searle & Co., which Monsanto purchased in 1985. Rumsfeld personally made at least $12 million from the transaction.

The above list is just a brief summary of the more high profile positions. Not listed are the hundreds of other GMO former company employees that are now employed in government positions. The cozy relationship our government has to GMO companies is also a two way streak of opportunities. For instance, the GMO companies make it plain to government employees that wonderful future employment opportunities exist to those employees that are amicable to their agenda. Some of the more high profile positions include:

- Josh King, former director of production for White house events, is now the director of global communication in Monsanto's Washington, D.C. Office.
- Clayton K. Yeutter, former Secretary of the USDA, former U.S. Trade representative who led U.S. negotiations in the U.S.-Canada Free Trade Agreement and helped launch the Uruguay round of the GATT negotiations, is now a member of the board of directors of Mycogen, whose majority owner is Dow. Mycogen is also the corporation that holds the patent on a technology to genetically alter plants to produce and deliver "edible vaccines."
- Terry Medley, former administrator of the USDA Animal and Plant Health Inspection Serve, former chair and vice-chair of the USDA Biotechnology Council, and former member of the FDA Food Advisory Committee, is now presiding as the director of regulatory and external affairs of Dupont's agriculture enterprise.
- Micky Kantor, former Secretary of the US Dept. of Commerce and former US Trade Representative, is now a member of the board of directors of Monsanto.
- Linda J. Fisher, a former Assiatnat Administrator of the EPA is now Vice-President of Public Affairs for Monsanto.
- William D. Ruckelshaus, the former chief administrator of the US EPA is now (and for the past 12 years) a member of the board of directors of Monsanto.
- Lidia Watrud, a former microbial biotechnology researcher at Monsanto, is now with the US EPA.
- Margaret Miller, a former laboratory supervisor for Monsanto, is now Deputy Director of Human Food Safety and Consultative Services in the US FDA.

Again, the list above is a very brief summary showing just the more high profile positions. There are many other employees that could be listed. Such shows a symbiotic culture where favoritism and rewards end up not only influencing, but drive public policy.

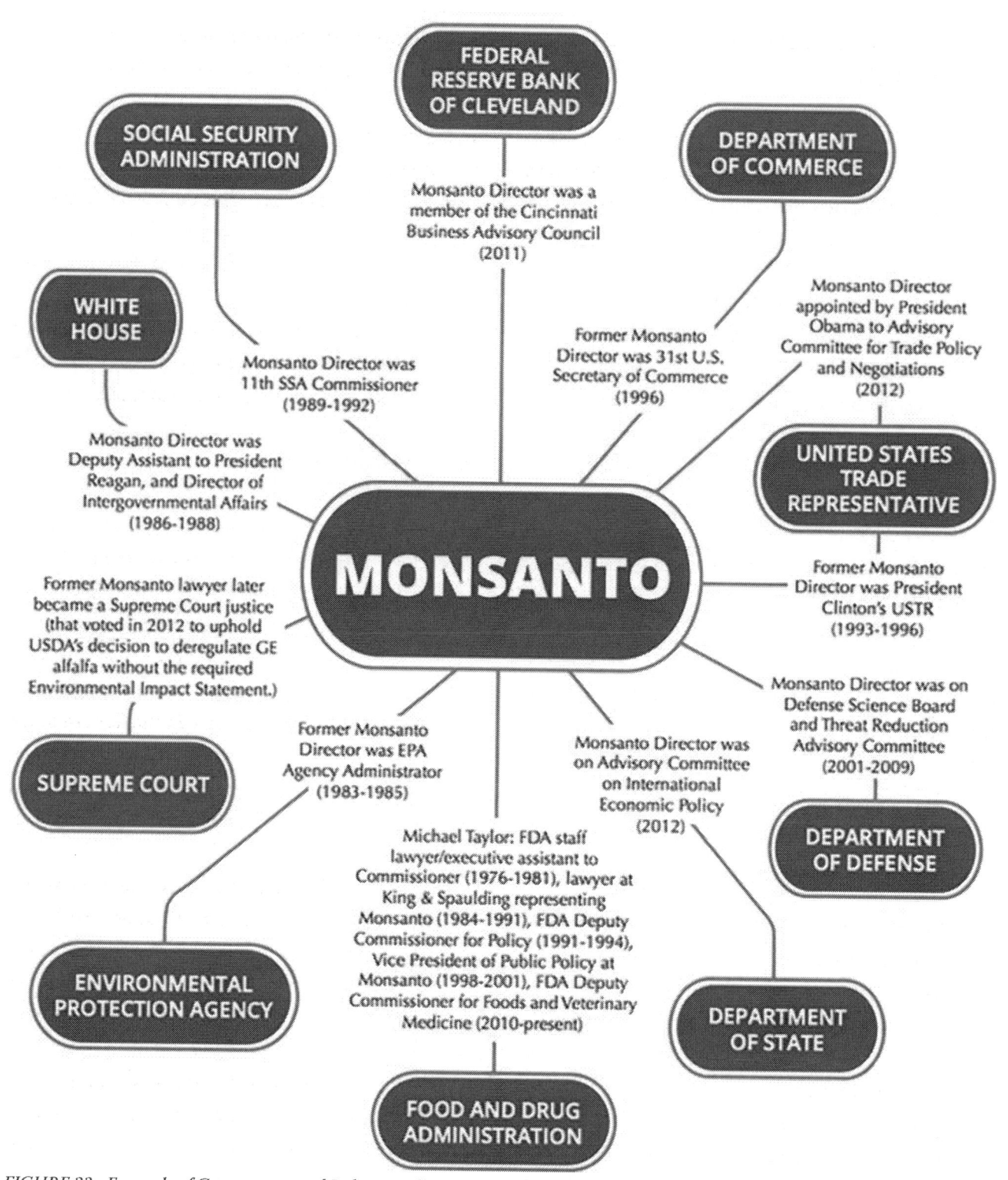

FIGURE 23. *Example of Government and Industry collusion which favors industry over the best interest of the public.*

Heavy Metals in Our Food Supply

Heavy metal poisoning is becoming a very serious issue in America today, as toxins in groundwater and soils, some of which have persisted there for decades, are increasingly turning up in both drinking water and the general food supply. All across the country, residues of arsenic, lead and other heavy metals have been detected in well water, crop soils and even foods marketed toward the health-conscious. Heavy metals are now being found in USDA certified organic foods, superfoods, vitamins, herbs and dietary supplements at alarming levels. Neither the USDA nor the FDA have set any limits on heavy metals in foods and organic foods, meaning that products can contain extremely toxic levels of mercury, lead, cadmium, arsenic, copper and even tungsten while still being legally sold across the USA.

The Natural News Forensic Food Lab, a very reputable independent laboratory headed by food researcher Mike Adams (AKA: the Health Ranger) has tested over 1,000 products using inductively coupled plasma mass spectrometry (ICP-MS) instrumentation. Below is a sample of some of the results obtained from the tests performed. A more comprehensive list of results are published at Labs.NaturalNews.com.

- Over 500 ppb Mercury in dried cat treats
- Over 10 ppm Tungsten in rice protein products
- Over 5 ppm Lead in ginkgo herb products
- Over 400 ppb Lead in cacao powders
- Over 500 ppb Lead and over 2000 ppb Cadmium in rice proteins
- Over 6 ppm Arsenic and over 1 ppm Lead in some spirulina products
- Over 100 ppb Mercury in dog treats
- Over 1,200 ppm Copper in children's multivitamins
- Over 200 ppb Lead in brand-name mascara products
- So much mercury contamination found in Maine lobster that the government halts fisheries.

In response to the publishing of these findings some companies have said that they believe heavy metals are actually good for you and that their customers should eat more heavy metals; therefore, they are not going to make any efforts to reduce heavy metals in their products. This response stands at odds with all known environmental science.

The industry desperately needs a standard—even a voluntary standard—to which products can be compared for their heavy metals composition. Neither the USDA nor the FDA have expressed any interest in promoting or enforcing such a standard.

What Levels Are Safe for Consumption?

What is clear to nearly all environmental scientists is that lower exposure to dietary heavy metals is better for your health. All heavy metals interfere with healthy cellular function. At what level they become "dangerous" depends on your genetics, your diet and your overall health.

Heavy metals bioaccumulate in the human body. When levels become high enough, they substantially interfere with healthy functioning of the brain, liver, kidneys, heart, skin, reproductive organs and other body systems.

- Learn more about heavy metals toxicity at http://labs.naturalnews.com/Heavy-Metals.html
- For expert clinical assistance regarding heavy metals detoxification, visit ACAM.org

Aren't Heavy Metals Safe Because They Naturally Occur In All Foods?

That is a myth promoted by some companies whose products are heavily contaminated with toxic heavy metals. Some of these companies are trying to convince consumers that heavy metals are not concerning by falsely claiming they are "naturally occurring." The claim is scientifically false and a prime example of scientific fraud perpetrated against innocent customers.

As proof of this, consider the fact that rice grown in China is often heavily contaminated with lead, cadmium and mercury while rice grown in the USA is incredibly clean, with virtually zero levels of those same heavy metals. If heavy metals were "naturally occurring" in all rice, then levels would be the same no matter where rice is grown. However, this is not the case. This proves that much of the rice being grown in China is grown in areas which are heavily contaminated by industrial pollution containing heavy metals.

The Natural News Forensic Food Lab has also verified huge concentration differences in seaweed products, depending on where they are grown. Wakame grown near New Zealand has almost zero heavy metals, while Wakame grown near China is heavily contaminated.

Arsenic in Our Food Supply

Arsenic exists in both organic and inorganic forms. The organic and inorganic forms of arsenic represent the total amount of arsenic present in any matter being studied, whether it be food, water, soil, air or other materials. Inorganic arsenic (iAs) is the form most consistently raised as a health concern in the scientific community, and therefore the form most focused on when addressing arsenic levels in the food supply. It is important to note that both organic and inorganic forms of arsenic are naturally occurring and are differentiated simply by the molecules in which the arsenic is attached. In very simple terms, organic arsenic contains carbon and hydrogen, and inorganic arsenic contains other metals and elements, such as oxygen and sulfur. It has only been in the past few years that arsenic speciation (differentiation between organic and inorganic arsenic) in food has been possible.

Arsenic is present in the natural environment, and is the 55th most common element in the Earth's crust. Because arsenic is ubiquitous in the environment, it is unavoidable to have traces in our food supply given the obvious need for soil, air and water to grow crops needed for direct human consumption and for feeding livestock. Mining, the use of arsenical pesticides, and the use of ground water (rather than surface water) are potential contributors to the levels of arsenic in the soil, and subsequently in food.

Relative Contribution of Inorganic Arsenic Sources

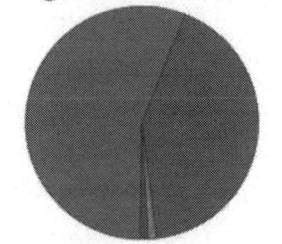

Food — 56%
Water — 42%
Soil — 1%
Air — 1%

FIGURE 24.

Arsenic is designated by the International Agency for Research on Cancer as a Class A human carcinogen. A U.S. Geological Survey (USGS) report on groundwater contamination, which includes well water and springs, found that arsenic is found just about everywhere. However, particular hotspots of contamination include New Jersey, Maryland and surrounding northeastern states, as well as Michigan, Idaho, California, Arizona and Washington. Other studies have also identified large swaths of the central South and Southeast.

The USGS study, published in 2000, shows concerning levels of arsenic greater than 10 micrograms per liter all across central Michigan and Idaho, as well as in California's Central Valley where the bulk of the nation's fresh produce is grown. Other highly contaminated areas, according to a USGS map, include Nebraska, South

Contribution of Inorganic Arsenic Intake by Foods

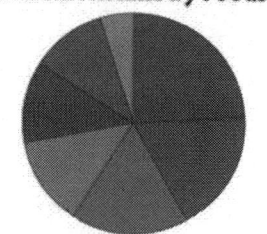

Vegetables — 24%
Fruit, Fruit Juices — 18%
Rice — 17%
Other — 13%
Beer, Wine — 12%
Flour, Corn, Wheat — 11%
Meat, Eggs — 5%

FIGURE 25.

Dakota, central Oklahoma and Illinois, and parts of Utah, Colorado and New Mexico.

Many public water supplies are also contaminated with arsenic, according to the report, and public utility filtration systems typically do not remove this toxin before delivering water to customers. This means that untold millions of people who drink unfiltered tap water are consuming arsenic daily.

Consumers Union, the public policy and advocacy arm of Consumer Reports, conducted laboratory tests of arsenic. Their testing, which included 200 samples of more than 60 different products, found inorganic arsenic at concerning levels in virtually every sample.

Exposure to Low-Level Arsenic is a Major Threat to Human Health

The U.S. Centers for Disease Control and Prevention (CDC) admits that even exposure to low levels of arsenic can cause major health damage over time. "Because it targets widely dispersed enzyme reactions, arsenic affects nearly all organ systems," says the CDC, noting that arsenic is also linked to gastrointestinal effects, renal damage, cardiovascular events, neurological damage, skin problems, anemia, leukemia, reproductive problems and cancer, including several types of skin cancer. "Arsenic can cause serious effects of the neurologic, respiratory, hematologic, cardiovascular, gastrointestinal, and other systems," adds the agency.

The Organic Trade Association research on arsenic revealed that the most commonly cited consequences of chronic exposure to low levels of inorganic arsenic include increased incidence of bladder, lung, kidney, and skin cancers, along with elevated levels of heart disease, skin hyperpigmentation, and skin lesions.

There is no standard for arsenic in foods. There is no arsenic testing requirements. Furthermore, there are no regulations placed upon food companies which mandate labeling or warning the public of arsenic dangers in their food products.

Other Toxic Food Considerations

The primary goal is to achieve and maintain a healthy, happy life, devoid of the chronic illnesses and conditions plaguing society today. However, this has become increasingly difficult considering the numerous problems with the current food system in the United States. That is why growing our own food, as well as using aquaponics, is so important.

The Food Industry

Understanding how the powerful $1.3 trillion food industry works today is critically important for making the necessary decisions to achieve and maintain a healthy life. For example, approximately 90 percent of the money Americans spend on food is spent on *processed* foods from large food manufacturers such as General Foods and Procter & Gamble. Their food marketers, employing some of the best and brightest minds to study consumer psychology and demographics, do a masterful job at targeting the right population and making it seem like their processed foods are the obvious, even healthy, choice. But not only are these processed foods "dead" and devoid of any natural nutrition, they can also be loaded with potentially carcinogenic substances.

The Media

Members of the media are subject to the same disinformation as the populace and thus are largely unaware of the problems in our current food system. A major funding source for the media, especially network television advertising revenue, comes from food and drug companies, making the information more unreliable, either promoting unhealthy and dangerous products, or failing to report on the issues with the trillion dollar industry. Celebrities and professional athletes often endorse products or participate in advertisements for processed food that they themselves may never or rarely consume, making the products seem more alluring.

The Government and the FDA

Like the media, our elected officials are consumers who are subject to the same disinformation and thus are often unaware of the health issues. Additionally, our politicians have been effectively controlled by the food and drug companies for so long that our government is now a large part of the problem, rather than being poised to be part of a solution. The Food and Drug Administration (FDA), although originally designed to protect consumers from unhealthy products, now often work diligently to protect the very companies it is supposed to regulate by keeping out competition and prolonging the economic life of the drug companies' government-sanctioned patents.

Sadly, government policy is now typically based on data supplied by the food or pharmaceutical industry, and lobbyist, rather than an unbiased source. A recent *USA TODAY* study found that more than half of the experts hired to advise the government on the safety and effectiveness of medicine and food have financial relationships with the companies that will be helped or hurt by their decisions. These experts are hired to advise the FDA on which substances should be approved for sale, and how studies should be designed. To date, the FDA has allowed more than 70,000 chemicals to infiltrate our food supply. Many of these dangerous ingredients are outright toxins, poisoning us, undermining our health and allowing cancer and diseases to enter our bodies.

Fish

Fish is high in protein, and full of essential nutrients and healthy fats. It is also easily digested, and assimilated, and full of essential nutrients and fats. Unfortunately, the vast majority of fish sold is imported from developing countries, much of which have proven to be contaminated with banned chemicals, poisons, carcinogens and high levels of antibiotics, often caused by being caught in contaminated waters.

However, imported fish is not the only fish that is likely to be bad for your health. According to a new U.S. Geological Survey study, scientists detected mercury contamination in every fish sampled in nearly 300 streams across the United States. More than a quarter of these fish were found to contain mercury at levels exceeding the criterion for the protection of people who consume average amounts of fish, established by the U.S. Environmental Protection Agency.

Exposure to mercury can damage your brain, kidney and lungs. Mercury is especially damaging to your central nervous system (CNS), and studies show that mercury in the CNS causes psychological, neurological, and immunological problems. Furthermore, mercury has an extremely long half-life in the human body that scientist believe is somewhere between 15 and 30 years. Additionally, non-dissipating medications, antibiotics, fungicides, pesticides, herbicides, untreated sewage, and other chemicals are also entering our waterways at alarming rates.

Farmed fish may not be the answer either. Studies show that farm-raised fish contain more polychlorinated biphenyl and over ten times the amount of dioxin.

Soy

Unfortunately, numerous so called experts and the media have extolled soy a healthy food. Supporters claim it as an ideal source of protein. It is also being recommended as a great tool to for lowering cholesterol, protecting against cancer and heart disease, reducing menopause symptoms, and preventing osteoporosis, as well as other things. However, studies based on sound research have found that soy products may increase the risk of breast cancer, cause brain cause damage, contribute to thyroid disorders, promote kidney stones, weaken the immune system and cause fatal food allergies.

One of the most disturbing of soy's ill effects on health has to do with its phytoestrogens that can mimic the effects of the female hormone estrogen.

Just moderate consumption of soy has been shown to alter a woman's menstrual cycle. This hormonal disruption is not healthy for men or women.

Processed Meat

Processed meats are not a healthful choice for anyone and should be avoided entirely, according to a recent review of more than 7,000 clinical studies examining the connection between diet and cancer! The report was commissioned by The World Cancer Research Fund (WCRF). Virtually all processed meats contain a well-known carcinogen: sodium nitrite. It's a commonly used preservative and antimicrobial agent that also adds color and flavor to processed and cured meats, but this additive can also cause the formation of nitrosamines in your system, which can lead to cancer. Hot dogs, deli meats and bacon are notorious for their nitrite content.

Besides nitrates, processed meats are typically far from being 100 percent beef contain some nasty filler and by-product materials, the least of which is chicken skin, chicken fat or other animal parts. It is also common for processed meats to contain MSG, high-fructose corn syrup, preservatives, and harmful artificial flavoring or artificial colors.

Charred Meat / Heated Foods

Heterocyclic amines, or HCAs, are compounds created in meats and other foods that have been cooked at high temperatures (above 212° F). The worst part of the meat is the blackened section, which is why you should always avoid charring your meat, or eating charred sections. Studies have shown that humans who consume large amounts of HCAs have increased risk of stomach, colon and breast cancers.

But HCAs aren't the only cancer-causing chemicals created during cooking. Polycyclic Aromatic Hydrocarbons (PAHs) are also extremely harmful. Many processed meats are smoked as part of the curing process, which causes PAHs to form. PAH's can also form when grilling. When fat drips onto the heat source, causing excess smoke, and the smoke surrounds your food, it can transfer cancer-causing PAHs to the meat.

Advanced Glycation End Products (AGEs) are also formed when food is cooked at high temperatures—including when it is pasteurized or sterilized. AGEs build up in your body over time leading to oxidative stress, inflammation and an increased risk of heart disease, diabetes and kidney disease.

Processed Food

When discussing processed food, the list of problems is lengthy, but a few specific issues should be discussed. The first is the contaminants in processed food. The FDA, in its long history of managing food safety, has established guidelines for acceptable quantities of a number of contaminants that it will allow in our food supply. Included on the list are various types of mold, flies, maggots, insect eggs or body fragments, rodent hairs and feces pellets, pus pockets, and larvae, to name just a few.

Another issue with processed foods is the hydrogenated oils used to lengthen the shelf life of products like crackers and cookies, which are also associated with diabetes and heart disease. They are also generally high in sodium, corn syrup and other unhealthy ingredients.

As with charred meats, researchers discovered that a cancer-causing and potentially neurotoxic chemical called acrylamide is created when carbohydrate-rich foods are cooked at high temperatures, whether baked, fried, roasted, grilled or toasted. The chemical is formed from a reaction between sugars and an amino acid (asparagine) during high-temperature cooking above 212°F. As a general rule, the chemical is formed when food is heated enough to produce a fairly dry and "browned" surface. Hence, it can be found in processed potatoes, grain products and even coffee.

Acrylamide is not the only hazard associated with heat-processed foods, however. The three-year long EU project known as *Heat-Generated Food Toxicants*1 *(HEATOX)*, identified more than 800 heat-induced compounds in food, 52 of which are potential carcinogens. For example, the high heat of grilling reacts with proteins in red meat, poultry, and fish, creating heterocyclic amines, which have also been linked to cancer.

Food Additives

More than 3,000 food additives, such as preservatives, flavorings, colors and other ingredients, are legally added to foods in the United States, most of which are unhealthy and often dangerous. Many of these additives have been linked to an increased risk of cancer, while recent studies have shown others are estrogen-mimicking xenoestrogens that have been linked to a range of hormonal health effects in males and females. Studies have also shown that a variety of common food dyes and preservatives cause some people to become measurably more hyperactive and distractible and can even do brain damage resulting in a significant reduction in IQ.

International Foods (China)

Generally speaking, quality control and employee health are not issues that Chinese industry or the Chinese government are overly concerned about. Without the regulation, China's products often contain plastics, pesticides, herbicides and other cancer causing chemicals.

According to Congressional testimony by Don Kraemer, Deputy Director of the Office of Food Safety at the FDA, *"The FDA has encountered compliance problems with several Chinese food exports, including lead and cadmium in ceramic-ware used to store and ship food, and staphylococcal contamination of canned mushrooms. While improvements have been made in these products, the safety of food and other products from China remains a concern for FDA, Congress, and American consumers."*

Artificial Sweeteners

All Artificial Sweeteners are toxic to the human body. They basically trick your body into thinking that it's going to receive sugar (calories), but when the sugar doesn't come your body continues to signal that it needs more, which results in carb cravings. Contrary to industry claims, research over the last 30 years—including several large-scale prospective cohort studies—has shown that artificial sweeteners stimulate appetite, increase cravings for carbs, and produce a variety of metabolic dysfunctions that promote fat storage and weight gain.

The most common artificial sweeteners are Aspartame *(NutraSweet, Equal)*, Acesulfame-K or Ace-K *(Sunett, Sweet One)*, Saccharin *(Sweet'n Low)*, and Sucralose *(Splenda)*. The list of side effects include but are not limited to headaches, chronic respiratory disease, dermatologic reactions, tachycardia, cancer and tumors, enlarged organs, and many neuropsychiatric disorders, including panic attacks, mood changes, visual hallucinations, manic episodes, isolated dizziness, memory impairment, nausea, temper outbursts, and depression.

Stevia isn't really an artificial sweetener because it's made from one of the extracts of the stevia plant (an herb); and is often marketed as a natural sweetener. However, Stevia can negatively interfere with absorption of carbohydrates and can further disrupt the metabolizing and conversion of food into energy.

Food Colorants

Americans are now eating 5 times more food dye than in 1955. There are serious hidden dangers of food colorants. Commonly, they can cause various tumors, allergic reactions, brain gliomas, immune system impairment, and hyperactivity. But with their widespread use, they can be found in baked goods, beverages, candies, cereals, treated fruit, ice cream, cosmetics and even dog food. So avoid foods with these common label ingredients: Blue #1, Blue #2,

Citrus Red #2, Green #3, Red #40, Red #3, Yellow #5 (Tartrazine), Yellow #6 or any other food dyes.

Preservatives

Preservatives are chemicals used to keep food fresh. Although there are a number of different types of food preservatives, antimicrobials, antioxidants, and products that slow the natural ripening process are some of the most common. Despite their function, preservatives can pose a number of serious health risks.

Cancer is a serious side effect associated with the use of preservatives. In fact, the National Toxicology Program reports that propyl gallate—a preservative commonly used to stabilize certain cosmetics and foods containing fat—may cause tumors in the brain, thyroid and pancreas. Similarly, InChem—an organization that provides peer-reviewed information on chemicals and contaminants—notes that nitrosames, including nitrates and nitrites, can lead to the development of certain cancer-causing compounds as they interact with natural stomach acids. Nitrosamines are found in a variety of foods, including cured meat, beer and non-fat dried milk.

Hyperactivity in children is another possible side effect associated with the use of preservatives. A study published in 2004 in "Archives of Disease in Childhood" noted a significant increase in hyperactive behavior in 3-year-olds who took benzoate preservatives. Children who were enrolled in the study also demonstrated a decrease in hyperactive behavior after they stopped taking benzoate preservatives. While benzoates can be found in a number of foods, they are often used to preserve acidic foods and beverages, like soda, pickles and fruit juice.

Some people may experience damage to their heart as a result of preservative use, reports InChem. Sodium nitrates can cause blood vessels to narrow and become stiffer. In addition, nitrates may affect the way the body processes sugar and may be to blame for the development of some types of diabetes, the Harvard School of Public Health notes.

Studies have shown and even the FDA has warned that other preservatives, such as Chlorphenesin and Phenoxyethanol, sulfites, BHA (Butylated hydroxyanisole) and BHT (Butylated hydroxytoluene) are toxic and in addition to the above named effects, can cause depression of the central nervous system, vomiting and diarrhea, severe allergic reactions including anaphylactic shock, stomach irritation, skin sensitivity, and even effects on the body's blood coagulation system.

Flavor Enhancers

Flavor enhancers are used in foods to enhance the existing flavor or modify flavors in the food without contributing to any significant flavor of their own. Food flavor enhancers are commercially produced in the form of instant soups, frozen dinners and most processed foods. Monosodium glutamate (MSG) is one of the most common flavor enhancers. The FDA has indicated flavor enhancers ingested in low doses are generally regarded as safe (GRAS); however, the list of dangerous side effects is extensive. Most are known to be carcinogenic to humans, but can also cause brain damage, neurological diseases, kidney problems, obesity, eye damage, headaches, depression, fatigue, disorientation, chest pain or difficulty breathing, and severe allergies.

High Fructose Corn Syrup (HFCS)

Food companies like to use high-fructose syrup, instead of traditional sweeteners because it costs less to make, is sweeter to the taste, and mixes more easily with other ingredients. The average American consumes nearly 63 pounds of high-fructose syrup a year. High-fructose corn syrup is commonly added to many processed foods and beverages. It maybe be listed on food labels as "corn sweetener," "corn syrup," or "corn syrup solids" as well as "high-fructose corn syrup".

Research shows that this liquid sweetener can upset the human metabolism, raising the risk for heart disease and diabetes. Researchers say that high-fructose corn syrup's chemical structure also encourages overeating. It forces the liver to pump more heart-threatening triglycerides into the bloodstream. In addition, fructose can deplete the body's reserves of chromium, a mineral important for healthy levels of cholesterol, insulin, and blood sugar. Additionally, almost half of tested samples of commercial HFCS contained mercury, according to a new study. Mercury was also found in nearly a third of 55 popular brand-name food and beverage products where HFCS is the first- or second-highest labeled ingredient. Mercury is toxic in all forms, and a poison to the brain and nervous system. At a minimum it causes extreme fatigue and neuro-muscular dysfunction

Food Labels

Food companies know that people are drawn to certain terms like "All Natural" and "Made with Whole Grains." They can make enormous profits if they can market their products as being healthy. "Natural" labeled food generated $22.3 billion in 2008, and 54 percent of all cereals are now labeled "whole grain," even processed, sugar-laden ones. Misleading food labels regularly dupe consumers with these keywords, questionable advertising practices, misleading packaging statements, and bold statements that appeal to people's dietary preferences and weight loss goals. Sometimes it is impossible to identify the ingredients in a product by reading the label. The FDA allows for ingredients that are present only in small quantities to be labeled as "artificial flavor", "natural flavor", or "spices". Only the manufacturer knows what all the ingredients are in the product. Unfortunately, the FDA does not regulate all food labels and cannot keep food manufacturers from using crafty wording.

Antibiotics in Food

Antibiotics are widely used in food-producing animals. Animals in factory farms are given hefty doses of antibiotics so that they can remain alive in stressful, unsanitary conditions, and to make them grow faster. Even many grazing animals not raised in farm lots are given antibiotics to help them grow faster. The practice of giving animals antibiotics greatly increases profit margins for livestock producers and pharmaceutical companies. As a result, according to the FDA in 2009 alone, a whopping twenty-nine million pounds of antibiotics were given to animals.

This use contributes to the emergence of antibiotic-resistant bacteria in food-producing animals. These resistant bacteria can contaminate the foods that come from those animals, and persons who consume these foods can develop antibiotic-resistant infections. Scientists around the world have provided strong evidence that antibiotic use in food-producing animals can have a negative impact on public health. Because of the link between antibiotic use in food-producing animals and the occurrence of antibiotic-resistant infections in humans, CDC encourages and supports efforts to minimize inappropriate use of antibiotics in humans and animals.

In addition to the emergence of antibiotic-resistant bacteria, common side effects from antibiotics include diarrhea, nausea, vomiting, fungal infections, and a variety of allergic reactions.

Fluoride/Chromium-6/Chlorine (Water Treatments)

Fluoridation is not legal or not used in the overwhelming number of countries including industrialized ones. In fact, 97 percent of Western Europe has chosen fluoride-free water. The United States is one of only eight countries in the entire developed world that fluoridates its water supply. It is added under the guise that it prevents and control tooth decay.

The fluoride added to our drinking water is actually chemical waste product. It is certainly not anything that should be ingested or included in our diet. The chemicals used to fluoridate water in the US are not pharmaceutical grade. Instead, they come from the wet scrubbing systems of the superphosphate fertilizer industry. These chemicals (90 percent of which are sodium fluorosilicate and fluorosilicic acid), are classified hazardous wastes and contaminated with various impurities.

Numerous studies have shown that fluoride is dangerous to our health. Five studies from China show a lowering of IQ in children associated with fluoride exposure. The Department of Health in New Jersey found that bone cancer in male children was far greater in areas where water was fluoridated. The U.S. Environmental Protection Agency (EPA) researchers confirmed bone cancer-causing effects of fluoride at low levels in an animal studies. Sadly, 23 human studies and 100 animal studies have linked fluoride to brain damage. Even the US Food and Drug Administration (FDA) has never approved any fluoride product designed for ingestion as safe or effective. Although they still allow it to be put into our drinking water.

Fluoride is Dangerous, but Chromium-6 (also known as hexavalent chromium) in your water supply may be far worse. Chromium-6 was detected in 31 of the 35 city water supplies tested and confirmed by a number of independent tests by various water utility companies. Hexavalent chromium is classified as "likely to be carcinogenic to humans" by the U.S. Environmental Protection Agency (EPA).

Chlorine is another noxious chemicals deliberately put in the water by public health officials. Chlorine is in the same chemical group as fluoride, which has been linked with cancer and osteoporosis. The chemical element chlorine is a corrosive, poisonous, greenish-yellow gas that has a suffocating odor and is 2-1/2 times heavier than air. It can result in arteriosclerosis, heart attack and stroke. Research has shown that individuals who consume chlorinated drinking water have an elevated risk of cancer of the bladder, stomach, pancreas, kidney and rectum as well as Hodgkin's and non-Hodgkin's lymphoma.

Pesticides/Herbicides

As discussed earlier in this chapter pesticides are widely used in producing food to control pests such as insects, rodents, weeds, bacteria, mold and fungus. Seven of the most toxic chemical compounds known to man are approved for use as pesticides in the production of foods. These toxins are referred to as Persistent Organic Pollutants (POP's). They're called persistent because they are not easily removed from the environment. The greatest risk to our environment and our health comes from the chemical pesticides. In spite of the dangers, the government maintains its approval of the use of toxic chemicals to make pesticides. And science is constantly developing variations of poisons.

Pesticides can be toxic to humans and animals. It can take a small amount of some toxins to kill, and other toxins that are slower acting may take a long time to cause harm to the human body. Even just using pesticides in amounts within regulation, studies have revealed neurotoxins can do serious damage. Several factors determine how your body will react including your level of exposure, the type of chemical you ingest, and your individual resistance to the chemicals. Some possible reactions are fatigue, skin irritations, brain and blood disorders, nausea and vomiting, liver and kidney damage, reproductive damage, breathing problems, cancer and even death.

Herbicides (weed killers) are mixtures of chemicals designed to spray on weeds, where they get inside the plants and inhibit enzymes required for the plant to live. The active ingredient in the most widely used herbicide is glyphosate, and some have the main component of Agent Orange. Until the introduction of GM crops about 20 years ago, herbicides were sprayed on fields before planting, and then only sparingly used around crops. The food that we ate from the plants was free of these chemicals.

In stark contrast, with herbicide resistant GM plants, the herbicides and a mixture of other chemicals (surfactants) required to get the active ingredient into the plant are sprayed directly on the crops and are then taken up into the plant. The surrounding weeds are killed while the GM plant is engineered to resist the herbicide. Therefore, the food crop itself contains the herbicide as well as a mixture of surfactants.

To accommodate the fact that weeds are becoming glyphosate resistant, thereby requiring more herbicide use, the EPA has steadily increased its allowable concentration limit in food, and has essentially ignored our exposure to the other chemicals that are in its commercial formulation.

As a result, the amount of glyphosate-based herbicide introduced into our foods has increased enormously since the introduction of GM crops. Multiple studies have shown that glyphosate-based herbicides are toxic and likely public health hazards.

New scientific studies link glyphosate to a host of health risks, such as cancer, miscarriages and disruption of human sex hormones. Other conditions with strong correlations are ADHD, Alzheimer's disease, anencephaly, autism, birth defects, brain, breast and other cancers, celiac disease, chronic kidney disease, colitis, depression, diabetes, heart disease, hypothyroidism, inflammatory bowel disease, liver disease, and many others.

Irradiation

Irradiation is the process of exposing fresh foods to low amounts of x-rays to sterilize and prolong its life. Commercialized food irradiation is done on a constant basis. Food growers and sellers say that food irradiation is safe. However, studies show the x-ray irradiation destroys vitamins A, E, and K. Some studies show that x-ray irradiation also destroys up to 85 percent of the vitamin C found in vegetables, meats and fruits. Irradiation also destroy the digestive enzymes in raw foods. Irradiation damages food by breaking up molecules and creating free radicals. When high-energy electron beams are used, trace amounts of radioactivity may be created in the food.

Some bacteria, like the one that causes botulism, as well as viruses are not killed by current doses of irradiation.

No one knows the long-term effects of a life-long diet that includes foods which will be frequently irradiated, such as meat, chicken, vegetables, fruits, salads, sprouts and juices. Studies on animals fed irradiated foods have shown increased tumors, reproductive failures and kidney damage. Some possible causes are: irradiation-induced vitamin deficiencies, the inactivity of enzymes in the food, DNA damage, and toxic radiolytic products in the food.

Bisphenol-A (BPA)

BPA is one of the biggest players in the wrapping industry. In 1999, more than 6 billion pounds of BPA was made, representing nearly $7 billion in sales. BPA is deeply imbedded in the products of modern consumer society, not just as the building block for polycarbonate plastic but also in the manufacture of epoxy resins and other plastics. BPA is routinely used to in plastic containers, water bottles, to line cans to prevent corrosion and food contamination. BPA has been used as an inert ingredient in pesticides, as a fungicide, antioxidant, flame retardant, rubber chemical, and polyvinyl chloride stabilizer. BPA contamination has become widespread in the environment in rivers and estuaries throughout the US. Unfortunately, BPA does not readily degrade in the environment.

What this all means is that we are constantly being exposed to BPA. A CDC (Centers for Disease Control and Prevention) study found 95 percent of adult human urine samples and 93 percent of samples in children had BPA. However, the FDA has approved the use of BPA and the EPA does not consider it cause for concern.

According to estimates, just a couple of servings of canned food can exceed the daily safety limits for BPA exposure in children. Even low-level exposure to BPA can be hazardous to one's health. There are more than 100 independent studies linking the chemical to serious health problems in humans. Some of the main problems are prostate cancer and breast cancer, diabetes and obesity, altered immune function, early sexual development in girls and disrupted reproductive function, learning and behavioral problems, including hyperactivity, abnormal heart rhythms and coronary artery disease, asthma, depression, diabetes, heart disease and reproductive disorders.

Growth Hormones

Today's hyper-productive animals are given injections and implants (in the case of cows) or genetic engineering (in the case of salmon), of artificially high levels of sex or growth hormones. This allows them to grow bigger, faster and produce better. Surprisingly little research has been done on the health effects of these hormones in humans, in part because it's difficult to separate the effects of added hormones from the mixture of natural hormones, proteins, and other components found in milk and meat.

However, many experts have valid concerns that these excess hormones in the food supply are contributing to cancer, early puberty in girls, and other health problems in humans. For years, consumer advocates and public health experts have fought to limit the use of hormones in cows, and some support a ban on the practice. Unfortunately, the FDA continues to approve of growth hormones in animals for human consumption, even though there has been minimal testing to determine whether or not there are any health consequences related to their use.

In 1993, the FDA approved recombinant bovine growth hormone (rBGH), a synthetic cow hormone that increases milk production when injected into dairy cows. Research has found that milk from rBGH-treated cows contains up to 10 times more IGF than other milk. Higher blood levels of IGF have been associated with more than a 50 percent increased risk of breast, prostate, and other cancers in humans.

IGF isn't the only hormone found in the food supply. Ranchers have been fattening up cattle with sex hormones, most notably estrogen, since the 1950s. Today most beef cows in the U.S, except those, that are raised 'organic', receive an implant in their ear that delivers a hormone, usually a form of estrogen (estradiol) in some combination with five other hormones. Even miniscule amounts of estrogen could affect prepubescent girls and boys. The majority of public health experts, health conscious nutritionist, and most registered dietitians urge consumers to stay away from rBGH-treated milk because of its potentially higher IGF levels.

Summary, Solutions, and Positive Approach

It is important that we not be duped by mainstream media, which receives millions of advertising dollars from food companies each year; or trust the government on this issue, since so much policy and government rhetoric is influenced by special interest groups, lobbying efforts, campaign contributions, political favors, and devious politics. Raising public awareness about the presence of arsenic in water and soils is beneficial. We also need to pressure the U.S. Food and Drug Administration (FDA) to set standards for arsenic in food, to prohibit the use of arsenic-containing drugs in livestock and poultry, and to limit the arsenic allowable in manure used on rice fields. A law requiring mandatory labeling of all GMO food is needed.

We need to remember that each dollar spent is a persuasive vote. We send a message every time we decide to purchase or not to purchase.

As individuals it is helpful to maintain a balanced and varied diet. Too heavy of a reliance on any particular food can have negative effects on one's health. By varying the types of food one eats, one minimizes the risk of any one food having undue deleterious impact on one's health.

The diets of infants, pregnant women and nursing mothers have significant influence at critical developmental stages of human growth. Individuals in these sensitive stages need extra protection against toxic food. We all need to take some responsibility in helping those that are less able in fighting for what is right and doing what we can to promote healthy food for all.

Obviously, aquaponics removes toxic food concerns. A comprehensive examination of plant species, crop options, associated considerations, and food growing recommendations are addressed in later chapters of this book. Being knowledgeable about our toxic food supply will enable you to take better care of your health, provide further motivation for growing your own food, and enable you to better market your aquaponic and home-grown harvest products.

CHAPTER 6

The Different Types of Aquaponic Systems

Overview

The three primary methods used in aquaponics are Raft/Deep Water Culture (D.W.C.), Nutrient Film Technique (N.T.F.), and the Flood-and-Drain System. Each of these methods is based on a hydroponic system design with accommodations for fish and filtration.

There are many different aquaponic system configurations. Regardless of the different setups there are two components common to every aquaponic system. These two components are the fish tank and a soil-free plant-growing device. Depending on the type of aquaponics system referenced, the plant growing device is either a conduit, grow bed, or container. The variables include filtration unit, plumbing components, the plant growing device, and the amount and frequency of water circulation and aeration. Generally speaking, systems that utilize a filtration unit to remove the solid fish waste are the N.T.F. system and the D.W.C. system. However, only Flood-and-Drain systems which have an undersized plant grow bed need supplemental filtration; as the grow bed serves as the filtration unit.

The Flood-and-Drain system is also referred to as an Ebb-and-Flow system or a Media-Filled Bed system. As previously mentioned, there are many different aquaponic arrangements, but they all fall within one of the three primary types of aquaponic system categories listed above.

Below is a brief introduction to these three main aquaponic methods. The latter chapters of this book provide detailed information and design options for each method. The chapters in between, and following the chapters with design details, will provide you with everything you need to know to be successful in aquaponics.

Nutrient Film Technique

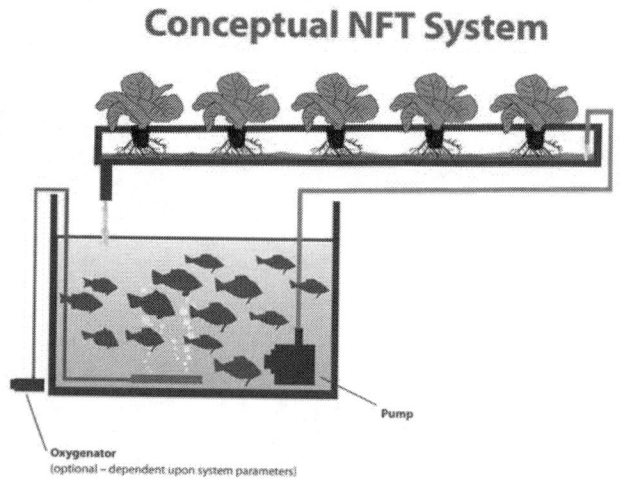

FIGURE 26. N.F.T. System.

FIGURE 27. Cross-Section of a N.F.T. System.

The Nutrient Film Technique (N.T.F.) is a commonly used hydroponic method. As a matter of fact the N.F.T. system is one of the most productive hydroponic systems. However, it is not as widespread in aquaponics. In N.F.T. systems, plants are grown in nutrient-rich water pumped down small, enclosed gutters (long narrow channels). The water flowing down the plant channel is only a very thin film, but it provides the plant roots with an abundant supply of essential nutrients, water, and oxygen. Plants are typically grown in small plastic cups, called net cups, allowing their roots to access the water and absorb the nutrients. In some N.F.T. systems, additional water filtration is necessary (discussed in more detail in later chapters).

CHAPTER 6: THE DIFFERENT TYPES OF AQUAPONIC SYSTEMS

Raft/Deep Water Culture (D.W.C.)

FIGURE 28. Deep Water Culture Raft System.   *FIGURE 29. Deep Water Culture Raft System.*

Deep Water Culture (D.W.C.), also referred to as a 'raft system', is done by floating a foam raft (typically a polystyrene board) on top of the fish tank water surface while allowing the roots to hang down into the water. However, another method is to grow the fish in a fish tank and pump the water through a filtration system, then into long channels, or a reservoir, where floating rafts filled with plants float on the water surface and extract the nutrients. Water flows continuously from the fish tank, through filtration components, through the raft tank where the plants are grown and then back to the fish tank. This method is the one most commonly practiced in large-scale commercial operations; but can also be done on a smaller scale.

Flood-and-Drain System

In the Flood-and-Drain System (also referred to as a media-filled bed system or an ebb-and-flow system) the media filled beds use containers filled with rock medium, expanded clay pellets or other similar media. This type of system can be run two different ways: (1) with a continuous flow of water over the rocks; or (2) by repeatedly flooding and draining the grow bed, in a flood and drain or ebb and flow cycle (most common and the recommended option).

This type of aquaponics system is most preferred for those desiring to grow a wide variety of plants, and/or taller plants such as tomatoes and bell peppers. It is also the simplest and most reliable design for a beginner.

With the media filled bed system, all waste, including the solids, is dispersed and eventually broken down within the plant grow bed. (NOTE: media grow bed and plant grow bed are the same, and used interchangeably throughout this book) In some systems, worms are added to the media-filled plant grow bed to enhance the break-down of the waste. The plant media bed serves as a natural water filter that gives the plants everything they need for vibrant growth, while returning filtered water back to the fish tank. This method uses the fewest components and no additional filtration, making it simple to operate. The media-filled bed is used for applications where greater plant diversity is desired, and whereby profit margin is not the primary goal. Not to say that a profit cannot be generated, but that in regards to looking at aquaponics from only a business viewpoint, N.F.T. and D.W.C. offer some revenue advantages to large scale operations. For instance, mass producing lettuce works best in N.F.T. and D.W.C. systems. However, a media-bed (flood-and-drain) system can still be extremely profitable if properly set-up and managed effectively.

Another benefit to a flood-and-drain system is that it can easily be scaled. In other words, it is relatively easy to continue enlarging it over time as you gain experience or develop a desire to produce more product i.e., fish and vegetables. Simply just add more fish tanks and grow beds.

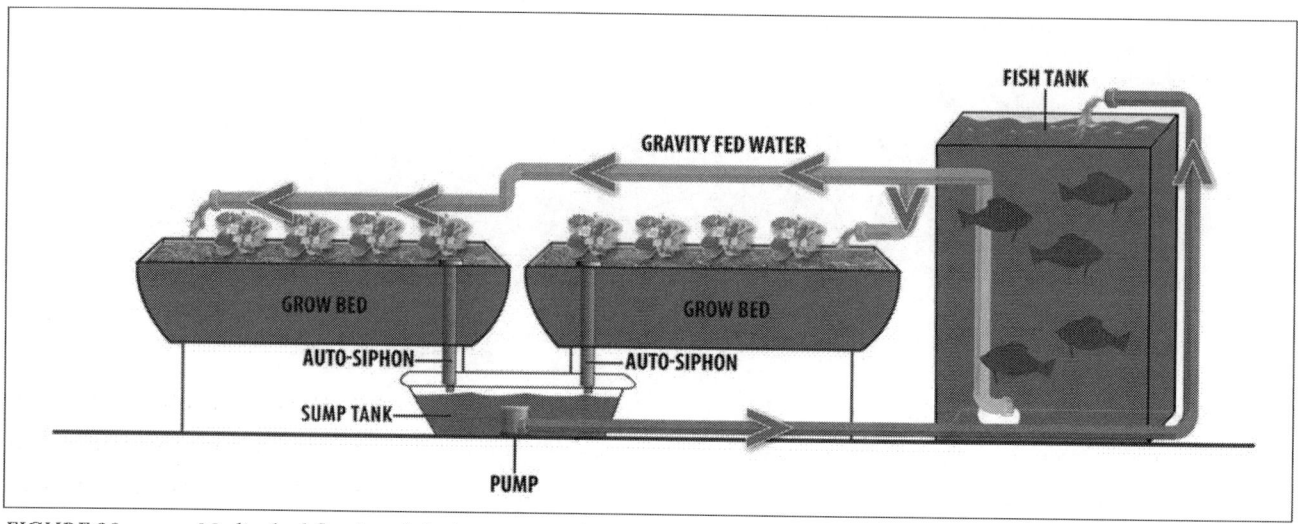

FIGURE 30a   *Media-bed flood-and-drain system with a sump tank.*

FIGURE 30b   *Basic Flood & Drain System. Note: Having the pump located in a sump tank, instead of in the fish task, is the ideal arrangement.*

CHAPTER 7

Aquaponics for You

Type of System Best Suited for You

The previous chapter provided a basic introduction into the three main types of aquaponics systems. These three types of aquaponic systems can be set up in a multitude of ways. As a matter of fact, it is rare to find two identical aquaponic systems. Nevertheless, the vast majority of systems fall within these three methods. In the following chapters of this book you will be provided more detailed information to equip you with the knowledge you need to be successful in whatever type of aquaponics system you decide to pursue, including set-up instructions and design details.

Aquaponics can be as simple or as comprehensive as you want to make it. For instance, you can take a small, flat piece of polystyrene board, cut holes in it, stick some plant cuttings (water cress) in the holes, and float it on the surface of an aquarium or a goldfish filled tote box to have a D.W.C. system. With minimal effort you will gain clearer water for your fish, and you will be able to harvest a surprising amount of edible plants. It is also worth noting that, although not as common, you can also mix different types of systems together to make one hybrid system. Some aquaponic operators have combined N.F.T. and D.W.C. into one system. A D.W.C. raft plant board can also be maintained on a flood-and-drain fish tank. Various multi-use aquaponic systems will be addressed later in this book.

Before deciding on the type and size of system to create, it is extremely important to identify your primary goal for getting into aquaponics before moving forward. As mentioned in the previous chapter Nutrient Film Technique (NFT) and Deep Water Culture (Raft System) is best suited for larger scale commercial applications where profit is the main objective. It is not ideal for all plants, nor is it the best approach for feeding your family, as diversity is limited. On the other hand, aquaponic professionals and hobbyists are all in agreement that the flood-and-drain media-based system is by far the most reliable and simplest method of aquaponics, especially for beginners, as plant and fish life-support components are more forgiving. It can be done very simply using a wide range of containers. The flood-and-drain media bed system also requires the least amount of maintenance. Best of all, it can support a much wider variety of plants. It can also be very profitable. This chapter was written to help you determine which type and size of aquaponics system is best suited for you.

Getting Reliable Information

Sound principles need to be applied when planning and sizing your own aquaponics system. Once the type of aquaponic system is determined, the most challenging aspect for many aspiring aquaponic operators is sizing the fish tank(s) and grow bed(s), as well as deciding how many fish and plants to cultivate. It can often take months, or even years, to research all this information, checking out various internet sites, participating in online forums, watching "how to" videos, reading books, and/or joining an aquaponics club. This can be a mind boggling process.

Aquaponics operators often suggest different numbers and methods based on their own experience, which adds to the confusion. Worse yet, there are many well-meaning aquaponic enthusiasts out there who give out bad advice and make recommendations that provide inefficient or unfavorable results. Getting sound advice from the right people can be costly. Obtaining consulting advice, taking university courses, or going through a training program is a considerable investment of money and time. Some programs can cost thousands of dollars and take months or years to complete. The good news is all the information you need to set up and operate a successful aquaponics system are provided in this book. Although we will dig deeper into many of these issues in following chapters, this chapter addresses very foundational issue that needs to be considered first.

Your Personal Endeavor

From the previous chapters it is obvious to see why aquaponics is certainly a worthwhile enterprise. Besides the fact that many well-renowned economists are predicting an economic crisis in the not-too-distant future, as well as an associated hyperinflation of food prices, raising your own healthy food is already extremely beneficial, considering all the harmful substances in our food these days: GMOs, pesticides, herbicides, fungicides, heavy metals, preservatives, coloring chemicals, etc. compared to homegrown organic food, which has a much higher nutrient density content than conventional grocery store produce, it is easy to understand the benefits. In addition, the satisfaction obtained from eating your own home-grown healthy food is priceless. Furthermore, as noted previously, aquaponics is much better for our world, environmentally, than traditional agricultural methods on so many levels.

Aquaponics can also be a wonderful family affair, with members sharing in the bettering of the family's well-being. Aquaponics is a wonderful activity for children. It teaches them many of the valuable lessons associated with the operation, such as the care of plants/fish, life cycle processes, water circulation, water characteristics, harvest, work ethic, responsibilities, fundamental business skills, organization, planning, etc.. Aquaponics is indeed a rewarding win-win for the entire family.

It has been determined that most aquaponics systems typically pay for themselves within 2.5 years. Obviously, this time could be less or more, depending upon the unique situation, but nevertheless, that is great news. All the while, you will be enjoying healthier food and making a positive contribution to your family, the community, and the environment as a whole.

There is no doubt that aquaponics is a positive endeavor. However, before moving forward it is important that you clearly define the reason(s) you want to get into aquaponics and what you desire to get out of it. Is it to be a means of supplementing your dietary needs, be the primary source of your diet, a means of providing for your family, or a method in which to acquire a bartering resource? Do you desire it to be a profitable business? If so, are you just after additional income, or do you desire it to be your primary source of revenue? Do you want it to be a family bonding affair and/or a means in which to teach children many valuable lessons? What best describes your desire to get into aquaponics: environmental reasons, social notoriety, health reasons, lower your food cost, or just as a hobby? Although, you may have several different reasons for getting into aquaponics it is critical that you clearly understand those motivations, and which one(s) are most important to you.

S.M.A.R.T. Goals

Therefore, before moving forward, it is important that you clearly define your goals and objectives. Only then can you create an effective plan that will produce the results you desire. Such is referred to as a strategic plan, and it should not be taken lightly. A strategic plan is your plan for success. It will define your mission, your present situation, where you want to be in three to five years, and help you get there efficiently. "If you fail to plan, you are planning to fail!"—Benjamin Franklin

Clearly defining what you desire to get out of aquaponics, putting together a realistic schedule, and defining how it is to be achieved is best accomplished through a process referred to as process called 'S.M.A.R.T.' goals. Writing out your S.M.A.R.T. goals creates a plan that promotes success in the least amount of time.

Defining your S.M.A.R.T. goals should be a process adopted early on in the aquaponic development process, preferably before investing any money; and your goals should be an integral part of ongoing operations. They can be refined as you grow and learn along the way. They also work well with all other endeavors in your life. In summary, the S.M.A.R.T. goals process is a clearly defined plan (in writing) that is consist of the following components:

S = Specific
M = Measurable
A = Achievable
R = Realistic
T = Time Bound

Specific — consider who, what, when where, why and how in developing the goal.
- What: What do I want to accomplish?
- Why: Specific reasons, purpose or benefits of accomplishing the goal.
- Who: Who is involved?
- Where: Identify a location.
- Which: Identify requirements and constraints.

EXAMPLE: A general goal would be, "Develop an aquaponics system." However, a specific goal would say, "Develop an aquaponic system in my backyard that will produce 'X' amount of vegetables and 'Y' amount of fish annually in order to provide for my family, earn 'Z' amount of revenue, ensure I have a realizable supply of pure organic non-adulterated food, etc..

Measurable — a numeric or descriptive measurement.
- How much?
- How many?
- How will I know when it is accomplished?
- You should be able to track and measure your progress along the way. Each measured success builds momentum and motivation by informing you of your current status as you make progress in the right direction.

Achievable — consider the resources needed and set a realistic goal.
- How: How can the goal be accomplished?

The goals that you set must also be attainable, which means that it *can* be achieved. Some things are just impossible to attain (world peace). However, most things can be achieved if you are strongly determined and put your heart and mind to it. History has proven this to be the case many times.

Big, long-term goals can be set if one is truly confident they can be achieved. However, one must also realize that success is achieved only by hard work, sacrifice, discipline and perseverance. Dream and plan for your long-term goal, but pursue many short-term goals along the way that are attainable at your current capability, skills and knowledge.

Relevant — make sure the goal is consistent with the mission (Result-Based, Results-oriented, Resourced, Resonant, Realistic)
- Does this seem worthwhile?
- Is this the right time?
- Does this match other efforts and needs in my life?
- A realistic goal means the goal is realistic when considering the other aspects and demands of your life. It is inherently related to the *attainable* aspect of the goal.

EXAMPLE: Starting and building a profitable business is certainly an attainable goal; however, if you are unable to carve extra time out of your schedule to pursue the goal, then it is not a realistic goal. Sometimes it is more important to first take control your current situation before taking on soemthing else. So even if a goal is ultimately attainable, it may not necessarily be realistic. It is up to you to decide whether you can make it happen or not. And if not, what can you do different in your life to make it happen?

Time-bound— set a realistic deadline.

(Time-oriented, Time framed, Timed, Time-based, Time-boxed, Time-bound, Time-Specific, Timetabled, Time limited, Trackable, Tangible)
- When?
- What can I do 6 months from now?
- What can I do 6 weeks from now?
- What can I do today?
- Without any timeframe, procrastination is likely; as there is no sense of urgency to take action to achieve the goal. When setting the timeframe, you should not be overly ambitious and set unrealistic timeframes. You also do not want to set too long a timeframe as that will make you complacent and lose momentum. *What you want is to set a timeframe that gives you the right amount of positive pressure to push you.*
- Set an overall timeframe for achieving the goal, as well as other shorter timeframes for achieving the short- and medium-term milestones and targets of the goal.

Finishing Strong

Personally, I have found that doing at least one thing every day to achieve each of my long-term goals works well for me. Life is full. Furthermore, we are often bombarded with distractions and unplanned interruptions. However, keeping to a plan of doing at least one thing each day (no matter how small) will help in remain on track and moving forward to the ultimate goal. That one thing could be as simple as making a telephone call, looking something up on the internet, making a purchase, organizing a file, etc. It may be a small step, but it is one less step that has to be taken on the journey to achieving the goal.

Whatever your desire, use S.M.A.R.T. goal setting, and you will not go wrong. It is easy to remember and very effective at the same time.

For more information and assistance on setting goals, refer to Appendix 11, Encouragement & Keys to Success.

"Failing to plan is planning to fail"
— ALAN LAKEIN

CHAPTER 8

Location, Location, Location

Selecting the Proper Aquaponics Location

The most successful aquaponics operations will meet the following "location" criteria: Location, location, location.

Plant/Crop Parameters

Plants need a minimum of six hours of sunlight a day to grow well. Artificial lighting can be expensive. Therefore, you should account for sunlight availability during the planning and designing stages. Also, winter and summer season sunlight paths need to be considered when laying out the system

Fish Parameters

Fish prefer to be in the shade. The fish tank should never be under direct sunlight, as it can cause increased temperatures and/or promote algae blooms in the tank. Algae blooms can cause a lack of oxygen which, in turn, could kill the fish. Placing the tank under a roof or covering the tank will address this issue. Also, growing floating plants on the surface of the fish tank will provide, shade, shelter, and hiding places for the fish. Hiding places make fish feel more secure and amounts to less stress on them.

Trees

The aquaponic system should not be set up under trees. Trees drop leaves, sticks, flowers, sap, and insects. Droppings from some tree species can alter pH levels or even be toxic to your fish. And obviously, trees block sunlight which would affect plant growth in the media beds.

If the aquaponics system is located outdoors, it is a good idea to have a protective cover or shelter to shield the water from debris and contaminants. Clean water is essential for the well-being of fish and plants.

Space Allowance and Accessibility

The aquaponics system should not be adjacent to a fence or wall, and the fish tank should not be located between the grow beds. Sufficient space for planting, harvesting and maintenance is a must. Easy or convenient access to the entire aquaponics system will save much time and hassle.

Also, keep in mind that when the grow beds are full of plants, it will be much more difficult to reach the back corners than it is when the grow bed is empty. In addition, plants typically grow over the edge of the grow beds, so take such in to consideration when setting up your system. As a general rule of thumb, it is a prudent plan for a minimum distance of 28-inches (70 cm) between grows beds.

It is also very important to have good accessibility to the fish tank for feeding, checking water quality and the health of the fish, as well as harvesting the fish. Since fish tanks are heavy, it is essential that they be properly supported or positioned on solid flooring i.e., secure ground, concrete, etc. Elevating fish tanks can be costly and complicated, as strong support structures are necessary.

Some operators try to maximize space by placing the fish tank directly under the grow beds. This is not a preferred approach as it typically limits access to the fish tank and puts the grow bed at heights inconvenient to maintain.

Power Considerations

An aquaponics system needs to have access to power for the pump(s) in order to keep the water circulating. Lighting is not always required, but certainly beneficial. Other items for power consideration are temperature management and supplemental grow lights. Power needs for lighting, grow lights, and temperature management is dependent upon the particular aquaponic system set-up, location, and the operator's goals.

Consideration should also be given to for the possibility of long-term power outages due to a storm event, or other infrastructure failure. It is prudent to have a readily available secondary power source, such as a generator, available in the event the power goes out for an extended time. This will ensure the life supporting pump(s) can continue operating and fish kill risk is minimized.

Some operators prefer to have their aquaponic system off the grid. As a result they obtain their electrical needs system via solar or wind power. Alternative power systems are covered in detail later in this book.

Heating and Insulation

Different fish species have their own optimal temperature ranges, but they all tend to prefer uniform water temperatures. Depending on the location of the aquaponic operation, and the type of fish being raised, it may be necessary to have a submersible heater (which requires electricity) to keep the water temperature within the optimal range. Some operators use heat wire around their water line(s).

Heating is not required if the fish species being raised has an optimal temperature range within the same temperature range as the location of the aquaponics system. However, even if located in the perfect climate, a heater can still be considered in order to maintain uniform water temperature so as to promote the growth of the fish and efficiency of the beneficial bacteria.

Plant growth is sped up due to increased availability of nutrients. Hence, even though the cost of electricity is higher, there can be increased yields which results in greater profitability.

If there is a potential for significant temperature fluctuation as a result of location, then insulation and submersible heaters to keep the water temperature constant and within optimal range is necessary; otherwise, the growth of fish can be stunted, or they may even die.

Ergonomically Correct

It is also helpful to plan a system that is ergonomically correct. In other words, the top of the media beds, top of the fish tank, equipment and devices should be designed to fit the human body and its correct functional abilities. Such will maximize productivity by reducing operator fatigue and discomfort. An ergonomically correct layout will not only be much more convenient and help optimize production, but it will make the entire aquaponics endeavor a much more enjoyable experience.

If the top of the fish tank has to be high, then consider building a platform to stand on. The fish tank can also be partially imbedded in the ground so that the top working surface is at a lower user-friendly height. Grow beds should be positioned so that bending over or reaching high positions are minimized. Keep in mind some vegetables will grow tall, and there will be a need to work within the grow bed media from time to time. Step ladders come in handy occasionally, for instance when harvesting the tops of tall vegetables such as tomatoes growing above the grow bed. However, the system should be designed so that the use of a step ladder is not a regular requirement.

Also, keep in mind that walking surfaces should be safe, free of tripping hazards, and have a non-slip surface, even when wet. The design height of all aquaponic components should can be based on a chart, or architectural standard (as will be presented in the system design chapters later in this book), but an even better approach is to construct the system specifically to the operator's body height—what is most convenient for the end user.

Storage Spaces

There are a lot of materials, tools, and supplies (piping, tubes, cables, fish food, gardening appurtenances, etc.) needed to keep an aquaponics system running smoothly. Easy accessible storage spaces nearby comes in very handy, as there will be frequent trips to this area.

Greenhouses and Net Houses

One of the benefits of having the entire system within the same greenhouse is that the fish tank is a wonderful source of thermal heating in the winter and thermal cooling in the warmer months. Many plants do better when they are in an environment with a higher humidity.

However, there is no hard and fast rule as to how the system must be set up. In some parts of the world everything can be located outside. Some operators have the fish tank and grow beds in separate buildings. Other operators may have part of the system outside and the other part inside. It really depends upon your location, space parameters, the type of plants and fish you desire to grow, and how important energy efficiency is to you. Several later chapters in this book discuss greenhouses in more detail (i.e. various types, construction, energy management, etc.).

A greenhouse is a metal, wood or plastic frame structure that is covered by transparent nylon, plastic or glass. The purpose of this structure is to allow sunlight (solar radiation) to enter the greenhouse and then trap it so it begins heating the air inside the greenhouse. As the sun begins to set, the heat is retained in the greenhouse by the roof and walls, allowing for a warmer and more stable air temperature during a 24-hour period. Greenhouses provide general environmental protection from wind, snow and heavy rain. Greenhouses extend the growing season by retaining ambient solar heat, but can also be heated from within. They can keep away animals and other pests, while seringe as some security against theft. Greenhouses are comfortable to work in during colder seasons and provide the grower with protection from the weather. Greenhouse frames can be used to support climbing plants or to hang shade material. Together, these advantages of a greenhouse result in higher productivity and an extended cropping season.

However, these benefits need to be balanced against the drawbacks of greenhouses. The initial capital costs for a greenhouse can be high depending on the degree of technology and sophistication desired. Greenhouses also require additional operating costs, because fans are needed to create air circulation to prevent overheating and overly humid conditions. Some diseases and insect pests are more common in greenhouses and need to be managed accordingly (use of insect nets on doors and windows), although the confined environment can favour the use of certain pest controls.

In some tropical regions, net houses are more appropriate than conventional greenhouses covered with polyethylene plastic or glass. This is because the hot climates in the tropics or subtropics raise the need for better ventilation to avoid high temperatures and humidity. Net houses consist of a frame over the grow beds that is covered with mesh netting along the four walls and a plastic roof over the top. The plastic roof is particularly important to prevent rain from entering, especially in areas with intense rainy seasons, as units could overflow in a matter of days. Net houses are used to remove the threat of many noxious pests associated with the tropics, as well as birds and larger animals. The ideal mesh size for the four walls depends on the local pests. For large insects, the mesh size should be 0.5 mm. For smaller ones, which are often vectors of viral diseases, the mesh size should be thicker such as mesh 50. Net houses can also provide some shade where the sunlight is too intense. Common shade materials vary from 25 to 60 percent sun block.

Rooftop Aquaponics

Flat rooftops are often suitable sites for aquaponics because they are level, stable, exposed to sunlight and are not already used for agriculture. However, when building a system on a rooftop it is crucial to consider the weight of the system, and whether or not the roof is capable of supporting it. It is essential to consult with an architect or civil engineer before building a rooftop system. In addition, be sure that materials can be transported both safely and effectively to the rooftop site.

Expansion and Transitioning into a Profitable Business

Ideally, when planning your aquaponics system your designed layout will be such that it provides you the flexibility to grow as you gain experience and are ready to scale up. It is easier to manage, operate, trouble-shoot problems when things go wrong with a smaller system compared to that of a large aquaponics operation. This is especially important when you are still relatively new to aquaponics. Furthermore, any errors with a smaller system typically do not hurt as much in regards to cost and time. However, even mistakes are wonderful opportunities to learn and improve, so they should not be viewed entirely as a bad thing, as they can be excellent teachers. Problems in a smaller system are just easier to recover from. Therefore, starting small and growing as you learn is usually the best approach. Just keep system expansion in mind when you are planning your design layout. Designating and preserving an area for future expansion, during your initial planning phase, will save you a tremendous amount of trouble, expense, and time later, compared to the complexity of relocating or rearranging an existing aquaponic system.

If your desire is to transition into a profitable aquaponic business operation, keep in mind that profit typically progresses with experience. The more you learn, the more you will earn.

FIGURE 31. A small media bed unit on a rooftop

FIGURE 32. Variety of vegetables growing on a rooftop in nutrient film technique systems

CHAPTER 9

Nitrogen Cycle

Nitrogen Cycle Explained

Nitrogen is an essential fundamental element necessary for all forms of life on Earth. It is an important constituent in both plant and animal cells. Living organisms need nitrogen to produce proteins, nucleic acids, and amino acids. Although Nitrogen gas (N2) is roughly 78 percent of the earth's atmosphere, it is basically unusable in its natural form. The vast majority of living organisms on earth can only use nitrogen when it is in its 'fixed' state: combined with carbon, hydrogen or oxygen.

One of the most important, yet least understood, aspects of aquaponics is bacteria, and its function in the nitrogen cycle. Basically, fish excrete ammonia. In a lake or ocean the vast volume of water dilutes this ammonia, which is eventually converted by bacteria in those habitats. In a fish tank ammonia must be dealt with properly as it is very toxic to the fish. Decomposing food also creates ammonia. Some effects of excessive ammonia include:

- Extensive damage to tissues, especially the gills and kidneys
- Impaired growth
- Decreased resistance to disease
- Death

Bacteria are the microscopic organisms that are involved in the conversion of fish waste into nutrients for the plants. It is important for aquaponic operators to understand how to create a healthy environment

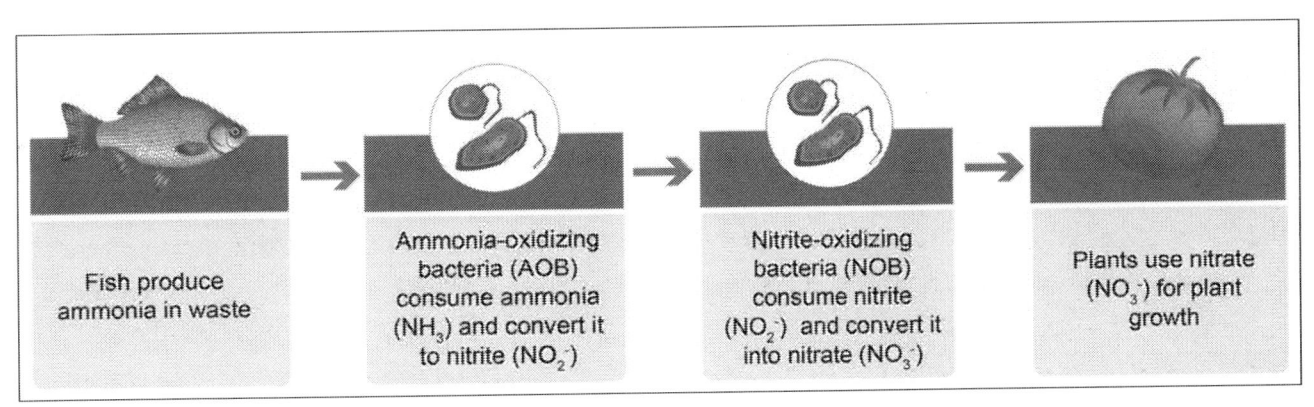

Fish produce ammonia in waste → Ammonia-oxidizing bacteria (AOB) consume ammonia (NH_3) and convert it to nitrite (NO_2^-) → Nitrite-oxidizing bacteria (NOB) consume nitrite (NO_2^-) and convert it into nitrate (NO_3^-) → Plants use nitrate (NO_3^-) for plant growth

FIGURE 33. *The nitrification process in an aquaponic system*

for the bacteria, so they will thrive within the system. A healthy colony of bacteria is essential for an aquaponics system to be successful. A thriving system will contain enough bacteria to break down and convert all the fish waste into nutrients for the plants.

The following factors outline how this process occurs within the aquaponics system. The fish in the tank are fed; they digest and break down the food and produce waste. Fish excrete ammonia through urine, feces (approx 17 percent) and their gills (approx 80 percent). The nitrification cycle is the process by which the ammonia produced by the fish is converted by one type of bacteria to nitrite and then by another type of bacteria to nitrate (the most plant-accessible form of Nitrogen in the cycle—and safest for the fish).

To elaborate further, in regards to more scientific explanation, the Nitrosomonas sp bacterium eats ammonia and converts it to nitrite. Nitrite is much less poisonous to the fish than ammonia, but it is by no means a good thing. It stops the fish from taking up oxygen and will cause damage to their gills. However, another bacterium (Nitrobacter sp.) eats nitrite and converts it to nitrate. Nitrate is a very accessible nutrient source for plants As a matter of fact, plants thrive off nitrate. In addition, the fish will tolerate a much higher level of nitrate than they will ammonia or nitrite. This is referred to as the nitrogen cycle.

When an aquaponics system has sufficient numbers of these bacteria to completely process the ammonia and nitrites, it is said to have "cycled." This process generally takes about a month; however, it can happen much quicker or much slower, depending on environmental conditions associated with the system. The bacteria will increase their numbers (reproduce) as the ammonia load increases, causing an ammonia "spike" when setting up a new tank.

The nitrogen cycle is the most significant process within aquaponics. It is responsible for the conversion of fish

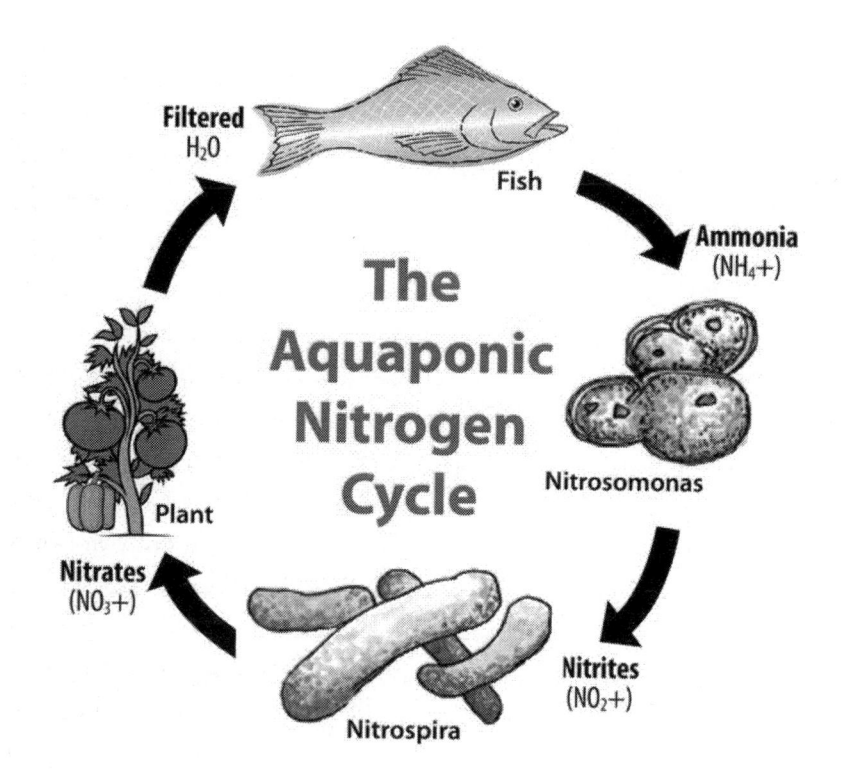

FIGURE 34.

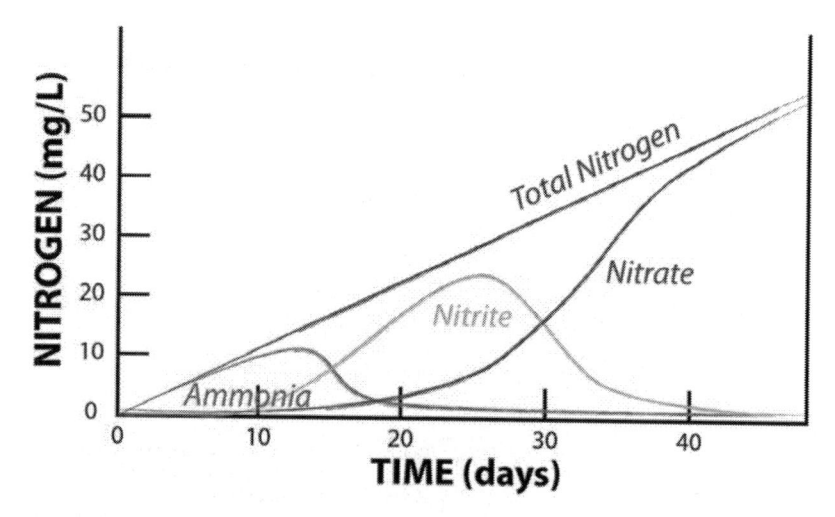

FIGURE 35.

CHAPTER 9: NITROGEN CYCLE

waste into nutrients for the plants. Without this process, the water quality would deteriorate rapidly and become toxic to both the fish and the plants in the system. Therefore, the water in aquaponics does not need to be treated chemically to make it "safe," nor does it have to be replaced. In aquaponics, a system is said to have "cycled" when there are sufficient quantities of bacteria to convert all the ammonia into an accessible form of nitrogen for the plants. The bacteria will arrive naturally to a system and colonize the media bed.

A pH of 6.8 to 7 is usually a happy compromise for most fish, plants, and the good bacteria, but again the pH can be slightly more or less depending upon the fish species The nitrogen cycle has a tendency to reduce pH; however, it is fairly easy to keep pH around 7 through the addition of calcium carbonate. Calcium carbonate increases pH but will stop dissolving when the pH level reaches around 7.4, meaning pH will stay fairly stable until all of the available calcium carbonate is depleted. See the section about pH in this book on using calcium carbonate.

With a NFT or DWC system a mechanical or biofilter may be needed to help remove excess waste. No filtration is needed for a properly setup and operating flood-and-drain system.

Maintaining a Healthy Bacterial Colony

The major parameters affecting bacteria growth that should be considered when maintaining a healthy biofilter in DWC and NFT systems and the grow bed of a flood-and-drain system are adequate surface area and appropriate water conditions.

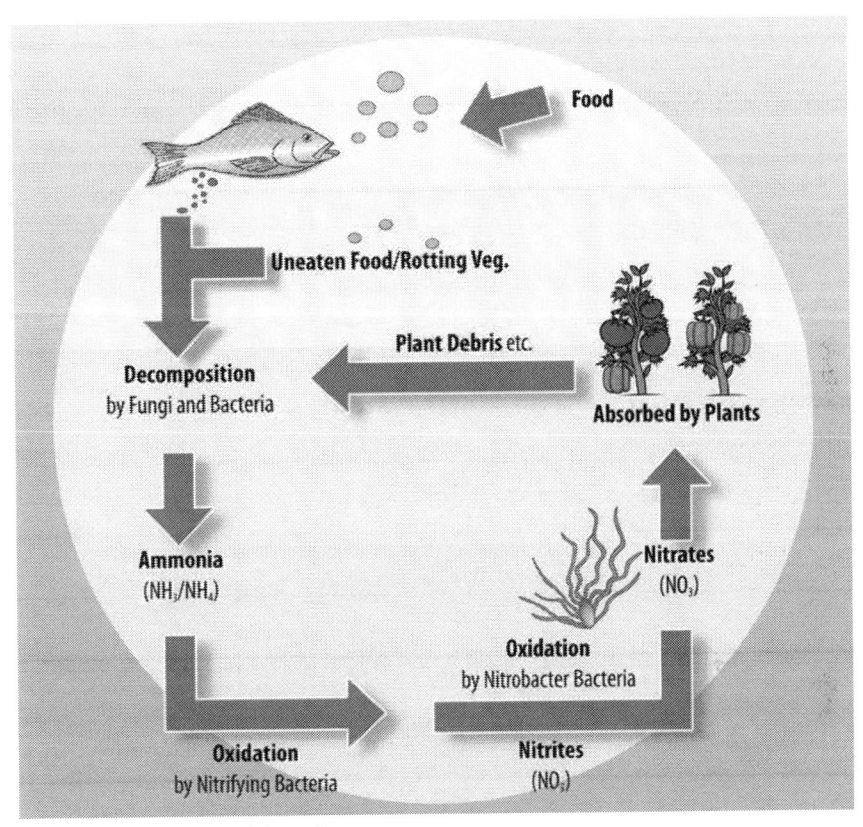

FIGURE 36. Bacteria life-cycle process

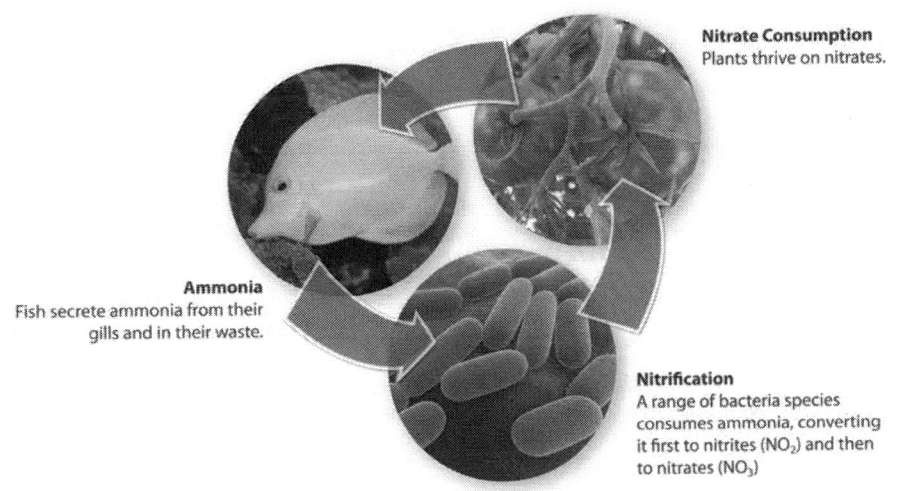

FIGURE 37.

Surface area bacterial colonies will thrive on any material, such as plant roots, along fish tank walls, inside pipes, and media in the grow bed or biofilter. The total area available for these bacteria will determine how much ammonia they are able to metabolize. Depending on the fish biomass and system design, the plant roots and tank walls can provide adequate area. Systems with high fish stocking density require a separate biofiltration component where a material with a high surface area is contained, such as inert grow media including gravel and tough or expanded clay.

Water pH and Bacteria

The pH is a numeric scale used to specify the acidity or basicity (alkalinity) of an aqueous solution. In aquaponics the pH level of the water has an impact on the biological activity of the nitrifying bacteria and their ability to convert ammonia and nitrite.

HMDPHM80 Digital pH/Temp. Meter

- Measures pH and Temperature
- One-touch automatic digital calibration
- Simultaneous temperature display
- Water resistant
- Cost: $45.17 (USA, year 2016)
- Cost and quality of pH/Temperature meters vary considerably from approx. $20 to $300+. The meter in Figure 38, or similar from another manufacturer, is sufficient for most aquaponic operations.

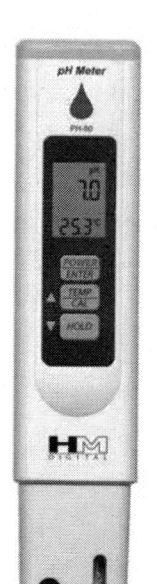

FIGURE 38.

The noted pH ranges for the two nitrifying groups below have been identified as ideal, yet the literature on bacteria growth also suggests a much larger tolerance range (6–8.5) due to the ability of bacteria to adapt to their surroundings.

NITRIFYING BACTERIA	OPTIMAL pH
Nitrosomonas spp.	7.2–7.8
Nitrobacter spp.	7.2–8.2

However, as mentioned earlier in this chapter, the most **appropriate pH range is 6–7 for aquaponics**, because this range is better for the plants and fish (discussed comprehensively in the Water Quality chapter of this book). Also, a loss of bacterial efficiency can be offset by having more bacteria, thus the biofilter or grow bed needs to be sized accordingly.

Water Temperature and Bacteria

Water temperature is an important parameter for bacteria and aquaponics in general. The ideal temperature range for bacteria growth and productivity is 63-93°F (17–34 °C). If the water temperature drops below 63°F (17 °C), bacteria productivity will decrease. Below 50°F (10 °C), productivity can be reduced by 50 percent or more. Low temperatures have major negative impacts on unit management during winter.

Dissolved Oxygen and Bacteria

Nitrifying bacteria need an adequate level of dissolved oxygen (DO) in the water at all times in order to maintain high levels of productivity. Nitrification is an oxidative reaction, where oxygen is used as a reagent; without oxygen the reaction stops. Optimum levels of DO are 4–8ppm (4–8 mg/liter). Nitrification will decrease if DO concentrations drop below 2ppm (2.0 mg/liter). Moreover, without sufficient DO concentrations, another type of bacteria can grow; one that will convert the valuable nitrates back into unusable molecular nitrogen in an anaerobic process known as denitrification.

TABLE 1. **Water quality tolerance ranges for nitrifying bacteria**

	Temperature	pH	Ammonia	Nitrite	Nitrate	DO
Tolerance Range	17–34 °C	6 — 8.5	< 3 mg/L	< 3 mg/L	< 400mg/L	4–8 mg/L
	63–93 °F		< 3 ppm	< 3 ppm	< 400 ppm	4–8 ppm

Ultraviolet Light and Bacteria

Nitrifying bacteria are photosensitive organisms, meaning that ultraviolet (UV) light from the sun is a threat. This is particularly the case during the initial formation of the bacteria colonies when a new aquaponic system is set up. Once the bacteria have colonized a surface (3–5 days), UV light poses no major problem. A simple way to remove this threat is to cover the fish tank and filtration components with UV protective material, while making sure no water in the hydroponic component is exposed to the sun, at least until the bacteria colonies are fully formed, and minimizing direct sunlight on the water as much as possible thereafter. Nitrifying bacteria will grow and thrive on material with a high surface area sheltered using UV protective material and under appropriate water conditions.

Balancing the Aquaponic Ecosystem

The term balancing is used to describe all measures an aquaponic operator takes to ensure the ecosystem of fish, plants and bacteria is at a dynamic equilibrium. It cannot be overstated that successful aquaponics is primarily about maintaining a balanced ecosystem. Simply put, this means there is a balance between the amount of fish, the amount of plants and the size of the biofilter, which really means the amount of bacteria. There are experimentally determined ratios between the size of the biofilter or grow bed, planting density, and fish stocking density for aquaponics. It is unwise, and very difficult, to operate beyond these optimal ratios without risking disastrous consequences for the overall aquaponic ecosystem. This section provides a brief, but essential, introduction to balancing a system. Biofilter or grow bed sizes and stocking densities are covered in much greater depth in the Biofilter and Flood-and-Drain chapters of this book.

Nitrate Balance

The equilibrium in an aquaponic system can be compared with a balancing scale where fish and plants are the weights standing at opposite arms. The balance's arms are made of nitrifying bacteria. Therefore, it is fundamental that nitrification is robust enough to support the other two components.

If the fish biomass and biofilter size are in balance, the aquaponic unit will adequately process the ammonia into nitrate. However, if the plant component is undersized, then the system will start to accumulate nutrients. In practical terms, higher concentrations of nutrients are not harmful to plants, but they are an indication that the system is underperforming on the plant side. On the flipside, too few fish will result in insufficient nutrients to adequately meet the plants' needs. This condition eventually leads to stressed plants and lower harvest yields.

The point is that achieving maximum production from aquaponics requires an appropriate balance between fish waste and vegetable nutrient demand, while ensuring adequate surface area to grow a bacterial colony in order to convert all fish waste. This balance between fish and plants is also referred to as the biomass ratio. Successful aquaponic systems have an appropriate biomass of fish in relation to the number of plants, or more accurately, the ratio of fish feed to plant nutrient demand is balanced. Although it is important to follow the suggested ratios for good aquaponic food production, there is a wide range of workable ratios, and experienced aquaponic operators will notice how aquaponics becomes a self-regulating system. Likewise, the aquaponic system provides an attentive operator warning signs in the form of water-quality metrics and the health of the fish and plants, as the system begins to slip out of balance,. All of these are discussed in detail in this book.

Feed Rate Ratio

Many variables are considered when balancing a system (note list below), but extensive research has simplified the method of balancing a unit to a single ratio called the feed rate ratio. The feed rate ratio is a summation of the three most important variables: the daily amount of fish feed in grams per day, the plant type (vegetative vs. fruiting) and the plant growing space in square meters. This ratio suggests the amount of daily fish feed for every square meter of growing space. It is more useful to balance a system on the amount of feed entering it than to calculate the amount of fish directly. Using the amount of feed makes it possible to calculate how many fish based on their average daily consumption.

Main Variables to Consider When Balancing an Aquaponic System

- Method of aquaponic production.
- Capacity (size) of operation.
- Type of fish (carnivorous vs. omnivorous, activity level).
- Type of fish feed (protein level).
- Type of plants (leafy greens, tubers or fruits).
- Proportion of crop (single or multiple species).
- Environmental and water quality conditions.
- Method of filtration (for N.F.T. or D.W.C. systems).

General Feed Rate Ratio:

- Leafy Green Vegetables: 40–50 grams of feed per square meter per day
- Fruiting Vegetables: 50–80 grams of feed per square meter per day

NOTE: The above feed ratio is provided with some reluctance and should only be considered as a very general rule of thumb. There are many other variables that can and will impact the feed rate ratio.

The point being that the proper feed rate ratio will provide a balanced ecosystem for the fish, plants and bacteria, provided the size of the biofilter or grow bed is adequate. Other variables may have larger impacts at different stages of the year, such as seasonal changes in water temperature. One must also be attentive to growth rate needs of the fish. The higher feed rate ratio for fruiting vegetables accounts for the greater amount of nutrients needed for these plants to produce flowers and fruits compared with leafy green vegetables.

Along with the feed rate ratio, there are two other simple and complementary methods to ensure a balanced system: health check, and nitrogen testing. They are described below.

Health Check of Fish and Plants

Unhealthy fish or plants are often a warning that the system is out of balance. Symptoms of deficiencies on the plants usually indicate that not enough nutrients from fish waste are being produced. Nutrient deficiencies often manifest as poor growth, yellow leaves and poor root development. In this case, the fish stocking density, feed (if eaten by fish) and the size of the biofilter (or grow bed) can be increased. Similarly, if fish exhibit signs of stress (gasping at the surface; rubbing on the sides of the tank; showing red areas around the fins, eyes and gills; or, in extreme cases, dying), it is often because of a buildup of toxic ammonia or nitrite levels. This often happens when there is too much dissolved waste for the media in the biofilter or grow bed to process. Obviously, any of these symptoms in the fish or plants indicate that the operator needs to actively investigate and rectify the cause.

Nitrogen Testing

This method involves testing the nitrogen levels in the water using simple and inexpensive water test kits (example shown below). If ammonia or nitrite are high > 1ppm (> 1 mg/liter), the media surface area in the biofilter or grow bed is inadequate and should be increased. Most fish are intolerant of these levels for more than a few days. An increasing level of nitrate is desired and implies sufficient levels of the other nutrients required for plant growth. Fish can tolerate elevated levels of nitrate, but if the levels remain high > 150ppm (> 150 mg/liter) for several weeks some of the water should be removed and used to irrigate other crops.

- 75 tests can be performed with this kit.
- Cost: $14.47 (USA, year 2016).
- Cost and abilities of water test kits vary considerably. The below kit, or similar from another manufacturer, is sufficient for aquaponic operations.

FIGURE 39. Seachem Multi-Test Nitrite and Nitrate Test Kit

If nitrate levels are low < 10ppm (< 10 mg/liter) over a period of several weeks, the fish feed can be increased slightly to ensure there are enough nutrients for the vegetables. However, increasing fish stocking density may also be necessary. It is worthwhile, and recommended, to test for nitrogen levels every week to make sure the system is properly balanced. Nitrate levels indicate the level of other nutrients in the water.

Fish stocking density, planting capacity, grow bed size and biofilter sizes are explained in much greater depth in their respective chapters of this book. The objective of this chapter was to provide an understanding of the nitrogen cycle, and to emphasize the importance, as well as the strategies, of achieving a balanced system.

PART II

Plants and Fish

CHAPTER 10

Plants/Crops — Keys to Success

Plants for Aquaponics

The most intensive aquaponic systems have been proven to achieve 20–25 percent higher yields than the most intensive soil-based culture. Although there is significant evidence which shows that, with the use of a greenhouse, yields can be anywhere from two to five times higher than traditional agriculture.

It is hard to find anything more rewarding than growing your own fresh, organic produce. With the exception of root vegetables, you can grow just about any type of plant in an aquaponics system. When deciding what plants to cultivate, all you have to ask yourself is what would you like to eat freshly out of your aquaponics farm, and what plants will grow best in your type of system. If you also desire to generate revenue from your aquaponics operation, then you will need to consider which plants are in highest demand and generate the most revenue.

Common plants that are being grown in aquaponics farms include: herbs, parsley, watercress, basil, sage, coriander, green leafy vegetables, (i.e. spinach, lettuce, etc.), and legumes such as peas and beans. Other plants, such as tomatoes, cucumbers, melons lettuce, chili, red salad onions, celery, broccoli, and cauliflower also do well. Some operators even grow dwarf citrus trees and flowers. For the most part, the sky is the limit.

It is also important to consider your choice of crops and fish together. For example, warm loving plants such as bell peppers would do better with Tilapias, which need warmer conditions than say Trout, a cold loving fish. If you are thinking of growing tomatoes, you will need to choose fish that do well in a densely populated tank, as tomatoes require a high level of nutrients.

It also makes the most sense to grow crops that are more expensive at the store, rather than those that can be purchased cheaply. In addition, some fruits and vegetables offered at the grocery store are known to usually have high concentrations herbicides, pesticides, heavy metals or other containments. It is always best to grow organic crops instead of purchasing their toxic counterparts at the store. More on this subject will be discussed later.

There are over 300 different aquaponic plants that have been tested which have proven to grow well in an aquaponic system. However, the type of plants that you can grow is dependent on your location and specific system parameters. Keeping this in mind, the following plants have been found to grow well in an aquaponics system.

Plants that do well in Aquaponics

- Amaranth
- Any leafy lettuce
- Arugula
- Basil
- Beans
- Begonias
- Black Seeded Simpson
- Bok choy
- Broccoli
- Cabbage
- Cantaloupe
- Celery
- Chard
- Chinese cabbage
- Chives
- Cilantro
- Collard
- Common Chives
- Coriander
- Corn
- Cucumbers
- Dill
- Eggplant
- Endive
- Flowers
- Ginger
- Impatiens
- Kale
- Lettuce
- Mint
- Most common household plants
- Mustard
- New Tomatoes
- Okra
- Pak Chov
- Parsley
- Peas
- Peppers
- Rapini
- Recao
- Redina Lettuce
- Spinach
- Squash
- Sweet potato
- Swiss Chard
- Taro
- Tatsoi
- Tomatoes
- Watercress
- Watermelon

NOTE: The major groups that do **NOT** grow well in an aquaponic system are root vegetables such as radishes, potatoes, carrots, etc.

Plants that have higher nutritional demands and typically only do well in a heavily stocked, well established aquaponic system:

- Tomatoes
- Peppers
- Cucumbers
- Beans
- Peas
- Squash
- Broccoli
- Cauliflower
- Cabbage

Plant Tips & Ideas

There are many resources available to obtain seedlings, starter plants, and cuttings. Sources range from the internet, mail order catalogs, local nurseries, garden centers, etc. It is important to educate yourself upon the source of your product to ensure that you obtain viable product from a reputable source. For instance, make sure you know that the item truly is organic, and that the company does not have a bad reputation for selling old seeds with poor germination rates. Read all seed packages for germinating and pollinating instructions. Beware of potential seed coatings that may not be safe for fish.

Winter plants can be grown during colder weather seasons. Flowers and potted trees can also be grown. Keep in mind that fish and bio-activity will slow down at lower temperatures, which typically correspond with plant activity.

It is normal for plant sprouts to shoot up quickly and then appear to stall for a few weeks. Don't fret, though. During this time, the plants are actually growing the root system out beyond the initial seed nutrients to support future growth. As the saying goes, patience is a virtue.

Duckweed can be grown directly on top of water surface. It multiplies rapidly with proper conditions. Some fish eat it naturally while other fish will take to it eventually. If your fish eat duckweed, be sure to keep some growing in a separate tank so the fish do not ravish your entire supply.

Be sure to grow only organic crops. Heirloom or organic seeds and plants obtained from a trusted reputable source are the best way to go. This will provide you the highest nutrient density and premium value produce.

Plant Design and Layout

A good layout of plant placement and crop design helps maximize production in the available space. Before planting, choose wisely which plants will be grown, bearing in mind the space needed for each plant and the associated growing season. It is prudent to plan the layout of the grow beds on paper in order to have a better understanding of how everything will fit together. Important considerations are: plant diversity, companion plants and physical compatibility, nutrient demands, market demands, and ease of access. For example, taller crops (i.e. tomatoes) should be placed in the most accessible place within the media bed to make harvesting easier.

In general, planting various crops and varieties provides a higher degree of security to the grower. If only one crop is grown, the risk for a serious infestation or epidemic increases. This can unbalance the system as a whole. As such, growers are encouraged to plant a diverse range of vegetables.

Staggered planting, as well as plant diversity, helps to maintain a balanced level of nutrients in the aquaponic system. At the same time, it provides a steady supply of plants to the table or market. Keep in mind that some plants produce fruit or leaves that can be harvested continually throughout a season, such as salad leaf varieties, basil, coriander and tomatoes, whereas some other crops are harvested whole. To achieve a staggered planting process there should always be a ready supply of seedlings. Maintaining a small side plant nursery for this purpose works well for many aquaponic operators.

Not only should the surface area be planned out to maximize space, but also the vertical space and time should be considered. Vegetables, such as cucumbers, are natural climbers that can be trained to grow up or down and away from the beds. Use wooden stakes and/or string to help support the climbing vegetables.

Planting vegetables with short grow-out periods (salad greens) between plants with longer-term crops (eggplant) will generate additional positive results. Lastly, another benefit of aquaponics is that plants can be easily moved by gently freeing the roots from the growing media and placing the plant in a different spot

Planting

Plants may be added to the grow beds as seeds or as seedlings. Most seeds require conditions that are moist (not drowned) and warm to germinate and sprout. Some seeds even require certain amount of light or darkness to germinate. Always be sure you follow seed germination instructions per each plant species. Also, wash off any dirt from plant roots before transplanting them in your system.

When beginning it is prudent to only plant about 1/4th of your grow bed area. Plant another 1/4th of the area after two weeks, and so on. Staggering plant growth will provide you with a continuous harvest (rather than all at once) and helps prevent against all plants dying off at the same time, causing a crash to the system. When the first group dies off, immediately replant it to keep the rotation going.

Taking Care of the Plants

Caring for plants in the aquaponics farm is much easier compared to a conventional garden. Unlike soil-based farming, you no longer have to battle against weeds which includes constant bending over which leads to back strain. If grown in a greenhouse, your crop is protected against the pests and wild animals that are often attracted to garden crops. Although bugs can still invade the aquaponics farm, they are far fewer due to its soil-less nature.

For successful pest management, it is helpful to educate yourself on any bugs found on your plants. Some bugs insects are beneficial, whereas others can be harmful to your crops. There are many environmentally friendly and aquaponic safe ways to combat insect problems. For instance, many small bugs, such as aphids, can be easily removed by spraying them with water. Alternatively, if the plant is small, you can even remove the plant from the media and soak it in the fish tank for a few minutes. This often removes the insects and provides food for the fish. If none of the above techniques work, you might want to consider using some natural, biodegradable foliar spray. You should never use non-organic pest control solutions as they are toxic to the fish, will contaminate the water within your system, and eventually make its way up the food chain to you and others. There are also a number of natural organic concoctions that can be made and sprayed on your plants that work very, or even better, than toxic pesticides. More on this subject later. Never spray your system with pesticides, herbicides, or fungicides, as they will harm or kill your fish. Furthermore, the heavy metals and other toxic pollutants from these toxic chemicals end up in your food supply, negatively impacting you and your family.

Another wonderful thing about aquaponics is that you no longer have to worry about the complexity of fertilizing your crops and whether you are over or under watering them. You are also removed from the problem of having to work with expensive fertilizers or chemicals as in hydroponics farming.

Crops grown in an aquaponics operation are completely organic. The essential nutrients needed by the plants are fully provided by the fish's waste. Bacterial colonies in the media bed convert the fish waste into nitrates and minerals that are readily absorbed by the plants for healthy growth. Supplements are not required in a well-managed aquaponics system. As matter of fact, supplements can be harmful your fish and should be avoided.

It is important, however, to ensure that the water is well-oxygenated and that the pH is maintained within the optimal range of your fish and plants. If the pH range is not optimal, the uptake of nutrients by plant roots as well as bacteria activity will be negatively affected, and waste will build up to toxic levels for the fish.

Plant Life

Leaves are where photosynthesis occurs. Photosynthesis is the process in which sunlight combines with carbon dioxide (CO_2) and water (taken up by the roots) to manufacture food (carbohydrates)

The Three Components of Plant Roots

for the plant; and oxygen is released as a by-product. Plants regulate small openings, stomata, on leaf undersides to allow CO_2, to enter and oxygen (O_2) to exit. When open, stomata allows water vapor to escape in the process of transpiration.

Roots anchor a plant in the ground and absorb water, nutrients, and air. Aquaponics allows absorption of water, nutrients, and air at an accelerated rate. Tiny root hairs increase the surface area for absorption. These root hairs are very delicate and must remain moist at all times. Large roots are similar to stems. They transport water and dissolved minerals (fertilizer) in the phloem.

Stems have a vascular system to transport water and nutrients throughout the plant. The xylem carries water and dissolved minerals from roots to leaves. The phloem transports food manufactured by leaves to the stems and roots. The vascular cambium is the growth zone that produces xylem on one side and phloem on the other. The stem also supports the plant and bean leaves, buds, flowers, and fruit

1. Primary Roots
These are the large diameter roots which first grow from the main body of the plant into the nutrient solution; and which grow so rapidly.

2. Secondary Roots
These are the smaller diameter laterals emerging from the primary roots. The large surface area of these roots enables them to absorb vital nutrients and water from the nutrient solution. By the way, plants don't always take-up nutrients and water from the solution at the same rate. There is a mechanism known as "active" uptake which enables plants to determine what they want to take-up from all that is provided in the solution.

3. Root Hairs
Root hairs are critical to plant health. Root hairs take-up atmospheric oxygen; which is as essential to plants as it is to humans. Root hairs are the lungs of the plant. Most plants cannot absorb sufficient oxygen from the solution for healthy growth so it follows that the roots of your plants should always have roots hairs present and above the solution level in the channel. It is essential to keep the solution from pooling so that roots are not submerged. This is achieved by having an appropriate channel slope, and proper flow of the solution (otherwise known as the fish wastewater). Atmospheric oxygen will always penetrate into the plant root zone of an N.F.T. system, so long as the roots are not completely submerged.

FIGURE 40.   *FIGURE 41.*   *FIGURE 42.*

Needs of the Plant

As a general rule, plants have a higher nutrient requirement during cooler months, and a lower requirement in the hotter months. Nevertheless, plants have five basic needs. Each need accounts for 20 percent of the plants ability to grow to its maximum potential. When all of these needs are met at the maximum potential, the result is maximum growth (20% x 5 = 100%). If one of these needs falls short (for instance, only 15 percent), growth is impaired. If two or more of these needs are not met, growth is hampered significantly.

Air 20%
- Temperature
- Humidity
- CO2 and O2 content

Light 20%
- Spectrum (color)
- Intensity
- Photoperiod*

Water 20%
- Temperature
- pH
- EC
- Oxygen Content

Nutrients 20%
- Composition
- Purity

Growing Medium 20%
- Air Content
- Moisture Content

* hours of light per day
** electrical conductivity — a measure of salinity

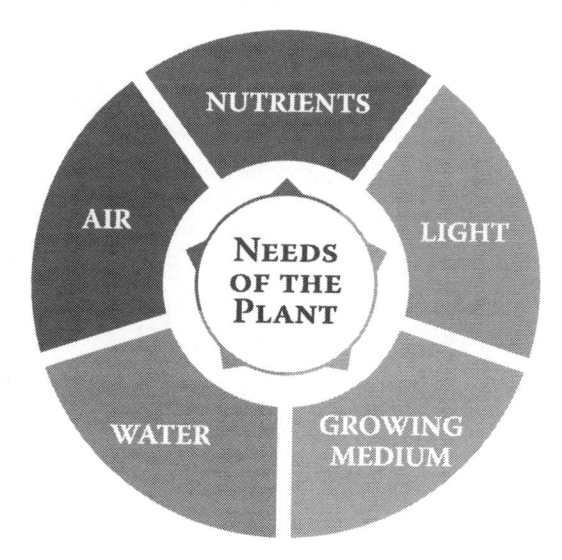

FIGURE 43. Needs of the plant

TABLE 2. **Mineral Elements Required by Plants**

ELEMENT	ABSORBED FORM	MAJOR FUNCTIONS
MACRONUTRIENTS		
Nitrogen (N)	NO_3^- and NH_4^+	In proteins, nucleic acids, etc.
Phosphorus (P)	$H_2PO_4^-$ and HPO_4^{2-}	In nucleic acids, ATP, phospholipids, etc.
Potassium (K)	K^+	Enzyme activation; water balance; ion balance; stomatal opening
Sulfur (S)	SO_4^{2-}	In proteins and coenzymes
Calcium (Ca)	Ca^{2+}	Affects the cytoskeleton, membranes, and many enzymes; second messenger
Magnesium (Mg)	Mg^{2+}	In chlorophyll; required by many enzymes; stabilizes ribosomes
MICRONUTRIENTS		
Iron (Fe)	Fe^{2+} and Fe^{3+}	In active site of many redox enzymes and electron carriers; chlorophyll synthesis
Chlorine (Cl)	Cl^-	Photosynthesis; ion balance
Manganese (Mn)	Mn^{2+}	Activation of many enzymes
Boron (B)	$B(OH)_3$	Possibly carbohydrate transport (poorly understood)
Zinc (Zn)	Zn^{2+}	Enzyme activation; auxin synthesis
Copper (Cu)	Cu^{2+}	In active site of many redox enzymes and electron carriers
Nickel (Ni)	Ni^{2+}	Activation of one enzyme
Molybdenum (Mo)	MoO_4^{2-}	Nitrate reduction

pH and Nutrient Uptake

The water pH has a direct impact upon optimal availability of all essential nutrients for a plant. At extremes in pH, many nutrients occur in forms unavailable for uptake by plant roots. The illustration below shows a relationship between pH and the availability of elements essential to plant growth for most species. Not all plant species will have the same pH/nutrient uptake characteristics. Some plants do better at a higher pH, whereas others absorb more nutrients at a lower pH.

Sowing

For the media-filled bed system (flood-and-drain system) there are a several ways to sow the seeds, dependent upon the type of plants to be grown. One method is to toss or plant the seeds evenly on the surface of the media-filled bed, and then thin them out, as needed, as they grow.

Another way is to germinate the seeds in a wet paper towel before transferring them to the grow media when their roots have grown at least 1 inch long. This also allows you to ensure that the plants are well-positioned in the media bed. This method works best for seeds that germinate very quickly, such as beans, peas, melons and cucumbers.

Seeds can also be sown in seed starting media (such as Rockwool, Peat sponges and Vermicompost) and then transplanted once they are established. This method works better for seeds that are harder to germinate (such as spinach and chard) or delicate plants (such as tomatoes and peppers).

Root cuttings or propagation can also be done on some plants. This works well if you have access to existing plants. It also saves money, as seeds don't have to be purchased.

Some aquaponic operators claim to have success planting crops twice as densely as a traditional soil-based farm. However, other operators find that their crops grow much bigger, wider, and fuller than traditional crops and thus, need more separation

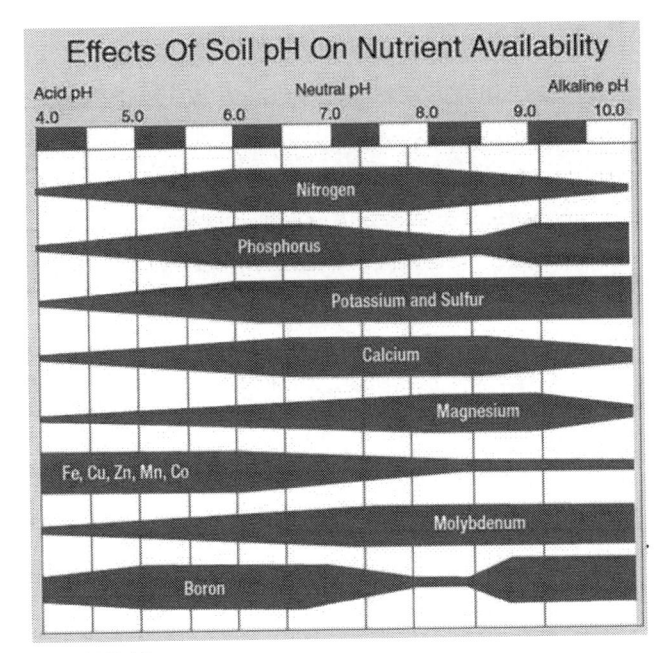

FIGURE 44.

chore that most operators like the best. Crops can be harvested for your family and friends or even sold as organic vegetables for lucrative profits. With a diversified crop (i.e. different types of plants and/or plants at different age stages) harvest is an ongoing process.

Different species of plants grow at different rates. Although the information on the back of the seed package will indicate how long it takes for the crop to ripen, with aquaponics plants typically grow much faster than via traditional methods.

Most Profitable Plants for Aquaponics

If growing plants for money, it is prudent to focus on growing those that are in demand are harder to obtain organically, and/or are the more expensive ones at the grocery store, and have potential for generating the highest profit margins. Even though not all high value plants are suitable for aquaponics, such as ginseng, there are many viable plant options to choose from that have tremendous revenue potential.

Regardless of whether you are growing plants for revenue or just for personal reasons some of the same principles can be applied. For one, it doesn't make sense to take time and space growing produce that can be inexpensively acquired at your local market (i.e. radishes). Also, some plants have a lengthy grow period from plant to harvest (months), whereas others can be harvested in a matter of weeks. Furthermore, some plants need lots of space, which will and can limit your available production space.

Most plants do much better in aquaponics than they do through traditional farming methods. For instance, research showed that Basil grew three times better in Aquaponics than it did in a conventional garden. Okra has been found to grow 18 times as well in Aquaponics than in does in soil gardens. To minimize risk, it is best to select plants that have a proven history of successfully growing in aquaponics. However, feel free to experiment with various species and varieties along the way. You may be able to improve upon the latest data available.

Keeping the above in mind, the following is a list of plants that can be grown in an aquaponics system which have the highest earning potential.

- **Bamboo:** Landscapers and homeowners are paying as much as $150 each for potted bamboo plants, and many growers are finding it hard to keep up with the demand. Why is bamboo so popular? It's a versatile plant in the landscape, as it can be used for hedges, screens or as stand-alone "specimen" plants. Bamboo is not just a tropical plant, as many cold-hardy varieties can handle sub-zero winters. If done right, through crop rotation and effective marking, it's possible to grow thousands of dollars' worth of profitable plants in a relatively small area.
- **Flowers:** If you are looking for a high-value specialty crop that can produce immediate income, take a look at growing flowers for profit. A flower growing business has almost unlimited possibilities, from fresh cut flowers to dried flowers — often called "everlastings", for their long life. It doesn't take much to get started growing flowers for profit either, just a few dollars for seeds. Even operators with small systems can produce incredible amounts of flowers and find lots of eager buyers at the local Saturday market held in most towns.
- **Ground Covers:** Due to high labor costs and water shortages, ground covers are becoming the sensible, low-maintenance way to landscape. Growers like ground covers too, as they are easy to propagate, grow and sell. Bringing profits of up to $20 per square foot, ground covers are an ideal cash crop for the aquaponics operator
- **Herbs:** Growing the most popular culinary and medicinal herbs is a great way to start a profitable business. The most popular culinary herbs include basil, parsley, chives, cilantro and oregano. Medicinal herbs have been widely used for thousands of years, and their popularity continues to grow as people seek natural remedies for their health concerns. Lavender, for example, has dozens of medicinal uses, as well as being a source of essential oils. Lavender is quite popular; hundreds of small nurseries grow nothing but lavender plants. The more commonly known herbs are easy to sell, but often don't bring in the higher revenues. Educate your customers on the benefits of other less common herbs and you can enjoy higher profit margins.
- **Ornamental Grasses:** Because ornamental grasses are drought-tolerant and low maintenance, landscapers are using more and more of them, as are homeowners. Because there are hundreds of shapes and sizes, they can be used for everything from ground covers to privacy screens. It's easy to get started growing ornamental grasses, as you simply buy the "mother" plants and divide the root clump into new plants as it grows. It's possible to grow thousands plants annually with a small system.

Highest Revenue Generating Plants

In regards to produce, table 3 is a list of popular aquaponic plants and their retail average retail cost (USA, year 2016) provided by two reliable sources. For the traditional gardeners and farmers out there, please note that produce producing plants that cannot be grown in an aquaponics system have been omitted from the following list.

TABLE 3. **Highest Revenue Generating Plants**

POPULAR AQUAPONIC PROVEN PRODUCE ONLY	AVERAGE PRICES (Commercial Vendors)			UNITED STATES DEPARTMENT OF AGRICULTURE (USDA) (year 2016 Retail Prices, USA)		
Vegetable	USD Value/ SF (2016) Grocery	USD Value/ SF (2016) Farmers Market	USD Value/ SF (2016) Organic	Avg. Retail Price (per lb. — fresh)	Avg. Retail Price (per 1 cup serving — fresh)	National Retail Price
Artichoke, Globe	$2.40	$2.64	$3.00	$2.21/lb	$2.27	$2.84/each
Arugula-Roquette	$20.00	$22.00	$25.00			
Basil	$5.28	$5.81	$6.60			
Bean, Bush	$2.00	$2.20	$2.50			
Beet	$2.50	$2.75	$3.17			
Broccoli	$0.80	$0.88	$1.00	$2.57/lb	$0.72	$0.97/lb
Broccoli, Chinese						
Brussels Sprouts	$4.80	$5.28	$6.00	$2.76/lb	$0.89	
Cabbage	$0.20	$0.22	$0.25	$0.58/lb	$0.25	$0.52/lb
Cabbage, Chinese Napa						
Cabbage, Savoy						
Cauliflower				$1.23/lb	$0.38	$1.09/lb
Celery				$1.11/lb	$0.40	$1.17/lb
Chard, Swiss	$3.28	$3.61	$4.10			$1.00/lb
Chives	$4.80	$5.28	$6.00			
Choi	$2.25	$2.48	$2.81			$0.98/lb
Cilantro	$10.33	$11.36	$12.91			$0.67/bunch
Corn	$0.66	$0.73	$0.83	$2.69/lb	$1.81	$0.33/ear
Cucumber	$4.77	$5.24	$5.96	$1.30/lb	$0.35	$0.67/lb
Dill	$6.40	$7.04	$8.00			
Eggplant	$2.25	$2.46	$2.82			$1.13/lb
Garlic	$1.60	$1.76	$2.00			
Grass, Lemon	$3.00	$3.30	$3.75			
Green Salad Mix						$2.15/lb
Greens, Mustard	$1.23	$1.35	$1.54	$2.57/lb	$0.94	$0.81/lb
Kale	$5.60	$6.16	$7.00	$2.81/lb	$0.77	$0.85/lb
Kohlrabi	$0.75	$0.83	$8.94			
Lettuce	$3.60	$3.96	$4.50			$1.97/each
Parsley	$4.03	$4.43	$5.04			
Peas, Edible Pod						
Peas, English	$1.50	$1.65	$1.88	$1.01/lb		
Peas, Snow	$1.80	$1.98	$2.25			$0.95/lb
Pepper, Bell	$4.50	$4.95	$5.63	$1.41/lb	$0.46	$1.48/lb
Pepper, Jalapeno	$1.80	$1.98	$2.25			$0.95/lb
Potatoes	$4.50	$4.95	$5.63	$0.56/lb	$0.18	$0.75/lb
Pumpkin	$2.70	$2.97	$3.38			
Rhubarb	$6.22	$6.84	$7.77			
Spinach, Spring/Fall	$1.00	$1.10	$1.25			
Squash, Summer, Yellow	$1.80	$1.98	$2.25	$1.64/lb	$0.85	$1.36/lb
Squash, Zucchini	$1.80	$1.98	$2.25			
Squash, Winter, Acorn	$7.95	$8.75	$9.94	$1.17/lb	$1.16	$0.88/lb
Squash, Winter, Butternut	$1.20	$1.32	$1.50	$1.24/lb	$0.89	$0.88/lb
Squash, Winter, Hubbard	$3.20	$3.52	$4.00			
Tomatillo	$1.20	$1.32	$1.50			$0.63/lb
Tomato, Cherry (small & medium)	$8.00	$8.80	$10.00	$3.29/lb	$1.35	$3.24/lb
Tomato, large	$6.94	$7.63	$8.68	$1.24/lb	$0.51	$1.38/lb

NOTE: The business side of aquaponics addressing economics, marketing, sales, and profitability is covered comprehensively in "Part VII — Making Money and Earning a Profit from Aquaponics" of this book.

PART II: PLANTS AND FISH

Out of Season Crops

Growing out of season produce is another means in which to potentially save money and generate higher revenues. Although some additional arterial lighting may be necessary to trick the plants into thinking, it is their season. The extra cost for lighting is typically negligible and is a very good investment in most cases compared to the cost of out of season produce; and the risk associated with consuming non-organic grocery store food.

The produce considered in or out of season varies considerably from one geographic location to another. Also, if you live in southern California, Florida, South America, or Australia, your options in December are far different and greater than someone living in Michigan, Ontario, or Finland.

If you are interested in growing out of season crops it is prudent to obtain seasonal crop statistics from your regional governing agricultural agency, local garden club, and/or your area college/university agriculture department. Keeping in mind that 'seasonal' is geographical dependent, the following fruit and vegetable lists can be used as general guidelines large portions of the United States.

Below Example Represents Central United States

(For your specific area, check with your regional agriculture department or local garden groups and nurseries)

TABLE 4. *Seasonal Fruit Planting Guide for most of the United States (not all are conducive to Aquaponics)*

SPRING	SUMMER		FALL		WINTER
Apricots	Apricots	Jackfruit	Black Crowberries	Key Limes	Cactus Pear
Cherimoya	Asian Pear	Key Limes	Catus Pear	Kumquats	Cardoon
Cherries, Barbados	Black Crowberries	Limes	Cherries, Barbados	Passion Fruit	Cherimoya
Honeydew	Black Currants	Loganberries	Apples	Pear, All	Clementines
Jackfruit	Blackberries	Longan	Cranberries	Persimmons	Dates
Limes	Blueberries	Loquat	Feijoa	Pineapple	Grapefruit
Lychee	Boysenberries	Lychee	Gooseberries, Cape	Pomegranate	Kiwifruit
Mango	Breadfruit	Mulberries	Grapes	Quince	Mandarins
Melon, Bitter	Cantaloupe	Nectarines	Grapes, Muscadine	Sapote	Oranges
Oranges	Casaba Melon	Olallieberries	Guava	Sharon Fruit	Passion Fruit
Pineapple	Champagne Grapes	Passion Fruit	Huckleberries	Sugar Apple	Pear
Strawberries	Cherries	Peaches	Jujube		Persimmons
	Cherries, Barbados	Persian Melon			Pummelo
	Cherries, Sour	Plums			Red Banana
	Crenshaw Melon	Raspberries			Red Currants
	Durian	Sapodillas			Sharon Fruit
	Elderberries	Sapote			Tangerines
	Figs	Strawberries			
	Grapefruit	Sugar Apple			
	Grapes	Tomatoes			
	Honeydew Melon	Watermelon			

86

The seasonal foods listed below in table 5 include all varieties of vegetables (roots, gourds, legumes, etc.), as well as herbs, lettuces and leafy greens. Not all are conducive to aquaponic gardening.

TABLE 5. **Seasonal Vegetable Planting Guide for most of the United States (not all are conducive to Aquaponics)**

SPRING	SUMMER	FALL	WINTER
Artichokes	Greens, Swiss Cha.	Beans, Chinese Lng	Okra	Beans, Chinese Lng	Mushrooms	Brussels Sprouts
Asparagus	Lettuce, Butter	Beans, Green	Peas	Black Salsify	Peppers, Jalapeno	Chestnuts
Asparagus, Purple	Lettuce, Manoa	Beans, Lima	Peas, Sugar Snap	Broccoli	Pumpkin	Collard Greens
Asparagus, White	Lettuce, Red Leaf	Beans, Winged	Peppers, Bell	Brussels Sprouts	Radicchio	Endive, Belgian
Beans, Fava	Lettuce, Spring Baby	Beets	Peppers, Jalapeno	Cardoon	Squash, Acorn	Kale
Beans, Green	Mushrooms, Morel	Corn	Potatoes, Yukon Gld	Cauliflower	Squash, Buttercup	Leeks
Broccoli	Onions, Vidalia	Cucumbers	Radishes	Diakon	Squash, Butternut	Squash, Buttercup
Cactus	Pea Pods	Edamame	Shallots	Radish	Squash, Chayote	Squash, Delicata
Chives	Peas	Eggplant	Squash, Chayote	Endive	Squash, Delicata	Squash, Sweet Dumpling
Corn	Peas, Snow	Endive	Squash, Crooknose	Endive, Belgian	Squash, Sweet Dumpling	Sweet Potatoes
Endive, Belgian	Radicchio	Garlic	Squash, Summer	Garlic	Sweet Potatoes	Turnips
Fennel	Ramps	Lettuce, Butter	Squash, Zucchini	Ginger	Swiss Chard	
Fiddlehead Ferns	Rhubarb	Lettuce, Manoa	Tomatillo	Jerusalem Artichoke	Turnips	
Greens, Collard	Sorrel			Kohlrabi		
Greens, Mustard	Squash, Chayote			Lettuce, Butter		
Greens, Spinach	Watercress					

87

Selecting the Right Crops for You

After you have a good idea as to what plants would do well in your geographical location, it's time to decide where to focus your efforts. Chances are you will grow multiple crops, integrating your resources to achieve the annual production goals you have established. Determining the ideal product mix is a task for each individual operator based on system parameters and his/her objective, but there are some basic guidelines which will facilitate the process.

Profitability Approach:

1. Known as the 'Profitability Approach', you can grow several crops annually that generate revenue. Estimate the annual income possible from each crop, and total up the amount of sales you think they will generate annually. Subtracting the result from your goal leaves the amount you need to earn in "new sales."

2. Finding the right product mix for your operation means considering not only what you can produce but, more importantly, what you can successfully sell. For example, selling vegetables can involve a lot of direct interaction with the purchaser. Customers at the farmers' market may want to know not only whether your tomatoes have good flavor, but also what culinary uses they might be best adapted. Prepare for this with a printed recipe or two.

3. Customers will engage you in discussions of your growing methods, offer their opinions about crop varieties, and what else you grow. If you find such extensive customer contact daunting, then perhaps you should focus more effort on selling your product to a wholesale buyer or find a few good chefs who will take most of your produce. You could also consider joining or even starting a CSA group. Time constraints will play a crucial role in your decisions, too. If you already have a crowded schedule, you may not be able to spend time running your operation and marshaling a local sales effort. Or maybe your leisure time does not coincide with the optimum sales schedule. Working a farmers' market is similar to having a part-time job. In these situations, selling via the internet may make more sense.

4. The internet has a highly developed community of avid gardeners, foodies, health conscious individuals, herb enthusiasts, and other potential customer groups. The possibilities are limitless. Internet sales also permit scaling your business to match your production capacity. When something is sold out, you can instantly remove it from your website, or even invite customers to put themselves on a "want list" for next year's crop.

5. If your work schedule and leisure time do not coincide with when the farmers' market is open for business then selling via the internet may make more sense. If so, look for products that lend themselves to this approach. You might be best-off developing a product that can be canned or dehydrated and shipped during their winter dormancy rather than having to cope with perishable seasonal vegetables. Again, using the internet is a great way to connect with potential customers.

6. When preparing your start-up plan, examine your own likes and dislikes, the ways you prefer to spend your leisure time, and how much enthusiasm you would truly have for selling your products. Also, consider if you would prefer to sell to others through one-on-one relationships, or prefer a more impersonal experience such as internet sales, or if you would most enjoy talking to a lot of people at a busy farmers market.

CHAPTER 10: PLANTS/CROPS — KEYS TO SUCCESS

7. You must also ask yourself how far you desire to go with your aquaponics operation. If adding value to a product is required to meet your sales goals, are you prepared to take that step? For example, say you realize that you can wring a higher dollar yield with more resources dedicated to canning a particular vegetable. Doing so, however, will require containers, labels, time to grow, harvest, and pack the finished product. Going the distance of actually creating a prepared food product on a large scale requires a commercial kitchen, permits, health inspections, and product liability insurance. Is this something that you really want to do? Do you have the capital for these endeavors? How long will it take to recoup that investment? These are questions that need to be answered before moving forward.

8. Besides the needs of your family and your personal aspirations, consider also the people living around you and your zoning laws. Keeping peace with the neighbors is important. Will your neighbors complain if you have a lot of traffic to and from your house? Will an occasional gift of fresh vegetables be enough to maintain the peace? Will customer parking going to be an issue?

Personal Savings Approach:

9. The 'Personal Savings Approach' is a plan with the objective being more about personal savings than profitability. Given the likelihood that you will probably grow several different crops, estimate the annual savings in grocery store food cost possible from each crop. Subtracting the result from your production cost leaves the amount you amount you will gain through your aquaponics crop selection.

10. Revaluate your crop selection to see if different species or more space may need to be dedicated to a particular species. Next, review the space you have available for expansion and determine how much more product can be produced. Dividing the annual goal of production benefit by the area of this space gives you the productivity you need to reach your production goal, in dollars per square foot. Armed with this number, you can proceed to create a crop plan.

11. Product uses is another determining factor that should be considered. For instance, does your selected tomato variety have only good fresh flavor, or can they also be used for many culinary options? Does plant selection provide you with long term food storage, such as canning, freezing, and dehydration options?

12. One should also consider the time involved for each crop. Some crops are higher maintenance than others. For instance, spinach and lettuce require very little time whereas tomatoes and bell peppers require higher maintenance as they must be supported. It is prudent to ensure that you know what time requirements will be placed upon you and you have the spare time to maintain the crop before making the crop selection.

13. It is also wise to know if the rest of your family is on board. You don't need ongoing resistance on the home front. Farming on any scale is, to some extent, a way of life, and running an aquaponics operation will likely alter some of your accustomed patterns and family activities. Thinking ahead of ways to mitigate potential problems and conflicts now will save a lot of frustration later. With that said, running an aquaponics operation can also bring a family closer together as each member works together towards a common goal that benefits all. Running an aquaponics operation often requires family members to turn off the electronic devices, put away the distractions, and spend time quality together.

These are all issues should be carefully considered during the planning phase. The more you think through these issues in the beginning, the greater your chance for optimal success and enjoyment.

Identifying the Best Plants to Grow

Every possible product your aquaponics operation might yield has its advantages and disadvantages. Thus, depending upon your particular situation, there are good and bad crop choices. To help sort them out, consider rating different possibilities according to these key criteria:

- **Productivity:** the amount of salable crop per square foot cultivated.
- **Market Value:** the price you can obtain for a unit of crop.
- **Estimated Demand:** the volume of a crop you can reasonably expect to sell.
- **Capital Requirement:** the amount of money and other resources that must be invested in order to bring in a crop to harvest.
- **Personal Zeal:** your familiarity with, or desire to learn about, the cultivation of a particular crop, as well as the enthusiasm you bring to its production and sale. It is much more difficult to be successful if you lack passion.

The ideal crop is one that generates a lot of produce per square foot over the course of a season, has high market value, is in great demand, and is one in which you enjoy growing. Ideally, it is also one with a capital requirement that you can afford or acquire the necessary funding to produce. In addition, growing a top quality product—one that is superior to others on the market— will give you a huge advantage over your competitors.

Time = Money

"Time is money" is an adage that is never more correct than when applied to the business of aquaponics. The operator gets paid for the time required to take a crop from start to salable size. This can range vary significantly, depending upon the type of plant in production.

Each of the crop's "life stages" offers a different level of risk and reward that the entrepreneur must take into account. For example, starting plants from seed is the cheapest way to obtain most species, but also takes longer and will result in the highest percentage of losses.

Production of Annuals

An annual is any plant that completes its life cycle in a single growing season. For the would-be aquaponics operator, product possibilities among the annuals include flower, herb, and vegetable plants. They are usually started from seed. Seedlings are generally grown under cover in order to get a good start, then transplanted to the aquaponics bed to maximize space and ensure that the grow bed is always being utilized by larger plants rather than seeds or sprouts. Larger plants also help filter the water before it returns to the fish tank.

Growing Flowers for Money

"Color," as annual flowers are known in the nursery business, can be a reliable source of income. Locally grown, fresh cut flowers are the fastest growing segment of the U.S. floral market. Approximately 90 percent of the fresh cut flowers sold in the U.S. come from abroad. While foreign growers are great at producing the top three floral crops (roses, carnations and chrysanthemums) at incredibly low prices, the flowers that you buy from your local florist, big box or supermarket are harvested 10 to 14 days before they reach your shopping cart. To achieve this, foreign producers can only grow a limited number of varieties and they have to use an incredible amount of chemical fertilizers, pesticides, and preservatives to make sure that their product looks fresh by the time it reaches the customer.

Most flowers are grown at a huge cost to the environment and the flowers have already lost up to two weeks of vibrant life due to transport and storage. Therefore, with the right kind of marketing local, fresh cut flowers can be a thriving business. Local growers produce a product that is fresher than their foreign competitors. In fact, bouquets can even be sold the same day they are cut. The local grower can also customize his or her offering based on seasonality and local preferences. In addition, the customers of local growers often develop a personal relationship with them. They value knowing who produces their flowers and how that person grew them. When you consider that the local grower does all of this and still keeps prices extremely competitive, you begin to understand why this movement has taken off. Furthermore, as you educate the community as to how your flowers are grown organically through sustainable farming practices, as compared to the way the vast majority of commercial flowers are produced, you will literally have customers begging for your product. Nobody in their right mind would want to give their loved one a flower contaminated with synthetic fertilizers, pesticides, sprayed with chemical preservatives, and produced in ways that cause a great deal of harm to the natural environment.

If you are interested in growing flowers for market, The Association of Specialty Cut Flower Growers (www.ascfg.org) is a great place to start. The ASCFG formed in 1988 to provide information on growing techniques, marketing strategies, and new developments to the field-grown and greenhouse-grown fresh cut flower producer. As of this writing, they have 450 members that specialize in this growing market. While some of these growers are large producers who supply tons of flowers to high-end florists and supermarket chains, the majority is small. These small, independent growers are making a living or supplementing their family income by growing flowers on farms as small as one acre. Many grow, harvest, package, market, and sell directly to their customers.

They can provide you with a wealth of valuable information that will save you time, expense, and help you maximize your profit.

Following are some helpful tips on how you can make money growing flowers:

1. Grow good flowers. That sounds obvious, but if you don't have a good product, it'll be harder to make a good profit. So make sure your flowers are grown organically under optimal conditions. Your flowers will be healthier, less susceptible to diseases and look better.

2. Do your research on the best varieties to grow. Spend time on the Association of Specialty Cut Flower Growers website learning what others are doing, talk to florist, search the Internet, and obtain some books on the subject of growing flowers for profit.

3. Educate your potential customers on the disadvantages of buying flowers through the common methods versus your local organic sustainable farming approach.

4. Look into setting up a web-based business or another online mechanism so that customers can easily buy your flowers.

Great Ways to Sell Flowers

- **Farmers' Markets and Flea Markets** — These markets most always draw big crowds. Setup a stand or table with your flowers, and show everyone that stops by why your flowers are the best.
- **Hotels and Restaurants** — Hotels and restaurants often like to make a table, entryway, or special room look nice by adding some pretty flowers. It is an easy way to spice up any establishment. Talk to local hotels and restaurants and

find out what flowers they are looking for. If you are able to grow those varieties, you could open yourself up to a lot of business, as these hotels and restaurants will want to keep their tables and entryways looking nice.

- **Fundraisers** — People will often be more willing to buy your products if they know some of the profits are going to a good cause. That's why fundraisers are an excellent way to sell your flowers. It makes your business look good, while making your wallet happy too.
- **Florists** — Florists have to get their flowers from somewhere, so why not your aquaponic operation? Make the rounds and talk to your local florists, and find out just what they're looking for. Find out what they routinely pay, and see if you can give them a better deal. If you can produce flowers that they can then sell and make good money, they will buy from you repeatedly.
- **Online** — People are more apt to buy form you if it is easy. They can also forward your website's link to to friends, co-workers, and family. Another advantage is that you can capture their contact info (phone number, mailing address, and email address) and reach out to them whenever you have a product to sell. A short, periodic, e-newsletter is also a great way to keep your business and products in the thoughts of your customers.

You really can make big money growing flowers. After you get up to speed on all the 'ins' and 'outs', and have your operation running efficiently, your profits will be grow faster than your flowers.

Growing Plants for Money

You'll be facing stiff competition from industrial-scale growers who can turn out plants by the millions, but if you carefully choose a market niche and grow quality plants, you can sell them at a nice profit. However, even common crops can be profitable if grow efficiently. In addition, many customers want variety. A frequent question growers hear is 'what else do you have for sale?

Again, the ideal situation would be to find plants that you enjoy growing, can be grown easily in your particular operation, can be grown as efficiently and inexpensively as possible, are ones that you and your family enjoy eating, are in high demand, and produce a high profit margin. However, it is rare for anything to be perfect. Therefore, you may need to pick the plants that will most closely coincide with your main purpose. Keep in mind your primarily objectives and go from there. Another thing to keep in mind, as eluded to previously, is that vast numbers of people are willing to pay more, wait, and go out of their way to purchase a product that is organic, grown locally, and produced without doing harm to the environment.

Homemade Organic Pest Control

Pesticides cause undesirable ecological and health impacts. Pest control methods that are relatively non-toxic with few ecological side-effects are sometimes called 'bio-rational' pesticides. The major categories of bio-rational pesticides include botanicals, microbials, minerals, and synthetic materials. However, not all bio-rational pesticides are safe to use in an aquaponic system or integrated into your food supply.

The first thing to keep in mind in regards to introducing any application to your aquaponic system is that of the health of your fish and the good bacteria that are driving your system. Never use non-organic pest control products. Furthermore, all pesticides (whether organic or synthetic) are also harmful to beneficial insects such as ladybugs, honeybees, praying mantas, butterflies and others that are good for your plants. Beneficial insects aid in pollination and/or eating the harmful insects. For instance, bees pollinate plants and ladybugs feed on aphids. A pesticide should only be used as a last resort.

Although there are organic pesticide products available on the market, the best and least expensive approach is to make your own organic bug spray. A homemade-pesticide-recipe to repel-garden-pests that is suitable to green living can be made from typical kitchen ingredients; even from leftovers. The following sections provide several methods for making your own organic, non-harmful pesticide.

Organic Bug Spray from Onion and/or Garlic Scraps

Simply save your onion skins, peels, and ends in a Ziploc sandwich bag. You can also include garlic leftovers. Keep the leftovers in the refrigerator. Once you have a full bag, place the onion pieces in a pail and fill with warm water. Soak for several days or up to a week. You can keep the pail in a sunlit location to steep, but this is optional. Finally, strain the onion bits out and store the onion water in spray bottles. You can spray this concoction in the plant area as well as directly on the plants.

How to Make Healthy Garlic Spray Pesticide

- 10 cloves minced garlic
- 2 tsp. mineral oil
- 2 1/2 cups water

Soak garlic in mineral oil for 1 day. Strain garlic out and add the water and soap. Mix well. Spray plants.

How to Make Healthy General Organic Pesticide

- 3 hot green peppers (canned or fresh)
- 2 or 3 cloves garlic
- 3 cups water

Puree the peppers and garlic cloves in a blender. Pour into a spray bottle and add the liquid soap and water. Let stand for 24 hours. Strain out pulp and spray onto infested plants, making sure to coat both tops and bottoms of leaves.

How to Make Healthy Pyrethrum Pesticide

Pyrethrum is a natural, botanical insecticide that is environmentally friendly and has several major advantages over chemically synthesized insecticides including its rapid breakdown in the environment and its lack of insect resistance.

Pyrethrum is extracted from a particular type of Chrysanthemum known as the African Daisy (Chrysanthemum cineraria folium). Oil is extracted from the seeds of these flowers.

Ingredients

- 1 tablespoon flower heads of either Chrysanthemum cinerariaefolium or Chrysanthemum roseum.
- 1 liter / 33fl.oz hot water or 1 Quart/32 fl.oz

Steps

- Obtain the flower heads.
- Mix the flower heads with the hot water; allow to stand for one hour.
- Strain off the flower heads.
- Add the soap powder and mix.
- Pour the mixture into a spray bottle.
- Spray lightly on the plants.

Warnings

- Spray with care—this will also impact beneficial insects.
- This is a controversial pesticide. Many web sites say it is harmless to humans and mammals and that it degrades quickly, but some studies indicate that it is very harmful to humans, dogs, cats, and that it lingers in dust. See Cox, Caroline, 2002, Insecticide Factsheet Pyrethrins/Pyrethrum, Journal of Pesticide Reform, Vol. 22, No. 1, 14–20.

CHAPTER 11

Fish — Everything You Need to Know

Importance of Fish

Fish play a very important role in the aquaponics system. They provide the vital nutrients for vibrant plant development and if you're growing edible fish, then they serve as a valuable source of low-fat, high quality protein, essential omega-3 fatty acids, vitamin D, Riboflavin, and other necessary nutrients. Growing a large population of fish may be a little daunting to some, especially those without any prior experience; however one need not be discouraged. Successfully raising fish in an aquaponic system is just a matter of following some basic fundamental principles. As you follow these simple, proven principles, you will not have any difficulty growing fish from fingerling size to ready-to-eat fish. The next chapter will explain further how you can easily breed and raise your own fish, so as to spare you the cost of having to purchase fingerlings.

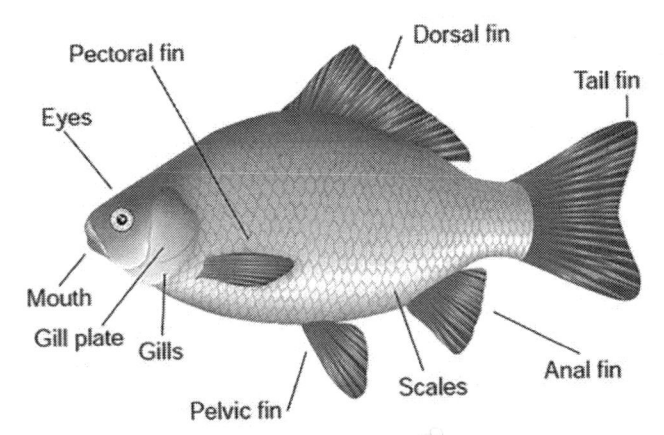

FIGURE 45. Illustration of the main external anatomical features of fish

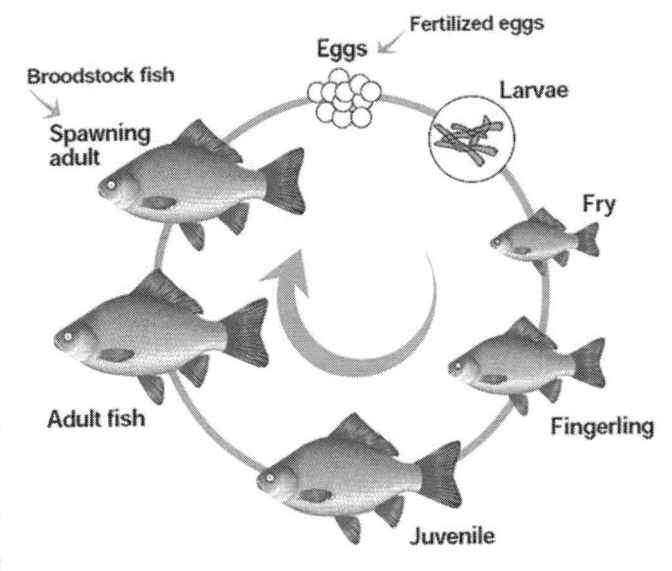

FIGURE 46. General life cycle of a fish

Deciding Upon the Fish Species to Rear

Reiterating an important point, you must clearly define your objectives for getting into aquaponics, and stay true to that vision. Not that you can't change your mind or alter your mission as you learn more along the way and gain experience, but it is important not to be diverted into a different direction due to the random external distractions that pop up in life.

Once you've define your goals, a plan can be established and implemented that will help you achieve optimal success. The following are some ideas that will, hopefully, help you further clearly define your purpose, and the rewards you desire to obtain from aquaponics.

Choosing a Fish Species

There are many different species of fish that can be used in an aquaponic system, depending on your growing environment, available supplies, and in most applications, your local climate.

Do you desire to rear the fish for consumption or for ornamental purposes? If edible fish is the objective, trout or tilapia are excellent choices. Goldfish and koi are good choices for ornamental fish.

If growing edible fish is your objective, and you live in a cooler climate, growing Trout year round may make the most sense. In warmer conditions, Barramundi or Jade Perch are often a good option. However, in most warm areas or controlled temperature environments (i.e. inside), Tilapia is the fish of choice.

Aquaponic Proven Fish

Ideally, the fish and plants selected for the aquaponic system should have similar needs in regards to temperature and pH. Realistically, a perfect match is rarely achieved, but the closer they match, the higher probability there is for success.

Warm loving fish and leafy crops, such as bush green beans and herbs, typically go well together. Fruiting plants such as tomatoes and peppers—due to their higher nutritional requirements—do the best when supported by a densely stocked fish tank(s).

Fish known to have excellent results in aquaponic systems are:

- Tilapia
- Blue Gill/Brim
- Sunfish
- Crappie
- Koi
- Fancy Goldfish
- Pacu
- Various ornamental fish such as angelfish, guppies, tetras, swordfish, mollies

Other fish commonly raised in aquaponics:
- Carp
- Barramundi
- Silver Perch
- Golden Perch
- Yellow Perch
- Catfish
- Bass

Best Fish Species for Profitability

When deciding on which species to produce in order to generate the maximum revenue, consider the following issues:

- **Choose a marketable species.** A good example of a species that is easy to produce, but can be difficult to market, is common carp. It is advisable to consider other, more widely accepted species. Of the sixty or so potential fish species used for food, channel catfish, tilapia, crawfish, rainbow trout, and salmon have large, established industries in the United States. Other species such as hybrid striped bass, and various sunfishes also offer considerable potential.

- **Know the complete production cycle.** Without complete production information, trying to raise some species can be a very risky venture. Although species such as walleye, shrimp, and lobsters have wide public appeal and are widely consumed, each has production peculiarities and challenges.

- **Raise a variety of species, if possible.** Many market outlets, for the smaller-scale fish producer, prefer buying a lesser quantity of numerous species. Production of more than one species may offer a competitive edge over single species operations. If production of a variety of species is not feasible, pooling resources with other producers may enhance species' availability.

CHAPTER 11: FISH — EVERYTHING YOU NEED TO KNOW

- **Raise a species that is in high demand.** When choosing what fish to rear, you will need to consider what are the types of fish are in high demand in the market where you are located and whether it suits your climate.
- In addition, you may also consider growing fish, such as goldfish or Koi, which can be sold as pets. Some Koi varieties bring in considerable money per fish.

Fish Details

In working with fish and fish feed it is easier to use metric units, and such is the industry standard; even in the United States. Following is a useful list of proven aquaponic fish with a few details about each species.

Warm Water Fish

BARRAMUNDI

The barramundi or Asian seabass (Lates calcarifer) is a species widely distributed in the Indo-West Pacific region from the Persian Gulf, through Southeast Asia, to Papua New Guinea and Northern Australia. Barramundi grown in the USA are fed a mostly vegetarian diet, and a small amount of fishmeal (which typically comes from partially from herring by-products). They are net protein producers, with a conversion of about 0.9 pounds. This high feed conversion ratio (feed-conversion efficiency), greatly increases return on investment (ROI), which makes it especially attractive to commercial producers.

Growing your own Barramundi excites guests and often generates a lot of stimulating conversation in the community. Barramundi provide a decent harvest at the end of the season and are one of the more majestic species of edible fish.

Barramundi is very popular in Thai cuisine. Barramundi grown in an aquaponic system has an exceptionally clean, crisp taste, with a varying amount of body fat.

TEMPERATURE RANGE: 26–29° C (79–84 ° F)
SIZE: Common between 25 and 100 cm (10 to 40 inches), but can grow to 200 cm (6.5 feet).

BLUEGILL

Bluegill are more sensitive to temperature, pH, and water quality conditions. They prefer a temperature range around 80°F (26.6°C). They can be raised on pellet food and are harvested in 12–16 months.

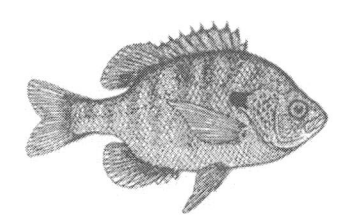

SCIENTIFIC NAME: Lepomis macrochirus
PRODUCTION POTENTIAL: Easy
MARKETING POTENTIAL: Moderate
MARKET SIZE: Three fish per pound
MARKET: Sport or food
TEMPERATURE REQUIREMENTS: *Growing*: 55–80°F, *Spawning*: 75-80°F, best at 80°F
FEED REQUIREMENTS: *Protein*: 32%, *Fat*: 8–10%
SPAWNING REQUIREMENTS: Spawn in spring, nest builders, 50,000 eggs/lb body weight

PART II: PLANTS AND FISH

CARP

There are many species of carp that could be very well suited to aquaponics. Unfortunately because of their reproductive capabilities, their tough nature, and ability to readily adapt in many areas of the world, carp have become noxious pests to native waterways and disrupt natural ecosystems. As a result, they are listed as an invasive species, not being permitted to be raised or farmed in many regions of the country; doing so could lead to hefty fines, high fees, and/or strict requirements for keeping them.

In most western cultures, carp have a fairly poor reputation as an eating fish. Never the less, carp is still the most widely cultured fish in the world as it's grown throughout most of Asia. They can also be considered as source of revenue in regards to selling them as an animal based food source, if local laws allow. Carp prefer a temperature range of 65–75°F (18–24°C). Carp are omnivorous and consume flake or pellet foods, bugs, plant roots.

SCIENTIFIC NAME: Cyprinus carpio
PRODUCTION POTENTIAL: Easy
MARKETING POTENTIAL: Moderate
MARKET SIZE: 3–5 lbs
MARKET: Food
TEMPERATURE REQUIREMENTS: *Growing*: 55–80°F, *Spawning*: Above 65°F
FEED REQUIREMENTS: *Protein*: 31–38%, *Fat*: 3–8%
SPAWNING REQUIREMENTS: 60,000 eggs/lb. body weight; eggs hatch in 2–7 days

CHANNEL CATFISH

There are many different species of catfish around the world that are well suited to aquaponics. However, channel catfish are the most widely farmed aquaculture species in the United States, so they are readily available in most areas. Catfish don't have scales so they need to be skinned, they are quick growing, and have a good food conversion ratio. Catfish can be harvested in 5–10 months.

They are a good fish for warmer climates. Catfish grow best at a temperature of around 80°F (26.6°C). They are very resistant to disease, but are more sensitive to dissolved oxygen levels. Therefore, it is wise to diligently monitor those levels regularly. Install a good aeration device if DO levels tend to average less than 5 ppm in your system.

SCIENTIFIC NAME: Ictalurus punctatus
PRODUCTION POTENTIAL: Easy
MARKETING POTENTIAL: High
MARKET SIZE: 0.75–2.5 lb.
MARKET: Food or sport
TEMPERATURE REQUIREMENTS: *Growing*: 80–85°F, *Spawning*: 72–82°F
FEED REQUIREMENTS: *Protein*: 28–30%, *Fat*: 6–12%
SPAWNING REQUIREMENTS: Annual spring spawners lay eggs in cavities at a rate of 3,000–4,000 eggs/lb body weight; eggs hatch in 7–8 days at 78°F.

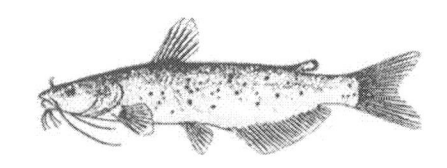

GOLDFISH

Although some people may group these fish with carp, most folks, pet shops, and other fish suppliers refer to them as goldfish. They are generally resilient, withstanding imperfect conditions, and are therefore often used in smaller aquaponic systems by beginners and hobbyist.

If you do not intend to rear aquaponics fish for consumption, rearing goldfish is an option worth consideration. The excess goldfish can be sold to your local pet shop or aquariums for profit.

In many areas they will breed in a tank, although they generally need plant cover within the tank to breed.

Goldfish prefer a temperature range of 65–75°F (18–24°C). Goldfish are omnivorous and consume flake or pellet foods, bugs, plant roots.

COMMON NAME: The comet variety is the most common type of goldfish, but there have been many other varieties developed such as black moors, calico, koi, and shubunkins.

SCIENTIFIC NAME: Carassius auratus

PRODUCTION POTENTIAL: Easy

MARKETING POTENTIAL: Moderate

MARKET SIZE: 1–6 inches for ornamental and 1–2 inches for feeder fish.

MARKET: Ornamental

TEMPERATURE REQUIREMENTS: *Growing*: 70°F, *Spawning*: Above 60°F

FEED REQUIREMENTS: *Protein*: 30–38%, *Fat*: NA

SPAWNING REQUIREMENTS: Instinctively spawns repeatedly from May to June, eggs hatch in 2–8 days, 50,000 eggs/lb. body weight.

An additional note on spawning goldfish. Their spawning period can be modified with artificial lighting. The primary method used is the egg transfer method. In this method, the broodstock spawn on spawning mats which are placed in shallow water along a pond's shore or in a separate tank. When mats are covered with eggs, they are then moved to rearing ponds or a tank.

HYBRID STRIPED BASS, SUNSHINE BASS, PALMETTO BASS, WIPER

A hybrid striped bass, also known as a wiper or whiterock bass, is a hybrid between the striped bass (Morone saxatilis) and the white bass (M. chrysops). Hybrid striped bass are known for aggressive feeding habits which makes them highly sought after by anglers. Often schooling by the thousands, these stocked fish will surface feed on baitfish-like shad. Their quality as a hard-fighting gamefish is closely followed by their delicious, firm, white, flaky meat. Many restaurants sell "striped bass" on their menus, but what you are really eating are farm raised hybrid striped bass.

SCIENTIFIC NAME: Morone saxatilis x M. chrysops or M. chrysops x M. saxatilis
PRODUCTION POTENTIAL: Moderate
MARKETING POTENTIAL: Moderate
MARKET SIZE: 1.5–3.5 lbs
MARKET: Food or sport
TEMPERATURE REQUIREMENTS: *Growing*: 77–86°F, *Spawning*: 61–68°F
FEED REQUIREMENTS: *Protein*: 45–50%, *Fat*: 15%
MOST COMMON PRODUCTION SYSTEMS: Ponds, recycle systems, and cages.
SPAWNING REQUIREMENTS: Instinctively spawns in late March through May. Cross made with either female white bass and male striped bass (sunshine bass), or with female striped bass and male white bass (palmetto bass). Female striped bass average 100,000 eggs/lb. body weight (25,000 eggs per ounce of spawn) and white bass females average 50,000 eggs/lb. body weight (100,000 eggs per spawn). Eggs hatch in two to five days. Survival of fry is low, less than 50 percent.

JADE PERCH

This native Australian fish has the highest levels of omega three oils of any fish species in the world. In fact, it's so high that growers are trying to breed a less oily fish because they've found people don't like the high content. They require warm water and an omnivorous diet. They are very well suited to aquaponics and grow quickly. Fingerlings are typically readily available in warmer areas.

KOI

Another species of carp, but better known as "Koi", are very common within many Asian communities. They are often found in ornamental ponds and public fish tanks throughout the world. For those who love Koi, an aquaponic system is a great proposition for stocking the fish. In terms of revenue, some individuals with distinct color patterns from certain varieties, can be quite lucrative bringing in tens of thousands of dollars each. Koi prefer a temperature range of 65–75°F (18–24°C). Koi are omnivores and consume flake or pellet foods, bugs, and plant roots.

LARGEMOUTH BASS

Largemouth Bass can be successfully grown in an aquaponic system, but it requires a vigilant patient grower to do so. They are much less tolerant to unfavorable water conditions than the Tilapia. They also take between 16-17 months to produce a table-ready fish. The longer it takes to grow, the higher probability that something will go wrong along the way. They require fairly delicate handling. They are stressed by bright light and cannot tolerate poor nutrition. They are one of the most sensitive fish to raise; and they are more sensitive to water temperature and quality, as well as pH and oxygen levels. The younger fingerlings must be trained to feed on pellets. Largemouth bass prefer temperatures between 77 and 86 degrees. Bass seldom feed at temperatures below 50 degrees and cannot survive for long at temperatures above 98 degrees.

When keeping bass in tanks, it is prudent to use some sort of cover or fencing/netting around the tank as they are prone to jump out.

SCIENTIFIC NAME: Micropterus salmoides
PRODUCTION POTENTIAL: Moderate
MARKETING POTENTIAL: High
MARKET SIZE: 1.0–2.5 lbs.
MARKET: Sport or food
TEMPERATURE REQUIREMENTS: *Growing*: 60–80°F, *Spawning*: 60–65°F
FEED REQUIREMENTS: *Protein*: 40%. These diets are normally fed from fingerlings to adults after fingerlings have been trained to accept commercial diets. *Fat*: NA
SPAWNING REQUIREMENTS: Instinct spawning runs from spring through summer when temperatures are around 75°F (24°C). Nest layers,;13,000 eggs/lb body weight. Eggs hatch in approximately four days.

MURRAY COD

Murray cod are a magnificent native Australian fish, known to grow to enormous sizes in their native habitats. Murray cod are widely grown in recirculating aquaculture systems; butcan also been grown in aquaponic systems. They are quick growing and a delicious edible fish. One disadvantage is that they must be kept at high stocking densities, and kept well fed otherwise they cannibalize each other.

SILVER PERCH

Silver perch, a native Australian fish, grow well under a variety of conditions. Perch are omnivorous and eat pellet foods, green scraps, and Duckweed and Azolla. They grow within a wide temperature range. However, they are not as fast growing as many other fish, taking 12–18 months for fingerlings to grow to plate size. They prefer a temperature range around 80°F (26.6°C).

TILAPIA

Tilapia is a species that is extremely popular in aquaponics systems. They are an ideal species for aquaponics for many reasons. They are easy to breed, fast growing, can withstand poor water conditions, consume an omnivorous diet, and are a delicious edible fish. They are very resilient as well, being less sensitive than other fish species to fluctuations in temperature, pH, dissolved oxygen levels, and a build-up in waste. They can also grow well in crowded tanks. Tilapias have especially high beneficial nutritional ratios of omega-6 to omega-3 fatty acids. Tilapias grow to plate size in about 6-9 months.

A potential downfall is that they may not be the most ideal species for aquaponic systems located in cooler climates as they require warm water. Every breed of tilapia is different, but most tilapia prefer temperatures of 77 to 86°F (25-30°C). They start to lose their resistance to disease and infections below 54°F (12°C). If you live in a cooler area and want to avoid large heating costs, you are much better off growing a fish species that will do well in your temperature range. Tilapias are also a declared an invasive species in many areas, and therefore are prohibited by U.S. Fish and Game laws, and/or other governing agencies.

There are over 100 different species of tilapia, each with unique characteristics and behavior. Members of the Oreochromis genus are a common choice for aquaponics or aquaculture. In terms of popularity, the Nile tilapia (*O. niloticus*) is the most widely cultured tilapia, followed by Blue tilapia (*O. aureus*), and Mozambique tilapia (*O. mossambicus*). All can be reproduced fairly easily by a beginner. Refer to the 'Chapter 32: Fish Breeding, Fish Reproduction, and Raising Your Own Crop of Fish' for details on reproducing your own Tilapia.

Tilapias start out as omnivores, but later become more like vegetarians. The Tilapia's digestive system is designed to eat algae, vegetation, other small fish, worms, and insects. The newly hatched, called 'fry', require a lot of protein for fast growth. You can actually stunt their growth by under feeding them, which is what some breeders do to keep their fish in the sellable fingerling size range. As they mature, they require less protein and are not interested in eating their own fry, which is why it's possible to raise fry with adult Tilapia in a single tank. Be sure to find a high quality tilapia feed that takes all this under consideration. Be diligent in checking out the protein and fat contents in the fish feed formula when shopping to ensure that it provides the necessary nutrients for each stage of the Tilapia's growth.

Tilapias are often fed pellet fish food, duckweed, and green plant products produced from the system. Some operators grow algae and duckweed in separate tanks to feed their Tilapia.

Commercial tilapia farms will usually feed their fish pellets made from fishmeal, grain, soybeans or other food products. In the wild, tilapia will eat vegetation, algae, plankton, insects, larvae, decaying organic matter, fish wastes, small fish and just about anything edible that they can get in their mouth. As stated, Tilapia can be vegetarians and survive just fine by eating only algae.

Tilapia can be cultured as mixed-sexes (males and females together) or as mono-sex (males only). Most large scale commercial growers prefer to raise only male tilapia because they grow faster than the females. Male tilapia populations can be produced by visual selection, hybridization, sex-reversal, and genetic manipulation. Tilapia with a minimum weight of 25 to 30 g can be separated by visual inspection of the genital papilla.

During the last 10 to 15 years, the most popular way to produce all-male populations is with hormone sex reversal of tilapia fry. Recently hatched tilapia fry obtained by harvest from spawning containers 18 days after brood fish are stocked or hatched from eggs taken from females are fed a powdered diet containing a male steroid for 20 to 28 days. Fry that would have been females if fed a steroid-free diet, will be functional males at the end of the hormone treatment. While all-male populations are hard to produce with sex reversal treatment, 95 to 98 percent males are commonly produced.

Tilapia reach sexual maturity at 3 to 5 months of age. Chapter 32 is dedicated to reproducing and raising your tilapia.

SCIENTIFIC NAME: Oreochromis niloticus, O. aurea, O. mossambicus, O. hornorum
PRODUCTION POTENTIAL: Easy except tilapia grow slowly at temperatures lower than 70°F and die when temperatures drop into the 50°F range.
MARKETING POTENTIAL: High
MARKET SIZE: 1.25–2.0 lbs
MARKET: Food
TEMPERATURE REQUIREMENTS: *Growing*: 80–87°F, *Spawning*: Greater than 72°F, Lethal: 55°F
FEED REQUIREMENTS: *Protein*: 25–30%, *Fat*: 6–8%
SPAWNING REQUIREMENTS: Maternal mouth brooders, spawn twice a month, 2,500 eggs/lb body weight, eggs hatch in 5-7 days. All male hybrids can be produced by crossing female O. niloticus and male O. aurea. or by crossing female O. niloticus and male O. hornorum. Stocking ratios for fingerling production is three females to each male.

Tilapia Feed Conversion Ratio Feeding Rates, Growth Rate, and Harvesting

The feed conversion ratio (FCR) or feed conversion efficiency (FCE), to define it simply, is a measure of an animal's efficiency in converting feed mass into increased body mass. Tilapia have an average feed conversion rate of 1.7. This 1.7 to 1 ration is incredibly efficient, especially compared to cattle which have a FCR of 8:1 (8-lbs of feed for every 1-lb of mass).

It takes between 7 and 9 months for a tilapia to reach harvest size of 1.25-lbs (0.57-kg or 567-grams) which produces two, 4-ounce (113-grams) fillets, or enough for a meal for two people. Commercial growers, using supplemental feedings and intensive management techniques, are able to achieve harvest of a 500-gram fish within five months. Keep in mind, though, that premium prices for your fish will only be achieved via being able to honestly market your fish as being grown through sustainable farming practices, fully organic (no artificial supplements & organic based feed products), in an animal friendly environment.

Tilapia can be harvested anytime they reach a marketable size of 1 to 2 pounds. Tilapia grow fast and usually reach market size in 6 to 12 months, depending on the feed and conditions. Large commercial operators generally harvest all of their fish at one time; whereas smaller operators typically harvest their fish over a period of time. Both options have their advantages. Harvesting all at once is more efficient, and the fish usually are sold to one buyer (i.e. a food company). For the small to mid-size operator, harvesting over time relieves system overcrowding, minimizes water quality challenges, and proves a steady supply of fresh fish for meals and/or an ongoing revenue stream.

Tilapia are most commonly processed into skinless, boneless fillets. The fillet yield is typically 30 to 37 percent of their total body weight, depending on fillet size and final trim. [1-ounce = 28.3 grams]

PART II: PLANTS AND FISH

TABLE 6. *Tilapia Growth and Feeding Rates*

MONTH	START WEIGHT (g)	END WEIGHT (g)	GROWTH RATE (g/day)	FEEDING RATE (% weight)
1	1	5	0.2	15–10
2	5	20	0.5	10–7
3	20	50	1.0	7–4
4	50	100	1.5	4–3.5
5	100	165	2.0	3.5–2.5
6	165	250	2.5	2.5–1.5
7	250	350	3.0	1.5–1.25
8	350	475	4.0	1.25–1
9	475	625	5.0	1

Cold Water Fish

TROUT

Trout are a good choice of fish for aquaponic systems located in cooler climates, where water temperatures can efficiently be kept cooler. Trout are best suited for water temperatures between 50°F (10°C) and 60°F (20°C). They have extremely fast growth rates and excellent food conversion ratios. They can be trained to be carnivorous, but will typically always eat smaller fish given an opportunity. They require high dissolved oxygen levels. They are sensitive to pH changes and water quality. Most aquaponic operators who raise trout feed them pellet fish food. They reach **plate size** in 12-16 months.

SCIENTIFIC NAME: Oncorhynchus mykiss
PRODUCTION POTENTIAL: Easy
MARKETING POTENTIAL: High
MARKET SIZE: 0/5–1.5 lbs.
MARKET: Food or sport
TEMPERATURE REQUIREMENTS: *Growing*: 50–60°F, *Spawning*: 50–55°F, *Lethal*: Greater than 70°F
FEED REQUIREMENTS: *Protein*: 45–50%, *Fat*: 12–15%
SPAWNING REQUIREMENTS: Spring and fall spawning strains, 500–10,000 eggs depending on size of female, eggs hatch in 24–31 days at 50–55°F.

TABLE 7. *Common Aquaponic Edible Fish*

LIFE SUPPORT	BASS	TROUT	TILAPIA
Thriving Temp.	74°–80°F 24°–26°C	55°–65°F 13°–18°C	74°–80°F 24°–26°C
Surviving Temp.	40°–90°F 5°–32°C	38°–68°F 4°–20°C	60°–95°F 16°–35°C
Carnivore or Omnivore	Carnivore	Carnivore	Omnivore
Oxygen Needs	Low	High	Low

Fish Overview — Common Aquaponic Edible Fish

Nearly all freshwater fish are edible, but their taste, production potential, and growing parameters must be taken into consideration. It is vital to take good care of your fish since they are the lifeline of your aquaponics system. Managing water quality and using only top-rated fish food helps ensure that your fish stay.

Another thing to consider is whether to buy fry or fingerlings. Fish fry is cheaper, but takes longer to mature; which means that they take longer to produce adequate nitrate levels for plants to absorb. On the other hand, fingerlings are more expensive but they produce more waste in the early stage optimization stages, which means an earlier start in growing vegetables.

Acclimatizing Fish

Acclimatizing fish into new tanks can be a highly stressful process for fish, particularly the actual transport from one location to another in bags or small tanks. It is important to try to remove as many stressful factors as possible that can cause fatality in new fish. There are two main factors that cause stress when acclimatizing fish: changes in temperature and pH between the original water and new water; these must be kept to a minimum.

The pH of the culture water and transport water should ideally be tested. If the pH values are more than 0.5 different, then the fish will need at least 24 hours to adjust. Keep the fish in a small aerated tank of their original water and slowly add water from the new tank over the course of a day. Even if the pH values of the two environments are fairly close, the fish still need to acclimatize. The best method to do this is to slowly allow the temperature to equilibrate by floating the sealed transportation bags containing the fish in the culture water. This should be done for at least 15 minutes. At this time small amounts of water should be added from the culture water to the transport water with the fish. Again, this should take at least 15 minutes, so as to slowly acclimatize the fish. Finally, the fish can be added to the new tank.

Stocking Density

The recommended stocking density is 1 pound of fish per 5–7 gallons of tank water (0.5 kg per 20-26 liters). However, a lighter fish stock density has been found to be more forgiving if things go wrong (pump malfunction, significant leak not immediately discovered, etc.).

Fish selection should take into account the following:

- Edible (i.e. Tilapia) vs. ornamental (i.e. Koi, goldfish, etc.).
- Water temperature based upon your climate (recommended).
- Diet of fish: Carnivore vs. omnivore vs. herbivore
- In order to maintain the perfect balance between fish and plant population density, it is best to progressively harvest the fish as soon as they are big enough.

FIGURE 47. Acclimatizing fish. Juvenile fish are transported in a plastic bag (a) which is floated in the receiving tank (b) and the fish are released (c).

Keep in mind that there are other factors affecting the number of fish that can be grown in an aquaponic system, such as the species of fish and feed rates. The more the fish are fed, the more waste they produce. It is important to monitor system parameters such as water flow-rates, oxygen levels, pump rates, and water temperature as they all play a critical role as well. All these issues are discussed in detail in their respective chapters of this book.

Harvesting and Staggered Fish Stock

A constant biomass of fish in the tanks ensures a constant supply of nutrients to the plants. This ensures that the fish eat the amount of feed calculated using the feed rate ratio. The next chapter, which covers feed conversion rates, will show how the feeding ration depends on the size of the fish, and small fish are not be able to eat enough feed to supply the full growing area with adequate nutrients. To achieve a constant biomass in the fish tanks, a staggered stocking method should be adopted. This technique involves maintaining three age classes, or cohorts, within the same tank. Approximately every three months, the mature fish (500 g each) are harvested and immediately restocked with new fingerlings (50 g each). This method avoids harvesting all the fish at once, and instead retains a more consistent biomass.

Table 8, using tilapia as an example, outlines the potential growth in one tank over a year using the staggered stocking method. The important aspect of this table is that the total weight of the fish varies between 10–25 kg, with an average biomass of 17 kg. This table is a basic guideline depicting optimum conditions for fish growth. In reality factors such as water temperature and stressful environments for fish can distort the figures presented here.

If it is not possible to obtain fingerlings regularly, an aquaponic system can be still managed by stocking a higher number of juvenile fish and by progressively harvesting them during the season to maintain a stable biomass to fertilize the plants. Table 9 on the following page shows the case of a system stocked every six months with tilapia fingerlings of 50 g. In this case, the first harvest starts from the third month onward. Various combinations in stocking frequency, fish number, and weight can apply providing that fish biomass stands below the maximum limit of 20 kg/m3. If the fish are mixed-sex, the harvest must

TABLE 8. **Potential growth rates of tilapia in one tank over a year using a progressive harvest technique**

MONTH	DEC	JAN	FEB	MAR	APR	MAY	JUN	JUL	AUG	SEP	OCT	NOV	DEC
STOCKING ROUND	WEIGHT (KG)												
1	1.5	3.75	6.0	8.25	10.5	12.75	15.0*						
2				1.5	3.75	6.0	8.25	10.5	12.75	15.0*			
3							1.5	3.75	6.0	8.25	10.5	12.75	15.0*
4										1.5	3.75	6	8.25
5													1.5
Total fish mass (kg)	1.5	3.75	6.0	9.75	14.25	18.75	24.75–9.75	14.25	18.75	24.75–9.75	14.25	18.75	24.75–9.75
Action							Restock harvest			Restock harvest			Restock harvest

NOTES: Fingerling tilapia (1.5 kg = 50 g/fish x 30 fish) are stocked every three months. Each fish survives and grows to harvest size (15g = 500 g/fish x 30 fish) in six months. The asterisk indicates harvest. The range during harvest/stocking months accounts for the range if not all 30 fish are taken at once, i.e. the 30 mature fish are harvested throughout the month. This table serves only as a theoretical guide to illustrate staggered harvest and stocking in ideal conditions.

TABLE 9. **Potential growth rates of tilapia in one tank over a year using a progressive harvest technique**

MONTH	DEC	JAN	FEB	MAR	APR	MAY	JUN	JUL	AUG	SEP	OCT	NOV	DEC
STOCKING ROUND 1													
Number of fish in tank	80	80	70	60	50	40	30	10					
Fish weight (g)	50	125	200	275	350	425	500	575					
Cohort biomass (kg)	4	10	14	17	18	17	15	5.8					
STOCKING ROUND 2													
Number of fish in tank							80	80	70	60	50	40	30
Fish weight (g)							50	125	200	275	350	425	500
Cohort biomass (kg)							4	10	14	17	18	17	15
Total tank biomass (kg)	4	10	14	17	18	17	19	15.8	14	17	18	17	15

NOTES: Tilapia fingerling are stocked every six months. Staggered harvest starts from the third month to keep the total fish below the maximum stocking biomass of 20 kg/m^3. The table shows the theoretical weight of each batch of harvested fish along the year if fish are reared in ideal conditions.

firstly target the females to avoid breeding when they reach sexual maturity from the age of five months. Breeding depresses the whole cohort. In the case of mixed-sex tilapia, fish can be initially stocked in a cage and males can then be left free in the tank after sex determination.

It is important to note that adult tilapia, catfish, and trout will predate their smaller siblings if they are stocked together. A technique to keep all of these fish safely in the same fish tank is to isolate the smaller ones in a floating frame. This frame is essentially a floating cage, which can be constructed as a cube with PVC pipe used as frame and covered with plastic mesh. It is important to ensure that larger fish cannot enter the floating cage over the top, so make sure that the sides extend at least 6-inches (15 cm) above the water level. Each of the vulnerable size classes should be kept in separate floating frames in the main fish tank. As the fish grow large enough not to be in danger, they can be moved into the main tank. With this method, it is possible to have up to three different stocking weights in one tank, so it is important that the fish feed pellet size can be eaten by all sizes of fish. Caged fish also have the advantage of being closely monitored to determine the FCR by measuring the weight increment and weight of the feed over a period.

Important Notes Regarding Fish

- Add fish only after the fish-less cycling process is complete, if applicable.
- Feed the fish as much as they can eat in 5-10 minutes, two times per day. Always remove uneaten feed after 30 minutes. Record total feed added. Balance the feeding rate with the number of plants using the feed rate ratio, but avoid over or under feeding the fish.
- Fish appetite is directly related to water temperature, particularly for topical fish such as tilapia, so remember to adjust feeding during colder winter months.
- A fingerling tilapia (50 g) will reach harvest size (500 g) in 6–8 weeks under ideal conditions. Staggered stocking is a technique which involves stocking a system with new fingerlings each time some of the mature fish are harvested. It provides a way of maintaining relatively constant biomass, feeding rate and nutrient concentration for the plants.

Other Aquatic Species for Aquaponics

There are other fish species which are quite suitable for aquaponics, which might be available in your local area. In Europe, many different species of carp are grown; within the United States species such as Bluegill are often available, while in Australia there are a number of other native species like Sleepy cod which would be suitable.

Other aquatic animals that are sometimes used in an aquaponic system are fresh water mussels, fresh water prawns, and fresh water crayfish. Mussels are a filter-feeder, and do a great job of helping to clean the water. They will grow in flooded grow beds, or can be incorporated into fish tanks. Crustaceans make a nice addition to an aquaponic system. There are a few different species available depending on your location and water temperatures.

Consideration for those that live in tropical areas is Redclaw, a fast growing native Australian species. For those in cooler areas there's Yabbies or Marron as a viable option.

Aquatic Animals to Avoid

Animals such as turtles and snails may work for general pond watching activity, but they are not good for an aquaponics system. Turtles eat fish while snails eat algae; they usually reproduce beyond control, can plug pumps filters, pipes, and will eat certain plants and plants roots. Growing snail shells will also deplete your system of carbonates causing a drop in pH level. Be sure to thoroughly research before adding new life to your aquaponics system. It is also important to keep domestic pets out of your system.

Keeping Fish Healthy

An aquaponic system fish tank has inputs of food, water, and oxygen. The fish tank has outputs of urine, ammonia, carbon dioxide, feces, and uneaten food.

Although each fish species is unique in regards to the ideal growing conditions, there are some generalities that are common for all fish. For instance, most fish need the pH to be somewhere between 6 and 8 (although the pH range of 6.8-7.0 is optimal for plants and fish) . Ammonia and nitrites are very toxic to fish. Nitrates are fairly safe for fish (and great for plants). Fish need oxygen (they will die within 30 minutes without it). Drastic temperature changes can cause health issues and even result in death. Fish are sensitive to light, and do best when not exposed to direct light. Fish need to be temperature matched to the water before releasing them into the tank.

TABLE 10. *Cause and symptoms of stress in fish*

CAUSES OF STRESS	SYMPTOMS OF STRESS
Temperature outside of range, or fast temperature changes	Poor appetite
pH outside of range, or fast pH changes (more than 0.3/day)	Unusual swimming behavior, resting at surface or bottom
Ammonia, nitrite or toxins present in high levels	Rubbing or scraping the sides of the tank, piping at surface, red blotches and streaks
Dissolved oxygen is too low	Piping at surface
Malnourishment and/or overcrowding	Fins are clamped close to their body, physical injuries
Poor water quality	Fast breathing
Poor fish handling, noise or light disturbance	Erratic behavior
Bullying companions	Physical injuries

Fish Maintenance
The following are general rules to keep in mind:
- Feed fish 2-3 times a day, but don't overfeed.
- Fish eat 1.5-2.0 percent of their body weight per day.
- Only feed fish what they can eat in 5-10 minutes.
- Fish won't eat if they are too cold, too hot, or stressed.
- Check water quality regularly.
- Add water or do partial water changes when necessary.
- In addition to checking water conditions of the fish tank, observe fish behavior and appearance. After becoming familiar with your fish, you will often be able to tell if there is something wrong or different.
- Some fish desire to be your friend and will regularly "greet you" at the tank.

Types of Fish Disease
Fish ailments can be separated into four general types:
- Bacterial Diseases
- Fungal Diseases
- Parasitic Diseases
- Physical Ailments

How to Prevent Fish Diseases
Precautions can and should be taken to reduce the possibility of your fish getting a disease. Following these precautions can also help keep fish diseases from spreading if they do occur:
- Buy only good quality, compatible fish, and then breed/raise your own fish thereafter.
- Quarantine new fish in a separate tank before adding them to your aquaponic system.
- Avoid stressing the fish with rough handling, sudden changes in conditions, or "bully" tank mates.
- Do not overfeed your fish.
- Remove sick fish to a separate tank for treatment.
- Disinfect nets used to move sick fish.
- Do not transfer water from the quarantine tank to the main aquarium.
- Do not let any metal that has the propensity to rust come in contact with your system's water.

Recognizing Disease
Diseases may occur even with all of the prevention techniques listed above. It is important to stay vigilant and monitor and observe fish behaviour daily to recognize the diseases early. The following lists outline common physical and behavioural symptoms of diseases.

External signs of disease:
- Ulcers on body surface, discoloured patches, white or black spots,
- Ragged fins, exposed fin rays,
- Gill and fin necrosis and decay,
- Abnormal body configuration, twisted spine, deformed jaws,
- Extended abdomen, swollen appearance,
- Cotton-like lesions on the body,
- Swollen, popped-out eyes (exophthalmia).

Behavioural signs of disease:
- Poor appetite, changes in feeding habits,
- Lethargy, different swimming patterns, listlessness,
- Odd position in water, head or tail down, difficulty maintaining buoyancy,
- Fish gasping at the surface,
- Fish rubbing or scraping against objects.

Abiotic Diseases: Most of the mortalities in aquaponics are not caused by pathogens, but rather by abiotic causes mainly related to water quality or toxicity. Nevertheless, poor water quality conditions can induce opportunistic infections that can easily occur in unhealthy or stressed fish.

Biotic Diseases: In general, aquaponics and recirculating systems are less affected than pond or cage aquaculture farming by pathogens. In most cases, pathogens are actually already present in the system, but disease does not occur because the fishes' immune system is resisting infection and the environment is unfavourable for the pathogen to thrive.

Healthy management, stress avoidance, and quality control of water are thus necessary to minimize any disease incidence. Whenever disease occurs, it is important to isolate or eliminate the infected fish from the rest of the stock and implement strategies to prevent any transmission risk to the rest of the stock. If any cure is put into action, it is fundamental that the fish be treated in a quarantine tank, and that any products used are not introduced into the aquaponic system. This is in order to avoid any unpredictable consequences to the beneficial bacteria. More details on this subject are available from regulatory agencies, such as the U.S. Fish and Wildlife Service, online research, and most local fishery organizations.

Beware of Fish Disease Treatments

Many recommended ailment treatments, although effective, are not conducive to an aquaponic system. For instance, antibiotics will kill beneficial bacteria and copper treatments can be harmful to your plants. These treatments should only be used in a quarantine situation, and the water from the quarantine tank should never be added to the aquaponics system. All recommended treatments need to be carefully scrutinized to ensure that they will not negatively impact the aquaponics system, as the vast majority of fish experts address disease and treatments based upon isolated conditions (i.e. aquaculture, aquarium, quarantine tank), and not within the scope of an aquaponics system.

Steps to Treating Sick Fish

If a significant percentage of fish are showing signs of disease, it is likely that the environmental conditions are causing stress. In these cases, check levels of ammonia, nitrite, nitrate, pH, and temperature in order to respond accordingly. If only a few fish are affected, it is important to remove the infected fish immediately in order to prevent the spread of the any disease to the other fish. Once removed, inspect the fish carefully and attempt to determine the specific disease and cause.

However, it may be necessary to have a professional diagnosis carried out by a veterinarian, extension agent, or other aquaculture expert. Knowing the specific disease helps to determine the treatment options. Place the affected fish in a separate tank, sometimes called a "quarantine" or "hospital tank", for further observation. Kill and dispose of the fish, as appropriate.

Disease treatment options in small-scale aquaponics are limited. Commercial drugs can be expensive and/or difficult to procure. Moreover, antibacterial and antiparasite treatments have detrimental effects on the rest of the system, including the biofilter and plants. If treatment is absolutely necessary, it should be done in a hospital tank only; antibacterial chemicals should never be added to an aquaponic unit. One effective treatment options against some of the most common bacterial and parasite infections is a salt bath.

Salt Bath Treatment: Fish affected with some ectoparasites, moulds, and bacterial gill contamination can benefit from salt bath treatment. Infected fish can be removed from the main fish tank and placed into a salt bath. This salt bath is toxic to the pathogens, but non-fatal to the fish. The salt concentration for the bath should be 1 kg of salt per 100 liters of water. Affected fish should be placed in this salty solution for 20–30 minutes, and then moved to a second isolation tank containing 1–2 g of salt per liter of water for another 5–7 days.

With bad white-spot infections, all fish may need to be removed from the main aquaponic system and treated this way for at least a week. During this time, any emerging parasites in the aquaponic unit will fail to find a host and eventually die. The heating of the water in the aquaponic system can also shorten the parasite life cycle and make the salt treatment more effective. Do not use any of the salt bath water when moving the fish back into the aquaponic system as the salt concentrations would negatively affect the cultured plants.

CHAPTER 12

Fish Feed

Fish Feed and Nutrition

Fish require the correct balance of proteins, carbohydrates, fats, vitamins, and minerals to grow and be healthy. This type of feed is considered a whole feed. Commercial fish feed pellets are readily available from numerous sources from online suppliers to local feed stores.

Protein is the most important component for building fish mass. In their grow-out stage, omnivorous fish, such as tilapia and common carp, need 25–35 percent protein in their diet, while carnivorous fish need up to 45 percent in order to grow at optimal levels. In general, younger fish (fry and fingerlings) require a diet richer in protein than during their grow-out stage. Proteins are the basis of structure and enzymes in all living organisms. Proteins consist of amino acids, some of which are synthesized by the fishes' bodies, but others have to be obtained from the food. These are called essential amino acids. Of the ten essential amino acids, methionine and lysine are often limiting factors, and these need to be supplemented in some vegetable-based feeds.

Lipids are fats, which are high-energy molecules necessary to a fish's diet. Fish oil is a common component of fish feeds. Fish oil is high in two special types of fats, omega-3 and omega-6, that have health benefits for humans. The amount of these healthy lipids in farmed fish depends on the feed used.

Carbohydrates consist of starches and sugars. This component of the feed is an inexpensive ingredient that increases the energy value of the feed. The starch and sugars also help bind the feed together to make a pellet. However, fish do not digest and metabolize carbohydrates very well, and much of this energy can be lost.

Vitamins and minerals are necessary for fish health and growth. Vitamins are organic molecules, synthesized by plants or through manufacturing, that are important for development and immune system function. Minerals are inorganic elements. These minerals are necessary for the fish to synthesis their own body components (bone), vitamins, and cellular structures. Some minerals are also involved in osmotic regulation.

Pelletized Fish Feed

There are a number of different sizes of fish feed pellets. Obviously, the recommended size of pellet depends on the size of the fish. Fry and fingerlings have small mouths and cannot ingest large pellets, while large fish waste energy if the pellets are too small. If possible, the feed should be purchased for each stage of the lifecycle of the fish. Alternatively, large pellets can be crushed with a mortar and pestle to create powder for fry and crumbles for fingerlings. Some operators use the same medium-sized pellets (2–4 mm) so that their fish continue to eat the same-sized pellet from the fingerling stage right up to maturity.

Fish feed pellets are also designed to either float on the surface or sink to the bottom of the tank, depending on the feeding habits of the fish. It is important to know the eating behaviour of your specific fish and supply the correct type of pellet. Floating pellets are advantageous because it is easier to identify how much the fish are eating. It is often possible to train fish to feed according to the food pellets available; however, some fish will not change their feeding culture.

Feed should be stored in dark, dry, cool, and secure conditions. Fish feed will attract rodents and other vermin, if not securely contained. Warm wet fish feed can rot, being decomposed by bacteria and fungi. These micro-organisms can release toxins that are dangerous to fish; spoiled feed should never be fed to fish. Fish feed should not be stored for too long, should be purchased fresh, and used immediately to conserve the nutritional qualities, wherever possible.

Avoid overfeeding your fish. Uneaten food waste should never be left in the aquaponic system. Feed waste from overfeeding is consumed by heterotrophic bacteria, which devours substantial amounts of oxygen. In addition, decomposing food can increase the amount of ammonia and nitrite to toxic levels in a relatively short period. Finally, the uneaten pellets can clog the mechanical filters, leading to decreased water flow and anoxic areas. In general, fish eat all they need to eat in a 30 minute period. If uneaten food is found, lower the amount of feed given at the next feeding.

Fish Feed

The following provides some general information about fish feed:

- Live foods are a good source of supplements and provide variety for carnivorous fish.
- Fish typically eat all their food within 3 minutes. If feed remains after 10 minutes it is a good indication that they are being overfed. Overfeeding will lead to food decomposition toxicity problems. Feeding a variety of different feed brands helps ensure proper nutrition.
- Most commercial fish feeds contain exact protein, carbohydrate, and other vitamin requirements for specific fish.
- Plant based proteins can include soy meal, corn meal, wheat meal, etc.
- Most commercial feeds are between 10 to 35 percent protein.
- Alternative feeds should be considered like duckweed, insects, worms, or black soldier fly larvae.
- Avoid fish meal based feeds as this source is not sustainable or derived from environmental friendly practices. Fish meal feed can also come from animals raised on growth hormones, antibiotics, pesticides, herbicides, and fungicide laced feed, which then enters your food supply.
- Most fish feed is GMO based, which can entail the presence of heavy metals. In turn, this is passed on up the food chain to you and your family in higher concentrations. Heavy metals wreak havoc on our health and well-being. Therefore, try to use only fish feed that is made entirely from non-GMO, USDA certified organic ingredients, is free of terrestrial animal parts, and contains no fish meal or soy. There are many suppliers of organic fish feed. An Internet search for organic fish feed will provide numerous options.
- Many online vendors will provide you with the nearest retail outlet that sell their feed, saving you the shipping cost. Depending upon your location, though, having it shipped to you may be the only option. Teaming up with other aquaponic or aquaculture operators in your area could enable

CHAPTER 12: FISH FEED

you to get better bargains via buying in bulk and sharing the shipping cost.

Feeding Fish — Parameters and Other Considerations

In some operations, feed costs can add up to half or more of the total annual operating costs. For this reason, it is important to use the proper feed and to take measures to be sure that the conversion of feed to fish flesh is efficient. The Feed Conversion Ratio (FCR) describes how efficiently an animal turns its food into growth. It answers the question of how many units of feed are required to grow one unit of animal. FCRs exist for every animal and offer a convenient way to measure the efficiency and costs of raising any given animal.

Fish, in general, have one of the best FCRs of all livestock. The FCR is expressed as the number of pounds of feed required to produce one pound of flesh. Depending on the species being grown, it may take 1.75 to produce 1 pound of fish. In good conditions, tilapias have an FCR of 1.4–1.8, meaning that to grow a 2.2 pound (1.0 kg) tilapia, 3-4 pounds (1.4–1.8 kg) of food is required.

Tracking FCR is not essential in small-scale aquaponics, but it can be useful to do in some circumstances. When changing feeds, it is worth considering how well the fish grow in regard to any cost differences between the feeds. Moreover, when considering starting a small commercial system, it is necessary to calculate the FCR as part of the business plan and/or financial analysis. Even if not concerned about the FCR, it is good practice to periodically weigh a sample of the fish to make sure they are growing well and to understand the balance of the system.

FIGURE 48. Weighing a sample of fish using a weighing scale

This also provides a more accurate growth rate expectation for harvest timing and production. As with all fish handling, weighing is easier in darkness to avoid stressing the fish. Following is a list of simple steps for weighing fish. Weighing fish of the same age, growing in the same tank, is more preferable than heterogeneous cohorts of fish as the measurement should provide more reliable averages.

Simple Steps for Weighing Fish

1. Fill a small bucket with 2.5 gallons (10 liters) of water from the aquaponic system.
2. Place the bucket on a weighing scale and record the weight (tare).
3. Scoop 5 average size fish with a landing-net, drain the landing-net of excess of water for a few second-ds then place the fish into the bucket.
4. Weigh again and record the gross weight.
5. Calculate the total weight of the fish by subtracting the tare from the gross weight.
6. Divide this figure by 5 to retrieve an average weight for each fish.

Repeat steps 1–6 as appropriate. Try to measure 10–20 percent of the fish (preferably no duplicates) for an accurate average.

Periodic weight measurements will give the average growth rate of the fish, which will be obtained by subtracting the average fish weight, calculated above, over two periods. The FCR is obtained by dividing the total feed consumed by the fish by the total growth during a given period, with both values expressed in the same weight unit (i.e. ounces and pounds, or gram or kilogram).

Total feed / Total growth = FCR

The total feed can be obtained by summing all the recorded amount of feed consumed each day. The total growth can be calculated by multiplying the average growth rate by the number of the fish stocked in the tank.

At the grow-out stage, the feeding rate for most cultured fish is 1.5–2 percent of their body weight per day. On average, a 3.5 ounce (100 gram) fish eats 0.04 to 0.07 ounces (1–2 grams) of pelletized fish feed per day.

For example, if you have 75 lbs (34kg) of fish in your tank, multiply 75 lbs x 1.5% (0.015) = 1.125lbs of fish feed daily (34kg x 1.5% (0.015) = 0.5kg).

If the fish still appear to be consuming all of the feed, then increase the amount to 1.75 percent and then to 2 percent (a multiplier of 0.0175 and 0.02, respectively). As mentioned in the previous chapter, in working with fish and fish feed, it is easier to use metric units; such is the industry standard, even in the United States.

Monitor the feeding rate and FCR to determine growth rates and fish appetite helps maintain overall system balance. However, don't just rely the calculations. Observe your fish eating to help determine the proper amount of feed needed above or below the recommended amount. If they rapidly devour all of the feed, they most likely need more. If the fish are showing little interest and much of the feed is being wasted on the bottom of the tank, then they are probably being overfed.

As a friendly reminder to an important and often overlooked issue mentioned earlier in this chapter, in selecting a proper feed, it is important to match the feed size to the size of the fish being fed. Smaller particle or pellet size is required for smaller fish, while larger pellets will be more efficiently used by larger fish.

The feed must be suitable to meet the nutritional needs of the species being cultured. A commercially manufactured feed is available for most species.

Feeding rates will vary by species and production system, but a few common factors contribute to determining these rates:
- Water temperature.
- Water quality.
- Size of the feed particles.
- Palatability of the feed to the fish.
- Frequency of feeding.
- Technique for feed delivery.
- Type of feed used (i.e., floating, sinking, etc.).

Feed can be delivered in a variety of ways, including feeding by hand, using automated stationary feeders, or by allowing the fish to feed themselves using "demand" feeders. The choice of feed delivery system used will be dictated by logistical needs, resources available, amount of feed to be delivered, number of fish to be fed daily, and the size, scope, and type of operation.

Keep carnivorous fry (newborns) separated from bigger fish so they are not eaten. Feed fry a diet of micro worms (nematode), brine shrimp, or soaked oatmeal (soft things). Feed fingerlings (between newborn and mid-grown) small fish flakes.

Extruded trout feed is a proven favorite of the North American Trout industry. Formulated to allow controlled growth and deliver excellent feed conversions, extruded trout feed can lead to more efficient production. It can be fed to numerous species including Trout, Perch, Bass, Sturgeon, Catfish, and Tilapia. The typical composition is shown in table 11:

TABLE 11. *Typical Composition*

GUARANTEED ANALYSIS
Crude Protein, min..............................40%
Crude at, min 12%
Crude Fiber, max 3%
Ash, max...12%

INGREDIENTS

Fish Meal, Soybean Meal, Wheat Flour, Stabilized Fish Oil, Wheat Midds, Poultry By-Product Meal, Blood Meal, Hydrolized Feather Meal, Corn Gluten Meal, Poultry Oil, Vitamin A Acetate, D-Activated Animal Sterol (D_3), Vitamin B12 Supplement, Riboflavin Supplement, Niacin, Folic Acid, Menadione Sodium Bisulphite Complex, Calcium Pantothenate, Pyridoxine Hydrochloride, Thiamine, Biotin, DL Alphatocopherol (E), L-ascorbyl-2-polyphosphate (C), Betaine, Zinc Sulfate, Copper Sulfate, Ferrous Sulfate, Manganese Sulfate, Ethylenediamine Dihydriodide, Ethoxyquin (Anti-Oxidant).

Practical Aspects of Feeding Fish

Hand-feeding Techniques

Hand and mechanized feedings are the two widely practiced techniques. Of these, hand feeding is the recommended one. Calibrated spoons and hand shovels should be used in order to ensure exact and uniform portions of feed.

Loss of appetite among fish is one of the most obvious symptoms of many different problems. It indicates, among others concerns, insufficient oxygen content of water or a developing disease in fish. Therefore, regular daily feeding is an excellent opportunity to observe fish, detect problems, and diagnose diseases.

Demand and Automatic Feeders

Demand feeders are those that release feed according to the appetite of fish. Because some fish species are very greedy, these feeders may allow unnecessary overfeeding of fish unless the portions are controlled.

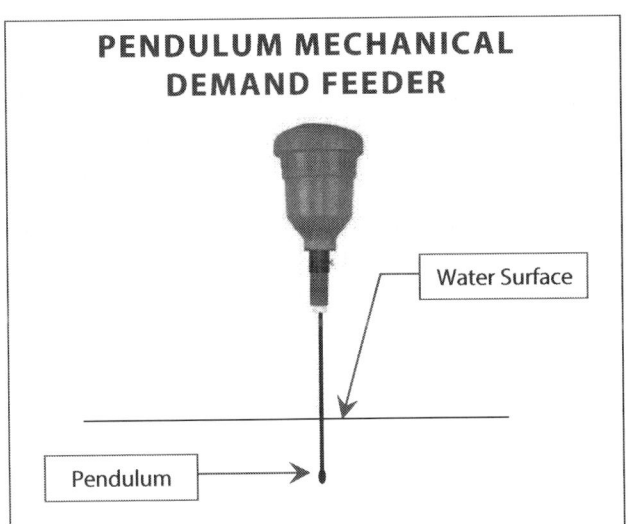

PENDULUM MECHANICAL DEMAND FEEDER

Batteries or an external power sources, are not needed. As fish bump the pendulum, feed drops into the water. Most of these feeders come with a unique twist-lock lid, which is wind- and varmint-proof. Furthermore, the better feeders are made with UV-resistant polyethylene. There are four common sizes, which accommodate #4 crumble to ¼-inch pellets.

Typical Cost (year 2015)

Demand Feeder 22 lbs (10 Kg)	$320.00
Demand Feeder 44 lbs (20 Kg)	$333.00
Demand Feeder 88 lbs (40 Kg)	$410.00
Demand Feeder 132 lbs (60 Kg)	$501.00

FIGURE 49. *Pendulum Mechanical Demand Feeder*

The advantage of mechanized and automatic feeders is that they save on labor. The most typical mechanized and automatic feeders are the demand bar feeder, used for fish size 50 g, and the clock-driven feeding belt.

Signs of Feeding Problems

Obvious signs of feeding problems are the increasing differences in individual sizes, growing aggressiveness, and cannibalism. Lack of sufficient feed manifests itself in bitten/damaged fish and dead fish.

PART II: PLANTS AND FISH

TABLE 12. *Estimating Feed Cost/lb Gain*

Cost/Ton	Cost/50 lb bag	Cost/lb 0.9	\multicolumn{11}{c}{Feed Conversion (lb feed required for lb gain)}
			1	1.1	1.2	1.3	1.4	1.5	1.6	1.7	1.8	1.9	2
$400	$10	$0.20	$0.20	$0.22	$0.24	$0.26	$0.28	$0.30	$0.32	$0.34	$0.36	$0.38	$0.40
$440	$11	$0.22	$0.22	$0.24	$0.26	$0.29	$0.31	$0.33	$0.35	$0.37	$0.40	$0.42	$0.44
$480	$12	$0.24	$0.24	$0.26	$0.29	$0.31	$0.34	$0.36	$0.38	$0.41	$0.43	$0.46	$0.48
$520	$13	$0.26	$0.26	$0.29	$0.31	$0.34	$0.36	$0.39	$0.42	$0.44	$0.47	$0.49	$0.52
$560	$14	$0.28	$0.28	$0.31	$0.34	$0.36	$0.39	$0.42	$0.45	$0.48	$0.50	$0.53	$0.56
$600	$15	$0.30	$0.30	$0.33	$0.36	$0.39	$0.42	$0.45	$0.48	$0.51	$0.54	$0.57	$0.60
$640	$16	$0.32	$0.32	$0.35	$0.38	$0.42	$0.45	$0.48	$0.51	$0.54	$0.58	$0.61	$0.64
$680	$17	$0.34	$0.34	$0.37	$0.41	$0.44	$0.48	$0.51	$0.54	$0.58	$0.61	$0.65	$0.68
$720	$18	$0.36	$0.36	$0.40	$0.43	$0.47	$0.50	$0.54	$0.58	$0.61	$0.65	$0.68	$0.72
$760	$19	$0.38	$0.38	$0.42	$0.46	$0.49	$0.53	$0.57	$0.61	$0.65	$0.68	$0.72	$0.76
$800	$20	$0.40	$0.40	$0.44	$0.48	$0.52	$0.56	$0.60	$0.64	$0.68	$0.72	$0.76	$0.80
$840	$21	$0.42	$0.42	$0.46	$0.50	$0.55	$0.56	$0.63	$0.67	$0.71	$0.76	$0.80	$0.84
$880	$22	$0.44	$0.44	$0.48	$0.53	$0.57	$0.62	$0.66	$0.70	$0.75	$0.79	$0.84	$0.88
$920	$23	$0.46	$0.46	$0.51	$0.55	$0.60	$0.64	$0.69	$0.74	$0.78	$0.83	$0.87	$0.92

Economics of Feeding Fish

Economics of production is related primarily to efficiency of the system and market price. Two of the most predictable and significant variable costs are fish and feed. Feed cost to produce a one-pound fish is related to conversion rate and cost of the feed.

Table 12 on the next page shows cost of feed per pound of fish flesh produced, using Rainbow Trout as an example. Trout will require from 1.2 to 1.7 lb of feed for one pound of gain for food size animals. Please note that the costs below are based upon buying in bulk, and are geographic specific. Prices will vary when buying in smaller quantities and your location.

Reducing Feed Cost

To offset cost, many operators report having success growing earth and blood worms in the media bed. Some operators maintain a nearby compost bin for worms or raise solder grubs. Others grow crickets, roaches, or other insects as feed. Some operators even grow maggots and feeder fish for food for their aquaponics fish.

If the fish tank is located outdoors, a low voltage submersible LED light can be integrated into the fish tank to attract bugs at night. If trying this method, be sure to shield the light so that the fish are not exposed to any direct lighting.

Duckweed is a popular aquaponics feed. Duckweed can be raised by those of the do-it-yourself crowd or purchased from a multitude of sources. Duckweed is addressed comprehensively later in this chapter.

It is perfectly acceptable to feed your fish store bought food or table scraps so long as it is corresponds to their diet. Research the eating habits of your fish, realizing that some fish only eat a plant based diet, while some species only eat smaller animals (fish, worms, insects, cut-up meat scraps, etc), and some fish eat both. This approach will greatly lower your feed cost.

Real World Commercial Aquaculture Data Farming Striped Bass

The table 13 on the following page is based on a real world hybrid striped bass commercial aquaculture businesses operating in the United States using a recirculating tank system. Labor, oxygen, and biocaronate line items used in commercial aquaculture operations have been omitted to better reflect a commercial aquaponic operation. Hybrid striped bass are generally harvested at a weight of 1.5 to 2.5 pounds when they are 18 to 24 months old. In controlled growing environments water temperature and quality can be controlled, not ponds where conditions are subject to change, growth is faster (harvest at18 months).

Reducing the above commercial aquaculture operation down to the typical size of a small backyard aquaponic operation (raising bass) in a 250 gallon fish tank, the economics would more closely resemble the following for one harvest cycle:

$200	Total Operational Cost.
$8	Cost per lb. (Cost is higher per lb. in smaller systems).
26.5	Pounds of ending biomass.
$13.54	Average retail market value for striped bass.
$5.54	Net profit per lb.
$146.81	Total Net Profit (250 gallon fish tank).

NOTE: Aquaponic economics, and cost-benefit analysis and earning revenue from aquaponics are addressed compressively in the chapters in section 'PART VII: Making Money and Earning a Profit from Aquaponics' later in this book.

In deciding what is the best species for you to grow, you should take a few factors into account; most importantly is what you want from your system. Another important factor depends on the type of fish that is available your area. You need to be able to buy fish to stock your system. Even if you plan to breed your own fish with a species such as

TABLE 13. **Bass Fish Production Cost/Profit**

INVENTORY & INPUT USE	
Beginning number of fish	250
Ending number of fish	212
Beginning biomass (grams of fish, 250 grams at 1 gram each)	250
Beginning biomass (lbs. of fish)	0.6
Ending biomass (lbs. of fish)	265
Max. standing biomass (lbs./gal.)	0.12
Feed used, lbs.	718
Kwh used	1,102

COSTS	
Fingerlings (each)	$1.56
Fingerlings (total quantity)	250
Fingerlings (total cost)	$390.00
Feed	$312.63
Electricity	$208.42
Total of above costs for this unit	$911.05
Cumulative cost per lb.	$3.44

NOVEMBER 2016 MARKET VALUE	
Market value for bass at current retail prices (price/lb)	$13.54

EST. NET PROFIT/2,500 GALLON FISH TANK	
Ending biomass (lbs. of fish)	265
Cumulative cost per lb.	$3.44
Market value for bass at current retail prices (price/lb)	$13.54
Net profit per lb.	$10.10
Total net profit/tank	**$2,676.50**

Tilapia that reproduces readily, you need to be able to get your broodstock in the first place.

In some areas of the United States, parts of northern California for example, it is illegal to have Tilapia, even in tanks isolated indoors. Therefore, it is important to check with your regional US Fish and Game office, or the governing regulatory agency in your region, to ensure that you are in compliance with all applicable laws in your area. Desiring to raise a species that is not common in your area or country will likely be very expense to obtain.

Tilapia and trout are among the most popular aquaponics fish. Other commonly raised aquaponic species include cod, perch, and striped bass. Goldfish are frequently raised in smaller systems that have a fish tank of 50 gallons or less.

Always keep in mind that in order to achieve maximum benefits with plants and fish, the aquaponics system must be sized correctly and fish population maintained at the optimal stocking density. Following is a Tilapia commercial aquaculture fish tank efficiency table based upon a 2016 recirculating aquaculture operation. Feed rates used in this table are lower because the feed was purchased in bulk. Check your local feed cost for an accurate cost of feed.

TABLE 14. **Fish Tank Efficiency**

Water volumne, galloons	2,500
Size stocked (grams)	1
Size harvested (grams)	567
Size harvested (lbs)	1.25
Survival rate	85%
Feed cost, per pound	$0.52

NOTE: Most backyard aquaponic systems have fish tanks ranging in size from about 250 to 1,000 gallons.

Feeding Summary

- A good rule of thumb is to feed your fish as much as they will eat in 5 to 10 minutes, 1 — 3 times per day. An adult fish will eat approximately 1 percent of its bodyweight per day. Fish fry (babies) will eat as much as 7 percent. Be sure that your fish are being fed enough. However, be cognizant of the fact that over feeding fish will negatively affect water quality, is wasteful, and is an unnecessary increase in cost.
- If your fish are not eating as they should, it is a good indication that they are stressed or unhealthy. Some factors that may result in fish not eating as they should:
  ◊ Living in conditions that are outside of their optimal temperature range.
  ◊ Water quality issues: Improper pH range, too much ammonia in the system, inadequate dissolved oxygen.
  ◊ Loud or irritating noises and vibrations.
  ◊ Direct lighting upon the fish tank.

Commercial Fish Feed Problem and a Possible Solution

About half of the world's seafood now comes from fish farms. From the environmental perspective, that is creating a major problem: millions of tons of wild fish like anchovies, sardines and mackerel are being caught in the ocean to feed farm-raised fish like salmon. Most anchovies and sardines don't end up on pizzas. Instead, they go to processing plants where they are turned into pellets to feed farmed fish. We are depleting the world's oceans to make a cheap protein.

Up to 90 percent of tiny harvested forage fish from the oceans go into pet food, poultry feed and fishmeal, never destined for human consumption. Small, filter feeding fish were traditionally used because they were inexpensive. But as wild fish stocks diminish, the cost of these forage fish increases. Meanwhile, the price for plants like corn and soy has decreased.

CHAPTER 12: FISH FEED

Now scientists and entrepreneurs are finding ways to create vegetarian diets for species like trout, which may lessen the strain on over-fished oceans. To avoid using wild fish in farmed fish diets, the United States Department of Agriculture has spent the past ten years researching alternative diets that include plants, animal processing products, insects and single-cell organisms like yeast, bacteria, and algae. "We have been hit over the head with the notion that farming carnivorous fish means that you have to catch fish in the ocean for its diet, but that's wrong," says Michael Rust, science coordinator of the NOAA Fisheries' Office of Aquaculture.

Recent advances in aquaculture research have shown that farmed carnivorous fish do not require any fishmeal or fish oil in their feeds. The USDA has proven that many species of fish can get enough nutrients from these alternative sources without eating other fish. If widely adopted by the aquaculture industry, this plant-based diet could significantly reduce the amount of wild fish that are harvested and turned into fish meal pellets.

The research comes at a pivotal time when a growing population will mean an increased demand for seafood. Americans consumed 4.5 billion pounds of seafood in 2012 and to meet that demand, 91 percent of it was imported.

Without increased aquaculture production, the world will face a seafood shortage of 50 to 80 million tons by 2030, according to the United Nations Food and Agriculture Organization. But that production needs to be sustainable by decreasing the use of wild fish according to the USDA.

Fish, like people, don't need specific foods but rather specific nutrients in order to stay healthy. In fact, all animals essentially need the same forty nutrients—a combination of amino acids, fatty acids, vitamins and minerals. With this in mind a pair of entrepreneurs started a company called 'TwoxSea'. They have found that aquaculture facilities were concerned about the environmental impact of traditional fish feed, future availability and cost of such; and as result were very open to replacing fishmeal with alternative feeds like corn, soy and algae. They now have several large commercial customers feeding their fish nothing but TwoxSea's nutritionally complete vegetarian feed. By all accounts, there is now hope that we can have a growing aquaculture industry that is sustainable and we won't have to rely on the ocean to get our fish.

Alternative Fish Feed

Fish feed is one of the most important and expensive inputs for any aquaponic system. It can be purchased or self-made. Purchasing a quality manufactured whole food fish feed is certainly the easiest way to go and ensures the nutritional needs of the fish are being met. Even so, below is an example of supplemental fish feed that can be easily produced domestically, which can help save money or used temporarily if manufactured feeds are not available.

Duckweed

Duckweed is the second smallest flowering plant in the world (watermeal being the smallest). It floats and

FIGURE 50. Duckweed growing in a container as fish feed supplement

grows directly on top of water. Not all fish take to duckweed immediately, but most herbivorous fish adjust quickly. Duckweed grows in calm waters. It produces oxygen for the water when in sunlight, and consumes oxygen from the water when shaded and at night.

Duckweed is a fast-growing floating water plant that is rich in protein and can serve as a food source for carp and tilapia. Duckweed can double its mass every 1–2 days in optimum conditions, which means that one-half of the duckweed can be harvested every day. Duckweed should be grown in a separate tank from the fish because otherwise the fish would consume the whole stock. Aeration is not necessary and water should flow at a slow rate through the container in which the duckweed is grown. Duckweed can be grown in sun-exposed or half-shaded places. Surplus duckweed can be stored and frozen in bags for later use. Duckweed is also a useful feed for poultry.

Duckweed consumes more nutrients than most plants. Duckweed can consist of up to a 45 percent protein, which even surpasses the protein concentration of soybeans. It also has all the essential amino acids. Duckweed can be obtained from a variety of sources online simply by performing an Internet search.

It is a useful addition to an aquaponic system, especially if the duckweed growing container is located along the return line between the plant grow devices (i.e. flood-and-drain grow beds, N.F.T. grow channels), and the fish tank. Any nutrients that escapes the plant's grow beds fertilize the duckweed, thereby ensuring the cleanest water possible returning to the fish.

Azolla (Water Fern)

Azolla is a genus of fern that grows while floating on the surface of the water, much in the manner of duckweed. The major difference is that Azolla is able to fix atmospheric nitrogen, essentially creating protein from the air. This occurs because Azolla has a symbiotic relationship with a species of bacteria, Anabaena azollae, which is contained within the leaves.

As well as providing a free source of protein, Azolla is an attractive feed source because of its exceptionally high growth rate. Like duckweed, Azolla should be grown in a separate tank with slow water flow. Its growth is often limited by phosphorus, so if Azolla is to be grown intensively an additional source of phosphorous is needed such as compost tea.

Insects

Insects are considered undesirable pests in many cultures. However, they have an enormous potential in supporting traditional food chains with more sustainable solutions. In many countries insects are already part of people's diets and sold at the markets. In addition they have been used as animal feed for centuries.

Insects are a healthy nutrient source because they are rich in protein and polyunsaturated fatty acids and full of essential minerals. Their crude protein content ranges between 13 and 77 percent (on average 40 percent) and varies according to the species, the growth stage, and the rearing diet. Insects are also rich in essential amino acids, which are a limiting factor in many feed ingredients. Edible insects are also a good source of lipids, as their quantity of fat can range between 9 and 67 percent. In many species, the content of essential polyunsaturated fatty acids is

FIGURE 51. Azolla *spp. growing in a container as fish feed supplement*

also high. These characteristics together make insects a healthy and ideal option for both human food, and feed for animals or fish.

Given their enormous number and varieties, the choice of the insect to be reared can be tailored to their local availability, climatic conditions/seasonality and type of feed available. The source of food for insects can include staple husks, vegetable leaves, vegetable wastes, manure, and even wood or cellulose-rich organic materials, which are suitable for termites. Insects also make a great contribution to waste biodegradation, as they break down organic matter until it is consumed by fungi and bacteria and mineralized into plant nutrients.

The culturing of insects is not as challenging as other animals since the only limiting factor is feed and not rearing space. Sometimes insects are referred to as "micro-livestock". The small space requirement means that insect farms can be created with very limited areas and investment costs. In addition, insects are cold-blooded creatures, this means that their feed conversion efficiency into meat is much higher than terrestrial animals and similar to fish. Lots of possible options and additional knowledge on insect farming as feed is available on the internet. Among the many species available, an interesting one to be used as fish feed is the black soldier fly.

Black Soldier Fly

The larvae of black soldier flies, Hermetia illucens, are extremely high in protein and thus, a valuable source for livestock, including fish. The lifecycle of this insect makes it a convenient and attractive addition to an integrated homestead farming system in favourable climate conditions. The larvae feed on manure, dead animals, and food waste. When culturing black soldier flies, these types of waste are placed in a compost unit that has adequate drainage and airflow. As the larvae reach maturity, they crawl away from their feed source through a ramp installed in the compost unit that leads to a collection bucket.

Essentially, the larvae devour wastes, accumulate protein and then harvest themselves. Two-thirds of the larvae can be processed into feed while the remaining one-third should be allowed to develop into adult flies in a separate area. The adult flies are not a vector of disease; adult flies do not have mouthparts, do not eat, and are not attracted to any human activities. Adult flies simply mate and then return to the compost unit to lay eggs, dying after a week. Black soldier flies have been shown to prevent houseflies and blowflies in livestock facilities and can actually decrease the pathogen load in the compost. Even so, before feeding the larvae to the fish, the larvae should be processed for safety.

FIGURE 52. Black soldier fly (Hermetia illucens) *adult (a) and larvae (b)*

Baking in an oven (170 °C for 1 hour) destroys any pathogens, and the resulting dried larvae can be ground and processed into a feed.

Moringa or Kalamungay

Moringa oleifera is a species of tropical tree that is very high in nutrients, including proteins and vitamins. Classified by some as a super food and currently being used to combat malnutrition, it is a valuable addition to homemade fish feeds because of these essential nutrients. All parts of the tree are choice edibles suitable for human consumption, but for aquaculture it is typically the leaves that are used. In fact, there has been success in several small-scale aquaponic projects in Africa using leaves of this tree as the only source of feed for tilapia.

These trees are fast-growing, drought-resistant, and easily propagated through cuttings or seeds. However, they are intolerant of frost or freezing and not appropriate for cold areas. For leaf production, all of the branches are harvested down to the main trunk four times per year in a process called pollarding.

Making Homemade Fish Feed

As mentioned above, fish feed is one of the most expensive inputs in aquaponics. Feed is also one of the most important components of the whole aquaponic ecosystem because it sustains both the fish and vegetable growth. Therefore, it is necessary that operators understand its composition. Also, if commercial pelleted feed is not available, it is important to understand how to make it.

Composition of Fish Feed

Fish feed consists of all the nutrients that are required for growth, energy, and reproduction. Dietary requirements are identified for proteins, amino acids, carbohydrates, lipids, energy, minerals, and vitamins. A brief summary of major feed components are listed below.

Protein

Dietary protein plays a fundamental role for the growth and metabolism of animals. A combination of more than 100 amino acids joined by peptides forms a protein; they are the building blocks of protein. Only some amino acids can be synthesized by animals while others cannot, so these must be supplied in the diet. Non-essential amino acids can be synthesized internally. However, this does not mean they are unimportant. It is just that the body is capable of producing a sufficient amount to meet the demands for growth and tissue repair. Essential amino acids cannot be synthesized by the body and must be acquired through outside sources.

For aquatic animals, there are ten essential amino acids (EAAs): arginine, histidine, isoleucine, leucine, lysine, methionine, phenylalanine, threonine, tryptophan, and valine. Therefore, feed formulation must find an optimal balance of EAAs to meet the specific requirements of each fish species. Non-compliance with this requirement would prevent fish from synthesizing their own proteins, and also waste the amino acids that are present. The ideal feed formulation should thus take into account the EAA levels of each ingredient and match the quantities required by fish.

Recommended protein intake of fish depends on the species and age. Tilapia and herbivorous fish the optimal ranges are 28–35 percent; carnivorous species require 38–45 percent. Juvenile fish require higher-protein diets than adults due to their intense body growth.

Besides any optimal amino acid content in the feed, it is worth stating the importance of an optimal dietary balance between proteins and energy (supplied by carbohydrates and lipids) to obtain the best growth performance and reduce costs and wastes from using proteins for energy. Although proteins can be used as a source of energy, they are much more expensive than carbohydrates and lipids, which are preferred.

In aquaponics, any increase in dietary proteins directly affects the amount of nitrogen in the water. This should be balanced either by an increase in plants grown in the system or the selection of vegetables with higher nitrogen demands.

Carbohydrates

Carbohydrates are the most important and cheapest energy source for animals. They are mainly composed of simple sugars and starch, while other complex structures such as cellulose and hemicellulose are not digestible by fish. In general, the maximum tolerated amount of carbohydrates should be included in the diet in order to lower the feed costs. Omnivorous and warm-water fish can easily digest quantities up to 40 percent, but the percentage falls to about 25 percent in carnivorous and cold-water fish. Carbohydrates are also used as a binding agent to ensure the feed pellet keeps its structure in water. In general, one of the most used products in extruded or pelleted feed is starch (from potato, corn, cassava, or gluten wheat), which undergoes a gelatinization process at 140-185°F (60–85 °C) that prevents pellets from easily dissolving in water.

Lipids

Lipids provide energy and essential fatty acids (EFAs), indispensable for the growth and other biological functions of fish. Fats also play the important role in absorbing fat-soluble vitamins and securing the production of hormones. Fish, as other animals, cannot synthesize EFAs, which have to be supplied with the diet according to the species' needs. Deficiency in the supplement of fatty acids results in reduced growth and limited reproductive efficiency.

In general, freshwater fish require a combination of both omega-3 and omega-6 fatty acids, whereas marine fish need mainly omega-3. Tilapias mostly require omega-6 in order to secure optimal growth and high feed conversion efficiency. Most diets are comprised of 5–10 percent lipids, although this percentage can be higher for some marine species. Lipid inclusion in the feed needs to follow optimal protein/energy ratios to secure good growth, to avoid misuse of protein for energy purposes (lack of fat/carbohydrates for energy purposes), and to avoid fat accumulation in the body (diet too rich in lipids).

Energy

Energy is mainly obtained by the oxidation of carbohydrates, lipids and, to a certain extent, proteins. The energy requirements of fish are much lower than warm-blooded animals owing to the reduced needs to heat the body and to perform metabolic activities. However, each species requires an optimum amount of protein and energy to secure best growth conditions and to prevent animals from using expensive protein for energy. It is thus important that feed ingredients be carefully selected to meet the desired level of digestible energy (DE) required by each aquatic species.

Vitamins and Minerals

Vitamins are organic compounds necessary to sustain growth and to perform all the physiological processes needed to support life. Vitamins must be supplied with the diet because animals do not produce them. Vitamin deficiencies are most likely to occur in intensively cultured cages and tank systems where animals cannot rely on natural food. Degenerative syndromes are often ascribed to an insufficient supply of these vitamins and minerals.

Minerals are important elements in animal life. They support skeletal growth and are also involved in osmotic balance, energy transport, neural, and endocrinal system functioning. They are the core part of many enzymes as well as blood cells.

Fish require seven main minerals (calcium, phosphorus, potassium, sodium, chlorine, magnesium, and sulphur) and 15 other trace minerals. These

TABLE 15. **Common feed ingredient sources of the most important nutrient components**

NUTRIENT COMPONENTS	FEED INGREDIENT SOURCES
Protein	*Plant-based sources*: algae, yeast, soybean meal, cottonseed meal, peanuts, sunflower, rapeseed/canola, other oil-seed cakes *Animal-based sources*: fishery by-products (fishmeal or offal), poultry by-products (poultry meal or offal), meat meal, meat and bone meal, blood meal
Carbohydrates	Wheat flour, wheat bran, corn flour, corn bran, rice bran, potato starch, cassava root meal
Lipids	Fish oil, vegetable oil (soybean, canola, sunflower), processed animal fat
Vitamins	Vitamin premix, yeast, legumes, liver, milk, bran, wheat germ, fish and vegetable oil
Minerals	Mineral premix, crushed bone

can be supplied by diet, but can also be directly absorbed from the water through their skin and gills. Supplementing of vitamins and minerals can be done according to the requirements of each species.

The production of feed requires a fine balance of all of the nutrient components mentioned above (protein, lipids, carbohydrates, vitamins, minerals, and total energy). An unbalanced feed will cause reduced growth, nutritional disorders, illness and, eventually, higher production costs.

Fishmeal is regarded as the best protein source for aquatic animals because of its very high protein content and it has balanced EAAs. However, it is an increasingly expensive ingredient, with concerns regarding sustainability. Moreover, fishmeal is not always available. Proteins of plant origin can adequately replace fishmeal; however, they should undergo physical (de-hulling, grinding) and thermal processes to improve their digestibility. Plant ingredients are, in fact, high in anti-nutritional factors that interfere with the digestion and the assimilation of nutrients by the animals, which eventually results in poor fish growth and performance.

The size of the pellets should be about 20–30 percent of the fish's mouth in order to facilitate ingestion and avoid any loss. If the pellets are too small, fish exert more energy to consume them; if too large, the fish will be unable to eat. A recommended pellet size for fish below 50 g is 2 mm, while 4 mm is ideal for pre-adults of more than 50 g.

The use of any raw ingredient of animal origin (fish offal, blood meal, insects, etc.) should be preventively heat treated to prevent any microbial contamination of the aquaponic system.

Homemade Fish Feed for Omnivorous and Herbivorous Fish

Two simple recipes for a balanced fish feed containing 30 percent of crude protein (CP) are provided in tables 16 and 17. The lists of the ingredients for each diet are expressed in weight (kilograms), enough to make 10 kg of feed. The first formula is made with proteins of vegetable origin, mainly soybean meal (see table 16). The second formula is mainly made with fishmeal (see table 17).

Step-by-Step Preparation of Homemade Fish Feed

1. Gather the utensils noted in table 18.
2. Gather the ingredients shown in tables 16 and 17. Purchase previously dried and defatted soybean meal, corn meal, and wheat flour. If these meals are unavailable, obtain whole soybeans, corn kernels, and wheat berries. These would need to be dried, de-hulled, and ground. Whole soybeans need to be toasted at 240°F (120 °C) for 1–2 minutes.
3. Weigh each ingredient following the quantities shown in the recipes above.

4. Add the dry ingredients (flours and meals) and mix thoroughly for 5–10 minutes until the mix becomes homogeneous.
5. Add the vitamin and mineral premix to the dry ingredients and mix thoroughly for another 5 minutes. Make sure that the vitamins and minerals are evenly distributed throughout the whole mixture.
6. Add the soybean oil and continue to mix for 3–5 minutes.
7. Add water to the mixture to obtain a soft, but not sticky, dough.
8. Steam-cook the dough to cause gelatinization.
9. Remove the dough, divide into manageable pieces, and pass them through the meat mincer/pasta maker to obtain spaghetti-like strips. The mincer disc should be chosen according to the desired pellet size.
10. Dry the dough by spreading the strips out on aluminium trays. If available, dry the feed strips in an electric oven at a temperature of 140–185 °F 60–85 °C for 10–30 minutes to gelatinize starch. Check the strips regularly to avoid any burn.
11. Crumble the dry strips. Break or cut the feed on the tray with the fingers into smaller pieces. Try to make the pellets the same size. Avoid excessive

TABLE 16. **Recipe for 10 kg of fish feed using vegetable-based protein, including proximate analysis**

FEED INGREDIENTS	WEIGHT (kg)	% OF TOTAL FEED	PROXIMATE ANALYSIS	%
Corn meal	1.0	10	Dry matter	91.2
Wheat flour	1.0	10	Crude protein	30.0
Soybean meal	6.7	67.2	Crude fat	14.2
Soybean oil	0.2	2	Crude fiber	4.8
Wheat bran	0.7	7.8	Ash	4.6
Vitamin and mineral premix	0.3	3	Nitrogen-free extract (NFE)	28.3
Total amount	10.0	100	-	-

TABLE 17. **Recipe for 10 kg of fish feed using animal-based protein, including proximate analysis**

FEED INGREDIENTS	WEIGHT (kg)	% OF TOTAL FEED	PROXIMATE ANALYSIS	%
Corn meal	1.0	10	Dry matter	90.9
Wheat flour	4.0	40	Crude protein	30.0
Soybean meal	1.5	15	Crude fat	10.5
Soybean oil	0.2	2	Crude fiber	2.1
Fishmeal	3.0	30	Ash	8.3
Vitamin and mineral premix	0.3	3	Nitrogen-free extract (NFE)	34.5
Total amount	10.0	100	-	-

TABLE 18. **Utensils**

COMPONENT	QUANTITY	SPECIFICATION
Weighing scale	1	Capacity 2–6 lbs. (1–3 kg), Divisions in ounces or grams
Grinder	1	Electric coffee-type grinder
Metal sieve	1	5–10 U.S. Sieve Size, 0.2–0.4 cm (2–4 mm) mesh
Mixing bucket	1	Capacity 3-gallon (10 liters)
Plastic bowl	1	Capacity 2-quart (2 liters)
Meat mincer / pasta maker	1	Manual or electric
Mixing spoon	1	Large size
Aluminum baking tray	10	Large baking tray

pellet manipulation to prevent crumbling. Pellets can be sieved and separated in batches of homogeneous size with proper mesh sizes.
12. Store the feed.

Storing Homemade Fish Feed

Once prepared, the best way to store fish feed is to put pellets into an airtight container soon after being dried and broken apart. Containers must be kept in a cool, dry, dark and ventilated place, away from pests. Keeping pellets at low levels of moisture (less than 10 percent) prevents them becoming mouldy and developing toxic mycotoxins. Depending on the temperature, the pellets can be stored for as long as two months.

Another way to keep pellets for long periods is to close them in a plastic container and store them in the refrigerator. Feed kept in this way can last for more than one year.

Feed must be used on a "first in, first out" basis. Avoid using any feed showing signs of decay or mould, as this could be fatal for fish.

PART III

Components of Aquaponics Used in Aquaponics

CHAPTER 13

Equipment & Component Overview

Overview of Aquaponic Equipment and Components

Following is a list of equipment that is available for an aquaponics system. Some components are necessary, whereas other items are optional, or may not even be applicable to your particular set-up. The list is provided for consideration. All of these items are available via the internet, and many can be obtained through your local home improvement store. Prices vary greatly, depending on the quality of parts and materials used and how sophisticated you want to go, i.e., automation vs. manual.

Only use food-grade plastics and materials typically used for potable water such as PVC. Some plastics will degrade from weather and break down or leak chemicals into your system. Avoid using copper or other metals in any part of the system exposed to water, as they can negatively affect the fish.

Some aquaponic operators use recycled items for system parts. This is a great way to lower costs; however, it is important to refrain from using materials previously used with chemicals or when their prior use is unknown. Even used parts with marine grade paint or galvanized coatings can leak chemicals and oxidized metal into your system. Inspect and thoroughly rinse all parts thoroughly before integrating them into your system. For additional protection certain parts can be encased with a safe marine grade epoxy coating.

Below is a list of items used in aquaponics. These items will be addressed in much greater detail in the following chapters, with the most prominent components (pumps, grow media, plumbing, fish tanks, liners, etc.) having an entire chapter dedicated to providing more comprehensive information.

Plumbing

Plumbing supplies which may include piping, elbows, bends, sleeves, pipe joint compound, gaskets, flares, reducers, tees, Y-joint, etc. will be needed. In 'Chapter 14: Plumbing', all aspects of plumbing will be addressed in detail.

Grow Bed

Grow beds can be made from any number of materials including wood, fiberglass, steel, concrete, bricks, etc. A liner can be installed to retain the water if necessary. The grow bed can also be made from totes, purchased directly from a vendor, or custom made. Grow beds are addressed in much greater detail in the Part V: Flood-and-Drain System Design and Layout chapters.

Grow Bed Stan

Keep in mind that a stand will most likely be needed for the grow bed. The stand should be firmly supported and strong enough to support the weight of the grow bed filled with media, water plants, and most likely the weight of the operator leaning against it.

Float Switches

Float switches are inexpensive devices used to control the pump depending on the water level (see figure 53 below). If the water level in the sump tank falls below a certain height, the switch will turn off the pump. This prevents the pump from pumping all of the water out of the tank. Similarly, float switches can be used to fill the aquaponic system with water from a hose or water main. A float switch similar to a toilet ballcock and valve can ensure the water level never falls below a certain point. It is very important to know that in certain types of loss-of-water events, such as a broken pipe, a float switch can actually make the flooding much worse, and this needs to be carefully considered in indoor applications and other situations where flooding could cause significant property damage or electrical shock dangers.

Hi-Low Water Level Sump Pump Controller (Dual Float Switch)

A Hi-Lo Pump Controller is a dual float device with a universal switch that works with all types of sump pumps and utility pumps. Its two sensors give you complete control of where your pump turns on and off. Using this type of controller with the right utility pump allows you to set the turn on level as low as 1/2" of water and turn off as low as 1/8" of water. When used with a sump pump the controller enables you to adjust the turn on and turn off levels to get the longest run time for the pump, which saves energy and lengthens the service life of the pump.

One example of a Hi-Lo pump controller switch is the HC6000 by HydroCheck. It is currently available on Amazon.com for $75.49, plus $6.34 for shipping in the U.S. (USD, year 2016), and has excellent reviews. This particular make/model Hi-Lo pump controller switch is presented as an example and hopefully a helpful starting point (resource) for the reader (see figure 54). There are also other Hi-Lo switches on the market with various features and degrees of quality, and cost.

Water Detector and Hi-Low Water Alarms (Optional)

There is a wide variety of water alarms available with a range of features, quality, and cost. They are available with battery, solar, and/or AC/DC power supply. Most units under $100 have sound alarms, but some also have a light 'on' alarm. More expensive units can be obtained that make a phone call, send an email, and/or a text message alert.

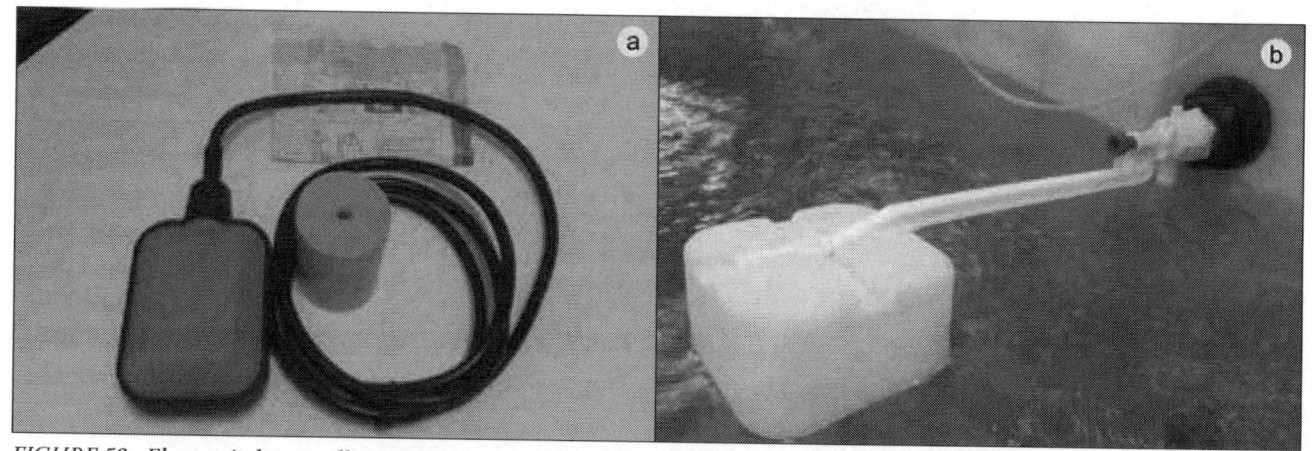

FIGURE 53. Float switch controlling a water pump (a) and a ballcock and float valve controlling the water main (b)

CHAPTER 13: EQUIPMENT & COMPONENT OVERVIEW

Water detection and hi-low water alarm devices are not mandatory in aquaponics but can provide the operator with a higher level of protection and peace of mind. They can also help prevent a disaster, saving the aquaponics operator from potential system-related loss (fish, pump burn out, plant) and/or property damage (flooding).

For instance, a water detector sensor alarm could be used where water could damage the surrounding environment (i.e. indoor applications). A hi-low water alarm can alert the operator 'immediately' of any water balance problems. Discovering the problem later during a routine operational check may be too late.

Again, these devices are entirely optional. They range in cost from under $10 to over a $1,000 in the U.S. (USD, year 2016). This expense needs to be weighed against risk factors, budget, and peace of mind.

Other Monitoring Alarms

Automatic monitoring devices for water temperature, water level, pumps, pH, blower, lights, air temperature, dissolved oxygen, or the entire system and system environment are available. Alarms can even call, text message or email you the status of conditions and alert you if there is a problem.

Water Quality Testing

Monitoring water quality in aquaponics is an important procedure. There are diverse water quality test kits and meters available, manual and automatic, in a wide price range. Water quality testing devices can be acquired which will monitor just one, several or all of the following components. Although all of these water parameters should be tested, budgetary constraints may determine how many tools are used to gather this data.

- pH
- Dissolved Oxygen
- Ammonia
- Water Temperature

- Works with all types of sump pumps and utility pumps
- Precise control of turn on and turn off levels
- Wide control range from 1/2" to 20' (Not timed based like other dual-float switches)
- Small profile fits in small or crowded spaces
- Sensors not affected by minerals and debris in water — Never have to be cleaned

- Part Number HC6000
- Product Dimensions 2.8 x 3.8 x 2.5 inches
- Voltage 120 volts; Wattage 2.4 watts; Amperage Capacity 14 A
- Cord Length 12 Feet
- UL Certified
- HC6000 is for indoor applications.

NOTE: For outdoor applications refer to HC6100 by HydroCheck.

FIGURE 54.

NOTE: Please refer to Chapter 31: Water Quality for a comprehensive examination on these components and associated monitoring equipment.

Inline Water Heater

Allows the operator to establish the desired water temperature. Most devices are fully equipped with a digital control mechanism that will provide precise all-season temperature regulation. They also come with a flow switch and safety thermostat and are easy to install with the option of water flow from either direction.

Ammonia Nitrogen Test Kit

Used for determining ammonia and nitrogen concentrations in water.

Water Hardness Test Kit

A portable testing kit to measure total hardness, calcium, and magnesium hardness in water.

Fish Tanks

Avoid tanks that have hollow pockets as they will collect fish waste that will go anaerobic and cause water quality problems. Also ensure that there is adequate circulation delivering dissolved oxygen throughout the tank. Please also refer to Chapter 15: Fish Tanks.

Sump Tanks

A "sump" is an area where liquid run-off accumulates. In an aquaponics system the sump (when necessary) is positioned at a point lower than the grow beds and is where the grow beds drain. A sump is not necessary if the grow bed drains via gravity directly into the fish tank.

Grow Lights

Grow lights are necessary when growing plants where adequate sunlight is not readily available. They are also beneficial during winter months and/or there is a desire for longer growing periods for plants.

Using grow lights to supplement the natural sunlight in a greenhouse can increase plant production by up to 40 percent. If you are growing indoors, you will need a high quality, full spectrum grow light. Metal halide lights and LED lights are available. Both offer an excellent light spectrum. The LEDs use less electricity but cost more to purchase. The metal halides use more electricity but cost substantially less to purchase.

Grow Media

There are many different types of growing media that can be used for a flood-and-drain aquaponic system. The ideal grow bed media will anchor the plants, will not decompose, will have enough space to provide a good circulation of air and water, will be beneficial for bacteria growth, and will have close to a neutral pH level so as not to impact the overall pH level of the system. Furthermore, oxygen should be able to move freely to lower levels of the grow bed and the plants. The most commonly used growing media includes gravel, clay aggregate, lava rock, packing foam, sponges, perilite, and vermiculite. See 'Chapter 18: Growing Media' for Plants for a comprehensive examination of grow media.

Liner

A liner is an impermeable geotextile used for water retention commonly referred to as pond liner, water garden liner, greenhouse cover material, hydroponics pond liner, aquaponics bed liner, and polyethylene tarp material.

Materials include LDPE (Low Density Polyethylene), PVC (Poly Vinyl Chloride), PVC with internal reinforcement, and HDPE (High Density Polyethylene). HDPE is frequently noted as the best, but it is the most expensive and hardest to work with because of its rigid nature. LDPE is most preferred by small to mid-size aquaponic operators. Avoid EPDM as it emits toxins harmful to the beneficial bacteria in your system, and the toxins will eventually make their way into your food supply. Also avoid vinyl liners (they are too stretchy) and any product that does not specifically identify what it is made of.

The liner thickness should range anywhere from 20 to 40 mils. Obviously the thicker the better in terms of protection against tears. However, the thicker the liner the more difficult it is to handle and work with during installation; thicker will also be more expensive.

The ideal liner will be UV resistant, fiber reinforced, have one side white for easier installation, food grade quality, and thick enough to not tear easily during installation or regular use. See 'Chapter 16: Liner Material' for a comprehensive examination of liners.

Vertical Gardening

Vertical gardening is a revolutionary new garden system that provides proven methods and techniques for growing up, not out, thus maximizing space. There are a variety of containers and systems available, with more being introduced at increasing rates. Vertical gardening aquaponics will be addressed in full detail at the www.FarmYourSpace.com website.

Pumps

The correct pump for your aquaponics systems is just as important as choosing the right fish, selecting bed material, and determining tank and grow bed size. Some folks simply pick up a pump at the local home improvement store order one online, or purchase whatever the retailer recommends before gaining an understanding of their system's parameters. Please do not buy your aquaponics pump this way. It is rare to find a sales representative or a retailer that knows aquaponics (growing fish and plants together). A pump needs to be selected based upon compatibility of pump specifications to the system parameters, such as how much water is in your system, how high the pump needs to raise the water (head), desired flow rate, and whether or not a siphon or timer system is being used. If you desire to have your system off the grid, then efficiency becomes a very critical factor as well. Even if you have your system on the grid, efficiency should be considered to minimize operating costs. Refer to 'Chapter 19: Pumps & Choosing the Right Pump' for more specific and detailed helpful information on pumps.

Aeration

Aeration devices are sometime needed, or just included for additional benefit, in order to raise dissolved oxygen levels. Oxygen is introduced into the aquaponics system naturally through the plant life, waves, cascades, and waterfalls. If natural aeration methods, such as plant life, waves, cascades, and waterfalls are insufficient then an aeration device is necessary. Aeration devices are also used if dead spaces in a tank are a problem such as bottom corners of square tanks, end of a long run, etc. Mechanical devices include air pumps, tubing, oxygen injectors and oxygen diffusers.

FIGURE 55. Vertical gardening

Net Pots

Net Pots are used in floating raft culture, N.F.T. and sometimes flood and drain systems. They are used to help hold and support plants. They can be purchased online or at hydroponic stores.

Heating and Cooling

Maintaining ambient room temperature for plants, as well as your comfort is important. There are multitudes of heaters, air conditioners, fans, misting devices, and evaporation systems that can be used to achieve this goal; however, it is important to ensure that whatever is used is not counterproductive to plant health.

Timing Controllers

Timing Controllers enable you to run your aquaponic system according to your preferences and needs. They can be set up to work a pump, multiple pumps, water heaters, lighting, fish feeding, humidity, and anything else you desire to automate. A time controller is really only necessary for pump operation, but having other components on an automatic timer controller is a nice convenience. It allows the operator to maintain the system on a consistent schedule and better enables one to make clearly defined adjustments to various system elements (flow rate, water temperature, feeding schedule, lighting preferences, etc.). Timing controllers come with a wide range of features and in many price ranges.

Dehumidifier

Depending upon the environment and the type of plants to be grown, a dehumidifier may be necessary. This is a rare item, but it is one that should be considered if trying to grow arid loving plants in a room where there is high humidity, or if there are extensive mold problems.

Ambient Lighting

Room and work lighting is also an important consideration. Just remember not to have any direct lighting impacting the fish tank.

Backup Energy

In case of an extended power outage it is essential to have a back-up energy source. Keep in mind that if your power is out for several days, others in your community will also most likely be without power, resulting in a run on generators from area rental companies and home improvement stores. Fish will perish quickly as oxygen becomes insufficient and waste begins to accumulate.

An electric generator is a device that converts mechanical energy to electrical energy, typically via burning some type of fossil fuel (gas, diesel, propane, etc.). Typically, the only items that must be provided power through an emergency event are pumps, aerators, and heaters. The remaining aquaponic components being without power through an extended outage will typically not result in catastrophic problems.

For small to mid-size systems, a pump can simply be plugged in to a power inventor box, which is in turn connected to any DC power source (typically just a 12-volt automobile battery). The AC output voltage of a power inverter device is often the same as the standard power line voltage, such as household 120AC. This allows the inverter to power numerous types of equipment designed to operate off the standard line power. Most often, if it is just a matter of a few hours, survival can be maintained by just keeping the pump in operation.

Alternative Energy

Alternative energy for aquaponics is referred to as any energy source that provides off-the-grid power supply. This back-up power can be used in case there is a disruption of service of the main public utility supply, or as a primary means in which to power the electrical components associated with an aquaponic system (i.e. pumps, lighting, heat, fans, etc.). Following is an example of one back-up alternative power supply which is relatively affordable.

Alternative Energy Supply Source (one example)
Sunforce 50048 60W Solar Charging Kit
Available on online for approximately $300.00 (USA, year 2016)

❶ Four 15W Solar Panels
❷ Plastic PVC Frame for Mounting
❸ 12V DC Plug, Alligator Battery Clamps and Mounting Screws
❹ Inverter
❺ 7 Amp Charge Controller

Features:
- Amorphous solar charging kit provides up to 60 watts of clean, free, renewable power
- Designed for back-up and remote power use
- Weatherproof, durable solar panels.
- Built-in blocking diode helps protect against battery discharge at night
- Complete kit includes four 15W amorphous solar panels, a PVC mounting frame, a 7-amp charge controller, 200-watt inverter, and wiring/connection cables

FIGURE 56.

CHAPTER 14

Plumbing

Overview of Plumbing

Plumbing your aquaponics system requires careful consideration of many different factors and will depend on your own design and the type of aquaponics method being implemented (i.e. N.F.T., D.W.C., and Flood-and-Drain system). Nevertheless, there are some basic principles that are applicable on almost every system which need to be applied. Therefore, issues pertaining to plumbing infrastructure are addressed in this chapter with the hope of providing you the best possible, user-friendly assistance.

Plumbing is an integral part of an aquaponics system and needs to be considered from the very beginning. In other words, one should consider plumbing to be just as important and as much as a priority as the fish tank and the grow beds. A common mistake is to develop the fish tank and grow beds (or grow channels), and then attempt to plumb in afterwards only to find a major problem. This chapter, as well as sections in the following design chapters of this book, will help you avoid those time consuming, and sometimes costly, mistakes.

Water Conveyance

It should be understood that water pressure and water volume rate are two separate and distinct issues. Water pressure is the force that water flows from a plumbing fixture. Water volume rate is the amount of water present to fill a tank or grow bed. Water pressure is typically noted in pounds per square inch (psi) or Kilopascal (kPa). Water volume rate is typically presented in gallons per minute (gpm), gallons per hour (gph), liters per second (L/s), or cubic meters per hour (m3h). Volume of water is normally described simply in terms of gallons, cubic feet, litters, or cubic meters.

Piping can be classified into two categories: (1) those delivering water to the grow beds, and (2) those removing it from the grow beds. As a general rule, drain pipes conveying water from the grow bed(s) to the fish tank are typically larger and function via gravity, whereas pipes or tubes conveying water from the fish tank to the grow bed are typically pressurized by the pump and can be smaller in diameter.

Pipe diameter size will depend on system parameters, but it is always better to go with a slightly larger sized diameter pipe than the minimum size needed. There are a number of engineering related calculators online that can help you work out the ideal pipe diameter to use and it is prudent to review these during the planning phase. The following are internet links to two such online calculators that can used to determine pipe size:

- http://irrigation.wsu.edu/Content/Calculators/General/Pipe-Velocity.php
- http://www.calctool.org/CALC/eng/civil/hazen-williams_g

Flow rate decreases as the length of pipe increases, so keep this in mind if conveying water or wastewater a considerable distance (i.e. an outdoor pond to a greenhouse is usually a considerable distance away, etc.). The line itself provides resistance to the water flow. Therefore, the length of the run is a major factor. The longer the run, the less gallons per minute that can flow through the line.

Length of run actually has a dramatic affect on the conveyance of fluid. As an example, a typical water line will lose approximately 33 percent of its water delivery capability when the length of the run is increased from 30 linear feet to 60 linear feet. As a specific example, 1-¼ inch diameter pipe can deliver approximately 21 gallons per minute over a run of 30 linear feet, yet only 14 gallons, approximately, per minute over a run of 60 linear feet. While the length of run is a major factor for water line size calculations, it becomes more of factor when the run is unusually long.

Table 19 is another helpful guide to determine the pipe size needed based upon the required amount of water that will be conveyed to empty the fish tank once per hour.

TABLE 19. **Water Flow Rate**

PIPE LENGTH (ft)	WATER FLOW RATE IN GLM — Pipe Diameter in Inches									
	0.5	0.75	1	1.5	2	2.5	3	4	5	6
5	23	66	140	407	868	1560	2520	5371	9659	15601
10	16	45	96	280	597	1073	1733	3694	6643	10730
15	13	36	77	225	479	862	1393	2968	5337	8620
20	11	31	66	193	410	738	1192	2541	4569	7380
40	7	21	46	132	282	508	820	1747	3142	5076
100	4	13	28	81	172	309	500	1065	1916	3095

PIPE LENGTH (m)	WATER FLOW RATE, m^3/hr — Pipe Diameter in mm									
	12	20	25	40	50	65	75	100	130	150
1	5.6	21.5	38.6	133.0	239.2	477.0	694.9	1481	2953	4302
2	3.9	14.8	26.6	91.5	164.5	328.1	478.0	1019	2031	2959
4	2.7	10.2	18.3	62.9	113.2	225.6	328.7	700.5	1396.7	2034.9
6	2.1	8.2	14.7	50.6	90.9	181.3	264.1	562.8	1122	1635
12	1.5	5.6	10.1	34.8	62.5	124.7	181.6	387.1	771.7	1124
30	0.9	3.4	6.2	21.2	38.1	76.0	110.7	236.0	470.5	685.5

Additional Pipe Sizing Considerations

Increasing the pipe to just one size larger makes a dramatic difference. Those not in the plumbing trade or engineering field do not realize that there is strong correlation between length and pipe diameter.

As an example, a 1-¼ inch diameter pipe is only 25 percent larger in diameter than 1-inch diameter pipe, but there is an area difference of 56 percent between these two slightly different pipe diameter sizes. Another example of this size/area relationship can be seen when examining the difference of the areas inside a 1-¼ inch diameter pipe compared to a 2- inch diameter pipe, which is about 77 percent.

When examining the flow-rate (gallons per minute), the differences are even more dramatic. Basing calculations of an average run of pipe of 50-feet in length, a 1-¼ inch diameter pipe cab will convey up to about 16 gallons per minute. On the other hand 1-inch diameter pipe only provides about 9 gallons per minute. Therefore, a 1-¼ inch diameter pipe provides almost 77 percent more gallons per minute than a 1-inch diameter pipe.

What does all this mean to the aquaponic operator? It means that for a nominal amount of money, increasing pipe diameter by just one size provides dramatic benefits. The photo below clearly illustrates this point.

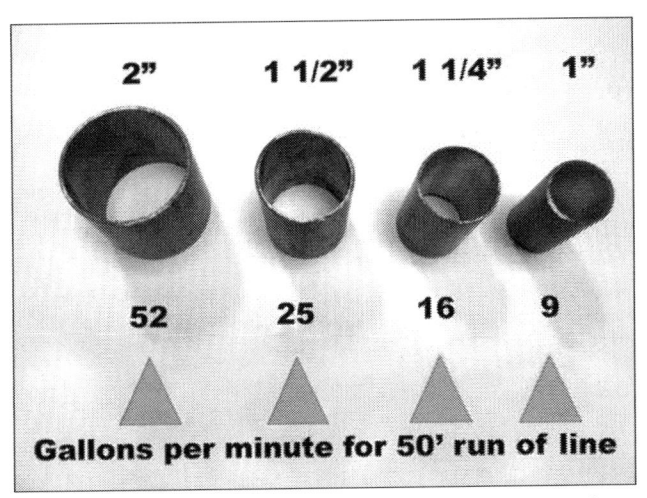

FIGURE 57. *Typical flow rates for common pipe sizes used in aquaponics.*

This is probably a good place to mention that the above noted plumbing principles apply to both tubing and pipes. There are minor head loss difference for each type of material being considered (i.e. PVC, steel, copper, polyvinyl, etc.). Furthermore, fitting and bends also impact conveyance of the fluid being delivered. However, for the vast majority of aquaponic systems, the head loss differences between material type and the number of fittings used are of such minor consequence that these differences can be ignored. Head loss needs to be considered, but not to the extent of examining each minor difference when planning and designing an aquaponic system. Pipe diameter, elevation difference, and length of pipe run are going to be the critical factors that need to be considered in the planning and design phases.

Over time, debris can build up on the inside of the pipes and this will negatively impact the flow rate. Pipes may need to be cleaned once or twice a year in order to ensure an unimpeded flow rate. The necessity to clean some sections of plumbing should influence your decision as to how you connect pipe and fittings (i.e. glue, installing a cleanout fitting(s) at certain plumbing points, etc.). Feeding a domestic garden hose through the pipe (with the water turned on) will sufficiently clean the pipe in most cases. A power washer does wonders, as well.

PVC Piping

The most commonly used plumbing materials used in aquaponics are PVC or UPVC pipes and irrigation tubing. Garden hoses are also used fairly often, especially when the fish tank is located a significant distance from the plant grow beds or channels.

Occasionally, there is an article questioning the safety of PVC, but after years of study by scientists all over the world, there is still very little evidence to prove that it is harmful; as such, it has been certified as safe for use in drinking water plumbing by government agencies throughout the world. In reality, all evidence suggests that the risk of harm from using

PART III: COMPONENTS OF AQUAPONICS USED IN AQUAPONICS

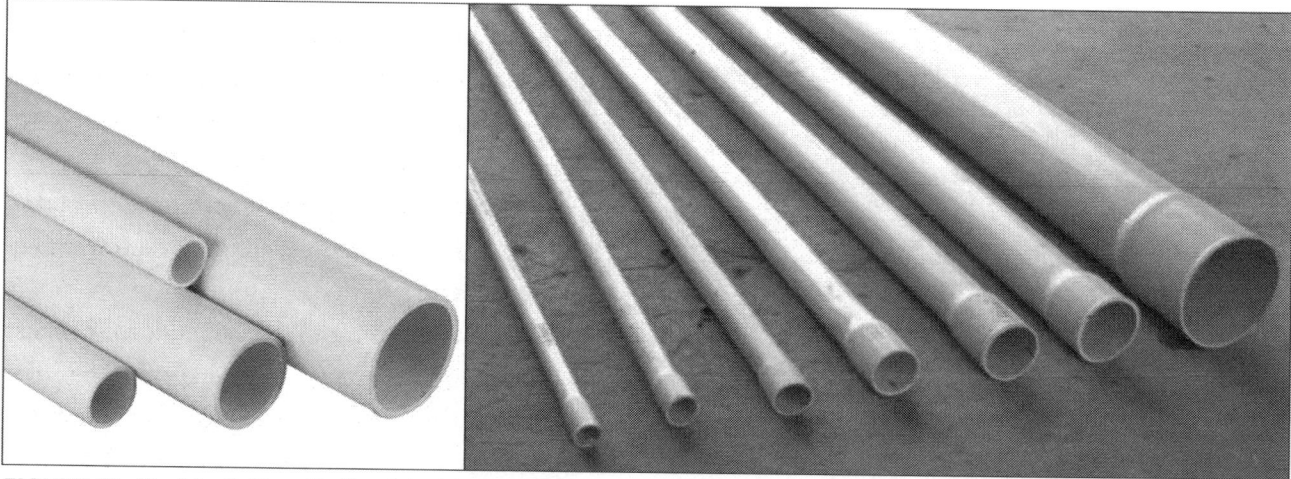

FIGURE 58. Straight (left) and bell end (right) PVC piping.

PVC pipes in food and water supply is so slight as to be negligible. Because of this, it is almost universally used for agriculture, aquaculture, hydroponics, and aquaponics.

Below are several beneficial features:
- PVC Pipe is almost universally available.
- PVC Pipe is usually extremely cost effective (low cost).
- PVC Pipe comes in standard sizes throughout the world.
- PVC Pipe has a wide range of adapters and connectors available.
- PVC Pipe is easy to use, cut and adapt.
- PVC Pipe is durable and long lasting. PVC Pipe is light weight.

Although other piping can be used in aquaponics (such as agricultural pipe, flexi-pipe, bamboo, hosepipe, etc.,) it important to make sure that it is safe for use in a system that grows food for human consumption, and make sure that it will not be harmful to fish or plants. As an example, it is prudent to avoid using metal piping, especially copper piping, as it can be highly toxic to fish. Also, if installing used plumbing materials, make sure that it was not used to covey any toxic or harmful substances in the past. If in doubt, it is best to pass on it, especially since PVC pipe piping and fittings are relatively in expensive.

Schedule 40 vs Schedule 80 PVC

If you've been shopping around for PVC you may have heard the term "schedule". Despite its deceiving title, schedule doesn't have anything to do with time. A PVC pipe's schedule has to do with the thickness of its walls. Maybe you've seen that schedule 80 pipe is slightly more costly than schedule 40.

Though the outside diameter of a schedule 80 pipe and a schedule 40 pipe are the same, an 80 pipe has thicker walls. This standard of measuring pipe came from a need to have a universal system for referring to PVC. Since different wall thicknesses is beneficial in different situations, the ASTM (American Society for Testing and Materials) came up with the schedule 40 and 80 system for classifying the two common types.

The main differences between Schedule 40 (Sch 40) and Schedule 80 (Sch 80) are:
- Water Pressure Rating
- Sizing & Diameter (Wall Thickness)
- Color
- Application & Use

Water Pressure for Sch 40 vs. Sch 80

Both schedule 40 and 80 PVC are used widely around the world. Each one has its benefits in different applications. Schedule 40 pipe has thinner walls, so it is best for applications involving relatively low water

CHAPTER 14: PLUMBING

pressure. In the vast majority of situations, schedule 40 is sufficient for aquaponics.

Schedule 80 pipe has thicker walls and is able to withstand higher PSI (pounds per square inch). This makes it ideal for industrial and chemical applications. To give you an idea of the size difference, 1" schedule 40 PVC pipe has a .133" minimum wall and 450 PSI, while schedule 80 has a .179" minimum wall and 630 PSI.

Sizing & Diameter

As mentioned earlier, both schedule 80 and schedule 40 PVC pipe have the exact same outside diameter. This is possible because schedule 80's extra wall thickness is on the inside of the pipe. This means schedule 80 pipe will have a slightly more restricted flow, even though it may be the same pipe diameter as an equivalent schedule 40 pipe. This means schedule 40 and 80 pipe do fit together and can be used together if necessary.

The only thing to be careful of is that the lower pressure handling schedule 40 pipe meets the pressure requirements of your application. A pipeline is only as strong as its weakest part or joint, so even one segment of schedule 40 pipe used where a higher pressure schedule pipe is needed can cause problems. However, it is rare to have such high pressures in aquaponics. An exception may be a back-up or supplemental pipe directly from another water supply source.

Schedule 40 and Schedule 80 Color

Generally, schedule 40 pipe is white in color, while schedule 80 is often gray in order to distinguish it from 40. PVC is available in many colors though, so be sure to check labels when purchasing.

Which Schedule PVC do I Need?

So what schedule PVC do you need? For aquaponics, home repairs, or irrigation projects, schedule 40 PVC is the way to go. Even schedule 40 PVC is capable of handling impressive pressure, and it is likely more than adequate for light to moderately heavy applications.

Using schedule 40 will save you money, especially if you plan on using large diameter parts. For large commercial, industrial, or chemical applications, it would be wise to use schedule 80 pipe and fittings. These are applications that will likely cause higher pressure and stress on the material, so thicker walls are imperative.

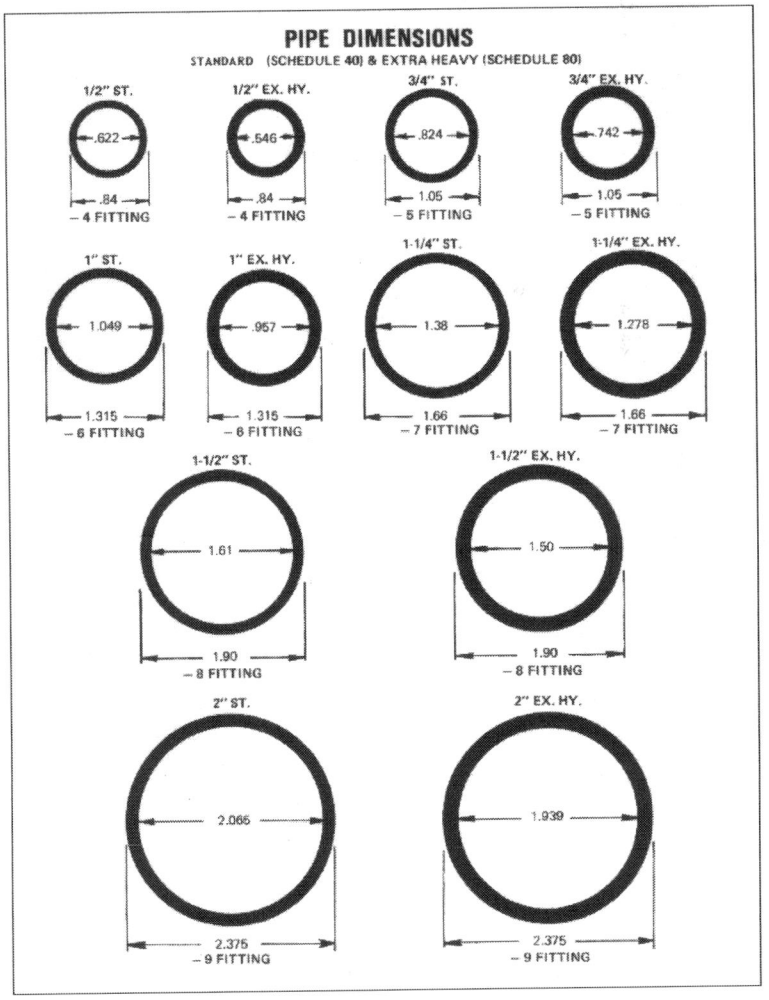

FIGURE 59. Dimensions of Schedule 40 and Schedule 80 PVC pipe.

PART III: COMPONENTS OF AQUAPONICS USED IN AQUAPONICS

Pipe Connections and Fittings

There are two commonly used PVC fittings in aquaponics—threaded connectors and slip connectors. Threaded connectors are ones that screw into one another and are designated as male and female. Slip connectors, as the name suggests, just slip into one another. In order to preserve pipe diameter, avoid restricting flow, and minimize clog points, it is prudent to use only female fittings. Female fitting fit over the outside of the pipe, whereas male fittings fit inside the pipe. Examples of common connectors are (see figure 60):

- 90° elbows
- 45° elbows
- 90° Tee fittings
- Ball Valves
- Bulkheads
- Reducers
- Couplings
- Wyes

FIGURE 60. Standard PVC fittings and connectors

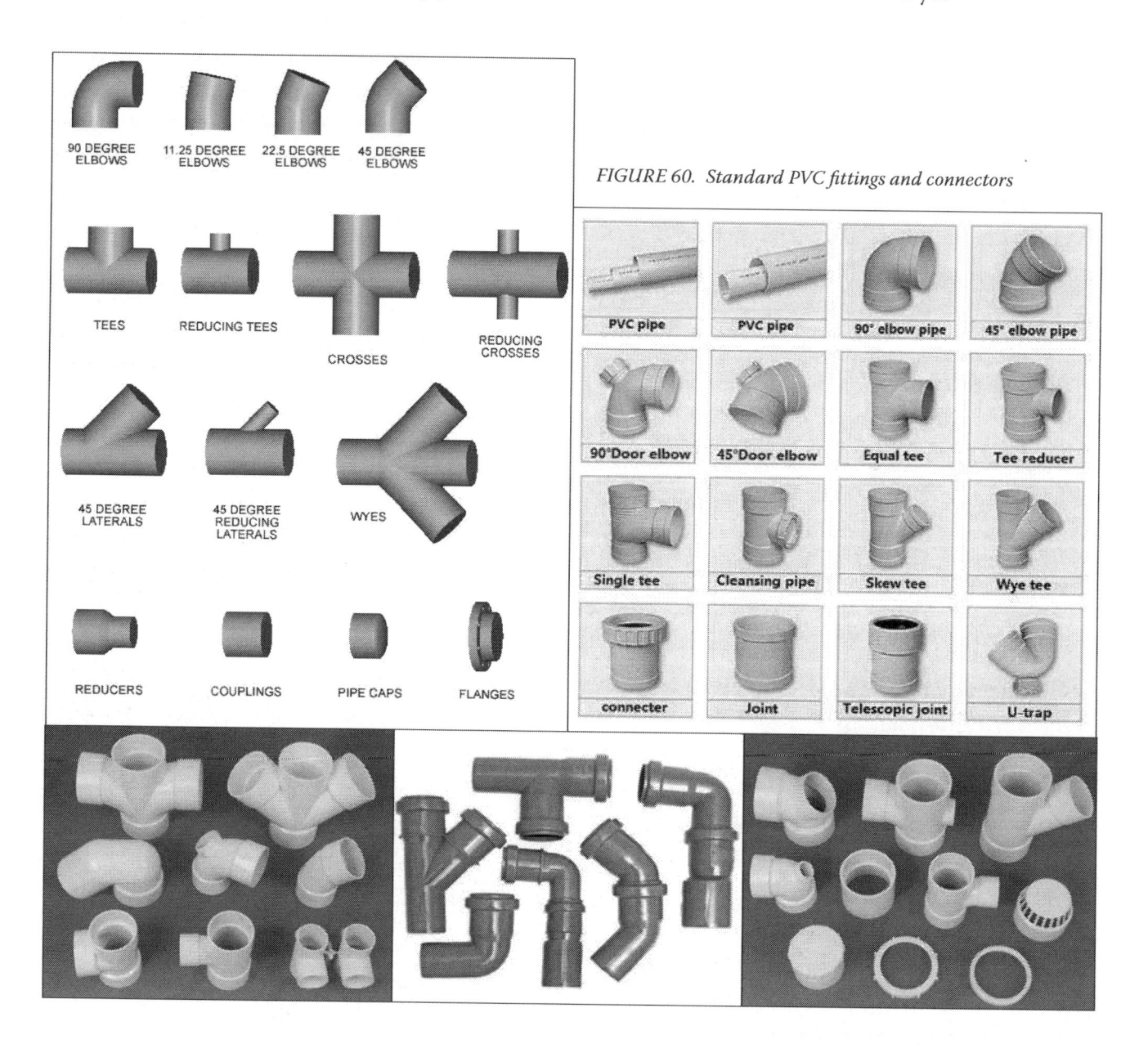

144

Irrigation Supplies

Irrigation materials are also commonly used in aquaponics. There are several good reasons as to why irrigation materials are integrated into aquaponics. They are inexpensive, require minimal effort in learning how to install them, they are safe for plants, fish and people, they are easy to install, and are effective conduits of aquaponic fluids.

However, not all irrigation materials are created equally. Some have thin walls, and are so light they are basically a cheap low grade inferior solution. As inexpensive as irrigation materials are, it is better to invest in the higher grade premium materials sold commercially to professionals from an irrigation supply store, rather than the typical residential stuff commonly sold at home improvement stores. Also, no more than what is typically used in a standard size aquaponics system, the extra cost for premium materials is nominal. The result for paying slightly more for premium irrigation supplies is that your aquaponic system will be much more reliable (less prone to leaks and not easily disrupted because of accidental damage). Premium irrigation materials are also much easier and faster to install than light weight residential materials.

It is also helpful to stick with the same name brand of irrigation materials, as they will be more interchangeable. Furthermore, it is helpful to shop at the same store so as to develop a solid working relationship with qualified staff. *Ewing Irrigation and Landscape Supply*, for instance, has expert staff on board eager to provide a wealth of information and assistance. And, no, I am not getting compensated for endorsing them. Nor do not have any friends or family working for Ewing, or benefiting from my praise of Ewing Irrigation and Landscape Supply Co. I say this as one consumer to another with the sole purpose of helping you. I have shopped for irrigation supplies for my nursery, aquaculture, hydroponics, and aquaponic operations for many years. I've been to many different supply stores (Ewing's competitors, several different large chain home improvement stores, Wal-Mart, etc.), and have tried many different products.

Bottom line: I am of the opinion that Ewing has the best quality products, and all the stores I have visited (as well as my phone calls), they have provided me with the most knowledgeable professional advice. Ewing is located throughout the southern half of the United States, as well as the west coast. They also have a very user-friendly website, and provide accredited education opportunities for those aspiring to learn more of development further professionally.

In summary, irrigation supplies offer a good alternative or supplement to PVC for plumbing of an aquaponics system. It is prudent to purchase premium quality irrigation products over cheaper residential type irrigation materials. It is also beneficial to shop at a store in which sound expert advice can be obtained, and build a positive relationship with the staff. Following are some of the more common irrigation materials used in aquaponics:

FIGURE 61. 90-degree elbow, tee fitting, coupling

FIGURE 62. ¼-inch tube tee, ¼-inch tube elbow, ¼-inch tube coupling

PART III: COMPONENTS OF AQUAPONICS USED IN AQUAPONICS

FIGURE 63. **Tubing**, *also called Poly Tubing, Poly Pipe, Supply Line, Trunk Line; all of which are common terms for this flexible polyethylene pipe (shown above). Common sizes are ½" (aka ⅝") or ¾" tubing. Emitters can be inserted into tubing or connected via ¼-inch diameter size micro tubing.*

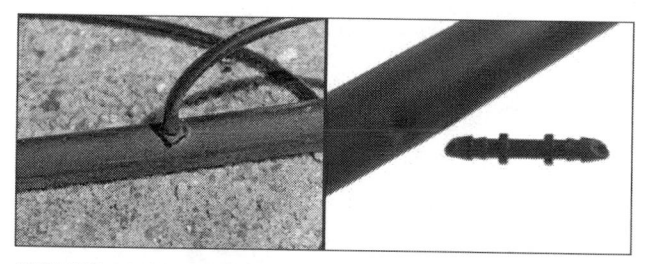

FIGURE 64. Two different types of ¼-inch tube to ⅝-inch tube connections

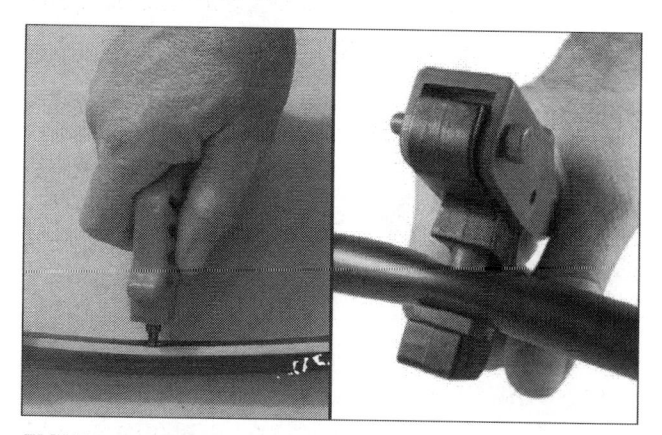

FIGURE 65. Two different kinds of hole punches used to install ¼-inch tubing in larger dia. Irrigation poly tubing.

Hoses

In addition to piping and irrigation tubing, hosing can also be used for water conveyance in aquaponics. The disadvantages of hoses are that they can kink, and some can be relatively expensive. Nevertheless, there is a wide variety of hose material types that can be used in aquaponics (see figures 66–70). Having clear hose will provide you the benefit of being able to watch your aquaponics system plumbing at work, which is both helpful and fun.

Use a hose clamp (figure 71) to obtain a secure watertight connection of the hose to pipes and fittings.

Fluid Dynamics

There are a few things to consider that influence how fluid actually flows through a pipe. The pipe and fittings causes friction to what would otherwise be smooth movement of the water, so the water in the very middle of the pipe is conveyed somewhat faster than the water flowing near the sides of the pipe. The difference is small, but it exists nonetheless.

In addition, comparing the flow rate through a straight length of PVC pipe, versus that of a pipe with a series of bends, the water flows more quickly through the straight pipe. There is no need to use complicated engineering equations; just recognize that there are some factors that will determine the flow rate through the aquaponics plumbing system, which in turn impacts what size diameter piping or tubing should be used. Simply put, in a given amount of time you can move a greater volume of water through a big pipe than you can through a small pipe. Also, if moving water over a significant distance (i.e. from an outdoor fish tank to a greenhouse grow bed), then it is especially important to caution on the side of using larger diameter piping or tubing instead of a smaller size, even if a reducer must be used at both ends (i.e. reducer at the pump and a reducer at the grow bed).

FIGURE 66.
Flex hose

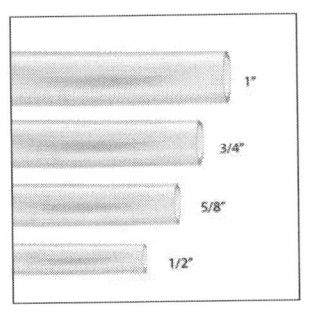

FIGURE 67.
Clear vinyl tubing

FIGURE 68.
Reinforced PVC hose

FIGURE 69.
Braided PVC hose

FIGURE 70.
⅝ to 1-inch diameter harden hose

FIGURE 71. Hose clamp

Many pumps suitable for an aquaponics system come with a ½-inch outlet. Rather than convey the pump outflow through a ½-inch irrigation line, it is better to install a reducer at the pump. For example, at the pump, install a short ½-inch irrigation tube about an inch in length, then add a ½-inch to ¾-inch reducer, and connect it to a ¾-inch poly irrigation tube to convey the fish tank waste water to the flood-and-drain plant grow bed (or grow channels if referring to a N.F.T. system).

Another thing to consider is gravity. Gravity will exert a constant downward pressure on the water in the aquaponics system. A pump used to lift water will be a fight against gravity. This means that the amount of water a pump can push will be reduced the more it has to lift the water. This will be discussed in more detail in the chapter covering pumps.

Gravity can also be used to provide a great advantage. It can be used to move water with no mechanical intervention. For instance, if a grow bed or grow channel is higher than the fish tank, then the overflow water can return to the fish tank directly through the use of gravity in a well designed system.

When considering plumbing, think about all the water that will be in your system. Such includes the water in the fish tank, grow beds or channels, sump if applicable, and in the pipes. The sum of that volume is the amount of water in the aquaponics system. As a general rule, for a flood-and drain system, the entire volume of water in the fish tank(s) should be moved every hour in order to maintain good water quality for the fish. This issue will be addressed in more detail in the water quality, pump, and flood-and-drain chapters.

As discussed in this chapter, and addressed in more detail later in the book, the 'head' of the system is how high in elevation the pump must move the water. Another item to consider is whether the system is to have a pump running continuously, or if it will have a timed flood and drain system. All of these factors together determine how much water needs to be moved around the time it takes to move it, and therefore plays an instrumental part in plumbing design and pump size selection. The pipes and/or tubing need to be large enough in diameter to convey the required volume of water.

Also, a well-designed system needs to have various controls and safety measures in place just in case there is any problem. Not to worry, as these issues can easily be determined, and they will be addressed in a user-friendly way later in this book. This chapter is meant to serve only as an introduction into these issues, and to emphasize that when selecting piping and/or tubing, it is better to pick slightly larger diameter plumbing component when possible.

Watertight Pipe Protrusion through Grow Beds and Fish Tanks

To achieve a watertight connection when passing a pipe or polyvinyl tubing through a fish tank wall, grow bed wall, or a N.F.T. grow channel a bulkhead fitting or a Uniseal can be used. Bulkheads are sturdier and work better for thick-walled applications, but also cost more than Uniseals.

The average price for a PVC bulkhead is approximately 25 percent more than a Uniseal.

Also, additional fittings are needed to connect the pipe to the bulkhead, whereas with a Uniseal, the pipe will slide through the opening (no additional fittings are needed).

For additional protection or if a drip leak is discovered after installation, with a Uniseal or a bulkhead fitting, heavily applied silicon caulking will often resolve the problem.

Bulkheads

Bulkheads come in a very wide variety of sizes and shapes, but can be easily assembled from parts readily available in most plumbing supply stores and home improvement centers, as well as online. Bulkhead fittings are identified by the size of the pipe it connects to, not the hole size. The bulkhead is a good, sturdy option for plumbing through a fish tank or grow bed. Bulkhead fittings are typically made of polyethylene, CPVC, PVC, polypropylene, Polytetrafluoroethylene (PTFE), and of various metallic materials. All types will work for aquaponics, so it makes sense to select the more common and inexpensive types, such as polyethylene, CPVC, or PVC.

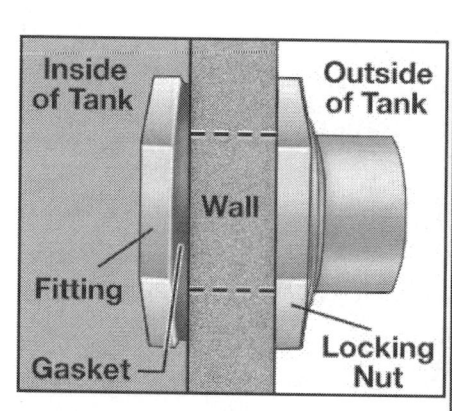

FIGURE 72. Bulkheads

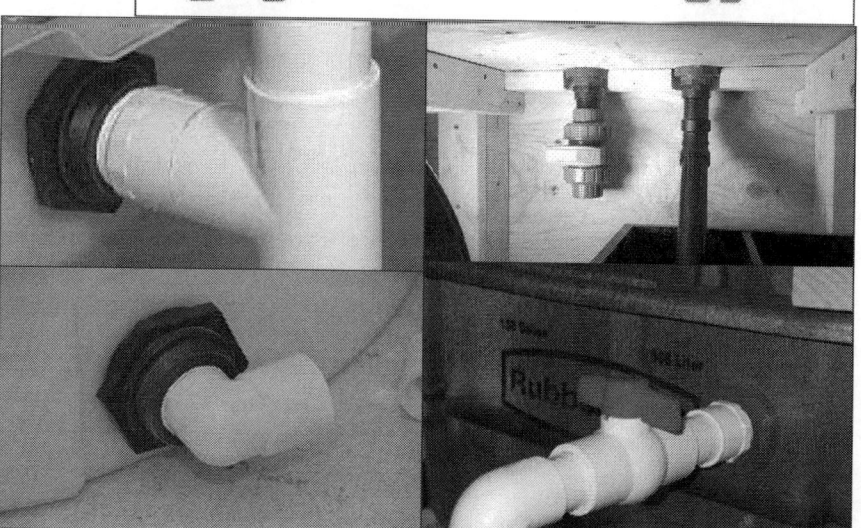

CHAPTER 14: PLUMBING

Grommets (Uniseal®)

Grommets (Uniseal®) are rubber rings that fit into the holes that have been drilled into the tank. Uniseal is a company trade name. A uniseal is a grommet, but they are so common that most people refer to all grommets as 'unseals'. It is similar to people referring to drywall as sheetrock. Sheetrock is a company name. Drywall (also known as plasterboard, wallboard, sheetrock, gypsum board) is a panel made of gypsum plaster pressed between two thick sheets of paper. It is used to make interior walls and ceilings. Since most people in aquaponics and plumbing refer to a grommet as a 'unseal', the same will often be done throughout this book so as to hopefully help the reader better relate to the message being provided.

A uniseal clamps around the hole making a watertight connection; the PVC pipe can then be slotted into the seal. The seals usually allow the pipe to be installed in only one direction, thus providing a watertight seal between the pipe and the connector. Unseals are inexpensive, costing anywhere from about $2 to $15 (USA, year 2016), depending upon quality, size, and where it was purchased. They make it easy to put a pipe through a tank, and they can also be used with rounded surfaces thus making them particularly useful for plumbing into barrels and other like rounded containers.

Uniseals will even allow you to plumb directly into five gallon buckets, brute trash cans, or any type of round surface. They accept standard Sch. 40 or Sch. 80 PVC and allow DIY projects that once may have been very costly to complete much more affordable.

Uniseals are used to attach pipe to just about any container in situation where bulkheads will not work, or not preferred. The most common use for uniseals is on curved surfaces such as storage drums, buckets, or even other pipes.

The advantage to using a uniseal is that it is inexpensive, and provides for a quick and easy installation. A hole is cut in the side of the tank or grow bed, and then the rubbery black uniseal is inserted. Next, a

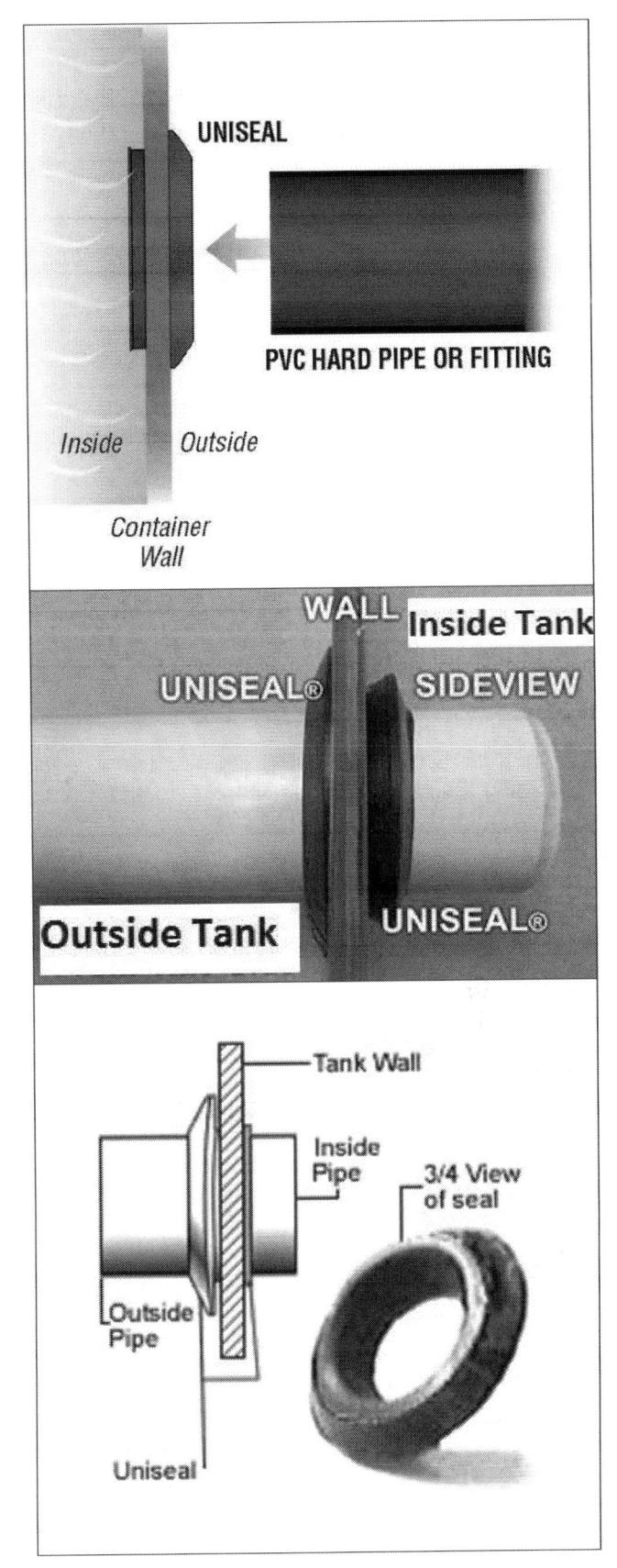

FIGURE 73. Grommets (Uniseal®)

slippery detergent film (i.e. dish soap) is applied to the exterior of the pipe. The pipe is then pushed through the seal completing the installation. When the pipe is pushed through the rubber-like uniseal from the outside, the uniseal becomes thin enough to allow the pipe to slip through. With this simple design, the uniseal solves many complex problems.

The disadvantage to using a uniseal is that it is prudent to replace them anytime you need to take out or reinstall the pipe, as they lose their watertight structural integrity. Also, they cannot be used in thick-walled applications. Lastly, in high pressure situations (for instance, in extremely tall tanks, but rare in aquaponics) a bulkhead would be a better choice.

Grommets (uniseals) have more than one dimension to consider. They are measured with both an inside and outside diameter; basically the thickness of the "wall" changes. Plus there is the thickness of the material the grommet is being used it as well.

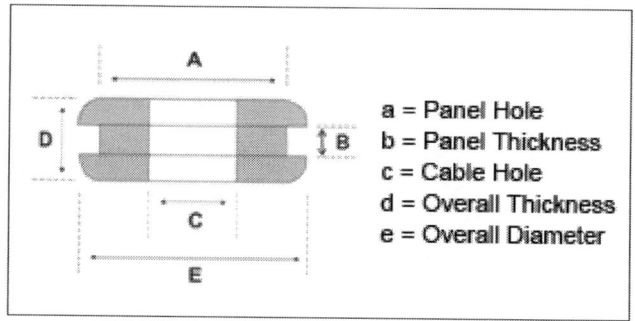

a = Panel Hole
b = Panel Thickness
c = Cable Hole
d = Overall Thickness
e = Overall Diameter

FIGURE 74. Grommet (Uniseal)

Uniseal Specifications

½" Uniseal
Fits Schedule 40 or 80, ½" pipe
Pipe ID — ½"
Hole saw size: 1 ¼" or 31.7mm (32mm)

¾" Uniseal
Fits schedule 40 or 80, ¾" pipe
Pipe ID — ¾"
Hole saw size: 1 ¼" or 31.7mm (32mm)

1" Uniseal
Fits schedule 40 or 80, 1" pipe
Pipe ID — 1"
Hole saw size: 1 ¾" or 44mm

1¼" Uniseal
Fits schedule 40 or 80, 1¼" pipe
Pipe ID — 1¼"
Hole saw size: 2" or 50.8mm

1½" Uniseal
Fits schedule 40 or 80, 1½" pipe
Pipe ID — 1 ½"
Hole saw size: 2½" or 64mm

2" Uniseal
Fits schedule 40 or 80, 2" pipe
Pipe ID — 2"
Hole saw size: 3" or 76mm

3" Uniseal
Pipe ID — 3"
Hole saw size: 4" or 102mm

4" Uniseal
Pipe ID — 4"
Hole saw size: 5" or 127mm

6" Uniseal
Pipe ID — 6"
Hole saw size: 7" or 177.8mm

Uniseal Installation Instructions

- Cut hole to the hole saw size indicated for the uniseal below.
- Ensure that the hole is clean with no sharp edges. Irregularities can cause a poor seal and leaks.
- Insert the uniseal into the hole with the wide flange on the outside of the container.
- Ensure the pipe end that will be inserted is clean of burs or sharp points. File the edges if needed.
- Insert the pipe into the uniseal. You can lubricate the pipe end with Windex. Most of it will be squeezed off during installation, but be sure to wash it out thoroughly before turning on your system.

Uniseal Grommets

1/2"	3/4"	1"
Fits 1/2" Schedule 40 PVC Tubing Drill 1 1/4" Hole	Fits 3/4" Schedule 40 PVC Tubing Drill 1 1/4" Hole	Fits 1" Schedule 40 PVC Tubing Drill 1 3/4" Hole

FIGURE 75.

Watertight Pipe Protrusion through a Liner

There are a number of devices that can be used to ensure a watertight seal where a pipe protrudes a liner. One item is a PVC liner boot. The PVC liner boot can either be built into the liner or made separate for field installation. A pipe coupler can also be used. A conduit flashing pipe boot (cone image below) also works well. The items can easily be obtained from your local plumbing supply store, hardware store, or home improvement center at a relatively low cost.

Pipe Maintenance

Periodic cleaning of pipes is a maintenance chore that should not be neglected. It will ensure that your aquaponics system continues to run efficiently, and that there are no unnecessary stress being placed upon pumps, plants, and fish. Pipe cleaning will remove blockage due to accumulation of algae, mosses, or other debris. Since each system is different, there is no hard and fast rule as to how often this should be done, but it is prudent to do it sooner and regularly rather than waiting until a problem occurs.

FIGURE 76. Several methods for making a watertight pipe protrusion through a tank.

PART III: COMPONENTS OF AQUAPONICS USED IN AQUAPONICS

A drain snake is the most standard pipe cleaning method used. Drain snakes are available as either a manual or power operated tool. Running water through the pipe during and after the process is recommended. However, it is best to detour the cleaned waste debris out of your system, or catch it in a bucket, rather than allowing it to enter back into your system. Never use chemicals.

A power pressure spraying machine is ideal for cleaning pipes. They can typically be rented from home improvement centers, as well as tool rental stores. However, most backyard aquaponic setups are too small to warrant the rental cost.

A built in debris trap located underneath the grow bed on the fish tank return line will greatly help reduce accumulation of pipe clogging debris. The following provides several debris trap examples:

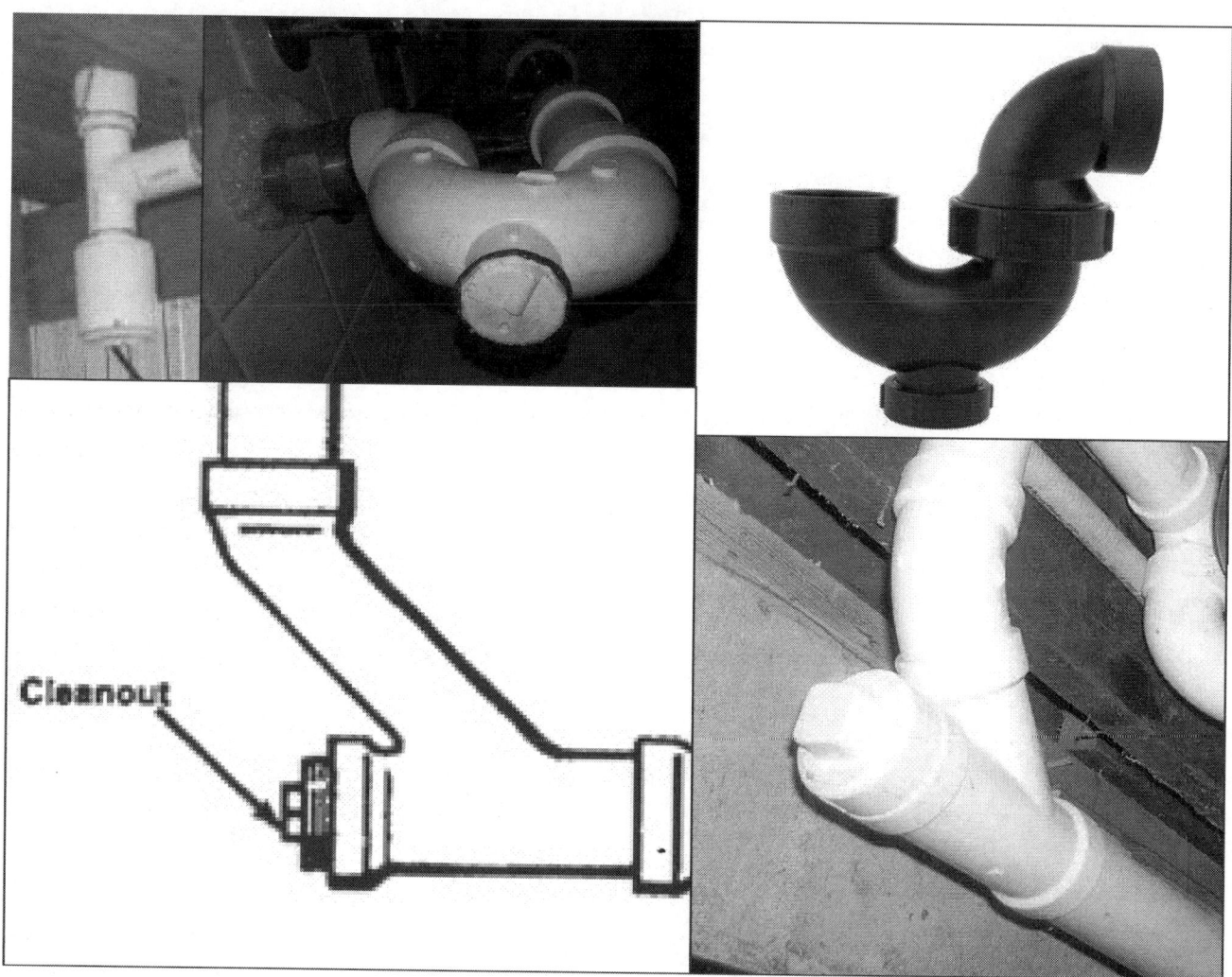

FIGURE 77. Debris traps

CHAPTER 15

Fish Tanks

The fish tank is a crucial part of an aquaponic system and so its size, safety, water quality, strength and long term usability must be carefully considered. As such, the fish tank is typically one of the most expensive components of an aquaponic system. The tanks can account for up to 20 percent of the entire cost of your aquaponic setup. However, there are some cost effective options, which we'll address in this chapter.

Fish require certain conditions in order to survive and thrive, and therefore the fish tank should be chosen wisely. There are several important aspects to consider including the shape, material type, placement, and color.

If the fish tank is to be located outside, it will be subject to environmental conditions. It needs to withstand direct sunlight (UV resistant) and temperature variations without cracking, warping, or leaching chemicals into the water. For best results, shade fish tanks to prevent algae growth and to reduce stress to the fish; they prefer dark hiding places. Avoid direct lighting on the fish tank.

If possible, in addition to shading the top, place at least one object within the water that will provide the fish a sense of security. Such is rarely done in medium to large scale aquaculture operations. However, an object in the water, which offers fish refuge, will go a long way in improving fish health via reduced stress, and is considered by many to be a more natural, animal friendly, and ethical means to rear fish.

Your choice of fish tank will also depend largely on your goal of aquaponics. If you are building a small size aquaponics system for hobby purposes, then you will be restricted to growing smaller or fewer fish. If you desire to rear fish for food, the fish tank should be large and sturdy enough to hold at least 50 gallons of water. The fish tank should be made of food-safe materials. However, fish tanks can be constructed from just about any structure or recycled materials, so long as they are lined with LDPE, PVC, HDPE pond liners, or other safe materials. Do not use EPDM liners or products in your system as EPDM releases toxic chemicals over time that are harmful to the living organisms within your system, and can also enter into your food supply.

As a general rule, use approximately one pound fish per six gallons of water, for fish stocking densities. This will determine the volume of water required for the size of the fish tank. Higher stocking densities per volume will cause the accumulation of toxic waste, and result in the loss of fish as there wouldn't be enough bacteria and plants to remove the nutrients fast enough.

Fish Tank Shape

Although any shape of fish tank will work, round tanks with flat bottoms are recommended. Square tanks with flat bottoms are perfectly acceptable, but require more active solid-waste removal. Tank shape greatly affects water circulation. There is a risk to having tank with poor circulation. Artistically shaped tanks with non-geometric shapes (featuring many curves and bends) can create dead spots in the water with no circulation. These areas can gather wastes and create anoxic, dangerous conditions for the fish. If an odd-shaped tank is to be used, it may be necessary to add water pumps or air pumps to ensure proper circulation and remove the solids. It is also important to choose a tank to fit the characteristics of the aquatic species reared because many species of bottom dwelling fish show better growth and less stress with adequate horizontal space.

As mentioned above, the aquaponic fish tank can be of almost any shape, but the most commonly used are circular. The round shape allows water to circulate uniformly and transports solid wastes towards the centre of the tank by centripetal force. A round tank is structurally stronger than any other shape, so given that the tank may be holding 250 plus gallons of water, the round shape allows for better structural integrity and less reinforcement. Therefore, round tanks don't have to be as heavy and bulky as square or rectangular shaped tanks. As a result, round tanks, such as the fiberglass tanks shown in the two images below (figure 78), are usually the least expensive tanks.

A round shape also allows for better water circulation. They have several distinct benefits for water quality and fish health, such as:

* Water circulation prevents thermal layering and thus helps to improve water quality.
* A good flow of water provides the fish with a current to swim against, which promotes health and muscle growth. This in turns makes raising fish for food better for consumption.
* Fish appear to naturally enjoy some moving water.
* Water movement aids in the exchange of gases between the air and the water (CO_2 exiting the fish tank, and oxygen entering the fish tank). Surface water constantly changing dramatically increases the rate of oxygen exchange.
* Increased oxygen through water movement benefits not only the fish, but also the de-nitrifying bacteria in the system, as well as inhibiting harmful bacteria that thrive in an anaerobic environment.
* In a rounded tank, solid waste has a tendency to gravitate toward the bottom-center of the tank. Strategic placement of the pump at this point will enable the collection and transport of this solid waste to the grow bed.

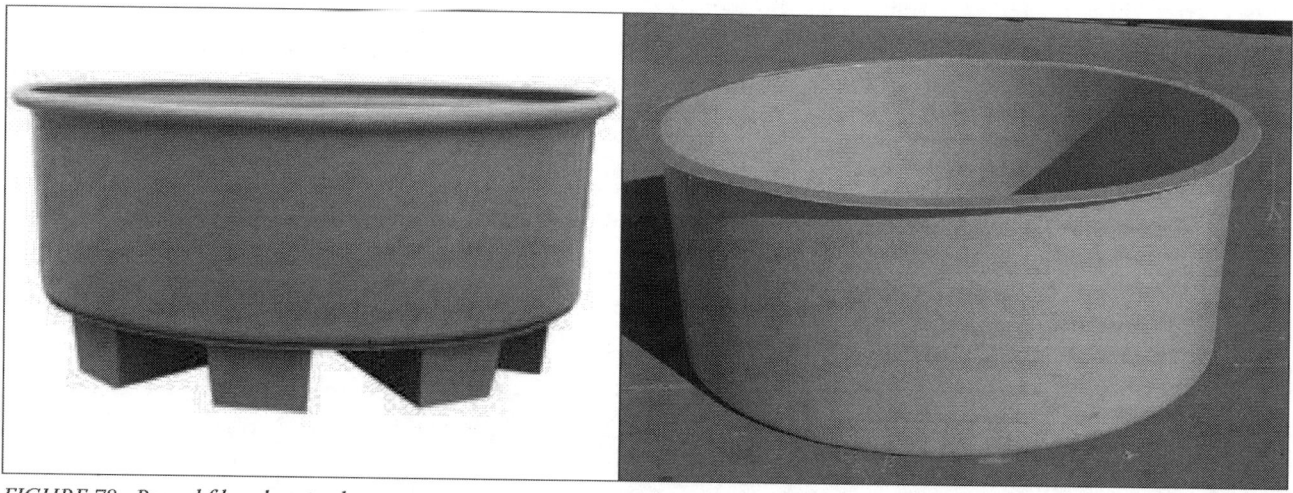

FIGURE 78. Round fiberglass tanks

NOTE: I highly recommend that you avoid bolt-up fiberglass tanks, similar to the ones in figure 79 which are crossed 'X' out. I have had lots of experience with these types of tanks, as well as many other types of fish tanks; they are highly problematic. There is a strong probability for leakage with bolt up tanks (eventually, if not initially). I have seen bolt up tanks leak way too many times in my career. I will never use or recommend a bolt up tank. Even if obtained for free, they will cost you time, and most likely product, when you have to eventually address a leak. Lastly, I believe bolt up tanks can be a safety hazard as well, should leaked water make its way to an electrical source (i.e. extension cord on the ground, etc.).

Square or rectangular fish tanks are also common and are especially useful if there is limited space. Care should be taken to ensure there is adequate water circulation so dead air pockets (low dissolved oxygen levels) do not form in the corners. Square tanks with rounded corners, such as the ones shown below, will assist in the water movement.

FIGURE 79. *Bolt together fiberglass round tanks. Not recommended due to higher probability of leakage.*

FIGURE 80. *Rectangular and square tanks.*

PART III: COMPONENTS OF AQUAPONICS USED IN AQUAPONICS

FIGURE 81. IBC (Intermediate Bulk Carrier) Tote

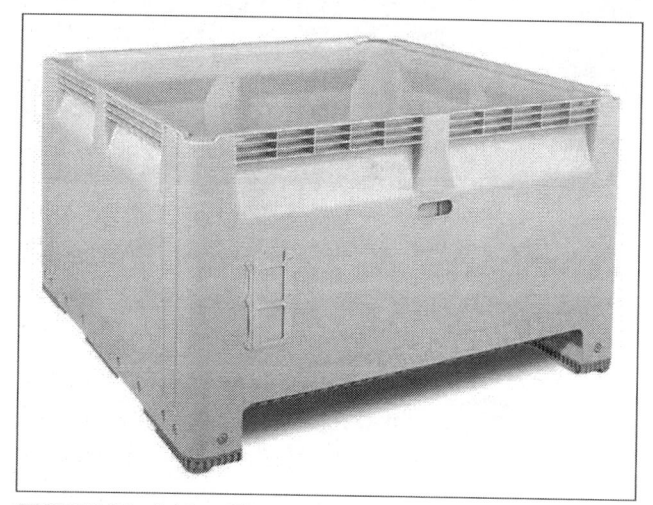

FIGURE 82. A Macrobin

IBC and Macrobin Tanks

IBC (Intermediate Bulk Carrier) totes (or high quality, food grade bulk shipping containers such as Macrobins) are commonly used, square aquaponics fish tanks. A used or reconditioned 275 or 330-gallon food grade IBC tote works great and can be acquired at a relatively low cost. Most standard size micro-bins can hold about 400 gallons of water.

Liner Tanks

Another practical and relatively inexpensive type of tank that works well for aquaponics is a DIY frame (wood, brick, concrete block, earthen, etc.) lined with a LDPE (Low Density Polyethylene) or HDPE (High Density Polyethylene) liner. The liner should range anywhere from 20 to 40 mils in thickness. Obviously, the thicker the liner, the better in terms of protection against tears. However, the thicker the liner the more difficult it is to handle and work with during installation, and the more expensive it will be.

The ideal liner would be UV resistant, fiber reinforced, have one side white for easier installation, be of food-grade quality, and be thick enough so as to not tear easily during installation or regular use. However, a liner that is made of LDPE or HDPE material, is 20 to 40 mils in thickness, and UV resistant is certainly satisfactory.

FIGURE 83. Liner tanks

Fish In and Predators Out

Regardless of what type of tank is used, it is important to ensure that small children and predators cannot enter the tank, and that fish cannot exit the tank. To keep fish in the tank, the water level can be lowered to approximately eight inches below the rim (this works for most species). However, doing so greatly decreases valuable volume capacity. A better method is to install netting or a fence around or over the fish tank. Keep in mind that it is also important to have quick access to the fish tank to check on the water quality, and for maintenance purposes.

I worked with a large aquaculture farm in California that raised several different types of fish in both ponds and 50-foot diameter tanks. One species raised in the tanks was largemouth bass. Although the surface water level was about a foot below the rim of the tanks, they had to keep a net stretched tightly over the bass tanks, because during feeding (when they dispersed feed on the surface of the water) the fish would get so excited that the water became extremely turbulent.; the fish would literally jump out of the tank if it weren't for the net.

Multipurpose Fish Tank

Planned correctly, a fish tank can serve several valuable purposes. The benefits of each should not be overlooked or taken lightly. Using a fish tank for more than just containing fish can significantly increase production, profits, and lower cost. And although I am using the 'tank' term, the same principles apply to a pond or any vessel used to raise fish. Furthermore, a fish tank's value increases when each of the following items are implemented.

- **Fish tank used to raise fish** or other aquatic species (obviously, raising fish or other aquatic animals is the main purpose of the fish tank).
- **Fish tank used for radiating heat and/or the cooling of the greenhouse.** Radiation is the process of heat transfer resulting from the temperature difference between elements. Standing

FIGURE 84. Net over fish tank

FIGURE 85. fence around fish tank

next to a hot fire (or standing outside on a hot sunny day), you can feel radiation heating the surface of your skin. Similarly, in a greenhouse, a fish tank radiates energy to the surrounding environment. In the summer, the tank helps cool the greenhouse, and in the winter the tank helps generate heat. This process helps to lower heating and cooling utility costs, and helps maintain a more consistent temperature balance within the greenhouse.

PART III: COMPONENTS OF AQUAPONICS USED IN AQUAPONICS

- **Fish tank used for a D.W.C. raft.** Although you may have decided upon a flood-and-drain or a N.F.T. aquaponic type system, if lighting and temperature conditions are sufficient, place a D.W.C. plant raft on the water surface of the fish tank. As mentioned in 'Chapter 11: Fish—Everything You Need to Know', fish like shaded areas best. Besides providing you additional plant product without sacrificing space, the raft helps the fish feel more secure, and as a result, lowers fish stress thereby improving their health. Just be sure to allow some spacing between the edge of the plant raft and the side of the tank for fish care and tank maintenance.

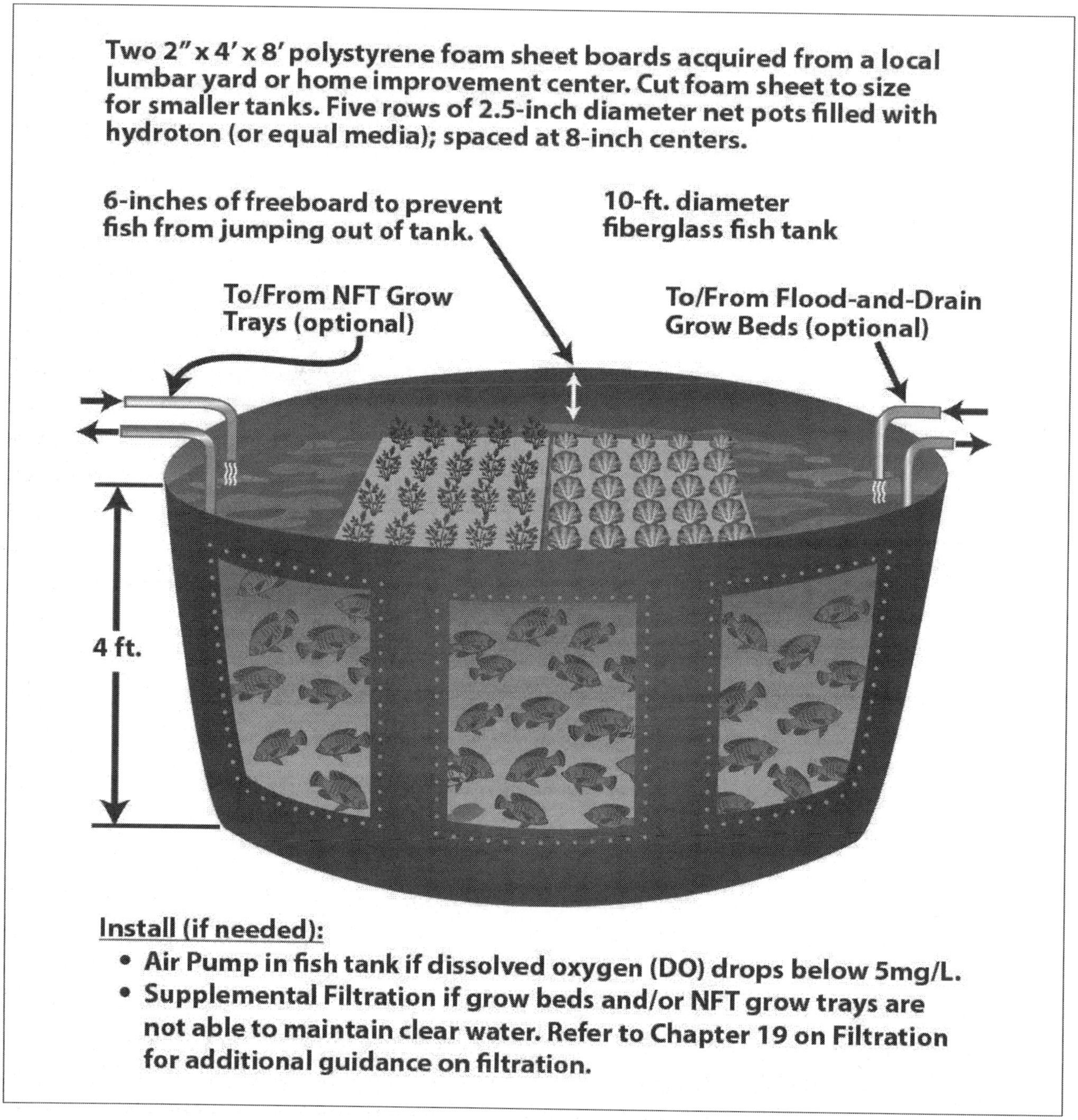

Two 2" x 4' x 8' polystyrene foam sheet boards acquired from a local lumbar yard or home improvement center. Cut foam sheet to size for smaller tanks. Five rows of 2.5-inch diameter net pots filled with hydroton (or equal media); spaced at 8-inch centers.

6-inches of freeboard to prevent fish from jumping out of tank.

10-ft. diameter fiberglass fish tank

To/From NFT Grow Trays (optional)

To/From Flood-and-Drain Grow Beds (optional)

4 ft.

Install (if needed):
- Air Pump in fish tank if dissolved oxygen (DO) drops below 5mg/L.
- Supplemental Filtration if grow beds and/or NFT grow trays are not able to maintain clear water. Refer to Chapter 19 on Filtration for additional guidance on filtration.

FIGURE 86. Fish tank providing multiple aquaponic uses (DWC, plus NFT grow channels and/or Flood-and-Drain media-bed supply feed)

CHAPTER 15: FISH TANKS

- **Fish tank used for multiple types of aquaponic systems.** Most aquaponic resources in the marketplace refer to the three different types of systems (D.W.C., N.F.T., and Flood-and-Drain). However, if planned properly, a system can be set up to be more than just one system, thus providing much greater flexibility and product yield.

Figures 86 and 87 show how a fish tank/pond can be utilized to provide "multiple-use" options, and thereby maximizing production without sacrificing space.

Multi-Use Earthen Fish Tank/Pond with Liner

Freeze Precautions:
- Bury hose/piping below frost line.
- Insulate, and/or use heat tape, where hoses or piping is above frost line.

2" x 4' x 8' polystyrene foam sheet board aquired from a local lumbar yard or home improvement center. Cut foam sheet to size for smaller tanks. Five rows of 2.5-inch diameter net pots filled with hydroton (or equal media); spaced at 8-inch centers.

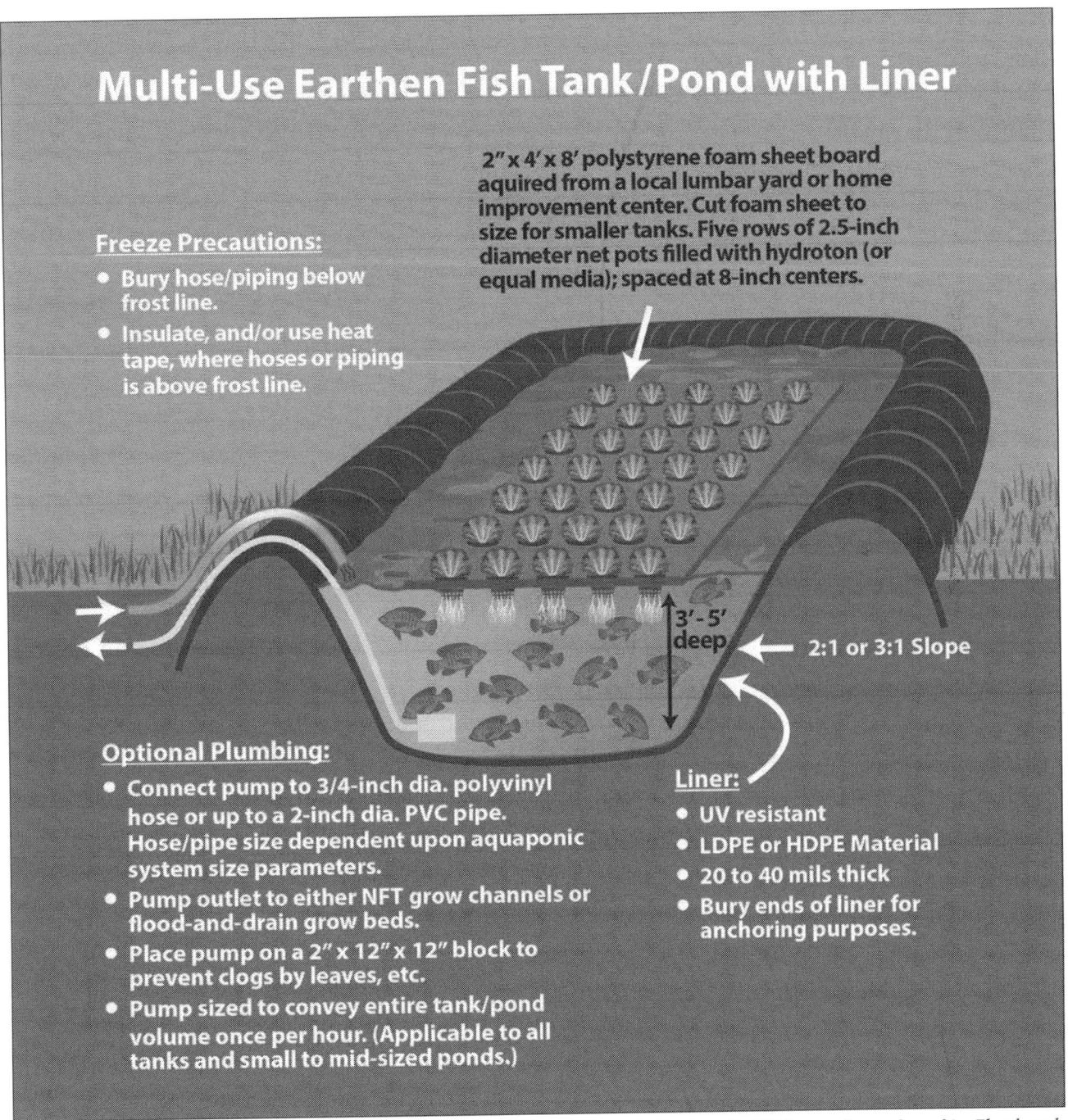

3'-5' deep

2:1 or 3:1 Slope

Optional Plumbing:
- Connect pump to 3/4-inch dia. polyvinyl hose or up to a 2-inch dia. PVC pipe. Hose/pipe size dependent upon aquaponic system size parameters.
- Pump outlet to either NFT grow channels or flood-and-drain grow beds.
- Place pump on a 2" x 12" x 12" block to prevent clogs by leaves, etc.
- Pump sized to convey entire tank/pond volume once per hour. (Applicable to all tanks and small to mid-sized ponds.)

Liner:
- UV resistant
- LDPE or HDPE Material
- 20 to 40 mils thick
- Bury ends of liner for anchoring purposes.

FIGURE 87. Fish pond serving multiple aquaponic uses (DWC, plus optional plumbing for NFT grow channels and/or Flood-and-Drain media-beds).

Birds of Prey

As mentioned in the introduction of this book, I served as the Engineering Manager over four large aquaculture farms, which produced 80 percent of the Nation's caviar and seven-plus tons of meat annually, which we also shipped all over the world.

True caviar is harvested from sturgeon when they are ten-plus years old. In my case, we raised white sturgeon fish from egg to harvest. By the time they were old enough to produce caviar, they typically weighed anywhere from 125 to 300 lbs (four to seven-plus feet in length). One of our farms was inside a steel building, but the other three were outdoors. All four farms had a water recirculation system, and the fish were raised in round tanks; most tanks ranged in size from 20 to 50 feet in diameter, with a depth of four to six feet. However, the larger tanks were ten feet deep. These tanks held a lot of fish. We had to keep all outdoor tanks with fish under about two years of age (less than 18-inches in length) covered with a net or shade cloth to prevent hawks and eagles from stealing them. The tanks were checked a least three times a day (feedings, maintenance, etc.), so employees had to regularly partially uncover the tanks to gain access. There were hundreds of tanks. Occasionally, an employee would fail to tightly seal the tank cover, after which we would sometimes find the remains of a partially eaten fish where a bird of prey took advantage of the opportunity for an easy meal.

If you live in a location where hawks and eagles are common, you may need to install a cover over your tank, if you find these awesome birds of prey are stealing your fish. It is doubtful that a scarecrow would work, as these birds will try to steal a fish off of a fishing line should they encounter a fisherman reeling in his catch.

Fish Tank Materials

Fish tanks are made via many different methods, and with a variety of materials. Some of the most common are food-grade plastic, fiberglass, and HDPE

FIGURE 88. A hawk with his catch.

FIGURE 89. An eagle stealing a fish off of a fishing line.

FIGURE 90. A shade covers over a fish tank helps regulate water temperature, shields fish from direct sunlight, and provides protection against birds of prey taking fish.

liner. However, concrete tanks and earthen structures are also used. All have their advantages and disadvantages. For practicality, in consideration of cost and ease of installation, plastic, fiberglass and HDPE liner materials make the most sense. The materials are light weight, easy to handle, and readily available throughout most of the world. Some creative aquaponic enthusiasts have even successfully converted swimming pools into viable aquaponics systems.

Wooden frames (or any other type of material for that matter) can also be built and turned waterproof via a HDPE liner, fiberglass, or an epoxy-based substance. These methods are certainly acceptable, but care needs to be taken to ensure that the frame has the structural integrity to withstand the lateral pressures, downward force, and stresses of the water within…and possibly several adults leaning against the tank. Water alone is extremely heavy by volume. Add to that several people pressing against the tank, and it is easy to understand that the tank must be fabricated using heavy duty construction methods.

Ponds can also be used in an aquaponics. Although care should be taken as an outside pond may contain and spread undesirable bacteria and/or sediment throughout the aquaponics system via the recirculating pumped water. This would compromise the water quality of the system and, at the least, require more frequent cleanings.

Concrete needs to be carefully considered and managed, as it can seriously affect water composition by causing imbalances in the pH. As water interacts with concrete, the water can dissolve various minerals present in the hardened cement paste or within the concrete aggregates. If the solution is unsaturated, dissolved ions such as calcium (Ca^{2+}) are leached out and can be detrimental to the aquaponics system as a whole. If concrete is used, it should be sealed with a food-safe, commercially available sealant

If constructing an aquaponic system with a previously used tank or materials, care needs to be taken to ensure that the tank is food grade, materials don't have any toxic residues, and there are no elements associated with what is being used that may be harmful to fish, plants, and human consumption. If the tank will be outdoors, it should also be of material that will not breakdown easily via UV light (UV resistant). If you are unsure about the material or tank, then it would be prudent to pass on it as potential problems are not worth the risk, could end up being a costly mistake, would be a major drain on your valuable time, and negatively impact on your peace of mind.

Ideally, the fish tank will be of an opaque material (not transparent), as direct sunlight will encourage algae growth. Algae is detrimental to an aquaponics system (primarily because algae are a prolific grower that also needs oxygen to multiply, thus depleting your system of the available oxygen needed by your fish, bacteria, and plants).

A window on the side of a tank can bring much pleasure, and also serves to make monitoring water quality and fish health much easier. It also makes aquaponics that much more fun for both adults and children. However, it is best if such a window is not located on a side of the tank that is exposed to direct sunlight or has a curtain to block it out.

Small aquariums that one finds in homes or businesses are able to use standard glass because they are micromanaged and have specific aquarium filters which address algae. Even so, a small aquarium can develop algae if in direct sunlight or if the filter is inadequate.

Fish Tank Size

There are several different issues that must be considered pertaining to fish tank size selection. Below is a summarized list:

Objectives, goals, and long-term plan.

What is your purpose for getting into aquaponics? What are your immediate and long-term production goals (i.e. amount of fish and vegetables you want to harvest each year)? Do you want to install your system and be done with it, or do you want to have the ability to enlarge it over time? Is your objective to just partially supplement your grocery store food with healthy, organic, home-grown food, grow enough for yourself others in your circle, or grow enough to barter and/or sell for a profit?

Location, facilities, and available space.

Will you need a greenhouse or green room to grow vegetables in the winter? Will your fish tank require temperature control during all or part of the year? Will your tank be indoors or outdoors? How much space do you have available for both the fish tank and plants?

Start-up and operational budget (i.e. feed cost, electrical cost for lights).

The larger the system, the greater the start-up and operational cost. How much are you willing to invest in creating an aquaponics system? What is your budget for feeding the fish, and potentially pay for increased utility cost (for temperature control and lighting)?

Maintenance and operational resources (i.e. your time, available assistance, etc.)

The larger the system, the more time and attention that will be required. It is great to have big ambitions, but if you don't have the time or resources to keep up with it, success will be hindered. Sometimes it is better to start off small and then grow in phases.

Type of aquaponic system (N.F.T., Deep Water Culture/Raft System, or Flood-and-Drain)

The tank to plant a grow area is not as critical with N.F.T. and Deep Water Culture as it is with a Flood and Drain System. As a matter of fact, with the exception of extreme circumstances, such as where the fish waste (plant nutrient solution) is so diluted that it is nearly non-existent, it is difficult to go wrong with tank size regarding N.F.T. and Deep Water Culture.

However, for a flood-and-drain system, a common rule of thumb is to use a 1:1 ratio when selecting the tank. In other words, the grow bed volume should be equal to (or larger than) the fish tank volume. This is by no means a hard-and-fast rule, but serves as an outline guide.

Regardless of the type of system or size of the fish tank, a certain level of water quality must exist. Adequate filtration must occur, whether it is through grow beds, a bio-filter, mechanical filtration, or a combination of these filtration methods. Water quality and specific details of system design are addressed later in this book

Redundancy

The optimal planned aquaponic system will have redundancy integrated within, where possible. Using a minimum of two fish tanks and two grow beds will have many advantages when it comes to addressing emergencies, harvesting operations, and maintenance issues.

Fish Tank Placement

A very common perspective of aquaponics is that of having the fish tank downstream, underneath, nearby by the grow bed (flood-and-drain), or grow channels (N.F.T.). Although there is nothing wrong with these layouts, they are by no means the only options. As a matter of fact, sometimes it is a big advantage to have the fish tank outdoors and located elsewhere on the property, a considerable distance from the greenhouse.

Another approach is for the grow beds to drain to a sump where the water is then pumped to the fish tanks. From the fish tank, the water can then either gravity feed back to the grow beds, or be pumped back if the elevation difference is not favorable.

The fish tank is the heaviest component in the aquaponics system. Each gallon of water weighs 8.34 lbs, which means a 300-gallon fish tank would weigh about 2,502 pounds (slightly less because of the fish). Since the fish tank will be too difficult to be move once it is filled, it is important to carefully consider its location. Furthermore, the fish tank should be located within or on the ground, or on solid flooring, since a structure to support all the weight would have to be enormous and very costly. In short, due diligence in determining fish tank location is paramount.

Fish Tank Safety Considerations

Regardless of the tank system being implemented or its size, remember that safety is a priority. If you have small children on your property then fencing, nets, and/or locked doors should be intact to prevent children from potentially drowning in the fish tank.

Water in contact with electrical components, electrical outlets, breaker boxes, and extension cords combine for a very dangerous situation. All electrical items should be securely protected from water and potential fish tank leaks. Thus, all electrical components, including extension cords, should be kept a safe distance off the floor.

In summary, take all necessary safety precautions to protect yourself and others. Be aware of all electricity and water combinations. Follow all National Electric Codes.

Fish Tank Selection

In summary, your fish tank choice will be influenced by space, budget, available time for maintenance, and aquaponic goals (desired yield). There is a wide variety of choices; deciding upon a fish tank can seem bewildering at first, but if you keep focused on what you desire to achieve out of aquaponics, then the size issue is easily resolved.

1. If space allows, a 250-gallon (1000 liters) or larger fish tank has been proven to create a more stable aquaponics system. Larger volumes are especially better for beginners because they allow more room for error. With larger volumes, changes to water quality happen more slowly than they do in smaller systems.

2. To raise a fish to a length of 12-inches ("plate size") a fish tank volume of at least 50 gallons is required.

3. The recommended stocking density is one pound of fish per five to seven gallons of tank water (.5 kg per 20-26 liters).

4. Determine fish tank volume from the stocking density rule above (one pound fish per five to seven gallons of fish tank volume or 1 kg per 40-80 liters).

CHAPTER 16

Liner Material

Liner Overview

A liner is an impermeable geotextile material used for water retention. They are commonly referred to as pond liner, water garden liner, greenhouse cover material, hydroponics pond liner, aquaponics bed liner, and polyethylene tarp material. However, liners are also used in many other types of industries in a wide variety of applications.

Liners are often used in aquaponics for flood-and-drain grow beds, D.W.C. tanks, fish tanks, and fish ponds. With the aid of structural support a liner has the ability to contain large volumes water, even against significant pressure. The advantage of using a liner over other materials is lower cost, ready availability, and ease of use.

There are many different types of liners available on the market. As matter of fact, there are well over 24 different types of common liner materials. However, only some materials are suitable for aquaponics.

Materials include LDPE (Low Density Polyethylene), PVC (Poly Vinyl Chloride), PVC with internal reinforcement, and HDPE (High Density Polyethylene). HDPE is frequently noted as the best, but it is the most expensive and hardest to work with because of its ridged nature. LDPE is most preferred by small to mid-size aquaponic operators.

Avoid EPDM as it emits toxins that are harmful to the beneficial bacteria in your system and the toxins will eventually make their way into your food supply. Also avoid vinyl liners as they are too stretchy as well as any product that doesn't specifically identify what is made of.

The liner thickness should range anywhere from 20 to 40 mils. Obviously, the thicker the better in terms of protection against tears. However, the thicker the liner, the more difficult it is to handle and work with during installation, and more expensive it will be.

The ideal liner will be UV resistant, fiber reinforced, have one side white for easier installation, be of food grade quality, and be of thick enough to avoid tears during installation or regular use.

Liner Installation Precautions

Whether installing a liner in a pond, tank, or grow bed care, needs to be taken to ensure that the piece is large enough to do the job. Allow excess to overlap grow bed rims and tanks or to be buried in the ground for anchoring purposes. In other words, measure twice, cut once.

Refrain from walking on or laying tools or supplies on the liner. Avoid installing liner material outdoors on windy days or make sure you have enough people to assist you. Also, don't let pets walk on the liner material either. Any of the above can cause a puncture and result in a major hassle when discovered after the system is filled with water.

Liner Modifications

Occasionally, two separate pieces of liner material need to be joined in order to have a piece large enough to do the job. The following are directions from the Colorado International Lining Company (www.coloradolining.com) for field seaming liner material.

RECOMMENDED GUIDELINES FOR FIELD SEAMING

Temperatures for seaming should be as follows:
- Minimum ambient temperature: 50° F.
- Liner surface temperature: 90-100° F.
- Solvent adhesive should be 70-80° F.

1. Before adhesive is applied, surfaces to be seamed must be dry and free of dirt and foreign materials. The contact surfaces of the panels should be wiped clean to remove all dirt, dust, or other foreign materials. A clean cloth or grout brush may be used to help clean.
2. Adhesion of one liner panel to another is accomplished by lapping the edges of panels a minimum of 3 -6" inches (6" up to 1' is more desirable).
3. If using a seaming board, it should be placed below the liner panels to be joined. Only the pull rope should be exposed through the seam.
4. To commence seaming, fold the upper sheet back and apply adhesive to both sheets at the overlap area using a 3" wide paintbrush. Only apply about a body width (shoulder to shoulder width) of adhesive at a time. The adhesive should be applied to reach the outer edge of the seam and be 2-3" wide (within the overlapped area). Use enough solvent to make the liner surface appear wet and shiny (not flooded or inundated). Once the solvent is brushed-on, the brush should be replaced in a can (not on the liner) and the seam be rolled flat with a seam-roller once both surfaces become tacky. Roller strokes perpendicular to the liner seam should be made to push out excess adhesive and air pockets. Some adhesive should barely start to ooze from under the top overlapped edge. Once 100 percent is burnished down, the next shoulder to shoulder width area can be adhered with the same repeated process. The seaming board shall be advanced as needed.
5. Always keep the solvent adhesive can sealed. Only remove the lid long enough to wet the seaming brush. Failure to do so will result in the adhesive losing its "solvent grab" through evaporation of volatiles and seam quality will deteriorate to unacceptable quality. One option is to have a separate paint style can with a slit carefully cut into the lid allowing the brush to fit snugly within. Once brushes begin to stiffen-up, they should be replaced.
6. Using too much PVC bonding solvent in a wide area could result in a poor quality seam and make repairs difficult.
7. About one hour after seaming is performed it is advisable to manually probe the seam. Any loose edges should be peeled back and re-glued or patched if necessary. All repaired areas should be identified with a crayon or marker for re-inspection after glue dries.
8. Seams typically require 24 hours to completely cure. They can reach 90 percent of their final cure strength in just a few hours.

CHAPTER 16: LINER MATERIAL

TOOLS SUPPLIES LIST FOR FIELD SEAMING

- PVC bonding solvent specially formulated for liners (i.e. LO-VOC X-15™ PVC Solvent, or equal). Purchase enough for your job. Read product label for coverage specifications.
- Typical solvent adhesive yield is: One gallon of adhesive should bond 75' — 100 lf of seam.
- Safety glasses or safety goggles.
- Rubber gloves.
- Well ventilated area.
- Knee pads.
- Cotton rags.
- Several 3" wide paint brushes
- One (1) whisk broom (or fox tail brush).
- One (1) 10" x 6' -8' long Douglas Fir clear board, rounded off on both ends as well as all
- edges, with a rope tied to one end.
- One (1) Stanley Knife.
- One (1) yellow crayon for marking liner surface.
- One (1) pair of scissors with rounded-off points.
- One (1) 2' plastic or wood wallpaper seam-roller per each field seaming crew.

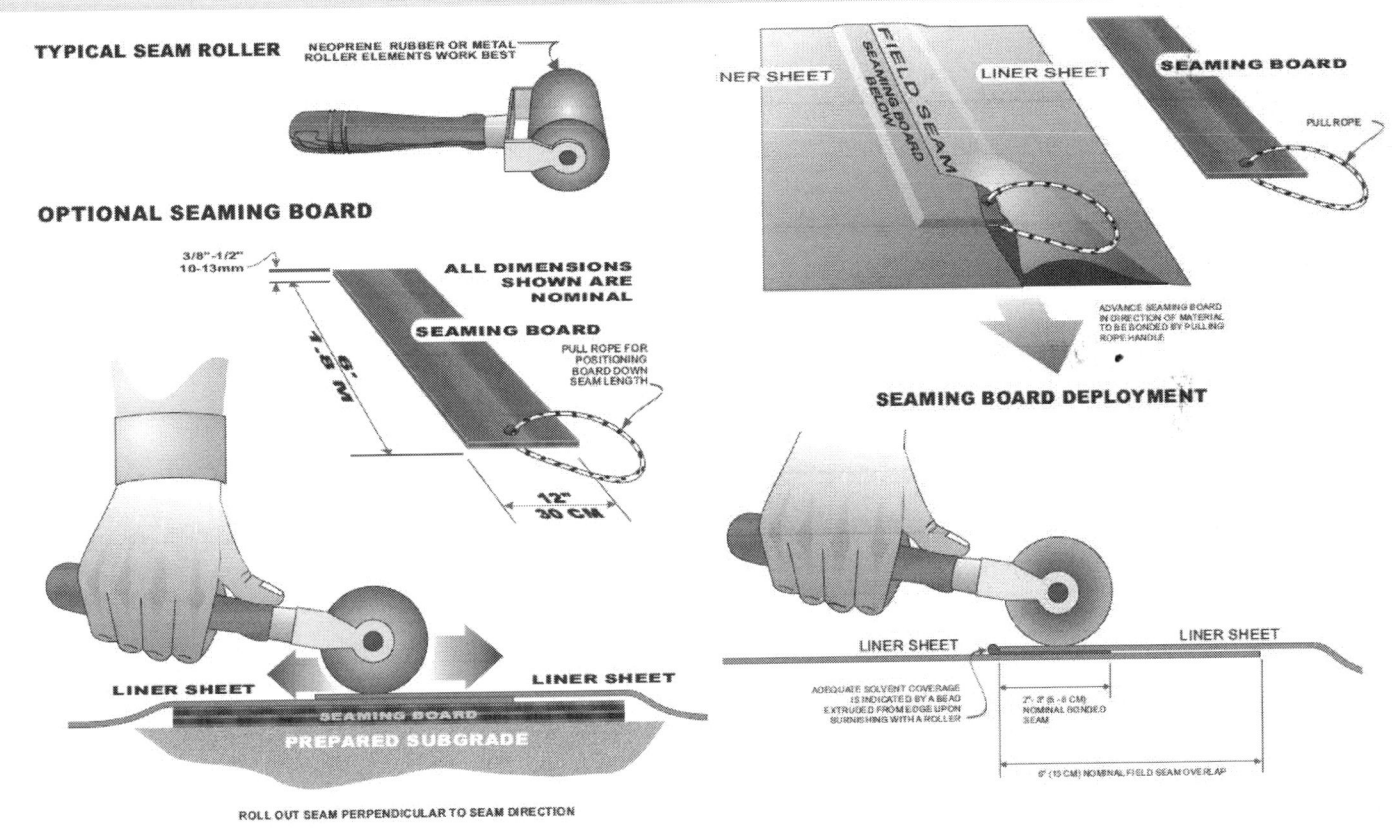

FIGURE 91. Joining two liners with a watertight seam.

PART III: COMPONENTS OF AQUAPONICS USED IN AQUAPONICS

Liner Repairs

To repair small rips or puncture holes, follow the instructions for sealing a seam above, but using the method described in the illustration (see figure 92).

Liner Plumbing Penetrations

Although this topic is addressed in greater detail within 'Chapter 14: Plumbing', along with several options, figure 93 is an illustration from Colorado International Lining Company which provides one approach of ensuring a watertight pipe penetration through a liner.

FIGURE 92. Patching a damaged liner.

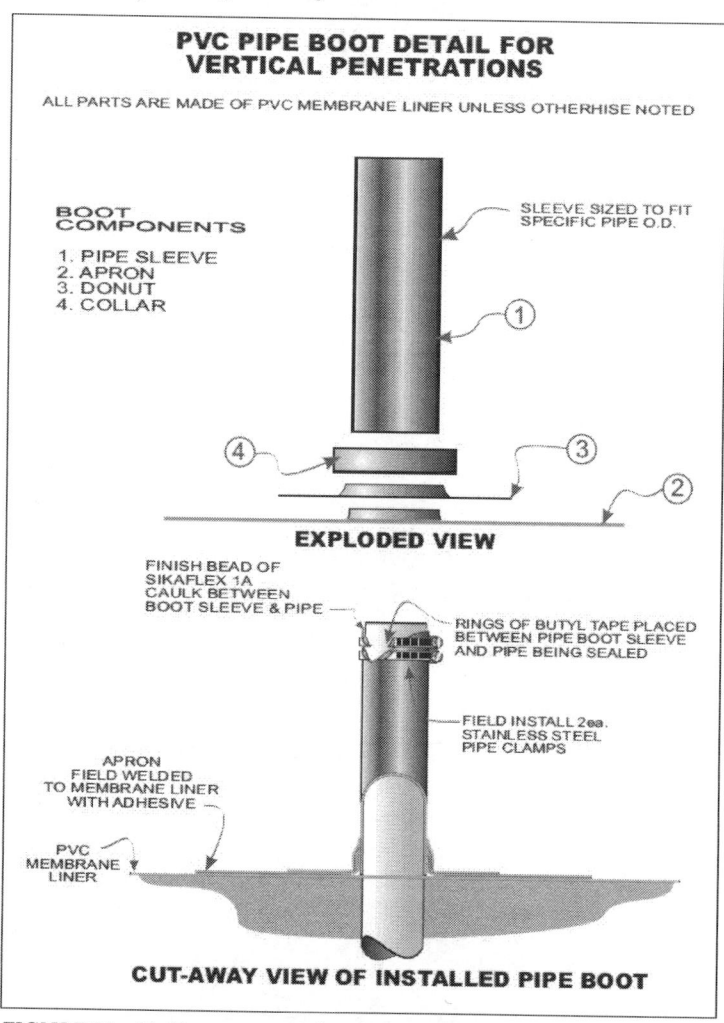

FIGURE 93. Making a watertight pipe protrusion through a liner.

CHAPTER 17

Making a Water Tight Container

Making a Water Tight Wood Frame Container

In addition to using a liner, as discussed previously, there are a variety of other waterproofing options available. The waterproofing materials discussed in this section are in the epoxy or resin categories, and adheres to the wood structure.

There are a variety of options available to seal a wood framed tank. Structural support is primarily achieved via a wood frame. Plywood and/or other boards are typically used for the walls and floor. The next step is to waterproof the container so that it will hold water.

The point of this section is not to endorse a particular product or method, but rather to provide comprehensive, unbiased, and reliable information about the primary methods that can be successfully implemented so that you can ultimately make your own educated decision. The following are other main methods in which to make a water-tight, sealed container (i.e. fish tank, grow bed, D.W.C. grow tank).

1. Two Part Epoxy Paint

A commonly used and reliable epoxy paint for plywood container builds is 'Sweetwater' epoxy available from Aquatic Ecosystems. It is non-toxic when cured and has good adhesion to lumber materials. This epoxy paint has solvents mixed in and is only 65-72 percent solids. It has strong, toxic fumes while curing, so a well-ventilated workspace is essential when using this product. It has a nice consistency and is very easy to work with and apply. Sweetwater can be used as a stand-alone product for waterproofing a wood container. It is important to have a well built structure when using this product as excessive flex could potentially contribute to leaks. Silicone will adhere well to Sweetwater so when sealing a tank with this method you should first apply the Sweetwater and then silicone (i.e. fish viewing windows, grommets, bulkheads, pipes, etc.).

Epoxy Paint Summary
PROS: Easy to apply, available in a range of colors, silicone will stick to it making window installation easy.
CONS: Toxic solvent requires well ventilated workspace, does not add structural strength to a container. Potential risk of failure in an inadequately supported tank with excessive flex.

2. Fiberglass Resin with Epoxy Paint Top Coat

Reinforcing a wood container with fiberglass is an excellent way to add structural strength to the build and impact resistance to a non-structural epoxy coating. An inexpensive way to apply fiberglass to a wood container is to apply a fiberglass resin to wet out the cloth. Lightweight fiberglass cloth and fiberglass resin are available from online vendors and are also generally easily found in home improvement centers such as Home Depot and Lowes, as well as in hardware stores such as Ace Hardware.

Fiberglass resin is generally a polyester resin and requires a small amount of hardener to be added as a catalyst. Effective application of fiberglass takes a little skill, but is fairly easy to learn with some practice. It is important to avoid bubbles in the fiberglass layer, which may ultimately pop and compromise the watertight seal. Polyester fiberglass resin is fairly inexpensive. It has a very strong toxic smell and requires a well ventilated workspace. In addition, it is not sufficient as a standalone waterproof barrier coating and will leach out chemicals into your system. Therefore, it is important to finish a fiberglass resin coated container with a non toxic, waterproof topcoat. Sweetwater epoxy (noted above) is an excellent product for this purpose. Using Sweetwater over a layer of fiberglass overcomes most of the potential cons of Sweewater alone.

Fiberglass Resin + Epoxy Paint Summary

PROS: A relatively inexpensive way to add structural strength and impact resistance to a wood container. The epoxy topcoat is available in a range of colors. Silicone will adhere to the epoxy topcoat.

CONS: Fiberglass application requires some practice and skill. Toxic fumes in the resin and paint require well ventilated workspace. Fiberglass resin is not sufficient as a stand-alone barrier coating.

3. Two-Part Marine Epoxy Resin

There are many different brands of epoxy resin available, but only true two-part marine epoxies should be used for wood frame aquaponic containers. These epoxies have a long established and successful history in waterproofing wooden boats. They differ from epoxy paints in that they are 100 percent solid. Several brands are available, but some tried and true options include:

- West Systems 105 (one of the more expensive products).
- US Composites (an inexpensive option).
- Max ACR (relatively affordable epoxy being marketed specifically for aquatic applications)

There are many other marine epoxy brands available that can be used in aquaponics. The three noted above cover the spectrum of price and have all been successfully used in waterproofing wood containers.

Marine epoxies come as two-parts, a resin and a separate hardener, which have to be mixed in a precise ratio. It is best to use slow hardeners when sealing a wood container with these products so as to be provided a longer working time and better penetration into the wood. It is strongly recommend that instructions be thoroughly read, and a 'how to' video on the internet be watched before beginning to work with these products. An instructional video will be posted on the www.FarmYourSpace.com website.

Marine epoxies can be used as a standalone product to provide a completely waterproof and non-toxic coating for wood containers. However, they are fairly brittle when cured, can be susceptible to stress fractures at seams, and damage from impact which will compromise the barrier coating. The best way to avoid these issues is to incorporate a layer of fiberglass cloth into the epoxy resin. Epoxy resins can be used to wet out fiberglass cloth in much the same way as polyester fiberglass resin, but offer several advantages.

There is not a strong smell. The cured resin layer is completely waterproof, non-toxic, and is slightly less brittle than polyester resin. With the exception of cost, epoxy resin is an all around better option than polyester fiberglass resin. As discussed above, wetting out fiberglass cloth does take a little practice, but with a little skill is an excellent way to add significant structural strength to a wood container.

When used with thickening agents such as colloidal silica, marine epoxy can also function as an excellent adhesive, particularly for slightly loose joints or joints where high clamping pressure cannot be applied. It can therefore also be useful as a waterproof adhesive during the construction and assembly of a wood container.

Silicone will adhere well to epoxy resin, so waterproof the wood container first and then silicone in your any viewing windows, bulkheads, grommets, etc. Epoxy resins can be tinted with a coloring coating agent but will usually still show wood surface grain and joints. Sweetwater epoxy, mentioned above, can be used as a topcoat if solid color is desired.

Two-Part Marine Epoxy Summary
PROS: Can add significant structural strength to a wood container, particularly when used with fiberglass. Minimal smell. Effective standalone waterproof barrier layer. Silicone will stick to it making window installation easy.
CONS: Expensive and it can suffer from stress fractures if used without fiberglass on an inadequately supported structure.

4. Pond Shield

Pond Shield is a 100 percent solid, two-part epoxy resin available from Pond Armor. It is different enough from the marine epoxies described above that it is discussed here as a standalone waterproofing option. Pond shield is non-toxic, is almost odorless, and is safe to apply indoors. Pond shield is an extremely thick epoxy which can make it somewhat challenging to work with. The black pond shield is the thickest and has a consistency slightly thinner than honey. It can be thinned marginally with denatured alcohol to make it more workable, but thinning also increases the risk of not getting the required thickness in a single coat. One 1.5qt kit of Pond Shield claims to cover 60sq ft at 10mil thickness. For many applications, a single coat is sufficient to get a 10mil coat that is completely waterproof.

Pond Shield is best suited for waterproofing well-supported structures with no flex. If there are concerns about the structural integrity of seams, it is best to fiberglass them in order to prevent the formation of potential stress fractures.

Pond Shield is ideal for sealing concrete and masonry tanks. While it adheres to wood, it holds even better to concrete and masonry. Therefore, one option when using it on a wood tank is to first line the inside with Hardiboard. If you instead choose to apply it directly to a wood tank, it is best to first do a light wash with 30-40 percent alcohol-thinned Pond Shield to get better penetration into the wood. Follow this with a single coat of un-thinned, or very slightly thinned, Pond Shield.

The thick consistency of Pond Shield can make it challenging to work with and may require touching up in some areas after the initial coat. Careful inspection and touch up is critical to the success of using this product.

Pond Shield Summary
PROS: Completely non-toxic and odor free. Available in a range of colors. Requires only a single coat and touch up. Silicone will stick to it making window installation easy.
CONS: Thick consistency can make it difficult to work with. While it remains somewhat flexible after curing, it is susceptible to stress fractures in poorly supported structures.

5. Liquid Rubber

These waterproofing products are elastomeric emulsions that remain highly flexible after curing. While there are a number of these products available, some examples of those that have been successfully used with wood containers are:

- Zavlar (named to Permadri Pond Coat in the US).
- Ames Blue Max.

These products have minimal odor and are easy to apply. Zavlar (Permadri Pond Coat) requires several coats to achieve a waterproof barrier of 40mil thickness. One gallon will cover 30 square feet at 40mil thickness.

The primary advantage of these products is that they're incredibly flexible, which allows them to stretch and resist fractures in inadequately supported structures. However, it is still a good idea to use drywall tape on seams while applying these products in order to reduce potential stress on the coating at these spots.

Zavlar (Permadri Pond Coat) will initially appear brown, but will dry to a black coating. However, this black coating will gradually turn back to brown after being submerged for a period of time. Ames Blue Max dries to a translucent blue color.

Zavlar (Permadri Pond Coat) will not cure when applied over silicone. Similarly, silicone will not adhere to cured Zavlar (Permadri Pond Coat). This incompatibility with silicone is a disadvantage of these products, but can be easily overcome. One strategy is to use butyl rubber, polyurethane caulk, or 3M 5200 instead of silicone when double waterproofing a viewing windows, grommets, and bulkheads.

As an additional measure of security, it is advisable to use a small amount of epoxy to first waterproof the area where a viewing window or bulkhead will be installed, such that epoxy layer bridges the seam between the Zavlar (Permadri Pond Coat) and silicone caulk. This way, in the event the Zavlar (Permadri Pond Coat) separates from the silicone caulk, the epoxy provides an additional waterproof layer that will prevent leakage.

Liquid Rubber Summary

PROS: Highly flexible coating resists fractures even in poorly supported tanks. Easy to apply. Minimal smell.
CONS: Limited choice of colors (brown in the case of Zavlar/Pond Coat, bright blue in the case of Ames). Incompatibility with silicone slightly complicates fish tank viewing window installation and backup waterproofing of grommets and bulkheads.

CHAPTER 18

Growing Media for Plants

Grow Media

Grow media is used in most aquaponic systems. It is even used in N.F.T. and D.W.C. aquaponic operations. In N.F.T. and D.W.C. operations net containers are typically filled with media to help anchor the plants.

However, it is used in abundance in flood-and-drain systems, as the grow beds are filled with media. The media is needed for anchoring of the roots to support the plants and to increase the surface area on which the beneficial bacteria cling to and live. With the help of bacteria and plants, the media also acts as a bio-filter. With the right media, and a balanced system, the grow bed becomes a very suitable habitat for the beneficial bacteria.

The most commonly used media to fill the grow beds in a flow-and-drain system are expanded clay pellets and gravel. The grow media anchors plants, is home to beneficial bacteria, and in some cases can even be used to raise worms. Some operators add worms to further aid in the break down accumulated solids and sediment. A surplus of worms to feed to the fish is a bonus for those raising carnivorous fish.

Factors to Consider when Choosing the Media

Particle size should be 8-16 mm (3/8–5/8 inches). If the particles are too small, the air space in between the particles is limited and the flow of water can be obstructed. Plant roots require oxygen and good water circulation. If the particles are too big, the surface area is greatly reduced and the plant roots would not be as well-anchored.

The pH of the media should be neutral. Expanded clay and most river stones are pH neutral. However, not all gravel is pH neutral. Some local crushed rocks or gravel contain limestone or other pH minerals, which are undesirable. Therefore, always know what you are getting before purchasing media.

For a flood-and-drain system, since gravel, plus the weight of water, is so heavy it is important to design and build strong supporting stands to withstand the weight. Expanded clay is lighter, but it is considerably more expensive than gravel. To reduce cost, it is perfectly acceptable to fill the bottom portion of the grow beds with gravel and the top with clay.

Growing Media Options

Table 20 below is a summary of several different growing media types, and their various pro's and con's.

TABLE 20. **Media Comparison Chart (Pros and Cons)**

	COST	WEIGHT	pH IMPACT	WATER RETENTION	DUST/ DEBRIS	AVAILABILITY
Gravel	Low	High	Possibly	Lowest	Needs Initial Rinse Only	Common
Sand	Low	Highest	Possibly	Low	Possibly	Common
Lava Rock	High	Medium	No	Medium	Needs Initial Rinse Only	Uncommon
Coir	Medium	Lowest	Yes	Medium	Yes	Common
Rockwool	Medium	Low	No	High	No	Uncommon
Hydroton	High	Medium	No	High	Needs Initial Rinse Only	Uncommon
Perlite/ Vermiculite	Low	Low	No	Medium	Needs Initial Rinse	Common

Expanded Clay Pellets

Expanded clay is lightweight and porous, allowing an abundance of air, water, and nutrients to reach the plants. It holds roots well and supports the beneficial bacteria necessary to convert the ammonia waste from the fish tank into nitrogen for the plants. Expanded clay mixes well with gravel. Expanded clay pebbles are often referred to as Hydroton, which is a really just a company's brand name for the product. It is also relatively easy to work with, although usually the most expensive of all mediums.

FIGURE 94. *Image shows a close-up view of Expanded Clay Pellets*

Expanded Slate

Expanded slate is mostly used as a traditional gardening soil additive, but it works well as a medium in aquaponics since it has a large surface area for the roots to grip; however, it is still lightweight. It is also less expensive than expanded clay. Expanded slate mixes well with pea gravel.

Gravel

Pea gravel contains small particles of gravel, most commonly produced by running gravel through a screen. Pea gravel and crushed gravel, another common aquaponics medium, provide additional substrate to mix with larger aquaponics media and serve as biofilter for the fish waste and home for the bacteria that nitrify it. Gravel is an excellent choice of aquaponics medium when mixed with other mediums that retain water, but it doesn't work well on its own because it doesn't hold water well. It also isn't large or heavy enough to provide the plant roots the support they need and can easily damage them. Crushed basalt is frequently used in place of gravel, although it provides both the same benefits and drawbacks. River stones are a slightly larger alternative (about the size of marbles) which provide many of the same benefits without the drawbacks.

Avoid gravel that has points or sharp edges as it can wreak havoc on hands, liners, and plant roots. Plant roots move about when exposed to running water through a flood-and-drain system. Gravel with sharp edges will damage or even cut the roots if gravel with sharp edges or points are used. Again, remember that it is heavy and stronger supports than usual will be needed to hold the weight of the gravel and water. As mentioned above, be sure to rinse the gravel well before adding it to your system.

Sand

Media beds filled with coarse larger type sand (not silt or fine sand) can make a good aquaponic substrate because sand retains and circulates water well. However, be sure to have a screen, which can be easily accessed for regular cleaning, so as to prevent sand from entering the fish tank and/or doing damage to the pump. Also, sand can sometimes be contaminated with clay particles.

Perlite

Perlite is a naturally occurring volcanic glass produced by hydrating obsidian. Perlite is a popular growing medium used by itself or as a mixture with other mediums. Perlite is commonly used with vermiculite (as a 50 — 50 mix), and is also one of the major ingredients of within a soilless mix. Perlite is also relatively inexpensive. Perlite does not retain water well, which means that it will dry out quickly between watering.

Unfortunately, perlite dust is bad for your health, so you should always wear a dust mask when handling it. Another drawback is that without proper screening at the drain outlet and/or via of a filter cage around the pump, the pump and emitters can become clogged and rendered useless.

Coconut Fiber

Coconut fiber (also called 'Coir') is rapidly becoming a popular growing medium. Coconut fiber is essentially a waste product of the coconut industry. It is the powdered husks of the coconut itself.

There are many advantages. It maintains a large oxygen capacity, yet also has excellent water holding ability, which is a real advantage for most systems that have intermittent watering cycles.

Coconut fiber is also high in root stimulating hormones and offers some protection against root diseases including fungus infestation. A large contingent of growers in the Netherlanders have found that a mixture of 50 percent coconut fiber and 50 percent expanded clay pellets is the perfect growing medium.

One word of caution, you must be careful not to purchase the commonly available, lower grade coconut fiber that has a high concentration of sea-salt and is very finely grained. This lower grade coconut fiber will lead to disappointing results for plants, and can negativity impact water quality, thus being detrimental to the fish. Also, coconut fiber tends to only last 3 to 4 years before it biodegrades.

Pumice

Siliceous material of volcanic origin, inert, has high water-holding water capacity, and high air-filled porosity.

Scoria

Porous, volcanic rock, fine grades used in germination mixes, lighter, and tends to hold more water than sand.

Polyurethane

Polyurethane is a relatively new material for aquaponics which typically holds 75 to 80 percent air space and has about a 35 percent water-holding capacity.

Composted Pine Bark

Composted pine bark combines an optimum water-retentive capacity together with good drainage. A medium size product does not need to be mixed it with any other media types. It is cost effective and it works.

Sawdust

The right kind of sawdust, with medium to course texture, is good for short-term uses, has reasonable water-holding capacity, and good aeration qualities. The problem is that pine bark from certain trees (i.e. western red cedar, etc.) can be harmful to plants and fish. Furthermore, the lumber industry sometimes uses chemical sprays on their materials. So unless you know for certain that your sawdust will not impact your pH and be harmful to fish or plants, it is best to stay away for it.

Vermiculite

Vermiculite is most frequently used in conjunction with perlite as the two complement each other well. Vermiculite retains moisture (about 200 percent–300 percent by weight), which is the opposite of perlite, so it is easier to balance your growing medium so that it retains water and nutrients well, but still supplies the roots with plenty of oxygen. A 50/50 mix of vermiculite and perlite is a very popular medium for drip type hydroponic and aquaponic systems. Vermiculite is inexpensive. It should never be used alone, as is it retains too much water and can suffocate the roots of plants if used straight.

Soilless Mix

There are many kinds of soilless mix's containing a vast assortment of ingredients. These mixes can be purchased or created by the do-it-yourself enthusiast. Many operators create their own mix using a wide variety of ingredients such as expanded clay pellets, sphagnum moss, perlite, and vermiculite.

Beware to take the proper precautions when using a soil mix as they can have some very fine particles that will clog pumps and drip emitters. If using any ingredient with fine particles, be sure to install a good filtration system. Panty hose and paint strainers serve as good filters when used on the return line and/or on a small cage around the pump. This will prevent fine particles from damaging the pump and interfering with emitters.

Most soilless mixes retain water well and have great wicking action while still holding a good amount of air, making them a good growing medium for aquaponic grow beds.

Sphagnum Moss / Peat Moss

Peat moss (also referred to as 'Sphagnum Moss') is the partially decomposed remains of formerly living sphagnum moss from bogs. Peat moss has long strands of highly absorbent, sponge-like material that hold and retain large amounts of water while simultaneously having good aeration. Sphagnum is usually purchased in dry, compressed blocks, and works best if soaked for approximately one hour before use.

A problem with this growing medium is that it can decompose over time and will shed small particles that can plug up your pump or drip emitters. Although, the biggest problem with peat moss is that it's environmentally bankrupt. Peat moss is mined, which involves scraping off the top layer of living bog. A sphagnum peat bog is a habitat for plants and animals; common as well as rare and endangered animals. Despite manufacturers' claims that the bogs are easy to restore, the delicate community that inhabits the bog cannot be quickly re-established. Yes, peat moss is a renewable resource, but it can take hundreds to thousands of years to form. Like all precious wetlands, peat bogs purify fresh air and even mitigate flood damage. Many conservationists, gardeners, and wetland scientists from all over the world have recommended a boycott of peat. Coir (a natural fiber extracted from the husk of coconut) has been touted as a sustainable alternative to peat moss as a growing media. Another viable peat moss alternative is manufactured in the USA (California) from sustainably harvested redwood fiber.

Rice Hulls

Rice hulls are a lesser known and underutilized substrate in most parts of the world, but they have proven to be as effective as perlite for the production of a wide range of crops. Rice hulls are a by-product of rice production and are typically inexpensive in rice production areas.

This free-draining substrate has low to moderate water-holding capacity, a slow rate of decomposition and low level of nutrients. However, rice hulls have a tendency to build up salt and decompose after one or two crops, so they should be replaced often.

Rockwool/Stone Wool

Made from rock that has been melted and spun into fibrous cubes and growing slabs, rockwool (also referred to as 'Stone Wool') has the texture of insulation and provides roots with a good balance of water and oxygen. Rockwool is suitable for plants of all sizes, from seeds and cuttings to large plants.

Rockwool is considered by many commercial growers to be the ideal substrate because of its unique structure. Rockwool can hold water and retain sufficient air space (at least 18 percent) to promote optimum root growth. Since Rockwool exhibits a slow, steady drainage profile, the crop can be manipulated more precisely between vegetative and generative growth without fear of drastic changes in pH.

Note that some Rockwool products require an overnight water soak before usage, as the bonding agents used to form slabs can result in high pH. Additionally, there has been a growing concern about disposing Rockwool after use because it never truly decomposes.

Oasis cubes

These lightweight pre-formed cubes are designed for propagation. A very popular medium for use when growing plants from seed or from cuttings. This product has a neutral pH and retains water very well. The cubes are meant to be a starter medium only. They typically come in three sizes, up to 2" x 2". They can be easily transplanted into practically any kind of hydroponic system or growing medium (or into soil).

CHAPTER 19

Pumps & Choosing the Right Pump

Pumps: Overview

Choosing the right pump is often a challenging decision for the new comers to Aquaponics who desire to build their own system. The pumps main purpose is to lift water to a certain height and move a specified amount of water over a unit of time (i.e. gallons per minute, etc.). The perfect mechanical device for this purpose is energy efficient, inexpensive, reliable, and will last a long time with minimal maintenance.

As a general rule, the pump should circulate all of the water from the fish tank in the aquaponic system at least every two hours, but preferably closer to every hour, 24 hours a day, seven days a week. Make sure the pump you choose will meet this standard of cycling half your system's water every hour. For example, if you have 100 gallons of water in your system, you will need a pump that can move at least 50 gallons per hour (.83 gallons per minute).

An efficient and reliable pump for your aquaponics system is essential. Without a good quality pump that is specifically right for your system, you simply will not succeed.

The flow rate of your aquaponics pump will either be measured in gallons per hour (gph), gallons per minute (gpm), liters per hour (lph) or liters per minute (lpm). **The flow rate will change depending on the head or the height the water must be pumped.** As the head increases the flow rate of your pump will decrease. Be sure to keep this in mind when deciding on your pump. The more head in your system, the more electricity will be required to move a specific amount of water. It takes one pound of pressure per square inch (PSI) to move water up 2.2 feet. The more head, the more pressure required to lift the water.

All pumps are different. One pump may take 100 watts to move five gallons per minute up 10 feet while another might only use 30 watts. The difference is how much energy is consumed and how much heat is produced. Daily power usage equals wattage of pump multiplied by 24 hours. Watt hours = wattage of pump x hours of operation.

Just as you need a powerful enough pump to do the job, make sure you do not get a grossly oversized pump. An oversized pump may move too much water out of your system or overly draw down the fish tank. It will also cause you unnecessary operating expense. Do not hastily jump in and find the first second-hand submersible pump that you see on the internet. You need the right pump for the specific operational characteristics of your system. Taking the time and doing your due diligence in selecting the right pump in the beginning will save you a lot of headache later. Your system must operate the way it is supposed to so you will not lose plants and/or fish.

When designing your system, strive to make it as efficient as possible in regards to pumping requirements. Keep the following in consideration:

- The higher the flow rate (amount of water being pumped), the more electricity will be required.
- The higher the water must be pumped (the higher the head), the higher the electricity consumption.
- A timer system will use less electricity than a siphon system as the pump will operate less.
- Keep in mind that the longer the pipe or hose run, the more turns, elbows, bends and fittings that are is the pump outflow line, the greater the friction head (pumping resistance), and thus, the harder the pump will need to work to overcome this resistance.

Steps for Selecting the Right Pump

1. Know the volume of your fish tank.
2. Determine how high (elevation difference) you need to pump the water. This is referred to a 'head', technically speaking. A pump simply capable of moving 40 gallons an hour could be suitable if the tank and the growing beds were adjacent on the same level, but an increase in height and distance between the tank and the growing beds will indicate you need a stronger pump. To calculate what size you need in regards to 'head', measure from the level of the pump intake up to the highest the water needs to be lifted which will be either over the rim of the growing beds, or over the rim of the filter-tank, depending on the system you will be using.

Evaluate the pump's efficiency, and don't be intimidated by this process. It is easy to learn about pump curves.

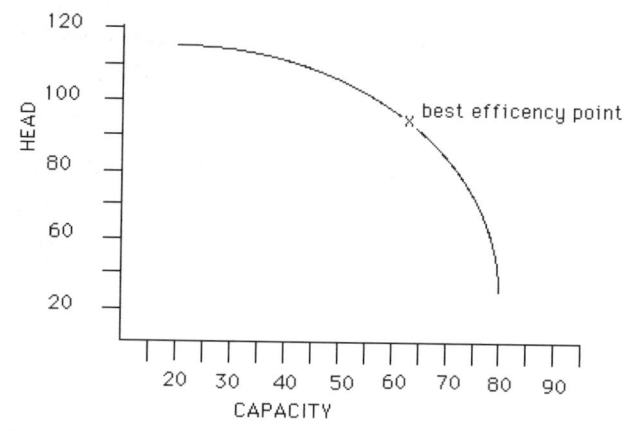

FIGURE 95. Pump efficiency curve.

How to Read a Pump Curve:

Figure 95 above shows the head of the pump with its capacity and optimal operating efficiency point. The head of a pump is read in feet or meters. The capacity units will be either gallons per minute, liters per minute, or cubic meters per hour. You can think of the above illustration in metric or imperial units, however you see fit.

The maximum head of this pump is 115 units. This is called the maximum shutoff head of the pump. Also note that the best efficiency point (BEP) of this impeller is between 80 percent and 85 percent of the shutoff head. This 80 percent to 85 percent is typical of centrifugal pumps. The pump manufacturer will typically provide you with the pump curve that will show the exact best efficiency point for the pump in consideration.

Ideally a pump would run at its best efficiency point all of the time, but we seldom hit ideal conditions. As you move away from the BEP, the shaft will deflect and the pump will experience some vibration. You'll have to check with your pump manufacturer to see how far you can safely deviate from the BEP and still get satisfactory operation (a maximum of 10 percent either side is typical)

The pump manufacturer typically states the maximum height the pump can efficiently move the water. Be wary of buying a model that does not supply that information. You may also want to consider using a pump that has a larger head than is strictly necessary for your aquaponics set-up. This will provide you with the option of diverting the water, should you ever have the need to do so, for maintenance or repair reasons.

Aeration of the water is extremely important and will affect the growth of both plants and fish. Water circulation through pumping action is typically the only way aeration is done in aquaponics, although some operators do use aerators. A good efficient pump may cost a little more to purchase, but plant and fish production can be maximized, and operating cost reduced; which can amount to a significant economic advantage over time.

Don't be fooled by advertisers boasting about horsepower, wattage, or voltage. Pump selection should never be made based upon these characteristics. Instead, selection of the pump needs to based entirely on flow rate, head, and pump efficiency. When these parameters closely match your system's parameters, then horsepower, voltage, and wattage will follow.

Customer reviews are helpful. Pumps that are extremely loud or produce a great deal of vibration can impede fish health via stress. User reviews are one of the best ways to determine if this may or may not be a problem with the pump you are considering.

Always have a spare pump on-hand.

Check your pump regularly to ensure that the inlet is not clogged or partially clogged. As the pump pulls in water, debris can sometimes cling to the inlet hole(s), inlet grate, or pump impeller. Some operators build a screen box for their pump to be housed in to help prevent clogging. Even so, it is wise to inspect and clean the screen box and pump regularly.

Maintain some type of back-up power system just in case your electricity is disrupted for an extended time. Imagine having to keep your pump working if the power went off for three weeks. Have an alternative plan in place and the necessary equipment readily available, just in case this happens. Refer to 'Chapter 22' for detailed information on back-up power, associated equipment and alternative energy options.

Shop around for the best pump at the best price. There are many places online where pumps can be purchased. Also remember that sometimes we get what we pay for, so be wary of exceptionally low priced pumps, as well as those on sale at a big discount store or have had the price reduced remarkably. Do your research.

Word of Caution

When buying a pump, or any other product, it is helpful to read product reviews by previous customers. Beware, though, as there are companies with large staffs, mostly located overseas, in which their business is influencing public opinion. For a fee, they will bombard various online retail outlets with positive comments, provide "Likes" or "Dislikes", give a thumbs up or down, and perform other such measures to sway opinion.

Some of the retail, media, and social websites they assail with this service is Facebook, Yelp, mainstream news articles, Yahoo, Google, Amazon, and various other online sites. Many manufacturers, suppliers, and even the federal government, use these companies to push their product or agenda. Coincidently, just before publication of this book I read a solicitation posting on Craigslist offering payment to individuals for posting positive reviews on certain Amazon product listing pages.

Therefore, read product and news article reviews with a skeptical eye. Sometimes it can be more helpful to read the negative product reviews. Obviously, those reviews noted as being from a 'verified purchase' are the most reliable.

CHAPTER 20

Filtration (Mechanical, Biofiltration, Natural)

Filtration

All cultured organisms, vertebrates or invertebrates, finfish or shellfish, produce waste as a result of the nutrition they receive. Fish excrete ammonia (NH^3), mostly from their gills, and it dissolves in the water in which the fish must live. This waste product is toxic to the fish and is an environmental stressor that causes reduced appetite, reduced growth rate, and death at high concentrations.

Maintaining water quality is imperative for fish health. Even with the perfectly designed aquaponic system, it is not a bad idea to have a spare or supplemental filtering mechanism available. If water quality is becomes a concern (regularly or approaches threshold limits on occasion), supplemental filtration can be used (regularly, intermittently to maintain water quality, or just switched on as needed in those rare occasions until water quality reaches optimal conditions). Some aquaponic operators make their own biofilter while others purchase a filtration system. Supplemental filtration is needed more often with D.W.C. and N.F.T. aquaponic systems, than with flood-and-drain systems.

Natural Filtration

Of course the best filtration one can have in an aquaponic system is the grow beds themselves. Sized correctly, kept in a favorable temperature range year-round, and plants maintained properly, the flood-and-drain grow beds serve as the best natural filtration system one could hope to have for the fish tank. Even if the system is primarily set up to be a D.W.C. or N.F.T. system, adding grow beds as a supplemental filtration mechanism can help clean the fish tank while also producing healthy organic vegetables.

Grow beds are addressed in much more detail in the chapters of 'Part V: Flood-and Drain System Design and Layout'. However, pertaining to our discussion here regarding the sole purpose of using grow beds for natural filtration it is prudent to consider the following guidelines:

- Size the grow bed using a 1:1 ratio of grow bed volume to fish tank volume.

The grow bed media should be at least 12-inches deep to allow for growing the widest variety of plants and to provide complete filtration. A grow bed with a freeboard of 1 to 2-inches above the top of the media is helpful in terms of practical application. This allows the operator to work within the media bed without spilling media or water on the floor. Therefore, a media bed 14-inches deep with 12-inches of media is recommended for optimal success.

Design the grow bed to be a working height that best suits the operator. An ergonomically correct height not only reduces strain on the back, arms and shoulders, but makes the aquaponics experience that much more enjoyable.

Turning over or stirring the bulk of the media within the grow bed every 4 to 12 months is beneficial. If using a liner, be sure not to get too close with your shovel or do anything else that may risk puncturing or tearing the liner.

Use a media guard around all plumbing fixtures. A media guard can be made from a wide variety of common hardware supplies, such as: window screen material, larger pipes, wood, aluminum, plastic, paint strainer, etc. A media guard will greatly facilitate cleaning and repair of plumbing fittings.

Mechanical Filtration

The easiest, but most costly, approach is to purchase a filtration unit. The size of filtration unit(s) is dependent upon the volume and operational characteristics of your system (i.e. numbers/size of fish, plant grow area/plant density, size of fish tank, etc.).

With so many variables, there is no easy formula for determining the size and number of filtration units needed, or how often they should be used. It is a trial and error process. One method however, is to purchase a filtration unit that is specified to address the volume of water contained within the fish tank.

For instance, assume you have one 400 gallon tank of fish within the normal recommended stocking density range (stocking density is addressed in chapter 11), then the following type of filtration unit would be a wise choice. It is designed to for aquariums up to 400 gallons.

The Fluval FX6 High Performance Canister Filter has very good reviews (figure 96 on the following page). At the time of this writing , it is being sold by Amazon, Petco, Big Als Pets, Pet Mountain, Marine Depot, and other retailers for around $300 USA dollars (2016 prices). The author is not providing a recommendation to any of these merchants, just listing them as options. Furthermore, there are other similar product options which will also work equally as well. The referenced Fluval FX6 is provided here only as one viable option. Tank volumes in excess of 400 gallons would need multiple Fulval FX6 units or a larger filtrer.

Biofiltration

Biological filtration is the use of beneficial bacteria to eliminate organic waste compounds from a body of water. It is differs from mechanical filtration in that it is a process whereby water is strained and suspended material is physically removed from the water.

Adequate filtration (used to trap and remove solid waste) is especially important for N.F.T. and D.W.C. systems. **The minimum volume of this biofilter container should be one-sixth that of the fish tank.** Without proper filtration, solid and suspended waste will build up in the grow channels and canals and will clog the root surfaces. Solid waste accumulation causes blockages in pumps and plumbing components. Finally, unfiltered wastes will also create hazardous anaerobic spots in the system. These anaerobic spots can harbor bacteria that produce hydrogen sulphide, a very toxic and lethal gas for fish, produced from fermentation of solid wastes, which can often be detected as a rotten egg smell.

Biofiltration is the conversion of ammonia and nitrite into nitrate by living bacteria. A lot of fish waste is not filterable using a mechanical filter because the waste is dissolved directly in the water, and the size of these particles is too small to be mechanically removed. Therefore, in order to process this microscopic waste an aquaponic system uses microscopic bacteria. Biofiltration is essential in aquaponics because ammonia and nitrite are toxic even at low concentrations, while plants need the nitrates to grow. In an aquaponic system, the biofilter is a deliberately installed component to house a majority of the living bacteria. Furthermore, the

CHAPTER 20: FILTRATION (MECHANICAL, BIOFILTRATION, NATURAL)

Fluval FX6 High Performance Canister Filter

FIGURE 96.

Product Overview:
- Super-capacity canister filter for aquariums up to 400 gallons
- Enhanced filter performance for superior aquarium water quality
- Includes all essential filtering media to streamline aquarium filtration

High performance canister filter boasts enhanced performance and setup ease. Fluval FX6 High Performance Canister Filter makes aquarium fishkeeping easier and more convenient. A 1.5-gallon media capacity filter is powered by an energy-efficient pump with a 925-gph pump output that draws just 43 watts. The result is cleaner and healthier aquarium water while reducing operating costs. Fluval FX6 Canister Filter comes with all the essential filtering media to provide complete 3-stage filtration.

Multistage filter features removable media baskets precision engineered to eliminate water by-pass for efficient filtration. Each media basket is lined with a mechanical foam insert for effective mechanical pre-filtering. Instant-release T-handles let you lift and separate the baskets quickly and easily, making routine maintenance simpler. For freshwater or marine aquariums up to 400 gallons.

DIMENSIONS	PUMP GPH	MECHANICAL AREA	FILTER GPH	HEAD HEIGHT	WATTS
17" dia x 21" high	925 gph	325.5 in²	563 gph	10.8 ft	43W

dynamic movement of water within a biofilter will break down very fine solids, which prevents. Biofiltration is unnecessary in the media bed systems (flood-and drain, ebb-and-flow) because the grow beds themselves serve as perfect biofilters.

The bacteria that do the work in a biological filter are part of the 'nitrogen cycle', which is compressively covered in 'Chapter 9: Nitrogen Cycle'. To summarize, fish constantly and continually excrete highly-toxic nitrogenous liquid waste from their bodies in the form of ammonia. If ammonia becomes too prevalent in the water, the fish will die. The ``nitrogen cycle'' (more precisely, the *nitrification* cycle) is the biological process that converts ammonia into other, relatively harmless nitrogen compounds. Several species of bacteria are involved in this process. Some species convert ammonia (NH_3) to nitrite ($NO2-$), while others convert nitrite to nitrate ($NO3-$). Thus, cycling the tank refers to the process of establishing bacterial colonies in the filter that convert ammonia -> nitrite -> nitrate.

Nitrates are one of the three forms of nitrogen found that plants use to grow and produce chlorophyll and proteins. It is a component of DNA, which transfers genetic information cell reproduction and plant reproduction.

By eliminating the organic waste compounds in the water, biological filtration detoxifies the water and makes it safe for fish. Additionally, by removing the organic waste compounds, algae is controlled because those compounds are the nutriment that algae require in order to grow.

A biological filter or "bio filter" is simply a home for the beneficial bacteria that perform the nitrogen cycle. The filter provides surface area that the bacteria can live on and a recirculating water pump ensures that water flows over the bacteria so they can obtain the nutrients and oxygen needed for survival.

The bacteria occur naturally in a fish tank or pond. They live on fish and other underwater surfaces. So if the bacteria occur naturally, then why is a bio filter necessary? It is because fish tanks have a much higher concentration of fish than would occur in nature, and the fish are usually fed a high-protein food. Hence, there is a higher concentration of organic waste, i.e., ammonia, than the naturally occurring bacteria can deal with. A bio-filter houses a higher concentration of beneficial bacteria, so that ammonia and the other nitrogen compounds can be reduced to levels which are safe for the fish.

Why do tanks/ponds turn green?

Ammonia is at the root of most green water (algae) problems. As discussed above, ammonia forms naturally in fish tanks and ponds, and is toxic to fish. Ammonia can be reduced or eliminated by the beneficial bacteria in a properly operating bio-filter, but if it is not, the water could become toxic were it not for algae. Algae help protect fish—up to a certain point. As ammonia levels rise, algae will colonize the pond or tank by taking up the ammonia as nutrient. However, too much algae makes for an unsafe environment for fish. Some algae produce toxic chemicals that pose a threat to fish. The toxins are released into the water when the algae die and decay. However, algae's biggest threat to fish is its ability to deplete dissolved oxygen in the water; thus, causing the fish to suffocate.

Types of Biological filters

There are two basic types of bio-filters: in-tank/pond and out-of-tank/pond. Out-of-tank/pond filters are divided into two types — pressurized and non-pressurized. The main function of any type of bio-filter (i.e. providing a home to bacteria), is the same regardless of design. The differences are in cleaning, space requirements, and add-on enhancements. Out-of-tank/pond filters work best for aquaponics.

A bio-filter can be a manufactured purchased unit or constructed as a Do-It-Yourself (DIY) project. Both types will be discussed later in this chapter.

The primary purpose of a bio-filter is to provide surface area for beneficial bacteria to live on, the size of the filter depends upon the amount of organic waste the bacteria have to deal with. In aquaponics, the amount of waste is a function of the fish population and uneaten feed.

If the bacteria colony is not thriving, ammonia will accumulate and algae will colonize the tank or pond. The most common problems with bio-filters are allowing it to be depleted in wastewater, improper cleaning, chlorine or chloramine in the water supply and copper leaching into the water from copper pipes. Winter weather (cold temperatures) can also reduce or kill off bio-filter bacteria.

How to Make a Make a Low-Budget Media Ball Biofilter

The filter is made up of a specially designed inert material such as ceramic or plastic. Plastic is les expensive. Commonly used biofilter mediums are Bioballs®, Kaldnes® Bio Media®, and BioMax Balls®, which are proprietary products available from aquaculture supply stores, aquarium stores, and online. There are many other equally realisable brands on the market, as well. These products are designed to serve as ideal biofilter material, because they are small, specially shaped plastic items that have a very large surface area for their volume. Other media can be used, including volcanic gravel, plastic bottle caps, nylon shower poufs, netting, polyvinyl chloride (PVC) shavings and nylon scrub pads; however the plastic bio media balls work best. Any biofilter needs to have a high ratio of surface area to volume, be inert and be easy to rinse. Bio filter balls have almost

CHAPTER 20: FILTRATION (MECHANICAL, BIOFILTRATION, NATURAL)

double the surface area to volume ratio of volcanic gravel, and both have a higher ratio than plastic bottle caps. When using suboptimal biofilter material, it is important to fill the biofilter as much as possible, but even so the surface provided by the media may be not sufficient to ensure adequate biofiltration. It is always better to oversize the biofilter, but secondary biofilters can be added later if necessary.

Biofilters occasionally need stirring or agitating to prevent clogging, and occasionally need rinsed if the solid waste has clogged them, creating anoxic zones. For cleaning, not all the media has to be removed. Removing and power washing approximately 75 percent of the media is usually sufficient. Leaving some media will help jump start good bacteria growth after the cleaned media is returned. Avoid using standard type cleaning solvents. Cleaning solvent residue remaining on the balls can kill the good bacteria and introduce undesired contaminants into the aquaponic system.

Another required component for the biofilter is aeration. Nitrifying bacteria need adequate access to oxygen in order to oxidize the ammonia. One easy solution is to use an air pump, placing the air stones at the bottom of the container. This ensures that the bacteria have constantly high and stable DO concentrations. Air pumps also help break down any solid or suspended waste not captured by the mechanical separator by agitating and constantly moving the floating bio filter media balls. To further trap solids within the biofilter, it is also possible to insert a small cylindrical plastic bucket full of nylon netting (such as Perlon®), sponges or a net bag full of volcanic gravel at the inlet of the biofilter. The waste is trapped by this secondary filter. The trapped waste is also subject to mineralization and bacterial degradation.

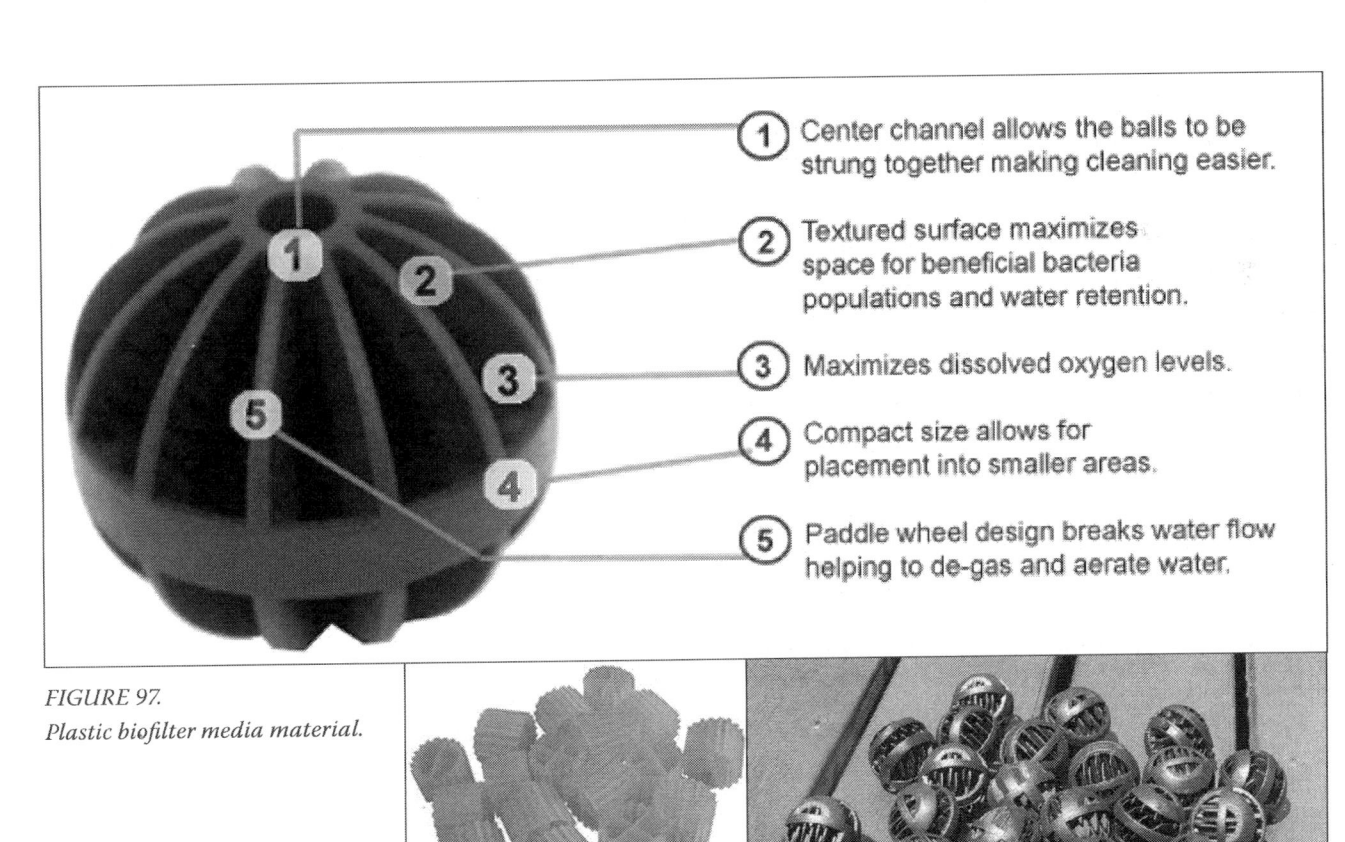

1. Center channel allows the balls to be strung together making cleaning easier.
2. Textured surface maximizes space for beneficial bacteria populations and water retention.
3. Maximizes dissolved oxygen levels.
4. Compact size allows for placement into smaller areas.
5. Paddle wheel design breaks water flow helping to de-gas and aerate water.

FIGURE 97.
Plastic biofilter media material.

How to Make a Low-Budget Polyester Woven Biofilter

A biological filter needs to be constructed out of non-corroding material such as plastic, fiberglass, ceramic or rock that has large amounts of surface area nitrifying bacteria cells can colonize. To make bio-filters more compact, material that has a large surface area per unit volume is usually chosen. This unit of measure is usually referred to as the specific surface area (SSA) of the biofilter media. Simply stated, the more surface area available, the more bacteria cells can be grown and the greater the nitrification capacity, which means that higher feed rates can be achieved. A biofilter with a higher SSA will be more compact than one with a lower SSA. Keep in mind, however, that some biofilter media with a higher SSA can become clogged with bacteria. Thus, there must be a balance between a high SSA and an operationally reliable biofilter media.

Below is an illustration of a homemade biofilter, followed by step-by-step instructions.

Steps to making a biofilter:

1. Determine if a biofilter is necessary. The necessity of a biofilter will vary greatly between D.W.C., N.F.T., and Flood-and-Drain systems. In some systems a biofilter may be an essential life support component that is in operation the all or the majority of time (D.W.C. and N.F.T.), and in other systems, such a flood-and-drain system, the biofilter may just be a supplemental or backup component to ensure the integrity of water quality for the fish should the grow beds fail to keep up with filtration demands (i.e. during winter months, maintenance purposes, etc.). In some systems additional filtration may not be necessary at all.

   There are two simple methods to determine if a biofilter (or any additional filtration) is needed. The first method is just a common sense approach. Take a look at the system (preferably during the design and planning phases) in regards to the volume of the fish tank, planned stocking density, plant grow areas, and consider whether or not the

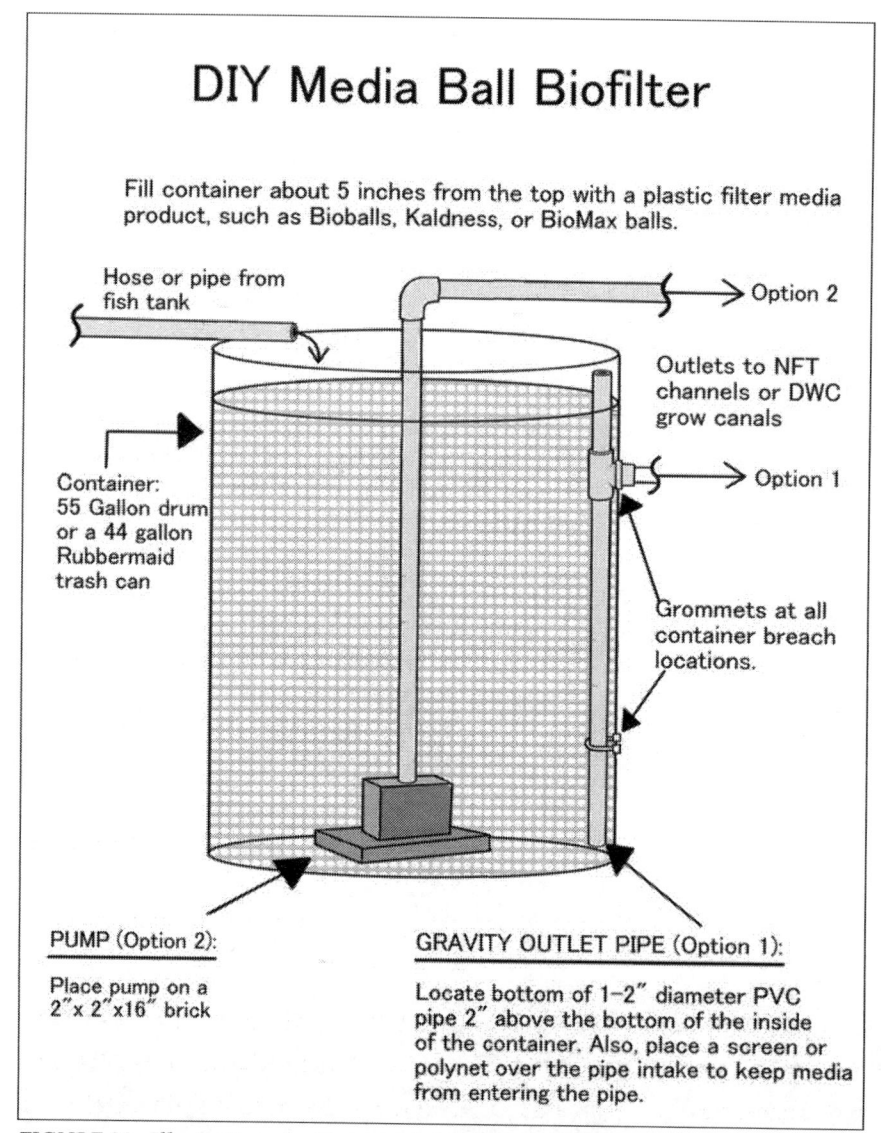

FIGURE 98. Illustration of a Homemade Media Ball Biofilter

CHAPTER 20: FILTRATION (MECHANICAL, BIOFILTRATION, NATURAL)

system has enough plant grow area to provide adequate filtration. In most cases, it is easy to determine whether or not additional filtration is necessary just by taking a close look at the system.

The second approach is based upon water quality observations. Problems with water quality and/or algae are typically good indicators that additional filtration is needed. This can typically be addressed by either adding filtration or increasing plant filtration area (i.e. more N.F.T. grow trays, larger flood-and-drain grow bed volume).

FIGURE 99. Above View of a Homemade Media Ball Biofilter (located between the fish tank and D.W.C. canals.

2. Determine the size biofilter needed. This is no easy task, though, as size will vary greatly dependent upon the type of system and the need for additional filtration. For a densely stocked tank it is ideal to size the pump so that the tank is pumped out every hour. For a flood-and-drain system, this volume should pass through the grow bed(s) every hour. If the beds are sized properly, and working properly (not partially dormant because of cold temperatures, etc.), then additional filtration will not be necessary. However, for systems where additional filtration is needed, then the biofilter can be sized accordingly:

- The ideal rate of filtration is 1/2 gallon of water per minute per square foot of biofilter bed. If your filtration is too fast it will result in insufficient exposure time to the bacteria.

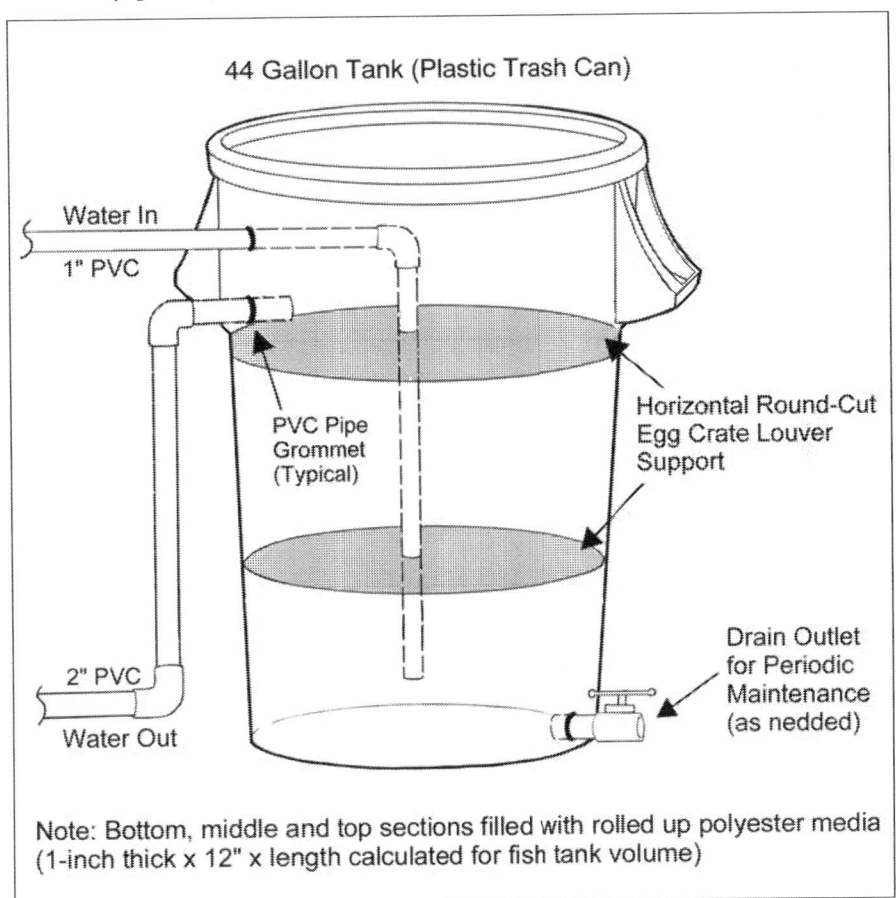

FIGURE 100. Homemade biofilter

- To calculate the size of your filter, figure the tank volume in gallons and divided by 60 (minutes). This equals the square footage of your biofilter material. This 'area' is the TOTAL area

189

PART III: COMPONENTS OF AQUAPONICS USED IN AQUAPONICS

of filter material put it in biofilter unit. It will be cut to size and rolled up.
- Acquire the supplies needed:
  ◊ One 44-gallon Rubbermaid (or equal) round trash can. See figure 101 below.

FIGURE 102. Standard 44-gallon plastic drum.

  ◊ A roll of 1-inch thick HVAC cleanable reusable filter (typically polyester). See figure 102 below.

FIGURE 101. HVAC cleanable reusable filter material.

NOTE: Any of the media materials described in Chapter 18 'Growing Media for Plants' would work. However, the advantage of using rolled up HVAC cleanable reusable filter material is that in addition to the micro-surface area it provides for bacteria adhesion and growth, it is easy to clean — just unroll it and blast it with water. Use a pressure sprayer to blast it clean if it becomes too clogged for a garden hose to do the job.

Another option is 1-inch thick dual sided poly fiber filter made of 100 percent polyester specifically for Koi ponds (see figure 103). It is washable or inexpensive enough that I can simply be replaced. It cost approximately $9.50 for a 12-inch by 120-inch roll (year 2016 USA dollars). It is available on line through eBay, Amazon, and various aquarium supply websites. Many local aquarium shops also carry it, or can order it for you.

FIGURE 103. HVAC cleanable reusable filter material.

- PVC plumbing pipe and fittings:
  ◊ 3/4" PVC Pipe (2)
  ◊ 2" PVC Pipe (1)
  ◊ 1" PVC Pipe (1)
  ◊ 2" 90 Degree Elbows (2)
  ◊ 2" Male Adapter, for 1/2 of bulkhead adapter
  ◊ 2" Female Adapter, for 1/2 of bulkhead adapter

CHAPTER 20: FILTRATION (MECHANICAL, BIOFILTRATION, NATURAL)

- ◊ 1" Male Adapter, (2), for 1/2 of bulkhead adapters
- ◊ 1" Female Adapter, (2), for 1/2 of bulkhead adapters
- ◊ 1" 90 Degree Elbows (1)
- ◊ 1" PVC Pipe to Garden Hose Fitting (2)
- ◊ 3/4" 90 Degree Elbows (8), for supports
- ◊ 3/4" Kris Cross Fitting (2), for supports
- ◊ 3/4" PVC fittings to get from the pump to a garden hose on the outside of the box.
- Egg Crate Louver from overhead fluorescent light fixture, for supports. Cut to fit horizontally in the 44-gallon trash can. See photos below for various examples.

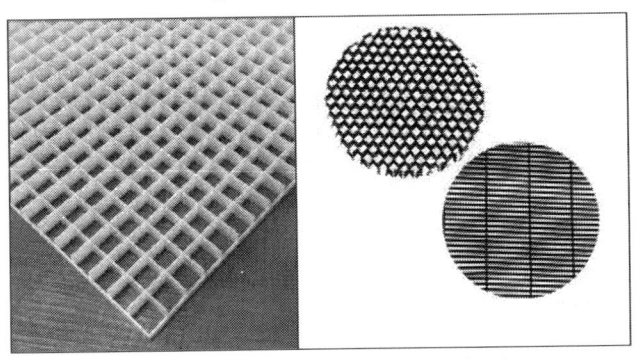

FIGURE 104. Dual sided poly fiber filter made of 100 percent polyester.

- Install the plumbing as shown in the DIY illustration below.
- Cut the HVAC cleanable reusable filter area previous calculated.
- Install ¼ of the previously determined HVAC cleanable reusable filter area in the bottom third of the 44-gallon can (this filter material should be rolled up like a newspaper).
- Place a round-cut egg crate louver support horizontally on top of the rolled-up reusable filter material.
- Install ½ of the previously determined reusable filter area in the middle third of the 44-gallon can (this filter material should also be rolled up) on top of the round-cut egg crate louver horizontal support.
- Place the second round-cut egg crate louver horizontally on top of the rolled-up reusable filter material that was placed in the middle section of the can.
- On top of the second horizontally placed round-cut egg crate louver support, install the remaining ¼ of the reusable filter area in the top third of the 44-gallon can (this filter material is also to be rolled up),.

Purchasing a Biofilter

Or those aquaponic operators would rather purchase a biofilter than make one, there are many options available. Below is one option that works well for aquaponic systems. The cost for the below biofilter is approximately $75 (USA, year 2016).

TetraPond Clear Choice Biofilter PF1
by Tetra Pond (figure 105)

- Works with fish tanks up to 500 gallons.
- Mechanical pre-filter sponges remove suspended debris to improve water clarity.
- Interchangeable .75" and 1" diameter intake fittings.
- Bio Ring media provide large surface areas for beneficial aerobic bacteria.
- Easy out-of-tank accessibility and simple maintenance.
- *ClearChoice* biofiltration removes these contaminants and ammonia using its advanced Trickle Flow and and Bio Ring technologies.

The *TetraPond ClearChoice BioFilter* works to keep your pond clean through biological and mechanical filtration. This easy-to-use system filters coarse debris and provides biological filtration media with massive surface area for beneficial bacteria to colonize and remove harmful pollutants. The *ClearChoice BioFilter* utilizes a unique Venturi system, which draws air into the in-flowing water, creating ideal

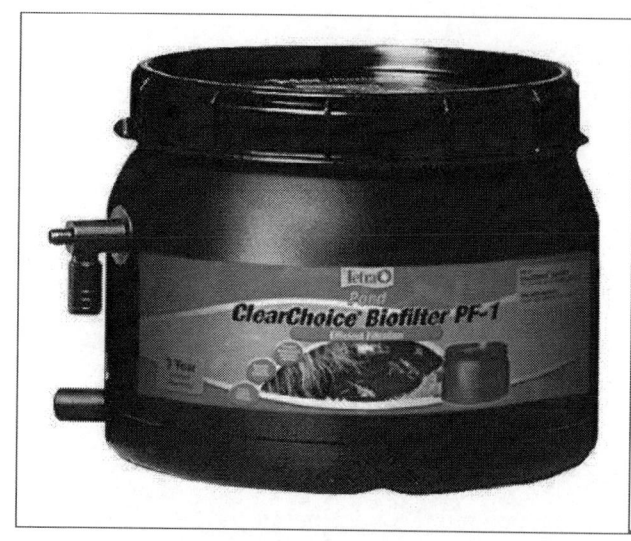

FIGURE 105. *TetraPond Clear Choice Biofilter PF-1.*

conditions for the beneficial bacteria to decompose waste. Take a look at the information below to see how the *ClearChoice BioFilter* could benefit your pond.

How the TetraPond Clear Choice Biofilter PF1 Works

STAGE 1: Pond water is pumped into the filter. It passes through the foam, which sieves coarse suspended debris from the water (Mechanical Filtration).
STAGE 2: The water from stage 1 then passes through the finer foam, which provides further filtration.
STAGE 3: The final purification process of the water occurs on the surface of the biological filtration media. Beneficial bacteria will naturally colonize on this massive surface area and convert the harmful pollutants in the water into relatively harmless nitrates, which are absorbed by the aquatic plants or removed in partial water changes.

The Tetra Venturi and Trickle-Flow System

The water inlet into your filter is fitted with an advanced Venturi system, which draws air into the in-flowing water, creating ideal conditions for the beneficial bacteria to decompose waste. This, when combined with the use of "Trickle Filtration" encourages the prolific growth of beneficial bacteria, which helps to keep the pond free of pollutants.

Recommended Pump

TetraPond recommends the *TetraPond Water Garden Pump* 325 GPH or 550 GPH. Do not use pumps with flows that exceed 550 gallons per hour unless a flow control valve is used.

CHAPTER 21

Greenhouses

Do you need a controlled plant-growing environment (a greenhouse)?

No matter how good your aquaponic system is, it will not function to maximum capacity if the environmental conditions are outside the comfort zone for the fish and plants. A controlled environment greenhouse can provide natural sunlight, proper temperature, humidity and ventilation. Greenhouses can be designed for every climate, from the tropics to northern climates with extreme winters. Greenhouses can also include bio-security features.

Bio-Security

Bio-security is a combination of equipment and procedures that you implement to keep your food safe. Essentially, bio-security is an implemented process that keeps pathogens and pests out of the greenhouse so that plants stay healthy and your food is never contaminated via air-borne pollutants. It is astonishing to learn about all of the air-borne contaminates that can settle on our fruit and vegetables when planted outdoors. Needless to say, the more we can do in regards to taking precautionary measures to prevent pollutants and toxins from coming into contact with our food, the better. Greenhouses provide an extra measure of security against toxic food.

Greenhouse Basics

A greenhouse is used to increase productivity of an aquaponics operation and ensure that produce can be enjoyed throughout the year. It can also be a means of giving plants a head start on spring outdoor planting.

The costs of building a greenhouse can vary from several hundred dollars to tens of thousands of dollars. Likewise, operating costs can be relative low or very high depending upon location, regional climate, type of greenhouse constructed, materials used, energy supply system(s), and how much additional heating, cooling, or lighting is required.

A greenhouse can be built using traditional methods, purchasing a kit, or installed by a contractor. Before you decide whether to build or buy, determine your needs in regards to preferred size, available space, and desired level of production. A greenhouse is a long-term investment that should provide the growing area and service desired while blending well with the home and landscape. With an open mind, examine as many greenhouse styles and equipment options as possible in person, books, and through the Internet before making a decision.

Greenhouse Kits

Kits are the most common method of establishing a greenhouse. Greenhouse kits are available in a wide range of costs and with a wide range of features. Prices range from a few hundred dollars to well over $25,000 depending on the size, style, accessories, and type of construction materials. Manufacturers can be found in the yellow pages of a telephone book, in advertisements in gardening magazines, by searching the World Wide Web, or by contacting the Hobby Greenhouse Association (8 Glen Terrace, Bedford, MA 01730-2048). Compile a list of manufacturers, and request information on different models. Compare costs and features for the style and size you are interested in. Generally, plastic-covered greenhouses are easier to assemble than glass houses. If any aspect of the assembly is beyond your skills and/or available time, many manufacturers and contractors will erect their products on-site for an additional fee.

Building Your Own Greenhouse

A kit, although common, is certainly not the only way to go. A greenhouse can be constructed easily and inexpensively by anyone able to use simple hand tools. Most of the construction materials can be purchased at building supply stores. Materials may also be available inexpensively at construction sites, through Craigslist or building supply salvage yards. If any aspect of the construction is beyond your skills, you may be able to hire a local carpenter or handyman to help. Plans for different styles of hobby greenhouses can be acquired online and through most libraries.

When you do decide to build a greenhouse, choosing the type of structure, covering, and environmental control equipment can be confusing. A step-by-step approach addressing your needs can help you organize the planning and implementation of your hobby greenhouse.

Greenhouse Needs

When deciding upon a greenhouse you need to consider what type of plants you prefer to grow, what seasons of year you will use the greenhouse, and how a greenhouse fits into your lifestyle. For example, you might want to use the greenhouse in one of the following ways:

- To start vegetable or flower seeds or propagate cuttings in the spring to plant in the garden. On a small scale, this can be accomplished with a structure as simple as an outdoor cold frame or hotbed. A freestanding greenhouse can also be used for this purpose but will probably be a simple and inexpensive model.
- To grow year-round tropical foliage in a conservatory setting. A greenhouse for this purpose will probably be more permanent and formal.
- To grow specialty flowers or ornamentals. Many greenhouses are constructed because owners develop an interest in specialty flowers or ornamentals that have unique requirements, such as orchids, African violets, or bromeliads. These greenhouses should be designed with the needs of the particular plant in mind.
- A greenhouse can be a part of your home in the form of a sun-room or porch. Sunrooms or porches usually have clear covering on one or more sides but not on the roof. A greenhouse can also be attached to the house, with an entrance to the living area. Plants, walks, furniture, a water pond, or a fountain may also be included and arranged formally or informally as an extended living room. Plants may "spill over" from the greenhouse through a sliding glass door into the living area of the home. This type of greenhouse can be used for relaxing, reading, or family gatherings.
- A greenhouse can be used a production facility providing year-round produce. In this case, it is also good to consider allocating a portion of the greenhouse as a work area

Different Types of Greenhouses

Greenhouse design styles vary widely and include Quonset, tri-penta, dome, gothic arch, slant-side, A-frame, gable roof, straight-side lean-to, curved-side lean-to, and slant-side lean-to (as shown in the illustration below). Some styles are more suited to flexible coverings like polyethylene, such as the dome, gothic arch, Quonset, curved-side lean-to, and tri-penta. Others work better with rigid coverings like glass or plastic, including the A-frame, gable roof, slant-side, and straight- or slant-side lean-to.

Likewise, some design styles are more efficient to heat and cool, such as the gable roof and Quonset. Others may look unusual and attractive but are difficult to construct or heat and cool, such as the dome and tri-penta. The A-frame style is easy to construct and is inexpensive, but the usable growing area is small and awkward, and the shape may not blend well with normal surroundings. The most commonly used styles are the gable roof, gothic arch, Quonset, and slantside lean-to. In addition to deciding what style your greenhouse should be, you need to decide whether it will be freestanding or whether it will be attached to your home. Freestanding greenhouses stand alone in the landscape. They can be constructed in a wider range of styles, can be larger, and can offer greater flexibility in location than attached greenhouses can. These greenhouses can be placed almost anywhere in the landscape where the ground is level and adequate light is available. The most widely used styles are the gable roof, gothic arch, Quonset, and slant-side. Attached greenhouses are attached to the home and may or may not have an entrance to the home. They can be designed to blend with the architecture and landscape of the home and are useful where space is limited.

An attached greenhouse may also cost less per square foot to build than a freestanding greenhouse. With an inside entrance, you can maintain the greenhouse without going outside during bad weather. Utilities such as electricity, water, and heat can be

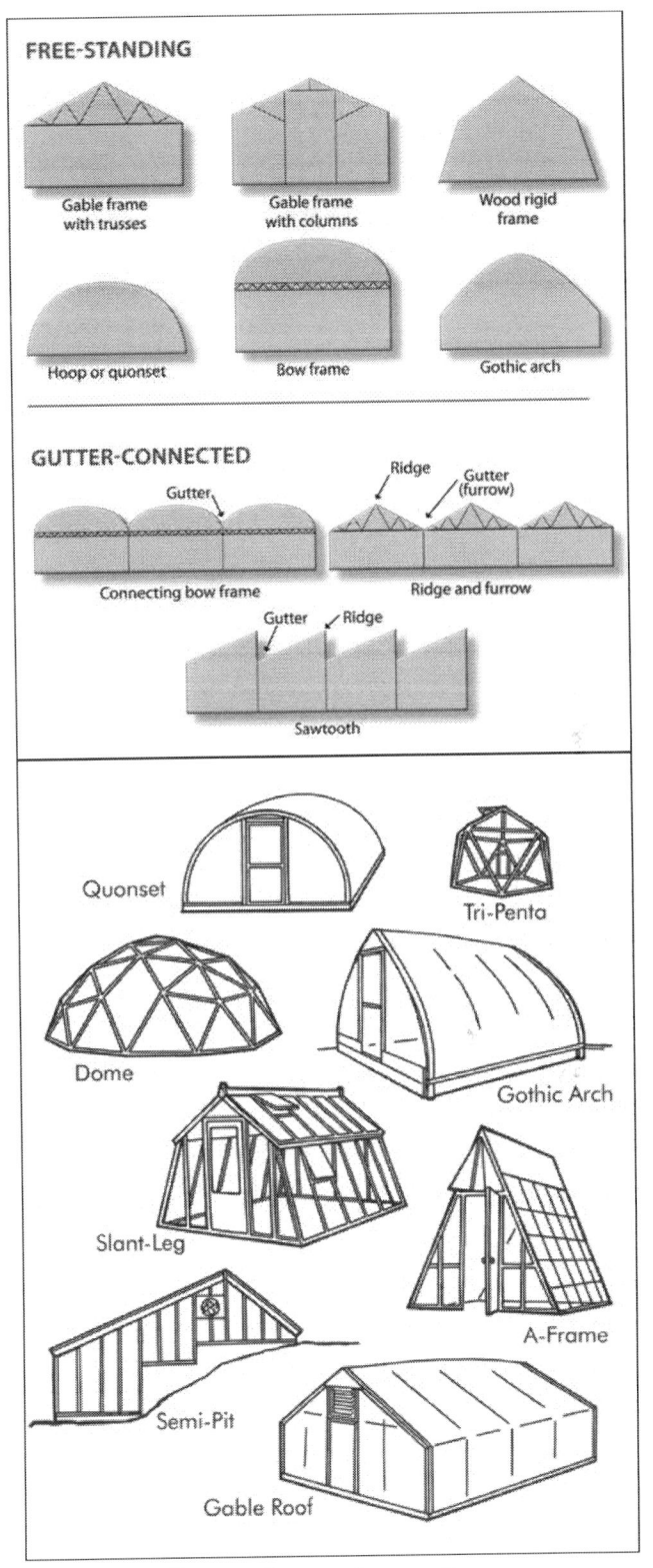

FIGURE 106. Freestanding greenhouse options.

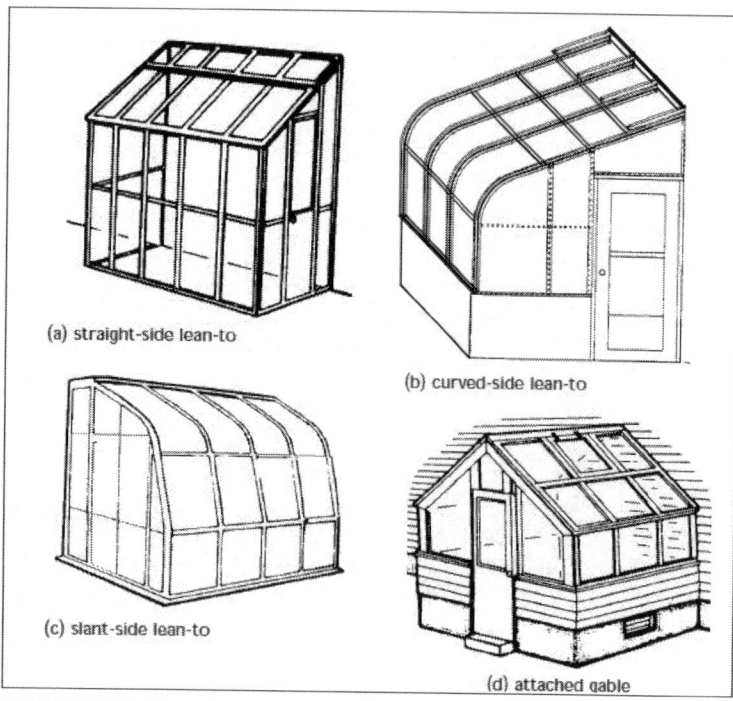

FIGURE 107. Greenhouse connected to a residence.

(a) straight-side lean-to
(b) curved-side lean-to
(c) slant-side lean-to
(d) attached gable

Greenhouse Location Importance

Often, there may be a limited choice of locations that will have adequate sunlight, adequate soil drainage, easy access for people and materials, access to utilities, and a pleasing blend with the landscape. For many homeowners, the appearance of the structure is most important, so compromises must be made to meet other location requirements. The exact size and shape of the property will have a direct impact upon the size of greenhouse that can be erected.

One of the most important location requirements is sun exposure. Many flowering plants require full sun to perform well. A freestanding greenhouse for these plants should be constructed with the long sides of the building facing southeast to southwest (figure 108). Vegetable and flower seedlings for transplanting outdoors in the spring need maximum sunlight, so choose a location that receives full sun. For plants requiring less light, the greenhouse can face northeast to southeast or northwest to southwest.

Exposure is particularly important for attached greenhouses. Consider these locations in order: first—south or southeast, second—east, third—southwest, fourth—west, and last—north.

Keep in mind that a western exposure can be too hot in summer, and a northern exposure usually does not receive enough light for most plants. Also, be aware that tall structures and trees near the greenhouse may block light for parts of the day.

Falling limbs can also be a major problem if the greenhouse is located too close to trees.

Other location considerations include ensuring that the site is level and that the soil drains well. Many locations may have to be graded to ensure that the foundation is level. Slope the soil away from the greenhouse to drain rainwater away. If drainage is

shared with the home if a greenhouse is planned for during home construction. If the greenhouse is attached later, you may need an additional heater because greenhouses lose heat five to ten times faster than an equivalent area of home. The cost of heating an attached greenhouse may be less than that of heating a freestanding greenhouse of the same floor area because one wall is not exposed; however, attached greenhouses usually receive less light for the same reason. Straight-side lean-to, slantside lean-to, and curved-side lean-to styles are ideally suited for small, easy-to-construct attached greenhouses (, although the Quonset, gothic arch, slant-side, and gable roof freestanding styles can also be attached. A solid foundation similar to the house foundation is often required or highly desirable for attached greenhouses. Joining and sealing the greenhouse to the house needs special attention. In some cases, it is best to consult an architect or building contractor to determine the method of attachment.

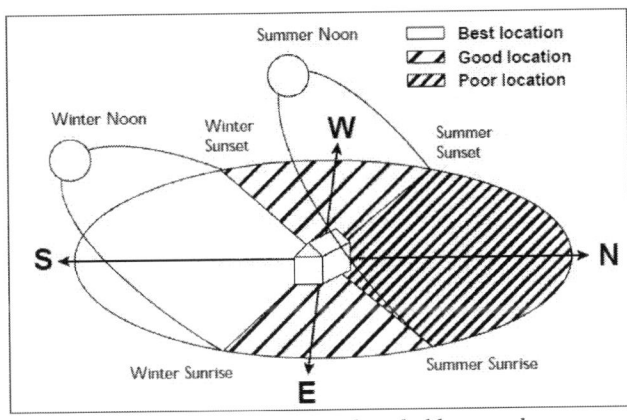

FIGURE 108. Location priorities for a hobby greenhouse based on ambient sunlight

a problem, consider installing drainage tile before constructing the greenhouse.

Avoid low areas with poor air circulation, especially those surrounded by woods or buildings. Cold, humid air can stagnate in these locations and increase heating costs. Conversely, avoid high elevations with direct exposure to strong winter winds. Convective heat loss through the covering material can increase heating demand. Wind can also wreak havoc on polyethylene film coverings.

The greenhouse should be convenient to a driveway to receive supplies and to haul away plants or garbage when needed. Walkways to and from the garden, house, and storage areas and access to water, fuel for heating, and electricity should all be considered.

The greenhouse should contribute to the appearance of the home and landscape and not be an eyesore either to the owner or neighbors. Consider room for expansion if you think you may be so inclined in the future.

Commercial Greenhouse Considerations

If you plan to have a commercial retail operation then it is important to be visible and easily accessible to customers. It is also prudent to plan for expansion. Therefore, make sure there enough room to grow until you can afford to purchase more property for the business. Make sure the property is zoned for business. Check to see if zoning is likely to change in the near future or if it has been disputed in the recent past. Also, check to see if there are any special restrictive clauses that may inhibit future expansion, as well as regulations on the size or height of signs would be needed to promote your business.

The business should be close enough to major roads for delivery and transport trucks to have easy access. Check to see if there are any weight limits or restrictions on large trucks that would need to access your business. For a retail greenhouse, locate the business so customers can see it from at least 200 feet and can get to the business easily and safely. A lot of shopping is done on impulse. The more customers have to cross major barriers or make a special effort to get to your business, the less likely they will do business with you.

Water quality and availability for a Commercial Greenhouse Operation.

Is city water available or will you have to dig a well? If you dig a well, how much water is available and how long will it last? Many greenhouses require about 6 acre feet of water per year for every acre of greenhouse production area. Regardless of the source, have a water quality test performed. This is an inexpensive, easy procedure that may save a lot of money in the future.

Private labs or your county Extension agent can help take the water sample and have it analyzed. Check the level of soluble salts and bicarbonates. Low soluble salts, a level below 0.75 mmhos/cm, is best because fertilizer is often added to the water during irrigation. When present in excess amounts, some salts are toxic to plants. Water bicarbonate level is important in plant production. A bicarbonate level of less than 100 ppm (parts per million) is recommended for growing most plants. You will also need a plan for water collection and a plan for water runoff from your greenhouse.

Proper Size Greenhouse for Your Needs

Available space and cost usually have a large impact on the choice of size for a greenhouse. Keep in mind, however, that a greenhouse that is too small may cost more to operate than the initial cost of building a larger one. The obvious problem with a small greenhouse is that it is too small to meet the needs of the owner. In addition, temperatures can fluctuate rapidly in a small greenhouse, and heat losses can be as high as they are in a larger greenhouse. Small greenhouses may also have limited headroom and be hard to work in. A taller, larger greenhouse obviously has more space, and it heats and cools more uniformly than a shorter, smaller one.

As a starting point for a small greenhouse, 100 square feet would be a minimum size, but 200 square feet is better. A house 9 to 14 feet wide by 20 feet long can be managed in a few hours per week. A larger greenhouse can also cost less to cover per square foot than a small greenhouse, so choose a size slightly larger than you think you need. On the other hand, ask yourself, do you have the time and dedication to maintain a large greenhouse, now and in the future?

Greenhouse Covering Materials

There are several covering materials to choose from, each of which has its own advantages and disadvantages. Common covering materials are glass, polyethylene film, fiberglass reinforced panels, and double layer structured panels.

Glass

Glass is the traditional greenhouse covering against which all others are judged. Good-quality glass is an attractive, very transparent, and formal (in appearance) covering material. It is very strong (tensile strength), but it is subject to shattering and can become brittle with age. Glass is also very expensive and because of its weight requires sturdier framing support than is required with other covering materials. Originally, glass panes for greenhouses were 18 by 16 inches, but larger sizes are more common now. Actually, larger panes are less fragile than smaller panes. Many greenhouses are covered with double-strength float glass (1/8 inch thick). Other greenhouses have larger glass panes of tempered glass (5/32 inch thick) which cost approximately two to three times more per square foot, depending on the pane size.

Polyethylene Film

Polyethylene film (PE) is a good choice for home-built greenhouses because less structural support is required and it costs much less than other materials in regards to initial capital costs; but over time (having to replace it every two to three years) it can become even more expensive than other material coverings. However, PE film only lasts about 2 years. Clear PE is used for growing most plants,

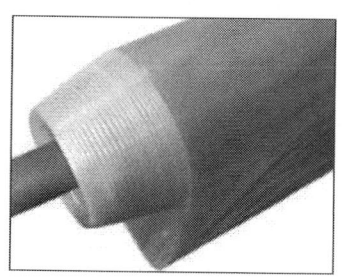

FIGURE 109. Polyethylene film

but white PE can be used to reduce light and heat for growing lowlight plants or for propagation. PE manufactured for greenhouse application comes in widths from 10 to 50 feet, and thicknesses ranging from 1 to 8 millimeters. Two layers of PE are frequently applied to greenhouses to reduce heating demand. Double-layer PE houses generally cost 30 to 40 percent less to heat than single-layered houses do. The two layers are kept air inflated by a 100 to 150-cubic foot- per-minute squirrel cage blower mounted to the inside PE layer. Purchase 4-millimeter PE for the inside and 4 or 6-millimeter PE for the outside. Use 6-millimeter PE for single-layer applications. PE can be installed on wood-frame greenhouses by nailing wood batten strips over the film into the foundation boards and rafters or arches. However, because PE must be replaced frequently, investing in special fasteners makes the job easier. Fastening systems are available for single- or double layer applications.

Facts about Polyethylene Film (PE)

- Can be applied in either single layer or double layer.
- Double poly film requires the installation of an inflator fan to provide air between the two layers.
- Different grades are available and are rated as either one, two, three or four year film.
- All high quality greenhouse film has a UV inhibitor co-extruded into the film. This will give you extra life before replacement is necessary.
- Many films are also available with an anti-condensate coating to prevent condensate drip on the inside of the greenhouse during humid conditions.

Fiberglass Reinforced Panels

Fiberglass reinforced panels (FRPs) are rigid plastic panels made from acrylic or polycarbonate that come in large corrugated or flat sheets. FRPs are available in 24- to 57-inch widths and up to 24-foot lengths. FRPs are durable, retain heat better than glass does, and are lightweight (less structural support needed). Large panels are flexible enough to be applied to a Quonset greenhouse. Light transmission may be better than glass simply because less structural support is needed, which creates less shadow. The prices of FRPs vary, depending on the guaranteed life span of the material. Inexpensive materials may be guaranteed for as little as 5 years; more expensive types may be guaranteed for as long as 20 years.

FIGURE 110. *Fiberglass reinforced panels*

Facts about Fiberglass Reinforced Panels (acrylic or polycarbonate):

- It is a hard plastic glazing material which approaches glass in its clarity and light properties. Transmission value of up to 92 percent.
- Relatively inexpensive. Requires no special extrusions for attachment.
- Easily installed. Lightweight and can be cut with utility knife or tin snips.
- Available with anti-condensate coating on inside surface to reduce condensate dripping.
- Virtually unbreakable which makes it extremely resistant to hail damage.
- UV protected outer surface to provide long life under high sun conditions.

Double-Layer Structured Panels

Acrylic or polycarbonate double-layer structured panels (DSPs) are made of two layers of plastic held apart by ribs spaced 1/2 to 1 inch apart. The double-layer construction increases structural strength and heat retention but decreases light transmission compared to single-layer materials. Panels are 4-feet wide and up to 39-feet long. DSPs made of polycarbonate cost slightly less than those made of acrylic material.

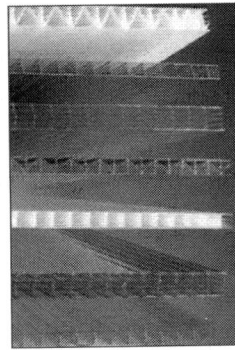

FIGURE 111. *Double-layer structured panels*

Facts about Multi-Wall Acrylic or Polycarbonate Panels

- Tough impact strength for resistance to hail.
- Provides insulation value similar to insulated glass. Reduces heating costs.
- UV protected outer surface to provide long life under high light conditions.
- Most come with a 10-year warranty against yellowing and loss of light transmission.

TABLE 21. **Greenhouse Covering Comparison**

COVERING	ADVANTAGES	DISADVANTAGES	LIGHT TRANSMISSION	U' FACTOR	INSULATION VALUE 'R'	ESTIMATED LIFETIME	COST PER SQ./FT (YR 2014)
Single (polyethane film)	Inexpensive; Easy to Install	Short Life	85%	1.2	0.83	1 to 4 years	$0.09
Double (polyethane film)	Inexpensive; Saves on heating costs; Easy to install	Short Life	77%	0.7	1.43	1 to 4 years	$0.17
Corrugated Polycarbonate	High transmittance; High impact resistance	Scratches Easily	91%	1.2	0.83	15+ years; 10 yr warranty	$1.30
Glass Double Strength	High transmittance; High UV resistance; Resists scratching	High cost; Difficult installation; Low impact resistance; High maintenance	88%	1.1	0.91	25+ years	$3.00
Glass (Insulated)	High transmittance; High UV resistance; Resists scratching	Very high cost; Difficult installation; Low impact resistance	78%	0.7	1.43	25+ years	$6.00
8mm Twin Wall Polycarbonate	High impact resistance; Saves on heating costs	Requires glazing system to install; Scratches easily	80%	0.61	1.64	15+ years; 10 yr warranty	$1.66
10mm Twin Wall Polycarbonate	High impact resistance; Saves on heating costs	Requires glazing system to install; Scratches easily	80%-	0.56-	1.79	15+ years; 10 yr warranty	$2.50
16mm Triple Wall Polycarbonate	High impact resistance; Saves on heating costs	Requires glazing system to install; Scratches easily	78%	0.42	2.38	15+ years; 10 yr warranty	$4.00

Summary of Greenhouse Coverings

Single (polyethane film):
- Advantages: Inexpensive, easy to install
- Disadvantages: Short Life
- Light Transmission: 85%
- 'U' Factor: 1.2
- Insulation Value 'R': 0.83
- Estimated Life: 1 to 4 years
- Cost per Sq./FT (year 2014): $0.09

Double (polyethane film):
- Advantages: Inexpensive, easy to install, saves on heating costs
- Disadvantages: Short Life
- Light Transmission: 77%
- 'U' Factor: 0.7
- Insulation Value 'R': 1.43
- Estimated Life: 1 to 4 years
- Cost per Sq./FT (year 2014): $0.17

Corrugated Polycarbonate:
- Advantages: High transmittance, high impact resistance
- Disadvantages: Scratches easily
- Tight Transmission: 91%
- 'U' Factor: 1.2
- Insulation Value 'R': 0.83
- Estimated Life: 15+ years, 10-year warranty
- Cost per Sq./FT (year 2014): $1.30

Glass, Double Strength:
- Advantages: High transmittance, high UV resistance, resist scratching
- Disadvantages: High cost, difficult installation, low impact resistance, high maintenance
- Light Transmission: 88%
- 'U' Factor: 1.1
- Insulation Value 'R': 0.91
- Estimated Life: 25+ years
- Cost per Sq./FT (year 2014): $3.00

Glass, Insulated:
- Advantages: High transmittance, high UV resistance, resist scratching
- Disadvantages: High cost, difficult installation, low impact resistance
- Light Transmission: 78%
- 'U' Factor: 0.7
- Insulation Value 'R': 1.43
- Estimated Life: 25+ years
- Cost per Sq./FT (year 2014): $6.00

8mm Twin Wall Polycarbonate:
- Advantages: High impact resistance, saves on heating cost
- Disadvantages: Requires glazing system to install, scratches easily
- Light Transmission: 80%
- 'U' Factor: 0.61
- Insulation Value 'R': 1.64
- Estimated Life: 15+ years, 10-year warranty
- Cost per Sq./FT (year 2014): $1.66

10mm Twin Wall Polycarbonate:
- Advantages: High impact resistance, saves on heating cost
- Disadvantages: Requires glazing system to install, scratches easily
- Light Transmission: 80%
- 'U' Factor: 0.56
- Insulation Value 'R': 1.79
- Estimated Life: 15+ years, 10-year warranty
- Cost per Sq./FT (year 2014): $2.50

16mm Triple Wall Polycarbonate:
- Advantages: High impact resistance, saves on heating cost
- Disadvantages: Requires glazing system to install, scratches easily
- Light Transmission: 78%
- 'U' Factor: 0.42
- Insulation Value 'R': 2.38
- Estimated Life: 15+ years, 10-year warranty
- Cost per Sq./FT (year 2014): $4.00

PART III: COMPONENTS OF AQUAPONICS USED IN AQUAPONICS

TABLE 22. *Characteristics of Structure Frames and Appropriate Coverings*

Frame Type	Frame Material	Support Strength	Light Transmission	Internal Access	Rigid Sheets	Plastic Film
Free Span Gable Frame with Trusses	Treated Wood Galvanized Steel Pipe	Good	Fair	Good	Yes	Yes
Gable Frame with Columns	Galvanized Steel Pipe	Good	Fair	Fair	Yes	Yes
Quonset	Galvanized Steel Pipe Aluminum Extrusion	Fair	Best	Good	Some	Yes
Gothic Arch	Treated Wood Laminate	Good	Good	Good	Yes	Yes
Rigid Frame	Treated Wood	Best	Good	Good	Yes	Yes
Gutter-connected Bow Frame	Galvanized Steel Pipe or Tubing	Good	Good	Good	Yes	Yes
Gable Frame	Galvanized Steel Pipe or Tubing	Good	Good	Good	Yes	No
Sawtooth	Galvanized Steel Pipe or Treated Wood	Good	Good	Good	Yes	No

NOTE: Diagrams of structures can be seen in 'Step 2' above (this Chapter).

Greenhouse Framing Materials.

To a large extent, this choice will be influenced by the type of covering material chosen, the desired permanence, and costs. Many homemade greenhouses are made of wood. Be sure to purchase treated lumber for exterior use. If you can afford it, treated lumber for the interior is ideal if you are going the wood route, especially I you are going house your fish tank inside the greenhouse. Metal-framed greenhouses are more permanent but are more difficult for do-it-yourself projects from scratch. However, greenhouse manufacturers offer many kits that are easy to erect and are reasonably priced. Aluminum or galvanized framing last a long time with little maintenance. Schedule 80 PVC pipe can also be used as supports for small Quonset-style greenhouses that are to be covered in PE, but PVC requires extra support to withstand strong wind.

Whatever the choice, the framing must adequately support the covering and any equipment and/or hanging baskets suspended from the framing. The framing must also be able to withstand wind, rain, snow, or ice common to the geographic location. The load exerted by hanging baskets can be greater than that caused by weather.

Wood preservatives for greenhouse framing and benches should be pressure-treated, salt type, or copper naphthenate. Avoid creosote (a cacogenic which is no longer legal) and pentachlorophenol preservatives—these are dangerous for people and toxic to plants.

The eave height for a gable roof greenhouse should be at least 5½ feet with a minimum gable height of 8 feet. However, gable heights of 10 to 15 feet are common. Roof pitch for a glass greenhouse should be 6-inch rise per foot (27 degrees) to prevent inside condensation from dripping on plants. Plastic-covered greenhouses require a steeper pitch of 7- to 8½-foot rise per foot to prevent drip.

Greenhouse Foundation

and Sidewall Materials

Simple Concrete foundation for a greenhouse. (poured concrete should be reinforced).

Greenhouses to be covered in PE usually do not require an extensive foundation, but the support posts must be set in concrete footings. The problem with a PE-covered Quonset house is usually not holding it up but holding it down. In strong wind, the shape of the house makes it function like an airplane wing, and it may lift off the ground. Figure 4 shows a simple poured concrete footing and a concrete block foundation.

Attached greenhouses and those covered with glass should have a strong concrete or concrete block foundation that extends below the frost line according to local building codes; and with steel reinforcement as required. A 2 to 3-foot-high sidewall can add considerably to the appearance of a greenhouse. Concrete block, stone, or brick are the most popular materials used, but shingle, clapboard, and asbestos rock have also been used. Choose the type that works best with the overall architectural scheme.

Greenhouse Flooring Options and Walkways

In many places where drainage is adequate, a solid floor is not necessary. Even so, most operators find that some type of flooring for walkways is helpful. Four inches of 3/4-inch crushed stone or pea gravel will help control weeds and provide a porous medium for water to evaporate in the summer to provide some cooling. If you want a solid concrete floor, install one or more French drains, and slope the floor toward the drains. Don't forget to plan for plumbing when planning the foundation.

Walkways can be constructed of concrete for easy movement of equipment and people, especially if a family member is disabled. Brick filled with sand, flagstone, or stepping stones can be used for decorative walks. Gravel under the benches keeps the walkways free of debris and reduces weeds. Walkways can be 2 to 3-feet wide in a small greenhouse. Adjust the width of walks if a cart or wheelbarrow will be used. Larger greenhouses often have 2 to 3-foot secondary aisles and a 4 to 6-foot main aisle. If a family member is in a wheelchair, use at least 4-foot-wide aisles and ramps for easy access. If you are going to be an official business, then you certainly need to be ADA compliant in all aspects.

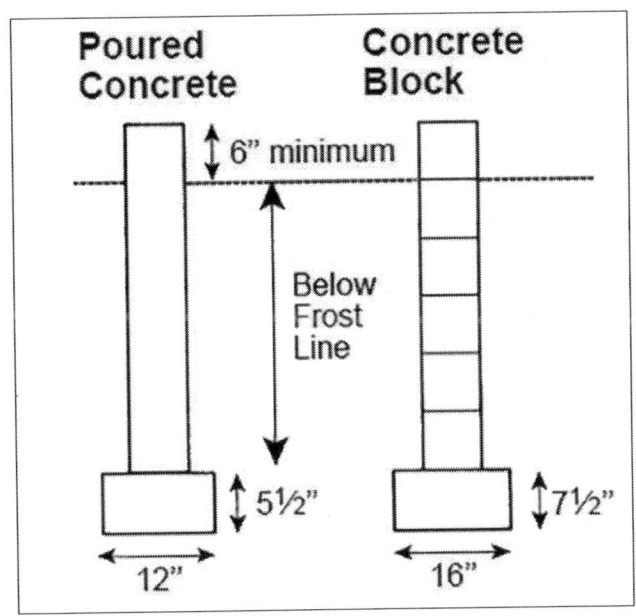

FIGURE 112. Concrete stem wall on concrete footing (left) and concrete masonry block stem wall (right) on concrete footing.

Greenhouse Benches (Sizes and Heights)

Benches can be constructed from a variety of materials and arranged in many different ways. If you plan carefully, 70 to 80 percent of the floor area can be devoted to growing plants. Make sure that the supports for benches are strong enough to hold the largest number of plants and the largest container size anticipated. Wood, metal pipe, or concrete block can be used as bench supports.

Also, make sure the bench surface is strong enough to support plants without sagging but that it is open to allow water drainage and air movement. Spruce or redwood lath and 14-gauge welded wire fabric or expanded metal make a strong, long-lasting, open bench top. Benches should be 2 to 3-feet wide with access from one side or 4 to 5-feet wide with access from both sides. If using solid-topped benches, set them back from the sidewall of the greenhouse by 6-inches to allow air movement. No setback is needed for open-top benches. Benches should be 24 to 36-inches high; for individuals in wheelchairs, bench height should be 30 to 36-inches, with little or no surface lip. Place bench supports 6 to 12-inches back from the surface edge to provide knee room.

Greenhouse Construction Tips

Whether you choose to buy a manufactured greenhouse or construct your own, there are some important construction points, which can save time and money later on. They are as follows:

1. If the site is at a latitude higher than 35°N and the greenhouse will be used in winter, it should be oriented longitudinally from east to west. Greenhouses with more light admitting area on one side than the other should have that side facing south. Both of these orientations provide more winter light and in general, steeper roof slopes will do the same.

2. Roof slopes of less than 25° do not allow snow to slide off readily and inside condensation will drip on plants rather than run to the sidewall.

3. Greenhouse roof beams or trusses should be engineered to support extreme local snow loads plus the anticipated load of wet hanging baskets.

4. Extremely small greenhouses are typically more difficult to heat and cool evenly. If possible, have a floor area of at least 500 square feet.

5. Endwalls, doors, utility service, and venting are significant expenses in greenhouse construction; it is usually less costly per square foot to build larger and longer greenhouses.

6. Efficient material and personnel movement should be designed into the greenhouse structure. Head space and aisle space must be adequate for free movement at a fast space.

7. The ability to add to facilities without major disruption should be planned in advance.

8. All structural members must be of aluminum, galvanized steel, properly painted where applicable, or treated lumber. The constant presence of water and the high humidity of greenhouses accelerates corrosion of metals and decay of wood.

9. Use the very highest quality paint possible; re-painting in a high-humidity greenhouse is extremely difficult and paint fumes can be damaging to crops and unhealthy to people.

10. Never use pentachlorophenol or creosote as greenhouse wood preservatives. The fumes are deadly to plants, considered carcinogenic to man, and will last for years. An acceptable wood preservative should contain copper naphthenate; one widely distributed brand is named Cuprinol.

11. An efficient work area should be provided and it must be easily accessible to all growing areas. If one plans to retail plants, customer comfort must be considered. People buying plants do not appreciate muddy walkways and clothes catching snags.

12. Air inlets must be provided for fuel combustion, especially in polyethylene houses. Heaters and boilers should be vented according to specifications and clearance from combustible materials must be observed.

13. Ventilation systems should have the ability to operate at several levels of capacity.

14. If possible, heat should be distributed near ground level so that it rises to the plants.

15. Greenhouse covering materials should be purchased from reputable supply firms.

16. Benching arrangements for plants should utilize every available square foot of space but still accommodate easy movement by workers. Benches must be at a height which workers find comfortable for working on plants.

17. Insects eventually become a problem in any greenhouse. Do not attach a greenhouse to your home. This arrangement makes it very dangerous to apply poisons for pest control.

18. Once established, weeds are difficult to eradicate from greenhouses. During construction, try to avoid bringing soil or materials onto the site if they have been located in a weed patch.

19. An alarm system to warn of power outages and temperature problems in the greenhouse after working hours should be installed.

It is relatively easy to obtain books through the library system you local library or via interlibrary loan), and via online sellers, which provide greenhouse construction details and plans, as well as associated cost estimates. Most helpful are the catalogs and technical bulletins published by greenhouse manufacturers. With a little research it is easy to plan heating and cooling systems for yourself. Do not blindly accept the plans a salesperson presents to you. Evaluate them critically and don't be afraid to question aspects which do not satisfy your common sense. You are the one who will live with it for years to come, not the salesperson

Heating Options for Your Greenhouse

Greenhouses are poorly insulated structures, and heat loss through the covering on cold, clear, windy nights can be considerable. Heat can be supplied using electricity, natural gas or propane, fuel oil, solar energy, wind energy, or kerosene. Electricity can be expensive. Kerosene should only be used during emergencies. Solar and wind are not always 100 percent reliable, but can still be used. You can also connect the greenhouse to your home heating system. Natural gas or propane are probably the most cost-effective ways of heating a hobby greenhouse. If natural gas is available in the home, plumbing into the existing line saves considerable cost compared to the cost of installing a new meter, construction of a separate gas line from the main, and incurring the monthly cost of a second set of utility fees and taxes. If natural gas is not available, check with your local propane gas companies for cost and availability. Determine if the company provides a free storage tank, if has to be rented, or purchasing is an option. In some instances purchasing a tank is a better long-term economical option. Consult the company to determine the tank size appropriate for the greenhouse.

A variety of gas unit heaters are available to heat smaller greenhouses. Some are designed to hang from a structural beam or rafter of the greenhouse; others sit on the floor. Costs for gas unit heaters range greatly in price depending upon the BTU output. Most heaters unit burn gas in a firebox. Heated air rises through the inside of a thin-walled heat exchanger on the way to the exhaust chimney. A fan draws air in from the greenhouse, across the outside of the heat exchanger, and out into the greenhouse. Thus, most of the heat is removed from the exhaust before it exits the structure. The exhaust chimney must be sufficiently tall to maintain an upward draft and extend above the greenhouse roof. An 8 to 12-foot chimney is usually sufficient. All open-flame heaters must be vented to the outside and given a fresh air supply for complete combustion. Fresh air should be provided by an unobstructed opening to avoid carbon dioxide buildup.

In larger greenhouses, a plastic tube system may be needed to distribute the heat evenly within the house. The system consists of a perforated polyethylene tube suspended overhead in the ridge and extending the length of the greenhouse. A fan connected to the tube blows warm air from the heater into the tube for greenhouse distribution. This system can also be used for circulating internal air when heating or cooling are not required. A variety of electric resistance heaters can also be used. Those available in department stores and home centers are usually only adequate for smaller greenhouses. Larger units can be purchased, but operating them can be costly.

Heater size for a given greenhouse and geographic location depends on the surface area of the greenhouse and the temperature difference between the inside and outside of the greenhouse. To determine the size heating system you need, calculate the total surface area of the greenhouse covering. Then determine the difference between the minimum expected outside temperature during winter in your area (see the USDA hardiness zone map, average annual minimum temperature) and the maximum night temperature you wish to maintain (generally 60 to 65 degrees F). Multiply the greenhouse surface area by the temperature difference by the heat loss conversion factor in Table 1 for the appropriate covering. The answer will be in BTUs per hour. Most heaters are rated in this unit. Many greenhouse supply dealers and various specialty construction companies can help you determine the proper size heater for your situation.

Virtual Grower, a free computer program developed by the U.S. Department of Agriculture — Agricultural Research Service (USDA–ARS) and available at www.virtualgrower.net, has enabled growers throughout the United States to predict heating costs for their greenhouses. You can use the program to help make decisions on growing temperature set points, use of alternative fuels and energy-saving investments. One of the significant uses of the program is the ability to predict the amount of energy needed to maintain a desired temperature at different times of the year. When combined with information on temperature's effects on crop timing, you can identify the most energy-efficient growing temperatures.

TABLE 23. **Heat Loss Factor for a Small Hobby Greenhouse**

GREENHOUSE COVERING	*HEAT LOSS (WINDY AREA)	*HEAT LOSS (CALM AREA)
Polyethene/Fiberglass	1.2	1.4
Glass	1.5	1.8
Double-Layer Plastic	0.8	1

*Heat Loss in BTU / Hour / Square Feet / Fahrenheit

Ventilation Options for Your Greenhouse

The purposes of ventilation are to exchange carbon dioxide and oxygen, to remove hot air, and to lower relative humidity. Hobby greenhouses can be vented by natural flow-through ventilation or by forced-air ventilation. Flow-through ventilation relies on side and top vents that pull cool outside air into the greenhouse through the side vent as warm air rises and exits through the top vent. The combined side and top vent area should equal about 20 percent of the roof area. Vents can be manually controlled, but this requires frequent temperature checks and vent adjustment according to outside conditions. Using an electric motor and thermostat for automatic vent control is much easier. Simple automatic systems open or close the vents based on a set-point temperature. More advanced systems open and close vents in stages based on multiple set-point temperatures. The advantage of natural flow-through ventilation is that it costs less than forced-air ventilation to operate; however, it may not be adequate to cool the greenhouse during the summer.

Forced-air ventilation systems consist of a louvered intake vent and electric fan(s) controlled by a thermostat. The fans pull cool air into the greenhouse from outside through an intake vent and push warm inside air out. Fans should be mounted in a waterproof housing with air-activated louvers to protect electrical components from inclement weather and to keep cold air out during the winter. Be sure to install a screen over the inside of the fans to prevent injury. Be sure that the distance between the fans and adjacent structures equals at least 1 1/2 times the fan diameter. The louver on the intake vent on the wall opposite the fans can be air-activated or motorized. The cost for a fan typically corresponds to the production rate measured in cubic-feet per-minute quantities.

Be sure that the fan capacity is large enough to exchange the air in the greenhouse at least once per minute. Recommendations for warm climates call for a fan capacity to remove 12 to 17 cubic feet per minute per square foot of floor area. If the greenhouse is attached to the east, west, or south wall of another building, solar heat will collect inside the greenhouse from this wall. If this is the case, add half the area of the attachment wall to the floor area when calculating the ventilation requirements. Work with a greenhouse supply or construction company to determine the proper size fan or vent for your situation.

Cooling Options for Your Greenhouse

One of the best ways to cool a greenhouse in the summer is to reduce light intensity. How much reduction to provide depends on the heat load in the greenhouse and the light requirements of the plants grown.

Greenhouse whitewash, shade cloth, screens of wood or aluminum, or Venetian blinds can be used for shading. Greenhouse whitewash is a special kind of latex paint that is diluted in water and sprayed on the covering surface. It is designed to be applied in the spring and gradually degrade by the action of rain and sun so that little remains by fall. Shade cloth is a black, green, or white woven fabric of polypropylene that is laid over the outside of the covering. Shade cloth can be purchased with various weave densities that result in 20 to 80 percent light reduction. For most smaller type greenhouse applications, 30 to 50 percent light reduction is usually sufficient.

Another inexpensive method for cooling a greenhouse is evaporative cooling, which relies on air passing through a porous pad saturated with water. The evaporating water removes heat from the greenhouse. There are two types of evaporative coolers: fan-and-pad systems and unit coolers (swamp coolers). Fan-and-pad systems consist of a cellulose pad at least 2 feet tall and extending the length of one wall, with water supplied from the top to keep the pad wet during operation. Fans are installed in the opposite wall to draw outside air through the pads.

Unit coolers consist of a metal box mounted outside the greenhouse and evaporative pads on three inside walls. These units may be more practical and attractive for small hobby greenhouses. A water connection, collection lines, and recirculating pump are built into the unit. A fan within the unit draws outside air through the pads and the cooled air into the greenhouse through a duct. A vent on the opposite side of the greenhouse provides an air exit. Unit coolers should provide about 15 cubic feet per minute for each square foot of floor area. Determining the evaporative cooling capacity for a given greenhouse and geographic location takes some effort. Work with a greenhouse supply or company that specializes in evaporative coolers to determine the proper size for your particular situation.

Generally, heating, ventilating, and cooling are controlled by thermostats located close to the center of the greenhouse at plant level. For accurate control, be sure the thermostat is shaded from direct sunlight. Mounting it in a plastic or wood box ventilated by a small blower works well.

TABLE 24. *Greenhouse Cooling Systems*

OUTDOOR TEMPERATURE	COOLING SYSTEM	DESCRIPTION	COOLING RESULT INSIDE GREENHOUSE
Fahrenheit: 95 Degrees or Celsius: 35 Degrees	No Cooling	1. Inside temperature of greenhouse can exceed 140 degrees F. 2. Plants will die within a short period of time.	Fahrenheit: 140 Degrees or Celsius: 60 Degrees
	Natural Ventalization	1. Roof vent opens to allow hot air to escape to outside. 2. Door or other opening must be left open to provide for incoming air from outside to replace exhausted air. 3. On hot summer days temperatures can rise 20 to 30 degrees (F) above outside temperature. 4. Recommended for mild climate areas only.	Fahrenheit: 115 Degrees or Celsius: 46 Degrees
	Fan & Shutter Cooling	1. Outdoor air is introduced through motorized inlet shutters. 2. Hot air is exhausted by exhaust fans. 3. Temperature inside house can be maintained within 10 degrees (F) of outdoor temperature with properly designed system. 4. Fans and shutters are controlled by thermostat.	Fahrenheit: 105 Degrees or Celsius: 41 Degrees
	Fan & Shutter Cooling with Shade Cloth	1. Outdoor air is introduced through motorized inlet shutters. 2. Hot air is exhausted by exhaust fans. 3. Shade cloth is placed over exterior of greenhouse or suspended from wires placed on inside. 4. Temperature inside house can be maintained within 3 to 4 degrees (F) of outdoor temperature with properly designed system. 5. Fans and shutters are controlled by thermostat.	Fahrenheit: 98 Degrees or Celsius: 37 Degrees
	Evaporative Cooling (Positive Pressure)	1. Outdoor air is cooled by Evaporative Cooler (located outside) and discharged into greenhouse. 2. Hot air is exhausted through outlet shutters which operate automatically on pressure differential. 3. Temperature inside house can be as much as 10 to 15 degrees (F) cooler than outdoor temperature with properly designed system. 4. Evaporative Cooler is controlled by thermostat. 5. System efficiency can be increased with the use of shade system. The fans will not have to work as hard to maintain the desired temperature.	Fahrenheit: 80 Degrees or Celsius: 27 Degrees
	Evaporative Cooling (Fan and Pad)	1. Outdoor air is drawn through pad cooling system located on one end wall of the greenhouse. This cool air enters into the greenhouse. 2. Hot air is exhausted by fans mounted on the opposite end wall of the greenhouse. 3. Temperature inside house can be as much as 10 to 15 degrees (F) cooler than outdoor temperature with properly designed system. 4. Fan & Pad System is controlled by thermostat. 5. System efficiency can be increased with the use of shade system. The fans will not have to work as hard to maintain the desired temperature.	Fahrenheit: 80 Degrees or Celsius: 27 Degrees

Utilities to Your Greenhouse

It is best to check with your local governing agency planner and/or local building codes before building a greenhouse on your property. Some county or city codes prohibit greenhouses or place restrictions on size, type, covering, or construction materials. Some agencies have restrictions on property location placement. If is also a good idea to check with the local electric company. In some areas, the utility company may request a utility pole and separate meter for the greenhouse. Be sure all electrical work is performed by a licensed electrician according to code.

Water can usually be plumbed from the home supply line as long as the volume and pressure are adequate. Install a backflow prevention valve in the water supply line to prevent the possibility of contaminating the water in your home.

Greenhouse Work and Storage Areas

Make a work area for potting and maintaining plants. This may be located inside or outside the greenhouse. The north wall is often a good location for a work area inside the greenhouse. The work area may also include a sink. Storage areas for soil and containers can be located outside the greenhouse but should be protected from the weather and not be an eyesore.

Light Required for Photosynthesis

Only part of solar radiation is used by plants for photosynthesis. The photosynthetically active radiation (PAR) contains the wavelengths between 400 and 700 nanometers, and falls just within the so-called visible spectrum (380-770nm). The total visible spectrum is perceived by us humans as white light, but with the aid of a prism, it is possible to see that the "white" light is actually separated into a spectrum of colors from violet to blue, to green, yellow, orange and red. Plants use the blue to red light as their energy source for photosynthesis.

Supplemental Lighting for Your Greenhouse

Why use supplemental lighting for indoor or greenhouse gardening? High intensity discharge (H.I.D.) lighting has traditionally been used only by commercial growers in large scale greenhouses. Business savvy professionals have discovered exceptional benefits of supplemental, artificial lighting for plant growth. From stronger, healthier seed starts, into faster maturing, vigorous plants that offer much higher yields and more spectacular flowering than can be achieved without supplemental lighting. H.I.D. lighting not only supplements sunlight, but can actually replace it during long winters where sunlight is in short supply. It is energy efficient and the cost of operating some light systems is comparable to using one of your kitchen appliances. For instance, Sunlight Supply, Inc. at https://www.sunlightsupply.com has several H.I.D options that can do just such. They also have the technical expertise to assist you with your project.

Plants need light for proper growth. The light spectrum range produced by artificial light (particularly H.I.D. light) enhances the natural light derived from the sun by many times over. The result, when combined with proper nutrients, is astonishing. This technology is now available to both the commercial market and individuals.

In summary, supplemental lighting has the following benefits:

- Significantly increase the health, strength, growth rate & yield of your plants.
- Supplement natural sunlight in your greenhouse virtually eliminates seasonal & geographical restraints. In addition, by extending the "day length" with supplemental lighting, growing success is greatly enhanced.
- Container plants that are outdoors on decks & patios during the summer can be moved indoors during the winter under H.I.D. or high

output fluorescent light allowing them to thrive year round.
- Indoor gardening — by using an H.I.D. light fixtures as a primary light source indoors, you can enjoy the gardening indoors throughout the year.

High Intensity Discharge lighting systems have revolutionized indoor gardening in the last two decades. They are the most energy efficient grow lights available, so they produce much more light for the amount of power consumed.

Linear Fluorescents

Traditional T12 and T8 fluorescent lighting is simply not powerful enough to light an area more than 8-10 inches below the bulb. With the introduction of the highly efficient T5 technology, T5 linear fluorescent fixtures can now put out a respectable 92.6 lumens per watt. T12 lamps typically put out about 30 lumens per watt. T5 fixtures are excellent for starting seeds and cuttings, but are also able to produce enough light for full term growth. Because of their minimal heat output, they can be placed very close to the plant canopy to maximize the light output.

High Wattage Self Ballasted Grow Lamps

These lamps have become quite popular in the recent past. *Sunlight Supply, Inc.* offers *EcoPlus* brand 105 & 150-watt mogul base fluorescent lamps that put out about 70 lumens per watt. They are excellent for starting seeds or to use over a small garden area.

Incandescent Lamps

These standard household bulbs do not emit enough light, or the proper spectrum to be used by serious gardening enthusiasts. They are not very efficient, using a considerable amount of power for the light they emit. They are typically only about 15 lumens per watt.

Which Light is Right for You?

Most gardeners use at least 25 watts per square foot of garden space. You may need less if your light is used to supplement natural sunlight, or if you are growing a plant that does not require as much light (i.e. such as lettuce). However, many gardeners prefer to double or even triple the recommended wattage to achieve faster growth rates. There really is no such thing as too much light at one time, but using a big light in a small space will sometimes result in high

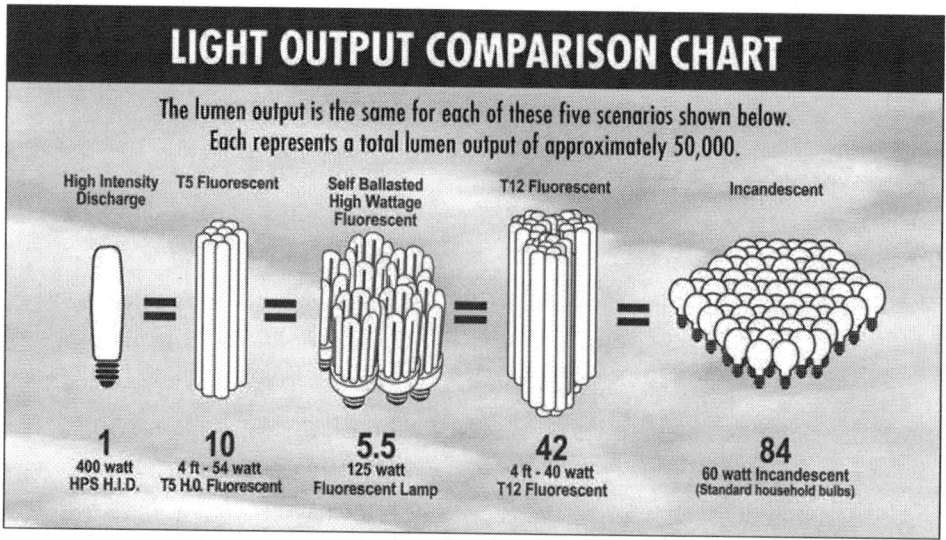

FIGURE 113.

temperatures that are difficult to control, or increase operational cost more than necessary. Keep in mind that plants need periods of darkness too. Most indoor gardeners use lighting from 12–18 hours per day.

Hanging Height for Greenhouse Lighting

A general guideline for the proper hanging height of an H.I.D. lamp would be 12"-48" depending on wattage (see below). Make sure to check for excessive heat at the top of your plants by placing your hand (palm down) over your plants. If the top of your hand is hot, you need to move your lamp up higher. If the light source is too close to your plants, it will burn them. Remember that as your plants grow you will need to adjust the height of your lamp. The latest air-cooled reflectors available on the market, like the Super Sun 2, allow you to place higher wattage bulbs closer to plants than was possible in the past.

When you raise the light up and away from your plants, light levels at your plants will be significantly reduced. For example: 1ft. = 1000 FC, 2ft. = 250 FC, 3ft. = 111 FC, 4ft. = 63 FC, 5ft. = 40 FC, & 6ft. = 28 FC. (*The acronym FC means foot-candle, which is a measurement of light intensity. This is a non SI unit, but it can be easily converted into more useful units of measure if you prefer including Lumens and Watts. FC refers to a unit of illuminance or illumination, equivalent to the illumination produced by a source of one candle at a distance of one foot and equal to one lumen incident per square foot.*)

Coverage Area Greenhouse Lighting

A fluorescent fixture can be placed much closer to plants than an H.I.D. fixture because it produces very little heat. You should place your fluorescent lights as close to the tops of your plants as you can without excluding the outside perimeter of your garden.

H.I.D. Average Coverage Area by Wattage

- 150/175 watts covers approx 2' x 2' area
- 250 watts cover approx. 3' x 3' area
- 400 watts covers approx. 4' x 4' area
- 600 watts covers approx. 6.5' x 6.5' area
- 1000 watts covers approx. 8' x 8' area

NOTE: Coverage area may need to be reduced if this is your primary light source.

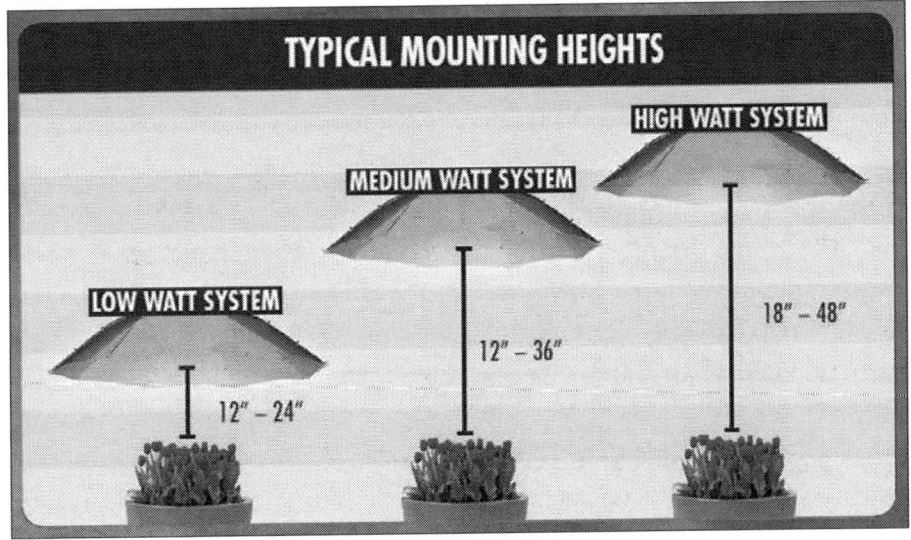

FIGURE 114.

High Pressure Sodium (hps) or Metal Halide (MH) Greenhouse Lighting

If you choose H.I.D. as your source of lighting, you have another choice: HPS or MH. *Sunlight Supply, Inc.* offers lights in MH and HPS. Metal Halide (MH) bulbs emit a light spectrum which appears blue-white to the human eye. This color spectrum is more conducive for vegetative growth, or starting seeds and cuttings. High Pressure Sodium (HPS) bulbs emit a spectrum which is more concentrated in red/orange light. This color is ideal for the fruiting and flowering stage of a plant's development. It is a good multi-purpose light as well.

Your style of gardening will determine which type of light is best for you. But whichever model you choose, you can be assured that your investment into the lighting technology used by the professional will be rewarded by increased, nutrient packed, yields and healthier plants.

Efficient Photoperiodic Lighting

When the natural day length (photoperiod) is short, many bedding plants and herbaceous perennials flower earlier when provided with artificial long days. Many long-day plants bloom faster when provided with a night length of less than 10 hours (when the day is at least 13 to 14 hours long). Therefore, in North America, and most of the northern hemisphere, photoperiodic lighting can be useful from around mid-September to early April. For Australia, and most of the southern hemisphere, from mid-March to early October, supplemental photoperiodic lighting is beneficial, if not essential.

Photoperiodic lighting is effective when delivered to extend the natural day or during the middle of the night. Day-extension lighting typically begins around sunset and ends once the total desired photoperiod is achieved. For example, if sunrise is at 7 a.m. and sunset is at 6 p.m., then a 15-hour photoperiod can be achieved when the lamps operate from 6 p.m. until 10 p.m. Night-interruption or night-break lighting is equally effective and is usually delivered by turning lamps on from 10 p.m. to 2 a.m. For those that have so-called electrical 'smart meters' off peak times are usually less expensive. If the supply of electricity is limited, then half of the crop could be provided with day-extension lighting from 6 p.m. to 10 p.m., and the other half with night-interruption lighting from 10 p.m. to 2 a.m. Regardless of lighting strategy, the minimum recommended light intensity is 2 $\mu mol \cdot m^{-2} \cdot s^{-1}$ (10 foot-candles, which equals around 100 lux) at plant height.

Many crops produce earlier when provided long days. Some crops even require long days to flower (critical step in fruit and vegetable production).

For most species, once a plant is induced to flower, the flowers will develop even under a natural photoperiod. Therefore, once flower buds are visible (often about 3 to 4 weeks after the start of long days), you can turn off lamps and long-day plants will proceed to flower. Long-day lighting is recommended until early April.

Pollination of Plants within a Greenhouse

A good harvest of many plants requires successful pollination. Birds, bees, butterflies and wind are some of the methods plants are pollinated in nature. Greenhouses create barriers that protect plants from adverse growing conditions, but they also present structural obstacles that prevent pollination. However, there are some intervention strategies that you can implement to ensure that your plants are successfully pollinated, even in protected greenhouses.

Manual Pollination

If you have a small greenhouse, manually pollinating plants is cost-effective and takes little time. Gently shaking plants or tapping flowers releases pollen from male flower parts to female structures. Some plants, such as squash, have separate male and female flowers so pollen must be transferred between blossoms. Other plants, such as tomatoes, have "perfect flowers," which contain male and female parts. Disturbing flowers on these plants distributes pollen within each bloom. By the way, tomatoes need pollinating every day for good fruit production, according to the University of Kentucky Cooperative Extension Service.

Device Pollination

You can buy battery-operated pollinating tools from greenhouse supply companies. Hand-held pollinator wands have vibrating heads that you touch to the base of flowers. Some operators get by using a battery-operated electric toothbrush to pollinate blossoms whereas other claim that such a method is too abrasive. Because pollinating devices are made specifically for this task, they generally cause less damage to flowers, fruits and the plant as a whole, according to the University of Florida IFAS Extension Service. Whether you choose manual or device pollination, it is recommended that you pollinate plants between 10 a.m. and 3 p.m.

Bee Pollination

Bees are the best pollinators. Bumblebees are more effective pollinators than other bees because of their longer tongues and wing vibration techniques, notes Clemson Cooperative Extension. Certain bumblebee species are raised commercially specifically for use in greenhouses. You purchase a box or hive of bees and place it in your greenhouse. Generally, you provide a supplemental food source, such as pre-packaged nectar, to ensure their survival. Because of the closed environment of a greenhouse, it is important not to use any kind of pesticide on plants (organic or synthetic), as it will kill the bees.

General Pollination Considerations for Greenhouse Plants

Greenhouse fans circulate air to keep plants dry, but for the most part, they do not create enough wind to pollinate flowers. Temperature and humidity affect pollen release, so pollination efforts are unsuccessful without the right environmental conditions. High humidity levels inside greenhouses cause pollen to stick together in clumps, which inhibits proper dispersal of individual grains. To overcome this, you can use fans to dry plants before pollinating them. Temperatures that are too high or too low stop plants from releasing pollen. Research the optimal pollen-release temperature ranges for your specific plants and adjust your greenhouse temperature accordingly.

Greenhouse "Do's" and "Don'ts."

Finally, consider the following points. They are intended to prevent problems and to make life with a greenhouse more enjoyable.

Do's

- Keep the greenhouse and surrounding areas clean and organized.
- Allot enough time to the greenhouse weekly to be successful.
- Learn more about greenhouses and growing plants by reading, conducting research online, and talking to others.
- Keep the greenhouse in a good state of repair.
- Discard weak, diseased, or badly insect-infected plants.
- Enjoy the greenhouse; arrange work intelligently so it doesn't become a chore.
- Experiment—try new methods.

Don'ts

- Don't accept sick plants form friends or family. You're asking for trouble if you do!
- Don't start with the most difficult plants. Gain experience with plants that are easier to grow before trying the difficult ones.
- Plants in a greenhouse are a responsibility. Properly care for them for good results.

Net House

In some tropical regions, net houses are more appropriate than conventional polyethylene plastic or glass covered greenhouses. This is because the hot climate in the tropics or subtropics increases the need for better ventilation to avoid high temperatures and humidity. Net houses consist of a frame over the grow beds that is covered with mesh netting along the four walls and a plastic roof over the top. The plastic roof is particularly important to prevent rain from entering, especially in areas with intense rainy seasons, as units could overflow in a matter of days. Net houses are used to remove the threat of many noxious pests associated with the tropics, as well as birds and some animals. The ideal mesh size for the four walls depends on the local pests. For large insects, the mesh size should be 0.5 mm. For smaller ones, which are often vectors of viral diseases, the mesh size should be thicker (i.e. mesh 50). Net houses can provide some shade if the sunlight is too intense. Common shade materials vary from 25 to 60 percent sun block.

FIGURE 115. *Net house structure to house a small aquaponic unit*

CHAPTER 22

Alternative Energy Options & Operating Off-The-Grid

Alternative Energy Overview

Renewable energy is clean, affordable, domestic, and effectively infinite. It produces no emissions and results in cleaner air and water for all. Energy prices are rising rapidly and fuel oil prices are breaking records on an almost daily basis. With the economy slowing down, the increasing cost of food, gas, utilities, and taxes consuming more of our budget, consumers are spending less on non-essentials. It is no surprise that many families and business owners are wondering how they will survive. One strategy is to lower energy consumption by becoming more energy efficient or by using alternative energy sources.

In recent years, a lot of attention has been given to biomass as an alternative fuel source (ethanol, aka corn). With many Americans having to sustainably curtail their food budget and hundreds of millions around the world starving, U.S. lawmakers have adopted an insane policy of burning up our food supply in the form of a corn-based ethanol fuel mandate. What is really crazy is that ethanol-laced fuel gets much worse mileage than gasoline; you have to buy more of it to get where you're going. It is a policy that has never made much sense, but adopted because of lobbyists, Big Agriculture, and political favors. Obviously, growing corn for ethanol reduces the available farmland to grow food crops. Furthermore, a recent Congressional Budget Office report concluded that the increased use of ethanol accounts for 10-15 percent of the increase in food prices, thus one of the reasons food prices are rising at an alarming rate.

This chapter will focus on solar and wind energy. These are alternative energy options that not only make good sense, but are relatively affordable and easy for most people and small businesses to put into place.

Economics of Electricity

The cost of electricity can be somewhat challenging to determine, as most utility companies have different rates for summer and winter months, and for peak and not-peak usage. With the development of so called 'smart meters', utility companies are able to determine exactly when and how much electricity you are using.

Before we see how much electricity costs, we have to understand how it's measured. When you buy gas, they charge you by the gallon. When you buy electricity, they charge you by the *kilowatt-hour (kWh)*. When you use 1,000 watts for 1 hour, that's a kilowatt-hour. See table 25 on the following page.

PART III: COMPONENTS OF AQUAPONICS USED IN AQUAPONICS

TABLE 25. **Electricity Costs**

DEVICE	WATTAGE	HOURS USED	kWh
Medium Window-Unit AC	1000 watts	one hour	1 kWh
Large Window-Unit AC	1500 watts	one hour	1.5 kWh
Small Window-Unit AC	500 watts	one hour	0.5 kWh
42" Ceiling Fan on Low Speed	24 watts	ten hours	0.24 kWh
Light Bulb	100 watts	730 hours (all month)	73 kWh
CFL Light Bulb	25 watts	730 hours	18 kWh

To get kilowatt-hours, take the wattage of the device, multiply by the number of hours you use it, and divide by 1,000. (Dividing by 1,000 changes it from watt-hours to kilowatt-hours.) Here's the formula to figure the cost of running a device:

wattage x hours used ÷ 1000 x price per kWh = cost of electricity

For example, let's say you leave a 100-watt bulb running continuously (730 hours a month), and you're paying 15¢/kWh. Your cost to run the bulb all month is 100 x 730 ÷ 1,000 x 15¢ = $10.95.

If your device doesn't list wattage, but it does list amps, then just multiply the amps times the voltage to get the watts. For example:

2.5 amps x 120 volts = 300 watts

(If you're outside North America, your country probably uses 220 to 240 volts instead of 120.)

Watts vs. Watt-Hours

- Watts is the *rate* of use at *this instant*. In other words, Watt is a measure of work.
- Watt-hours is the total energy used over time.

- We use *watts* to see how hungry a device is for power. (i.e., 100-watt bulb is twice as hungry as a 50-watt bulb.)
- We use *watt-hours* to see how much electricity we used over a period of time. That's what we're paying for.

So, just multiply the *watts* times the *hours used* to get the *watt-hours*. (Then divide by 1,000 to get the kilowatt-hours, which is how your utility charges you.) Example: 100-watt bulb x 2 hours ÷ 1,000 = 0.2 kWh.

The national average rate for electricity is useless for two reasons:

1. **Electricity rates vary** *widely.* They vary not only by region (i.e., an average of 7.5¢ in Idaho vs. 36¢ in Hawaii), but they also vary from the same utility. As a matter of fact, rates can range from 12¢ to 50¢ per kWh from the same provider. The only way to know what you're actually paying is to check your bill carefully.

2. **Electric rates are usually** *tiered,* meaning that excessive use is billed at a higher rate. This is important because your *savings* are also figured for the highest tier you're in. For example, let's say you pay 10¢/kWh for the first 500 kWh, and then 15¢/kWh for use above that. If you normally use 900 kWh a month, then every kWh you save reduces your bill by 15¢ (technically, once you get your use below 500 kWh, then your savings will be 10¢ kWh, but you get the point).

For simplicity in determining a 'rough' cost for an item, use a rate of 15¢ per kWh. This isn't a "typical" rate, since there's no such thing as typical when it comes to electricity rates. And it's certainly not average. It's just a *reasonable expectation*. Your own rate could be dramatically higher or lower than 15¢ per kWh

Table 26 shows the typical cost of electricity. As of 2014, where I live the in northern California, the cost of electricity averages about $0.14/KWH, with annual increases planned for the next four years. What is NOT shown are the fees and taxes that are also imposed upon us in addition to our base electrical charges. It is easy to see how even a small solar powered system can help alleviate some of this financial burden, whether it be used for aquaponics or for other purposes.

TABLE 26. **Electricity Costs**

COST PER kWh	12-HOUR DAYS Day	12-HOUR DAYS Month	18-HOUR DAYS Day	18-HOUR DAYS Month
$0.05	$0.60	$18.00	$0.90	$27.00
$0.06	$0.72	$21.60	$1.08	$32.40
$0.07	$0.84	$25.20	$1.26	$37.80
$0.08	$0.96	$28.80	$1.44	$43.20
$0.09	$1.08	$32.40	$1.62	$48.60
$0.10	$1.20	$36.00	$1.80	$54.00
$0.15	$1.80	$54.00	$2.70	$81.00
$0.20	$2.40	$72.00	$3.60	$108.00
$0.25	$3.00	$90.00	$4.50	$135.00

Electrical Definitions

- Watt'—a measure of work.
- Kilowatt (kW or kw)—a unit of power, equal to 1,000 watts.
- Kilowatt-hour (*kWh* or *kW·h*)—a measure of electrical energy equivalent to a power consumption of 1,000 watts for 1 hour.
- Ampere (amp)—a measure of electricity in motion.
- Volt—a measure of electricity under pressure.

Grid Connection

Depending on local regulations and laws, so-called net-metering systems can be installed that allow your electricity meter to run 'backwards' when your power generator is producing more than you need. In some locations, excess power delivered to the grid will result in additional reimbursements. In many states, electrical utilities are mandated to purchase a specific percentage of their energy from 'green' or renewable sources and welcome special arrangements with energy producers.

Solar Energy Fundamentals

Solar power is one of the fastest growing sectors in the U.S. The price of solar panels has dropped by 30 percent since 2010 and costs continue to fall. This is not just a trend in America, the rest of the world is also moving in the same direction. As a matter of fact, the United States was a net exporter of solar products in 2010 by $2 billion.

While plants convert sunlight into biomass production, photovoltaic (PV) panels convert it into electricity. The conversion efficiencies of PV panels have increased over the years to as high as 17 percent at maximum light intensity. Some experimental PV cells have achieved efficiencies of 40 percent.

PV panels should be mounted for maximum light interception. In the Northern hemisphere, panels can be attached to south-facing roofs, other support structures, or on a tracking device that follows the position of the sun across the sky.

PV panels generate DC power that can be converted to AC power to operate lights, pumps, and other household/small business equipment. PV systems can be interconnected with the local electrical grid, ensuring that electrical power is always available. In a grid-connected system, excess power from the solar installation can be sent to the grid. Interconnection requirements vary from state-to-state and utility-to-utility.

Off-grid PV systems require some form of electrical storage to provide power during periods of little or no sun. Typically, banks of batteries are installed for this purpose. Off-grid systems are best suited to applications where there is no nearby electrical grid or for standalone systems, such as aquaponics.

A significant portion of sunlight reaches the surface of the Earth as heat radiation. This energy can be used to heat water. Typically, not much water is needed for washing and cleaning purposes, but aquaponic operators can use warm water to heat the greenhouse. The most solar energy that can be collected is during the middle of the day, so storing the warm water for use during the night is a good strategy to reduce the use of heating fuel. The fish tank can serve as an efficient means of regulating temperature in a greenhouse or other enclosed area. The rise in heating prices makes long-term storage of warm water more attractive.

There are also many other technologies for converting incoming solar radiation into heat. The most common systems are flat plate collectors that allow water or other fluids to flow through a panel that is oriented toward the sun. Very simple flat plate collectors are often used for heating swimming pools, but can and are used for many other applications. Flat plate collectors are very efficient. Other systems for converting sunlight into heat include evacuated tube collectors and parabolic reflectors. These products are capable of generating higher temperatures, but are significantly more expensive and are often dependent upon a tracking systems to maintain an optimal orientation.

Other solar technologies are passive. For example, big windows placed on the sunny side of a building allow sunlight to heat-absorbent materials on the floor and walls. These surfaces then release the heat at night to keep the building warm.

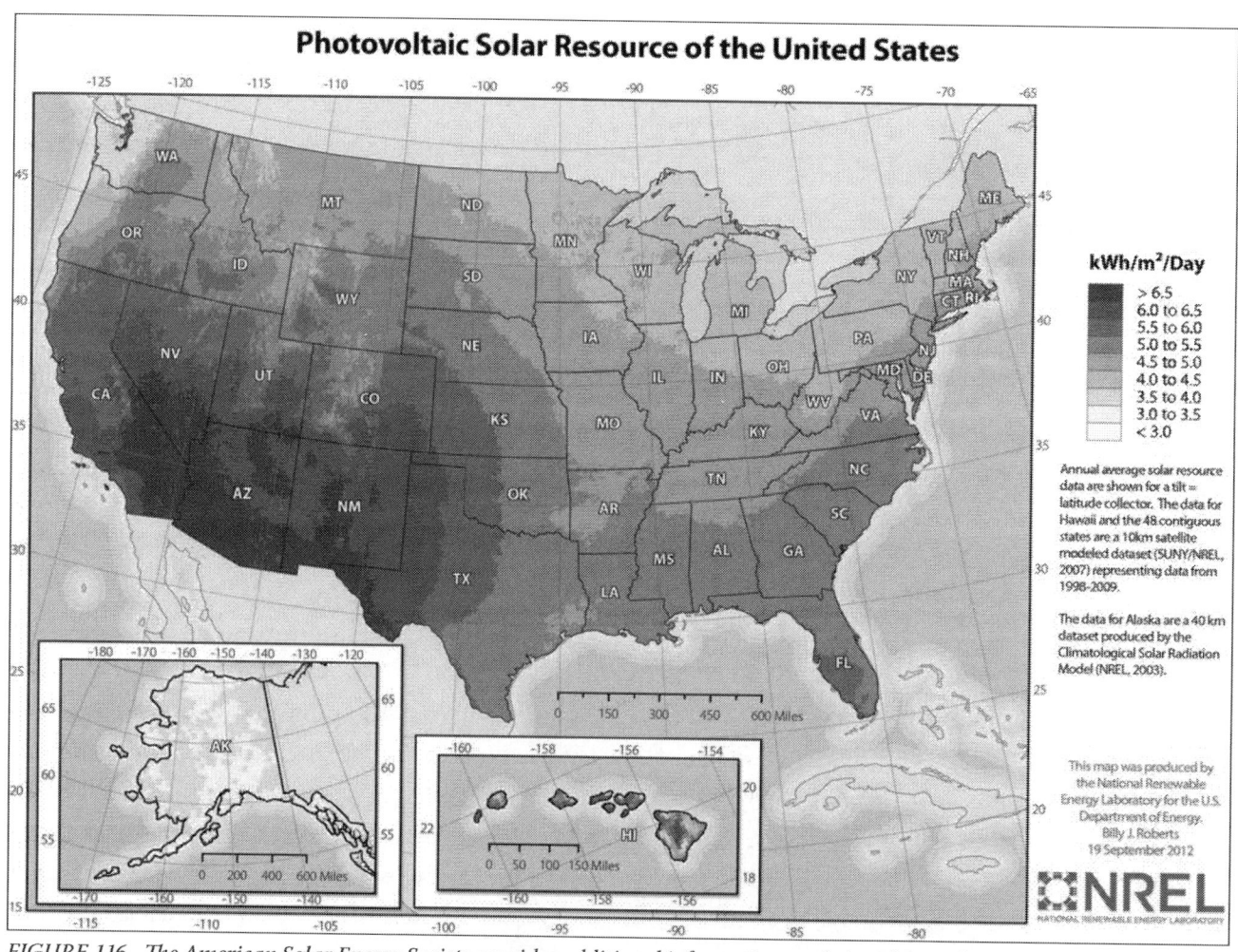

FIGURE 116. The American Solar Energy Society provides additional information on their web site: (http://www.ases.org/)

Benefits of Solar Energy Summarized
- Feel good about saving energy and the environment.
- Helps promote a model for sustainable living.
- Save tens of thousands of dollars over the life of your solar energy system.
- Allows you to put current utility bill money to better use.
- Use all your appliances without feeling guilty.
- Helps the U.S. get one step closer to energy independence.
- Doing something good for future generations.
- Solar energy is an inexhaustible fuel source that is pollution and noise free.

Average Daily Solar Radiation per Month at Your Location
Figure 116 shows the general trends in the amount of solar radiation received in the United States and its territories.

Steps to Integrating Solar Energy
1. Review your power needs.
2. Look for ways that you can conserve power (i.e. more efficient lighting, temperature control, etc.).
3. Calculate the savings over time, and consider how those funds could be better spent.
4. Calculate the cost of installation.
5. Remember that it is an investment, and that there is a good probability that you can sell the system (thus receiving a portion of your money back) in the future if you ever decide to go a different direction.
6. Obtain free advice from solar energy equipment vendors regarding your operation.
7. Educate yourself on solar energy equipment.
8. Compare cost.
9. Installation.
10. Maintenance (cleaning the panels periodically).

Solar Energy Grants, Loans, Tax Incentives, and Other Resources
- Energy Technology Inc. has solar powered equipment, supplies, and accessories for homeowners and small businesses: http://personal-solar.com
- Solar Resource Calculator at: http://solar.ucsd.edu/SolarApp.html
- The Solar and Energy Loan Fund (SELF) provides low cost financing options and energy expertise to help homeowners and small business owners lower energy bills and implement alternative energy systems, such as solar equipment:
- http://cleanenergyloanprogram.org/how-it-works/homeowners
- The USDA Rural Business and Cooperative Programs has a wide array of programs, as well as grants and loans, for increased energy efficiency and alternative energy systems: http://www.rurdev.usda.gov/Energy.html
- Check with your electric utility company to see if they have other resources as well as provide solar energy conversion benefits.
- With tax incentives, solar electricity typically pays for itself in five to ten years.

Wind Energy
The United States has some of the best wind resources in the world, with enough potential energy to produce nearly 10 times the country's existing power needs. Wind energy is now one of the most cost-effective sources of new-generation power technology, often having a significant competitive advantage over the coal, gas, and nuclear power technologies. Its cost has dropped steadily over the past few years, as wind turbine technology has improved. Currently, over 400 American manufacturing plants build wind components, towers and blades.

Using wind to pump water and generate power is not a new idea. Before the start of rural electrification in 1936, wind energy was widely used across the U.S. During the last ten years, technological improvements and rising energy prices have significantly increased the number of wind energy installations. In many cases, large installations have occurred on farmland, but often the farmers are not the main users of the generated energy, nor do they own the equipment. Many farmers only receive a land lease payment for the area used by wind turbines.

The success of wind energy installations depends on site-specific wind conditions. Wind maps have been compiled for all regions of the U.S. and these maps are useful for a first approximation of the average wind speed at a given location (links provided at the end of this section). However, local topology, vegetation, and building structures significantly affect the average wind speed. Where possible, use local wind speed measurements to determine whether a site is appropriate for wind generation. Currently, an average wind speed of 9 mph for small wind generators and 13 mph for large generators (measured at 100 ft above the ground) is considered necessary for the economical use of wind power.

Even small wind generators can be used to operate a fairly large size aquaponic system. Such a system can be operated off-grid, but a connection to the grid is prudent. Grid-connected-systems have the advantage of having power available when the wind system is not functioning at full capacity, and does not require batteries for electrical storage.

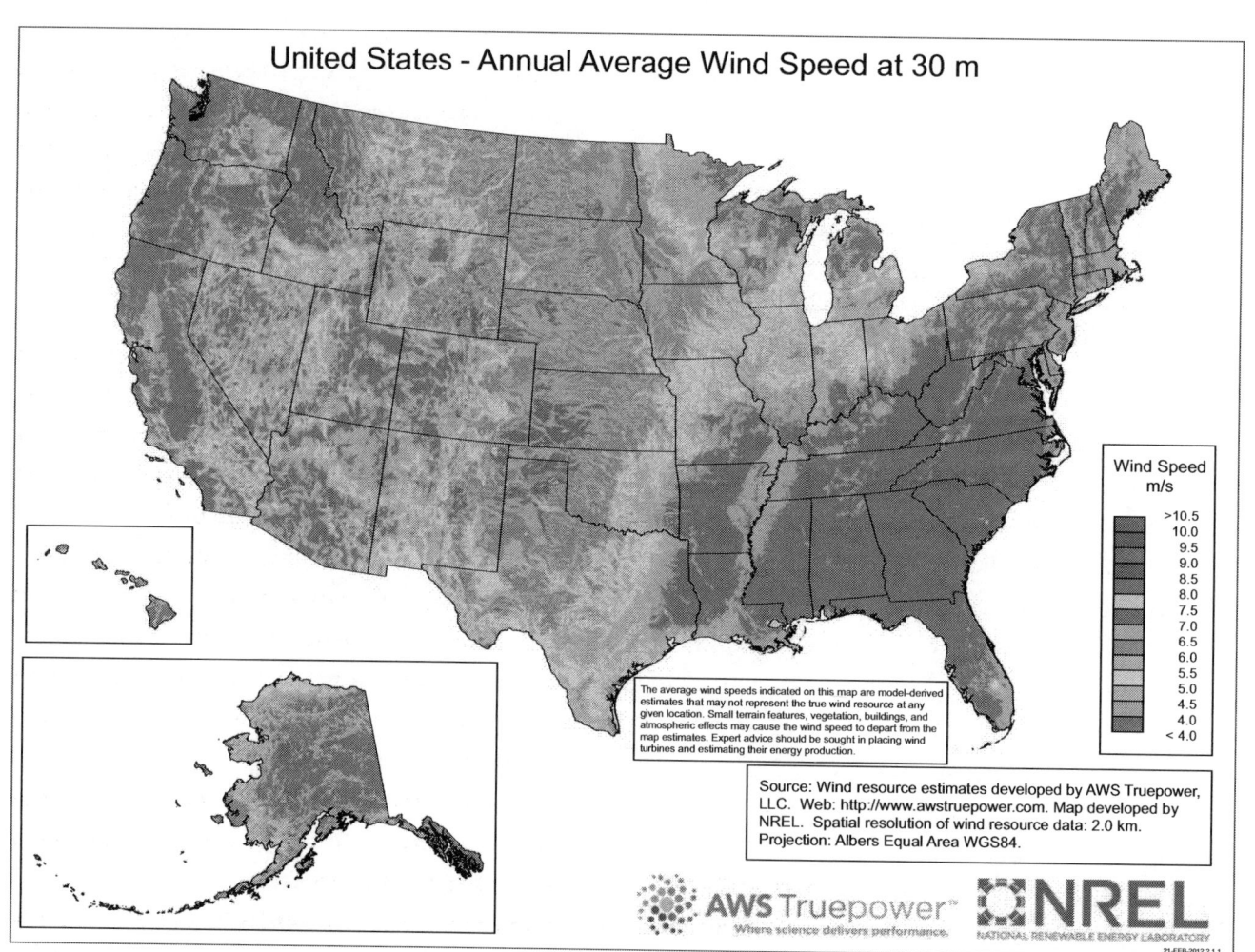

FIGURE 117.

Wind Energy Helpful Resources
- Very interesting current wind map of USA: http://hint.fm/wind
- A collection of helpful worldwide wind and climate maps: http://www.climate-charts.com/World-Climate-Maps.html
- The U.S. Department of Energy provides an 80-meter (262.5-ft) height, high-resolution wind resource map for the United States with links to state wind maps. States, utilities, and wind energy developers use utility-scale wind resource maps to locate and quantify the wind resource, identifying potentially windy sites within a fairly large region and determining a potential site's economic and technical viability. http://www.windpoweringamerica.gov/wind_maps.asp

Grant, Loan and Rebate Programs

Local utilities, as well as state and federal organizations, offer a variety of grants, loans, and rebate programs for alternative energy installations. Each of these programs comes with its own set of requirements and often entail cost-sharing. Nevertheless, these programs can reduce the investment costs and/or reduce the pay-back period. Many of these programs are announced on web sites, requiring some effort to learn about them. In some states, energy regulating commissions, such as the Board of Public Utilities or state energy agencies, have programs for renewable energy systems. Your local utility and county extension service, state departments of agriculture, the USDA, and the NRCS are good places to start investigating the various opportunities.

Renewable Energy Certificates

Some states administer renewable energy certificate (REC) programs that allow certified producers of eligible renewable energy to sell these certificates that represent proof that 1,000 kWh of electricity was produced. Thus, in addition to reducing your electric power consumption from the utility grid (i.e., by lowering your monthly electricity bill or receiving payment for excess electricity you exported to the grid), the RECs generated by your system can provide additional income when sold (i.e., to a power company that was mandated to deliver a certain percentage of its total output as renewable energy). While prices for RECs fluctuate, REC programs provide additional financial incentives for renewable energy production (http://www.eere.energy.gov/greenpower/markets/).

Energy Conservation

Before you consider a solar or wind system for your aquaponic operation, the first step in any renewable energy project is ensuring that the existing system is functioning efficiently. The reason is quite simple: the cost of implementing energy efficiency measures is less than the cost of installing renewable energy technologies to compensate for inefficient use of conventional energy sources.

Energy Conservation Resources
- An excellent energy conservation guide full of valuable ideas and practical techniques has been produced by the U.S. Dept. of Energy and can be found at http://energy.gov/sites/prod/files/2013/06/f2/energy_savers.pdf.
- The State of California Consumer Energy Conservation Department has a list of energy conservation and efficiency tips for your home, office, business, vehicle, and other areas, located at: http://www.consumerenergycenter.org/tips/index.html
- A useful reference source is the book titled "Energy Conservation for Commercial Greenhouses", published by NRAES.
- Some of the many ways to achieve better energy performance in greenhouses include using thermal curtains, where possible, and checking that

they seal properly (i.e., form a continuous barrier), verifying that environmental control systems are doing what they are supposed to, and sealing glazing leaks through unintended openings in your walls and roofs.
- Most local utility companies have a brochure and/or webpage showing you many ways on how you can better conserve energy in your home and business.

Living or Operating Off-The-Grid

A growing trend, especially within the United States, is being "off-the-grid". This term can conjure up many different meanings for people, ranging from a peaceful community of folks who still travel by horse-and-buggy to groups preparing for the end of the world and are of anti-everything. While these groups can fall under the basic definition of off-the-grid, the general term is simply people who are not connected to a public utility. Their reasons for choosing to live in such a way varies from one end of the spectrum to the other. For the most part, off-the-grid folks are not connected to their local electrical system; they are a stand-alone-system.

Off-The-Grid Alternative Energy Source

Electricity, however, is just the main aspect of being off-the-grid; it can also relate to all energy sources. For instance, off-the-grid homes are autonomous; they don't depend upon the municipal water supply, sewer, natural gas, cable or internet services, or similarly related utility services. What the author would consider a true off-the-grid house is one that is able to operate completely independent of all traditional public utility services.

Although there are many different types of alternative energy sources (such as geothermal), the ones that are most applicable to the topic-at-hand are wind, solar, biofuel, ethanol, and water (well, creek, or pond).

Of these, solar polar is by far the most commonly used alternative energy source in aquaponics.

The initial set-up costs for living off-the-grid are a bit high, turning many away from considering alternative energy. However, most find that the initial price tag is compensated through time as they are free from the monthly financial drain as well as government mandated taxes and regulations. In addition, many have peace of mind knowing that they are generating a non-polluting energy source and that they won't have the deal with as many outages and shortages. Lastly, they take comfort in knowing that they are no longer financially supporting and sanctioning the fossil and nuclear fuel industries.

Alternative Food Source

Many who live off-the-grid have taken to growing their own food as well. Compared to taking the jump into an alternative energy source, growing your own food will require constant dedication and a lot of hard work; the end results are always worth it, though. Through aquaponics yields are much greater than traditional gardening for equal areas. Growing your own food has many advantages, such as:

- **Receiving the nutrients your body requires.** Much of the food we eat today has been modified and changed with chemicals and preservatives. Anyone that does even ten minutes of research online will quickly learn that what we are purchasing from the grocery store just isn't healthy to consume. Most of the chemicals used in farming were approved by the Environmental Protection Agency without any research into how these chemicals could harm individuals. Currently, the EPA considers 60 percent of all herbicides, 90 percent of all fungicides, and 30 percent of all pesticides as carcinogenic (cancer-causing). In 1987, a study by the National Academy of Sciences found that pesticides may contribute to an additional four million cancer cases in the United States alone. Imagine what that number might

GOING IT ALONE
With enough renewable technologies you will only need the grid to help pay your bills

······ Electricity ······ Water
······ Heat ······ Waste

- WIND TURBINE
- SOLAR PANEL
- Hot water cylinder
- SOLAR THERMAL COLLECTOR
- CONTROL BOX
- ELECTROLYSIS OF WATER
- HYDROGEN PRODUCTION
- Hydrogen
- Fuel cell
- Water use appliances (bathroom/kitchen)
- Central heating
- Rainwater recycling
- ELECTRICITY GRID
- HYDRO TURBINE
- Hot water accumulator
- Electric appliances
- Drinking water
- Biomass stove
- Inverter
- Batteries
- Toilet waste to septic tank or for composting
- Water from river
- Water back to river
- Water filter
- Rainwater recycling
- Water pump
- GROUND SOURCE HEAT PUMP
- Water tank
- Water from spring, stream or river

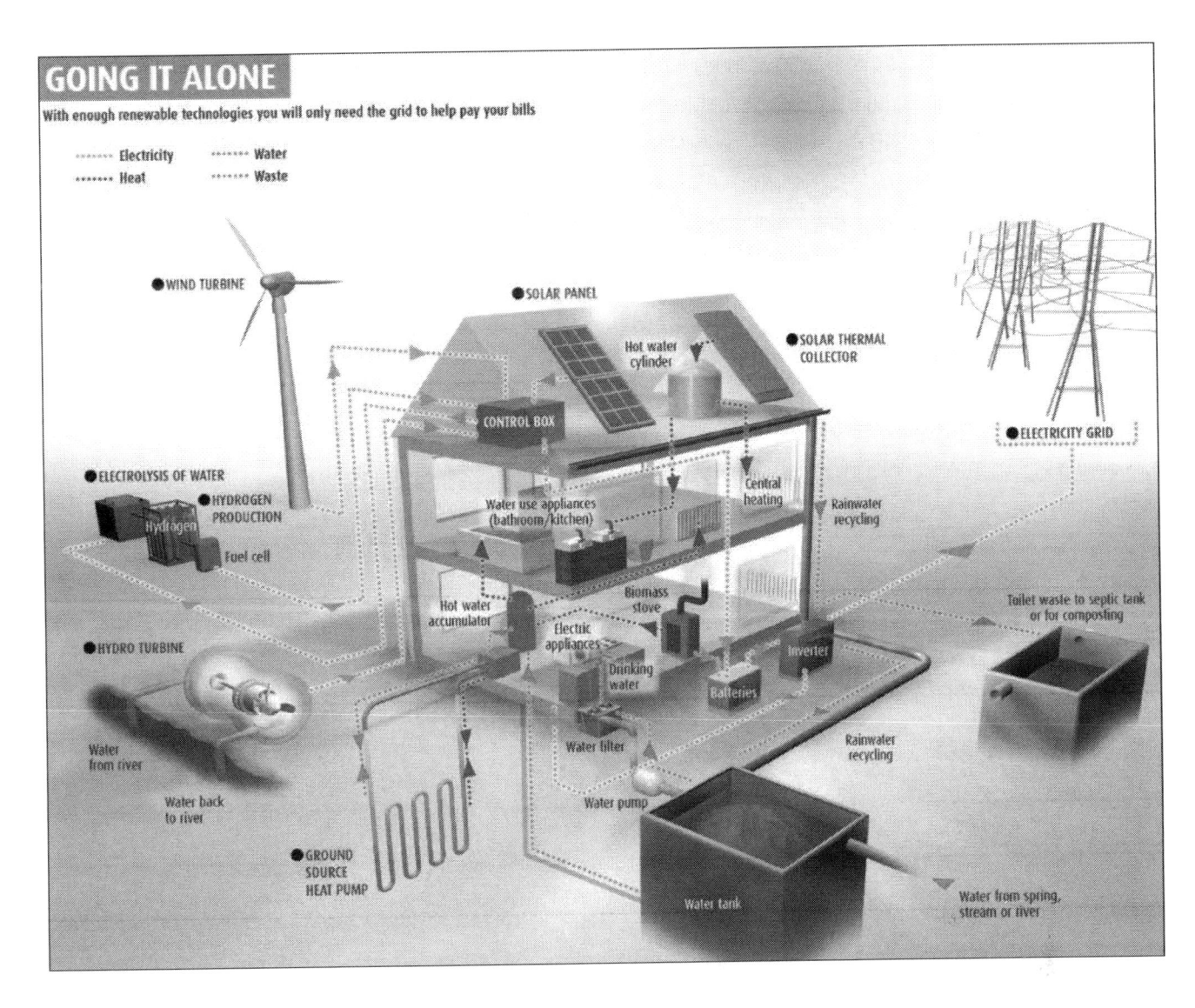

be today! Our body craves particular minerals, nutrients, and vitamins; growing and eating your own organic crops helps meet this need. Bonus: the food tastes better than anything you'll buy from the store.

- **It will save you money!** The math is simple: if you are no longer spending money every week at the grocery store, but are growing your own food to consume, then you are saving money that will rapidly add up over time. If you decide to sell some of it to neighbors, family, and friends, then you just created an additional cash flow as well.
- **Stops soil erosion.** According to The Soil Conservation Service, over three billion tons of topsoil are eroded each year from the U.S.'s croplands; or about seven times faster than it is being built-up by mother nature.
- **Better water quality.** Due to most crops across the United States being sprayed with an array of chemicals (such as pesticides), at least 38 states have reported that their groundwater has been contaminated. A possible outcome of drinking such water is cancer. Growing your own crops, free of pesticides and other chemical agents, helps prevent the polluting of your own groundwater (which you may be drinking if you are truly off-the-grid).
- **Saves energy.** Most crops you buy at the store today are grown on mega-farms. In order to keep up with production, many farmers are forced

to use petroleum (more than any other single U.S. industry). This energy is used to create the synthetic fertilizers as opposed to growing healthy crops.

- **Emergency Supply.** Should a disaster ever strike your family or community, whether that be from a storm or something worse, you will have a ready-supply of food to consume. Many know what a grocery store looks like right before a major storm system hits; shelves are emptied in a hurry leaving you with few options if you didn't arrive first (and if you did, you may have left with multiple bruises and a higher blood-pressure). The old Boy Scout motto of "always be prepared" is a good one to remember here.

While it takes a lot of work, time, and effort, the benefits of being off-the-grid and supplying your own power and food far out-weight the negative. Upfront costs for setting up are quickly reimbursed, and there is a peace-of-mind that simply cannot be purchased when you are no longer dependent upon local utility companies, stores, and certain government regulations.

Friendly Reminder: As mentioned previously in this chapter, one should be sure that making alternative energy changes and/or implementing a farm (i.e. adding solar panels, harnessing the wind, growing a garden, etc.) doesn't violate any local laws, will not void your home owners insurance, will be tolerated by your local utility companies, and is conformance with your homeowner association (if applicable).

PART IV

N.F.T. & D.W.C. Design and Layout

CHAPTER 23

Nutrient Film Technique

Nutrient Film Technique

Nutrient Film Technique (N.F.T.) is a commonly used hydroponic method. In fact, N.F.T. systems are one of the most productive hydroponic systems. However, it is not as widespread in aquaponic systems. In N.F.T. systems, plants are grown in nutrient rich water pumped down small enclosed gutters (long narrow channels). The water flowing down the gutter, or channel, is only a very thin film, but provides the plant roots with an abundant supply of essential nutrients, water, and oxygen. Plants are typically grown in small plastic cups, called net cups, allowing their roots to access the water and absorb the nutrients.

It is important to note that N.F.T. is only suitable for certain types of plants, generally leafy green vegetables such as lettuce, which do not grow tall. Larger plants have roots which spread too much and become invasive to the system, causing numerous problems. Also, larger plants like bell peppers and tomatoes become too top heavy to be supported by the non-anchored structure. These top heavy plants can be grown in a N.F.T. system, but it is typically a high labor endeavor as the plants must be continually attached to some sort of support apparatus as they grow so they will not fall over. Typically, the plants are supported via twine or wire hanging from the greenhouse ceiling. N.F.T. aquaponics shows potential, but it is used less than the other two major methods referenced in this chapter (Deep Water Culture and Media Filled Beds Systems).

Pros

- Materials are readily available.
- Relatively low cost of construction materials.
- More precise growing conditions.
- No concerns for pH changes related to media.
- Fish tank can be outdoors and a considerable distance from the grow beds.
- Very high yields of crops which can be grown in a N.F.T. system (plants grow quickly).
- Plants grown at a good working height (no bending over).

Cons

- Requires more filtration to remove fish waste.
- Doesn't allow for many different crop options. In other words, plant diversity choices are limited.
- Susceptible to power outages and pump failures.
- Grow channel tubes tend to clog regularly and require frequent monitoring.
- An aquaponic N.F.T. system is not recommended for beginners.
- For a hydroponic N.F.T. system (not aquaponics), the water solution should be replaced following each harvest to alleviate pH balancing issues and to eliminate the build-up of residual contaminates found in synthetic fertilizers.

Author's Opinion on N.F.T.

- N.F.T. works well for commercial growing hydroponic operations growing large volumes of lettuce or other similar small plants, but does not appear to be the best option for aquaponics. The additional filtration needed for fish, limitations of plant diversity, and the necessity for very close system monitoring is a real turn off.

- Nevertheless, since N.F.T. seems to always be listed as an option for those looking into aquaponics, I wanted to address it in detail within this book. Therefore, this chapter provides everything one, who is considering this option, needs to know about developing and managing an aquaponic N.F.T. operation. Grow channels can also be successfully added to Flood-and-Drain system

N.F.T. System Layout

If the Nutrient Film Technique coincides with your goals, then read on the learn more. N.F.T. uses a continuous running pump to deliver nutrient rich water to the high-end of a grow tray and gravity does the rest. The grow tray is placed at an angle to allow the water to flow downhill, whereby the solution is constantly flowing over the roots.

The nutrient solution flows in a thin film over the roots, ensuring that they are watered and fed, but not completely soaked. The thin, nutrient rich film solution only makes contact with the lower roots. The upper parts of the roots remain dry and are exposed to air where they derive the necessary oxygen.

Nutrient Film Technique is most commonly used for crops that grow fast and don't need a lot of support, such as lettuce. With N.F.T., the roots are not able to anchor the plants and it is typically a challenge to support the plants that grow more than 14-inches in height. If you want to grow plants such as tomatoes, it is doable, but you need to give considerable consideration as to how you plan to support the plants (i.e. string from the ceiling, trellis, etc.).

N.F.T. Grow Tray

As mentioned above, Nutrient Film Technique uses tubes or channels on an incline instead of flat plant growing container. This makes it easier to set it at an angle and to make sure that the nutrient solution flows directly to the roots without wasting any of it.

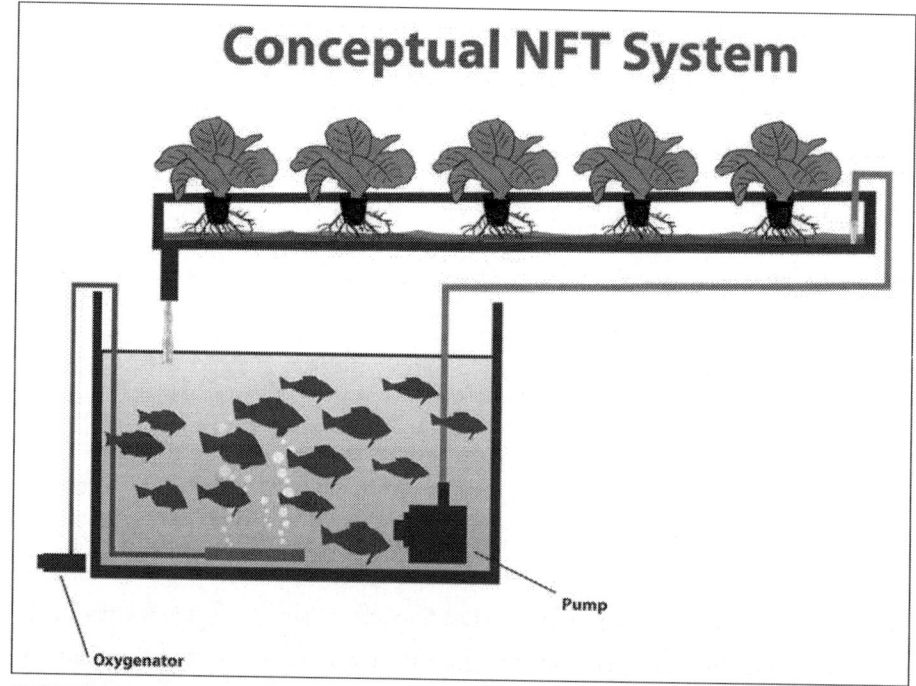

FIGURE 118. Cross-section of a nutrient film technique (N.F.T.) system.

CHAPTER 23: NUTRIENT FILM TECHNIQUE

FIGURE 119. N.F.T. Grow Channels.

Many DIY enthusiast use 4-inch diameter round PVC pipe, 2-inch x 3-inch gutter downspouts, or 4-inch x 4-inch white PVC decorative fence post sleeves. Channels can also be built using basic materials such lumber and water-proof plastic lining or fiberglass.

Holes are drilled in the grow tray. Net pots, although not always necessary, are sometimes placed in the grow tray holes to help hold the plants in place.. Holes and net pots range in size, depending upon the operator's goals, with 2.5 to 3-inch holes being the most common size. The spacing of the holes is dependent upon the operator's goal, with no industry standard being practiced. Some operators prefer to have many holes so that they can rearrange plants as they grow, or so they can plant their crop close and harvest leaves as the plants grow and become crowded (i.e. spinach leaves). Many holes, even if they are not all filled with plants, provide the operator with much flexibility. Other operators prefer to create a system with holes spaced in accordance with the recommended plant spacing distances of the particular plant species in which they will be growing—allowing each plant to grow to its maximum size in the same hole.

The main disadvantage to using a round PVC pipe as the grow tray is that the film will not evenly coat the roots. The roots in the middle will have access to a greater depth of solution, while the roots closest to the edges would only have a shallow depth. This can cause uneven growth and weakness in the plants. By using a flat-bottomed channel, this problem is eliminated.

For the record, the hard core N.F.T. purists believe that using a round PVC pipe as the grow tray is not truly qualify as a N.F.T. system, but is really some type of hybrid set-up. Of course, this view is subject to debate, but the main point is that a true N.F.T. system is one that conveys a thin film of water along the channel.

With that said, you can either place plant seedlings (starter plants) directly into the holes of your grow tray

FIGURE 120. N.F.T. Grow Channels.

(PVC pipe or channel), or for greater stability, grow the plants in net pots and place the net pots in the holes. Commercial operators and many hobbyists do not use a growing medium with N.F.T., but let the roots fall directly through the net pots and on to the film. If you choose to use a growing medium, use it sparingly and make sure that the roots have plenty of room to fall through to the bottom of the pot. Depending on what you are growing, sometimes it is necessary to regularly check and trim the roots in order to prevent out of control growth that can clog the system.

The N.F.T. channels need to be enclosed with holes, and not open. An open channel will result in unwanted algae growth and become a maintenance burden. Therefore, the channels need to be enclosed on the top side with holes for the plants being the only opening.

Another important consideration is the length of your grow tray. As the nutrient solution flows over the roots, it decreases in nutrient concentration and oxygen levels. "Short run" trays offer an advantage over longer ones because they ensure that plants at the end of the line receive nutrient solution with the same composition as those at the beginning of the line. Operators using long run channels will often rotate the plants periodically. This, of course, is more labor intensive, but is one method to balance things out if plants at the end of the long line tend to grow more slowly or are not producing as much. Beware, though, as it is common for the roots to spread out wider than the hole. This can cause a problem as there is a higher risk of damaging the plant during relocation. It also takes a lot more time to pull the roots carefully out of one hole, and carefully stuff them in another hole.

Unlike a Flood-and-Drain system grow bed, N.F.T. grow trays are light. There is rarely a reason to fasten the grow channels to the stand. It is nice to have the flexibility of sliding channels on the stands one way or another as needed for maintenance, as plants grow and need more space, as well as during harvest. Having the ability to slide the grow trays on the stand really comes in handy.

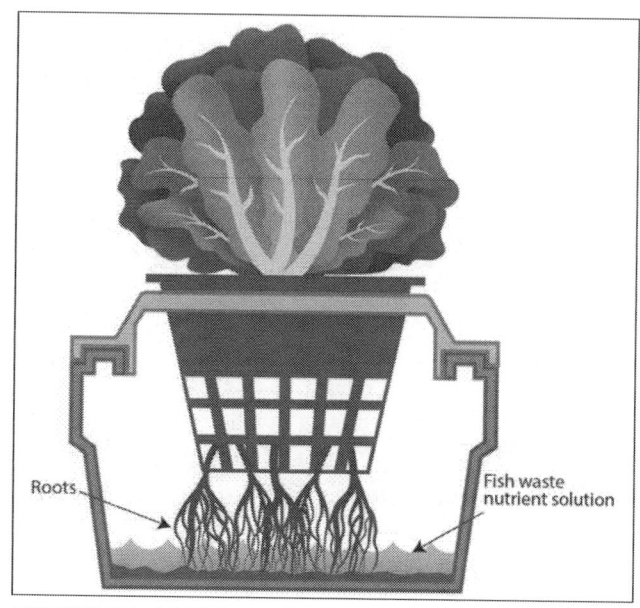

FIGURE 121. NFT Grow Channel Cross-Section

The grow trays should be positioned at an ergonomically correct height so that working with the plants can be done without strain and bending over too much. An ergonomically correct design makes the N.F.T. aquaponic experience that much more enjoyable. The design of the grow bed should be at a working height that best suits the operator. In most cases, the height would be about the same as a typical kitchen counter, with 36-inches being an industry standard. Extremely short or tall people, or those with special needs, may want to modify that height to better suit their needs.

Nutrient deficiencies can occur when grow channels are too long. The grow channel should not be longer than 30 to 40 feet (10 to 15 meters). However, having an additional nutrient feed line stationed along the grow channel can help eliminate this problem.

Unfortunately, some N.F.T. systems are incorrectly set up so that the grow channel is on too steep of an angle. This causes the nutrient rich water to pass through the grow tray at an undesired velocity, which in turn inhibits nutrient uptake by the plants, puts unnecessary stress on the plants, and can even damage the roots.

CHAPTER 23: NUTRIENT FILM TECHNIQUE

FIGURE 122. Phase 1: Planting

FIGURE 123. Phase 2: Growth

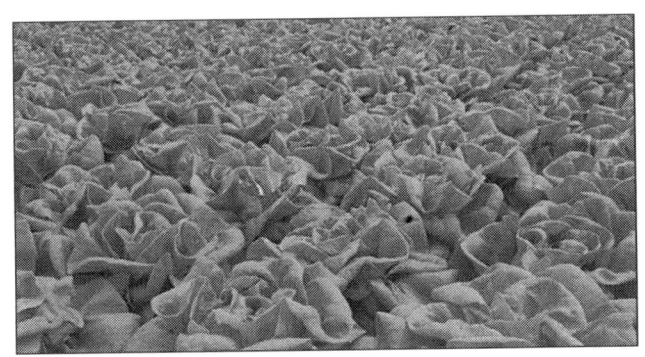

FIGURE 124. Phase 3: Harvest

How to Obtain Grow Channels

1. **Do-It-Yourself (DIY)** — Make your own using from rain a gutter (i.e. same type of gutter used on residential homes).

2. **Hydroponic Supplier** — Sold by most hydroponic suppliers. The cost, sizes, configurations, and quality vary greatly from one supplier to another.

   - American Hydroponics Inc.
     1-800-458-6543
     www.americanhydroponics.com

   - FarmTek
     1-800-327-6835
     www.FarmTek

   - CropKing
     1-330-302-4203
     www.cropking.com/catalog/N.F.T.-systems-supplies

3. **PVC Pipe** — <u>Not recommended</u>, as a round pipe does not allow for the conveyance of a thin film of fluid needed for optional plant growth.

The recommended slope for a N.F.T. system is typically a 1:30 to 1:40 ratio. That is for every 30 to 40 inches of horizontal length, one inch of drop (slope) is recommended. Better yet, the ideal approach is to construct the N.F.T. systems so that the slope can be adjusted periodically while the plants are growing. With a fixed slope, as plants get bigger, their root systems can cause the fish waste water (nutrient solution) to pool and dam up the water flow. A N.F.T. that has an adjustable grow channel slope can be tilted at a steeper angle if needed to prevent such flow constrictions. It is also important to ensure that there is a consistent slope along the channel. If there are sags or low spots, water will pool up in those areas as well.

The recommended flow rate for a N.F.T. system is typically between 1/4 to 1/2 gallon per minute (1 to 2 liters) for each grow tube (channel), or as otherwise stated, between 15 to 30 gallons per hour (60 to 120 liters per hour). While the plants are just seedlings the flow rate can be substantially less, and then increased towards the optimal flow rate (stated above) as the plants get bigger. Flow rates higher than these can result in nutrient deficiencies and root damage.

PART IV: N.F.T. & D.W.C. DESIGN AND LAYOUT

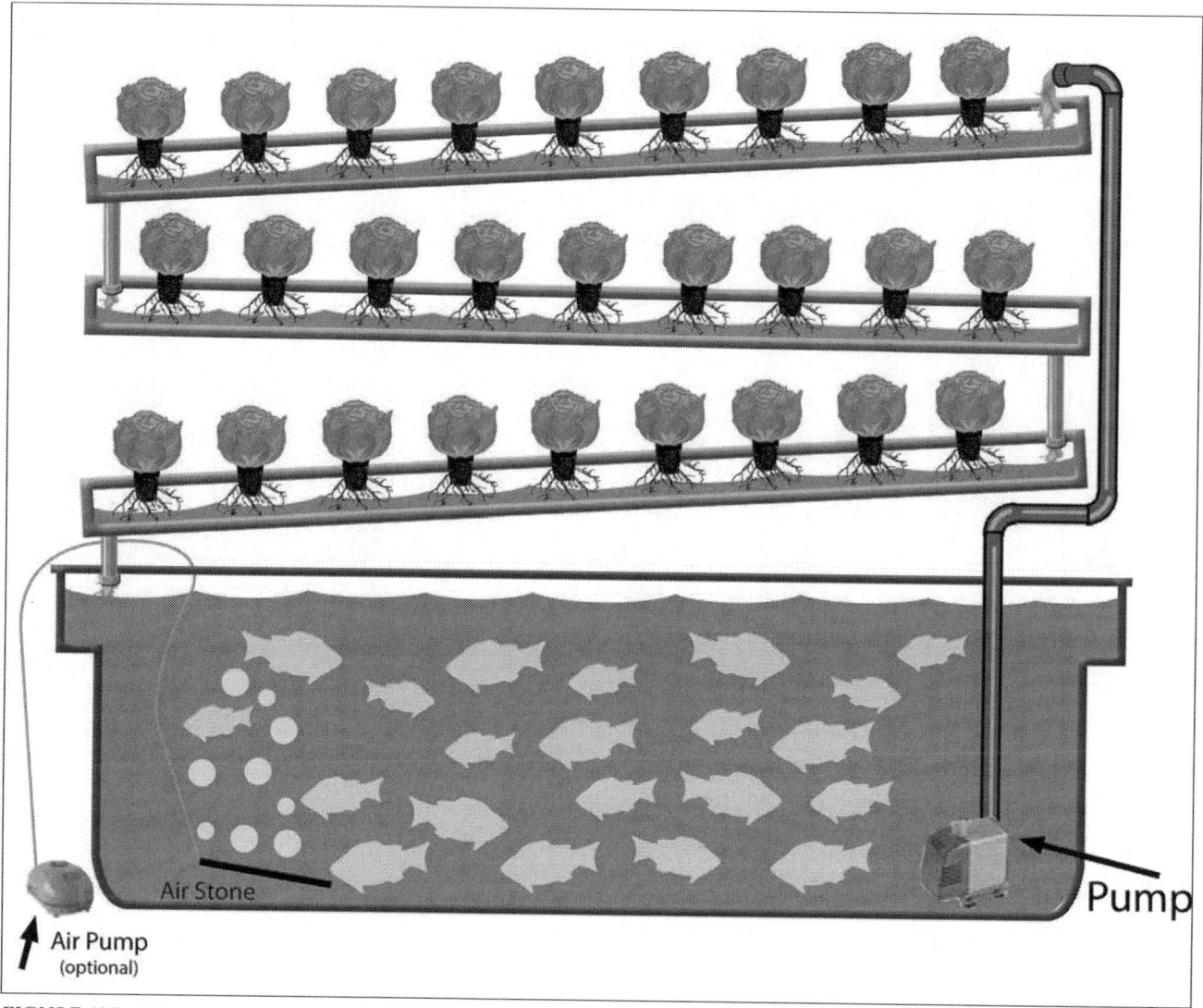

FIGURE 125. NFT System

FIGURE 126. Small N.F.T. "Hydroponic" Package Unit (Double) — Fish could be integrated into this unit with a water filter and air pump. Available through most hydroponic retailers.

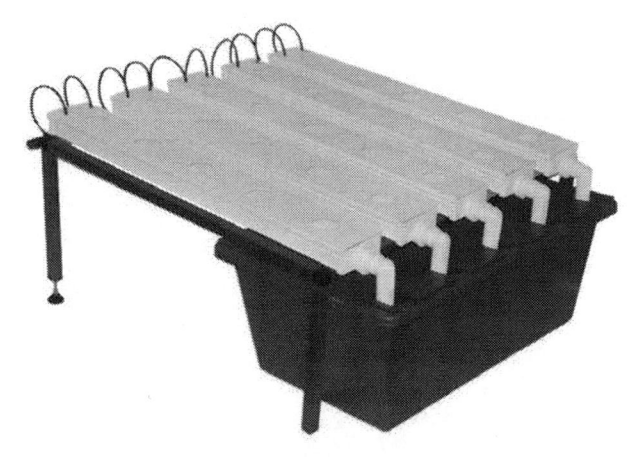

FIGURE 127. Small N.F.T. "Hydroponic" Package Unit (Single) — Fish could be integrated into this unit with a water filter and air pump. Available through most hydroponic retailers.

N.F.T. Reservoir and Pump

Unlike other systems, the N.F.T. does not use an automatic timer connected to the water pump because the pump runs constantly. This can be a big problem in the case of power outages, blockages, or system failures, so it is imperative that the water pump have a backup-ready power supply available.

Some N.F.T. systems need an air pump, otherwise known as an air stone or other air distribution devices in the reservoir, which is connected to an air pump outside of the reservoir. N.F.T. operators choose to install air pumps for one or more of the following reasons:

1. They believe that it will help the plants by oxygenating the roots even more.
2. Their water level is so deep inside the channel that it is more of a deep water system than a N.F.T.—a hybrid system—so an air pump is needed.
3. To maintain dissolved oxygen levels within the fish tank at life support levels for fish above the minimum of 3 ppm, and preferably above 6 ppm.

In all cases, the air pump will not hurt the plants, but unless the roots are completely and always under water, and dissolved oxygen levels are in an acceptable range, then the air pump will likely just be an unnecessary cost. It is best to monitor the dissolved oxygen levels within the fish tank before purchasing an air pump. Such may prevent having to incur unnecessary cost, if the system parameters are within a tolerable range without an air pump. (NOTE: Dissolved oxygen, and how to measure it, is explained later in this book).

FIGURE 128. Homemade N.F.T. System (Hydroponic)

N.F.T. Commercial Hydroponic Operations

FIGURE 129. Commercial N.F.T. Operation.

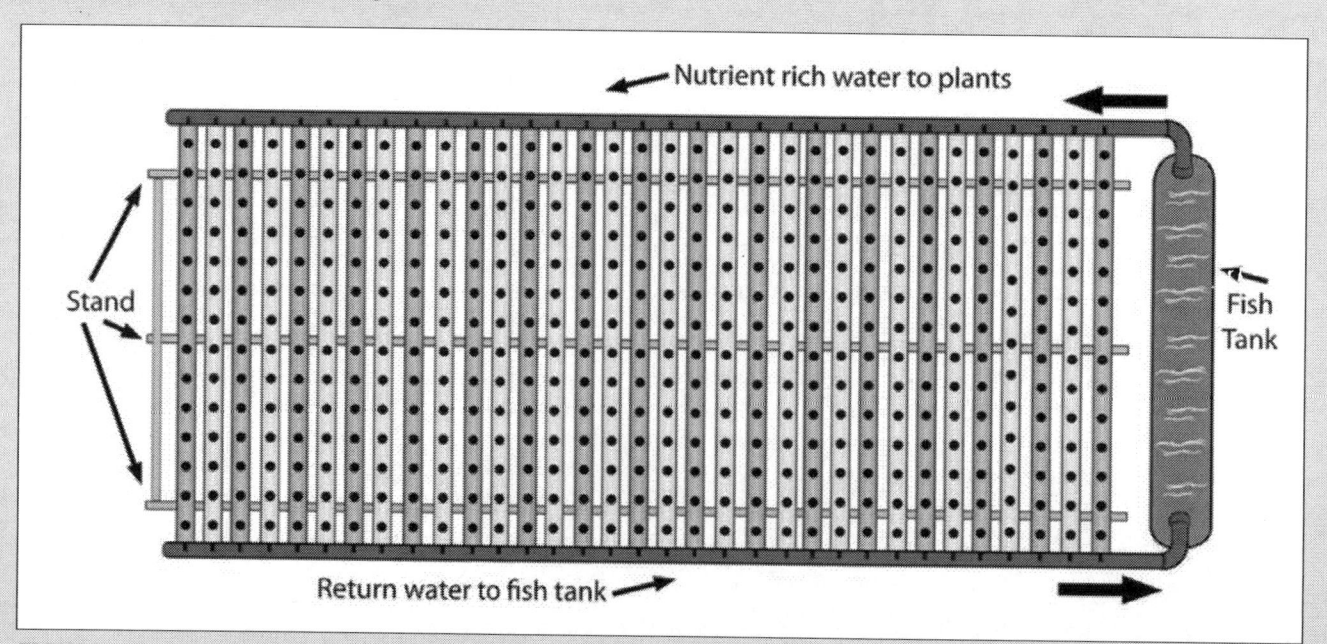

FIGURE 130. Note: Filtration and/or air pump may be necessary for fish tank. Dependent upon stocking density and DO levels (5 ppm DO min.)

CHAPTER 23: NUTRIENT FILM TECHNIQUE

N.F.T. Net Pots or Not

N.F.T. growers often use net pots to grow their plants in. Net pots allow for very fast flooding and filling with the rich nutrient solutions (aquaponics—fish tank waste). They are available in many sizes. The most common are: 2", 3", 3.75", 5" and 6", but there are sizes in between and even larger.

To keep expenses down, 16-oz disposable cups can be used for plant containers instead of the commercially available net pots (see figure 132). Using a soldering iron has proven to be the best way to make the holes in the cup.

N.F.T. — Medium or Not

Some operators use growing mediums in the net pots. A growing medium is used to help support the plants. Mediums used are expanded clay pellets (also referred to Hydroton— a company's brand name of the product) lava rock, stones, rock wool, and basically any other growing medium that will not fall through the holes. This would eliminate growing mediums such as sand, perlite, and smaller size particle growing mediums. Coco peat can be used if you only use the longer strands of the coco peat and not the smaller particles that would fall out of the holes, thereby clogging the system.

Plant roots can sometimes be damaged by using sharp-edged growing mediums that can cut into the plant roots as the water vibrates and moves the roots, or as the operator checks on or relocates the plants to another hole. This can be very detrimental to the plants. Expanded clay pellets are relatively smooth and round which is better for plant roots. One still needs to be careful relocating plants, though, as the dangling unprotected roots can be easily damaged.

FIGURE 131. Net Pots

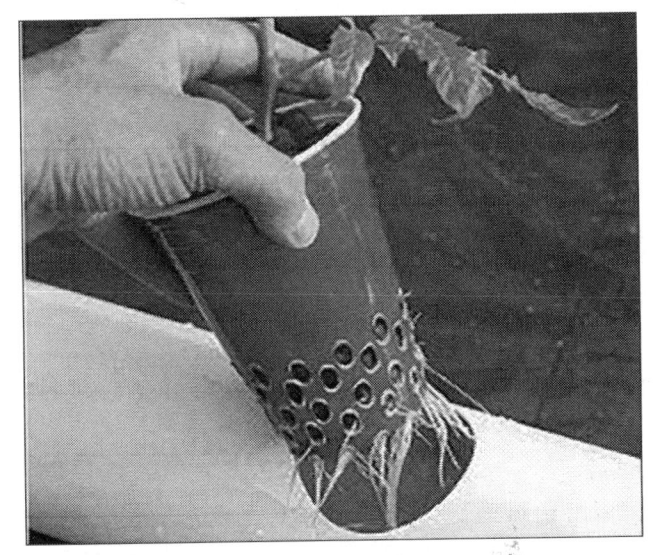

FIGURE 132.

FIGURE 133. N.F.T. with medium (left) and without medium (right)

Human Urine as Fertilizer
'Hydroponics' & Soil Gardening

For those who are considering a 'hydroponic' system (no fish), but the thought of adding synthetic fertilizers to your food supply is a concern, there is a better option. It is also free and effective — human urine.

Yes, that's right, human urine has been proven to be an effective agricultural fertilizer. Researchers say our liquid waste not only promotes plant growth as well as industrial mineral fertilizers, but also would save energy when used on sewage treatment. In one of several proven experiments, environmental scientists at the University of Kuopio in Finland conducted an extensive study which showed that human urine worked as well as traditional fertilizers.

Fresh urine is high in nitrogen, moderate in phosphorus and low in potassium and can act as an excellent high-nitrogen liquid fertilizer or as a compost accelerator. The nitrogen-phosphorous-potassium ratio (N-P-K ratio) of human urine is typically about 11 – 1 – 2.5.

Urine also contains salt — sometimes quite a lot of it if you have a diet high in sodium. Because of both the salt and high nitrogen levels, urine should generally be diluted 10:1 before use on garden crops. Greater dilution – 20:1 or more – is appropriate for more tender plants, seedlings and potted plants which are more susceptible to salt build up.

Fresh pee can have a pH anywhere from 5 to 9 depending on a person's diet, but it tends to move toward neutral as it ages and breaks down when applied outside. Therefore, the pH of human urine is not really a concern for most garden uses.

Many gardeners have known for years that human urine was an effective, environmental-friendly, and cheap fertilizer. People have been collecting their urine in containers and jugs to spread around their plants for years. Some men even perform the direct application when the urge hits. Ladies may need to be a little more creative if desiring to implement the direct approach.

However, like any fertilizer, too much is too much and can burn up the plants. Fertilizer contains salts—sometimes quite a lot of it if you have a diet high in sodium. Salt draws moisture out of plants. When you apply excess fertilizer to plants, the result is yellow or brown discoloration and root damage. Urine fertilizer burn symptoms may appear within a day or two of over application.

Because of both the salt and high nitrogen levels, urine should generally be diluted 10:1 before use on garden crops. Greater dilution – 20:1 or more – is appropriate for more tender plants, seedlings and potted plants which are more susceptible to salt build up.

Human feces are not sterile. This solid waste can be infected with salmonella and E.coli. Steps should be taken to ensure that your urine isn't contaminated by that solid waste. Also, if you're on any medication that will be present in your urine stream, don't use that urine as it could damage the plants and/or be absorbed into the edible components. Also, don't use your urine if you have a urinary tract infection as the germs from the infection will also contaminate your urine.

If you prefer to have a more natural and cheap alternative to synthetic fertilizers, then human urine is the way to go. Just make sure you have good aim.

CHAPTER 23: NUTRIENT FILM TECHNIQUE

N.F.T. Layout

So far, all images and text have referenced a N.F.T. system that is compacted, meaning where the tank and grow channels are adjacent to each other. However, there is no reason that a N.F.T. system (or a Flood-and-Drain system) cannot be spread out. In other words, the grow channels and the fish tank can be located a considerable distance apart. For instance, the fish tank could be located outside, and the grow channels can be kept in a greenhouse. In addition, by using a two pump system, elevation obstacles can even be eliminated. Following are two examples of how a N.F.T. could conceptually be laid out where the fish tank is not directly under the grow channels.

In the illustration below, the tank is located at a higher elevation than the grow channels. Gravity is utilized to convey feeds water (fish waste nutrients) grow channels, through the grow channels, and then to the sump or reservoir. The pump is located in a reservoir or sump, whereby the return water is forced uphill and back into the fish tank. As mentioned previously in Chapter 20 regarding filtrations options and systems, a supplemental water filter may also be necessary. A cover to shade the fish tank—or better yet a D.W.C. raft in the fish tank—would be ideal, so as to provide the fish shade and security. A D.W.C. raft would also provide increased plant production and further help clean the water.

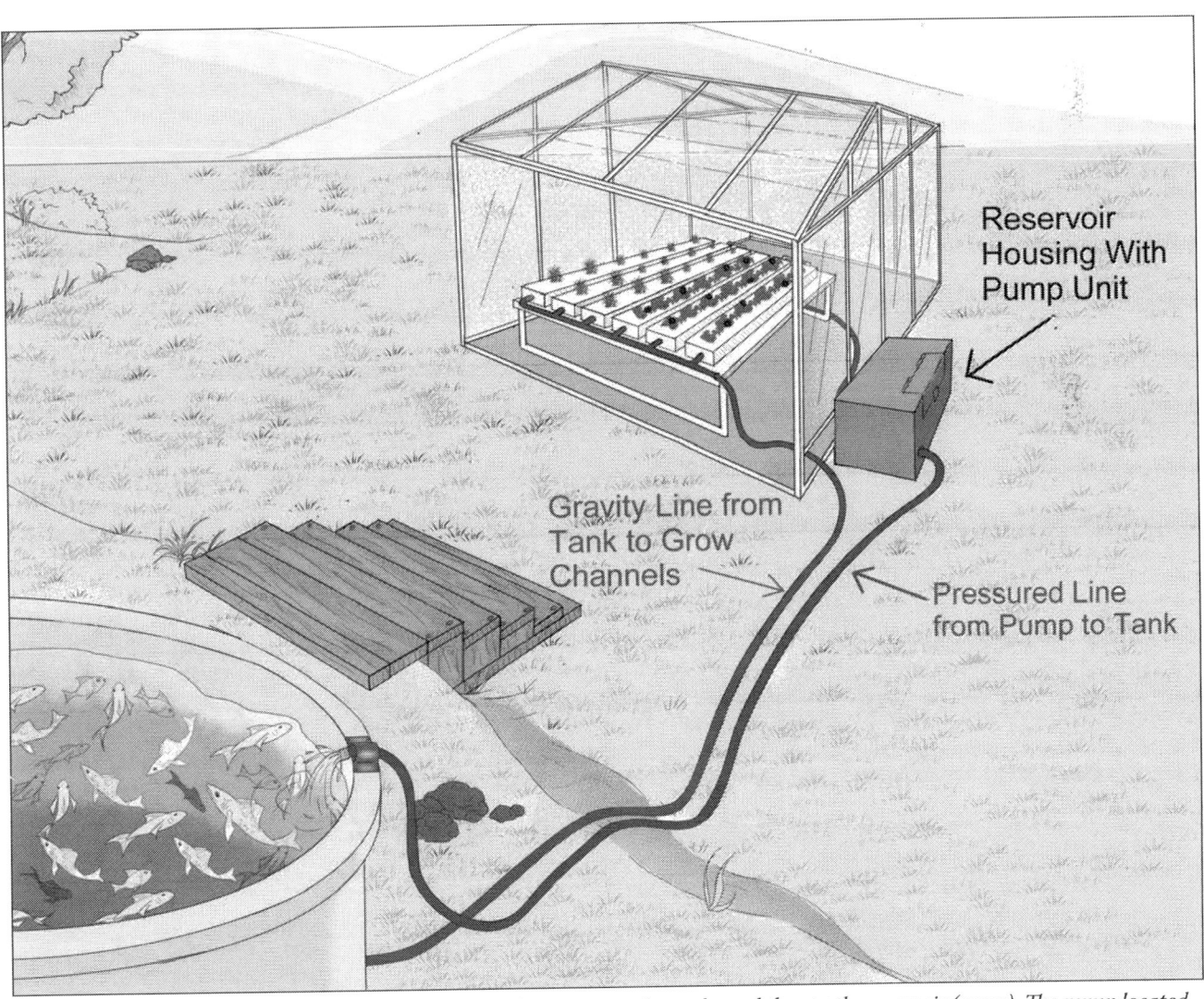

FIGURE 134. Water with fish waste nutrients gravity flows to grow channels, and then to the reservoir (sump). The pump located in the reservoir is used to return the water back into the fish tank.

PART IV: N.F.T. & D.W.C. DESIGN AND LAYOUT

Greenhouse with either NFT Grow Channels or Flood-and-Drain grow bed(s)

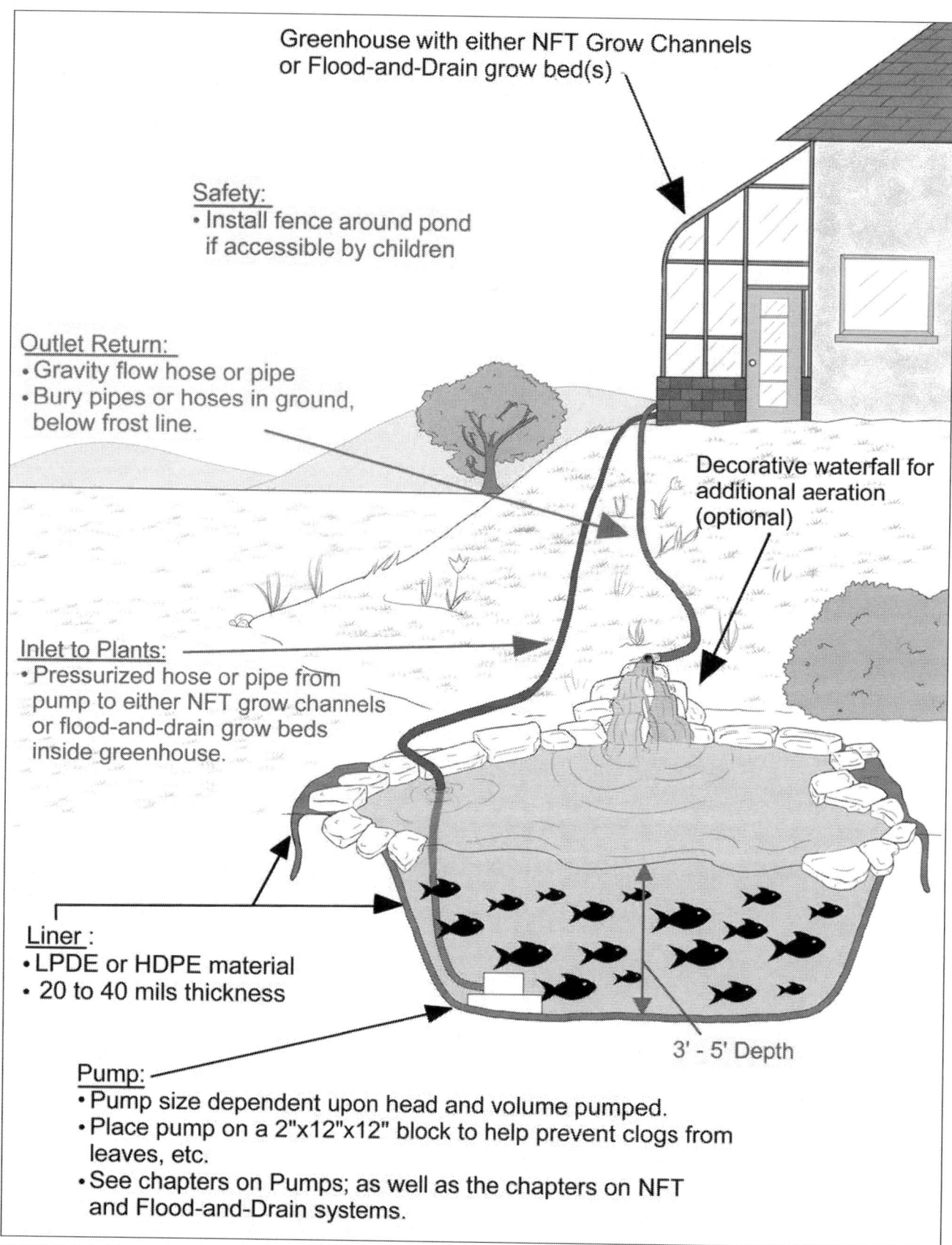

Safety:
• Install fence around pond if accessible by children

Outlet Return:
• Gravity flow hose or pipe
• Bury pipes or hoses in ground, below frost line.

Decorative waterfall for additional aeration (optional)

Inlet to Plants:
• Pressurized hose or pipe from pump to either NFT grow channels or flood-and-drain grow beds inside greenhouse.

Liner:
• LPDE or HDPE material
• 20 to 40 mils thickness

3' - 5' Depth

Pump:
• Pump size dependent upon head and volume pumped.
• Place pump on a 2"x12"x12" block to help prevent clogs from leaves, etc.
• See chapters on Pumps; as well as the chapters on NFT and Flood-and-Drain systems.

FIGURE 135. The water is being pumped upward from the fish tank to N.F.T. grow channels and/or Flood-and-Drain grow beds.

N.F.T. Fish Tank

Fish, fish care, water quality management, and pumps are thoroughly addressed later in this book. This section will focus on the fundamentals associated with the fish tank in regards to its relationship with N.F.T. grow channels. As mentioned above, most N.F.T. systems — whether referring to the past or present and for hobby and commercial uses — are hydroponics operations. However, fish can be integrated into the system. It just takes some additional effort.

As mentioned in the early chapters of this book, the problem with hydroponics is that synthetic fertilizers with heavy metals and other containments are used to supply nutrients for the plants. These contaminants enter into our food supply, which definitely not a good thing. Furthermore, water chemistry must be closely monitored, and after awhile, the water solution needs to be replaced. Fish will allow the system to operate more organically.

Unlike a Flood-and Drain system, the size (volume) of the fish tank to plant grow space is not that critical. However, grow channels do not filter the water as efficiency as a flood-and-drains system. Therefore, a supplemental water filtration system is typically needed. Some N.F.T. operators make their own bio-filter. However, the easiest approach is to purchase one or more filtration units. The quality of filtration units is dependent upon the size and operational characteristics of your system (i.e. numbers/size of fish, grow channels/plant density, size of fish tank, etc.).

The science of how many filtration units are needed is basically a trial and error process. Assuming that you have at least one 250-gallon tank of fish within the normal recommended stocking density range (stocking density is addressed in Chapter 11), then the following type of filtration unit would prove sufficient:

Please refer to the Chapter 20 'Filtration (Mechanical, Bio-filtration, Natural)' of this book for information on how to make a low-budget bio-filter in this book, rather than purchase a ready-made filtration unit.

Fluval FX6 High Performance Canister Filter

FIGURE 136.

Product Overview:
- Super-capacity canister filter for aquariums up to 400 gallons
- Enhanced filter performance for superior aquarium water quality
- Includes all essential filtering media to streamline aquarium filtration

High performance canister filter boasts enhanced performance and setup ease. Fluval FX6 High Performance Canister Filter makes aquarium fishkeeping easier and more convenient. A 1.5-gallon media capacity filter is powered by an energy-efficient pump with a 925-gph pump output that draws just 43 watts. The result is cleaner and healthier aquarium water while reducing operating costs. Fluval FX6 Canister Filter comes with all the essential filtering media to provide complete 3-stage filtration.

Multistage filter features removable media baskets precision engineered to eliminate water by-pass for efficient filtration. Each media basket is lined with a mechanical foam insert for effective mechanical pre-filtering. Instant-release T-handles let you lift and separate the baskets quickly and easily, making routine maintenance simpler. For freshwater or marine aquariums up to 400 gallons.

DIMENSIONS	PUMP GPH	MECHANICAL AREA	FILTER GPH	HEAD HEIGHT	WATTS
17" dia x 21" high	925 gph	325.5 in^2	563 gph	10.8 ft	43W

The Fluval FX6 High Performance Canister Filter mentioned on the previous page has very good reviews. At the time of this writing, it is being sold by Amazon, Petco, Big Als Pets, Pet Mountain, Marine Depot, and other retailers for around $300 dollars. The author is not providing a recommendation to any of these merchants, just providing them as options. Furthermore, there are other similar product options which will also work well. Tank volumes in excess of 400 gallons will need multiple Fulval FX6 units.

N.F.T. Design and Set-Up Instructions

A N.F.T. system has many advantages for a commercial hydroponic system, but as mentioned at the beginning of this chapter, the author is of the opinion that it is not the best option for aquaponics. However, for those desiring to give it a try, following are some options as to how to set-up a N.F.T. aquaponic system.

Option #1:

Purchase a readymade hydroponics system that has a plant grow area which best suits your preferences (your preferred size). Replace the hydroponic reservoir tank with a fish tank. Add a Fluval FX6 water filtration unit (or equal) to the fish tank. Add an air pump to the fish tank if needed.

Fish tanks, pumps, and water quality are addressed later in this book. A complete N.F.T. hydroponic system of varying sizes can be purchased via:
- American Hydroponics Inc.
  1-800-458-6543
  www.americanhydroponics.com
  (612 N.F.T. Herb and Lettuce System)
- CropKing
  1-330-302-4203
  www.cropking.com

Option #2:

Purchase all the materials and components needed, and then build the system yourself using the information presented in this chapter. This approach will allow you to have a competed system at approximately 40 percent less than the cost of buying a complete N.F.T. hydroponic system.

Materials and components to build your own system can be purchased via the two above referenced vendors, from FarmTek (1-800-327-6835, or www.FarmTek), as well as most hydroponic vendors.

Do-It-Yourself N.F.T. System Design

Using a DIY approach, below are design details for a Nutrient Film Technique aquaponic system. Obviously, a N.F.T. system can be created in a wide variety of sizes and configurations. The example system described below is sized for a typical residential lot with the hope that it will suffice for the majority of readers. It is assumed that the reader has some basic 'handyman' skills, or can easily acquire the knowledge and resources to build this system with the design instructions provided. If a larger system is desired, then double everything stated in this example.

1. **Build a stand** to support the grow channels. The stand would support a six (6) grow channels, 12-ft in length purchased from FarmTek. The stand can be made out of whatever type of material you prefer, including metal, PVC piping, or lumber. A fixed slope stand would be 32" on the low end and 36" on the high end.

   Another approach is to use sawhorses as the stand, and fasten boards to them to ensure that the grow channels have a slope of approximately 1:35 (1-inch rise for every 35-nches in horizontal length).

CHAPTER 23: NUTRIENT FILM TECHNIQUE

FIGURE 137. N.F.T. Grow Channels.

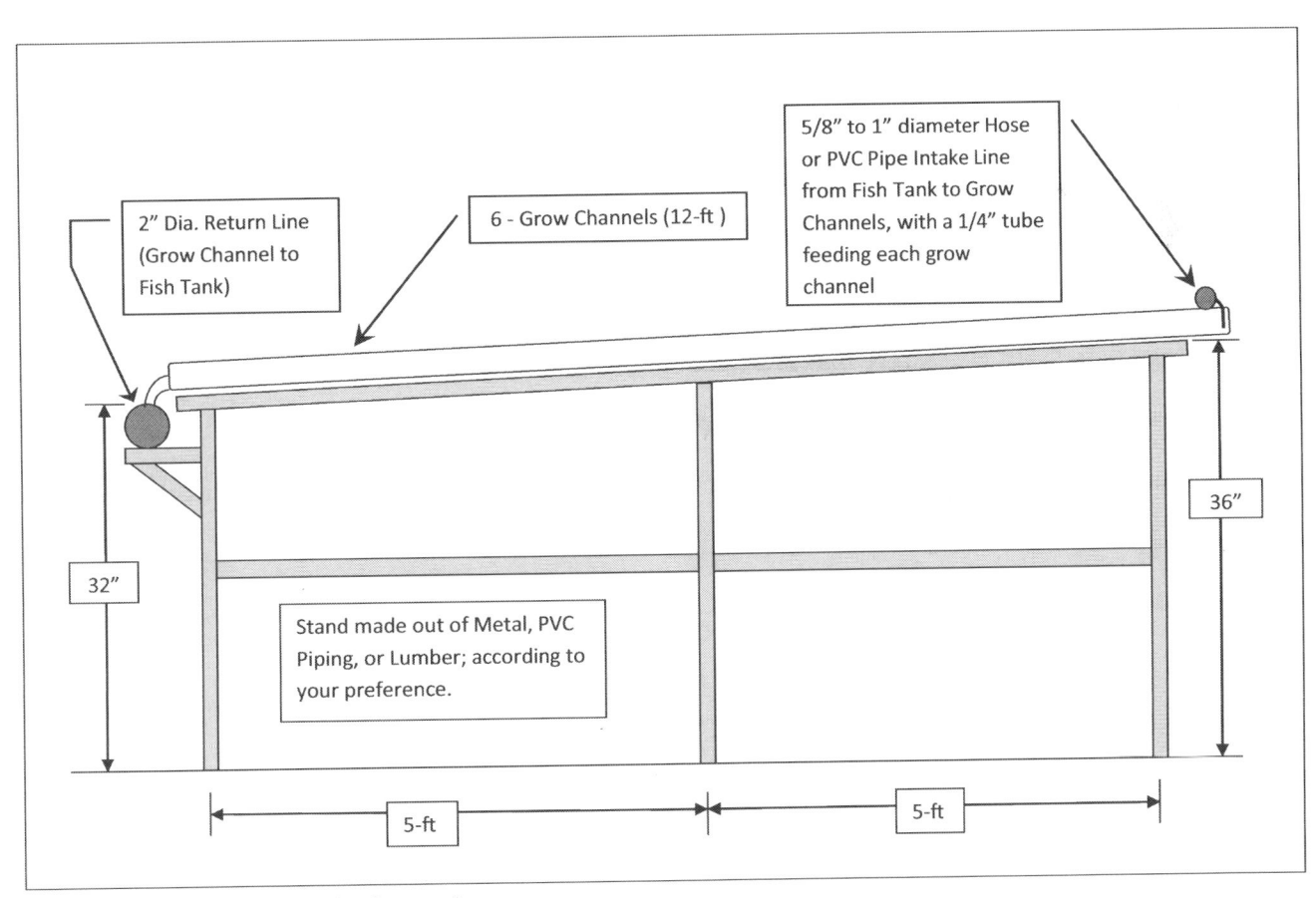

FIGURE 138. N.F.T. Construction Instructions.

2. **Construct a Fish Tank** from a used or reconditioned 275 or 330 gallon food grade IBC tote. Make sure that you can achieve gravity flow from the N.F.T. grow channel outlet collection point to the fish tank. In some instances, the fish tank may need to be placed somewhat below grade, or the grow channels raised in order to achieve gravity flow from the grow channels to the fish tank.

   Different fish tank options are addressed in 'Chapter 15: Fish Tanks'. Any number of those options could be used. However, for the construction of the system used in this example, an IBC tote is being specified. IBC totes are a common item and can be acquired at a reasonable cost. A used or recondition IBC tote (food grade), is sufficient and will cost less than a new one.

   If using an IBC tote, it may be necessary to install approximately 25 percent of it below a finished floor grade so that the N.F.T. grow trays can gravity feed back to the tote. If this is not possible, then a wider tank that is not as tall may be a better option than an IBC tote.

FIGURE 139. IBC tote.

3. **Pump Selection** — As stated earlier in this chapter, the recommended flow rate for a N.F.T. system is typically between 1/4 to 1/2 gallons per minute (1 to 2 liter per minute) for each grow channel. For a N.F.T. system described in this example, with six grow channels, a 2-gallon per minute pump (2 gpm) is needed. Such a pump can be acquired through many online retailers, at home improvements stores, etc. As of January 2016, the prices ranged from $40 to $160 (USA currency).

4. **Install Plumbing** — 'Chapter 14: Plumbing' addresses this subject in more detail. For now, know that most 2 gpm pumps are equipped with a ½-inch outlet. Therefore, the best approach is to use a standard black 5/8 or 3/4-inch irrigation tube sold at home improvement and irrigation supply stores. Avoid using a 1/2-inch diameter tube so as to reduce friction head loss. Depending upon the pump you select, a ½-inch to 5/8-inch or ½-inch to ¾-inch reducer fitting may be necessary to make the connection at the pump. Install the 5/8 or 3/4-inch tubing from the pump to head of the grow channels. Install one 1/4-inch black polyvinyl tube from the 5/8-inch tube into each one to the grow channels.

   Use a garden hose, PVC piping, or irrigation tubing for the drain line from the grow channels to the fish tank.

5. **Water Filter** — Install the Fluval FX6 High Performance Canister Water Filter described earlier in this chapter, an equal product, or a biofilter.

6. **Air Pump** — Depending upon the density of the fish in the tank, an air pump may be required. Fish density, water oxygen levels, and monitoring such are addressed in more detail later in this book. For now, know that dissolved oxygen (DO) levels need to remain above 5 ppm.

CHAPTER 23: NUTRIENT FILM TECHNIQUE

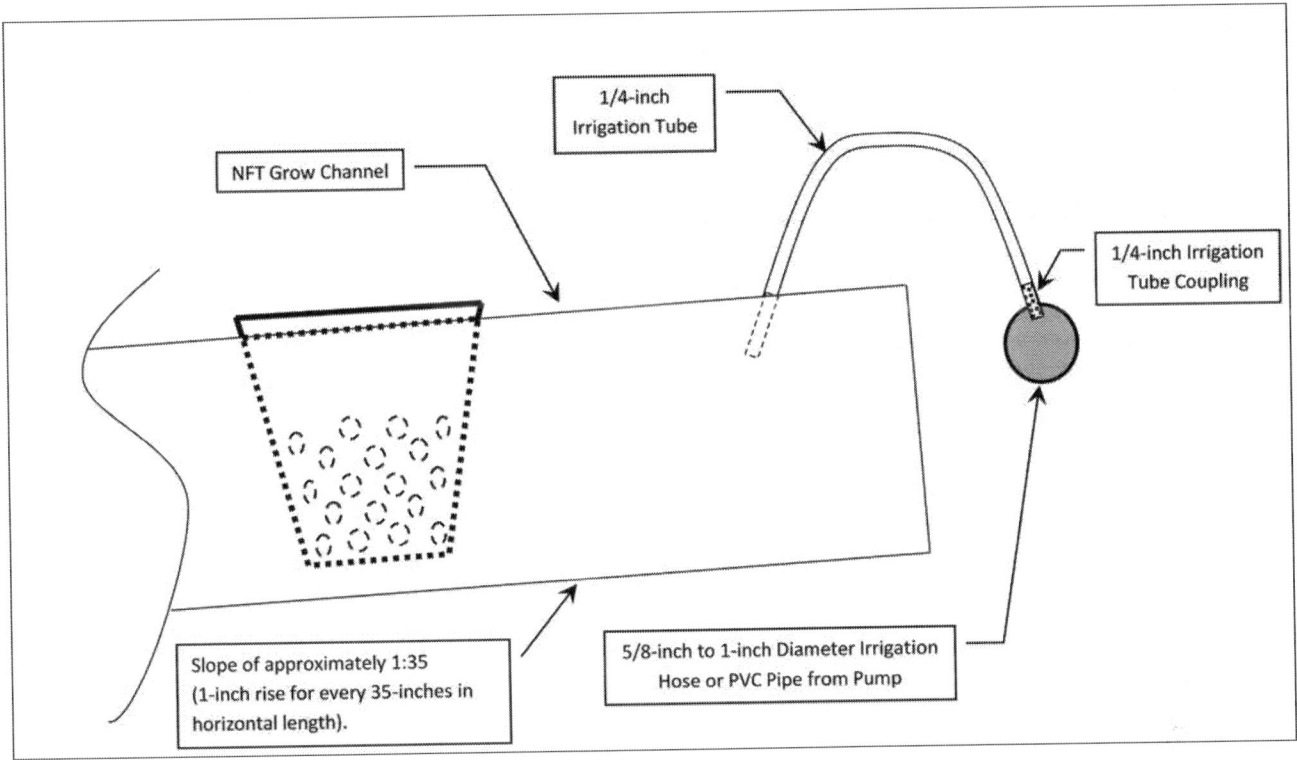

FIGURE 140. N.F.T. channel intake and slope instructions.

Air pumps can be acquired online through pet stores, Amazon, aquaculture vendors, or your local pet or aquarium shop. The cost varies depending upon unit capacity, but is still very affordable. The primary things to look for is: (1) a low operating noise level (most are quiet, but can be noisy), and (2) the ability to provide enough oxygen for the size of your tank (read the unit specs to ensure it can handle your tank size).

The air pump selected in this example is a Tetra Whisper Air Pump AP300 (see figure 141), and cost approximately $60. It is for tanks up to 300-gallons. Even if you are using a 330-gallon tote, this air pump will suffice as you will be getting oxygenated water from the grow channel return.

Tetra Whisper Air Pump AP300

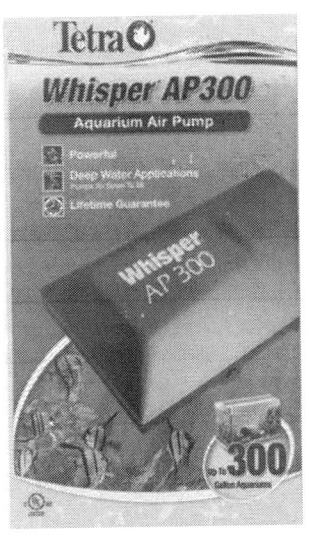

FIGURE 141.

Model DW96-2
- Great for all tanks up to 300 US gallons
- 115 Volts
- 60Hz
- 7.5 Watts
- 74" electrical cord

243

7. **Add plants.** Refer to Chapter 10 for information on plants and plant selection.

   Small plants may need to be watered until the roots are able to reach the water solution film at the bottom of the grow channel.

   For best results, install plants in net pots filled with round pebbles (1/2 to 2/4-inch diameter) or Hydroton. **Ideally plants will be at least 2-inches tall when installed in the net pots.**

8. **Add fish.** Refer to Chapter 11 for information on fish, stocking density, and fish selection.

9. **Maintain, Harvest, and Enjoy!**
   Refer to Chapter 31 for information on maintaining water quality.

Important !!!

For an N.F.T system, it is prudent to have a back-up pump, water filter, and air pump on hand. A back-up power supply (i.e. a generator, inverter box for a car battery, solar, etc.) is also a good idea. These items will prevent you from losing your fish and plants should there be a mechanical failure or lengthy power outage.

CHAPTER 24

Deep Water Culture / Raft System

Deep Water Culture / Raft System

Deep Water Culture (D.W.C.), also referred to as a 'raft system', is done by floating a foam raft (typically a polystyrene board) on top of the fish tank's water surface while allowing the roots to hang down into the water. However, another method is to grow the fish in a tank and pump the water through a filtration system, and then into long channels, or a reservoir, where floating rafts filled with plants float on the water surface and extract the nutrients. Water flows continuously from the fish tank, through filtration components, through the raft tank where the plants are grown, and then back to the fish tank. This method is the one most commonly practiced in large-scale commercial operations, but can also be done on a smaller scale. As a matter of fact, D.W.C. systems come in all sizes, from a 20-gallon aquarium tank to over one acre in size.

The beneficial bacteria live in the raft tank and throughout the system. One of the greatest benefits of D.W.C. is that the extra volume of water in the raft tank provides a buffer for the fish, reducing stress and potential water quality problems.

In a commercial system, raft tanks can cover large areas. In temperate zones (not tropical or subtropical) the system is typically installed within a greenhouse. Plant seedlings are transplanted on one end of the raft tank. The rafts are pushed forward on the surface of the water over time and then the mature plants are

FIGURE 142. Deep Water Culture (D.W.C.) / Raft System.

harvested at the other end of the raft. Once a raft is harvested, it can be replanted with seedlings and set into place on the opposite end. This optimizes floor space, which is especially important in a commercial greenhouse setting.

Pros

- Great for commercial setups.
- Very high-yield of both fish and plant crops.
- Typical for a small system — 100 lbs of fish, 925 heads of lettuce (annual yield).
- Typical for a large system — 7,500 lbs of fish, 194,400 heads of lettuce (annual yield).
- On some systems, the fish tank is mostly covered (shaded) with the rafts, so algae growth is inhibited.

Cons

- Requires more extensive filtration methods.
- Plant diversity is limited. Usually grows a specific, low profile crop like lettuce or basil. It is difficult to grow tall plants because roots cannot anchor and vertical support braces are generally not practical. Typically D.W.C. has to be done inside a greenhouse (although they can be located outside in temperate climates).
- On some systems, the fish tank is covered with the rafts, so checking on the fish requires additional effort.

Raft System / Deep Water Culture (D.W.C.) System Design Layout

Deep water culture (D.W.C.) is a method of plant production by means of suspending the plant roots in a nutrient-rich, oxygenated water solution. In hydroponics, nutrients are added to the water. In aquaponics, the fish tank water supplies the nutrients.

An air pump moves air from outside the tank to an air distribution device. In smaller systems, this is typically done by the use of an air stone. In larger systems, an air hose, tube, or small PVC pipe may be used. The hose, tube, or pipe will have many small holes that allows the air to be released into the tank. The air oxygenates the nutrient rich water. When sufficiently oxygenated, the plant roots can remain submerged indefinitely.

Roots readily absorb large quantities of oxygen and rich nutrients with deep water culture. Furthermore, the plants do not have to expend energy forcing roots to grow through soil. This leads to rapid growth throughout the life of the plant. (See figure 143.)

Without a lot of effort, plant diversity is limited to smaller type plants, such as lettuce. It is difficult to grow tall plants as the roots are not able to provide any structural support. Even if a taller type plant can be braced laterally, the increasing weight of the plant as it grows causes it to sink deeper into the water, which in turn will eventually result in stem rot. Therefore, taller type plants, such as tomatoes, would need to be suspended from above. Quite frankly, most people find that trying to grow taller plants just takes more time and isn't worth the trouble. Plus, there is some cost involved in suspension and bracing materials. Therefore, plants with shorter gowth are about the only type that is grown in a Deep Water Culture system.

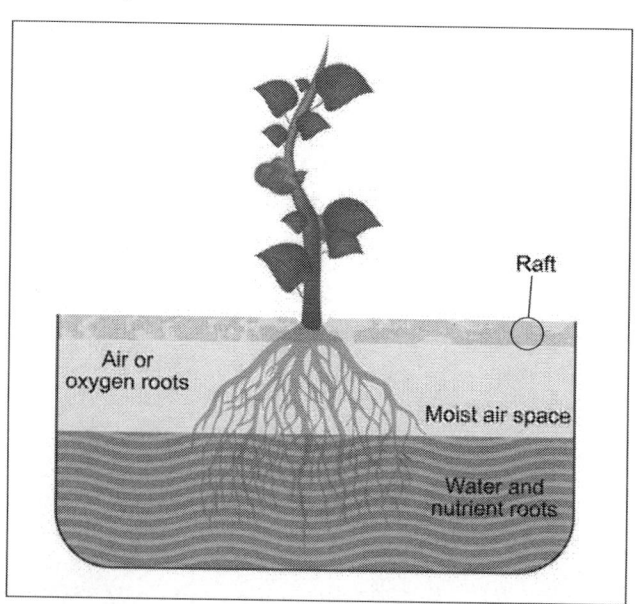

FIGURE 143.

CHAPTER 24: DEEP WATER CULTURE / RAFT SYSTEM

There are many different D.W.C. configurations and they also come in many sizes. The following pages show just a few of D.W.C. systems.

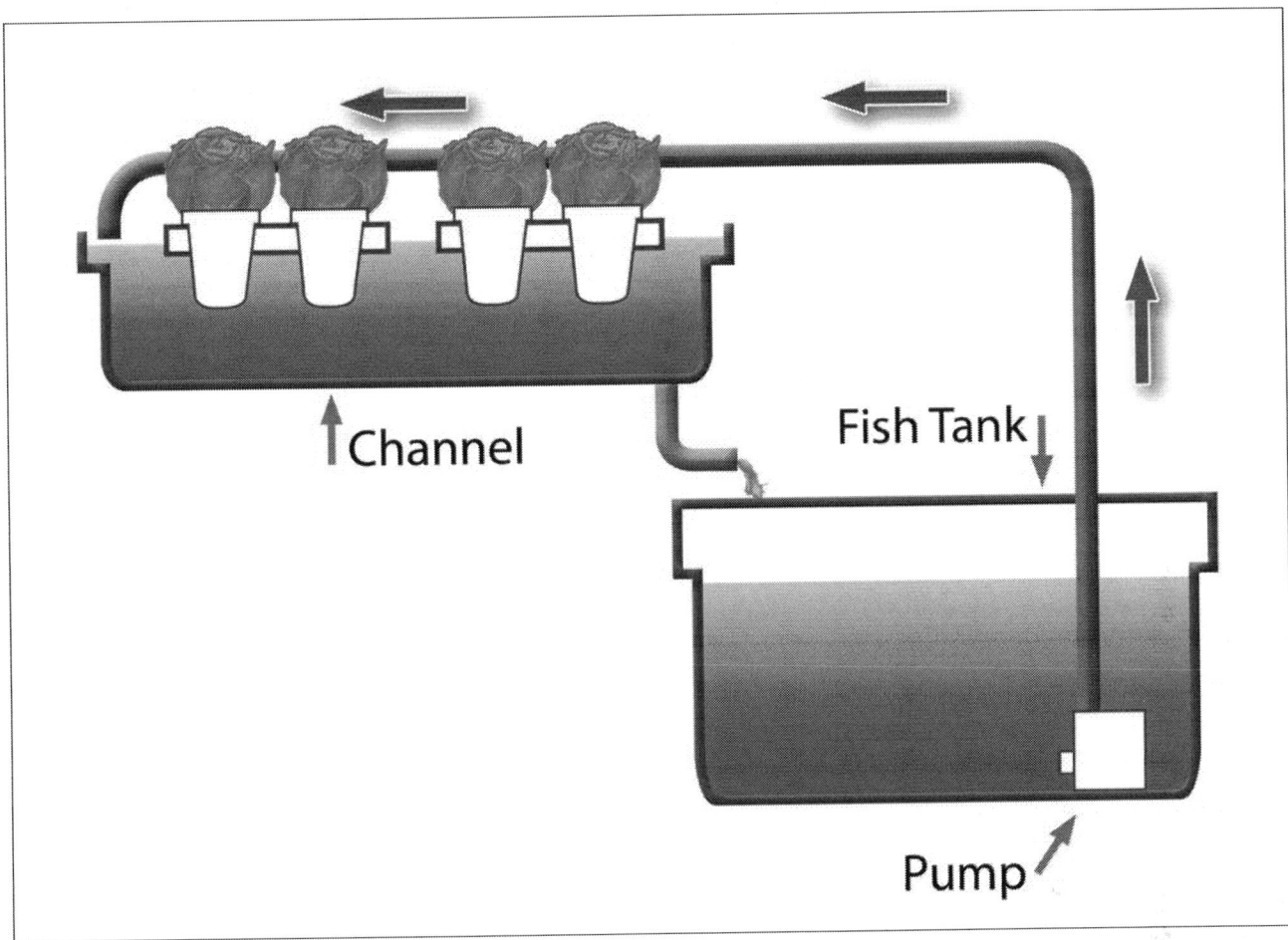

FIGURE 144. *Deep Water Culture (D.W.C.) / Raft System.*

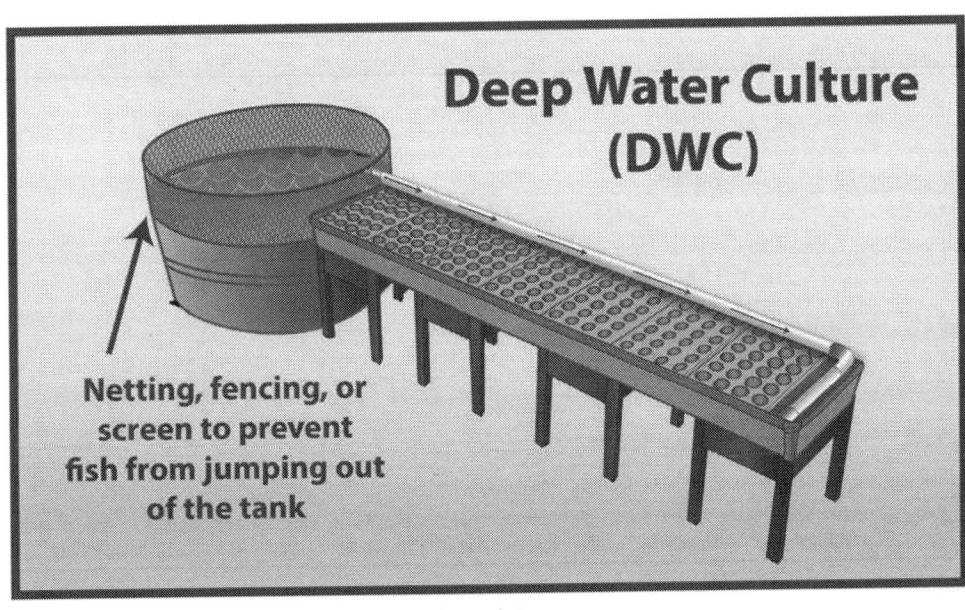

FIGURE 145. *Deep Water Culture (D.W.C.) / Raft System.*

247

PART IV: N.F.T. & D.W.C. DESIGN AND LAYOUT

FIGURE 146. Commercial Deep Water Culture (D.W.C.) / Raft System.

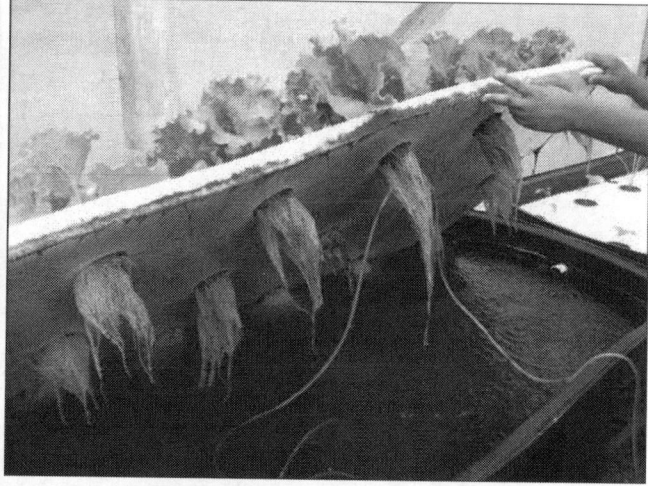

FIGURE 147. Commercial Deep Water Culture (D.W.C.) / Raft System.

FIGURE 148. Commercial Deep Water Culture (D.W.C.) / Raft System.

FIGURE 149. Raft / D.W.C. System Results

FIGURE 150. Raft System in a Pond

FIGURE 152. Roots of curly kale (Brassica sp.) growing in a deep water culture unit

FIGURE 151. Raft System at University of the Virgin Islands (UVI)

248

D.W.C. Grow Canals

D.W.C. grow canals, construction and planting Canals can be of variable lengths, from a few feet to hundreds of feet (one to tens of meters). Unlike N.F.T. grow channels, the length of D.W.C. grow canals is typically not an issue because the large volume of water enables adequate nutrient supply. More importantly, optimal plant nutrition in very long canals should always allow for adequate water inflow and re-oxygenation to ensure that nutrients are not depleted and that roots can breathe. As far as the width is concerned, it is generally recommended to be the standard width of a sheet of polystyrene, but it can be wider. However, narrower and longer canals enable a higher water speed that can beneficially hit the roots with larger flows of nutrients. Furthermore, it is advantageous to have the plants within arm's reach so as to be accessible to the operator. The recommended minimum depth for a grow canal is 1-foot (30 cm) to allow for adequate plant root space. Similar to fish tanks, canals can be made out of any strong, inert material that can hold water. For small-scale units, popular materials include fabricated IBC plastic containers or fiberglass. Much larger canals can be constructed using wood lengths or concrete blocks lined with food-grade waterproof sheeting (liner). If using concrete, make sure it is sealed with non-toxic, waterproof sealer to avoid potential toxic minerals leaching from the concrete into the system water.

The retention time for each canal in a unit is 1–4 hours, regardless of the actual canal size. This allows for adequate replenishment of nutrients in each canal, although the volume of water and the amount of nutrients in the deep canals is sufficient to nourish the plants over longer periods. Plant growth will definitely benefit from faster flow rates and turbulent water because roots will be hit by many more ions, whereas slower flows and almost stagnant water would have a negative impact on plant growth. If performing D.W.C. on pond or above a large tank, then using an air pump to assist in providing water movement of nutrients and oxygen to the plans is recommended.

Aeration for D.W.C. units is vital. In densely planted long canals, the oxygen demand for plants can cause DO levels to plummet below the minimum. Any decomposing solid waste present in the canal would exacerbate this problem, further diminishing DO. Thus, aeration is required in these situations. The simplest method is to place several small aeration outlets in and periodically along the canals.

Do not add any fish into the canals that could eat the plant roots (i.e. herbivorous fish such as tilapia and carp). However, some small carnivorous fish species, such as guppies, mollies, or mosquito fish, can be used successfully to manage mosquito larvae, which can become a significant nuisance to the operator, his/her family, and neighbors in some areas.

The polystyrene sheets should have a certain number of holes drilled to fit the net cups (or sponge cubes) used for supporting each plant. The amount and location of the holes is dictated by the vegetable type and the distance desired between the plants, where smaller plants can be spaced more closely.

Seedlings can be started in soil blocks or a soil-less medium. Once these seedlings are large enough to handle, they can be transferred into net cups and planted into the D.W.C. unit (see figure 152 below).

FIGURE 153. Transplanting to a Deep Water Culture (D.W.C.) / Raft System.

The remaining space in the net cup should be filled with hydroponic media, such as volcanic gravel, rock wool, or light expanded clay aggregate (LECA), to support the seedling (more on this subject in the sections below). It is also possible to simply plant a seed straight into the net cups on top of the media. This method is sometimes recommended if vegetable seeds are accessible because it avoids the transplant shock during replanting. When harvesting, be sure to remove the whole plant, including roots and dead leaves, from the canal. After harvest, the rafts should be cleaned but not left to dry so as to avoid killing the nitrifying bacteria on the submerged surface of the raft. Large scale units should clean the rafts with water to remove dirt and plant residues, and immediately repositioned in the canals to avoid any stress to the nitrifying bacteria.

D.W.C. Plant Containers

Similar to a N.F.T. system, D.W.C. growers normally use net pots in which to grow their plants. Net pots allow for very fast flooding and filling with the rich nutrient solutions (aquaponics-fish tank waste). They are available in many sizes. The best sizes for D.W.C. are the 2" and 3" sizes.

Net containers can be purchased at most hydroponic retail stores and online. Please refer to the reference section of this book for a list of vendors.

D.W.C. Grow Medium

Grow medium is used to help support the plants. Mediums used are expanded clay pellets (also referred to Hydroton— a company's brand name of the product) lava rock, stones, rock wool and basically any other growing medium that will not fall through the holes. Smaller size particle growing mediums, such as sand and perlite, are not acceptable as they will fall through the container holes. Coco peat can be used if you only implement the longer strands of the coco peat and not the smaller particles that would fall through of the holes.

Expanded clay pellets are relatively smooth and round, which is better for plant roots. Please note that roots can easily be damaged if pulling the container out of the hole. Therefore, the only time a container or plant should be removed from a hole in D.W.C. is during harvest.

D.W.C. Raft

Rafts can be purchased or easily made at a low cost. Rafts are typically about 2-inches thick and can be cut to any size under 4-ft x 8-ft.

A company called Beaver Plastics sells the following raft, named 'Lettuce Raft', through several online distributors, farm supply companies, and retail hydroponic stores.

Another option is to purchase a hydroponic D.W.C. unit, such as the figure 155 from Farmtek, and convert it to aquaponics by replacing the reservoir with a larger fish tank, adding a water filter, and air pump.

DIY Aquaponic Rafts

DIY Aquaponic Rafts are generally made from construction grade, 2-inch thick, 4-ft x 8-ft polystyrene foam sheets or blue-board purchased at a local home improvement store or lumber yard (see figure 154).

Since most crops grown on rafts have to be of the smaller variety, such as lettuce (NOT tall and top heavy plants like a tomato plant), it is recommended that hole spacing be arranged to maximize the available area on the raft. A layout that has alternating holes and rows (as shown in the image below) provides optimal production of your product. Rows and holes spaced in parallel (side-by-side) are also common. However, an alternating pattern allows for a denser crop (more crop product—larger harvest).

Using lettuce as an example for determining hole spacing, determine how far apart to space your crops. Go by the adult size of the plant variety you plan to grow. Generally, lettuce is spaced 6-inches apart in rows that are spaced about 12-inches apart.

CHAPTER 24: DEEP WATER CULTURE / RAFT SYSTEM

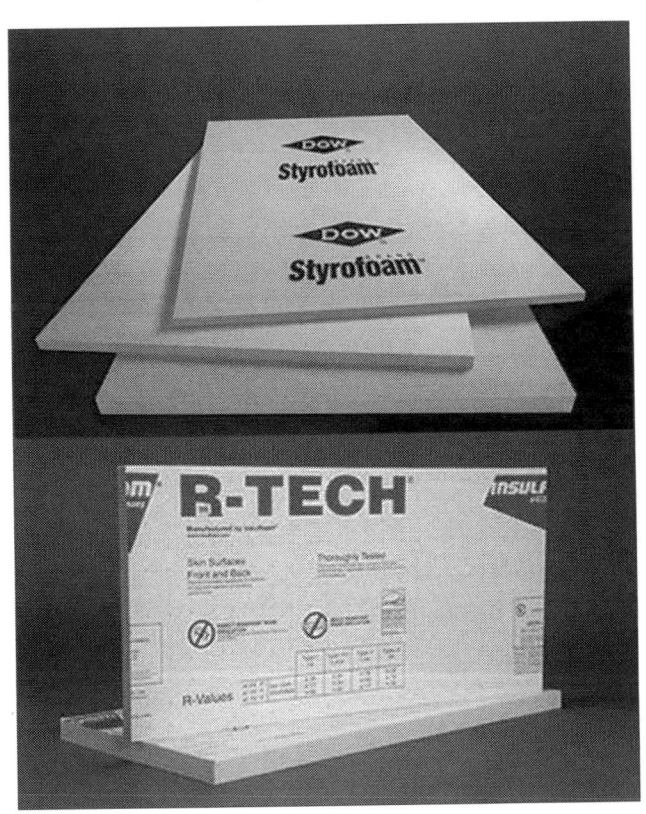

FIGURE 154. Raft material options.

If you know that you'll be harvesting leaves from the individual plants before they reach maturity, you can place the plants even closer together.

Therefore, for lettuce, the holes should be spaced 6-inches apart. If we use an alternating pattern, we can space the rows 6-inches apart as well. Again, growing a dense crop maximizes space and production.

Since a 3-inch size net pot appears to work best in D.W.C., a 3-inch diameter hole would be drilled every 6-inches on one row, with the next row 6-inches over, in an alternating pattern, similar to what is shown in the image on the following page (figure 156).

Float Station
FarmTek

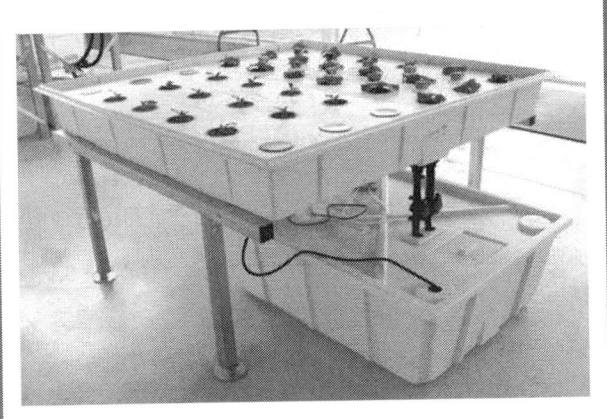

Can be used as a standard raft system to grow lettuce, swiss chard, pac choi, and more.

Constant recirculation provides constant mixing of solution to help prevent nutrient settlement, maximum aeration within the tank as to not disturb root zones, more stable water temperatures, and easy nutrient solution maintenance and testing.

Each system includes: efficient and potable water safe magnetic drive recirculation pump, two outlet air pumps with stones and tubing, 16-gauge, galvanized, square steel tubing frame with height adjustable feet for leveling, float tray, nutrient tank, bulkhead fittings, spacers, supply, and drain tubing with flow control valves.

ITEM #	DESCRIPTION	PRICE (USA, YR 2016
113503	Propagation Float Station 40 Gallons	$449.65 /EA
113504	Propagation Float Station 100 Gallons	$930.75 /EA

Farmtek; 1-800-327-6835; www.farmtek.com

FIGURE 155.

FIGURE 156. D.W.C. raft with net pots filled with Hydroton (clay pellets) and starter plants.

FIGURE 157. A saw tipped drill bit works well for drilling the holes. A 2-½ or 3-inch standard hole saw size works best. The average retail price is $13.00 (USA, year 2016)

Deep Water Culture Design Plans

As mentioned above, D.W.C. systems come in many different configurations and sizes. With the belief that most readers desire a system that will help feed their family, and can be constructed on a standard residential lot, the following instructions are provided.

Objective One — Consider Your Goals:

1. Produce just enough product for your family, or,
2. Produce enough for your family and extra for others (bartering) or,
3. Produce enough to have a side business.

Objective Two — Consider Your Resources:

1. How much space you have available?
2. How much space do you want to dedicate for a system?
3. How much time are you willing to dedicate to this venture?

The above will help you in determining the size of D.W.C. unit to install. Below, as well as in other places within this chapter, are some examples of the different types of D.W.C. systems and the various sizes for consideration.

The outdoor D.W.C. system in figure 158 also has an 'optional' pump so that water can be transferred to a media filled grow bed or a N.F.T. grow channel, if desired. Ideally, the media grow beds or N.F.T. grow channels would be located in a temperature controlled greenhouse. The return line is not shown on the below illustration. Depending upon geographical location, stocking density, and environmental/climate conditions, an air pump and/or additional filtration may be necessary to maintain suitable water quality for the fish.

DWC Filtration & Air Pump Considerations

As with NFT, depending upon fish stocking density a biofilter and/or an air pump may be required. Fish density, water oxygen levels, and monitoring such are addressed in more detail later in this book. For now, know that dissolved oxygen (DO) levels need to remain above 5 ppm.

An air pump may also be necessary if the tank does not have some circulation of water and/or oxygen to the plants is not adequate. Monitor plant heath. A plant that is over-watered will droop, just like an under watered plant. Leaves may also turn a yellow color or get brown patches and drop from the stem prematurely.

Multi-Use Earthen Fish Tank/Pond with Liner

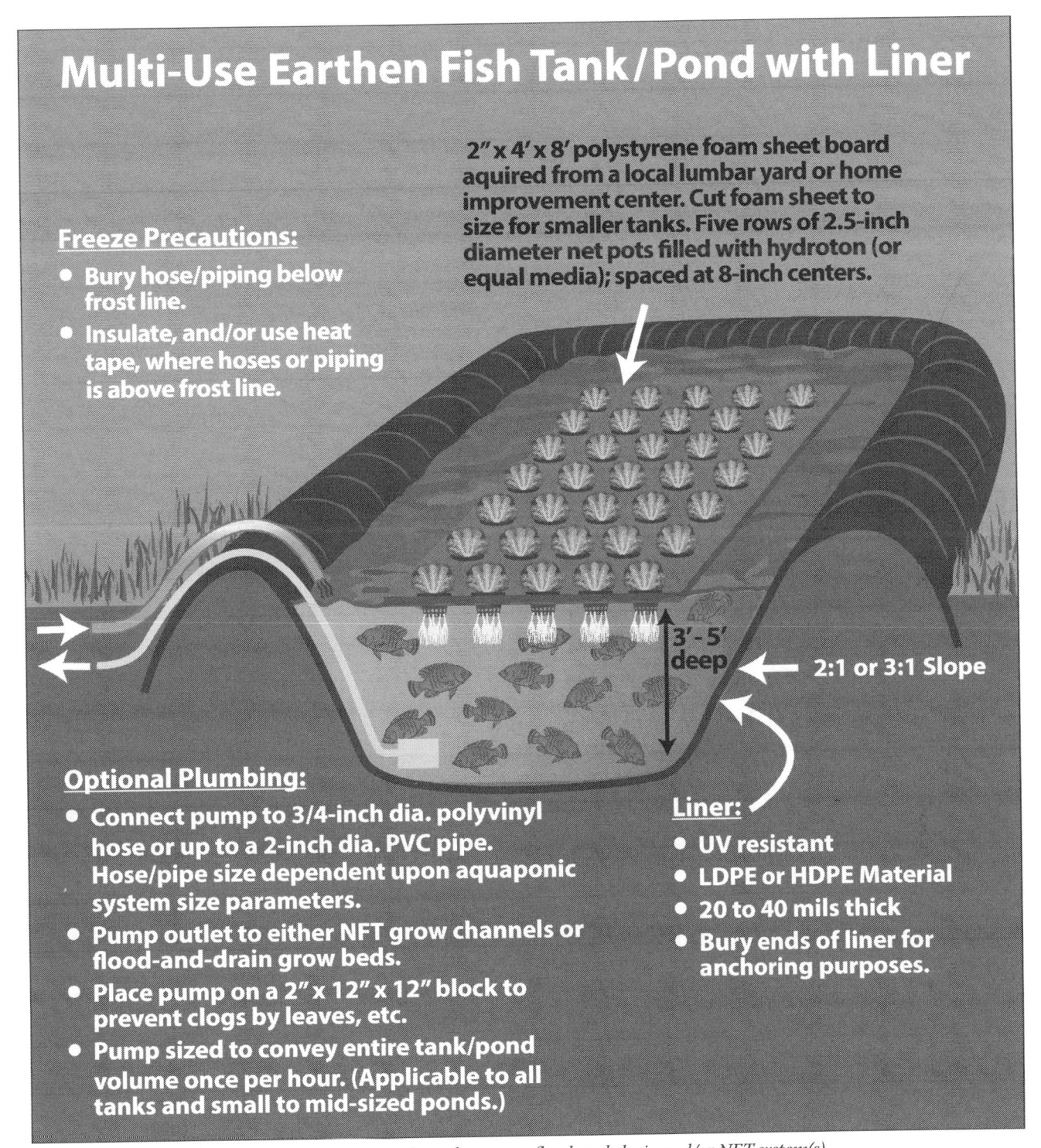

Freeze Precautions:
- Bury hose/piping below frost line.
- Insulate, and/or use heat tape, where hoses or piping is above frost line.

2" x 4' x 8' polystyrene foam sheet board aquired from a local lumbar yard or home improvement center. Cut foam sheet to size for smaller tanks. Five rows of 2.5-inch diameter net pots filled with hydroton (or equal media); spaced at 8-inch centers.

3'-5' deep

2:1 or 3:1 Slope

Optional Plumbing:
- Connect pump to 3/4-inch dia. polyvinyl hose or up to a 2-inch dia. PVC pipe. Hose/pipe size dependent upon aquaponic system size parameters.
- Pump outlet to either NFT grow channels or flood-and-drain grow beds.
- Place pump on a 2" x 12" x 12" block to prevent clogs by leaves, etc.
- Pump sized to convey entire tank/pond volume once per hour. (Applicable to all tanks and small to mid-sized ponds.)

Liner:
- UV resistant
- LDPE or HDPE Material
- 20 to 40 mils thick
- Bury ends of liner for anchoring purposes.

FIGURE 158. *Outdoor DWC with optional plumbing shown to a flood-and-drain and/or NFT system(s).*

Another type of D.W.C. system can be achieved by using IBC containers and homemade bio-filter. An example of such a set is shown in figures 159 and 160.

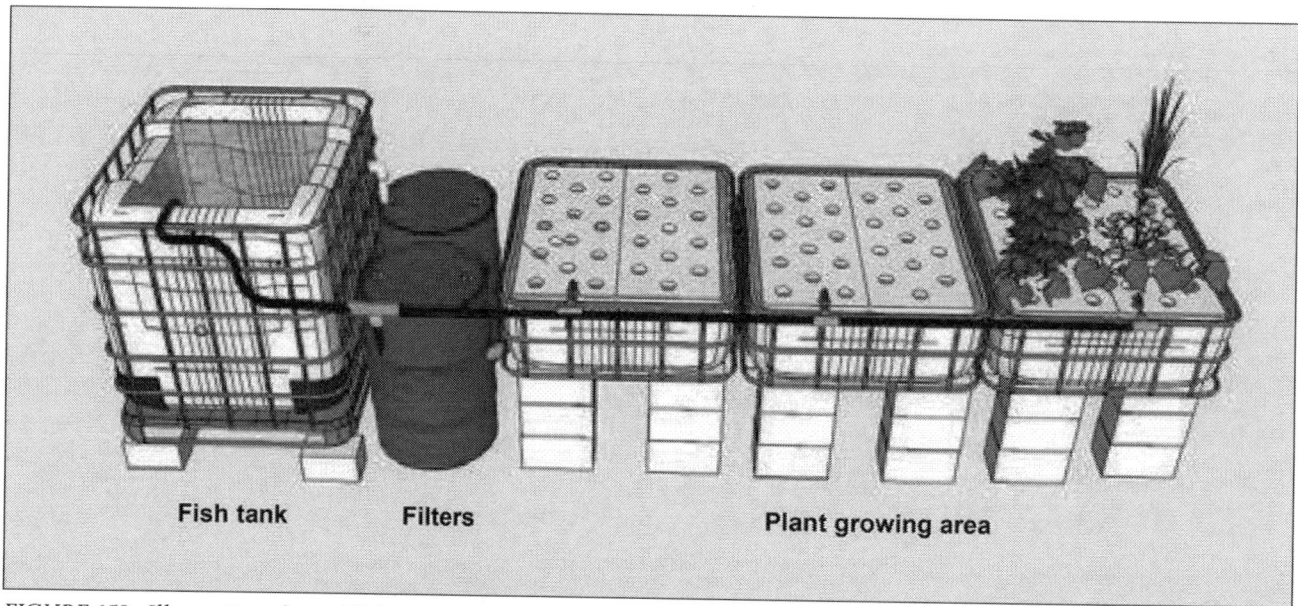

FIGURE 159. *Illustration of a small deep water culture unit*

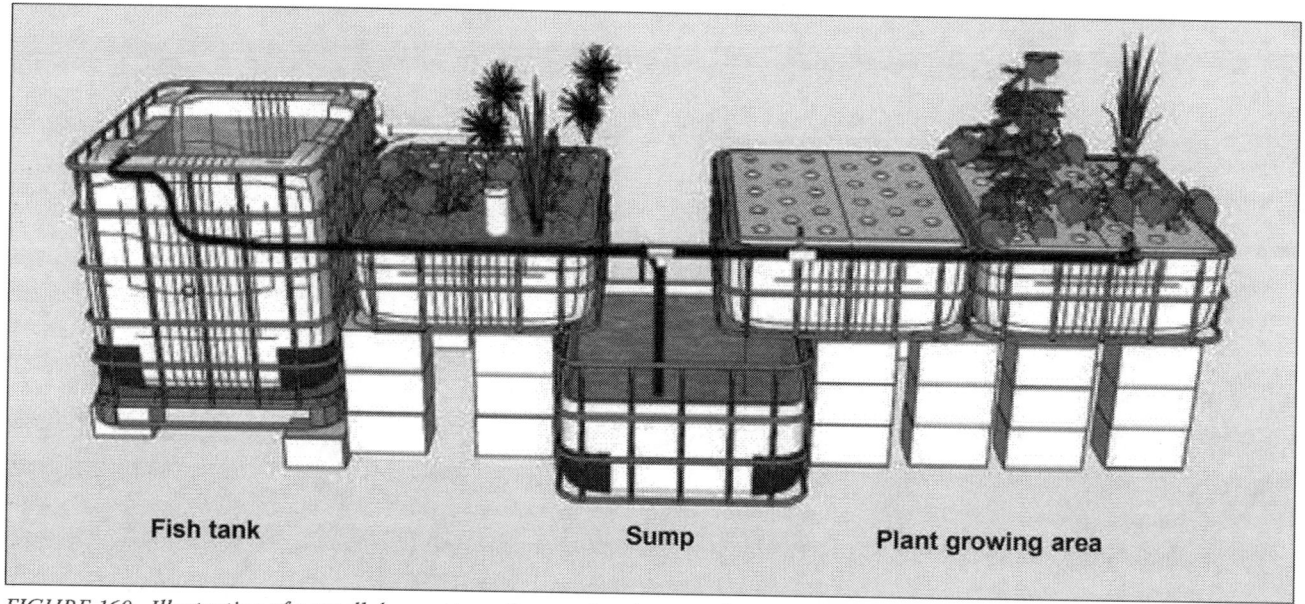

FIGURE 160. *Illustration of a small deep water culture unit using a media bed as filtration*

CHAPTER 24: DEEP WATER CULTURE / RAFT SYSTEM

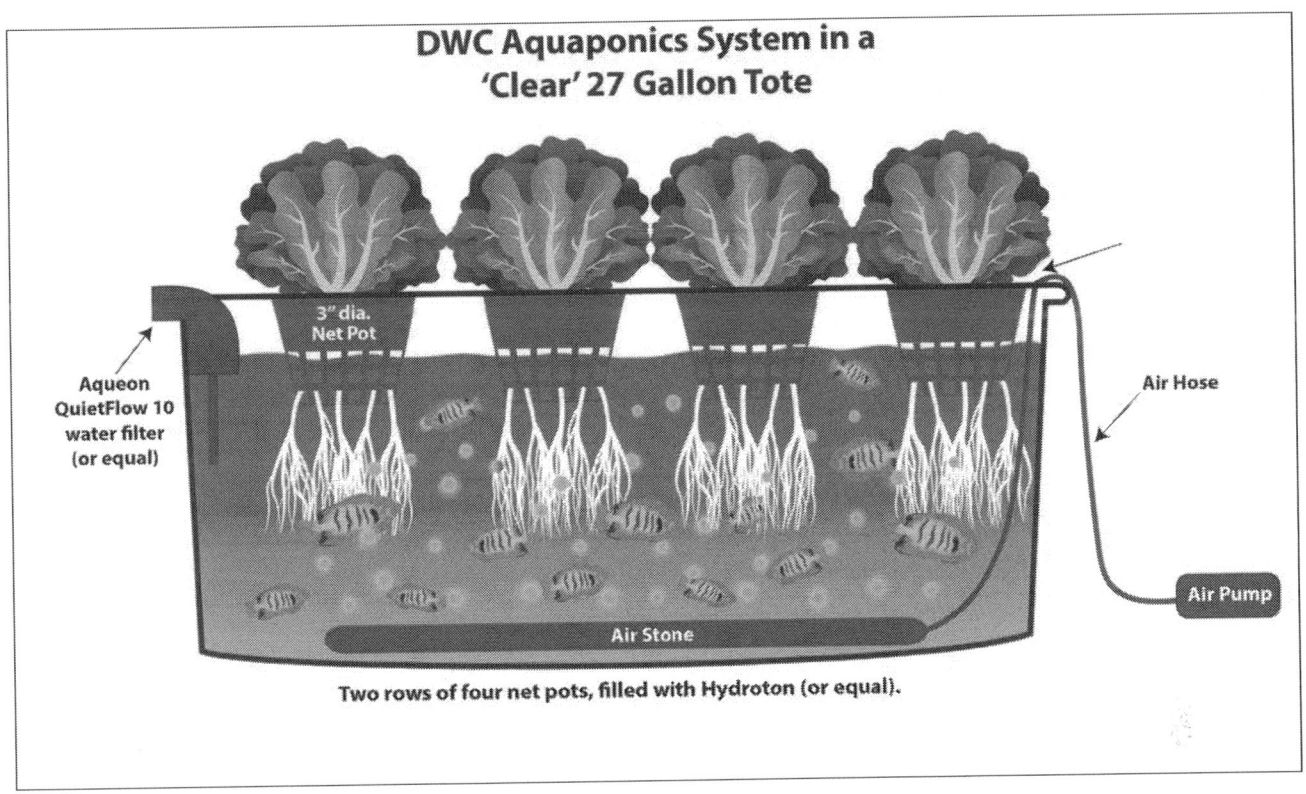

FIGURE 161. A 27-gallon clear tote aquarium DWC system

Special Case D.W.C.

Low fish density, no filters Aquaponic D.W.C. units can be designed that don't require external additional filtration (note illustration below). These units carry a very low stocking density of fish, approximately 2 to 3 lbs per square yard (1–1.5 kg of fish per m3) of fish tank, and then rely mainly on the plant root space and interior area of the canals as the surface area to house the nitrifying bacteria. Simple mesh screens capture the large solid waste, and the canals serve as settling tanks for fine waste. The advantage to this method is the reduction in initial economic investment and capital costs, while at the same time eliminating the need for additional filter containers and materials, which can be difficult and expensive to source in some locations. However, lower stocking densities will lead to lower fish production. At the same time, many aquaponic ventures make the vast majority of their profits on the plant yield rather than the fish production, essentially only using the fish as a source of nutrients. Often, this method requires nutrient supplementation to ensure plant growth. If considering this method, it is worth assessing the desired fish and plant production, and considering the relative costs and gains.

The main difference between the two designs (high fish stocking vs. low fish stocking) is that the low-density design does not use either of the external filtration containers, mechanical, or biological. Water flows by gravity from the fish tank straight into D.W.C. canals, passing through a very simple mesh screen. Water is then returned either to a sump and pumped back to the fish tanks, or directly to the fish tanks without a sump. Water in both the fish tanks and canals is aerated using an air pump. The fish waste is broken down by nitrifying and mineralizing bacteria living on the plant root surface and the canal walls.

Another option used with low stocking densities can include an internal bio-filter within the fish tank, consisting of a simple mesh bag of bio-filter material near an air stone. This can help to ensure adequate bio-filtration without adding external filtration. Finally, increasing the overall water volume without increasing the fish stocking density, basically using large fish tanks for few fish, can help to mitigate water quality issues by diluting wastes and ensuring adequate time for the farmer to respond to changes before the fish become stressed. Please note that this may dilute the available nutrients and hinder vegetable growth.

The lower fish density also means that the water flow rate can be lower. A smaller pump can be used, reducing the cost; ensure that at least half of the total fish tank volume is exchanged per hour.

The major advantage of using the D.W.C. low stocking density approach is that it is a simpler unit to manage. This system is easier to construct and has a lower start up cost. The fish are less stressed because they are grown in more spacious conditions. Overall, this technique can be very useful for initial projects with low capital, and for growing high-value fish such as ornamental fish or specialty crops such as medicinal herbs; the lower production is compensated with higher value.

However, a serious disadvantage is that these units are hard to scale-up. Fewer plants and fish are grown in a given area, so they are less intensive than some of the systems previously outlined. In order to produce a large amount of food, these systems would become prohibitively large. Essentially, filtration, be it from a bio-filter or grow bed, are what allow aquaponics to be so productive within a small area.

Furthermore, fish production cannot function independently from the hydroponic component; plants must be in the canals at all times. The plants keep the water clean for the fish. If it were ever necessary to harvest all the plants at once, which can occur during a disease outbreak, season changes, or major climate events, the reduced bio-filtration previously provided by the plants, would cause high ammonia and fish stress. On the other hand, with bio-filtration fish production can continue during crop reductions.

PART V

Flood-and-Drain System Design and Layout

CHAPTER 25

Flood-and-Drain System Instructions & Procedures

Flood-and-Drain System (Media-Filled Bed System)

In the Flood-and-Drain System (also referred to as a media-filled bed system and an ebb-and-flow system), plant grow beds are containers filled with various types of media through which the fish tank wastewater passes through. This style of system can be setup two different ways: 1) with a continuous flow of water through the grow bed(s), or 2) by flooding and draining the grow bed(s.

This chapter addresses the Flood-and-Drain System aquaponics from a Do-It-Yourself (DIY) perspective. The objective is to provide you with the information needed to be successful in developing a system that operates at optimal capacity. Although this chapter primarily refers to a Flood-and-Drain system many of the principles covered also apply to NFT and DWC/Raft systems as well. Additionally, even though the term flood-and-drain is used extensively throughout this chapter, most everything discussed applies to a continuous flow system as well. If there is a difference between the flood-and-drain system and a continuous flow system, the difference will be noted.

As discussed in the first several chapters of this book, aquaponics is a great way to grow your own organic produce, free of pesticide and chemicals. You can have fresh clean fish free of mercury, lead, and many other contaminates common to fish purchased at grocery stores and restaurants. Aquaponics can be done easily in your backyard, greenroom of your home, or in a small greenhouse.

Once established, the symbiotic relationship between the fish, plants and nitrifying bacteria does most of the work, practically self-regulating and requiring minimal maintenance. Aquaponics gardening saves tons of time and effort compared to the traditional soil-based gardening.

There are also many other benefits in addition to the health and economic advantages. With aquaponics the joys of gardening and rearing fish are combined. Moreover, you will find aquaponics also enriches your social and family life and is also a positive way to stay active. It is an edifying endeavor to share with children and can help you bond closer with your significant other.

After your system has been set up and running efficiently, all you need is just a few minutes a day to tend to the plants and fish. You can even integrate many automatic controls to reduce your time further. Automated temperature and water quality control devices, as well as grow lights and dispersion of fish feed set via a timer, can further eliminate the time needed to manage the system. Although, like most everything else, the more you put into it, the more you will get out of it.

Beginner Basics

The flood and drain configuration is the simplest and most reliable aquaponics design making it the best suited for beginners. It requires minimal care and can be constructed using a wide range of containers, tanks, materials and sizes. Media also provides better plant support than the other types of aquaponic systems and is more closely related to traditional soil gardening. The cost of building the system can be relatively inexpensive, because there are fewer components and so many material options.

Flood-and-Drain System for the Beginner

The media-filled bed provides the ideal environment for beneficial bacteria to thrive and serve as a reliable natural bio-filter where all of the ammonia-rich fish waste is converted to soluble nitrates on which plants thrive. Any remaining solid waste decomposes into mineral nutrients that also benefit the plants. This design eliminates the need for expensive filtration to keep the water clean and free of toxins for the fish. It is also worth mentioning that some aquaponic operators have success growing worms in their media grow beds.

A properly set up system has good and even distribution of the incoming nutrient-rich water throughout the grow bed, thus delivering nutrients to all the plant roots. The periodic flood-and-drain method produces better results than the continuous flow method for these purposes. The draining of the aquaponics system increases aeration to the media, hence increasing oxygen supply to the plant roots and bacteria living in the media-filled bed. The circulating water picks up oxygen molecules as it makes its way through all of the cavities within the media bed. The returning water further aerates the fish tank water via turbulence as it strikes the tank water surface.

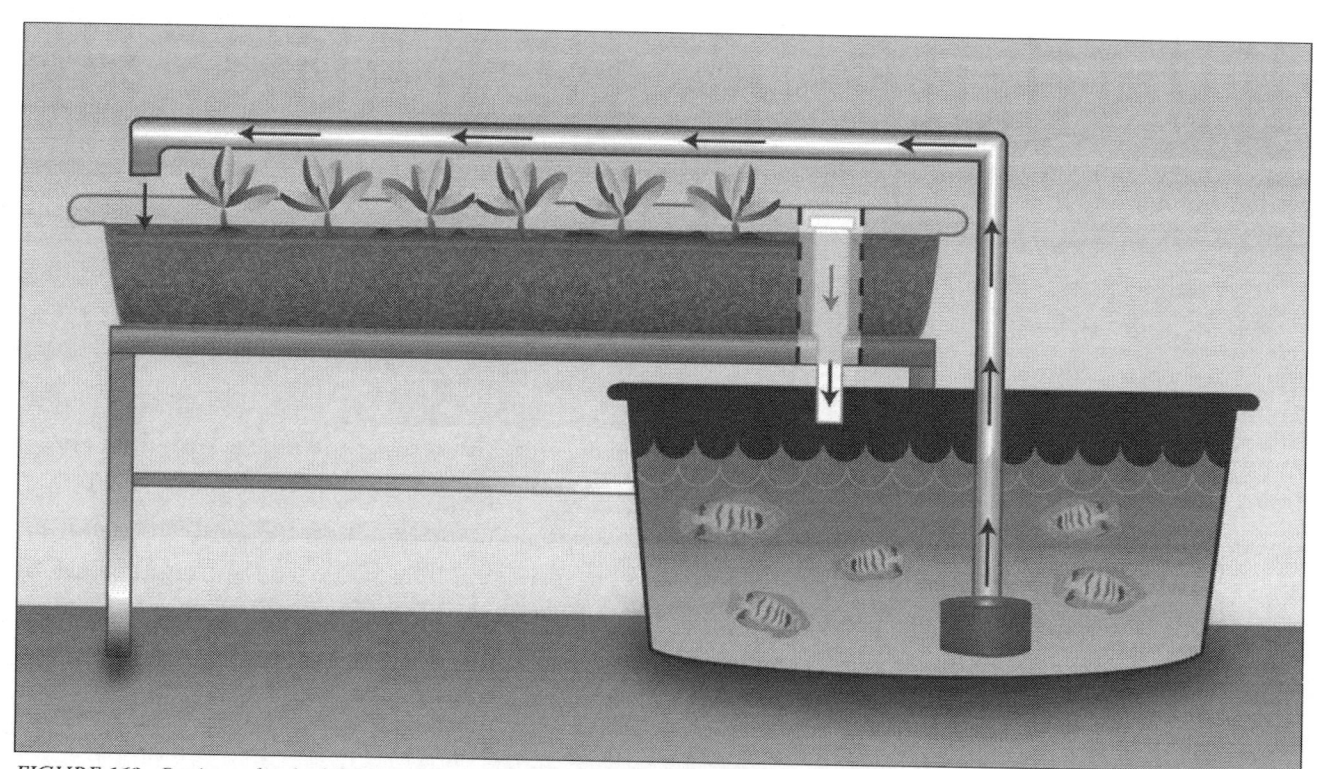

FIGURE 162. Basic media-bed flood-and-drain system without a sump tank.

CHAPTER 25: FLOOD-AND-DRAIN SYSTEM INSTRUCTIONS & PROCEDURES

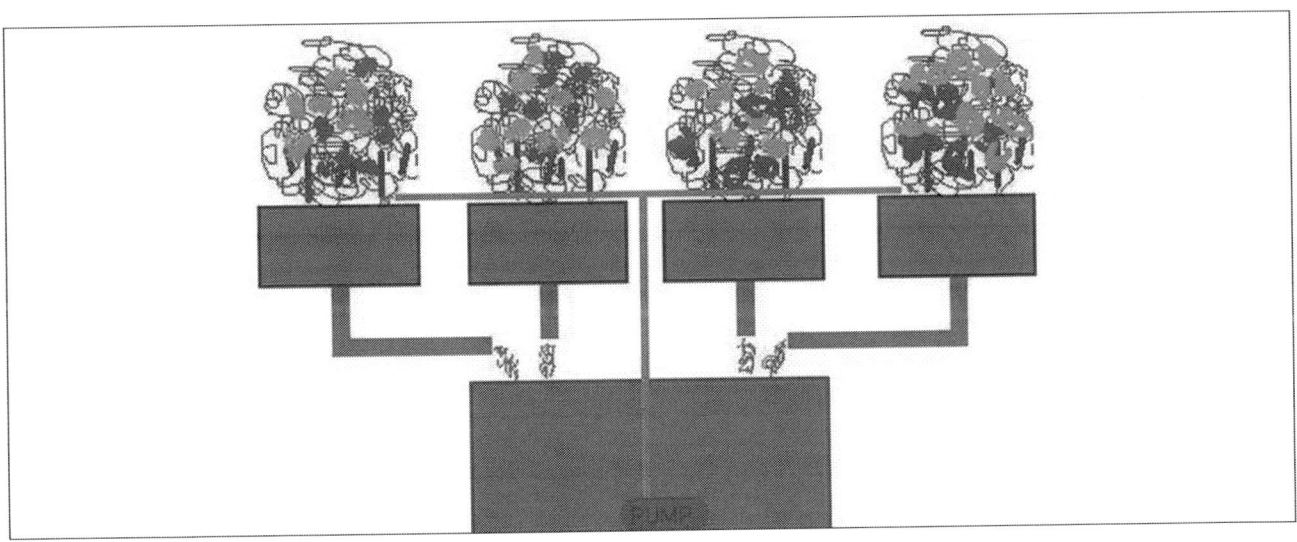

FIGURE 163. Simple growbed over fish tank setup

Flood-and-Drain System Fundamentals

A basic flood-and-drain aquaponics system consists of a media-filled plant grow bed placed above the fish tank so water is able to return to the fish tank via gravity flow. A submersible pump is typically installed in the fish tank to pump water up to the grow bed. This design layout is the most energy efficient configuration as it requires the use of only one (1) pump. Another benefit is that the media grow bed and the fish tank can easily be set up so that they are at an ergonomically correct height for the operator (easy to maintain without a lot of bending over). It is a relatively simple design.

The flood-and-drain system can be set-up many different ways. The grow bed can be set up where it partially overhangs the fish tank, set-up a distance from the fish tank, or even in a completely separate building from the fish tank. For instance, the grow beds can be in a greenhouse or even outdoors if you live in a year-round warm weather climate. The fish tank can be located near the grow beds, outside, or in a separate building such as a shed, barn, or garage. The variety of possible configurations are only limited by the owner's local climate conditons and available space.

There are also no rules regarding the size of a flood-and-drain system. Some systems are so small they are located in an apartment, whereas other systems can be larger than an acre, or operate out of a large warehouse-type building. It all just depends upon your goals and available resources. If the space is available, a system can be set up and built upon over time, with the addition or phasing in of more tanks and media beds as experience and resources allows.

With the media-filled bed system all waste, including solids, is dispersed and eventually broken down within the plant bed. As mentioned previously, in some systems, worms are added to the gravel-filled plant bed(s) to enhance the break-down of the waste.

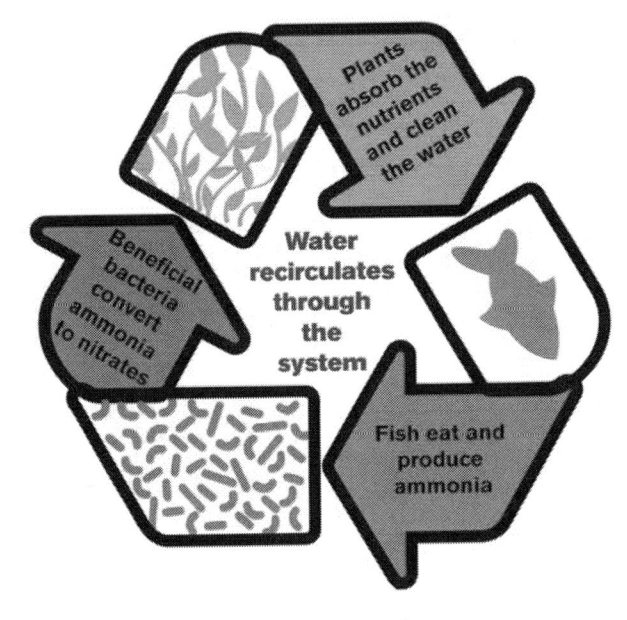

FIGURE 164. Aquaponics nutrient cycle.

261

PART V: FLOOD-AND-DRAIN SYSTEM DESIGN AND LAYOUT

The grow bed serves as a natural water filter that gives the plants everything they need for vibrant growth. It is important to note, however, that production of measurable, edible produce from a media-filled bed operation is typically lower than the other two methods described below. However, a much wider variety of plants can be grown in this type of system; therefore, the media-filled bed is typically used for applications where a greater amount of plant diversity is desired. Large commercial operators typically use deep water culture (or Raft System) and the Nutrient Film Technique to grow lettuce quickly and repeatedly.

Pros:

- It is very basic to build and maintain
- It has an easily customizable design with much flexibility in materials and configurations.
- Self-Cleaning — The grow media filters out the fish waste by performing three (3) filtering functions:
    ◊ Removal of solids from fish tank
    ◊ Mineralization (the release of plant-available compounds such as ammonium during decomposition)
    ◊ Bio-filtration

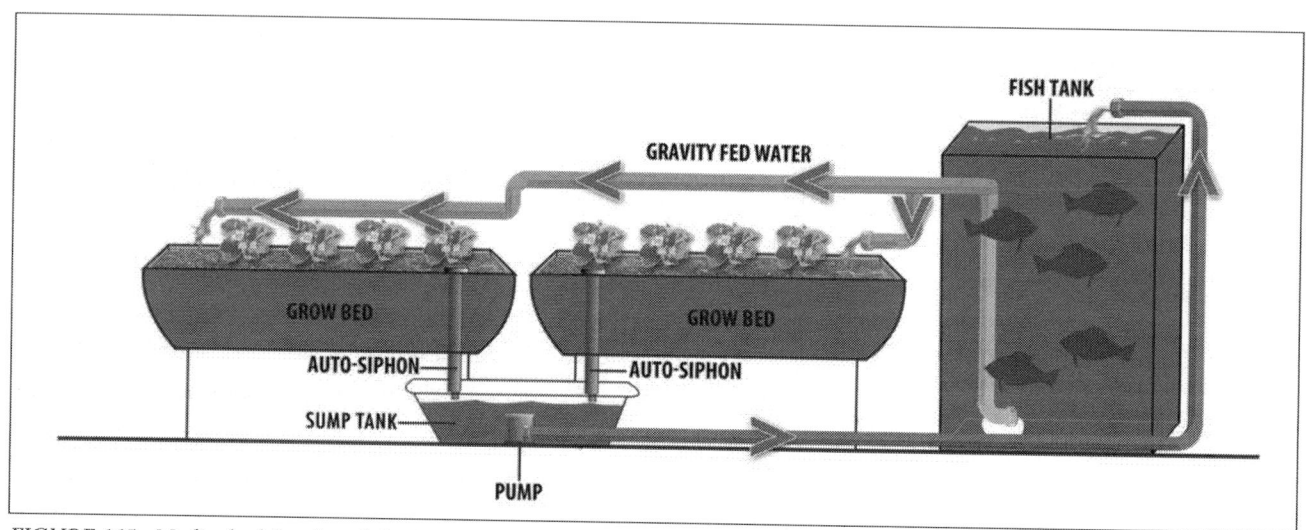

FIGURE 165. Media-bed flood-and-drain system with a sump tank.

FIGURE 166. Basic Flood & Drain System. Note: Having the pump located in a sump tank, instead of in the fish task, is the ideal arrangement.

CHAPTER 25: FLOOD-AND-DRAIN SYSTEM INSTRUCTIONS & PROCEDURES

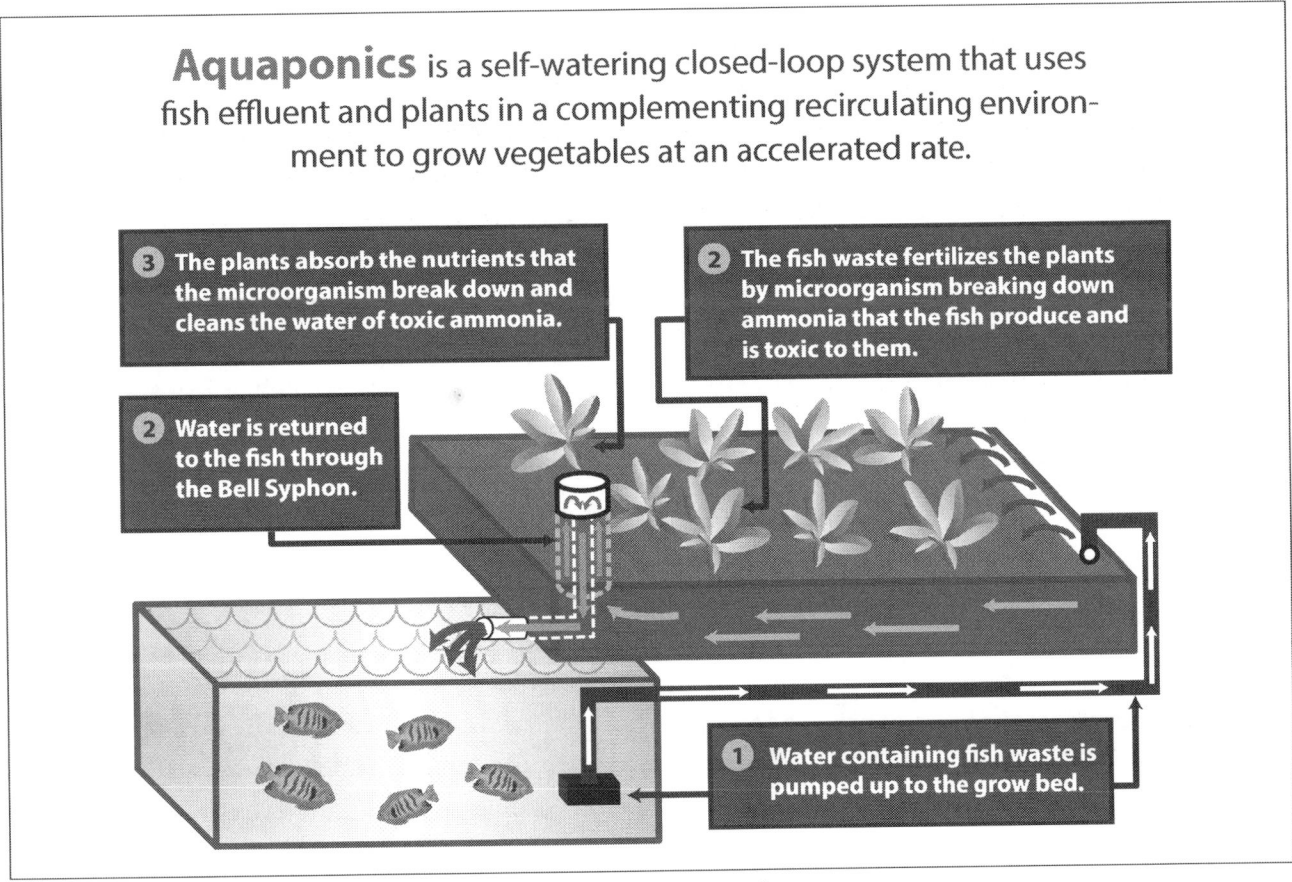

Aquaponics is a self-watering closed-loop system that uses fish effluent and plants in a complementing recirculating environment to grow vegetables at an accelerated rate.

3 The plants absorb the nutrients that the microorganism break down and cleans the water of toxic ammonia.

2 The fish waste fertilizes the plants by microorganism breaking down ammonia that the fish produce and is toxic to them.

2 Water is returned to the fish through the Bell Syphon.

1 Water containing fish waste is pumped up to the grow bed.

FIGURE 167. *Basic media-bed flood-and-drain system without a sump tank.*

- Media provides better plant support and is more closely related to traditional soil gardening, because there is a media in which to grow plants.
- It does not require a sump tank (low space used for collecting undesirable liquids and/or increasing water volume), although, some operators do integrate a sump set-up in their system.

Cons:
- Adding more grow beds or fish tanks can disrupt system balance. Expansion requires maintaining the grow bed to fish tank ratio. For instance, if you decide to just increase your grow beds there will be too much water drained in every flood cycle from your fish tank, unless you use an indexing valve. Low water levels could mean more stress for your fish.
- The upfront media cost can be expensive.

Do-It-Yourself Aquaponics

Even with the help of the internet, quality aquaponics plans and Do-It-Yourself (DIY) guides are hard to find; thorough books, plans, and other helpful resources are in short supply. Specific information can be obtained, but at a substantial cost. Training, workshops, and educational programs provided by professional consultants, successful aquaponic companies, and educational institutions can cost thousands of dollars, and it can take months or even years to acquire the desired knowledge. While it is possible to purchase ready-made aquaponics kits, they are expensive, and technical support can be non-existent.

In consideration of the issues noted in the above paragraph, one of the best options is to design and build your own aquaponics system using the basic fundamentals. If you educate yourself, create a plan, and adequately prepare, it is not as dauntingly difficult

PART V: FLOOD-AND-DRAIN SYSTEM DESIGN AND LAYOUT

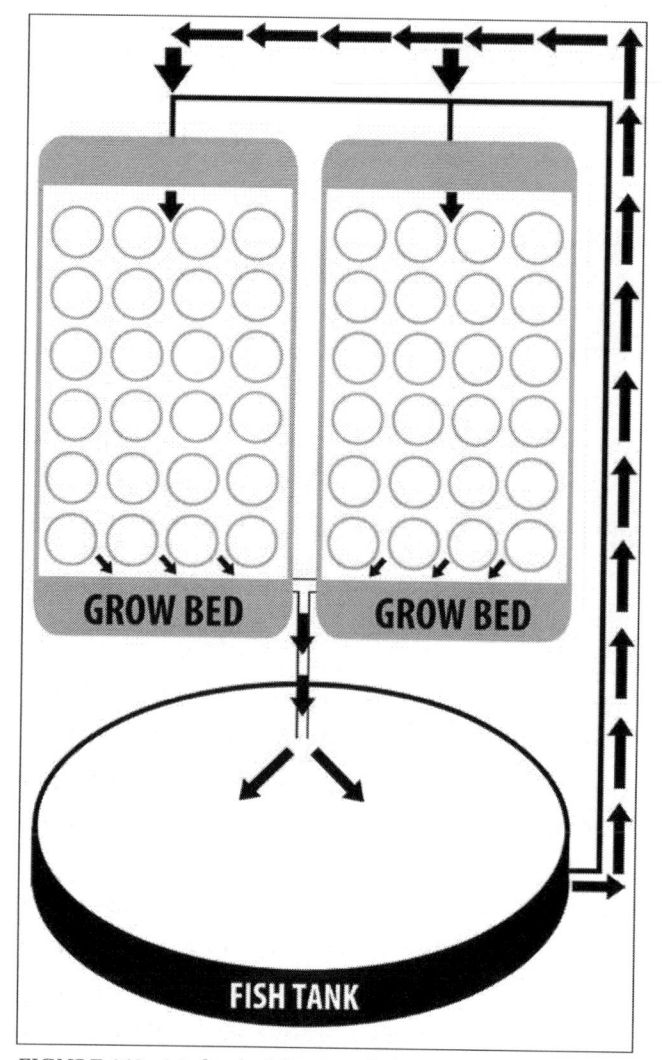

FIGURE 168. *Media-bed flood-and-drain system showing one simple flow pattern option.*

it is important to understand the types of aquaponic designs available, which suits you best, and if are getting what you really need. If you are purchasing a system understand what you will be getting for your money. Like any endeavor, it is always prudent to determine your cost and time involvement up front. The first step would be to make sure you read and consider 'Chapter 7: Aquaponics for You'.

The design of an aquaponics system can be as simple or as complicated as you make it. As mentioned previously, the Flood-and-Drain System is the most common and simplest aquaponics design. This type of system has been adopted by most backyard aquaponics operators, those just desiring to feed their family, and smaller commercial applications (small businesses). The media-filled bed aquaponics design can also be subdivided into two configurations — continuous flow system and the flood and drain system (also known as ebb-and-flow cycle), based on the way the water is being re-circulated.

Flood-and-Drain Configuration vs. Continuously Flooded Configuration

There are two different configurations of recirculating water between the fish tanks and media-filled grow beds that must be considered, namely the flood and drain (also known as ebb and flow) and the continuously flooded systems.

The flood-and-drain configuration is the method of choice among most aquaponic operators. This configuration has been proven to provide good and even distribution of the incoming nutrient-rich water throughout the grow bed. The periodic flooding and draining of the aquaponics system also increases aeration to the media as well as the water returning to the fish tanks. Hence, it helps increase the oxygen supply to the fish, plant roots and bacteria living in the media-filled bed. Furthermore, since the pump does not have to work around the clock, there is a small energy savings compared to a continuous flow system.

as it may at first seem.. Building an aquaponic system yourself also allows you to gain much valuable hands on experience with all details of your system, before fish and plants are added. Furthermore, you can save yourself costly mistakes if you start small and grow with knowledge, versus starting big as with some kits.

Aquaponics Types and Design Alternatives

If you desire to get into aquaponics, you need to first consider your long-term goal and what type of system will best enable you to successfully achieve that goal.

Whether you are planning to purchase a ready-made aquaponics system or build it yourself,

System Setup

System design is critical. Although aquaponics is a fairly straight-forward process, it is possible to set it up wrong. An example of an incorrect set up is when a great deal of the system water is in the grow beds with little in the fish tank. It is important that proper planning be done in advance of beginning any work to avoid such problems.

The grow beds need to be set up so the maximum water flood level is at least 1-inch (25mm) below the top of the grow bed media. This helps prevent algae, as most algae needs water and light to grow. It also helps guard against accidental splashing of water out of the system. A separate sump container is sometimes used which allows for a constant water height in the fish tank. A back-up overflow pipe can also be installed to ensure water never exceeds the designed maximum water level height of the grow bed should the main drain outlet become clogged. Either way, grow bed drainage should be directly routed back into the fish tank. This ensures good circulation and returns a full dose of oxygen to the fish.

Another approach implemented by some aquaponic operators is to install a ball valve on each grow bed inlet to control the speed of fill time. However, since this approach will produce backpressure on the water pump, it is important to provide a water return back into the fish tank from the water pump in order to eliminate undue stress on the pump.

Take precautions to ensure the water pump will not completely drain the fish tank should another part of the system fail to return water due to an unforeseen problem or leak. A low water alarm in the fish tank or a backup overflow pipe installed from the grow bed to the fish tank will provide an additional measure of safety against such problems.

If you have the pump in the fish tank, be sure to raise it off the bottom an inch or two; this will help prevent clogging. Furthermore, always make sure the pump is secure and will not move around. The author's preferred design is to keep the pump out of the fish tank to provide a more natural environment for the fish without the hum and vibration of the pump; however, I am not aware of any studies that indicate the pump is actually stressful to the fish or impedes normal, healthy growth.

Be sure to provide sufficient support for the grow bed and fish tank. For the grow bed, account for the weight of the media, water, the operator's leaning body weight, full grown plants, and plant supports. Keep in mind that 100 gallons of water is about 800 pounds. In short, due diligence is needed in planning and constructing the supports of all raised structures (grow beds, stands, stairs, piping, grow light supports, etc.).

Shade fish tanks to prevent algae growth and to reduce stress to the fish. Remember, fish prefer dark hiding places. Including at least one object in the water, in addition to shading the top, will help them feel more secure. Avoid direct lighting on the fish tank.

For initial system material cleansing, route the water through a clean towel/sock free of fabric softeners or other laundry residuals to catch any silt, chemical residue, or other contaminants that may be associated with the materials used to build the system.

Design Layout Overview

An internet search for aquaponics will quickly show the diversity of aquaponic systems out there in regards to types, sizes, and setup configurations. As a matter of fact, it would be a challenge to find two identical systems. Even aquaponic operators who purchased package units typically have different combinations of fish and/or plants growing in them and may integrate various components into their system, such as a greenhouse, artificial lighting, temperature control devices, automation controls, etc. Much of this depends on local climate conditions, the owner's growing parameters (available resources), and the operator's preferences.

There is no one type of aquaponic setup arrangement that is perfect for everyone. Your climate, goals, local regulatory parameters, and available resources (space, time, family, etc.) are driving forces to consider when planning an aquaponic system.

Although there are many different types of systems, various setup configurations, and a multitude of package units for sale, some operators have much more success than others, even when using similar setups in generally similar climate conditions. Yes, one can construct an aquaponic system that produces fish and plants using a multitude of ways. However, there is a big difference in just producing product and consistently producing product at an optimal level. The difference can be measured in hundreds of pounds of fish, bushels of vegetables, and/or thousands of dollars in profit a year. Therefore, it is very important to not just set up a system that works and produces, but to setup a system properly that will provide maximum results.

Among other critical factors, this chapter will address design specifications essential for achieving the most productive aquaponic systems proven to generate the highest yields of product on a consistent basis. Keep in mind that the system instructions provided in this chapter will only attain optimal effectiveness if the principles (fish tank to grow bed ratio, water quality conditions, fish care, etc.) discussed elsewhere in this book are followed. The various layouts presented in this book and on the www.FarmYourSpace.com website can be uniformly sized according to your space parameters and goals.

Flood-and-Drain System Design Layout

The following layouts have proven to be the most effective systems in regards to generating the highest yields of product on a consistent basis. System details are provided elsewhere in the book; and those basic principles need to be followed as well in order to maximize success.

Basically, there are two types of schematics for a Flood-and-Drain system, with many variations of each. The two types of Flood-and-Drain schematics are

- Pump located in the fish tank;
- Use of a sump with the pump located in the sump.

Both have advantages and disadvantages. Having the pump located within the fish tank simplifies the process, as you only have two major components—the fish tank and the grow bed. However, a pump located within the fish tank has a slightly higher probability of becoming clogged due to fish waste and uneaten food. However, this is usually a rare event and potential clogging can be minimized by placing the pump on a brick, putting a screen around the pump, and/or through periodic tank maintenance cleanings with a standard swimming pool cleaning net or pool leaf rake.

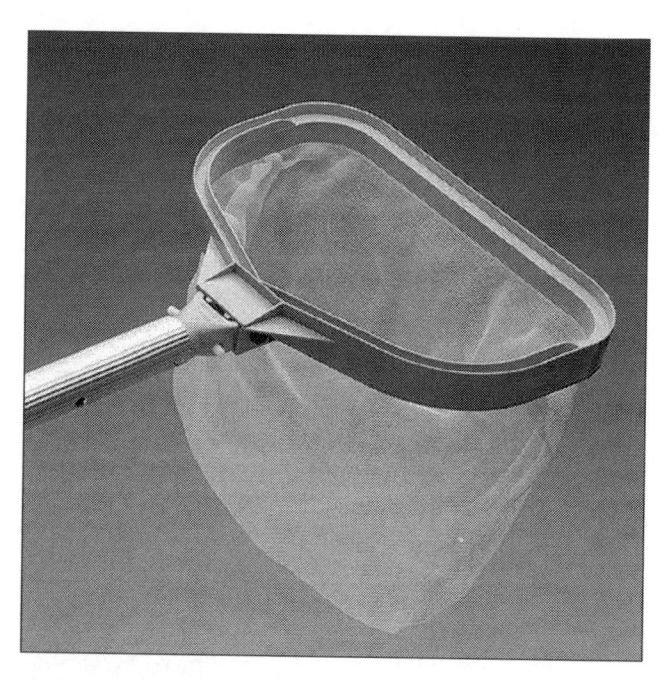

FIGURE 169. A swimming pool cleaning net serves as an excellent fish tank cleaning tool.

CHAPTER 25: FLOOD-AND-DRAIN SYSTEM INSTRUCTIONS & PROCEDURES

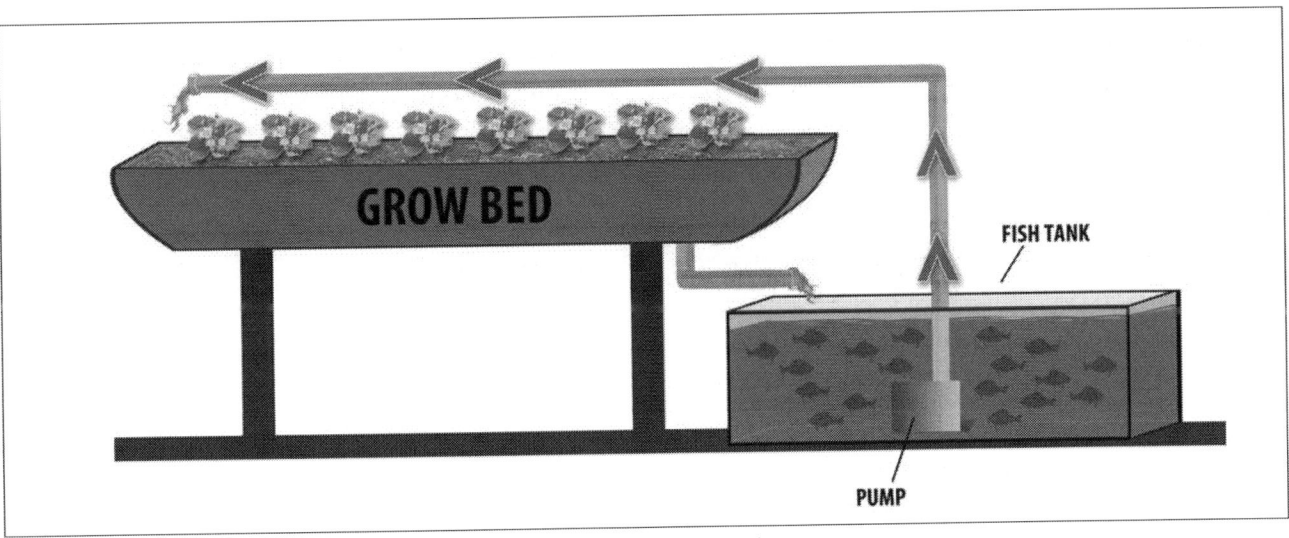

FIGURE 170. Flood-and-Drain System with Pump Located in the Fish Tank

Basic Set-Up Fundamentals:
- Fish Tank at the lowest point.
- Pump in the Fish Tank — Pump up to the grow bed(s).
- Grow bed(s) at the highest point.
- Simple gravity flow back to the Fish Tank.

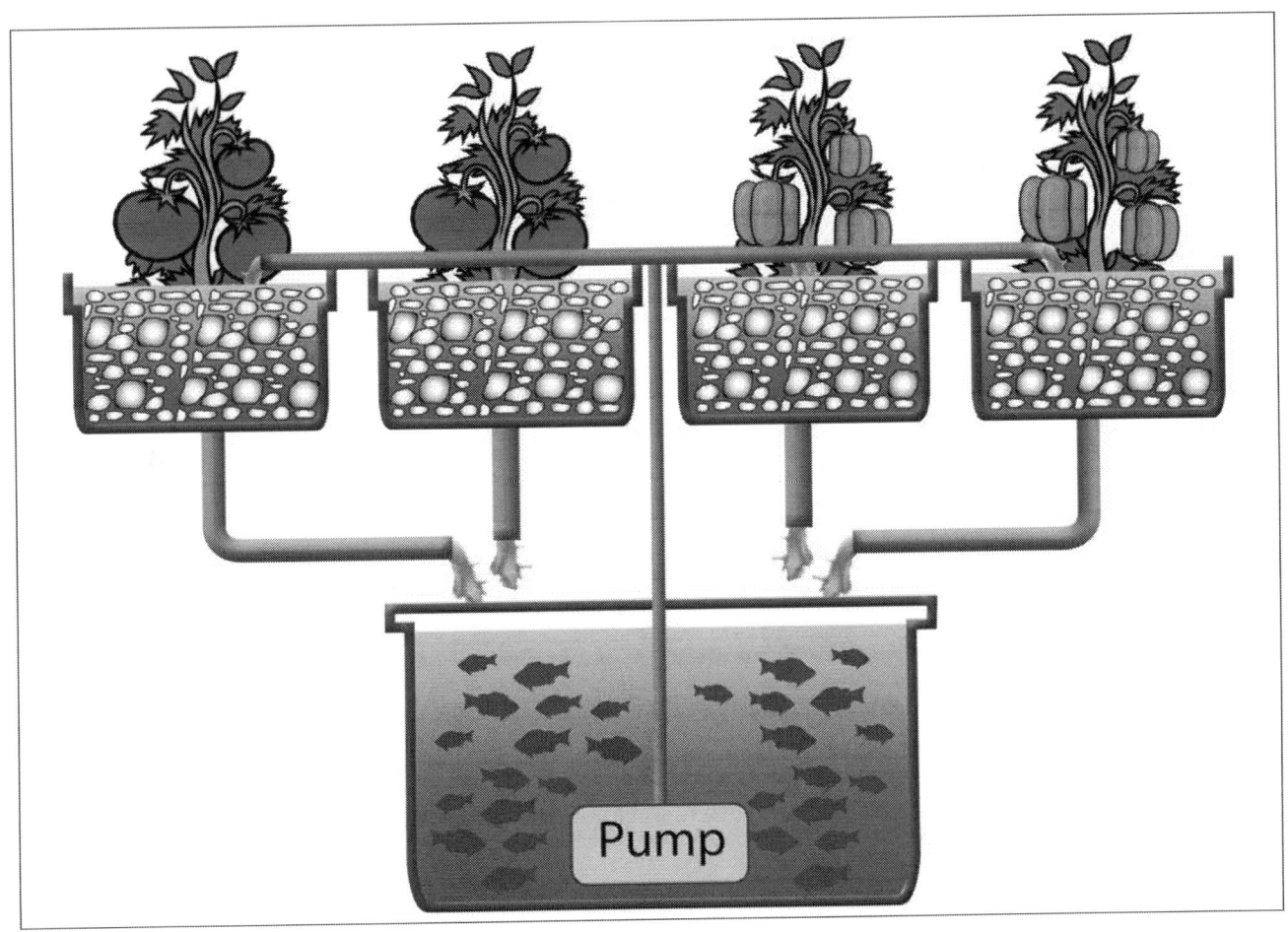

FIGURE 171. Simple Grow Bed Over Fish Tank Setup

Some aquaponic operators prefer to use a sump because they do not want the pump in the fish tank, as they believe the buzz and vibration of the pump causes needless stress to the fish; they feel it is not being friendly to our animals. This author is still studying the issue, but as of now I am not aware of any studies or data that suggest that a pump in the fish tank stresses the fish, harms them in any way, or interferes with optimal growth. However, placing a mechanical pump in the fish tank is unnatural and could place the fish under an element of unnecessary stress. Chronic stress does negatively impact the health of any creature.

Sump Set-Up Fundamentals:
- Sump located at the lowest point
- Simple gravity flow to a Sump Tank
- Water Pump in the Sump — Pumps to the Fish Tank
- Pump in sump controlled by timer or a float

Water is a much better conductor of sound than air. If you have ever gone swimming in a lake when motor boats or jet skis are in the general proximity, you can appreciate this fact. The engine of these crafts may sound nearby underwater, even though they may be hundreds of yards away. Although we plan to harvest our fish for consumption and market, we do not want them to suffer in any way during their life or be exposed to unnecessary stress. As soon as I am able to obtain factual data on this issue, I will post it on the website: www.FarmYourSpace.com

As of now, the author's preferred and recommended approach is to use a sump tank so as to keep the pump out of the fish tank.

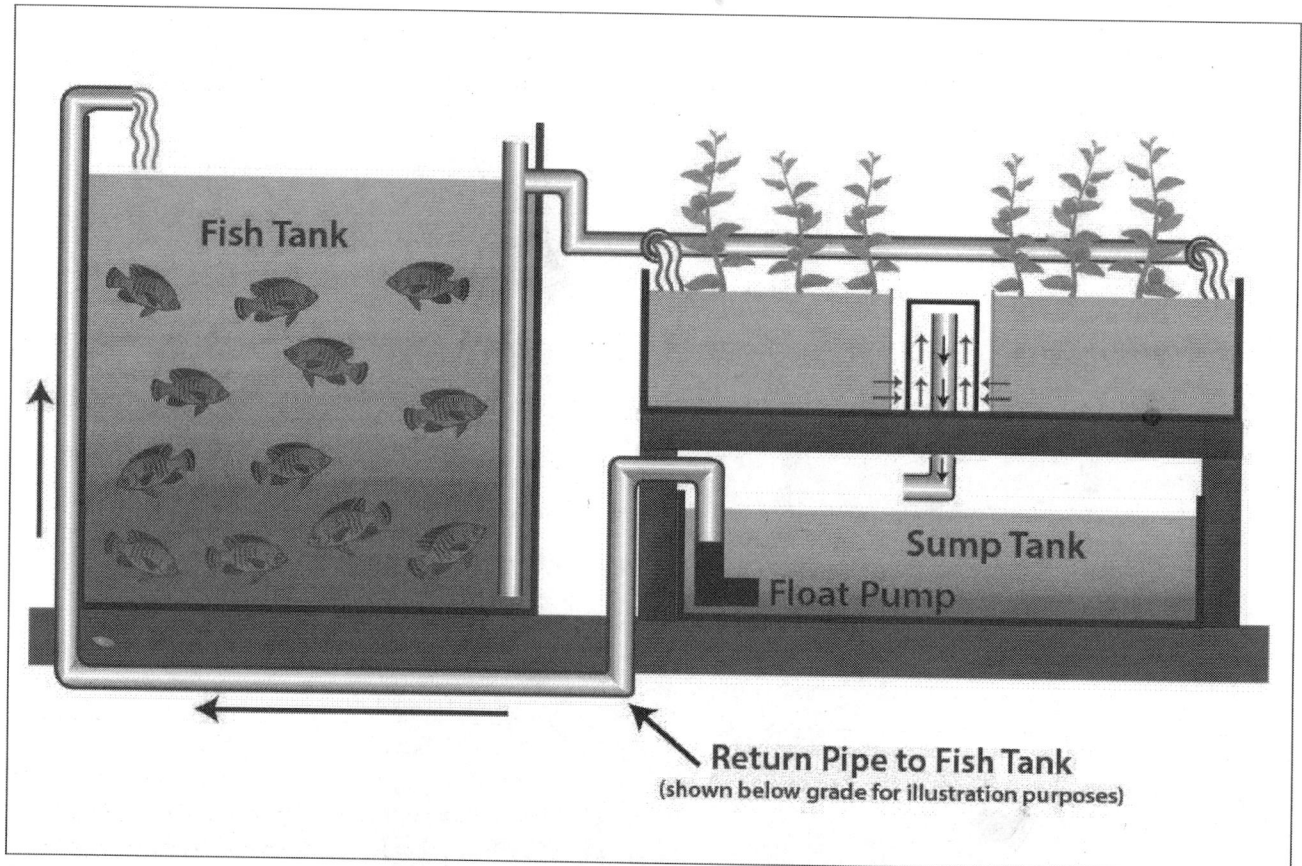

FIGURE 172. *Flood-and-Drain System with the use of a sump (pump located in the sump). Pipe shown under floor for illustration purposes only.*

CHAPTER 25: FLOOD-AND-DRAIN SYSTEM INSTRUCTIONS & PROCEDURES

Gravity flow from the Fish Tank to the Grow Bed(s) using a Solids Lift Overflow (SLO)

Several previous illustrations throughout this book have eluded to the method of conveying water from the fish tank to the grow bed via gravity flow, but it has not yet been discussed. This option works well and has the advantage of reducing the probability of the pump becoming clogged with fish waste and uneaten food which accumulates at the bottom of the fish tank where a pump might otherwise be located. The disadvantage of this setup is the fact that elevated fish tanks full of water need heavy duty structural support, or the fish tank needs to be taller and narrower to maintain the recommended tank to grow bed volume ratio. Nevertheless, the solid lift overflow method works great, and it certainly needs to be considered as an option.

When the fish tank is positioned above the grow bed(s), gravity can be utilized to efficiently convey the outflow water and fish waste to the grow bed(s). If an outflow is simply located out the side of the fish tank, the water will indeed flow out of the tank and directly into the grow beds. However, the problem is that since the nutrient-rich, solid waste is heavier and at higher concentrations closer to the bottom of the fish tank, using just an outlet in the top half of the fish tank does not move these rich nutrients, and it never makes it to the grow bed(s). So in essence, just having an outlet in the upper half of the fish tank deprives the grow bed(s) and plants of nutrients and exacerbates undesired water quality problems for the fish, since the waste continues to build up in the fish tank.

To resolve this problem, a Solids Lift Overflow (SLO) can be implemented. The SLO draws the water

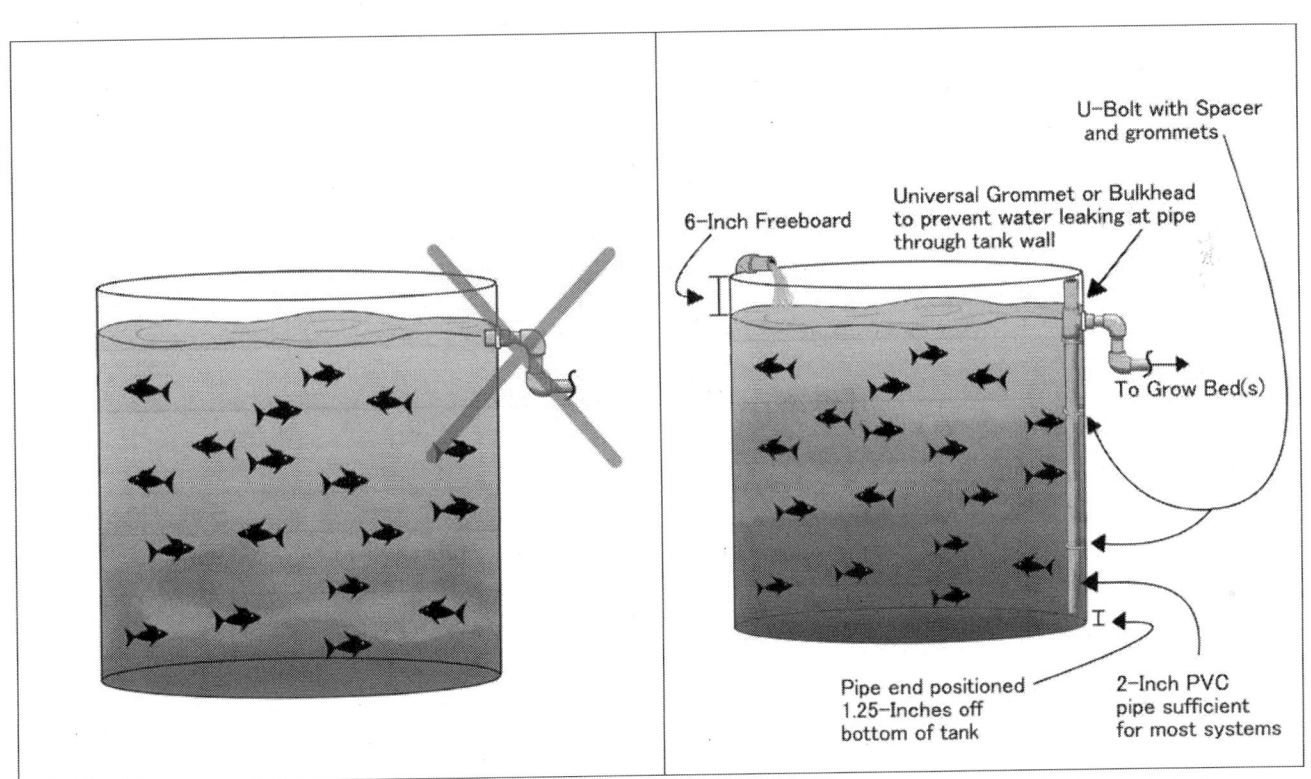

FIGURE 173. *Draining from the surface will not remove the condensed fish waste water from the fish tank.*

FIGURE 174. *U-bolts with grommets for the vertical pipe may not be appropriate for all tank material types. Also, strategically placing a few bricks against the bottom of the pipe can help support it. Just make sure the pipe inlet is not obstructed. Also, placing a screen or net over the pipe inlet will prevent smaller fish from getting pulled into the pipe.*

and the solid waste from the bottom of the tank, as desired, thereby cleaning the fish tank and delivering the waste (nutrients) to the grow bed and plants where needed.

The SLO is typically made up of a 2-inch diameter pvc pipe and fittings. Ideally, it would be positioned at the center of the fish tank. However, it can be secured easier against the inside wall of the fish tank, so that is where it is mounted in most aquaponics systems using a SLO. Locating the SLO intake pipe approximately 1.25-inches off the bottom surface of the fish tank has proven to be most effective in removing the solids without running the risk of clogging the pipe. The vertical pipe is fitted with a T-connector at the best location within the top half of the fish tank, whereby the pipe then passes through the tank wall via a uniseal grommet or bulkhead fitting and then on to the grow bed(s). Using a Tee fitting instead of a 90-degree elbow prevents a siphon from forming and ensures water never flows backward into the fish tank.

Operation Method Options (and Recommendation)

The next consideration is how you will flood and drain the grow beds. There are three main methods that are commonly used:

- Constant Flood
- Timed Flood-and-Drain
- Flood and Drain with Siphons
- Mechanical -Pump in grow bed. (This is a practical option if the fish tank is located at a higher elevation than the grow bed)

In order to make the best possible decision as to which one of the above methods is best for you, consider the following:

- **Simplicity** — The more complicated it is — the more likely it is to break!
- **Short** — Shorter pipe runs help reduce temperature variations, reduces friction loss, and is easier to maintain.
- **Cleanings** — Consider how you will clean your system during the planning and design phases.
- **Accessibility** — Plan a way to access all system infrastructure. It will be difficult to make repairs, clear possible clogs, and clean pipes if there is no access to them.
- **Redundancy** — Plan and build in redundancy where possible. Using a minimum of two fish tanks and two grow beds has many advantages. An overflow pipe from the grow bed to the fish tank is helpful in case there is ever a problem with any of the grow bed outlet infrastructure. Always keep fish health and security first.
- **Water Conveyance** — Ensure water can flow easily through the system. Complicated joints and connections or under-sized piping will slow your water down.
- **Solid Waste Management** — Consider how you will move and deal with solids.
- **Future Expansion and Upgrades** — Keep in mind the possibility of future expansion and system upgrades. In the future, you may want to add more fish tanks and/or grow beds or upgrade system components. Expansion and upgrades will go much smoother, take less time, and be less expensive if you plan for such in the beginning.

The most basic system will encompass a grow bed and fish tank. This type of system works well as a starter or introduction into aquaponics. However, considering the above, in most cases it makes a lot of sense to implement a system where the fish tank is the highest point in the system and water gravity flows down to the grow beds and then on to a sump from where it is returned to the fish tank. (see figure 175.)

This method has several benefits:

- **Safety:** The Fish tank will never be drained if there is a plumbing problem.
- **Solid Waste Management:** Solid waste is efficiently removed from the tank. Pumping is not impacted by solid waste.

CHAPTER 25: FLOOD-AND-DRAIN SYSTEM INSTRUCTIONS & PROCEDURES

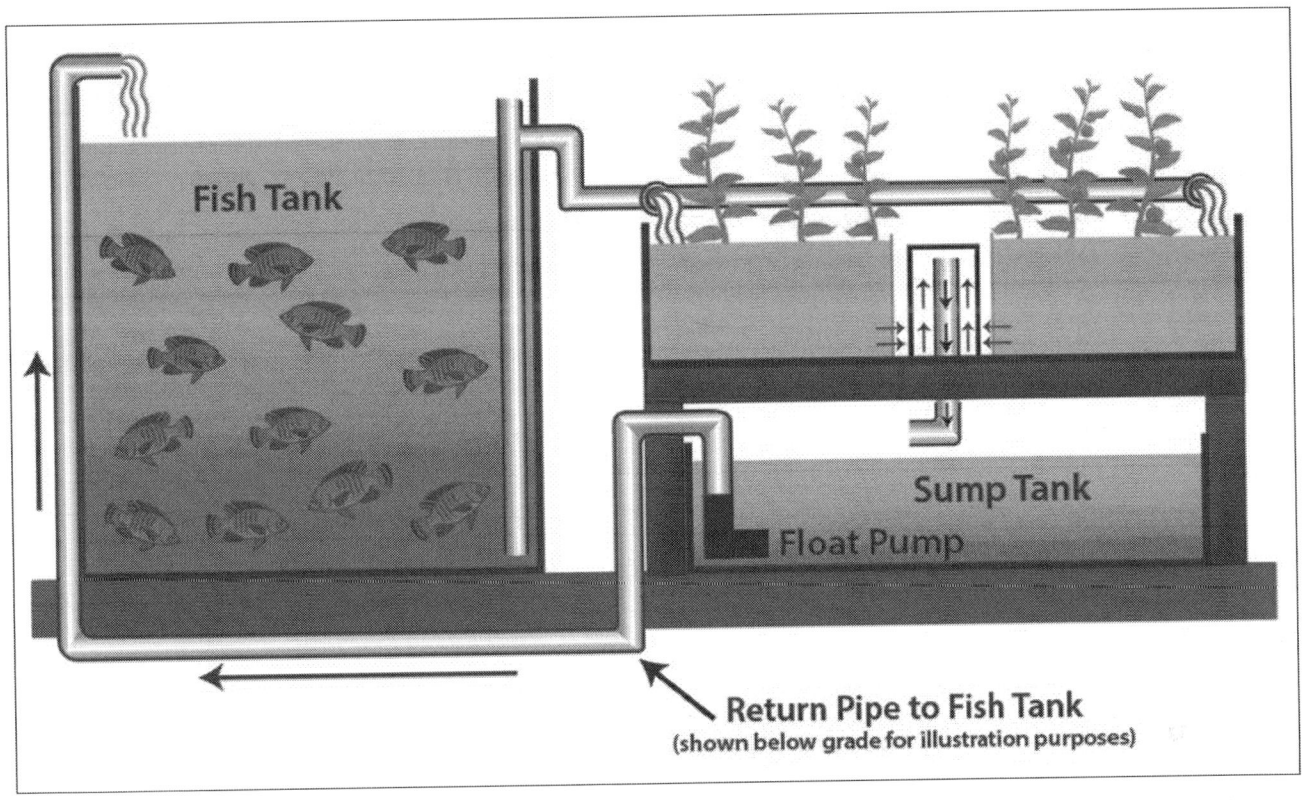

FIGURE 175. Media-Bed (Flood-and-Drain) Aquaponic System with a Sump Tank an Auto-Siphon Drain (i.e. Bell Siphon). See Chapter 26 for plans and instructions on how to build this unit. Pipe shown under floor for illustration purposes only.

- **Efficient:** This set-up uses only one pump (even with multiple fish tanks and grow beds) and can be arranged very efficiently.
- **Flexible:** More fish tanks and grow beds can be added relatively easily.
- **Water Back-Up:** The sump adds water capacity to the system.
- **Fish Comfort:** The fish are not negatively affected by the constant humming of a pump.

This type of set up takes more space than other methods, and there is a slightly higher cost for the sump and additional plumbing. However, if your situation allows, I recommend this type of system.

NOTE: Design plans and assembly instructions for the above illustrated system are included in 'Chapter 28: Design Plans and Construction Details DIY Flood-and-Drain System'.

Sizing a Flood-and-Drain System

After identifying your specific goals for getting into aquaponics, you will have a better idea regarding the size of system needed. Since the grow bed takes up the most square area space, it is the best place to begin planning the layout of your aquaponic system.

A National Garden Association study found traditional gardening methods yield 0.9 pounds per square foot (0.9 lbs/SF or 4.4 kg/m^2) on average. That accounts for one growing season. As discussed in earlier chapters, aquaponics plants grow much faster and produce greater yields than traditional farming methods. With the use of a greenhouse for year-round production, aquaponics produces yields of 7 pounds per square foot (7 lbs/SF or 34.18kg/m^2).

A properly setup and maintained system with a grow bed of 25 square feet will produce enough vegetables and soft fruits (tomatoes, peppers, squash) to supply one adult throughout the year, which amounts to 175 lbs of produce annually. That figure is based upon year-round yields without seasonal interruptions, typically a greenhouse operation in areas that are not tropic or subtropical. A family of four would need approximately 100 square feet to produce around 700 lbs of vegetables and certain fruits annually. Using the number of people you want to feed times the 25 square feet required for each person will provide the size of grow bed needed to supply 100 percent of your produce. Of course, one can have a smaller aquaponic system to simply reduce the amount of vegetables needed from the grocery store.

For a "real-world" example using a family of four (assuming all family members are over the age of 12) a 100 square feet grow bed that measure 5-ft x 20-ft. or 6-ft x 16-ft would work perfectly. These sizes are easy to manage as one can easily reach to the center of the grow bed from both sides. It is difficult to maintain the center portion of the grow bed if it is wider than 6-ft. Also, the above referenced sizes will fit comfortably inside a 10-ft x 24 ft or a 12-ft x 20-ft size greenhouse, respectively.

A second example would be sizing the grow bed for a family of eight (8) or planning to have excess to sell and barter. Using 200 square feet as an example, one 6-ft x 33-ft grow bed would work, but such a long grow bed would make it difficult to get an even distribution of nutrients throughout the grow bed. Yes, it can work, but careful planning must be done prior to assembly to ensure there is a relatively even dispersion of nutrients throughout the grow bed. Also, such a long grow bed is typically not the ideal configuration for most people in regards to available space. A better approach would be the use of multiple grow beds instead of just one. Furthermore, since it is difficult to reach more than three feet, walking or service aisles should be planned between and outside of any grow beds larger than three feet wide. This will allow easy planting, maintaining, and harvesting of plants. Two grow beds measuring 6-ft x 17-ft, three beds measuring 6-ft x 11-ft, or four beds measuring 6-ft x 8.5-ft would be a better option than one large bed.

For commercial installations a 1,000 square feet of grow bed area will produce enough product to provide 40 people all the vegetables they need for an entire year. Using the 25 square feet per person rule: 2,000 SF = 80 people; 3,000 SF = 120 people; 4,000 SF = 160 people; 5,000 SF = 200 people.

Grow Bed Size Parameters

The size of the grow bed(s) area was addressed above in regards to sizing the aquaponic system, because the grow bed takes up the most area. However, the size of the grow bed and the volume of the fish tank volume must be considered together, because the plants need the fish waste to thrive. The bigger the grow bed and plant mass, the more fish waste required; therefore, there have to be enough fish to support the plants.

Ideally, it is best to have at least two separate grow beds so when the roots in one bed are being harvested or cleaned out, there will be another grow bed working to clean the waters. This also provides longer harvest times throughout the year, as the crops in the separate grow beds are typically at different stages of growth (one young and the other closer to harvesting), if planned correctly.

The length of the bed is not as crucial as the depth and width. For most hobbyists, beginners, and smaller aquaponic operators a width of about 30 inches is common and seems to work well. This allows one to easily reach across the grow bed.

The ideal grow bed is 12 inches deep to ensure root support for most plants and enough bio-filter volume for bacteria growth. This also gives the bacteria enough time to break down fish solid wastes as they 'filter' through the media-filled beds, preventing them from over-accumulating. The area of grow bed is determined by your space constraints, desired crop production, and size of the fish tank.

Bed depth significantly greater than 12 inches is unnecessary, will be heavier, and will require more

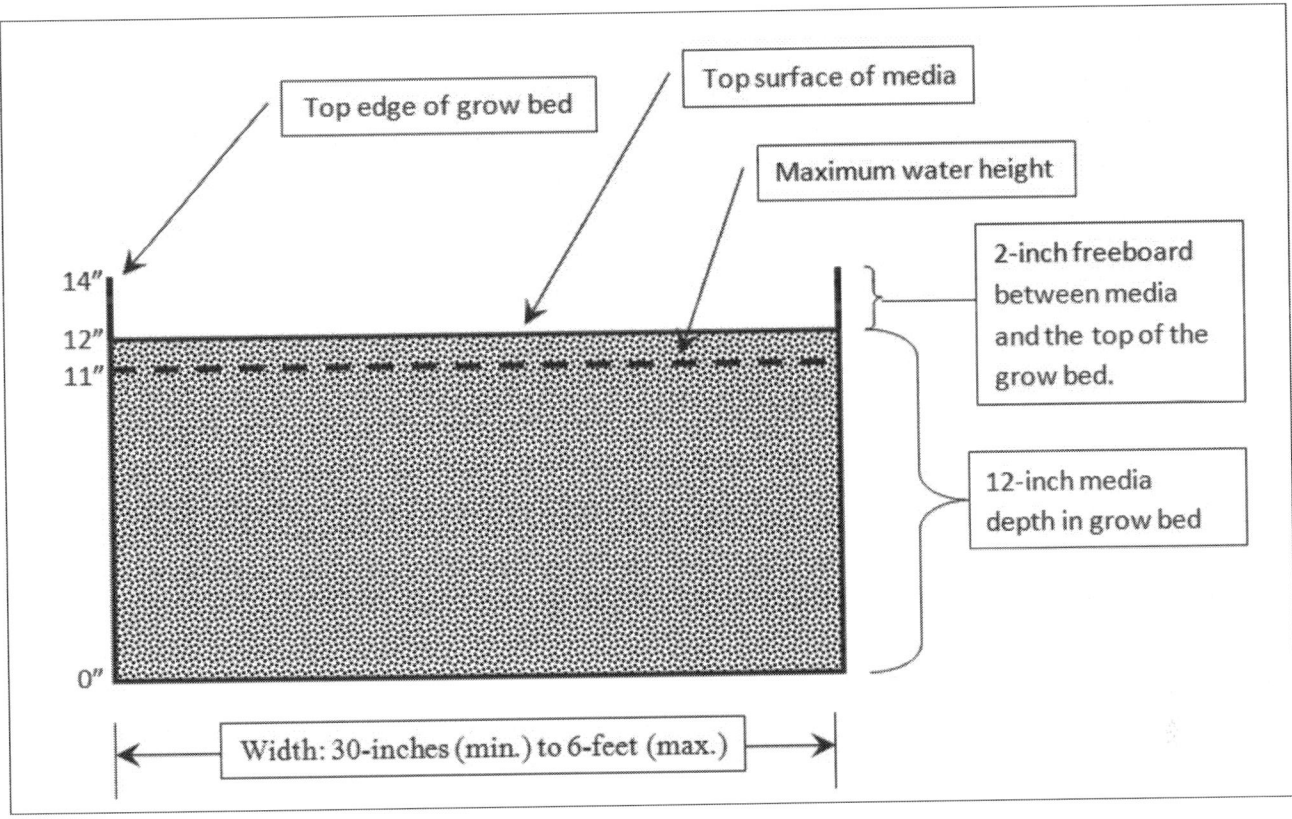

CHAPTER 25: FLOOD-AND-DRAIN SYSTEM INSTRUCTIONS & PROCEDURES

FIGURE 176. Recommended cross-section of grow bed.

water to be pumped. The media-filled bed should not be flooded to the top. It works best if the water level is never allowed to exceed a height 1 inch below the top of the media surface. This prevents algae growth on the top surface of the media and helps keep mold from developing on the bottom leaves of the plants. It is also preferable to have 2 inches of freeboard. Freeboard is the distance between the top of the gravel surface and the lip of the grow bed. This will help keep the media in the tank during planting, plant care, grow bed maintenance, and harvesting of the plants. It also provides additional protection against significant water loss, if the grow bed drain system becomes clogged.

Grow Bed to Fish Tank Ratio

In general the recommended grow bed volume to fish tank volume ratio is 1:1. However, it can be 2:1, even as large as 3:1, if the fish tank remains densely stocked. However, a fish tank stocked dense enough to support a 3:1 grow bed configuration has to be closely monitored to ensure the fish are not overly stressed, and no detrimental water quality problems arise. In addition, as the fish are harvested in a 3:1 configuration equal replacements of fingerling fish must be added to the system replacing the harvested one, so as to maintain adequate flow of essential nutrients to the bacteria and plants. Technically a 3:1 configuration will work, but it takes a great deal of effort to ensure water quality is maintained, fish health is not compromised, and system balance is not disturbed during the harvesting of fish or if there is a disruption to grow bed(s) production for maintenance, etc. Therefore, with all of the potential problems that can occur, a 1:1 configuration (grow bed being in equal volume to fish tank) is ideal.

To compare the volume of the tank and media bed, convert both to cubic feet. To determine cubic feet of the grow beds and fish tank multiply the dimensions (length x width x depth of media).

Only consider depth of media for the grow bed. For instance, a 14 inch deep grow bed would have media 12 inches deep. In this case, you would consider only the 12-inch dimension. Convert all dimensions to feet by dividing inches by 12. For example, to determine the size of a 48-inch grow bed in feet, divide 48 by 12 to determine it is 4 feet. To convert gallons to cubic feet, by multiply by 7.48; consequently, 1 cubic foot = 7.48 gallons.

Grow Bed Functions

An aquaponics media-filled grow bed is simply a suitable container filled with media such as gravel, hydroton, lava rock, any number of other substances, or a combination of materials (See Chapter 18: Growing Media for Plants). The grow bed performs four separate functions in aquaponics:

- **Support for Plants** — The grow bed with media provides a means for plants to use their roots to anchor themselves.
- **Water Filtration** — The media filters (removes) solid waste in the water. This waste gets trapped in the grow bed.
- **Mineralization** — This is the process whereby the solid waste breaks down so that the plants can use it. In more scientific terms, mineralization is the breakdown of organic matter into individual elements (macro nutrients like potassium, calcium, sulfate, phosphorous, magnesium and micro nutrients like iron, copper, molybdenum, zinc, etc) into plant-accessible forms.
- **Biological Filtration** — The media provides extensive surface area for the beneficial bacteria to colonize.

The media-filled grow bed performs all of these functions easily in a very space-efficient and cost-effective way.

Grow Bed Materials

In a flow-and-drain system plants are cultivated in the media-filled grow bed. There are a number of things to be consider during the grow bed planning process. Obviously, a grow bed needs to be waterproof. It also needs to be strong enough to hold the weight of the media, water, plants and potentially the weight of the operator leaning on or against it. Depending on the size of the grow bed this could equate to a fairly significant amount of pressure.

Grow beds can be purchased, or made from a number of different types of materials, such as plastic, fiberglass, concrete, bricks, etc. However, the most common and cost-effective is wood, lined with a water-proof plastic liner or thick epoxy compound. In cooler climates, depending upon the set up, adding insulation to the sides of the grow beds to keep them warm in the winter and cool in the summer is a good idea, especially if the aquaponics system is outdoors.

Some operators line their beds with plastic, while others use fiberglass they have mixed and prepared themselves. Both have advantages and disadvantages, and anyone can easily learn how to do both. YouTube videos, other internet resources, library books, and product vendors provide a great deal of "how to" information. A plastic liner is the least expensive and easiest to install; however, it is more susceptible to rips or puncture holes than a thicker alternative material. Creating your own fiberglass mix requires proper personal safety precautions. One can also purchase a fiberglass or plastic tank, although size options are limited, and they cost substantially more than a plastic liner.

Basically, any watertight, food-safe, fish-safe container can be used, depending on the size of the aquaponics system.

- Small systems can use food containers, plastic containers from the DIY store or IKEA, and wooden boxes with a suitable liner.

CHAPTER 25: FLOOD-AND-DRAIN SYSTEM INSTRUCTIONS & PROCEDURES

- Medium sized systems can use cut-up IBC totes, Rubbermaid-type water troughs, animal feed troughs, and concrete mixing troughs.
- Larger sized systems can use multiple units from above, hand built and lined wooden grow beds, and, of course, commercially available aquaponics grow beds.

A grow bed can be made from a wide variety of materials, but care should be taken to ensure it is made of materials that will not leak unwanted chemicals into the water or affect the pH of the water. If the grow bed is being used outside, it will also need to be UV stabilized to ensure it will not degrade in the sunlight and leach chemicals into the system.

Water Entry and Exit Placement in Grow Beds

There are a number of ways the inflow of water into the grow beds can be configured; however, the main objective is to ensure that waste from the fish tank is spread throughout the grow bed as uniformly as possible. A concentrated area of waste can form an anaerobic spot which will prohibit plant growth, deprive plants in other locations of obtaining maximum nutrient benefits, and impede water quality process. Good relative placement of your inflow and outflow plumbing will allow for good water movement within the grow bed, provide an equal spread of nutrients, and help optimize water quality. The following are a few common ways of setting up a grow bed.

In figure 177, water enters the grow bed at one single point and exits via the outflow pipe in the opposite end of the grow bed. With this set up there is potential for solid waste to have limited movement within the grow bed. This is however a common setup in many aquaponic systems and can still provide

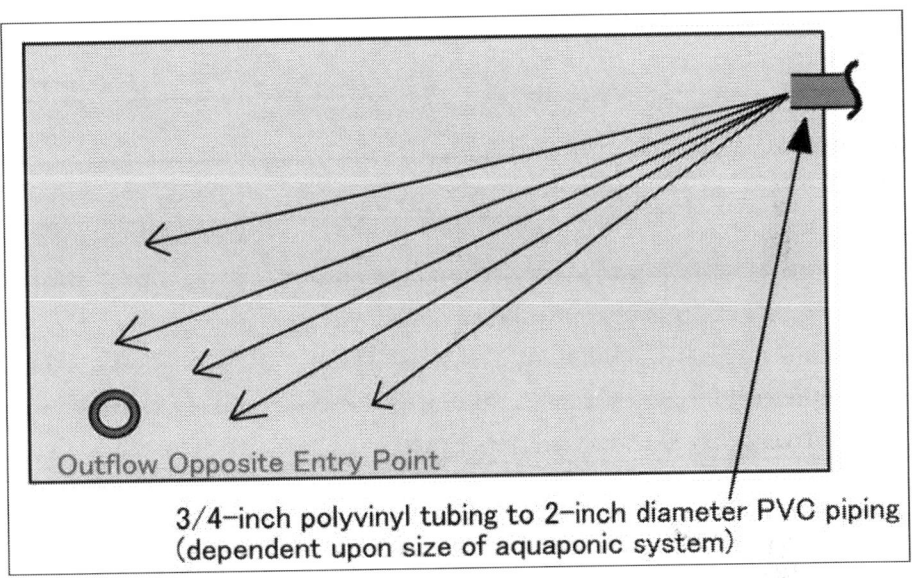

FIGURE 177. Media filled grow bed with one inflow and outflow in opposite corner

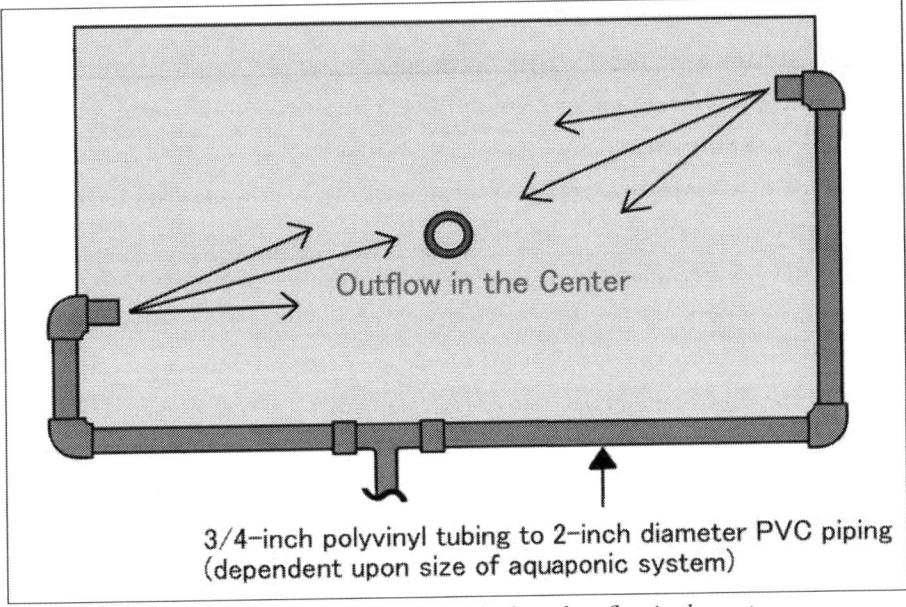

FIGURE 178. Media grow bed with two inflow feeds and outflow in the center

successful results. It is also very easy to set up.

In figure 178, the outflow is placed in the center of the grow bed with two intakes located at opposite ends of the bed. This allows for a more uniform dispersion of waste water throughout the grow bed. It is also a fairly easy set up that reduces the potential for waste build up and anaerobic (no oxygen) spots.

In figure 179, water enters the grow bed on all sides. This is accomplished by drilling evenly spaced holes in irrigation tubing or PVC piping. In this set up water is spread evenly around the entire grow bed and ensures maximum dispersal of waste water (nutrients). Although requiring slightly more work than the two previous methods discussed, it is still relatively simple to set up. Over time the holes can clog, so periodic inspection is needed. Clogs can easily be undone by poking the end of a Phillips screwdriver in the clogged hole.

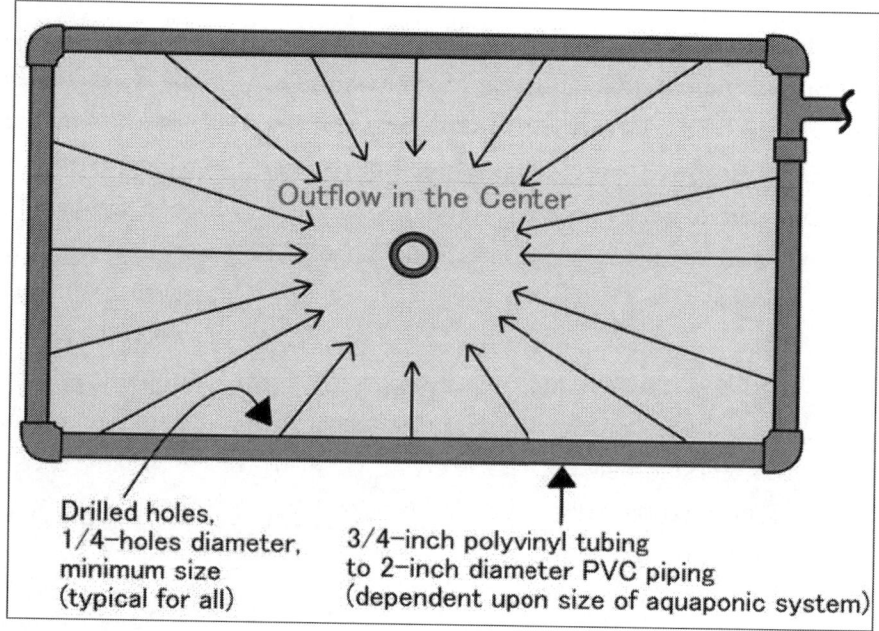

FIGURE 179. Media grow bed with uniform inflow and centralized outflow

Emergency Overflow

When planning a system it is best to consider incorporating a back-up overflow pipe or hose as extra protection should something go wrong. This will simply be a hole strategically placed immediately below the top edge rim in the side of the grow bed and the fish tank. This hole will allow water to flow out of those containers, if the water should rise too high.

If the grow bed is placed above the fish tank then the water should flow back either to the sump or to the fish tank. If the fish tank is at the highest point, then the water should flow either to the grow bed or to the sump. The objective is to keep the water in the system, rather than have it spill out of the system. It is also prudent to place a screen at both ends of the overflow so that neither fish nor media can enter the overflow pipe or hose.

Natural, Mechanical and Bio Filtration

Some aquaponic operators prefer to integrate a filtration unit into their system. A biological filter is positioned between the fish tank and the grow bed. The water is routed from the fish tank into the biological filter then to the grow bed(s), NFT Channels, or DWC tank, and then back to the fish tank. Biofilters can be purchased or crafted by the DIY enthusiast and come in all types of configurations and sizes.

Biofilters, when used, are most often used in a Flood-and-Drain aquaponic system, where the fish waste production is more than the grow beds can process. Biofilters allow for more bacteria growth to ensure adequate conversion of the toxic ammonia fish waste and suspended solid waste into beneficial nitrates and dissolved solids for plants and to prevent toxicity to the fish.

In a balanced flood-and-drain system, the grow bed (with bacteria and plants) serve as the biofilter; therefore, additional filtration is typically not necessary. In other words, with a properly planned flood-and-drain aquaponics system additional or supplemental filtration (mechanical, natural, or bio-filtration) is not needed. To learn more about mechanical, natural, and bio filtration, refer to 'Chapter 20: Filtration (Mechanical, Biofiltration, Natural)'.

Pumping from the Fish Tank, Pond, or Sump

The pump is a critical component in a flood-and-drain aquaponics system. An entire chapter of this book is dedicated to pumps (Chapter 19: Pumps & Choosing the Right Pump). It is important that the chapter on pumps not be overlooked, or taken lightly, as it contains a vast amount of helpful information. For the purposes of addressing the pump in a flood-and-drain system, several issues will be reiterated in this chapter.

Two very important specifications must be considered when selecting a pump.
- **Flow rate** — tells how much water the pump can move per minute or per hour.
- **Working head specification** — the height at which the pump will pump the water.

The flow rate and working head are shown on the pump curve provided for most pumps. However, sometimes the manufacturer will just provide a range for both of these parameters rather than a pump curve. Typically, the greater the head (the higher the pump must push the water) the lower the flow rate.

When selecting a pump for your system, it is absolutely essential that you ensure both of these parameters (flow rate and working head) meet the operational needs of your system. Your pump needs to pump the entire volume of your fish tank every hour. If the elevation of your pump intake is 3.5 ft below the pump outlet elevation, then you will need a pump with a working head of 3.5 ft at your required flow rate. For instance, if you have a 500-gallon fish tank and must pump water up 3.5 ft in elevation, your pump must be able to pump 500 gallons per hour (gph), or otherwise stated as 8.3 gallons per minute (gpm), at 3.5 ft of head. To convert gph to gpm, divide by 60. Literature provided with most pumps has the flow rate stated in gph or gpm.

The pump works best when it is working at full capacity and not being throttled (restricted) between the pump and discharge outlet by a partially opened valve, undersized fitting, kink in a hose, undersized pipe or hose, etc. When the pump works without impediment there is less strain on the motor, pumping efficiency is optimized, and pump longevity is maximized.

If the pump is being used to draw solid waste from the fish tank, it is worth checking to ensure it can do this easily. Some pumps draw water from the sides, but not from underneath. With some pumps a few carefully drilled holes under the grill of the pump can make it easier for solid waste to be pushed through the pump to the grow beds. In other instances, where the pump has regular pumping problems, placing the pump on a paving stone slightly off the bottom of the tank (or pond) and/or installing a screen around the pump may be necessary.

One last point: always keep a backup pump in stock. If the pump breaks down it is important that you have the ability to promptly replace it. Even if several of your local home improvement stores usually have the pumps you need for your system in stock, it is still prudent to have your own backup pump on hand.

Plumbing / Piping / Hoses / Fittings

Piping is addressed comprehensively in 'Chapter 14: Plumbing'. It is highly recommended that you review Chapter 14 during the planning phase of your aquaponic system.

Steps for Planning and Building a Flood-and-Drain Aquaponic System

Following are step-by-step instructions for a re-circulating, media-filled aquaponic system after you have identified your specific goals for getting into aquaponics:

- Determine the area you have to work with, including space for storage of equipment and materials, walkways, and a harvesting area. Keep in mind future expansion.
- Determine your total desired grow bed area in square feet (or square meters).
- From grow bed area, determine the fish weight required (pounds or kg) using the ratio rule 1 lb (.5 kg) of fish for every 1 sq ft (.1 sq m) of grow bed surface area, assuming the beds are at least 12" (30 cm) deep.
- Determine fish tank volume from the stocking density rule above (1-pound fish per 5—7 gallons of fish tank volume or 1 kg per 40-80 liters).

Grow Bed and Fish Tank Relationship

- Start with a 1:1 ratio of grow bed volume to fish tank volume. The ratio can be increased up to 2:1 in a mature system (minimum 4—6 months) if desired after becoming confident in operating your system.
- Media bed supports must be strong enough to withstand the lateral and downward forces of the media, leaning (working) human body weight, water, and plants.
- The system needs to be made of only food-safe materials. They should also be safe for regular handling and not alter the pH of your system. Be wary of media or other items (tools, sharp gravel, etc.) that can rip liners or cut hands.

- When your fish are young and small, keep the number of plants in proportion to the size of the fish and their corresponding feed rate/waste production.

Grow Bed

The grow bed media should be 12 inches (30 cm) deep to allow for growing the widest variety of plants, to provide complete filtration, and to optimize bacterial colonization. A grow bed with a freeboard of 2 inches above the top of the media is helpful in terms of practical application. This allows the operator to work within the media bed without spilling media or water on the floor. Therefore, a grow bed 14 inches deep with 12 inches of media is recommended for optimal success.

Design the grow bed to be a working height that best suits the operator. An ergonomically correct height not only reduces strain on the back, arms and shoulders, but makes the aquaponics experience that much more enjoyable. Ergonomic studies show that the optimal standing work counter (surface of the grow bed) is about 4 inches (10 cm) below the operator's elbow.

Turning over or stirring the bulk of the media within the grow bed every 4 to 12 months is beneficial. If using a liner, be sure not to get too close with your shovel or do anything else that may risk puncturing or tearing the liner.

Use a media guard around all plumbing fixtures. A media guard can be made from a wide variety of common hardware supplies such as window screen material, larger pipes, wood, aluminum, plastic, paint strainers, etc. A media guard will greatly facilitate cleaning and repairing plumbing fittings.

Fish tank

A 250-gallon (1000 liters) or larger fish tank has been proven to create a more stable aquaponics system, where space allows. Larger volumes are better for beginners, because they allow more room for error.

With larger volumes, changes to water quality happen more slowly than in smaller systems. Keep the optimal 1:1 size ratio in mind, though (volume of grow bed equal with fish tank volume). To raise a fish to a length of 12 inches or "plate size", a fish tank volume of at least 50 gallons is required

Plumbing and Pumping

Flooding and draining the grow bed works much better than continuous flow. The draining action allows more oxygen to infiltrate through the grow bed. The least complicated way to achieve a reliable flood-and-drain system is using a timer. Although more complex, using a siphon is also a viable option. Advantages and disadvantages of both methods are presented later in this book.

If using a timer, running it for 15 minutes on and 45 minutes off has provided excellent results for standardized setups (12-inch deep media, 1:1 fish tank to grow bed size ratio, and rightly sized pump).

The right size pump will move the entire volume of the fish tank through the grow bed(s) every hour. Therefore, if the pump is setup to run for 15 minutes every hour (see above), and you have a 100 gallon tank, you need at least a 400 gallon per hour (gph) pump. Larger tanks may require additional "on" time. Moving the entire fish tank volume every hour is possible with most small to medium size tanks, but much more challenging with large tanks or ponds. With these large tanks or ponds reasonable compromises need to be made to move the most realistic quantities of water without causing problems.

When choosing a pump, go by the height that you have to lift the water above the pump (head) and flow rate (gph). It is prudent to select a pump with a slightly larger head than the minimum needed in case you need to divert water elsewhere during repairs or while conducting a maintenance task. Do not get hung up on horsepower. All this is explained in much greater detail in 'Chapter 19: Pumps & Choosing the Right Pump'.

Media

- Try to select a media that is inert, i.e., will not alter the pH of the system.
- Avoid media materials that decompose.
- Try to select media that is of the preferred size (1/2" — 3/4" is optimal).

The most commonly used, and proven effective, media types are lightweight expanded clay aggregate, lava rock, expanded shale, and gravel. If gravel is your preference, avoid limestone and marble as they can alter system pH. It is prudent to rinse the media prior to adding it to your system. Refer to Chapter 18: Growing Media for Plants for a comprehensive understanding of media.

Be sure to use non-chlorinated water in your system. This can be achieved by using well water, capturing rain water, running tap water through a filter to remove chlorine, or allowing tap water to sit in a holding tank for several hours to "off gas" prior to adding it to your system.

Ideally, only non-fluoridated water will be added to the aquaponic system. See 'Chapter 31: Water Quality' of this book for information about water quality, the harmful effects of fluoride, and ways to remove it.

Temperature

- It is best to select fish that will thrive at the water temperature your system will naturally gravitate towards.
- It is easier and much less expensive to heat water than it is to cool it.
- Heat is attracted to the system when surfaces are black.
- It is much less costly to retain heat through good insulating techniques than to rely heavily on a heating system.

Oxygen

It is imperative that dissolved oxygen levels for fish be above 3 ppm and preferably above 6 ppm. Use an air pump if achieving these essential dissolved oxygen levels is a problem. The air pump can be set on a timer to supplement oxygen on occasion to maintain desired levels. It will take some trial and error to find the right "sweet spot" in regards to how frequently the air pump needs to be turned on. Running it all the time is more costly and usually not necessary. There are automated dissolved oxygen sensors that will turn the air pump on and off as necessary, but such systems are costly and usually only used in large commercial aquaculture applications or public aquariums.

pH

Generally, targeting a pH of about 6.8 to 7.0 works best. This is a good compromise between the optimal ranges of the fish, the plants, and the bacteria. pH must be tested at least once per week. It is optimal to test daily; however, 3 — 4 times per week is usually sufficient. Review 'Chapter 31: Water Quality' for a comprehensive understanding on pH.

During cycling there is usually a rise in pH. Be cautious about using aquarium industry products to lower pH, because they often contain sodium which is unhealthy for plants. The best methods for lowering pH, in order of preference, are as follows:

- Hydroponic acids (i.e. nitric or phosphoric acids). The nitrate or phosphate is also beneficial to most plants.
- Other acids, such as vinegar (weak), hydrochloric (strong), and sulphuric (strong). However, be very careful about adding these products. They should be used as a last resort since directly adding these acids to your system can be stressful for your fish.

Avoid adding anything to your system containing sodium as it is harmful to plants. Sodium can also build-up in system concentrations over time. Do not use citric acid, as this is anti-bacterial agent and will kill the beneficial bacteria in the media.

After cycling your system pH usually keeps dropping and requires the operator take measures to raise it to the desired range. High pH is generally due to the water source such as hard ground water, or because there is a base buffer in the system such as eggshells, oyster shell, shell grit, incorrect media, etc. The best method for raising (buffering) pH if it drops below 6.6 is calcium hydroxide — "hydrated lime" or "builder's lime", potassium carbonate (or bicarbonate) or potassium hydroxide ("pearlash" or "potash"). If possible, alternate between calcium hydroxide and potassium hydroxide each time you need to raise the pH in your system. These substances are also good for plants. Natural calcium carbonate products such as eggshells, snail shells, and sea shells will raise the pH, but they take a long time to work. The problem is most operators add these ingredients, check pH two, four, or even six hours later and nothing has changed, so they add more. Then suddenly, the pH spikes, because they have added too much.

Stocking Density

The recommended stocking density is 1 pound of fish per 5 — 7 gallons of tank water (.5 kg per 20-26 liters).

Fish selection should take into account the following:

- Edible (Tilapia) vs. ornamental (Koi, goldfish, etc.).
- Water temperature based upon your climate (recommended).
- Diet of fish: Carnivore vs. omnivore vs. herbivore

In order to maintain the perfect balance between fish and plant population density, it is best to progressively harvest the fish as soon as they are big enough. Although the 1:1 fish tank to grow bed volume size ratio is optimal, it is better to have a larger grow bed to fish tank ratio than vice versa. Also, a lighter fish stock density has been found to be more forgiving if things go wrong.

However, bear in mind that there are other factors affecting the number of fish that can be grown in an aquaponic system, such as the species of fish

CHAPTER 25: FLOOD-AND-DRAIN SYSTEM INSTRUCTIONS & PROCEDURES

and feed rates. The more the fish are fed, the more waste they produce. It is important to monitor system parameters such as water flow rates, oxygen levels, pump rates and water temperature, as they all play a critical role as well. All these issues are discussed in detail in their respective chapters of this book.

Introducing Fish into the System

Make sure your system is fully cycled before adding fish. Cycling is discussed in comprehensive detail within 'Chapter 30: Starting and Managing Your Aquaponic System'. Ensure the pH is at the desired level for fish, plants, and bacteria. As a general rule of thumb a pH of 6.8 to 7.0 is typically ideal for most plants, fish and bacteria. Ensure the fish tank water temperature is within the recommended range for your fish.

Feeding Fish

A good rule of thumb is to feed your fish as much as they will eat in 5 minutes, 1–3 times per day. An adult fish will eat approximately 1 percent of its bodyweight per day. Fish fry (babies) will eat as much as 7 percent. Be sure your fish are being fed enough. However, be cognizant of the fact that over feeding fish will negatively affect water quality, is wasteful, and is an unnecessary increase in cost.

Fish not eating as they should, is a good indication they are stressed or unhealthy. Some factors that may result in fish not eating as they should include

- Living in conditions outside of their optimal temperature range
- Water quality issues: Improper pH range, too much ammonia in the system, inadequate dissolved oxygen

FIGURE 180. Step-by-step procedure of transferring a seedling into a media bed unit. Removing the seedling from the nursery tray (a); digging a small hole in the medium (b); planting the seedling (c); and backfilling the medium (d).

- Loud or irritating noises and vibrations
- Direct lighting upon the fish tank

Plants

It is prudent to avoid plants that prefer an acidic or basic soil environment. Plants can be started for aquaponics the same way they would for a soil garden — by seed, cuttings or transplant.

If plants are not thriving, it is probably due to the pH being out of the 6.8 to 7.0 range or inadequate nutrient (fish waste) delivery to plants (very rare problem). Adding red worms to the grow bed once the system has been fully cycled can benefit plants and the system as a whole.

Starting the System— "Cycling" with Fish

- Start out by adding only 1/2 as many fish as it would take for the fish tank to be fully stocked.
- Test daily for elevated ammonia and nitrite levels. If either becomes too high perform a partial water exchange.
- During the start-up cycling period, feeding the fish only once a day will help control ammonia levels.
- See 'Chapter 30: Starting and Managing Your Aquaponic System'.

Fishless Cycling

There are several sources of ammonia including synthetic (pure ammonia and ammonium chloride) and organic (urine and animal flesh). Add the ammonia to the fish tank a little at a time until a reading from your ammonia kit of ~5 ppm can be obtained. Record the amount of ammonia this took and add that amount daily until the nitrite appears (at least 0.5 ppm). If ammonia levels exceed 8 ppm, stop adding ammonia until the levels decrease back to 5 ppm. Once nitrites appear, cut back the daily dose of ammonia to half the original volume. If nitrite levels exceed 5 ppm stop adding ammonia until they decline to 2.0ppm. Once nitrates appear (5 — 10 ppm), and both the ammonia and the nitrites have dropped to zero, fish can be added to the system.

System Maintenance (after "cycling" is complete)

Ammonia and Nitrite levels should be less than 0.75 ppm. If ammonia levels rise suddenly, check to see if there is a dead fish in your tank. If Nitrite levels rise undesirably, something has likely occurred that has damaged the bacteria environment in the system. If either of these circumstances occur, stop feeding the fish until the levels stabilize, and, in extreme cases, do a 1/3 water exchange to dilute the existing solution. Nitrates can rise as high as 150 ppm without causing problems. If nitrates exceed 150 ppm, it would be prudent to harvest some fish, add additional plants or expand the system by adding another grow bed.

CHAPTER 26

Flood-and-Drain System Drain Options, Plans, and Instructions

Media Guard

Regardless of what type of drain option selected, they all need a media guard. The media guard can be a screen or a wide diameter piping with small holes or gaps drilled or cut into its sides. Its purpose is to allow the water to flow, but to block the media from getting into the plumbing. The media guard is very important as it performs a couple of functions. First, it allows you access to the outflow plumbing in case you should need to change anything. Secondly, it helps to stop roots and media from getting into the plumbing.

Some operators prefer to have the media securely fastened in place. Others prefer to have it sitting freely so that it can be rotated to break loose attached roots or removed easily for cleaning. Regardless of your preference, the media guard should be constructed in such a way as to allow air to enter at the top so that it does not form a siphon.

Drain System Options for a Flood-and-Drain System

There are two standard systems used to achieve the flood-and-drain effect. One or the other can be used, but not both within the same system. They are:

1. **An outlet drain located within the grow bed, with a timer based pump in the fish tank set on a defined periodic pumping schedule.**
2. **An auto siphon in the media grow bed with the pump in the fish tank or within a sump tank.** There are several types of auto siphons being used in aquaponic media bed systems. They are Bell, J-bend, loop, or pivot siphons.

These drain system options are explained in more detail below.

1. Outlet Drain with a Timer Based Pump System

A timer-based system is one common operational method as it is easy to setup and maintain. Basically, a timer is attached to the pump so that the pump is powered on for a specified time, depending on the media being used. A schedule of 15 minutes every 30-45 minutes tends to work well for most setups. During the pumping phase, the water is pumped into the grow bed until the water level reaches the level of the 'overflow drain'.

The overflow outlet drain high-point is set at 1-inch below the surface of the grow bed media. The

PART V: FLOOD-AND-DRAIN SYSTEM DESIGN AND LAYOUT

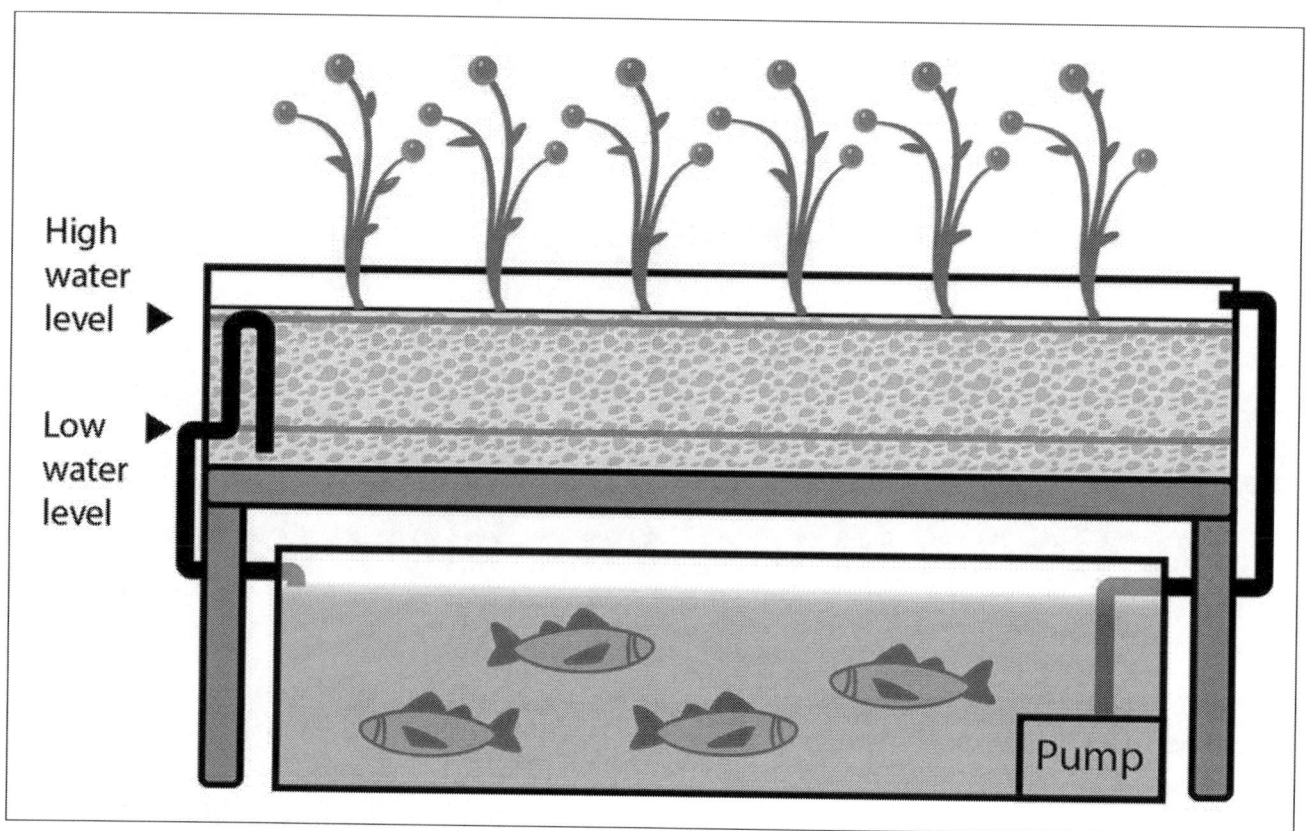

FIGURE 181. Flood and Drain Style Grow Bed

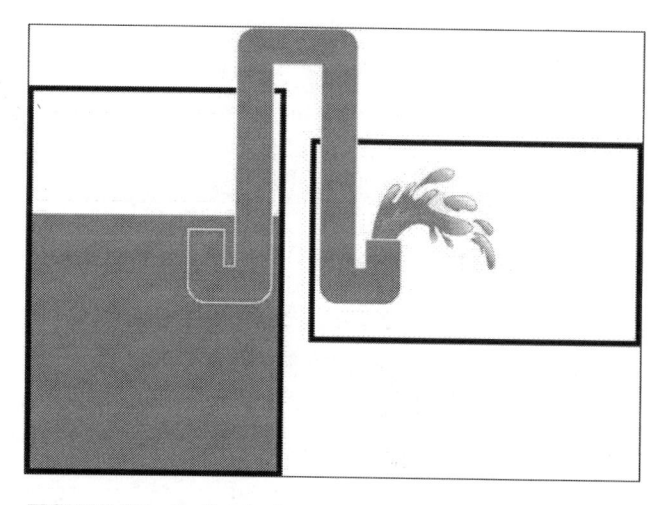

FIGURE 182. Outlet drain system concept

objective is to keep the top layer dry. This helps prevent the bottom leaves of the plants from becoming moldy, and to avoid an algae bloom from developing on the wet media surface.

After 15 minutes or so of flooding, the pump is switched off via of an automatic timer system. Water then drains back through the 'water-in-pipe' at the base of the grow bed and back to the fish tank. Figures 181 and 182 is an illustration of an outlet drain system using a timer based pump.

284

CHAPTER 26: FLOOD-AND-DRAIN SYSTEM DRAIN OPTIONS, PLANS, AND INSTRUCTIONS

J-Bend (Carlson Surge Device) / U-Bend System

A J-Bend is a tube bent like the letter "J" and used upside down. The long end is the drainage outlet. The short end is where the water will enter and seal off the opening. Once the water level is high enough to cover the top of the upside down "J", the siphon is started. These are used inside the container with the long end protruding through the container. A small U-Bend water trap at the bottom of the outlet pipe may be helpful if difficulties arise using this type of siphon.

U-Bend Siphon

Three U-Bends are assembled together to create a siphon-activated raised tank "no holes overflow" to transfer water outside a container without drilling holes through it and without spilling over the top of the container. It must be primed with water to remove all internal air and both ends at the same height to prevent air leaking inside. It is important that 'no holes overflows' be checked regularly to ensure that no air has accumulated inside the top, thus preventing normal operation. This is not usually visible when it occurs.

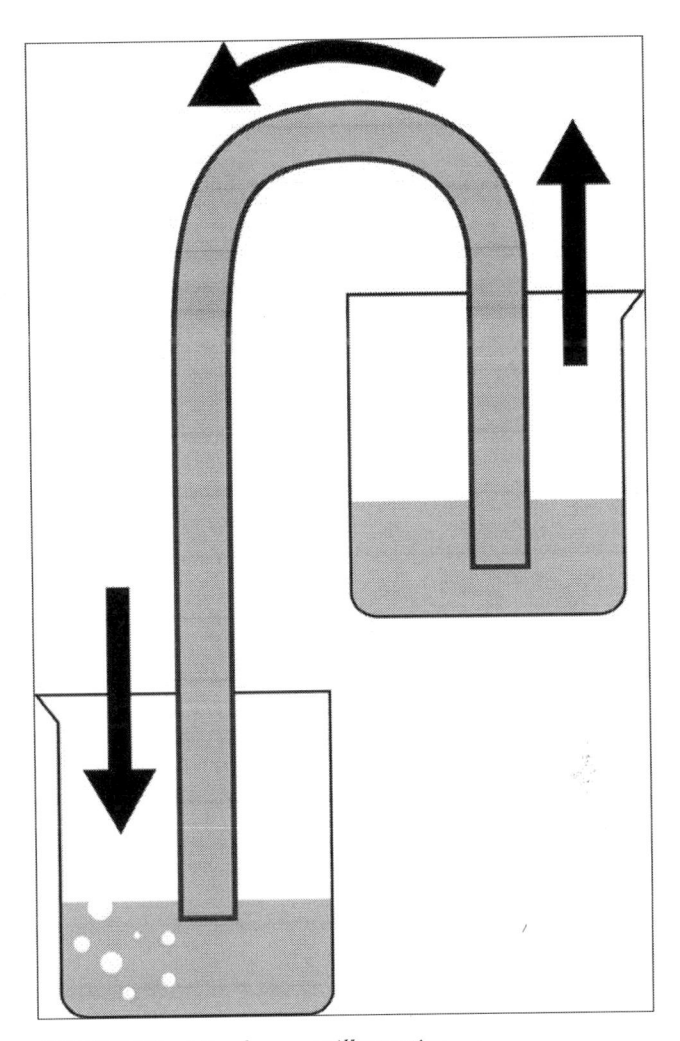

FIGURE 183. J-Bend concept illustration

FIGURE 184. U-Bend

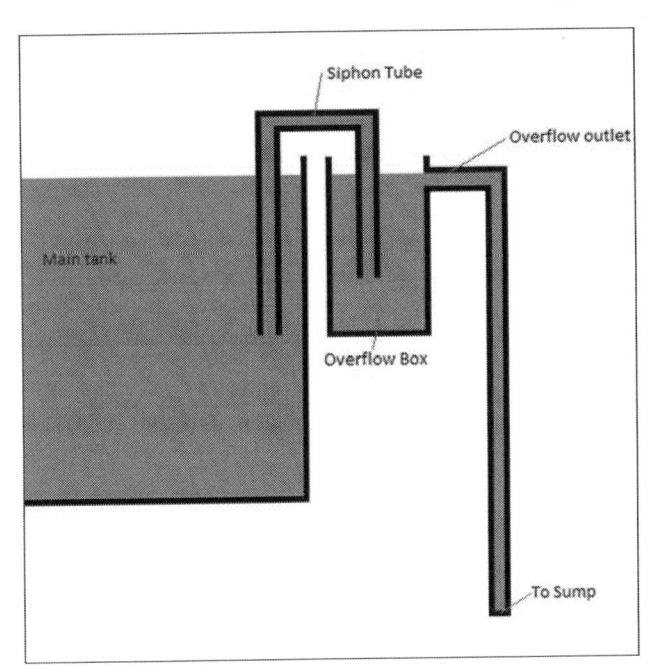

FIGURE 185. U-Bend concept illustration

285

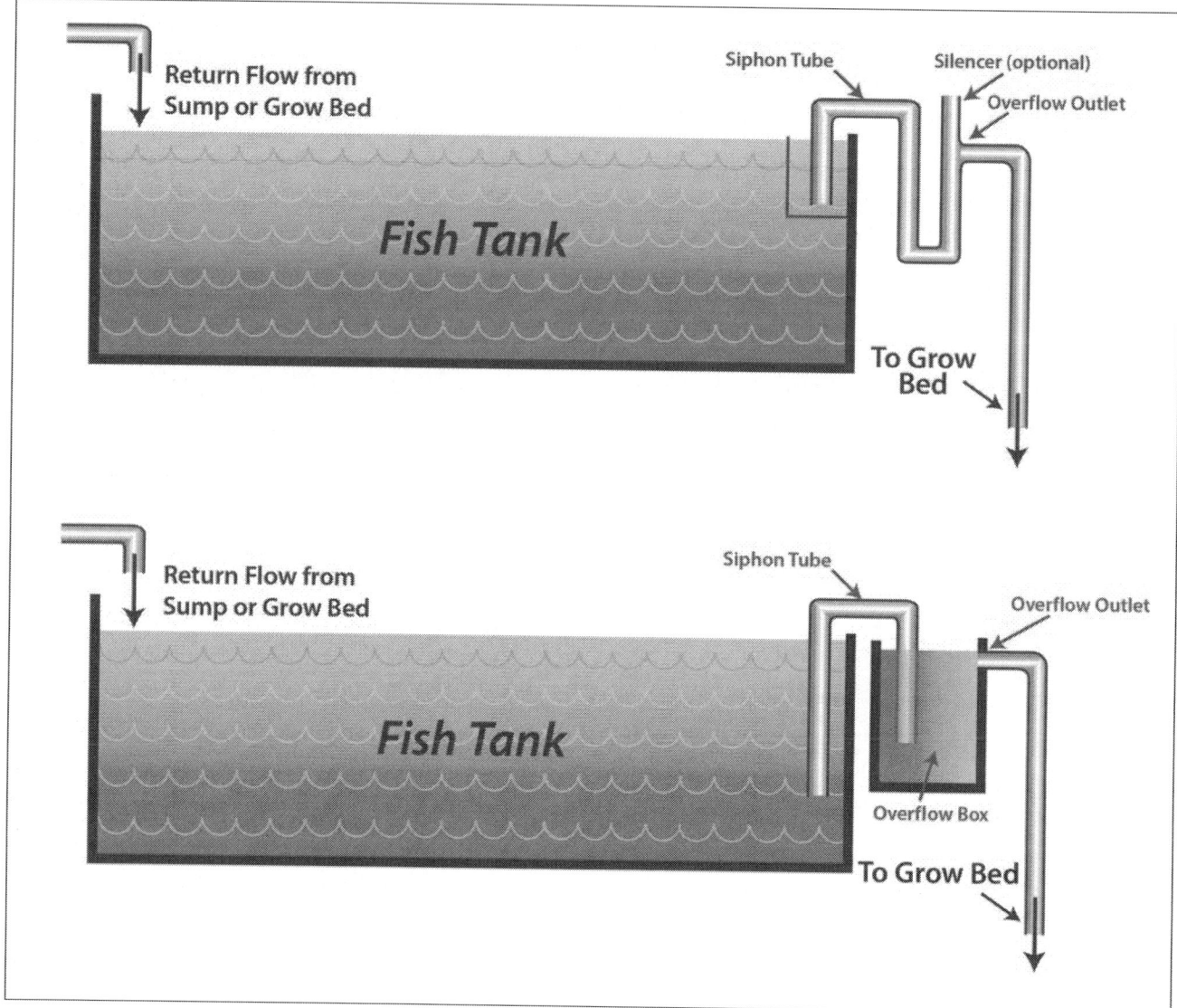

FIGURE 186. U-Bend examples

An "overflow box" is an upside down U-Bend/U-Tube with both ends in separate boxes, and works on the same principal. Figures 184, 185, and 186 are examples. A clear hose facilitates visual inspection.

Loop Siphon System

NOTE OF CAUTION: It is with much reservation that a loop system is included within this book. The author has not had any success using a loop system. However, it is included as others claim it as a reliable tank draining option.

This system features a loop of tubing with the outlet end usually pointed down. The open input end is where the water will enter and seal off the opening. Once the water level reaches the height of the top of the loop, it starts to push air and water down the output side, which engages the siphon affect. It is supposed to work the same way as a J-Bend, but the tubing forms a complete circle. A loop siphon can be used inside or outside the container.

The tube diameter also needs to be varied according to the size of the grow bed and the volume of water that needs to be drained. As a rule of thumb,

CHAPTER 26: FLOOD-AND-DRAIN SYSTEM DRAIN OPTIONS, PLANS, AND INSTRUCTIONS

FIGURE 187. Loop Siphon System

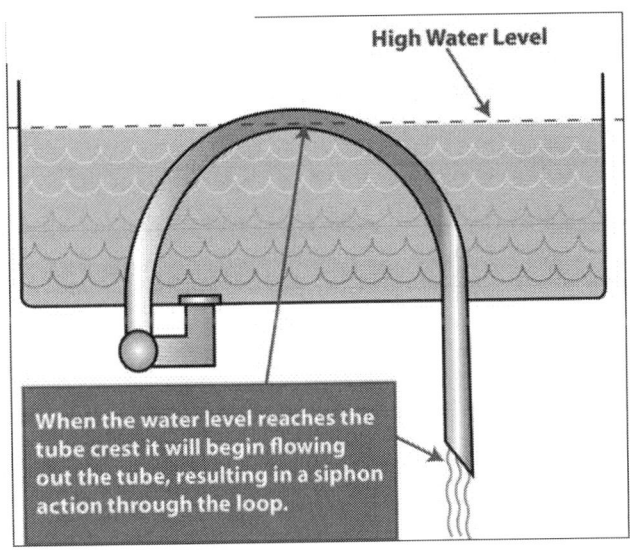

FIGURE 188. Loop Siphon System

it should be larger than the piping that brings the water to the grow bed so that it can drain faster than the rate that the grow bed is being filled.

As mentioned above, clear tubing provides the operator with the joy and confidence of watching the system at work. If using clear tubing, makes sure it is not exposed to direct sunlight, otherwise algae will grow on the inside of the tube.

2. Auto Siphon in Media Grow Bed with Pump in the Fish Tank or a Sump Tank

The second drain option is to use an auto siphon in the media grow bed with a pump in the fish or sump tank. The drain line can be tubing, garden hose, or PVC piping. Preferably, this will be larger in diameter than the pump line. Although not necessary, using clear tubing on the siphon drain line (or on a segment of the drain line) will allow you to observe water flowing through the tubing to ensure it is working properly. While a clear hose cost more, it also provides more confidence being able to watch the entire system at work.

The maximum height of water in siphons will match the maximum height of water in the container, so you need to adjust your siphon size and position it accordingly so it works at the proper water level, which is about 1-inch below the surface of the media. Another important item to note is that if siphons trap air at the top, they may not function properly, so it is best if they are tube shaped.

Aquaponic operators use all types of auto siphons, also referred to as bell siphons. With all siphon systems, typically the pump constantly propels water from the fish tank into the grow bed. As the water level in the grow bed rises, it fills the interior of the siphon, which is located within the grow bed. When the water level reaches a specific height, it overflows into a pipe within the siphon.

A low-pressure area is created within the siphon, triggering the drawing off of water from the grow bed, back to the fish tank. When the grow bed is almost completely drained, air enters the siphon and the draining action stops. The grow bed then begins to fill up slowly.

PART V: FLOOD-AND-DRAIN SYSTEM DESIGN AND LAYOUT

The design of the siphon is critical to the successful functioning of the siphon and its timely draining effect. The www.FarmYourSpace.com website has helpful instructional videos on these types of siphons, and well as other useful information.

NOTE: Some bell siphon setups have a funnel fitting at the top of the standpipe. Others have just a straight standpipe. Both configurations will work, therefore both types are shown within this book. The author favors a straight standpipe (without the funnel at the top), as it is easier to construct, and the funnel has not proven to provide any substantial claimed advantage.

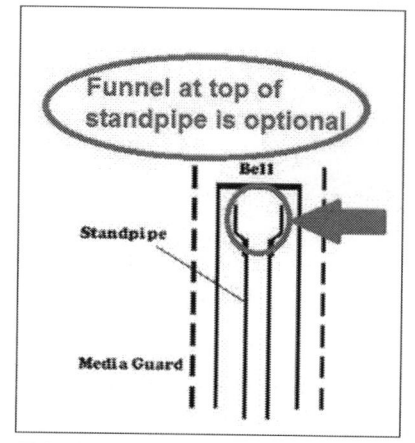

FIGURE 189. Auto siphon.

FIGURE 190. Top of an auto siphon.

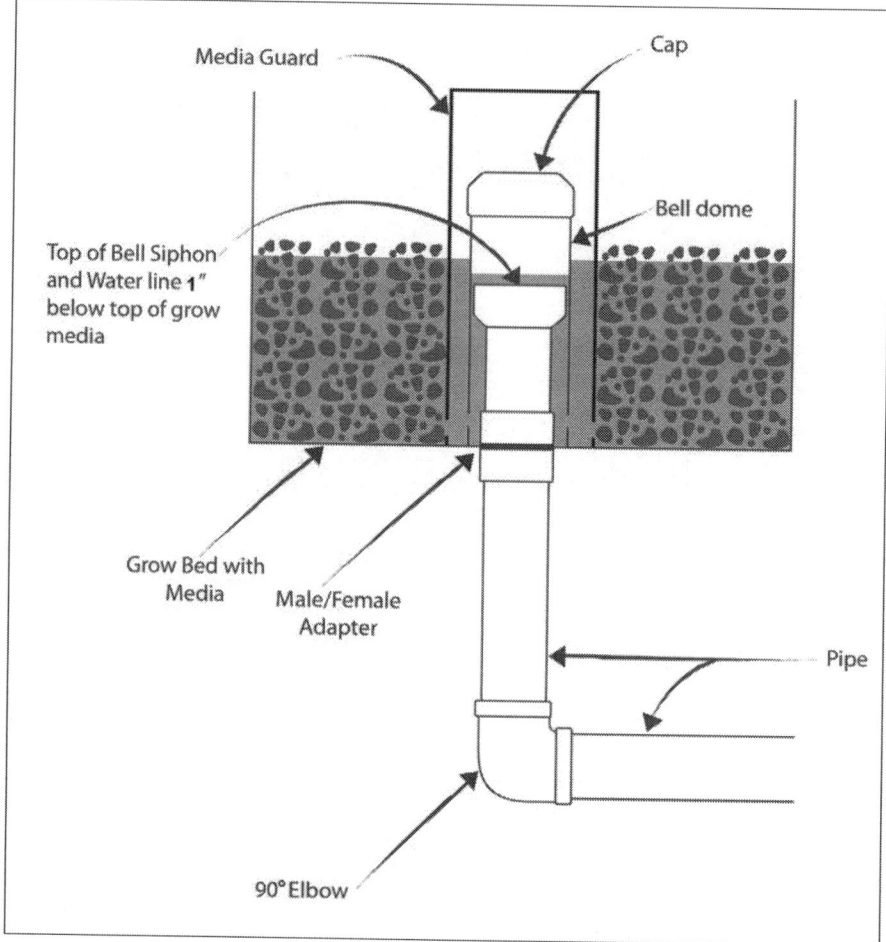

FIGURE 191. Auto siphon.

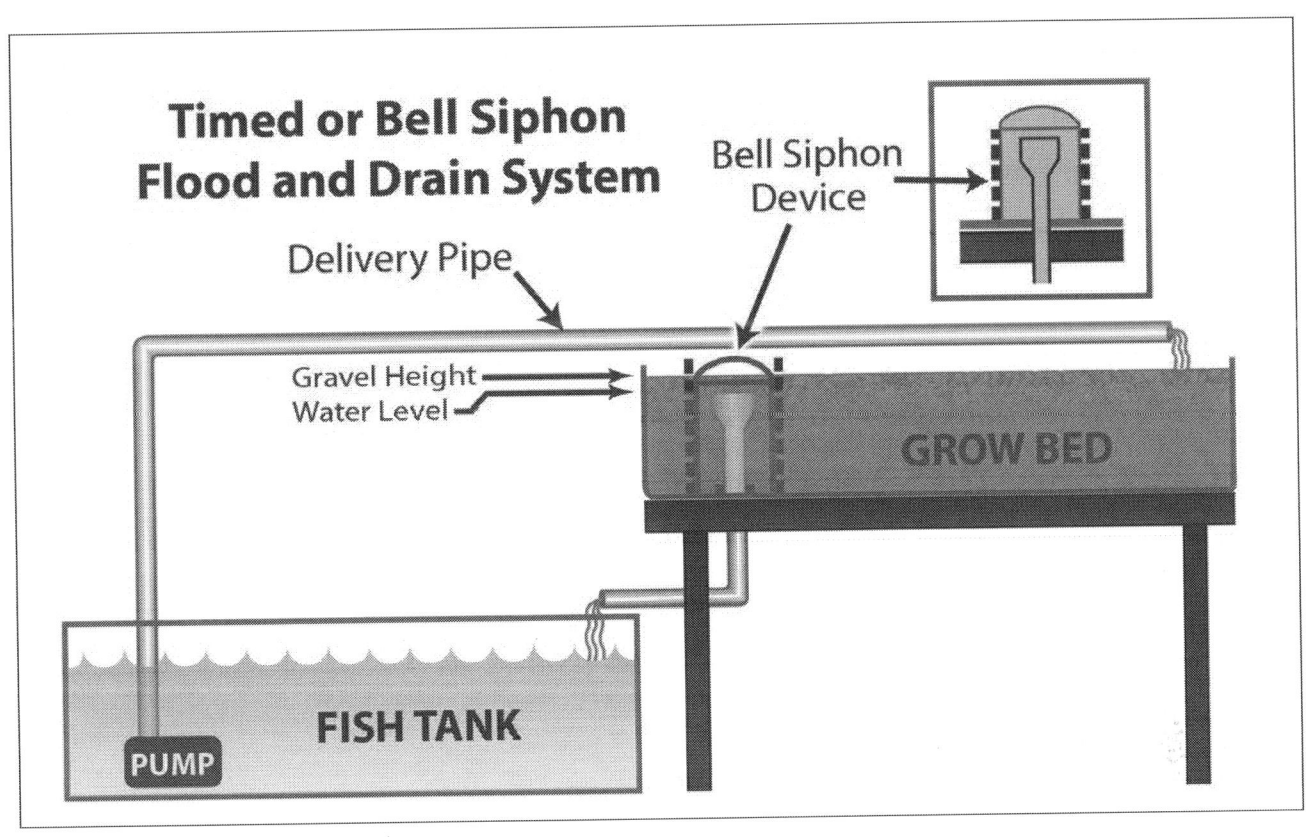

CHAPTER 26: FLOOD-AND-DRAIN SYSTEM DRAIN OPTIONS, PLANS, AND INSTRUCTIONS

FIGURE 192. Media-bed system without a sump.

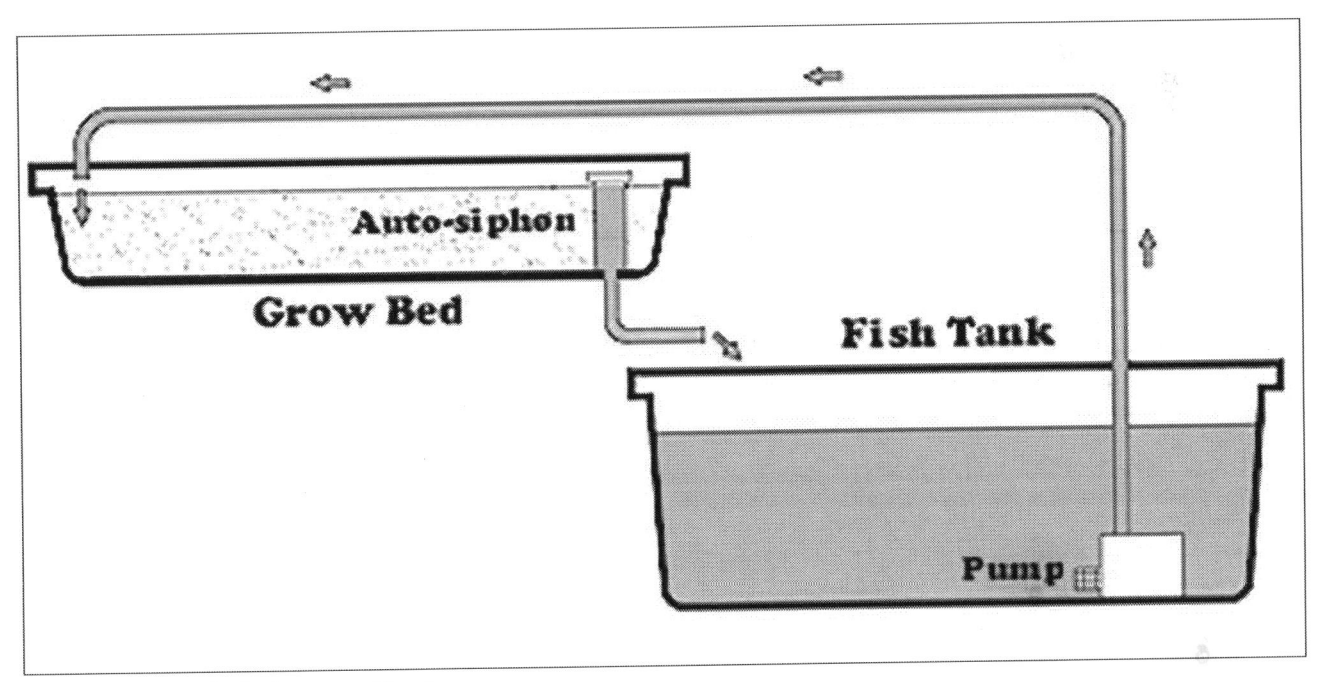

FIGURE 193. Media-bed system without a sump.

CHAPTER 27

How to Make a Auto Siphon (Bell Siphon)

Bell Siphon

A bell siphon is an ingenious way to drain the grow bed in a flood, and drain aquaponics system without using a timer to turn the pump on and off. Bell siphons enable you to effectively run a media filled aquaponics system by using the laws of gravity. To keep things simple, the bell siphon acts as a way to drain your system faster than it fills up. This process provides the plant roots with abundance of water, nutrients, and oxygen; thereby maximizing plant growth.

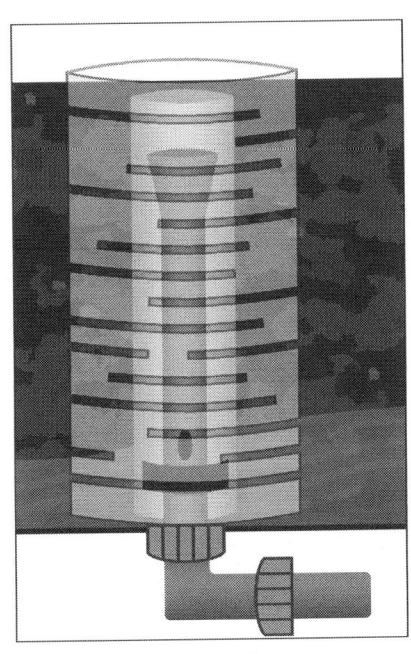

FIGURE 194. Auto-siphon

Bell Siphon Fundamentals

- As the water level rises in the grow-bed, water is forced through the teeth on the bottom of the bell and up between the walls of the standpipe and bell.
- As the water level exceeds the height of the standpipe and the drain begins to fill, a siphon is created.
- Most of the water in the grow-bed is then drained by the siphon until the water level reaches the height of the teeth and tip of the snorkel.
- Air is then forced through the snorkel and, as a result, the siphon is broken resulting in the grow-bed beginning to fill again; the cycle then repeats itself.

Benefits of Using a Bell Siphon:

- **No more timer** — This eliminates this cost of the timer and will lengthen the lifespan of your pump.
- **Pump runs continuously** — Pumps that run continuously suffer less wear and tear. This will result in a longer pump life.
- **More aeration** — If you are diverting part of your water stream back into your fish tank for aeration, using a siphon to drain your beds will increase the aeration you deliver to your fish tank (since

PART V: FLOOD-AND-DRAIN SYSTEM DESIGN AND LAYOUT

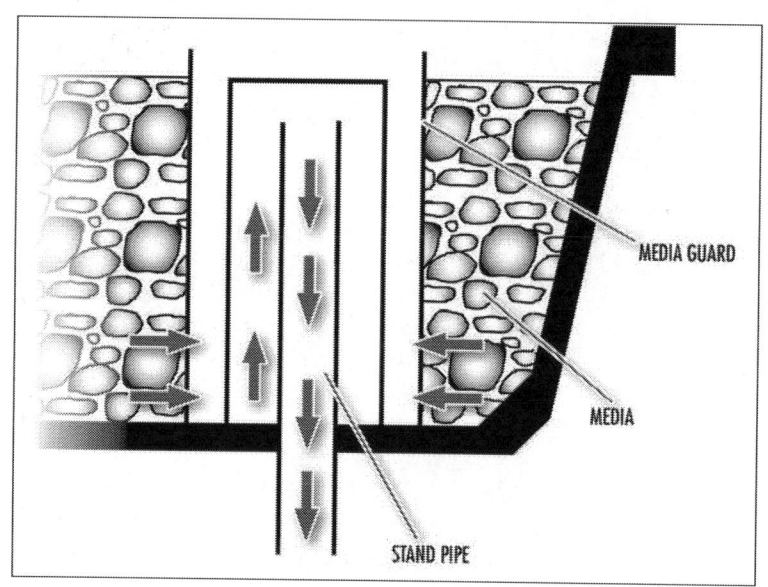

FIGURE 195. Auto-siphon illusrtation

aeration based on diverting a portion of the water stream only works while your pump is running).

As mentioned above, with a bell siphon, the pump is working all the time so the grow bed is filling continuously, but at regular intervals the siphon is activated and pulls all the water from the grow bed. The bell siphons work constantly, day and night, powered by nothing but simple physics (otherwise known as hydrodynamic water head pressure).

As the grow bed fills up with water, it reaches the designated set point whereby the water will start to drain. With just a drainage pipe, the media grow bed would fill and drain at the same rate. The water pressure build up creates a vacuum and drains the grow bed faster than it fills.

Sizing Bell Siphons and Drains

Before constructing a bell siphon, you need to decide which size of drain is appropriate for your grow-bed. The appropriate size of the bell siphon depends on the size of the individual grow-bed. In general, the larger the grow-bed, the greater the volume of water it can hold, and a larger standpipe and bell siphon is necessary to drain it. The recommended ratio of bell siphon size to drain is 2:1; the diameter of the pipe used to build the bell siphon should be twice that of the standpipe (i.e., if the standpipe is 1-inch in diameter, the bell siphon should be made using a 2-inch diameter pipe). Table 27 below shows some general sizing parameters for square and rectangular grow beds.

Figure 196 shows four sizes of bell siphons with accompanying standpipe and drain assembly. From left to right: 1/2-inch standpipe with 1-inch bell pipe, 1-inch standpipe with 2-inch bell pipe, 1.5-inch standpipe with 3-inch bell pipe, 2-inch standpipe with 4-inch bell pipe. Note that the top of the standpipe should be even with the bottom rim of the bell pipe when fully assembled in the growbed.

TABLE 27. **Measurements of Bell Siphon Components for Various Size Tanks**

Bell Pipe Diameter	2-inch	3-inch	4-inch
Standpipe/Drain Size (Diameter)	1-inch	1.5-inch	2-inch
Snorkel Tube Size (Dia.) and Material Type	7/16 OD × 5/16 ID Vinyl Tubing	5/8 OD × 1/2 ID Vinyl Tubing	7/8 OD × 1/2 ID Vinyl Tubing
Grow Bed Dimensions	1' x 4' x 4'	1' x 4' x 6'	1' x 4' x 8' to 1' x 6' x 10'
Volume of Grow Bed	16-ft^3, 120-gallon	24-ft^3, 180-gallon	32-ft^3, 240-gallon to 60-ft^3, 448-gallon

NOTE: Diameters(Inside Diameter = ID; Outside Diameter = OD) in inches

CHAPTER 27: HOW TO MAKE A AUTO SIPHON (BELL SIPHON)

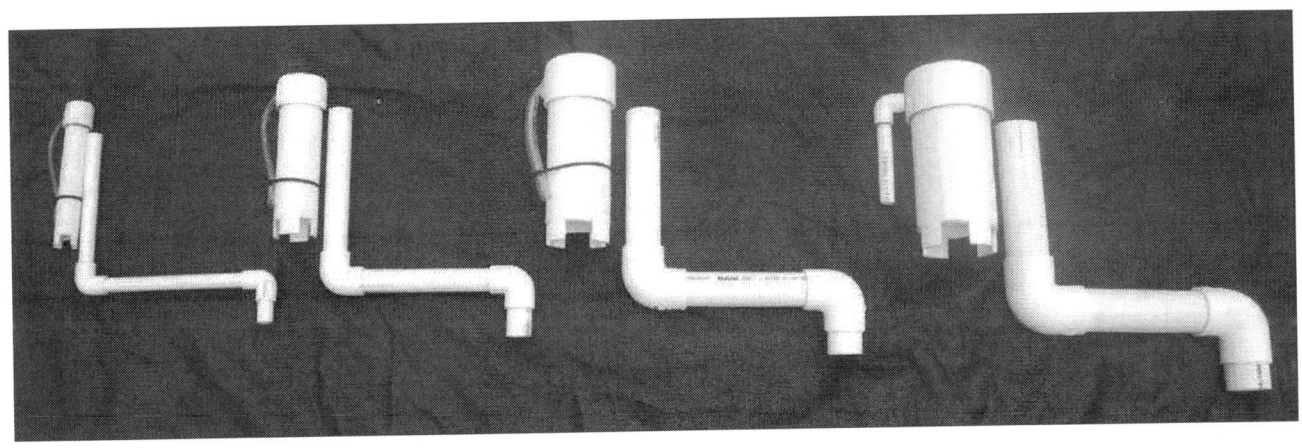

FIGURE 196. Auto siphons of different sizes.

How to Build a Bell Siphon

The height of the bell siphon needs to be appropriate for the size of your standpipe. Creating a space of about 1 to 2-inches between the top of the standpipe and the top of the bell siphon works best. The bell siphon is connected to the bulkhead at the bottom of the grow bed.

Step 1

Install a bulkhead fitting that will hold the standpipe in the grow-bed and drain the water into the fish tank. Bulkhead and Uniseal fittings are addressed comprehensively in 'Chapter 14: Plumbing'. Some aquaponic operators place 100 percent silicone caulking around their bulkheads as additional protection against leaks.

Step 2

Adjust the height of the standpipe to be 1-inch below the surface of the grow media. As discussed in the previous chapter, the grow-bed should be 14-inches deep with 12-inches of grow media. Therefore, the standpipe (where the water enters) would be 11-inches above the bottom surface of the grow bed. This keeps water from reaching the surface of the media where algae can grow.

NOTE: The top of the standpipe should be level with the bottom edge of the cap on the bell siphon. Coincidently, this is also 1-inch below the surface of the media. See figure 197.

Figure 197 shows the components of a bell siphon and companion standpipe with a bulkhead fitting. The dotted line (top large arrows) represents the maximum water level in the grow bed (located 1-inch below the surface of the media). Note that it is even with the top of the standpipe as well as the bottom of the bell cap. The lower dashed line and large arrows represents the minimum water level or the bottom of the grow bed. The dashed line is even with the top of the bulkhead fitting and the bottom of the teeth.

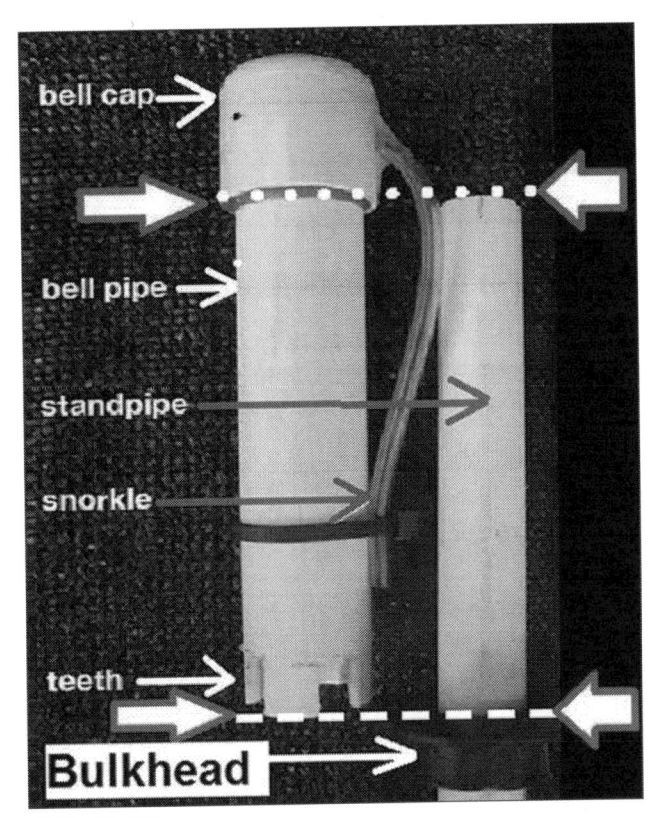

FIGURE 197.

PART V: FLOOD-AND-DRAIN SYSTEM DESIGN AND LAYOUT

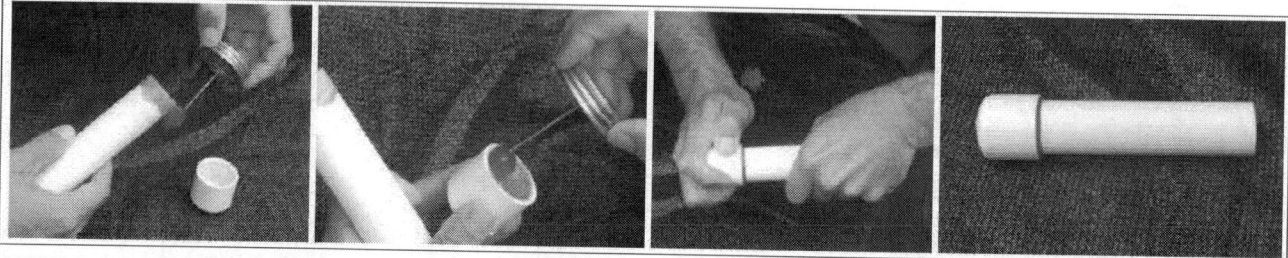

FIGURE 198. Attach the bell cap to the bell pipe with PVC primer and glue. After priming the inside surface of the cap and the outside surface of the pipe, apply glue to the rim of the pipe and the inside of the cap, as shown. Then, push the cap onto the pipe and twist the cap a quarter turn to seal the joint. Make sure the seal connecting the cap to the bell tube is airtight.

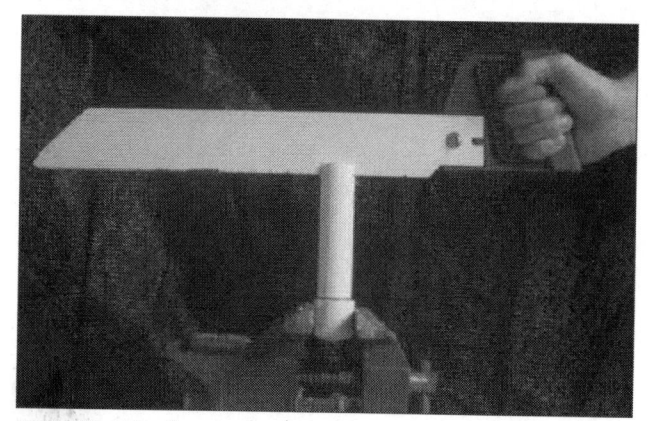

FIGURE 199. Cut notches to make teeth on the bottom of a bell tube. This can be done with various tools and methods.

Step 3

Prime and glue a PVC cap onto the end of the bell pipe (see figure 198). Next, cut the standpipe to the appropriate length (1-inch below the media surface — be sure to allow for the bulkhead fitting), and cut notches (or "teeth") into the bottom of the bell pipe. This can be done by securing the capped end of the bell pipe in a vise, and making two sets of two straight cuts perpendicular to each other across the open end of the pipe using a saw (see figure 199). Additional cuts should then be made on the lateral surface of the bell pipe, at the apex of the first cuts, to loosen the teeth (see figure 200). Pliers can be used to gently break away the material between the teeth, revealing the spaces through which the water will flow (see figure 201).

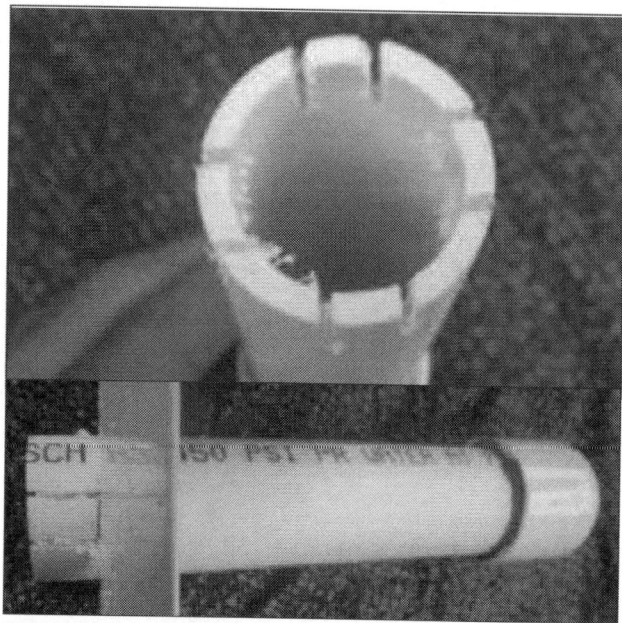

FIGURE 200. Make lateral cuts to weaken the spacers between the teeth.

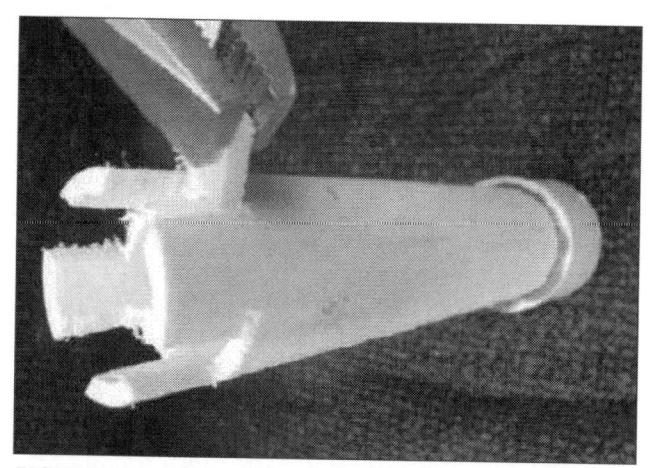

FIGURE 201. Remove the spacers between the teeth with pliers. If cut properly, the spacers break off cleanly and with ease.

Step 4

Once the teeth have been cut in the bottom of the bell pipe, the snorkel that will ultimately break the siphon must be made. Depending on the size of your bell siphon, use a drill bit or hole-saw with a diameter approximately the same size as the tubing or pipe (see figure 202), and drill a hole into the side of the cap that makes up the end of the bell (see figure 203).

Next, push the tubing or pipe through the hole so that the tubing penetrates the inside of both the cap and pipe wall of the bell and extends 1/4 inch inside the bell cap (figure 204). Using a bead of 100 percent silicone, seal the gap surrounding the tubing at the entrance to the bell and allow to dry completely. It is important to create an airtight seal, because if air enters the top of the bell through the space around the snorkel during use, the siphon will not start properly.

After the seal has dried, gently train the snorkel along the length of the bell pipe and secure the snorkel in place with a cable tie. Next, cut the free end of the snorkel so that it ends above the teeth. If the snorkel is cut too long (i.e., the open end of the snorkel is lower than or even with the height of the teeth), the siphon will not break properly.

An alternative approach to snorkel design is to drill a threaded "tap" hole in the bell pipe and screw a 90° hose barb fitting into place. This way, the snorkel tube can extend directly down along the bell pipe toward the teeth without having to make a sharp turn. The absence of a silicone seal and the lessened stress on the vinyl tubing should extend the life of the bell siphon. An example of this design can be found in figure 196 (the 4-inch bell siphon was built using a 1/2-inch, 90° PVC elbow and straight pipe).

Step 5

Once the bell siphon is completed, a "media guard" needs to be constructed. This is a porous tube placed around the standpipe and bell siphon before adding the media to the grow bed. The function of the media guard is to prevent clogs to the standpipe and bell

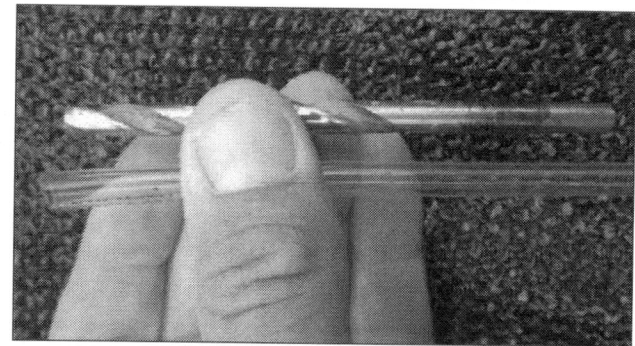

FIGURE 202. *Choose the appropriate size (diameter) drill bit or hole-saw for making the snorkel. The hole drilled in the bell pipe should be only slightly larger than the diameter of the tubing itself, to ensure a tight seal.*

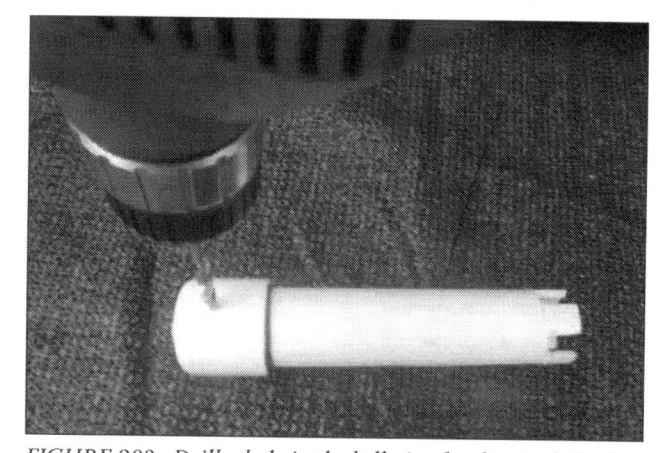

FIGURE 203. *Drill a hole in the bell pipe for the snorkel tube.*

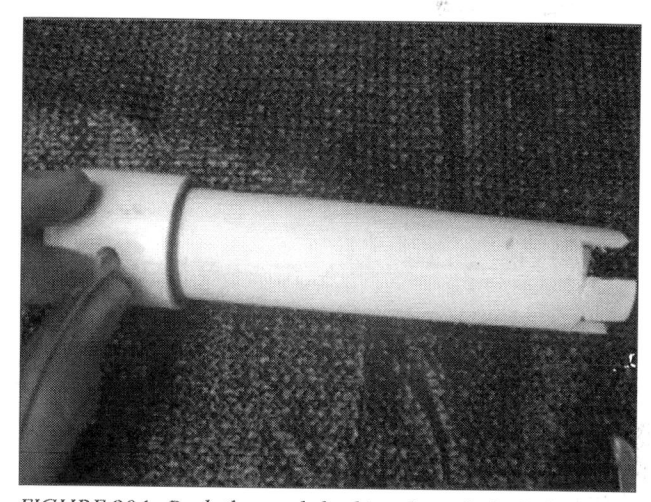

FIGURE 204. *Push the snorkel tubing through the bell pipe cap assembly.*

PART V: FLOOD-AND-DRAIN SYSTEM DESIGN AND LAYOUT

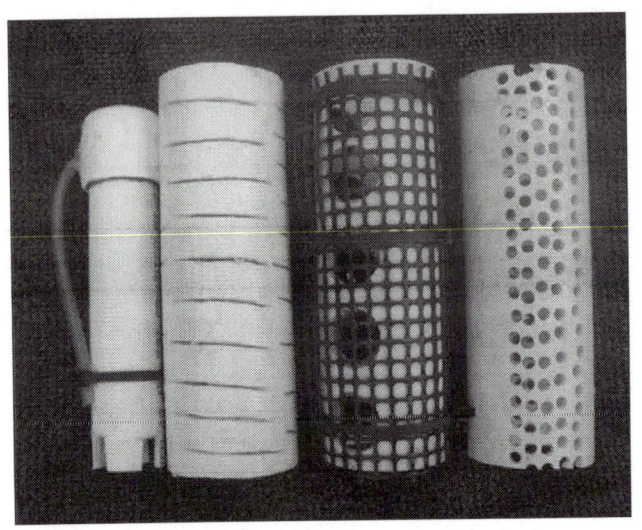

FIGURE 205. *Different styles of gravel guard*

siphon while allowing water to easily flow through. The medial guard also allows easy access and maintenance to the bell without having to remove or dig through the media. Although a media guard is not necessary for bell siphon function, it is a prudent investment of time and materials for the added benefit of ease of maintenance and protection. Particles from the grow media can easily obstruct or constrict the flow to the bell siphon (see figure 205).

Step 6

Following construction of the bell siphon and media guard, the completed auto siphon array can be assembled over the standpipe in the grow bed. Place the bell pipe over the standpipe so that the teeth rest evenly on the bulkhead or on the bottom of the grow bed. Next, place the media guard over the bell siphon. Refer to 'Chapter 18: Growing Media for Plants' for media options. If using cinder, gravel, or clay balls media, be sure to thoroughly rinse them before adding them to the grow bed. Carefully add the media around the base of the media guard so as not to disturb its placement. Once a firm base of medium has been added around the gravel guard, continue filling the rest of the grow-bed to a height of about 1-inch above the top of the bell siphon (a total of 12-inches of media depth). This added height of media ensures that the ebbing water in the grow-bed is not exposed to sunlight before the system flushes, which helps prevent algal growth in the grow-bed.

Step 7

The last step in the bell siphon assembly is the installation of the drain extending from the bottom of the standpipe on the underside of the bulkhead beneath the grow-bed (see figure 206 below). Add a 90° elbow to the bottom of the standpipe. Extend this elbow with a length of straight pipe and add an additional 90° elbow fitting. The bends in the drain created by the elbows assist the bell siphon in starting and stopping the flow of water by providing some resistance to the water exiting the grow bed.

FIGURE 206. *The drain assembly on the underside of the grow bed. Note the two sequential 90° elbow fittings, which help start and stop the siphon and direct the flow of water back to the rearing tank beneath the grow bed.*

Bell Siphon Assembly Summary

The height of the standpipe in the grow bed should be level with the bottom of the cap on the bell pipe. This relationship of standpipe to bell pipe height is important in ensuring that the volume of air resident in the top of the bell pipe is sufficient to start the siphon. The "double-double rule" is that the diameter of the media guard should be at least double the diameter of the bell pipe, which is double the diameter of the standpipe.

The drain assembly (consisting of the plumbing on the underside of the grow bed extending from the bottom of the standpipe) should contain two 90° elbow fittings in series connected by a length of straight pipe (see photo above). This arrangement is necessary to restrict the flow of water moving through the drain, and it assists both the starting and stopping of the siphon. An alternative approach is to add a reducer fitting to the bottom end of the standpipe, which acts in a similar way.

How fast water flows into the grow bed will determine the duration of cycling in the system. In other words, the faster water is added to the grow-bed, the faster it will fill up, and the shorter the duration between flushes. In general, flow-and-drain cycles should be about 15–20 minutes, regardless of the size or volume of the grow-bed. If possible for your size system, adjust the flow water into the grow bed (via proper pump size selection) so that the bell siphon starts, drains, and stops every 15–20 minutes. This also removes the fish tank volume once every hour. As a friendly reminder, the depth of the growth media should be around 12-inches for optimal filtration and plant growth.

CHAPTER 28

Design Plans and Construction Details DIY Flood-and-Drain System

Building Your Own Flood-and-Drain Aquaponic System

The aquaponic system design details provided in this chapter are for a system that can easily be constructed using common materials. If budget is an issue, these materials can gradually purchased over time. Perhaps used lumber can even be acquired at a lower price in your area. Craigslist can be a good resource for used building supplies, where applicable.

Another benefit to this system is that the sump tank is also located directly under the grow bed. Therefore, the sump tank does not take up any additional space. This configuration will allow for efficient drainage of the grow bed during operation, and if/when the grow bed ever needs to be emptied for

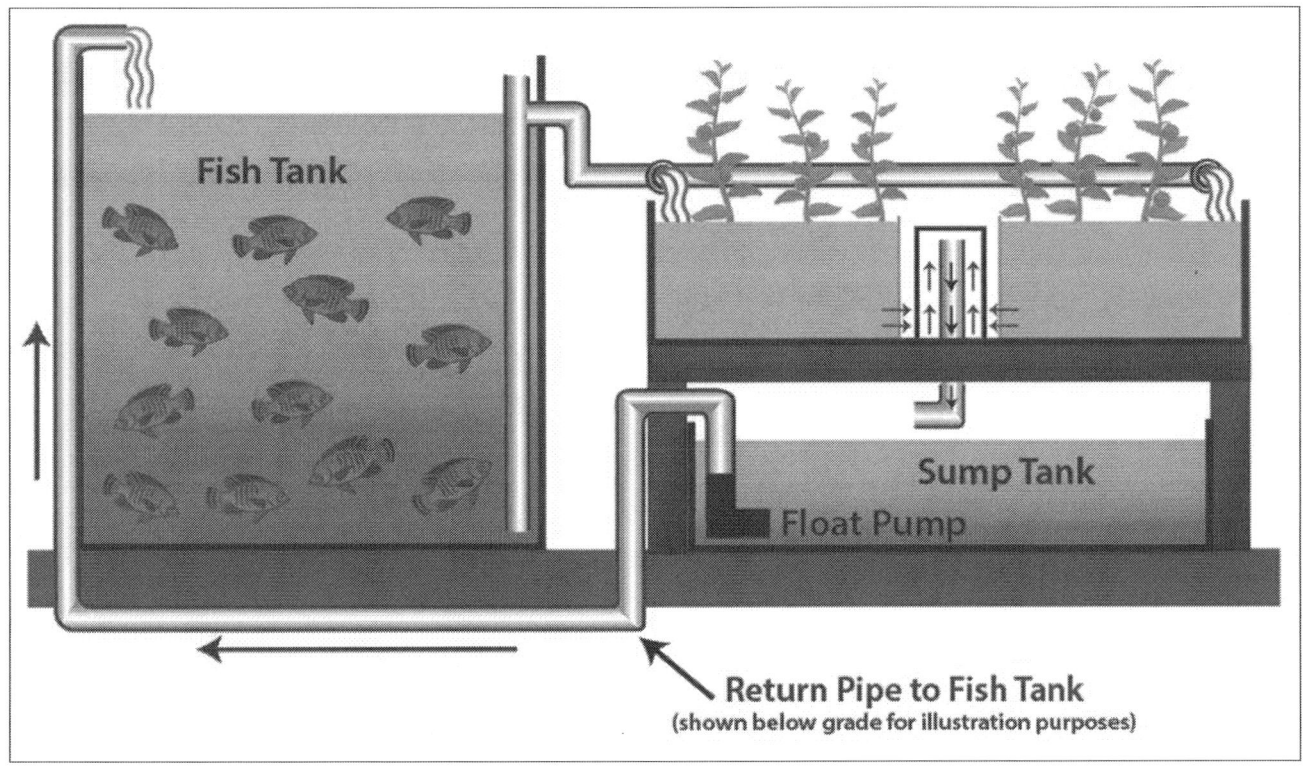

FIGURE 207. *Illustration of aquaponic system referenced in this chapter.*

299

maintenance, repairs, or for swapping out a different type of media. The sump tank size and location will also allow it to be utilized as a temporary back-up fish tank in case there is an emergency problem with, or the need to switch out, the main fish tank.

The size of this system is appropriate for beginners and can easily be enlarged over time. Furthermore, this system will easily fit in most backyards and still leave the home owner with space to spare. Best of all, this low-maintenance system will provide a substantial amount of healthy food at a very low operational cost. It will also serve as a wonderful conversation topic during social encounters, and is fun to show off to others. It can also be a very fun and educational family enterprise; providing bonding opportunities, and teaching youth many valuable lessons.

System Annual Production Output

Under normal operating conditions this system will produce an average of 77 lbs (35 kg) to Tilapia per year. This relatively small system will produce 144 lbs (65 kg) of vegetables annually. These figures are based upon proven values for tank space volume for Tilapia and grow bed square footage area of vegetables of typical flood-and-drain aquaponic system remaining in active operation throughout the year. Full year production is accomplished via location in a temperate climate or with the use of a greenhouse in colder climate regions and supplemental lighting up to a few hours each during winter months.

Construction Flexibility Options

Several fish tank options are provided in this design layout so as to give the end some end product and cost flexibility alternatives. Several options are also provided in regards to waterproofing the grow bed and sump tank. The hope and intention of these construction options will provide the end user some flexibility in constructing the level of quality aquaponic system that best fits his/her preferences and budget.

System Development Notes

1. **Fish Tank:** This system is based upon 300 gallons of water in the fish tank. The actual size of the fish tank will be listed greater than 300 gallons to accommodate freeboard (the vertical distance between the maximum water level and the top of the tank).
2. The **grow bed** and **sump** are each designed to hold 300 gallons of water.
3. **Ratio:** The volume of fish tank to grow bed volume has the desired 1:1 ratio (approx.) in order to provide optimal system operational conditions.

System Weight:

4. One (1) cubic foot of water = 7.48 gallons = 62.43 lbs.
5. One (1) gallon of water weighs 8.3 lbs.
6. The fish tank will have an approximate weight of 2,575 lbs.
7. The grow bed can weigh up to 3,100 lbs (water+media+plants+growbed+weight of operator leaning against the bed).
8. The sump tank can weigh up to approximately 3100 lbs.
9. The weight of the finished grow bed, sump, and appurtenances thereto could potentially weigh up to 6,500 lbs (approx. 136 lbs/SF).
10. The total weight if this operational aquaponic system is acceptable for most concrete slabs-on-grade floors. Further structural evaluation is needed if considering placement of this system on any ancillary structures (i.e. second floor, deck, etc.).

Material Notes:

11. **Bell Siphon:** Refer to Chapter 27 on 'How to Make a Bell Siphon' for making a Bell Siphon for this system.
12. Use **wood screws** for assembly.
13. **Adaptor:** Installing an adaptor at the ½-inch diameter pipe outlet to allow the return line to

be either 5/8-inch diameter irrigation tubing or a 1-inch diameter PVC pipe will help to reduce friction head on the return line.

14. **Growing Media:** Using a growing media of expanded clay pebbles (common trade name: Hydroton) would provide the best results, but is also the most costly option. Cost can be reduced by using a growing media mix of expanded clay pebbles and smooth pea gravel (no sharp edges on the gravel). Be sure to rinse the pea gravel thoroughly before placement. Other growing media options are available as well (refer to 'Chapter 18: Growing Media for Plants'). This system will need approximately 52 cubic feet of media.

15. **Lumber:** Treated lumber is more costly, but will provide the greatest longevity and is a wise choice given the ambient area moisture from the fish tank, sump, and grow bed. However, using treated lumber is not essential.

Material Information

The total material and supplies estimated cost to build this system is $1,000 (USA year 2016 prices). However, considerable savings of several hundred dollars can be achieved by purchasing a used fish tank, used and/or lumber that is not treated.

Only one pump is needed for this system. However, it is prudent to purchase two pumps so as to have a backup pump readily available in case the system goes out.

Plumbing Material Cost: 2-inch diameter PVC piping and fittings, auto siphon (Bell Siphon) parts, Return Line from sump pump to fish tank, plumber's

TABLE 28. **Lumber Sizes Need**

NOMINAL SIZE	ACTUAL SIZE	QUANTITY
1" x 12" x 12'	¾" x 11¼" x 12'	4
1" x 12" x 4'	¾" x 11¼" x 4'	*4
1" x 3" x 12'	¾" x 2.5" x 12'	4
1" x 3" x 4'	¾" x 2.5" x 4'	*4
2" x 4" x 12'	1.5" x 3.5" x 12'	2
2" x 4" x 38"	1.5" x 3.5" x 38"	**9
4" x 4" x 37.25"	3.5" x 3.5" x 37.25"	***8
15/32" x 4-ft x 8-ft Plywood	————	1
15/32" x 4-ft x 4-ft Plywood	————	1

*The most practical and economical approach is to purchase two standard 8-ft long boards and cut them in half to obtain the required 4-ft long boards.

**The most practical and economical approach is to purchase five standard 8-ft long boards and cut them to obtain the required four 38-inch length 2"x4" boards.

***The most practical and economical approach is to purchase four standard 8-ft long boards and cut them to obtain the required four 37.25-inch long 4"x4" boards.

PART V: FLOOD-AND-DRAIN SYSTEM DESIGN AND LAYOUT

TABLE 29. **Lumber Shopping List & Estimated Cost**
(Table 28 Summarized for User-Friendly Shopping Purposes)

LUMBER	QUANTITY	USA 2016 AVERAGE COST (EACH)	USA 2016 AVERAGE COST (TOTAL)
1" x 12" x 12'	4	$23.25	$93.00
1" x 12" x 8'	2	$19.50	$39.00
1" x 3" x 12'	4	$5.75	$23.00
1" x 3" x 8'	2	$4.50	$9.00
2" x 4" x 12'	2	$8.27	$16.54
2" x 4" x 8'	5	$5.67	$28.35
4" x 4" x 8'	4	$10.57	$42.28
15/32" x 4-ft x 8-ft Plywood	2	$26.27	$52.54
Box of 3-Inch Long Wood Screws	1	$8.47	$8.47
Box of 2-Inch Long Wood Screws	1	$2.75	$2.75
Box of Staples for Staple Gun –Used to Staple Liner (Not Needed if Using the DIY Fiberglass Option for Sump Tank and Grow Bed)	1	$3.00	$3.00
		Subtotal	$317.93
		Estimated Sales Tax (various per location)	$27.02
		Total Estimated Lumber Cost	**$344.95**

TABLE 30. **Fish Tank Options**

OPTIONS	TANK TYPE	APPROX. COST (USA, YR 2016)
Option #1	330 gallon IBC Tank	$150 (used)
Option #2	48" diameter x 44" deep Fiberglass Tank	$780

TABLE 31. **Sump Tank & Grow Bed Waterproof Options**

OPTIONS	TANK TYPE	APPROX. COST (USA, YR 2016)
Option #1	20 to 40 mil HDPE Liner – 6.5-ft x 14.5-ft for Grow Bed – 6.5-ft x 14.5-ft for Sump Tank	$82–$147
Option #2	Fiberglass (DIY Inside Boards)	$75–$125

CHAPTER 28: DESIGN PLANS AND CONSTRUCTION DETAILS DIY FLOOD-AND-DRAIN SYSTEM

TABLE 32. **Plumbing List**
(Summarized for User-Friendly Shopping Purposes)

ITEM	QUANTITY
2-Inch Dia. PVC Pipe (Schedule 40) typically sold in 10-ft lengths	Purchase two 10-ft pipes.
2-Inch Dia. PVC Tee-Fitting	2
2-Inch Dia. PVC 90-Degree Bend Fitting	4
2-Inch Dia. PVC 30-Degree Bend Fitting	1
PVC Cement (Glue), 8-Ounce Can	1
Plumber's Tape (Teflon Tape), Roll of Tape	1
Sump to Fish Tank Return Line: 5/8-Inch Dia. Polyvinyl Tubing (Irrigation Tubing), or 1-Inch Dia. PVC pipe. Be sure to purchase fittings (Adapter from pump outlet to tube or pipe, and 90-Degree Bends).	Approximately 20-ft of 5/8-Inch Dia. Irrigation Tubing or 1-Inch Dia. PVC Pipe needed (plus associated fittings)
Pump (Brand: United Pump UP-580 Submersible (or equal), 580 GPH. Only one pump is needed, but it is prudent to have a back-up readily available; therefore two pumps are recommended ($60 each at USA 2016 prices)	1
2" x 12" x 12" Concrete Block	1
U-Bolts to secure SLO 2-Inch Dia. Pipe within Fish Tank (2.5" x 3" U-Bolt)	2
Grommet for U-Bolts (Size to Fit Thread Dia. Of U-Bolt)	4
Grommet for SLO 2-Inch Dia. PVC Pipe Protrusion through Fish Tank (2-Inch Dia.)	1
Silicon Caulking, Tube	1
Pipe Clamp to Secure 2-Inch Dia. PVC Pipe to Top Edge of Grow Bed	6
Pipe Clamp to Secure 5/8-Inch Dia. Irrigation Tubing or 1-Inch Dia. PVC Piping to Wood Frame	4
Bell Siphon Assembly (Refer to the Chapter on How to Make a Bell Siphon for Details)	1
Grow Media (refer to note #14 above regarding the grow media)	52 cubic feet

tape (Teflon Tape), PVC cement (glue), and misc. items. Approximate Cost = $185 (USA 2016 prices).

Safety Precautions

- A simple tank leak or plumbing problem could be deadly situation if the water comes into contact with electricity. Secure all electrical conduits. Do not leave extension cords on the ground.
- Ensure all electrical outlets in the area are of the GFCI type, or replace them with such if they are not.
- If the aquaponic system is to be constructed on a concrete slab, then it is recommended that a non-slip epoxy coating be applied on walking paths near the system.
- Be sure that remove all tripping hazards from the area.
- If applicable to your situation, be sure to child proof the fish tank and any other potential hazards.

Design Plans

Build according to the following drawings.

PART V: FLOOD-AND-DRAIN SYSTEM DESIGN AND LAYOUT

DIY AQUAPONIC SYSTEM PLAN VIEW

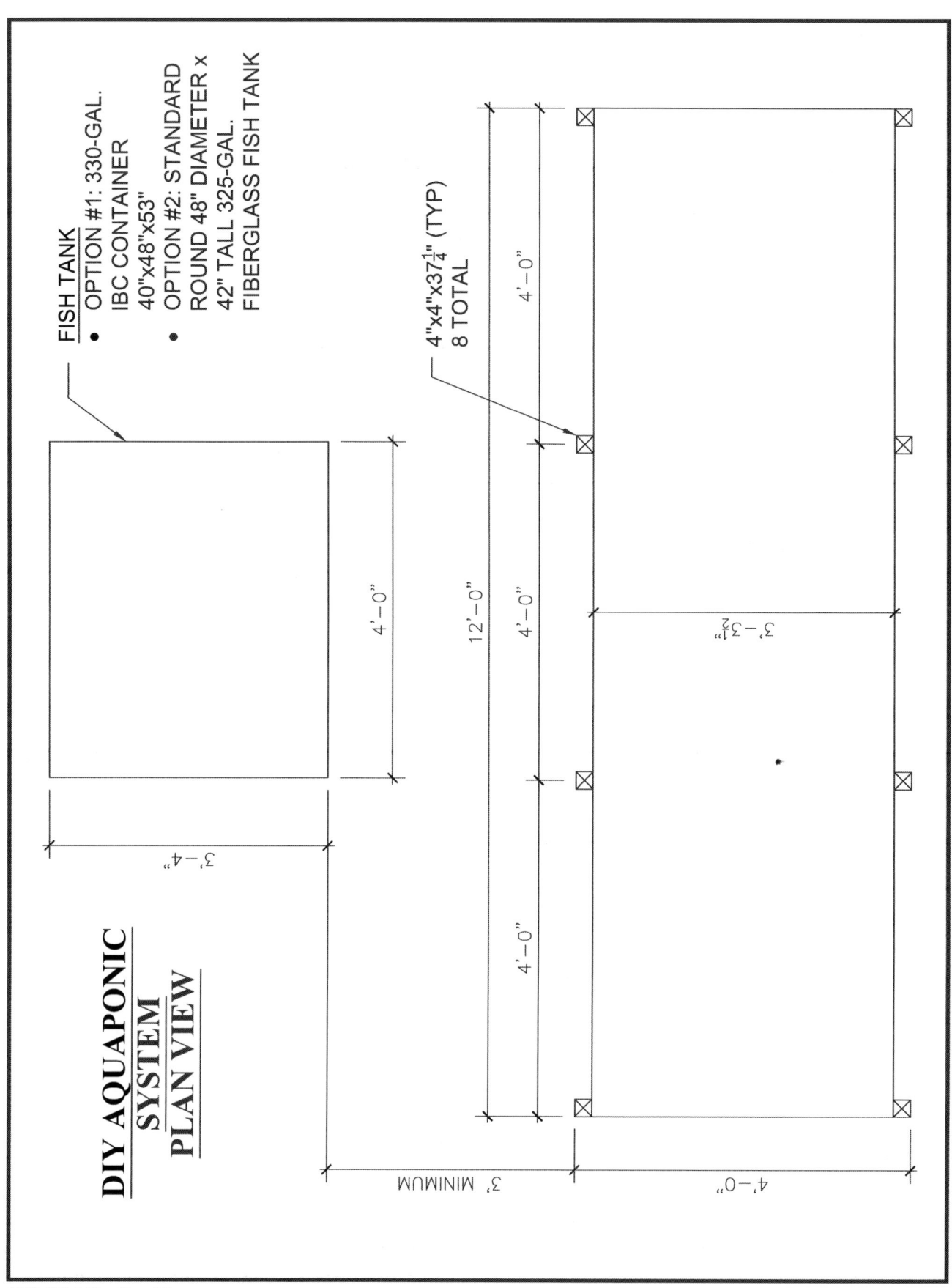

FISH TANK
- OPTION #1: 330-GAL. IBC CONTAINER 40"x48"x53"
- OPTION #2: STANDARD ROUND 48" DIAMETER x 42" TALL 325-GAL. FIBERGLASS FISH TANK

4"x4"x37 1/4" (TYP) 8 TOTAL

4'-0"
4'-0"
4'-0"
12'-0"
3'-4"
3'-3 1/2"
3' MINIMUM
4'-0"

FIGURE 208. Plan View.

304

CHAPTER 28: DESIGN PLANS AND CONSTRUCTION DETAILS DIY FLOOD-AND-DRAIN SYSTEM

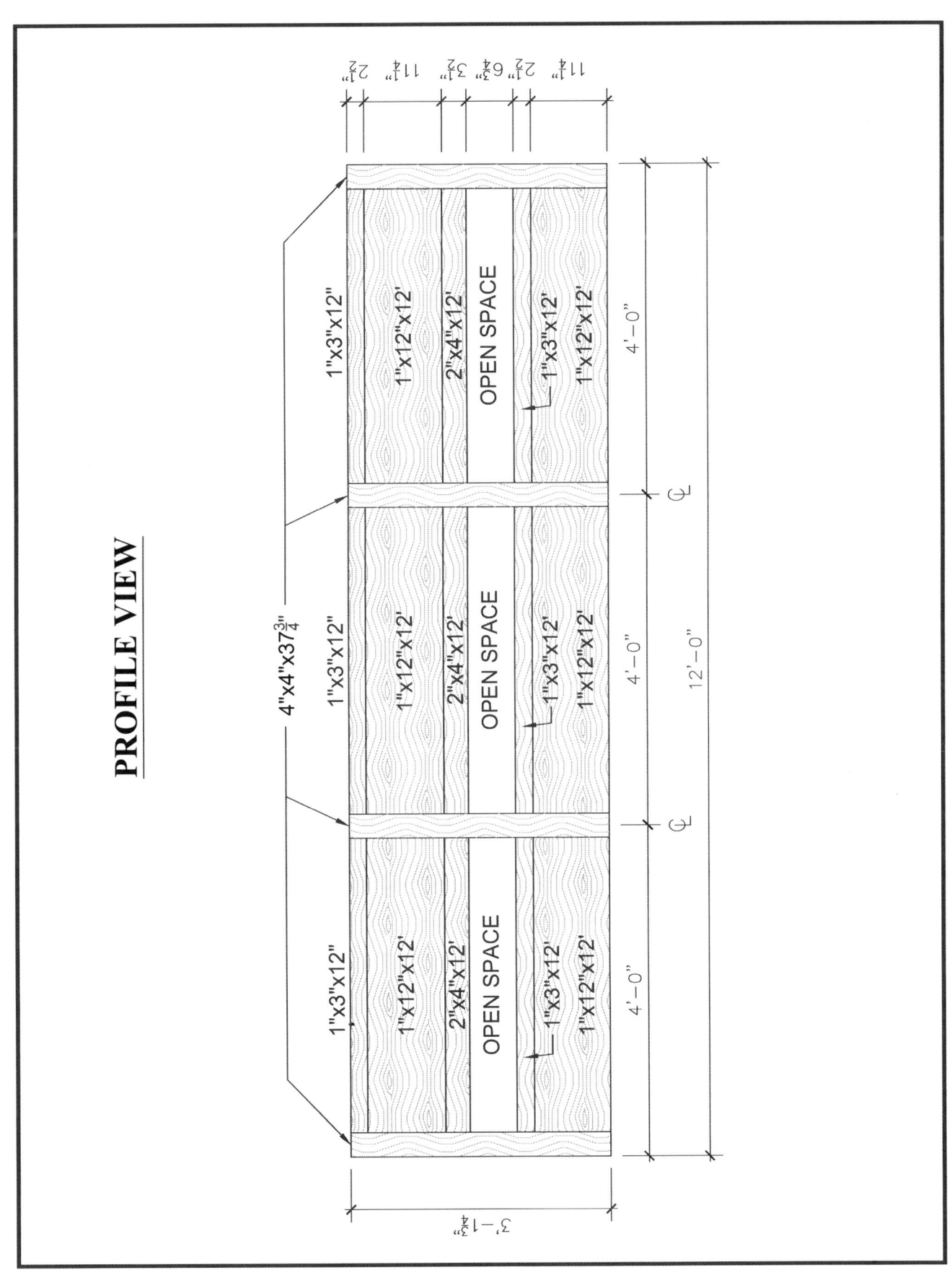

FIGURE 209. Profile View.

PART V: FLOOD-AND-DRAIN SYSTEM DESIGN AND LAYOUT

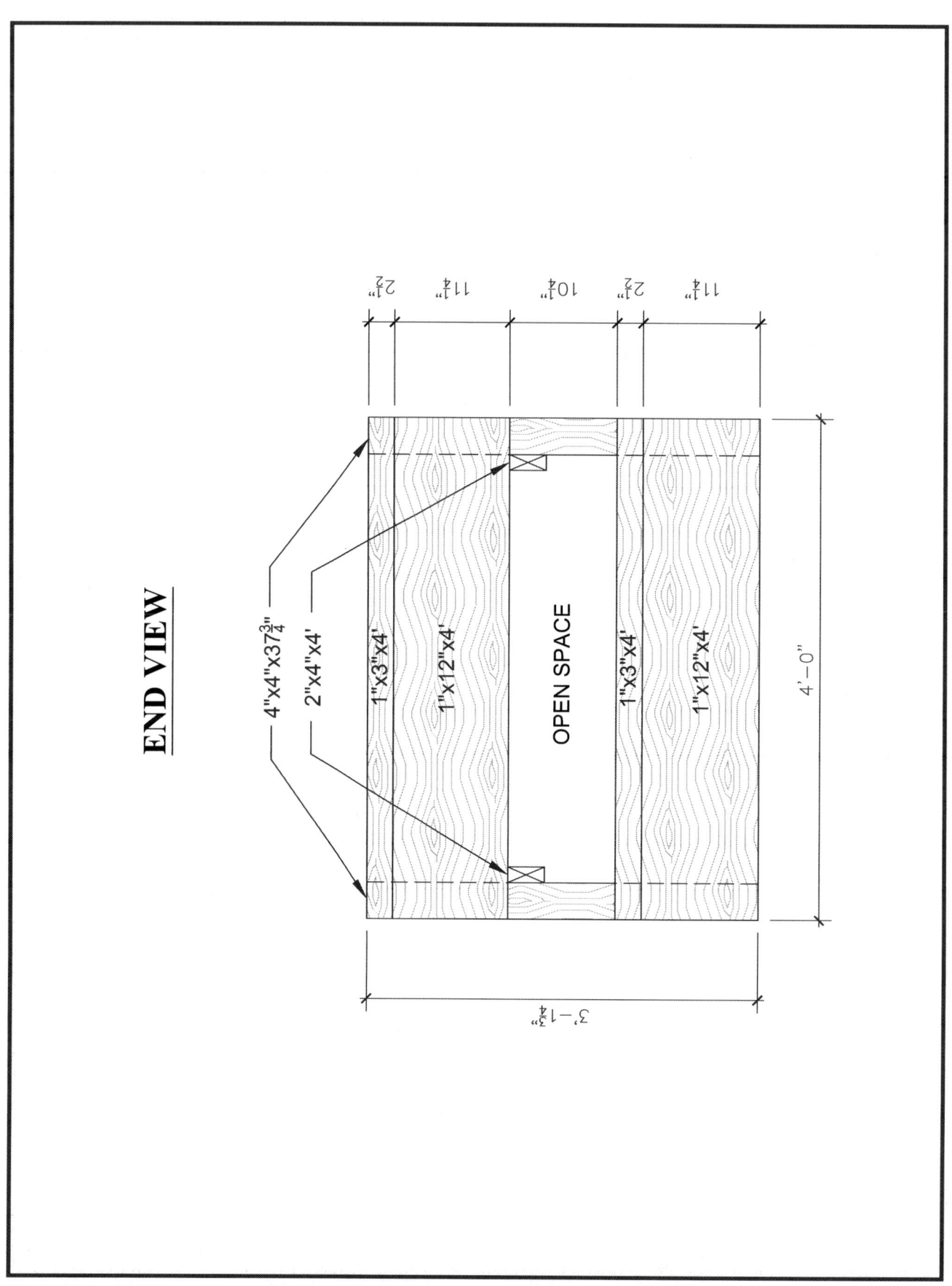

END VIEW

FIGURE 210. End View.

CHAPTER 28: DESIGN PLANS AND CONSTRUCTION DETAILS DIY FLOOD-AND-DRAIN SYSTEM

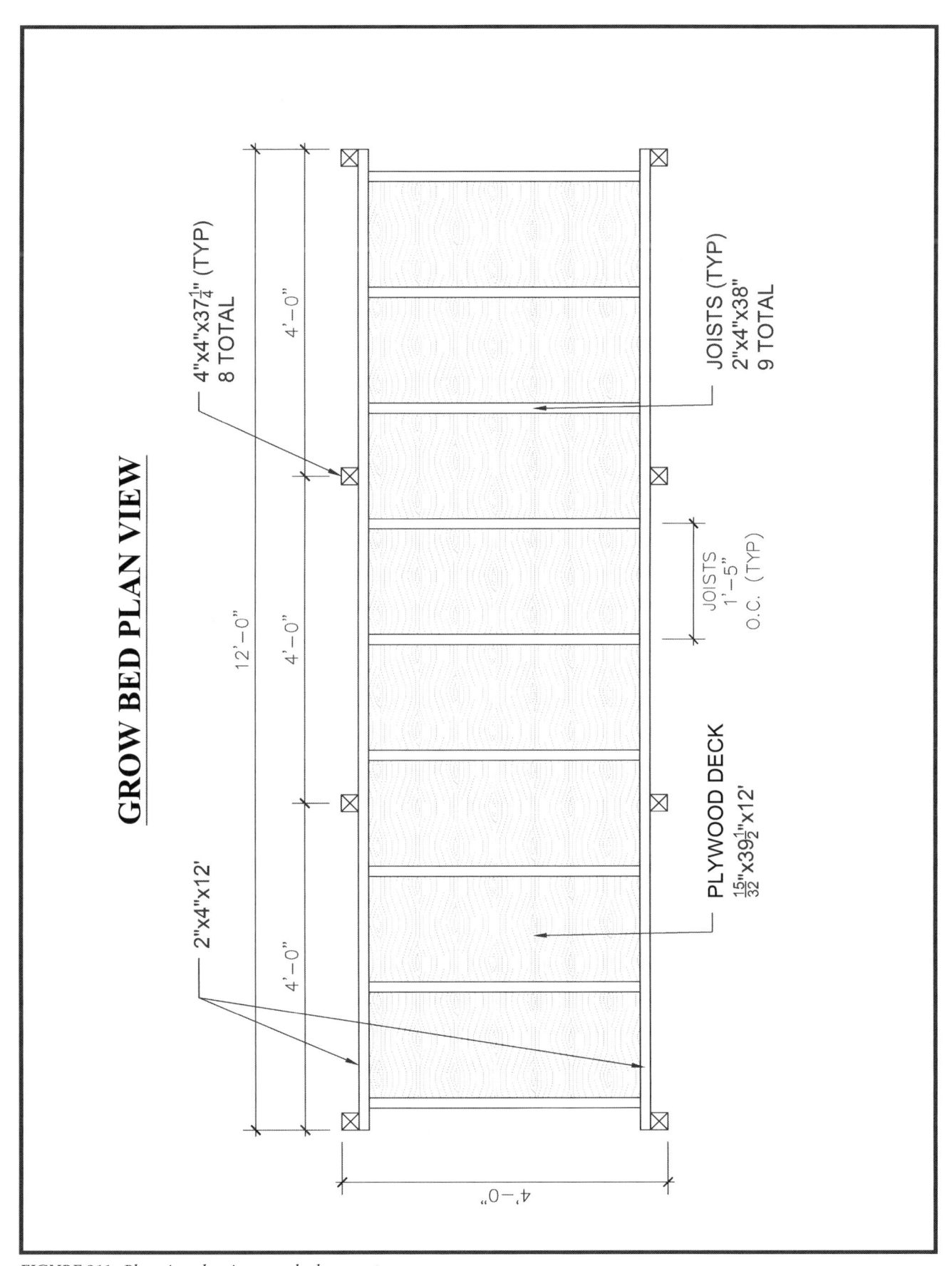

FIGURE 211. *Plan view showing grow bed supports.*

PART V: FLOOD-AND-DRAIN SYSTEM DESIGN AND LAYOUT

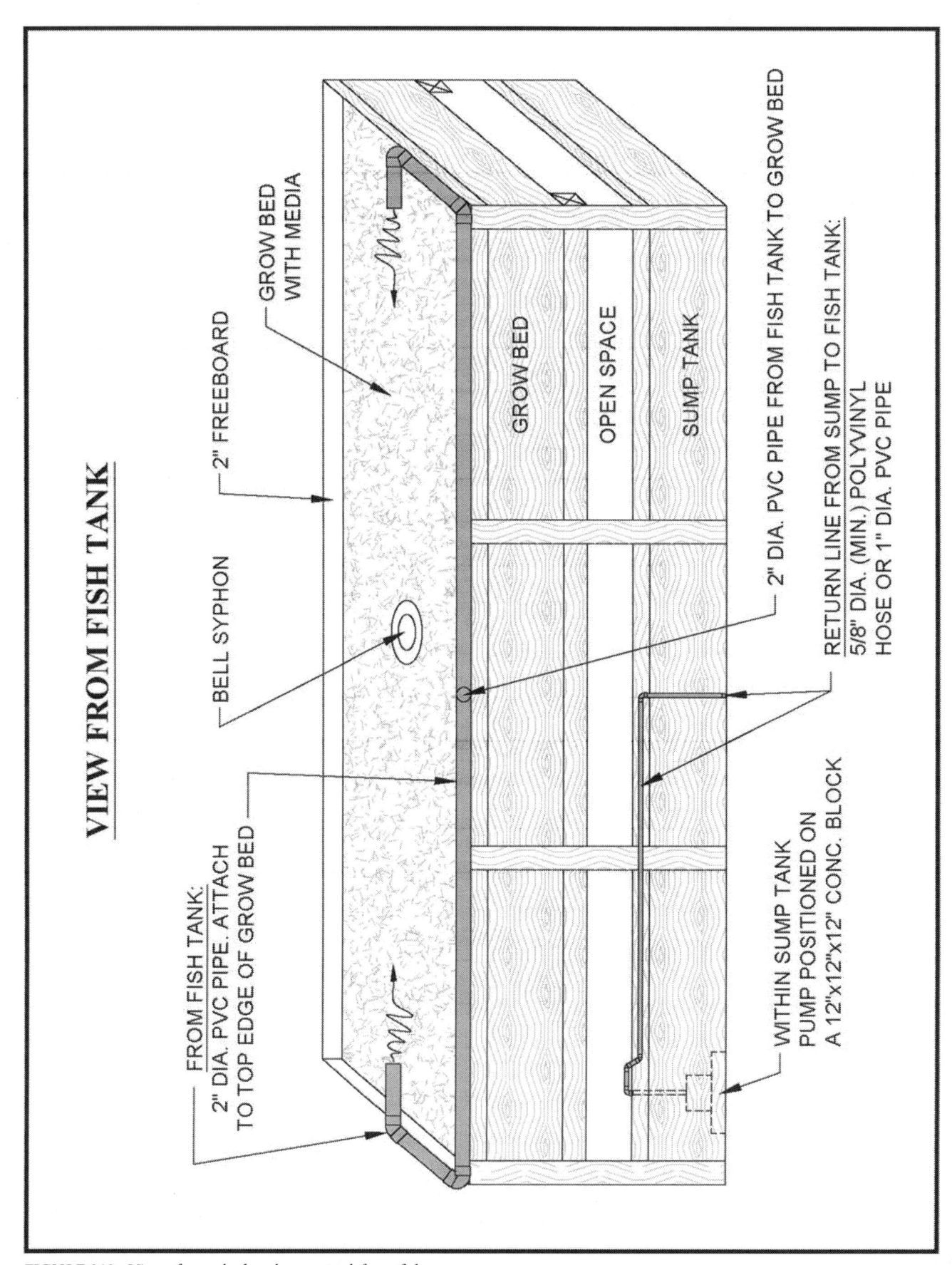

FIGURE 212. View of grow bed and sump tank from fish.

CHAPTER 28: DESIGN PLANS AND CONSTRUCTION DETAILS DIY FLOOD-AND-DRAIN SYSTEM

FISH TANK (PLUMBING PLAN)

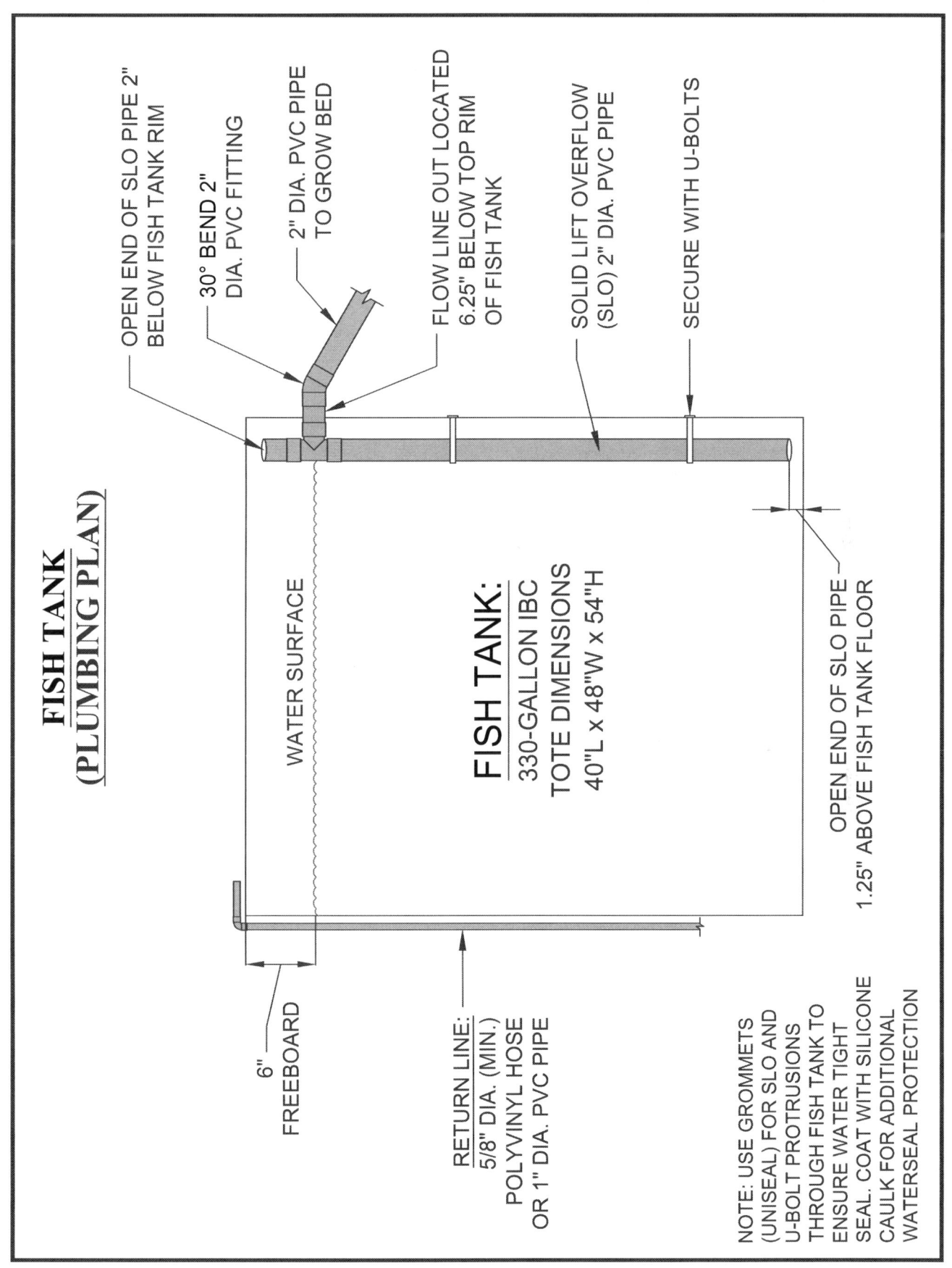

- OPEN END OF SLO PIPE 2" BELOW FISH TANK RIM
- 30° BEND 2" DIA. PVC FITTING
- 2" DIA. PVC PIPE TO GROW BED
- FLOW LINE OUT LOCATED 6.25" BELOW TOP RIM OF FISH TANK
- SOLID LIFT OVERFLOW (SLO) 2" DIA. PVC PIPE
- SECURE WITH U-BOLTS
- WATER SURFACE
- **FISH TANK:** 330-GALLON IBC TOTE DIMENSIONS 40"L x 48"W x 54"H
- OPEN END OF SLO PIPE 1.25" ABOVE FISH TANK FLOOR
- 6" FREEBOARD
- RETURN LINE: 5/8" DIA. (MIN.) POLYVINYL HOSE OR 1" DIA. PVC PIPE

NOTE: USE GROMMETS (UNISEAL) FOR SLO AND U-BOLT PROTRUSIONS THROUGH FISH TANK TO ENSURE WATER TIGHT SEAL. COAT WITH SILICONE CAULK FOR ADDITIONAL WATERSEAL PROTECTION

FIGURE 213. Cross-section view of fish tank.

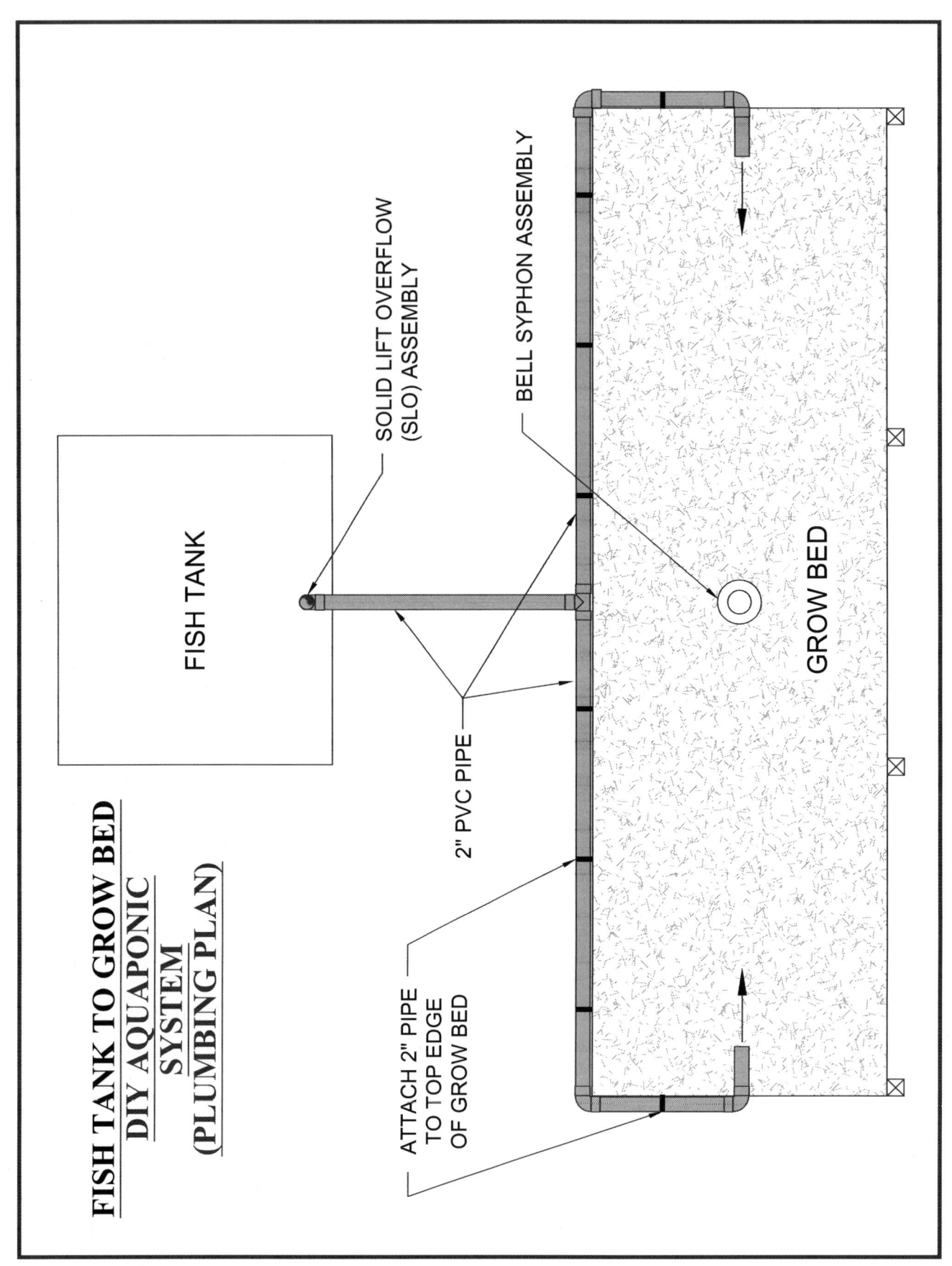

FIGURE 214. Fish tank to grow bed plumbing plan.

CHAPTER 28: DESIGN PLANS AND CONSTRUCTION DETAILS DIY FLOOD-AND-DRAIN SYSTEM

GROW BED TO FISH TANK DIY AQUAPONIC SYSTEM (PLUMBING PLAN)

FISH TANK

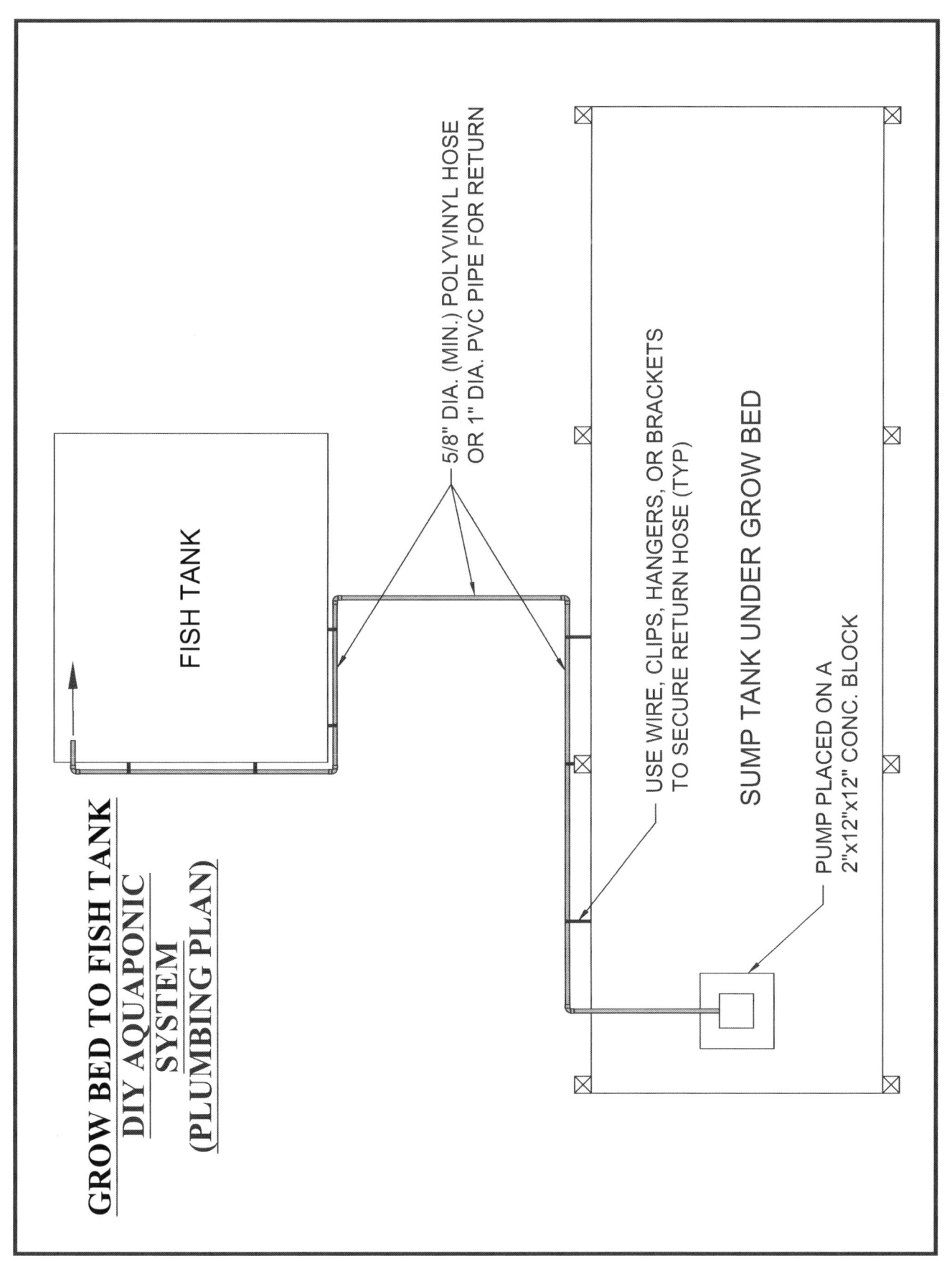

5/8" DIA. (MIN.) POLYVINYL HOSE OR 1" DIA. PVC PIPE FOR RETURN

USE WIRE, CLIPS, HANGERS, OR BRACKETS TO SECURE RETURN HOSE (TYP)

SUMP TANK UNDER GROW BED

PUMP PLACED ON A 2"x12"x12" CONC. BLOCK

FIGURE 215. Sump tank to fish tank plumbing plan.

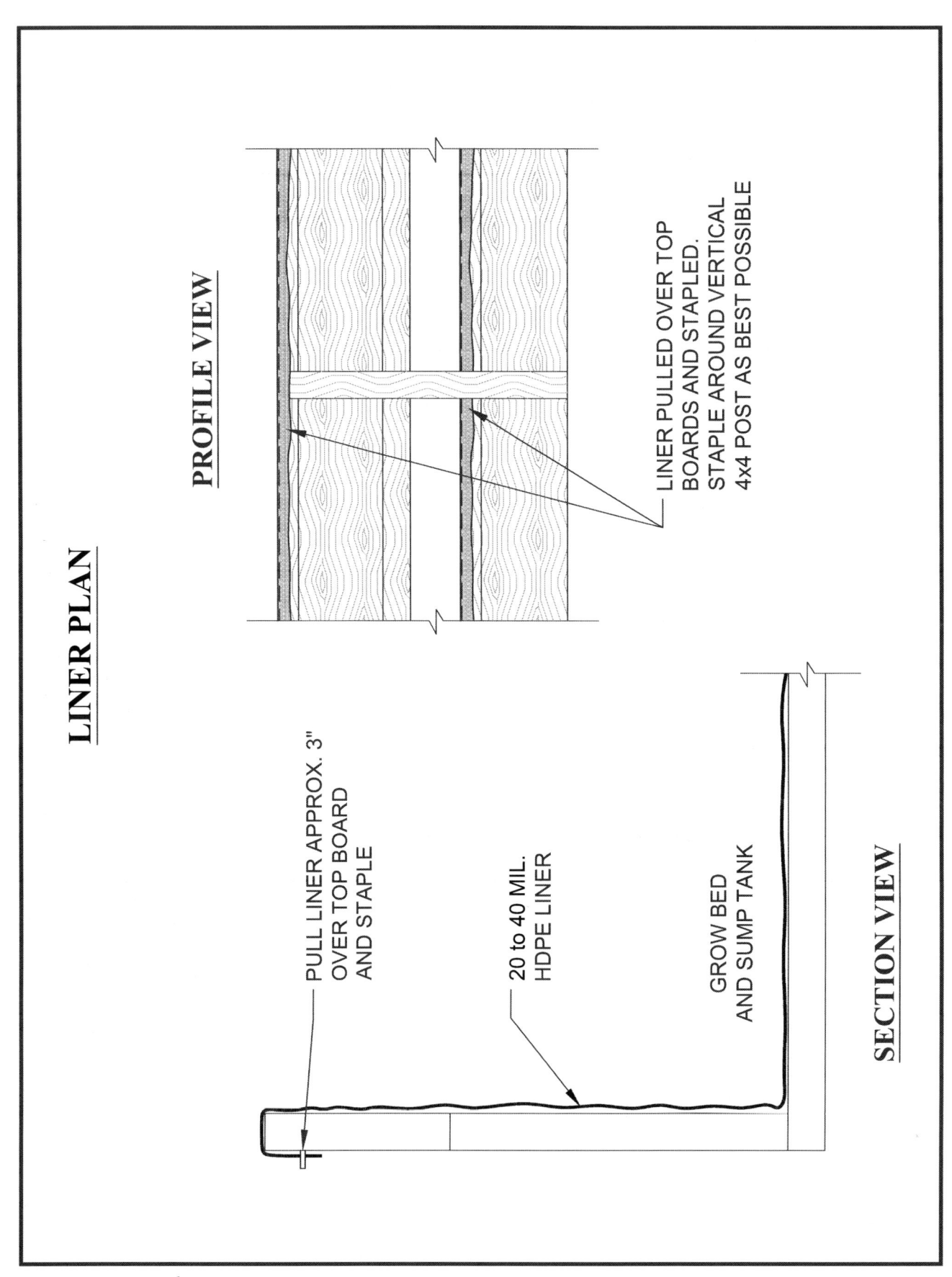

FIGURE 216. Liner plan.

CHAPTER 29

Design Plans and Construction Details Using IBC Containers Flood-and-Drain System

This chapter will provide step-by-step guide instructions and design details on how to build the media bed aquaponic system, shown below, using Intermediate Bulk Carrier (IBC). This is the same system in which a cost-benefit analysis is provided later in 'Chapter 35: Cost-Benefit Analysis (Aquaponic Economics)'.

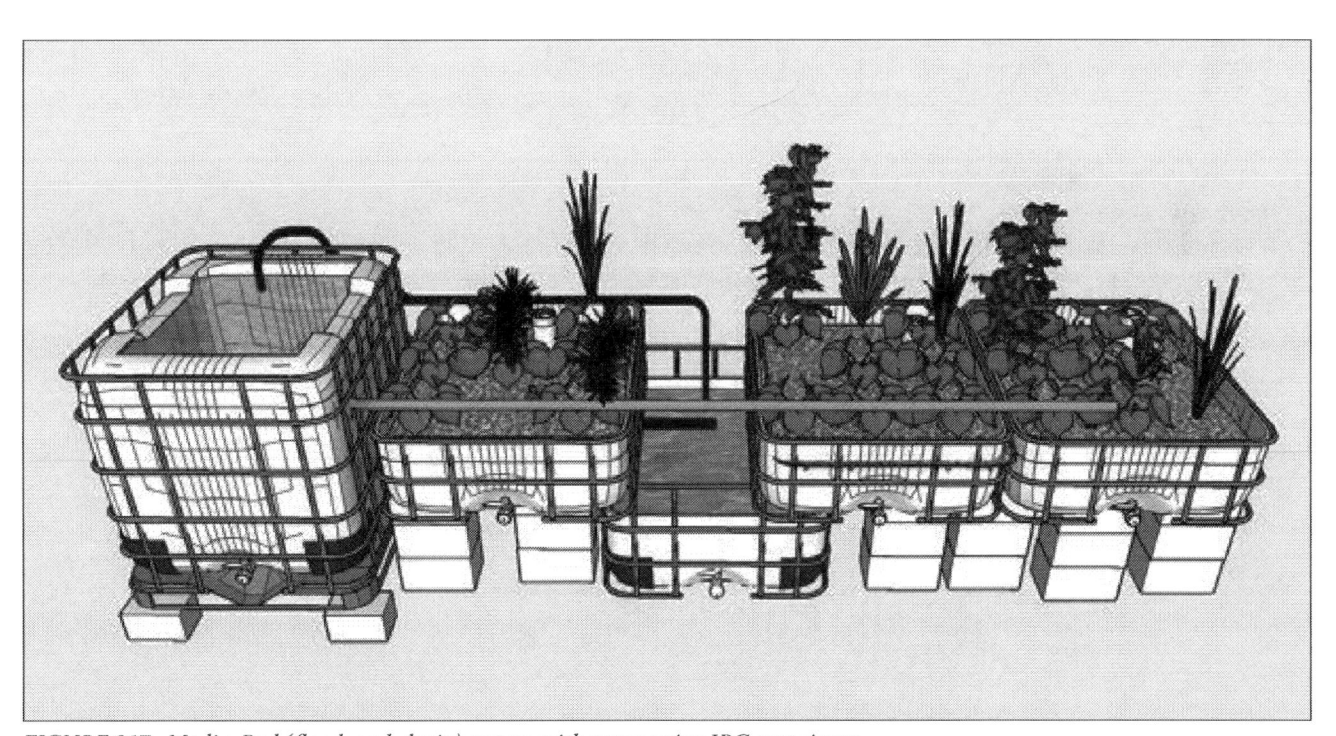

FIGURE 217. Media-Bed (flood-and-drain) system with sump using IBC containers.

PART V: FLOOD-AND-DRAIN SYSTEM DESIGN AND LAYOUT

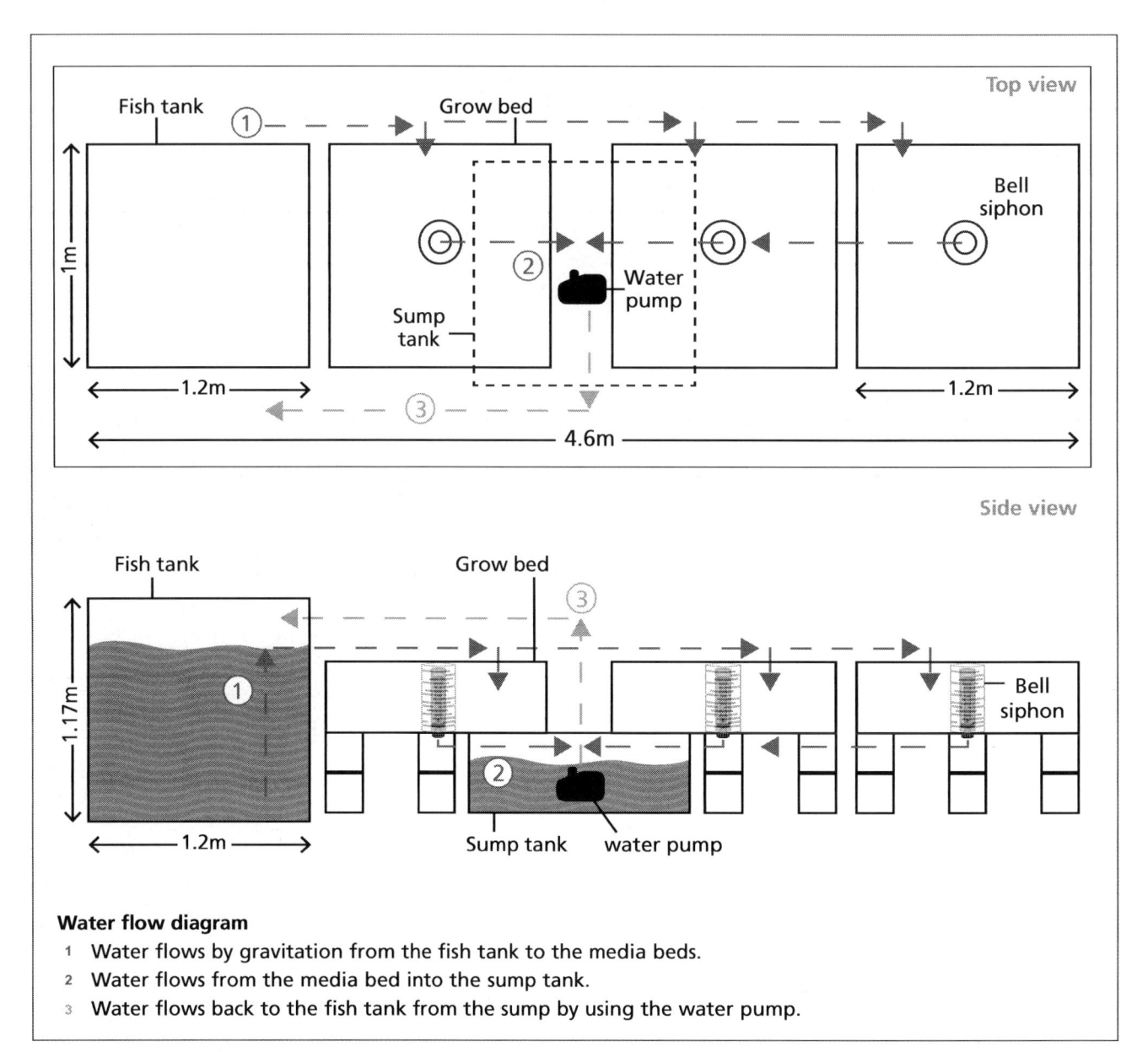

Water flow diagram
1. Water flows by gravitation from the fish tank to the media beds.
2. Water flows from the media bed into the sump tank.
3. Water flows back to the fish tank from the sump by using the water pump.

FIGURE 218. *Overall dimensions and schematic views.*

CHAPTER 29: DESIGN PLANS AND CONSTRUCTION DETAILS USING IBC CONTAINERS FLOOD-AND-DRAIN SYSTEM

Tools Needed

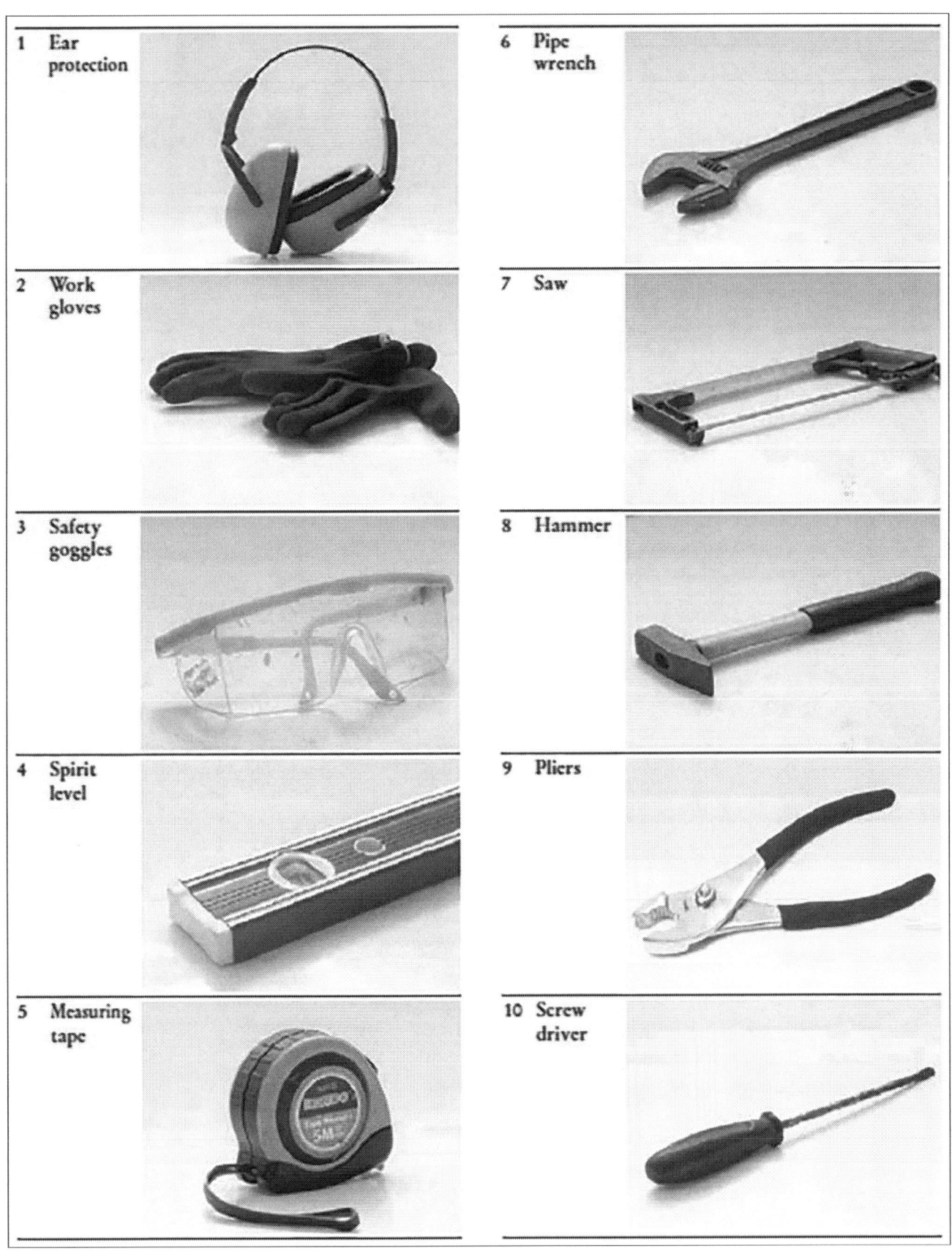

1	Ear protection	6	Pipe wrench
2	Work gloves	7	Saw
3	Safety goggles	8	Hammer
4	Spirit level	9	Pliers
5	Measuring tape	10	Screw driver

FIGURE 219.

PART V: FLOOD-AND-DRAIN SYSTEM DESIGN AND LAYOUT

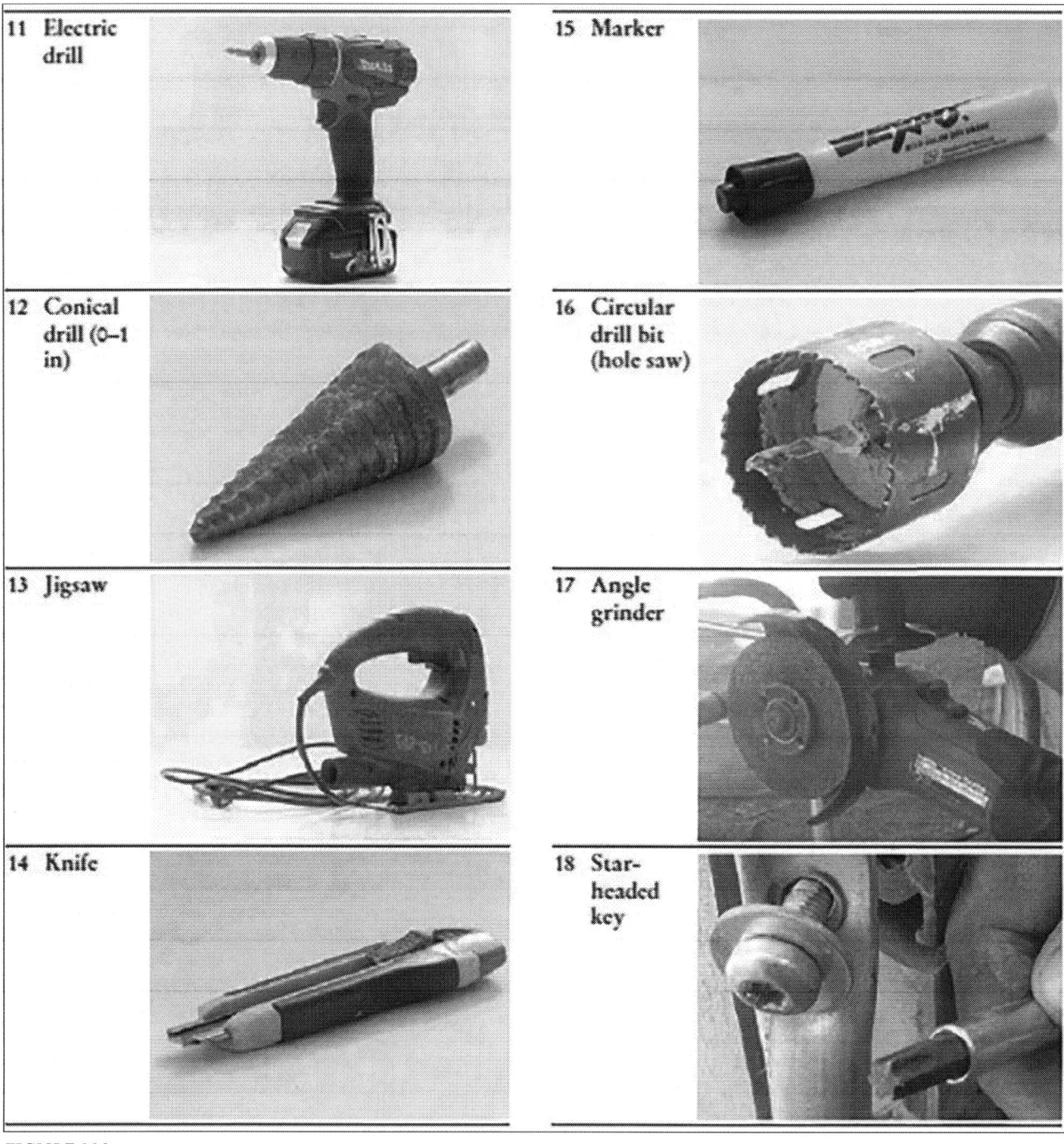

11	Electric drill
12	Conical drill (0–1 in)
13	Jigsaw
14	Knife
15	Marker
16	Circular drill bit (hole saw)
17	Angle grinder
18	Star-headed key

FIGURE 220.

316

TABLE 33. **List of Material Items Needed**

NO.	ITEM DESCRIPTION	QUANTITY
1	IBC Tank (275 gallon or 1000 liter)	3
2	Concrete Blocks (6"x8"x16")	48
3	Lumber, 1"x4" x8' (1cm x 8cm x 2.4m)	3
4	Pea Gravel Media	13-Cubic Feet (375 liters)
5	Expanded Clay Media	13-Cubic Feet (375 liters)
6	Shade Material for Fish Tank	10 Square Feet (2 Square meters)
7	Teflon Tape (Plumber's Tape)	2 Rolls
8	Cable Ties	15
9	*Electric Box Upgrade	1
10	*Residual-Current Device (RCD)	1
11	Ecological Soap or Lubricant to slide pipe through Uniseal*	1
PVC PIPE AND FITTINGS		
12	PVC Pipe; 2-inch or 50mm Diameter	25-Feet (7.5 Meters)
13	PVC Elbow; 2-inch or 50mm Diameter	5
14	PVC Coupler, Straight; 2-inch or 50mm Diameter (if using male to female slide over fittings, couplers will not be needed)	6
15	PVC Connector (T-Connector); 2-inch or 50mm Diameter	2
16	PVC Endcap/Stopper; 2-inch or 50mm Diameter	4
17	PVC or Metal Tap (1-inch or 25 mm); Male to Female Fitting	3
18	Rubber Grommet (Uniseal*; 2-inch or 50mm Diameter	1
BELL SIPHON		
19	PVC Pipe; 4-inch or 110mm Diameter	1-Foot (0.9 Meters)
20	PVC Pipe (3-inch or 75mm) with Flared End + PVC Endcap (3-inch or 75mm) + Rubber Washer (3-inch or 75mm)	3
21	PVC Pipe; 1-inch or 25mm Diameter	2-Feet, 7.5-Inches (0.8 Meters)
22	PVC Bulkhead (Barrel Connector); B-Type; 1-inch or 25mm Diameter	3
23	PVC Adapter (Reducer); (1-1.5 inch or 25-40 mm)	3
24	PVC Female Fitting; 1-inch or 25mm Diameter	3
25	PVC Elbow, Female; 1-inch or 25mm Diameter	3
26	Polyethylene Hose or Tubing; 1-inch or 25mm Diameter	30-Feet (9 Meters)

*NOTE: Replace existing electrical box with a residual-current device (RCD). This is a type of circuit breaker that will cut the power to the system if electricity grounds into the water. The best option is to have an electrician install one at the main electric junction. Alternatively, RCD adaptors are inexpensive and readily available at most hardware or home improvement stores. Also, replace current electrical junction box with an outdoor junction box. One each is listed in the above table, but use as many RCDs and outdoor junction boxes as needed in the aquaponic area. These simple precautions can save lives and protects fish.

TABLE 34. *Images of Material Items Needed*

ITEM DESCRIPTION	PHOTOGRAPH
IBC Tank	
Shade Cloth for Fish Tank	
Plastic Netting; Prevents Fish from Jumping Out	
Concrete Block	

CHAPTER 29: DESIGN PLANS AND CONSTRUCTION DETAILS USING IBC CONTAINERS FLOOD-AND-DRAIN SYSTEM

ITEM DESCRIPTION	PHOTOGRAPH
Lumber	
Active Aqua, AAPW550 550-GPH (2000 LPH) Submersible, Pump — 6 Foot Cord (Cost: $23 USD, year 2016); or different brand pump with equal specifications. ((NOTE: Due to variations in plumbing, media, or other conditions between the directions provided in this chapter and your system, should you find that the grow beds are not going through the 'flood-and-drain' cycle, then install a timer or float switch on this sump pump.))	
Teflon Tape (Plumber's Tape)	
Cable Ties	

ITEM DESCRIPTION	PHOTOGRAPH
PVC Pipe	
PVC Pipe (3-inch or 75 mm) with Flared End + PVC Endcap (3-inch or 75 mm) + Rubber Washer (3-inch or 75 mm)	
Polyethylene Hose or Tubing (1-inch or 25 mm Diameter)	
Rubber Grommet (Uniseal®) (2-inch or 50 mm)	
PVC Adapter (Reducer) (1-1.5 inch or 25-40 mm) ((examples of two different types shown))	

CHAPTER 29: DESIGN PLANS AND CONSTRUCTION DETAILS USING IBC CONTAINERS FLOOD-AND-DRAIN SYSTEM

ITEM DESCRIPTION	PHOTOGRAPH
PVC Female Fitting (1- inch or 25 mm Diameter)	
PVC Elbow (1-inch or 25mm Diameter) ((examples of two different types shown))	
PVC Elbow (2-inch or 50 mm Diameter)	
PVC Coupler, Straight (2-inch or 50 mm Diameter)	

ITEM DESCRIPTION	PHOTOGRAPH
PVC T-Connector Fitting (2-inch or 50 mm Diameter)	
PVC Endcap/Stopper (2-inch or 50 mm Diameter)	
PVC Bulkhead (Barrel Connector); B-Type 1-inch or 25mm Diameter	
PVC or Metal Tap (1-inch or 25 mm Diameter) Male to Female Fitting	

Construction Instructions (Step-By-Step Details)

1. Preparing the Fish Tank

1.1 — Remove the two horizontal steel lengths attached to the top surface of the IBC tank holding the inner plastic container in place. The steel lengths are fixed with 4 star-headed screws. Remove these four screws (figure 221, photo-1) using a star headed screwdriver (figure 221, photo-2) or star-headed key (figure 221, photo-3). Once the steel lengths are removed, pull out the inner plastic tank. If there is no star key, cut the screws with an angle grinder.

FIGURE 221.

1.2 — After pulling out the tank, draw a rough square shape on the top surface of the tank, 2-inches (5 cm) from, parallel along each of the 4 sides of the tank (figure 222, photo-4). Then, using the angle grinder (figure 222, photo-5), cut along the square shape and remove the cut piece from the top (figure 222, photo-6). Once removed, wash the inside of the container thoroughly with soap and warm water and leave to dry for 24 hours (figure 222, photo-7). The cut out top piece removed can be used as a fish tank cover.

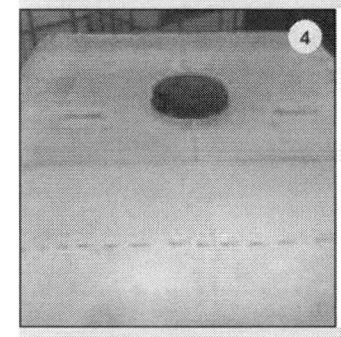

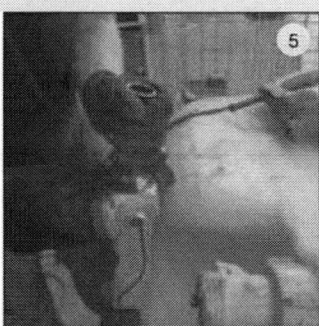

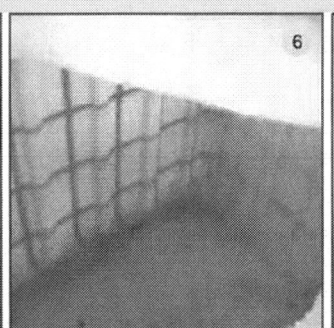

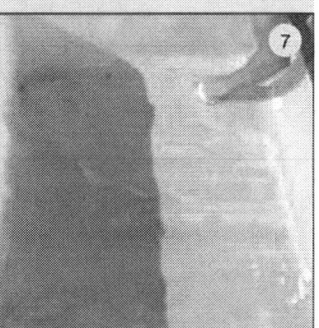

FIGURE 222.

2. Installing the Fish Tank Exit Pipe

2.1 — On one side of the IBC tank, mark a point 4.75-inches (12 cm) from the top and 4.75-inches (12 cm) from the side of the tank (figure 223, photo-8), and drill a hole at that point using the 57 mm circular drill bit (figure 223, photo-9). Insert a 2-inch (50 mm) uniseal (figure 223, photo-10) inside this hole. NOTE: The circular drill bit size should be 2 ¼ -inches (57 mm) and not 2-inches (50 mm) (see figure 223, photo-8).

FIGURE 223.

2.2 — The fish tank exit pipe is made of 2 lengths of PVC pipe 2-inch (50 mm) combined using a PVC elbow 2-inch (50 mm) and PVC coupler/straight connector 2-inch (50 mm) (figure 224, photo-11). The length of PVC 2-inch (50 mm) along the bottom surface of the tank is cut with horizontal slits 2–3 mm wide by using the angle grinder (figure224, photo-12) to allow solid waste to enter the pipe but to prevent fish from doing so. The open end of the PVC length along the bottom surface of the fish tank is sealed with a PVC endcap/stopper 2-inch (50 mm). Slot a short length of PVC (50 mm) through the uniseal 2-inch (50 mm) and attach to a PVC elbow 2-inch (50 mm) on the inside end (figure224, photo-11) and then attach the other (vertical) pipe length to the elbow that is now connected to the uniseal 2-inch (50 mm). Finally, drill a ¾ to 1-inch (2–3 cm) diameter hole into the PVC elbow 2-inch (50 mm) attached to the uniseal 2-inch (50 mm) (figure 224, photo-13). This small hole prevents any air seal forming inside the pipe, which would drain all the water out of the fish tank in the event of power cut or if the pump stopped working. This is also called an accidental siphon. This step is not optional.

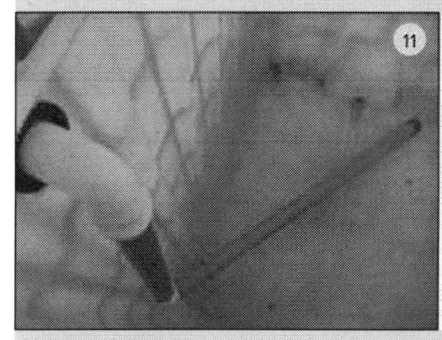

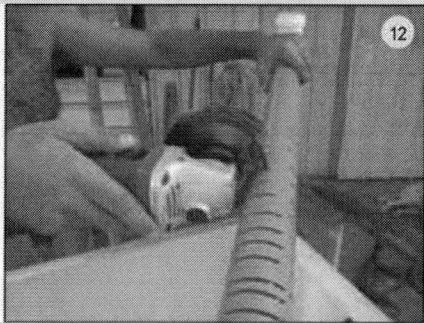

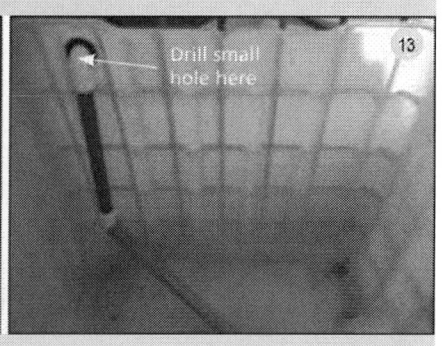

FIGURE 224.

3. Preparing the Media Beds and Sump Tank

3.1 — To make the 3 media beds and 1 sump tank, the 2 other IBC tanks are needed: the first to make the sump tank and 1 media bed, and the second to make the two remaining media beds. Take the 2 IBC tanks and remove the 4 steel profiles and pull out the plastic containers as shown before in figures 221, photos-1–3.

4. Making Two Media Beds From One IBC

4.1 — First, stand the plastic inner container upright (figure 225, photo-14) and mark, using a metre stick and pencil, two bisecting lines 12-inches (30 cm) from both sides of the tank (as seen in figure 225, photo-15). Make sure to mark the exact lines (shown in the figure 225, photo-15). Take the angle grinder and carefully cut along both bisecting lines marked out to create two uniform containers with a depth of 12-inches (30 cm) (figure 225, photo-16). Then, take both containers and wash them thoroughly using natural soap and warm water and leave them out to dry in the sun for 24 hours.

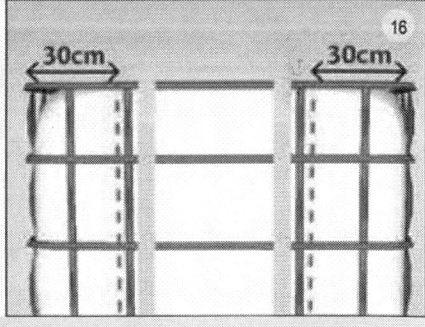

FIGURE 225.

5. Metal Supports For Both Media Beds

5.1 — Take the IBC metal support frame and cut out two support frames by following the same bisecting lines shown in figure 225, photo-14 using the angle grinder (figure 226, photo-17). When cutting the two 12-inches (30 cm) sides of the support frame, make sure to keep the two horizontal steel profiles intact as they will provide excellent support to the sides of the beds once they are full of water and medium (figure 226, photo-18).

FIGURE 226.

5.2 — Then, take both support frames and lay them out on the floor. Take the wood lengths (4 lengths of 41-inches (104 cm), 1 length of 16 ½–inches (42 cm) and 1 length of 19–inches (48 cm) and place them on top of the support frame as shown in figure 227, photo-19. These wood lengths keep the media bed horizontal, which is vital for the functioning of the bell siphons. Next, take the washed media beds and place them on top of the support frame and wood lengths (figure 227, photo-20). Finally slot in the remaining wood lengths in between the plastic media bed and support frame on both sides of each bed to provide further support (figure 227, photo-21).

FIGURE 227.

6. Making A Sump Tank and One Media Bed From an IBC

6.1 — Take the remaining IBC, place it upright and mark out, using a metre stick and pencil, only one 12-inches (30 cm) bisecting line as seen in figure 228, photo-22. Then, take the angle grinder and cut the inner plastic container and metal support frame at once by following the bisecting line (see figure 228, photo-22). Remove the 12-inches (30 cm) container (third media bed) from the remaining 27 ½–inches (70 cm) container (sump tank) (figure 228, photo-23). Wash out both containers thoroughly with natural soap and warm water and leave in the sun for 24 hours.

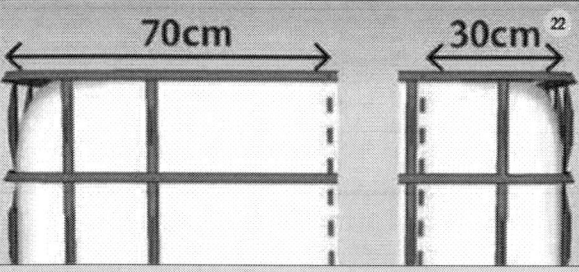

FIGURE 228.

6.2 — For the third media bed, follow the same steps regarding the wood lengths as detailed above for the first two. Finally, take the sump tank container and drill two holes 1-inch (25 mm) diameter) using the conical drill bit as shown in (figure 229, photo-25) where the 1-inch (25 mm) pipes will be inserted into both of these holes later, the pipes will drain water from each media bed.

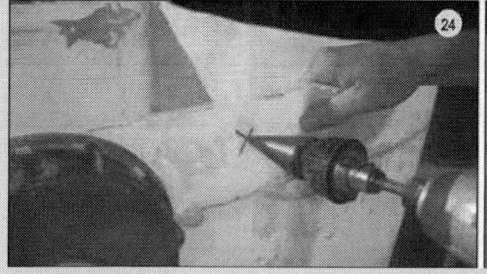

FIGURE 229.

CHAPTER 29: DESIGN PLANS AND CONSTRUCTION DETAILS USING IBC CONTAINERS FLOOD-AND-DRAIN SYSTEM

7. Preparing the Bell Siphons

As explained in 'Chapter 26: Flood-and-Drain System Drain Options, Plans, and Instructions' of this book, bell siphons are simple mechanisms used to automatically flood and drain each media bed. Refer to Chapter 27 and/or this section for information on how to make a bell siphon. The following materials are needed to make one siphon. Multiple by three, since a total of three bell siphons will be needed for this IBC aquaponic media bed system.

- 14-inch (35 cm) media guard using 4-inch (110 mm) diameter PVC pipe
- 10 ½-inch (27 cm) bell using [PVC pipe 3-inch (75 mm) with flared end + endcap/stopper 3-inch (75 mm) + rubber washer 3-inch (75 mm)]
- 6-inch (16 cm) standpipe (using a 1-inch or 25 mm PVC pipe)
- Bulkhead Fitting (Barrel Connector) (1-inch or 25 mm)
- PVC Adapter (Reducer) 1 ½ — 1-inch (40–25 mm)
- PVC Female Fitting; 1-inch (25mm) Diameter
- PVC Elbow, Female; 1-inch (25mm) Diameter

7.1 — First, create the bell. Take a 10 ½- inch (27 cm) section of 3-inch (75 mm) diameter PVC pipe and cut out 2 pieces as shown in figure 230, photo-26 using the angle grinder. Then, drill a hole with a 25/64 (10 mm) drill bit about 5/8–inch (1.5 cm) from the two cut pieces as shown in figure 230, photo-26. Finally, seal one end of the bell using the PVC endcap/stopper 3—inch (75 mm) and rubber washer 3-inch (75 mm).

7.2 — Next, make the media guards from the 14-inch (35 cm) length of PVC pipe 4-inch (110 mm) and cut 5 mm slots along their entire length using the angle grinder (figure 230, photo-27).

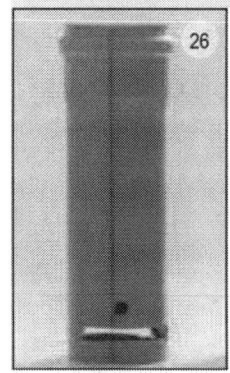

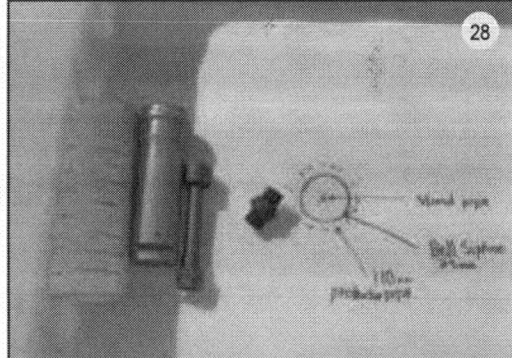

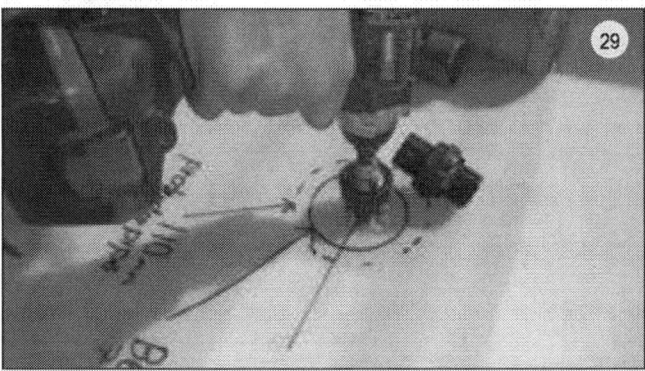

FIGURE 230.

PART V: FLOOD-AND-DRAIN SYSTEM DESIGN AND LAYOUT

7.3 — Now, take each media bed and mark their centre points in-between the two wooden lengths below as shown in figure 230, photo-28. Drill a hole 1-inch (25 mm) in diameter at each center point (figure 230, photo-29) and insert the bulkhead fitting (barrel connector) 1-inch (25 mm) with the rubber washer placed inside the media bed. Tighten both sides of the bulkhead (barrel connector) using a wrench (figure 230, photo-30).

7.4 — Screw the PVC adaptor 1–inch (25 mm) onto the bulkhead fitting (barrel connector) 1–inch (25 mm) inside the media bed and then slot the standpipe into the PVC adaptor 1–inch (25 mm). After, attach the second PVC Adaptor (Reducer) 1½ — 1-inch (40–25 mm) to the top of the standpipe (figures 231, photos-31–33). The purpose of this adapter is to allow a larger volume of water to initially flow down the standpipe when the water has reached the top. This helps the siphon mechanism to begin draining the water out into the sump tank.

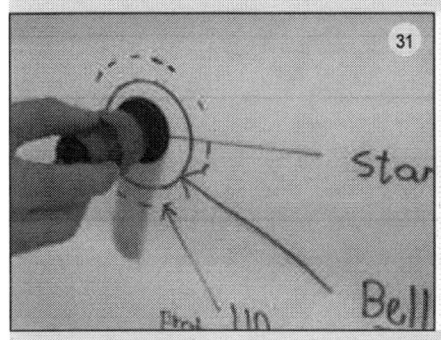

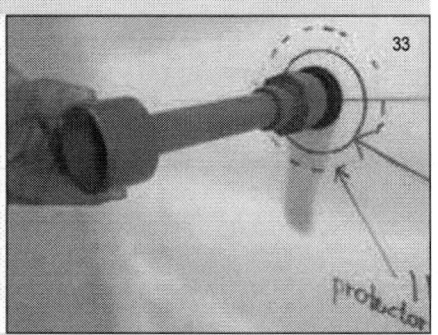

FIGURE 231.

7.5 — Place the bell siphons and the media guards over the standpipes (figures 232, photos-34–36).

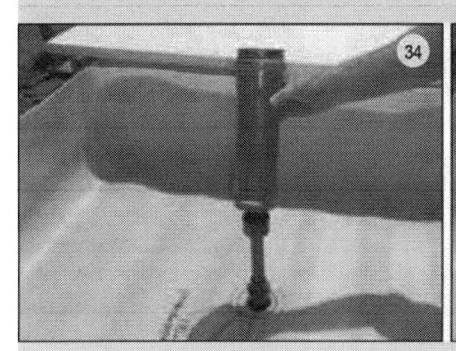

FIGURE 232.

7.6 — Finally, connect the PVC elbow 1 inch (25 mm) to the other end of the bulkead fitting (barrel connector) underneath the media bed, which allows the water to flow out of the media bed (figures 233, photos-37–39).

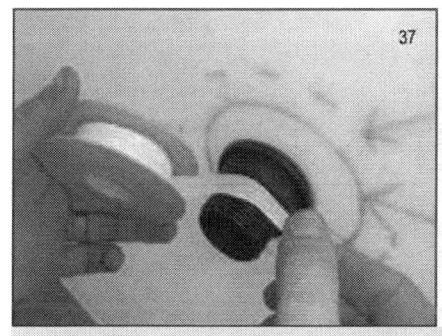

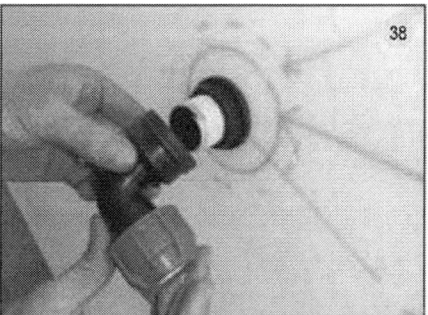

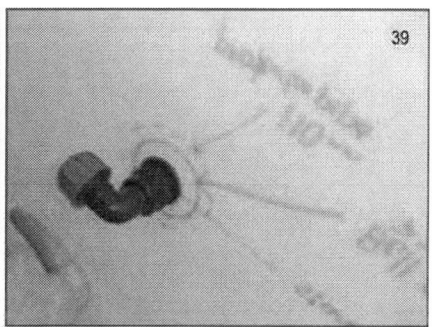

CHAPTER 29: DESIGN PLANS AND CONSTRUCTION DETAILS USING IBC CONTAINERS FLOOD-AND-DRAIN SYSTEM

FIGURE 233.

8. Assembling the Media Beds and Sump Tank

8.1 — First, place the sump tank and brace it with six concrete blocks from each side (12 blocks in total) as shown in figures 234, photos-40 and 41. Make sure the blocks do not cover the holes already drilled into the sump tank (figure 234, photo-42).

FIGURE 234.

8.2 — Place the remaining blocks and the fish tank according to the distances described in figure 235, photo-43. The fish tank should be raised up about 6-inches (15 cm) from the ground. This can be done by using concrete blocks as shown in figure 235, photo-43. Place the three media beds (including the metal support frames and wood lengths) on top of the blocks (as shown in figure 235, photo-44). Make sure the grow beds are secured on top of the blocks and horizontal by verifying with a spirit level. If not, slightly adjust the layout of the blocks underneath.

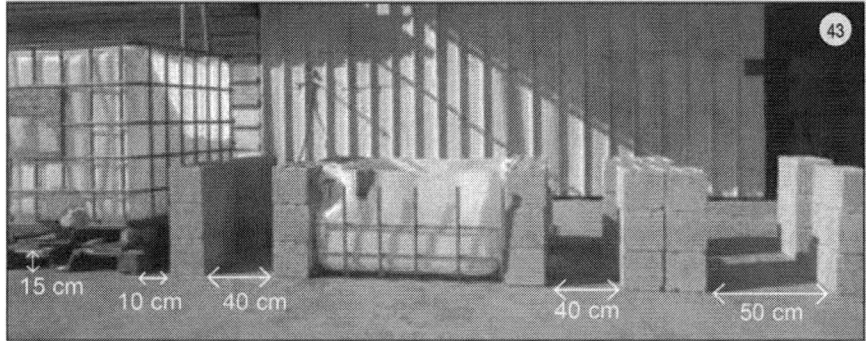

FIGURE 235.

329

9. Plumbing the System: Fish Tank to the Media Beds (Distribution Manifold)

9.1 — The plumbing parts needed for this section are as follows:

- Bulkhead Fitting (Barrel Connector), B-type (1-inch or 25 mm) × 3
- PVC Tap (1-inch or 25 mm) × 3
- PVC Endcap/stopper (Diameter: 2-inch or 50 mm) × 3
- PVC Elbow (Diameter: 2-inch or 50 mm) × 2
- PVC T-Connector (Diameter: 2-inch or 50 mm) × 2
- PVC Coupler (Diameter: 2-inch or 50 mm) × 3
- 60-inches (150 cm) of PVC Pipe (Diameter: 2-inch or 50 mm) × 1
- 34-inches (85 cm) of PVC pipe (Diameter: 2-inch or 50 mm) × 1

9.2 — Go back to the "preparing the fish tank" (2.2) instructions. The last instruction shows a length of PVC (50 mm) slotted through the uniseal 2-inch (50 mm) diameter and exiting the fish tank. Take another PVC elbow 2-inch (50 mm) and connect it to the pipe slotted through the uniseal (figure 236, photo-45). Then, using a PVC straight coupler 2-inch (50 mm) and another PVC elbow 2-inch (50 mm), connect the fish exit pipe to the distribution pipe 2-inch (50 mm) at the same height as the top of the media bed (figure 236, photo-46).

FIGURE 236.

9.2 — On each media bed, a valve is used to control the water flow entering the bed. To include a valve, first take a PVC endcap/stopper 2-inch (50 mm) and drill a hole (25 mm diameter). Insert a bulkhead fitting (barrel connector) 1-inch (25 mm) into the hole and tighten both ends using a wrench. Then, wrap Teflon tape around the threads of the male end of the barrel connector and screw the tap valve 1-inch (25 mm) onto the bulkhead (barrel connector) (figures 237, photos-47–50). There is one valve for each media bed for a total of three valves.

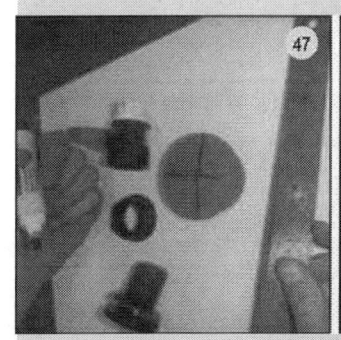

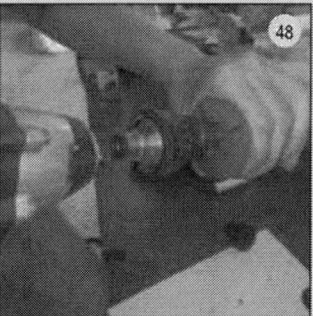

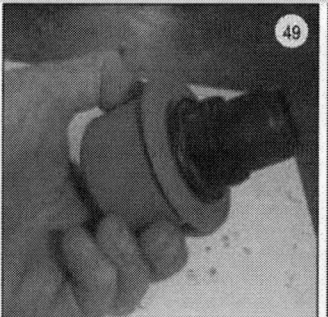

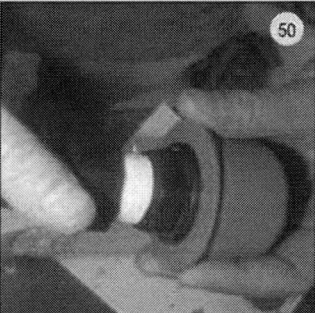

FIGURE 237.

CHAPTER 29: DESIGN PLANS AND CONSTRUCTION DETAILS USING IBC CONTAINERS FLOOD-AND-DRAIN SYSTEM

9.4 — From the PVC elbow 2-inch (50 mm) attached to the fish tank exit pipe, follow the pipe layout shown in figure 238-51 that allows water to flow into each media bed. Materials include: PVC pipe 2-inch (50 mm), PVC elbow 2-inch (50 mm) and PVC T-connector 2-inch (50 mm). Next, attach the pipe caps fitted with the valves to the PVC T connectors and PVC elbow connectors from the distribution pipe as in figure 238, photo-51, using one for each media bed. Use a PVC straight coupler 2-inch (50 mm), if necessary.

FIGURE 238.

10. Plumbing the Unit: Media Beds to the Sump Tank (Drain Pipe)

10.1 — Figures 239, photos-52 and 53 show the media beds marked as A, B and C. For media bed A, attach a drain pipe of 24-inch (60 cm) length of PVC pipe 1-inch (25 mm) to the elbow connection underneath the media bed (figure 240, photo-54), which exits from the bottom of the bell siphon standpipe. Next, slot the 24-inch (60 cm) length of pipe into the closest drilled hole on the side of the sump tank allowing the water to flow directly into the sump.

10.2 — Attaching media beds B and C (figure 239, photo-53): Under media bed C: attach a PVC elbow connector (25 mm to 1 inch) to the end of the bulkhead (barrel connector) (figure 240, photo-54). Then, take a 80-inch (2 meter) length of polyethylene pipe 1-inch (25 mm) diameter and attach it to the drilled holes at the side of the sump tank (figure 239, photo-53 and figure 240, photo-55).

10.3 — Do the same with media bed B using 40-inch (1 meter) length of polyethylene pipe 1-inch (25 mm) diameter (figure 240, photo-55). Now, the water exiting media beds B and C will flow through separate polyethylene pipes 1-inch (25 mm) into the sump tank.

FIGURE 239.

Finally, it is advisable to fix the pipes underneath the beds to the metal frame using cable ties to relieve any pressure on the pipe fittings (figure 240-54).

11. Plumbing the Unit: Sump Tank to the Fish Tank

11.1 — Take the submersible pump and attach a polyethylene pipe 1-inch (25 mm) diameter using a

FIGURE 240.

PVC straight connector 1-inch (25 mm) diameter, or any other connector that can attach the specific pump to the 1-inch (25 mm) diameter pipe (figure 241-56). Take a length of the polyethylene pipe 1-inch (25 mm) diameter that is long enough to reach the inside of the fish tank from the submersible pump (figure 241-57). Attach one end to the submersible pump and the other into the top of the fish tank (see figure 241-57–60). It is recommended to use the fewest connectors, especially elbows, between the pump and fish tank which will decrease pumping capacity.

11.2 — Place the electric box in a safe place higher than the water level and shaded from direct sunlight. Make sure it is still waterproof after plugging in the water and air pump plugs (figure 241-61).

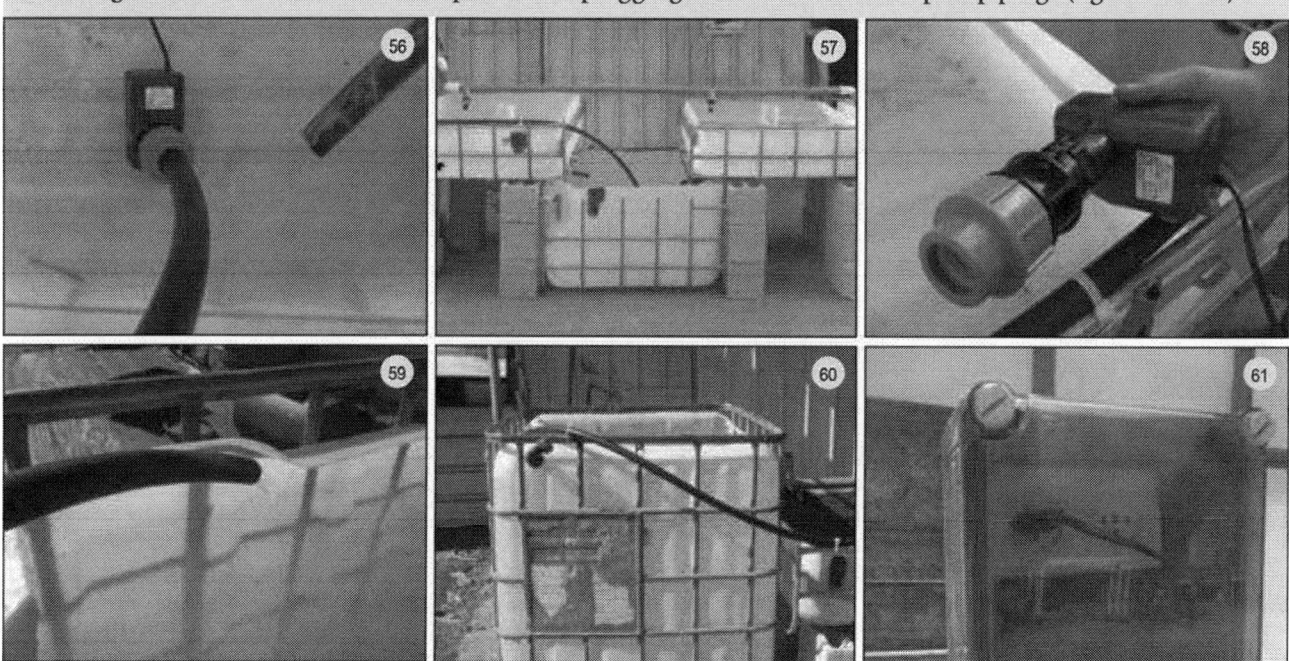

FIGURE 241.

12. Adding the Medium and Running the Unit

12.1 — All parts of the system are now in place. Before adding the media fill the fish tank and sump tank with water and run the pump to check for any leaks in the system. While checking for leaks,

CHAPTER 29: DESIGN PLANS AND CONSTRUCTION DETAILS USING IBC CONTAINERS FLOOD-AND-DRAIN SYSTEM

remove the standpipe and bell siphon so the water flows straight into the sump tank. If leaks appear, repair them where they arise by tightening the plumbing connections, re-applying Teflon to the treaded connections and making sure all taps are in their ideal position (figures 242-62–67).

12.2 — Once the water is flowing smoothly through all components of the unit, re-assemble the siphon bell and standpipes. Now mix and rinse the growing media outside the system. Flowing the rinsing of the media, fill the media beds to a depth of 12-inches (30 cm) (figures 242-68–69).

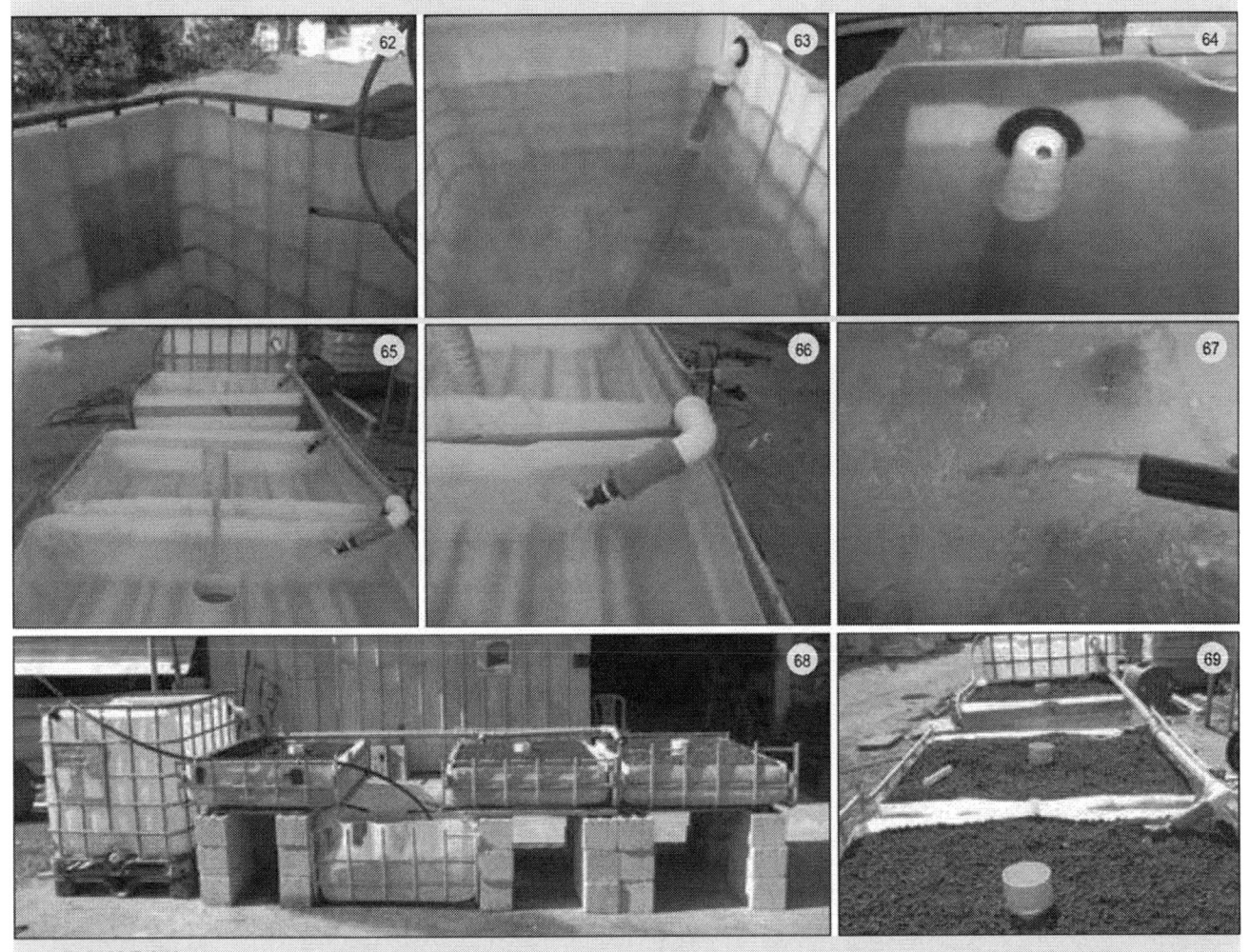

FIGURE 242.

12.3 — Verify that the grow beds are filling and draining appropriately (a thin layer of water in the grow bed is acceptable). Due to variations in plumbing, media, or other conditions associated with the system provided in this example and your system, should you find that the grow beds are not going through the 'flood-and-drain' cycle, then install a timer or float switch on the sump pump. A timer or float switch can easily be purchased at a well equipped hardware store, home improvement center, or online for less than $25 (USD, year 2016).

12.4 — Following successful construction of the system refer to 'Chapter 30: Starting and Managing Your Aquaponics System', of this book. Enjoy an abundance of healthy organic food.

PART VI

Operation and Maintenance

CHAPTER 30

Starting and Managing Your Aquaponics System

Successfully Activating Your Aquaponic System

When starting the system, you must be very patient. Don't expect your system to be running smoothly immediately. It can take up to five weeks, or even longer in some cases, before all living components of your system are in a thriving state of development. This is a good time to gain experience in testing the water with a test kits and/or getting familiar with your automatic monitoring systems. Practice testing the tap water immediately out of the faucet and test it again after it has been in the system every couple of days. Also practice recording the water quality readings, adjustments to your system that you have made, and organizing everything associated with your system (i.e. supplies, paperwork, tools, etc.). This will make things go much more smoothly over the long run, saving you time and trouble in the future.

Most municipal tap water has chlorine added. This water needs to be dechlorinated. When starting, the water can be dechlorinated in the system over a few days. After your system it running, it is important to dechlorinate the water before adding it to your system by letting it sit in another container for a day or two (the more water, the longer it will take it to dechlorinate). If your tap water contains chloramines, it will be necessary to obtain a chemical neutralizer, available in most fish shops or online, to properly address this issue (just follow the directions on the container).

After you put your system together (pipes, water containers, media) and fill it with water, add some quick growing plants (such as beans), and run the system as if it were fully stocked. This will provide you with an opportunity to troubleshoot any problems, and should be enough time to let chlorine dissipate from the system. It will also allow you time to evaluate whether or not you need to rearrange your system for more optimal plant growth in regards to lighting, etc. Ensure all equipment (i.e. timers, backup power system, pumps, etc.) and piping is functioning properly. By the way, you can always replace the trial-and-error, fast growing plants (i.e. beans) with more desirable plants once you are comfortable the system is running smoothly.

Starting the 'Cycling' Process

After the system has been setup, it is important that it is 'cycled' properly. Just planting the seeds in the pots the moment you place the fish in the tanks will result in poor, if not fatal, results. The bacteria that is essential in establishing the balanced ecosystem and synergy between the fish and plants are lacking at the beginning stage. 'Cycling' is the term used to describe the process of establishing the vital system bacteria.

Without the establishment of these bacteria, the ammonia-rich fish wastes cannot be broken down and transformed into soluble nutrients (nitrates) that can be absorbed by plants for growth. As a result, the fish wastes will build up to toxic levels, and the fish will perish. The plants are also unable to grow adequately due to the lack of nutrients. Therefore, bacterial populations need to be established within the aquaponics system so that plant growth and fish life can be supported.

The two main methods for cycling (or establishing the bacteria) is by raising fingerlings (feeder fish) and fishless cycling. However, the first thing needed is a freshwater test kit to monitor the ammonia, nitrite, and nitrate levels as well as pH throughout the cycling process. This is the only way to identify problems, determine if corrective action is needed if any of these elements get out of range, and to know when the system is fully cycled.

'Cycling' with Fingerlings

Using fingerlings is a traditional method to establish bacteria populations naturally. It can take up to two months. You can start up your system by using either the fingerling of your intended fish species or cheap feeder fish such as goldfish or minnows.

The fish in the tanks start producing ammonia-rich waste, which in turn attracts the air-borne nitrosomonas bacteria to populate the surface of the water. These bacteria convert the toxic ammonia into nitrites, which are still toxic to fish. However, the presence of nitrites will attract the nitrobacter bacteria. The nitrobacter bacteria then convert the nitrites to nitrates, which are harmless to the fish but are essential nutrients for the plants.

During this process it is important to monitor the water quality closely, as well as the state of health of the fish. At this initial stage of getting your system cycled, toxic ammonia and nitrites are accumulating in the tanks. This can be very stressful to the fish, and even be toxic if not maintained properly.

When cycling the system it is best to feed the fish small amounts of food to avoid ammonia spike and algal bloom. Cease feeding the fish if you observe an algae bloom or the ammonia/nitrites levels increasing to a high level of 1 ppm; as this signals that the bacterial population is insufficient to transform the wastes. Only resume feeding when the ammonia and nitrite levels return to 0.5 ppm and below. Start increasing the feeding rates after two months, when a thriving bacterial base has been established.

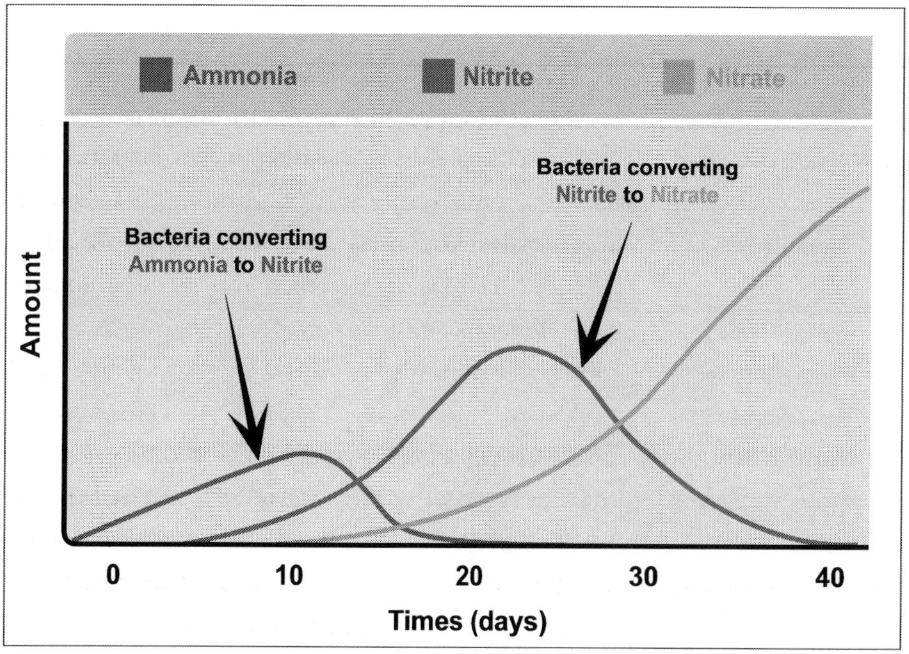

FIGURE 243. Cycle process.

Fishless 'Cycling'

Fishless cycling makes use of other sources of ammonia instead of fish. Compared to the traditional method of cycling described above, fishless cycling is faster as higher doses of ammonia can be generated in a shorter period of time. You can also control precisely how much ammonia to add without being concerned about fish survival.

Fishless cycling can be accomplished within 10 days. You can then fully stock up your fish tank once the aquaponics system as it will be fully cycled. This is contrary to the traditional method which requires you to gradually increase the stocking density.

Two Primary Methods to Add Ammonia to the System:

1. **Pure Liquid Ammonia:** It is relatively inexpensive and can be found in most home improvement and cleaning supply stores. It is also available online.
2. **Ammonium Chloride:** Also known as Sal ammoniac. Sold in some hardware stores, but most commonly purchased online. It is more costly than liquid ammonia.

Once you have decided which source of ammonia to use, simply add to the water in incremental doses to start the cycling process; and then monitor the ammonia, nitrites and nitrate levels over time until the parameters within the water reach the desired quality.

Knowing When to Start Plants

Check temperature, pH, ammonia, nitrite, and nitrate levels, with a water test kit (available online, at most local aquarium and pet stores). Keep a log of date, reading levels, and any changes made to the system, including anything added, removing fish, plants, grow bed adjustments, equipment changes, modifications to lighting, and feed data. Without a proper recording everything it is extremely difficult, if not impossible, to operate the system at optimal efficiency. Keeping accurate records will also provide you with the information needed to quickly troubleshoot system problems. The more data recorded, the higher the success rate.

Each aquaponics system will level out at a different pH reading. This is determined by water source and constituents within the water, and media type; as well as system configuration and components.

If nitrite level is not zero, run the system another week and recheck. If the readings remain unsuitable for over a week with no changes to the system, the system may be imbalanced or materials used may be negatively affecting it.

When reading levels are good, add plants if you have not already done so. Many operators proclaim to have greater success when they gradually add more fish over time, rather than all at once. It is prudent to add plants and fish at a slow rate to make sure that the water nutrient levels are stable and to give the system time to adjust. Always be sure to temperature match the water before releasing fish into the tank. Once the system has "cycled", the water should be clear. Larger systems can adapt and absorb changes easier than a smaller system.

The only way to determine when is the cycling process has reach the desired outcome is to monitor the water quality daily with the water test kit. Cycling is completed when the ammonia and nitrite levels drop to zero and the nitrate levels increase to 5-10 ppm. Once theses readings are achieved fish can be added to the tank (if done via fishless cycling) and plants can be planted.

The cycling process can also be sped up by 'seeding' the system (i.e. adding bacteria) from existing colonies. Sources of bacterial include nitrifying bacteria product found in commercially available cycling kits and disease-free Aquaponics filters.

How to Introduce New Fish into Your Tank

Moving into a new home can be a shock for any creature, but for fish it is particularly stressful. Not only will a new tank be completely unfamiliar for your new fish, but the water will likely be of a different quality. Subtle changes in water temperature, pH, and other factors make a big difference in a fish's life. To keep your fish safe and healthy, there are certain measures you should take when introducing fish to a new tank. Following are some tips on how to make an easy transition:

1. Test your water quality before adding your new fish. Your chlorine level should be at zero, and your pH should match that of the fish shop's where your pet is coming from.
2. Gradually add your system water to the transport tank to equalize the temperature.
3. Feed the fish in your aquarium before adding any new fish. This will make the existing fish less aggressive.
4. Before adding your new fish, dim the lights. This will create a less stressful environment.
5. When the transport tank water has relatively the same water parameters (i.e. temperature, pH, DO) as your fish tank, then relocate the fish to the new fish tank. Only relocate the fish. Do not relocate the transport water.
6. If possible, add more than one fish at a time, as this reduces the chance of one fish being picked on and harassed.

System Startup Checklist

1. Define your short-term and long term goals. What is your objective for getting into aquaponics? Where would you like to be with your aquaponics endeavor in five years?
2. What type of plants and fish do you want to raise? Are they compatible with your proposed aquaponic system environment/climate? Is the demand/need for the plants & fish you have in mind worth your effort or would it make more sense to raise produce that has a higher profit margin, is expensive to purchase at the store, and/or difficult to obtain organically in your area?
3. Decide on type and size of system to build.
4. Draw design plans. Take into consideration a plan for future potential expansion.
5. Research where to get parts, equipment, and fish.
6. Buy and assemble components
7. Start plants from seed or find source for seedlings. Be sure to obtain organic (non-GMO seeds).
8. Fill system with water and circulate (at least a week).
9. Add plants to system and enjoy their growth.
10. If using a fishless cycle, begin nitrification process.
11. Add fish to the system. Start out with about 20 percent of the final maximum capacity stocking density.
12. Monitor water quality, and perform partial water changes as needed.
13. Maintain system. Stay organized and keep good records.

Tips and Tricks for Aquaponics

1. Rinse media well before putting in the system, otherwise it can produce very cloudy dirty water introduce pathogens.
2. If using gravel media, always test the pH of the gravel media before adding it to your system. This will let you know if your system will have a tendency to be more on the acid or base side, so that you can be better prepared in regards to taking appropriate measures in maintaining the ideal pH.
3. For smalle systems vitamin C and an air pump can be used to more quickly bubble out chlorine and chloramine from tap water.
4. Worms, such as red wigglers, in media beds can be used to breakdown solids and reduce anaerobic zones.

5. Never use cleaning products, pesticides, algaecides, fertilizers or like substances in fish tanks or grow beds.
6. Spray plants with diluted vinegar and water solution if you discover aphides.
7. Avoid direct sunlight on fish tanks. You can limit light on the fish tank to avoid algae growth. Although Tilapia will eat algae.
8. Never change more than ⅓ of water at a time. More than that will destroy the good bacteria in the system.
9. Cover outdoor plants during a frost, and shade sensitive plants from the scorching summer sun.
10. Make sure you have backup power available for pumps and aerators.
11. Prepare for system failure. Be proactive by having at least one spare pump and battery backup system on hand.
12. It is prudent to keep spare dechlorinated water readily on hand. It is recommended that you store at least 10 percent of your system water volume in a separate container. If using municipal water be sure to let it sit for a minimum of 24 hours before adding to your system so it can properly dechlorinate. Water will need to be added to the system on a regular basis to offset evaporation. Adding chlorinated water directly to your system will kill off much, if not all, of the beneficial bacteria.

Aquaponic System Operation and Maintenance

Following is a preview of the tasks that can be expected in aquaponics gardening. It may seem like a lot of information to keep up with, but really, once the system has been established and you learn what to do, it only takes a small amount of your time. Furthermore, there are automated monitors and controls that you can integrate into your system which will simplify the process even more.

Visual Inspection

A daily visual inspection is necessary to ensure that the system is operating as it should. A walkthrough inspection would consist of observing the filters, water clarity, pumps, aerators, and plants for any problems. The water circulation between the grow beds and fish tanks should be flowing normally. Water levels should be adequate, and topped off if necessary, especially during the summer.

Water Quality and Temperature Monitoring

An entire chapter in this book is dedicated to water quality (Chapter 31). The Water Quality chapter provides more detail than what is covered in this section. However, this section will provide a brief overview on water quality and temperature.

Maintaining water quality is essential to the success your aquaponics operation. Water pH should be checked at least 3–4 times a week. Generally speaking, the pH levels should be maintained as close to neutral as possible, preferably in the range of pH 6.8–7. This pH range is a compromise to the optimal range for most plants, fish and bacteria. However, this range may need to be adjusted to suit plants and fish that do not fit this norm.

If the pH falls below the optimal range, you can alternate between using calcium hydroxide and potassium carbonate or potassium hydroxide. Both calcium and potassium are important minerals for healthy plant growth. To bring down the pH levels, you can use some of the same pH regulators that are used in hydroponics. Be sure to avoid using pH regulators that contain sodium, as they are detrimental to the life within your system. In addition, do not use products containing citrus to adjust pH.

The water introduced into the system should be free of chlorine and chemicals. Bacteria colonies will be killed off by chlorine, chemicals can be harmful to all life within the system and enter into your food chain. If using municipal water, it is important to let it stand in a separate tank for 2 to 3 days so the chlorine can vaporize off before being added to your system. Aerating the water in the tank can speed up the process.

Also monitor the water temperature using a submersible thermometer to ensure that water in the fish tank is kept stable and within the optimal range of your fish. Depending on your location, and the particular species of fish you are raising, you may need additional heating and insulation to help maintain temperatures.

Feeding of Fish

Feeding your fish is often viewed as one of the most enjoyable daily tasks by aquaponic operators. Fish need to be fed 1-2 times a day, but they don't necessarily have to be fed at regular timings. The process can even be automated with the use of auto-feeders. Some of these feeders can be acquired fairly inexpensively.

Most fish used in aquaponics can be fed with commercial feeds, worms that you even grow yourself, or with some species, scraps of food. Some operators have separate tanks that they use to grow algae in which they net out periodically to feed their Tilapia. Beware of commercial fish feed that is not organic if you are desiring to maintain a food supply that is completely free of heavy metals and other chemicals used on crops, as GMO feed will be a gateway of such into your food chain.

When fish are not eating, it typically signifies that they are stressed. It could be that they are too cold, too hot, there is too much direct lighting, oxygen is lacking, and/or there are water quality problems. If this happens, it is important to check all system parameters to identify the problem, and correct it right away.

Routine Aquaponic System Management Practices

Below are daily, weekly and monthly activities to perform to ensure that the aquaponic system is running well. These lists should be made into checklists and recorded. That way, multiple operators always know exactly what to do, and checklists prevent carelessness that can occur with routine activities. These lists are not meant to be exhaustive, but merely a guideline for management activities based upon a typical aquaponic system. Following this section are two 'go-by' checklist forms which will hopefully be helpful.

Daily Activities

Check that the water and air pumps are working well, and clean their inlets from obstructions.
- Check that water is flowing.
- Check the water level, and add additional water to compensate for evaporation, as necessary.
- Check for leaks.
- Check water temperature.
- Feed the fish (2–3 times a day if possible), remove uneaten feed and adjust feeding rates.
- At each feeding, check the behaviour and appearance of the fish.
- Check the plants for pests. Manage pests, as necessary.
- Remove any dead fish. Remove any sick plants/branches.
- Remove solids from the clarifier and rinse any filters.

Weekly Activities
- Perform water quality tests for pH, ammonia, nitrite and nitrate before feeding the fish.
- Adjust the pH, as necessary.
- Check the plants looking for problems. Address, as necessary.
- Check the fish tank from the bottom up.

- If using a biofilter, check it.
- Plant and harvest the vegetables, as required.
- Harvest fish, if required.
- Check that plant roots are not obstructing any pipes or water flow.

Monthly Activities

- Stock new fish in the tanks, if required.
- Clean out the biofilter, or other filters, if not operating efficiently.
- Clean the bottom of the fish tank using fish nets.
- Weigh a sample of fish and check thoroughly for any disease.

System Maintenance Overview

1. Feed the fish daily, monitor fish health.
2. Test water quality (every other day for the first month, then about once a week, and when you suspect problems.
3. As needed clean out filter screens, filter tanks (if using), tubing, water pump, growbed media, etc.
4. Check plant health, trim back, harvest or take cuttings.
5. Check plants for bugs or nutrient deficiencies.

Aquaponic Work Tools

Following are two very helpful tools for the aquaponic system operator:

1. Aquaponic System Data Tracking Worksheet (table 35)
2. Daily Maintenance Checklist (table 36)

Plant Care

To be a committed aquaponics operator, it is important to check on the plants regularly (preferably daily). Attention to pruning, plant structural support, and harvesting, when necessary, will provide optimal results. With aquaponics, gardening chores are much simpler as they are confined to waist-level activities, require no weeding, watering, or compost compared to soil-based traditional gardening chores.

TABLE 35. *Aquaponic System Data Tracking Sheet*

DATE	pH	AMMONIA	NITRATE	NITRITE	NOTES

TABLE 36. *Aquaponics System Maintenance Checklist*

Month_____

	MONDAY	TUESDAY	WEDNESDAY	THURSDAY	FRIDAY	SATURDAY	SUNDAY
Week 1							
Week 2							
Week 3							
Week 4							
Week 5							

Weekly Tasks

	WEEK 1	WEEK 2	WEEK 3	WEEK 4
Check pH				
Check Ammonia				
Add Water				
Check for Insects				

Monthly Tasks

	MONTH
Clean Out Pumps and Plumbing	
Check Nitrate Levels	

Harvesting of Fish and Stocking of New Fish

With the right set-up, and when done correctly, aquaponics allows for the regular harvesting of fish. You do not have to harvest the entire tank of fish at the same time. In fact, it is not advisable to do so. As a matter of fact, the closer you keep the system operating at optimal stocking density, the better the plants will do. If growing Tilapia or another species that does not typically eat the young, you can stock the tank with new fingerlings when the fish population has gone below the optimal stocking density.

Fish Health Check and Addressing Fish Illness

It is prudent to regularly net some fish to check on their state of health. Also, consider having a separate tank to isolate fish that appear to be sick, so as to prevent the spread diseases among the fish population.

Some aquaponics operators add salt to their aquaponics system periodically for disease control as salt acts as a natural anti-bacterial agent on the fish body. Due to various risk to the overall health of the system, this practice needs to be done slowly over time, with water quality and plant health closely monitored throughout and following the process. Only pure sea salt or swimming pool salt should be used. Table salt should never be used.

It is important to add only predetermined calculated amounts of salt as different fish species have varying levels of tolerance. Too much salt can stunt plant growth, or even be fatal to the plants and fish. Whenever salt is added to the system, it is important that a refractometer be used throughout the process so salt concentration in the system can be properly monitored. Adding salt should be a last resort, and only done after identifying the problem.

Addressing Emergencies

Emergency Response Processes to Have In Place

- Ready Phone List
  - ◊ Plumbers
  - ◊ Electricians
  - ◊ Staff
  - ◊ Fire/Emergency
- Trained Staff
  - ◊ Logical Troubleshooting Procedures
  - ◊ Triage
  - ◊ Know when and who to call for addition help.
  - ◊ Safety
- System Designed to Fail Reliably

Necessary Response Times for Emergencies

- **High** (fast response time — minutes)
  - ◊ electrical power
  - ◊ water level in tank
  - ◊ dissolved oxygen — aeration system/oxygen system
- **Medium** (moderate response time — hours)
  - ◊ temperature
  - ◊ carbon dioxide
- **Low** (normally slowly changing — days)
  - ◊ pH
  - ◊ alkalinity
  - ◊ ammonia-nitrogen
  - ◊ nitrite-nitrogen
  - ◊ nitrate-nitrogen

TABLE 37. **Potential Emergencies**
(not all situations are applicable to all different types of aquaponic systems)

TYPE OF PROBLEM	CAUSES
Beyond your control	Flood, tornadoes, wind, snow, ice, storms, electrical outages, vandalism/theft
Staff error, diffusers plugged	Operator "errors", overlooked maintenance causing failure of back-up systems or systems components, alarms deactivated.
Tank water level	Drain valve opened, standpipe fallen or removed, leak in system, overflowing tank.
Water flow	Valve shut or opened too far, pump failure, loss of suction head, intake screen plugged, pipe plugged.
Water quality	Low dissolved oxygen, high CO2, supersaturated water supply, high or low temperature, high ammonia, nitrite, or nitrate, low alkalinity.
Filters	Channeling/plugged filters, excessive head loss
Aeration system	Blower motor overheating because of excessive back-pressure, drive belt loose or broken or disconnected, leaks in supply lines.

Aquaponic Safety

Safety is important for both the operator and the system itself. The most dangerous aspect of aquaponics is the proximity of electricity and water, so proper precautions should be taken. Food safety is important to ensure that no pathogens are transferred to human food. Finally, it is important to take precautions against introducing pathogens to the system from humans.

Electrical Safety

Always use a residual-current device (RCD). This is a type of circuit breaker that will cut the power to the system if electricity grounds into the water. The best option is to have an electrician install one at the main electric junction. Alternatively, RCD adaptors are available, and inexpensive, at any hardware or home improvement store. An example of an RCD can be found on most hairdryers. This simple precaution can save lives. Moreover, never hang wires over the fish tanks or filters. Protect cables, sockets, and plugs from the elements, especially rain, splashing water, and humidity. There are outdoor junction boxes available for these purposes. Check often for exposed wires, frayed cables, or faulty equipment and replace accordingly. Utilize "drip loops" where appropriate to prevent water from running down a wire into the junction.

Food Safety

Good agricultural practices (GAPs), should be adopted to reduce as far as possible any food-borne illnesses, and several apply to aquaponics. The first and most important is simple: always be clean. Most diseases that affect humans would be introduced into the system by the workers themselves. Use proper hand-washing techniques and always sanitize harvesting equipment. When harvesting, do not let the water touch the produce; do not let wet hands or wet gloves touch the produce either. If present, most pathogens are in the water and not on the produce. Always wash produce after harvesting, and again before consumption.

Second, keep soil and faeces from entering the system. Do not place harvesting equipment on the ground. Prevent vermin, such as rats, from entering the system, and keep pets and livestock away from the area. Warm-blooded animals often carry diseases that can be transferred to humans. Prevent birds from contaminating the system however possible, including through the use of exclusion netting and deterrents. If using rainwater collection, ensure that birds are not roosting on the collection area, or consider treating the water before adding it to the system. Preferably, do not handle the fish, plants, or media with bare hands; disposable gloves are always the safest bet.

General Safety

Often aquaponic systems, farms, and gardens in general, have other general hazards that can be avoided with simple precautions. Avoid leaving power cords, air lines, or pipes in walkways as they can pose a trip hazard. Water and media are heavy, so use proper lifting techniques. Wear protective gloves when working with the fish and avoid the spines. Treat any scrapes and punctures immediately with standard first-aid procedures — washing, disinfecting, and bandaging the wound. Seek medical attention, if necessary. Do not let blood or body fluids enter the system, and do not work with open wounds. When constructing the system, be aware of saws, drills, and other tools.

Keep acids and bases in safe storage areas, and use proper safety gear when handling these chemicals. Always keep all dangerous chemicals and objects properly stored and away from children. Ensure that children cannot fall into fish tanks.

Safety Summary Overview

- Take all necessary safety precautions where electricity and water combinations occur.
- Follow all National Electric Codes.
- Use RCD on electric components to avoid electrocution.
- Use only low voltage equipment and sources 24 VAC or 12 or 24 VDC.
- Shelter any electric connections from rain, splashes, and humidity using correct equipment.
- Implement a maintenance schedule & regular system checks.
- Ensure that everyone working with the aquaponics system is properly trained.
- Have strict policies, procedures, and supervision involved for all working with the system; especially for entry level staff and visitors.
- Follow OSHA guidelines for all aspects of the operation.
- Do not contaminate the system by using bare hands in the water.
- Avoid trip hazards by keeping a neat workstation.
- Wear gloves when handling fish and avoid spines.
- Wash and disinfect wounds immediately. Do not work with open wounds. Do not let blood enter the system.
- Be careful with power tools and dangerous chemicals; wear protective gear.
- Adopt GAPs to prevent contamination of produce. Always keep harvesting tools clean, wash hands often, and wear gloves. Do not let animal faeces contaminate the system.

Troubleshooting: Problems and Solutions

Plants Troubleshooting: Potential Problems & Solutions

Problem: *Plants are Dying*
Reason: Potentially due to any number of reasons such as: fish/plant ammonia ratio out of balance, not enough food for plants, oxygen deficiency, water quality, minerals, temperature, pH, bugs, lighting, fungus, etc.
Solution: Perform an internet search by typing in plant symptoms, and/or consult plant expert.

Problem: *Aphids Eating Plants*
Reason: Natural phenomenon in the real world. Not as common in greenhouses.
Solution: Spray the plants with water and/or buy a bag of ladybugs from local nursery. Ladybugs eat aphids. Release the ladybugs at night, when they are not active and can easily be handled.

Problem: *Caterpillars Eating Plants*
Reason: Natural phenomenon in the real world. Less common in greenhouses.
Solution: Use garlic spray on plants. Put caterpillars into the fish tank as free fish food.

Problem: *Plants are Discolored*
Reason: Typically, it is a sign of a mineral deficiency.
Solution: Perform an internet search by typing in plant symptoms, and/or consult a plant expert.

Problem: *Plants Not Growing or Have an Extremely Slow Growth Rate*
Reason: The pH may be too high. Inadequate nutrients in water.
Solution: Check pH, increase feed if fish eat it all too quickly or reduce number of plants in system.

Water Quality Troubleshooting: Potential Problems & Solutions

Problem: *The pH over 7.5 or pH is less than 6.5 (or out of your systems optimal range)*
Reason: Acidity of water is too low or too high.
Solution: Refer to section in this book regarding pH and how to adjust it. Remember though, quick pH level changes are hazardous to fish. Since most systems naturally settle into their own operating pH level over time, adjusting and maintain pH beyond this natural range can be an ongoing challenge. If a system component (like media, water content) is already buffering the system, additives will only change pH temporarily. It is best to find long-term solutions such as adjusting the media type, or go with plants and fish that best suit your natural pH parameters.

Problem: *Low Water Level*
Reason: Surface evaporation and plant transpiration reduce water in system. The larger the plant, the more water they absorb and transpire. Warmer temperatures, low humidity, and turbulent water will result in greater evaporation. Ensure there are no water leaks.
Solution: Top off system with dechlorinated water.

Problem: *Water is Green*
Reason: Too much algae. May be due to water having too many nutrients and/or too much light hitting the water surface.
Solution: Make sure you have the correct tank to grow bed volume ratio, you are not over feeding the fish, and the fish tank is properly shaded. Also note that some algae eating fish will only eat certain types of algae.

Problem: *Water is Dirty*
Reason: Usually a result of fish not eating all of their food, and/or the system is under-filtered.
Solution: Ensure that you are not over feeding. Use a towel or sock as a filter on the return water to help filter out sediment that is not captured in the grow bed.

Problem: *Water is Cloudy*
Reason: Overfeeding, system is out of balance, algae growth, and/or the system is under-filtered.
Solution: Ensure that you are not over feeding. Also make sure that all water parameters (i.e. temperature, pH, dissolved oxygen, etc.) are in the desired range.

Problem: *Top of Tank Has Iced Over*
Reason: Conditions are too cold.
Solution: Adjust temperature of water and/or room temperature. If fish are still alive, add air bubbles to fish tank to maintain oxygen levels.

Problem: *Water is Foaming*
Reason: Detergents or other chemicals may have been introduced into the system.
Solution: Perform 50 percent dechlorinated water changes every day until the foaming is gone. It is important to prevent fish shock as much as possible in this situation, but it is also critical to eliminate the contaminants. Be careful not to eliminate most of your beneficial bacteria, as they keep the system cycling properly Refrain from cleaning media.

Fish Troubleshooting: Potential Problems & Solutions

Primary Ingredients for Fish Health:
1. Know and maintain the optimal water parameters for your fish species (i.e. temperature, pH, dissolved oxygen, etc.).
2. Feed fish at a rate appropriate to the biomass of your fish and stage of growth. Observe feedings to ensure fish are being fed enough or that there is not a lot of waste. Adjust feedings appropriate with growth rate and with changes of stocked density.

3. Shade fish tank as appropriate. No direct lighting on fish tank.
4. Remove or minimize stress (i.e. vibration, loud noises, banging, etc.). The author prefers aquaponic system designs that do not have the pump inside the fish tank.

Problem: *Dead Fish*
Reason: Unknown.
Solution: Remove dead fish from the tank immediately to prevent decaying ammonia buildup. Try to identify the reason for death. Observe other fish to ensure there is not major problem with the system.

Problem: *Fish Seriously Sick, Almost Dead, No Hope of Recovery*
Reason: Unknown.
Solution: Euthanasia. Don't flush, suffocate (by allowing fish to die out of water), freeze, or use antacid; this is extended torture and inhumane). The best approach is to remove the fish from the tank and flatten fish's head hard and fast with a brick.

Problem: *Fish Swimming at Surface of Water Unusually or Gulping Air at Surface ('piping')*
Reason: Lack of dissolved oxygen. However, piping can also be a symptom of fish being infected with gill parasites.
Solution: Immediately add an air bubbler. Manually splash water but try not to stress fish.
Water/air exchange introduces oxygen into the water. Test to see if pH levels are low and increase slowly if necessary. Remember, changing pH as little as 0.2 quickly can be dangerous to the fish.

Problem: *Fish Acting Unusual, (i.e. not behaving normally, swimming sideways, floating upside down, etc.)*
Reason: Possibly a problem with the feed you are using. Could be the result of a disease, injury, or poor water quality. Odd behaviors could indicate breeding activity.
Solution: Monitor activity. Try a different feed. Do an internet search with the specific symptoms and type of fish. Consult a fish expert.

Problem: *Fish Jumping Out of Tank*
Reason: It could be that the water level is too high, they are trying to catch insects, a result of poor water quality, or water parameters are not within the proper range for your fish species.
Solution: Reduce water level if necessary, cover tank with wire mesh/netting, test water parameter conditions, test water quality, and/or filter water intake with a rag or sock to catch sediment if water is not clear.

Problem: *Fish Has Creamy White Film Slime*
Reason: Could be the result of low pH level, breeding behavior, or disease.
Solution: Check pH level, try to determine if it is potentially disease related (fish not acting healthy), and do an internet search with the specific symptoms.

Problem: *Fish has been determined to have Ich, Ick, or White Spot Disease*
Reason: A protozoa (one-cell animal parasite) called Ichthyophthirius multifiliis. Typically takes hold as a result of stress (environment change/fighting) which often negatively impacts the immune system of the fish making it more susceptible to such problems.
Solution: Easily treated if caught in time. It is best to treat the fish in a separate quarantine tank. Other solutions include:
Adding one-tablespoon sea salt (non-iodized salt) per 1-gallon water in the system. In addition, gradually raise water temperature to the upper limits of your fish species optimal temperature range.
Purchase one of the following products online, or from your local aquarium or pet shop: Coppersafe, Quick-Cure Ich-Ease, Aquari-sol, Cure-Ick, or Super Ick Cure. Follow directions.

Problem: *Fish Disappeared*
Reason: Critter, thief, or the fish flopped out of tank.
Solution: Cover tank with wire netting and secure it, erect a scarecrow to the tank, close and lock greenhouse at night, and/or monitor aggressive fish.

Problem: *Algae on Fish*
Reason: Algae does not grow on fish; it grows on fungus on your fish. It can also be the result of cyanobacteria and/or high nitrates in the tank. Too much light on the fish tank can exacerbate this problem.
Solution: Ensure tank conditions are in the optimal range. Filter the intake coming from your grow bed with a towel or sock (something extra to filter the water before if enters back into the fish tank). Some operators claim to have had success adding a little espin salt to the tank (be sure to research this process thoroughly in advance to going this route).

Problem: *Fish Are Not Eating in Colder Weather*
Reason: Fish eat less the colder the water temperature. Water temperature impacts fish metabolism. Fish digestive enzymes are also not as active when temperature drops.
Solution: Feed less and/or raise the temperature (depending upon the circumstances and species). Every situation is unique.

Problem: *Power Outage*
Reason: Power grid is unreliable.
Solution: Always have a backup system readily available.

Problem: *Water Pump Stops*
Reasons: Lack of electricity, defective or clogged pump.
Solution: A system crash can occur in less than 24 hours. Switch to backup power system (if power outage). Replace pump right away. It is prudent to have multiple pumps in service, and at least one back-up pump readily available.

Problem: *System is Leaking Water*
Reason: Parts are old and cracking, sealants/tape defective.
Solution: Temporarily seal holes in pipes with Teflon plumbers tape, replace worn parts, shutoff water flow to parts of system that are leaking and repair, use a silicon product to seal leaks.

Problem: *Red Worms Appeared*
Reason: Probably introduced by natural elements, feed, or someone else.
Solution: Do nothing, as red worms provide additional benefit, and can be good free fish food.

Bell Siphon Troubleshooting: Potential Problems & Solutions

NOTE: The vast majority of bell siphon problems can be solved by using a wider and taller siphon pipe. Others can be solved with a cheater air break tube, cheater bottom bends, or larger crenellations. A few problems are because the siphon pipe is too large in diameter, have extra holes where they do not belong, the airtight siphon pipe cap was not installed (or is not airtight), the cheater air break tube needs to be shortened, or it has too many bottom bends.

Semantics: The standpipe is the inner drain pipe. The siphon pipe is the outer pipe with top cap. A shroud is the protective pipe with many holes (or other permeable material) that is around the unit to allow maintenance while preventing media from smothering it.

Problem: *Bell Siphon is Not Working Properly*
Reason: Poor design.
Solution: Most bell siphon issues can be solved with a wider and/or taller siphon pipe.

Problem: *Bell Siphon Will Not Start*
Reason: Standpipe is not level; water flows over one edge instead of entire edge surface.
Solution: Cut the top of the standpipe so it is level.

CHAPTER 30: STARTING AND MANAGING YOUR AQUAPONICS SYSTEM

Problem: *Bell Siphon Does Not Stop Draining*
Reason: Siphon is not strong enough to break water tension at bottom of siphon pipe.
Solution: Make one crenellation/hole at bottom of siphon pipe higher for an air break. Make siphon pipe taller if it is too close to the top of the standpipe to increase vortex area. If this approach does not work, try using a wider siphon pipe.

Problem: *Bell Siphon Never Engages; Standpipe Only Drains Water Output to Equal Water Input*
Reason: Siphon pipe cap is not airtight or cap is missing. Another reason for this problem may be that the standpipe (output pipe) is too wide to create a water lock and begin the siphon action.
Solution: For the first reason, seal the cap on top of siphon pipe. For the second reason reduce width of standpipe or add an adapter to the top of it that reduces the opening. Some operators install a bend at the bottom of the standpipe to resolve the problem.

Problem: *Bell Siphon Takes a Long Time from Dribble to Fully Starting the Siphon*
Reason: Siphon pipe does not have enough headroom to initiate a vortex over the standpipe
Solution: Make siphon pipe taller.

Problem: *Bell Siphon Does Not Drain Fast Enough*
Reason: Bell siphon was improperly designed and is thus too small for the amount of water it needs to address.
Solution: Reduce incoming water flow with ball valve in or use a larger bell siphon.

Problem: *Bell Siphon Stops Draining Midway*
Reason: Water pressure has equalized inside and outside the siphon at the standpipe height.
Solution: Use fewer crenellations in siphon pipe or use smaller diameter siphon pipe.

CHAPTER 31

Water Quality

Fish Life Support

As already addressed in previous chapters of this book, growing organic vegetables and fresh fish from your own aquaponic system has many advantages. These benefits come with lower costs and less maintenance compared to traditional hydroponics and aquaculture. Expensive hydroponic nutrient solution for the plants and expensive aquaculture filtering equipment to clean the waters for the fish are NOT needed with aquaponics.

To summarize key points of these previous chapters, even small systems can produce such a high production yields that excess vegetables and fish can be given as gifts, bartered, and/or sold for a profit. Most importantly, you can rest assured that the fish that you are eating is free of mercury and other toxins, unlike the fish caught from many of our polluted oceans, rivers, and lakes. The produce cultivated will be 100 percent organic and free of harmful pesticides, herbicides, and fungicides widely used on conventional crops.

Because fish, bacteria, and plants are totally dependent upon the water surrounding them and are unable to escape this environment, the quality of the water is of great importance. Several water quality parameters are important in fish culture, including dissolved oxygen (DO), temperature, mineral content, and the concentrations of several waste products excreted by the organisms. The presence of fish has a direct impact water chemistry. For example, fish excrete waste products that are high in ammonia and other nitrogen compounds, which tend to pollute the water. Likewise, plants favorably influence water chemistry by adding oxygen and by removing nitrogen compounds. Consequently, the relationship between the species in your aquaponic system and the surrounding water is a dynamic and crucially important one.

As emphasized early in this book, an aquaponic system operating at an optimal level is only achieved when there is harmonious balance of all the life support components of three groups of organisms: fish, plants and bacteria. Each organism has a specific tolerance range, and optimal point, for each water quality parameter (i.e. DO, pH, temperature, nitrate, nitrite, ammonia, etc.). These primary water quality components' tolerance ranges are relatively similar for all three organisms, but they are not the same. Therefore, a compromise is necessary. Studies have shown that the best balance is achieved when water quality parameters are maintained, as shown in table 38 on the following page.

PART VI: OPERATION AND MAINTENANCE

TABLE 38. **General Water Quality Tolerances for Fish, Aquaponic Plants, and Nitrifying Bacteria**

ORGANISM TYPE	TEMP (°C)	pH	AMMONIA (mg/liter)	NITRITE (mg/liter)	NITRATE (mg/liter)	DO (mg/liter)
Warm water fish	22–32	6–8.5	< 3	< 1	< 400	4–6
Cold water fish	10–18	6–8.5	< 1	< 0.1	< 400	6–8
Plants	16–30	5.5–7.5	< 30	< 1	-	> 3
Bacteria	14–34	6–8.5	< 3	< 1	-	4–8

TABLE 39. **Compromise of Ideal Water Quality Parameters for Fish, Plants, and Bacteria**

	TEMPERATURE	pH	AMMONIA	NITRITE	NITRATE	DO
Tolerance Range	18–30 °C	6–7	< 1 mg/L	< 1 mg/L	< 5-150 mg/L	> 5 mg/L
	64–86 °F		< 1 ppm	< 1 ppm	< 5-150 ppm	> 5 ppm

NOTE: It is important that the beginner not be intimidated by, or overwhelmed, the thought of maintaining a balanced aquaponic system. If constructed and maintained according to the basic principles and parameters stated in this book, operating a successful aquaponic system is a relatively simple and very enjoyable endeavor.

Fish Population Density

Aquaponics stocking density can be referred to the number of fish kept in the fish tank(s). Although, it is usually defined as the weight of fish held in regards to the volume of water. The accepted approach for stating fish stocking density is the kilograms (weight) of fish per liters (volume) of water.

The more fish, the greater yield and the higher profit. However stocking too many fish will lead to ill health of the fish, and can also be detrimental to the plants. The fish will not grow as well and will start dying, and there will be a buildup of wastes which will be toxic to both fish and plants. Finding the perfect balance of stocking as many fish as possible without adverse impacts is key.

Some species are able thrive in a denser population while others require considerably more space. Some species are territorial and expend a lot of energy fighting if packed too close together. Whereas others prefer to be in a denser pack, shoaling, and schooling. It is important to select a fish species that is not territorial. Territorial fish forced to live close together expend a great deal of energy, have high stress levels, and lower survival rates. In addition, it is not ethical to raise animals under stressful conditions. As a rule, smaller fish are more likely to live out their lives in schools, although some large fish will school together. The fish described in this book are species common to aquaponics and do fine when stocked at the recommended densities. However, if you are thinking about raising fish that are not listed in this book, it would be prudent to ensure that they are not a territorial species.

What is the recommended stocking density? It will depend upon your bio-filter (grow bed) capacity, the feeding rate, ammonia production rate, oxygen consumption rate, and aeration rate. Either gas exchange or ammonia toxicity are what will most likely be the limiting factor that will determine your stocking density. An experienced operator is usually able to stock a tank at 1 lb of fish per 5 -10 gallons of water, or one to two fish per 10 gallons of water (38 liters). Inexperienced operators are better off following a stocking density of one fish per 22.5 gallons (85 liters) of water.

However, it is more accurate to talk about stocking densities in terms of kilograms of fish per cubic meter or liters of waters, as this is the industry standard and doing math via the metric system is a much easier process. For most aquaponics set-ups, using 66 to 88 pounds of fish to 264 gallons (30 to 40 kg of fish per 1000 liters) will work fine.

Large, commercial recirculating fish growing facilities will keep fish at densities up to approximately 3-lbs/ft^3 (50 kg/m^3) in aerated systems and up to 9.5-lbs/ft^3 (150 kg/m^3) or higher in direct oxygen injected systems. Aquaponic hobbyists rarely approach these high fish densities in their smaller-scale systems, however this does not mean that the standards of fish tank design should not be adhered to. In order to ensure fish health and well-being via water quality optimization, most hobby aquaponics practitioners keep fish at densities well below 1.5-lbs/ft^3 (20 kg/m^3). Commercial aquaponic operators will often reach higher aerated fish stocking densities as this is required to recover capital and operating costs, and so the design and efficient operational characteristics of the system are critical.

Lower stocking densities are much more manageable and have higher success rates as there is more room for error. Lower stocking densities are recommended for those that are new to aquaponics. The amount of food you put into the tanks will also dictate how well the system runs. If you feed too much, there will be an accumulation of waste that the bacteria may not be able to handle.

Ideally, you will harvest the fish as soon as they're big enough. This avoids ending up with a stocking density which otherwise strains the system. Since individual fish grow at different rates, you should start taking out the bigger ones as soon as they are large enough to eat, rather than waiting until you can harvest all the fish at one time. This is actually works best for smaller operations, as you will have an ongoing manageable harvest rate where you can eat and/or sell them without having to freeze storage a large quantity of fish at one time.

Temperature Range

What is the temperature range of your aquaponics location? Ideally, it should match the optimal temperature of your aquaponics fish. If not, you will incur higher upfront capital cost and more expensive ongoing operational cost maintaining the temperature at the preferred range of the fish, via heating, cooling, and insulation adjustments. Also, a temporary power failure can end up being fatal to your system if it has to be kept at a significant different temperature than the exterior environment.

Fish prefer a specific temperature range, and do best when raised at their optimal temperature. As mentioned above, the easiest, most energy efficient and cost effective option is to select a fish species that has an optimal temperature range that matches your location's natural temperature. However, the ultimate decision comes down to your personal preference and if you are willing to incur the cost of additional cooling, heating, and insulation. This is essential if your environment has widely fluctuating temperatures. Placing a fish tank in the garage, a shop, or green house will help maintain a stable water temperature.

Fish cannot regulate their body temperature like humans do. They are dependent on the water temperature for their body temperature. To maintain fish health the water temperature should never be adjusted, or allowed to fluctuate, more than 3°F per day.

Rates of Growth

Obviously, fast growing fish provide food and revenue quicker than slow growing fish. However, fast growing fish have higher metabolic rates, consume more food, and produce more wastes. This is not necessarily a bad thing. It just means that growing conditions need to be monitored more closely, and adjustments made to accommodate these higher growth rates. Higher waste production can be beneficial if the plants being grown require higher levels of nutrients (such as tomatoes).

Balance of Water Constituents

One also needs to be careful not to assume that fish data that is based upon a particular species is applicable to all fish species. Each species, and even varieties of certain species, will have unique characteristics, growth rates, and require specific growing conditions.

The fish tank encompasses a direct relationship between ammonia levels, water temperature, and pH, and if their inter-relationships are not understood it can lead to impending disaster.

Dissolved oxygen levels also plays a very important role. The following need to be well understood by the aquaponics operator:

Ammonia

All fish give off ammonia. Ammonia is generated from their gills and waste. Uneaten fish food also is converted into ammonia as it breaks down. If not properly addressed the buildup of ammonia becomes toxic to fish. Ammonia levels as small as 5 ppm is typically deadly to fish. Fortunately, beneficial bacteria within the grow beds converts ammonia into nitrates that the plants ingest, thus addressing the problem. This is known as the Nitrogen Cycle. When first starting up, it can take a new aquaponics system some time to fully cycle or balance. It can take as long as six to eight weeks in winter months or just a few weeks in the summer (or warmer conditions) before the system is said to have "cycled" and all is in balance.

The Nitrogen Cycle will be addressed in more detail later within this book. For now, know that nitrite is the product of ammonia being digested by bacteria in the water, on the skin of the fish, and in the bio-filter (if the system has one). Nitrite is poisonous to fish at 225 ppm. Nitrite is broken down by bacteria to form nitrate. Nitrate is a natural fertilizer, used by plants.

An overstocked tank with fish is much more susceptible to fatal flaws if any of these parameters are pushed beyond their tolerable limits. Ammonia levels will depend greatly on the temperature of the fish tank water and the pH of the water, fish density, and how much uneaten fish food remains floating in the system. It gets more complicated with warmer water and a pH that is out of balance. Nevertheless, it is easy to monitor these conditions, and there are a number of things that can be done to ensure the system remains healthy and successful. Ideally your ammonia level should be near zero but there will always be traceable amounts being emitted constantly by your fish. It gets more complicated if you stock your system with a heavy fish load.

Fish can tolerate higher levels of ammonia the cooler the water. The same goes for dissolved oxygen. Cold water can store more dissolved oxygen than the same volume of warm water.

Understanding the relationship between ammonia and water temperature provides you with the ability to better manage your aquaponic system and avert catastrophic danger.

pH Range

The pH is the measure of the hydrogen ion (H+) concentration in the water. The pH scale ranges from 0-14 with a pH of 7 being neutral. Pure water has a pH of 7, but additives (i.e. chlorine, fluoride, etc.) can alter the pH. A pH below 7 is acidic and a pH of

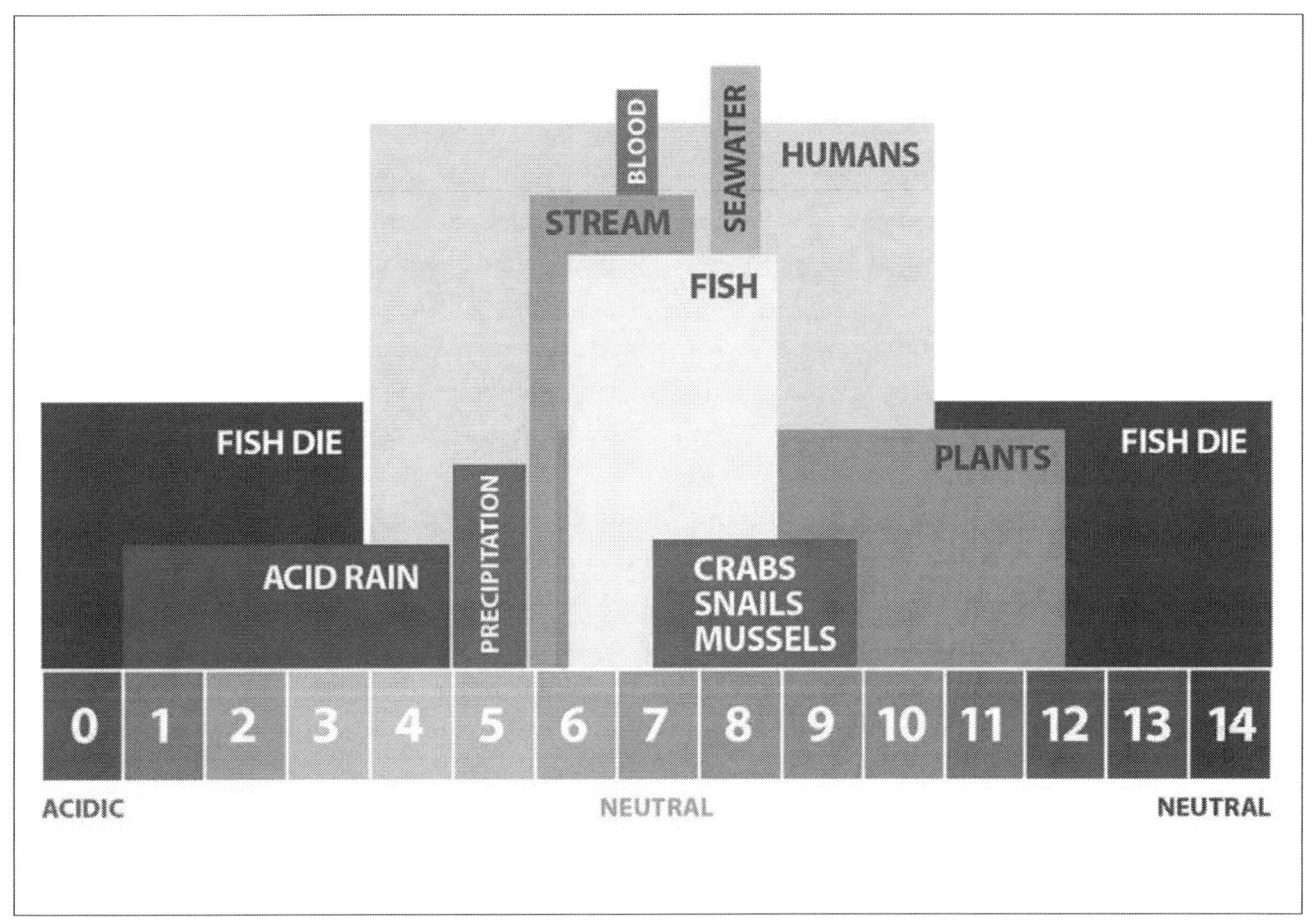

FIGURE 244. *Typical pH range of various living organisms.*

above 7 as basic. Different species of fish have varying pH requirements. However, the typical pH range for aquaponics is between 6 and 7.2. A pH of 6.8 to 7 is usually a happy compromise for most fish, plants, and the good bacteria; but again the pH can be slightly more or less depending upon the fish species.

The pH levels of tap water can significantly exceed the required pH range required. As a matter of fact, such detrimental differences are fairly common. Therefore, the pH of tap water needs to be properly adjusted before adding the fish.

Adjusting pH: Adjust pH Slowly After Fish are Present

When fish are already in the tank, it is important to adjust the pH slowly. If pH is lowered too much too fast, it will stress the fish, possibly the plants, and can even be fatal. Therefore, pH should be lowered over the span of a week or more when fish are present.

pH Fluctuations

The pH will change in response to system input (rain, fish, plants, topping off the tank periodically). Temperature will also cause your pH to vary. When testing pH, measure at several points in the day for an average, or measure at the same time and temperature each day in order to obtain reliable data.

In the beginning, when you have water without a bacteria colony, pH will fluctuate significantly. This is normal. When adding water from most municipal taps, the chlorine will off gas, which will cause the pH to change.

Lowering pH

Again, the lower the pH, the more acidic the water. The following are some safe additives that can be implemented to lower the pH in an aquaponics system.

- Hydrochloric Acid one or two caps per 250 gallons
- Acetic Acid (Vinegar)
- Sulphuric Acid
- Maidenwell media or Diatomite
- Iron sulfate fertilizer

Hydrochloric acid (swimming pool acid) is most frequently used to lower pH to the optimal level. Some gravel, such as limestone, will lower pH. Injecting CO_2 directly into the water has also been reported to lower pH.

Citric acid should never be used. Citric acid is an antibacterial agent that can be fatal to the good bacteria living within the media.

Raising pH

Higher pH readings are called "base". The following are some additives that can safely be used in an aquaponic system to raise the pH:

- Dolomite Lime — Calcium Magnesium Carbonate
- Calcium Hydroxide (hydrated/builder's/slaked/ hydrated limes)
- Potassium Carbonate (bicarbonate)
- Potassium Hydroxide (pearl ash/potash)
- Snail Shells
- Sea Shells
- Egg Shells

If using shells, it is best practice to boil, bleach, or use hydrogen peroxide on them first in order to kill all bacteria. Containing these chemicals or ingredients in a nylon stocking, paint strainer sack, or other breathable bag will allow for easy removal once the desired pH range is achieved.

pH Cycling

When the term 'cycling the aquaponic system' is used, it simply means that the nitrifying bacteria (good bacteria) are present and the system is stable. Regular monitoring and recording pH enables one to not only evaluate the current status of the system at a specific point in time, but shows any pH trends in one direction or another. Keeping pH within the desired range also prevents "nutrient lockout" enabling plants to readily take up the right nutrients and grow properly.

pH Adjustment Notes:

1. Adjusting pH fast can be hazardous to fish. A matter of fact, changing pH by only 0.2 too quickly can be dangerous.
2. If your system has a pH crash, identify the problem and remedy what caused it rather than just trying to adjust the pH when starting over.
3. Small crushed particles work faster while larger particles work slower to maintain system pH.
4. Keep "pH increase" equivalents on hand: sodium bicarbonate (baking soda), limestone, and calcium carbonate (egg shells, snail shells, sea shells).
5. It is easier to increase pH than it is to decrease.
6. Keep "pH decrease" equivalents on hand: vinegar, Hydrochloric Acid (one or two caps per 250 gallons), Acetic Acid (Vinegar), Sulphuric Acid, Maidenwell media, or Diatomite .
7. Iron sulfate fertilizer.
8. If possible, keep a back up water tank (or barrel) nearby filled with dechlorinated water bucket for topping off the tank periodically as water is lost through evaporation and plant transpiration, or even in an emergency (i.e. pipe leak, etc.). Preferably, this back up supply will hold at least 10 percent of the fish tank volume.
9. Use a mixture of plants so they do not all die off at the same time, causing the system to crash.

pH Impacts Upon Fish Breeding

There is a relationship between pH and aquatic life breeding cycles. As pH changes, so will the number of fish eggs that are produced, if they can be produced at all. Keep this in mind if you desire to reproduce your fish.

Calcium Carbonate

What is calcium carbonate? Calcium is a mineral that is found naturally in foods. Calcium is necessary for many normal functions of the body, especially bone formation and maintenance. Calcium can also bind to other minerals (such as phosphate) and aid in their removal from the body. Calcium carbonate supplements are used to prevent and to treat calcium deficiencies. Natural forms of calcium carbonate are eggshells, snail shells, and seashells.

Even though calcium carbonate works to lower pH and natural forms don't do any harm, be cautious in using it, especially natural products. It can take a long time to dissolve and affect the pH. Therefore, after adding a little, wait and check pH after two hours before adding any more. Adding too much at one time or over a short period can cause the pH to spike, and could result in doing more harm than good.

How much should you add to your system to raise the pH? That question does not have a specific answer. It all depends on the type of calcium carbonate you are using and its form (i.e. natural egg shells, bag of ground powder, etc.), the size of your system, fish density, if you are adding other substances to your system, the type of media you are using, temperature, and how far your pH is off the optimal range. It is always better to be safe than sorry. Raise the pH using small amounts of calcium carbonate over time, little by little every two hours. With some experience, you will come to know your system, and have a better understanding of how much to add whenever the pH needs to be adjusted.

Dissolved Oxygen (DO)

The limit of dissolved oxygen in fish tank water is termed "saturation level" and the level is measured in parts per million (ppm). DO is essential to the health of your fish, and can be impacted by the temperature of your water. Beneficial nitrifying bacteria, which break down the metabolic waste of aquatic creatures, are equally dependent on oxygen.

Oxygen is introduced into the aquaponics system naturally through the plant life, waves, cascades, and waterfalls. If these natural methods are not sufficient, then an aeration device is necessary. Poor aeration your tank, and/or an extremely high density of fish in your tank can lead to inadequate DO.

The amount of oxygen that water can hold is temperature dependent. Warm water holds less oxygen than colder water. In other words, the colder the water, the more oxygen the water can hold.

Water is 800 times denser than air and contains 95 percent less oxygen. Fish expend a considerable amount of energy breathing but are able to extract 80 percent of the available oxygen from water, as opposed to humans, who extract only 25 percent of the oxygen they breathe in. The saturation level of oxygen in water is in the region of seven to eight parts per million. Under normal conditions, there is never too much oxygen in a fish tank, unless the operator purposely attempts to super saturate the water. Operators can super saturate aquarium water by diffusing 100 percent liquid oxygen into the water, but there is no real value to doing so.

As a matter of fact, too much oxygen is also harmful to fish. Hyperoxia is the state of water when it holds a very high amount of oxygen. At this state, water is described as having a dissolved oxygen saturation of greater than 100 percent. This percent can be 140-300 percent. If fish are exposed to such water, their blood equilibrates, bubbles form in the blood, and can block the capillaries. In severe cases, death

PART VI: OPERATION AND MAINTENANCE

occurs rapidly as a result of blockage of the major arteries. The remedy is either to remove the fish to normally equilibrated water or to provide vigorous aeration to strip out the excess gas.

Most aquaponic fish, including Tilapia, prefer dissolved oxygen levels above 5 ppm. Ideally, dissolved oxygen levels will never drop below 3 ppm.

The fish will suffocate from low DO. Symptoms of low DO include fish swimming close to the water surface grasping for air, fish lying on the bottom of the tank, and red gills. However, waiting till then is not a great idea, so it is a good idea to check oxygen level regularly so preventative measure can be taken before the health of your fish is jeopardized. Most importantly, be sure you have operational features built into your system so that there is always adequate dissolved oxygen present should your grow bed and system setup be inadequate. If your system does not naturally generate enough DO, then simply adding an aeration device or air pump will ensure that there is sufficient supply for the fish to thrive.

To measure DO, use a DO meter. This is typically the most expensive piece of measuring equipment used in aquaponics. Although many operators get by without them, they do so at a risk. If you have the money to spend, they are nice to have on hand.

It is important to have a backup aerator (battery or generator operated) that is independent of the main power supply. This ensures that air can be continually pumped into the water to keep the fish alive, in case of a power failure.

Temperature has a very important effect to dissolved oxygen. As the temperature of the water goes up, the water loses the ability to hold the dissolved oxygen and the concentration goes down. When the water cools, it regains the ability to hold higher amounts of oxygen. Figure 245 below simplifies this rule. Knowing this relationship, one can deduce that hypoxia tends to occur in the warmer months of the year, namely during the summer.

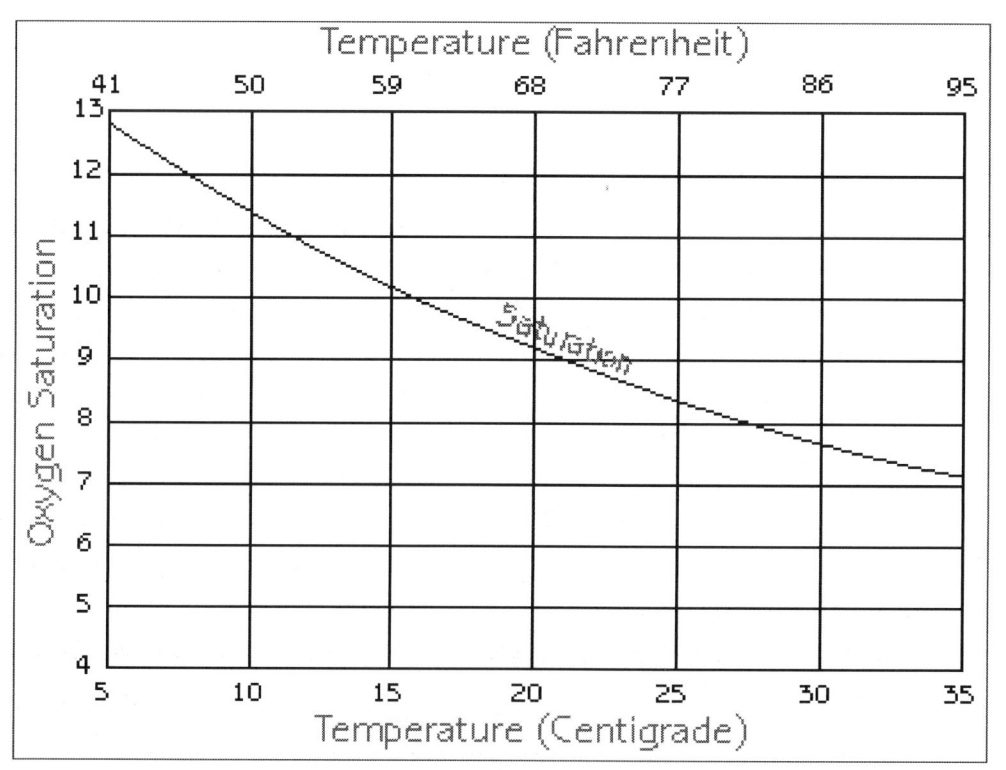

FIGURE 245. Oxygen and temperature relationship.

CHAPTER 31: WATER QUALITY

TABLE 40.

BENCHMARK	OXYGEN DEFICIENCY	ALGAE BLOOM	DISEASE
Fish Behavior	Swimming Near Surface Gasping for Air	Abnormal Swimming	Abonorma Swimming
Time of Fish Kill	Nighttime into Morning	Brightest Part of Day	Anytime
Size of Dead Fish	Large Fish Die First	Small Fish Die First	Small Fish Die First
Microscopic Algae Abundance	Algae Dying	One Dominant Algal Species	No Effect
Dissolved Oxygen Concentration	Less than 3 ppm Oxygen	Supersaturated Oxygen Concentrations	No Effect
Water Color	Brown or Gray or Black	Dark Green or Brown or Golden	No Effect

Aeration Sizing

The aeration rate is measured in cubic feet per minute (CFM). CFM is the measurement of the volume of air flow being introduced into the system.

Air pressure is the pressure required to deliver the correct amount of air flow for proper aeration. Air pressure is measured in terms of pounds per square inch. Professional operators typically strive to ensure that 1 cfm per 300 gallons is being achieved.

This is typically not a problem when the system is being properly circulated. However, aeration devices, oxygen injectors, or oxygen diffusers are necessary if the fish tank does not have adequate circulation (dead spaces), and/or there are high stocking densities.

As figure 246 illustrates below, dissolved oxygen increases throughout the day and decreases throughout the night. The most critical time to aerate is just before dawn.

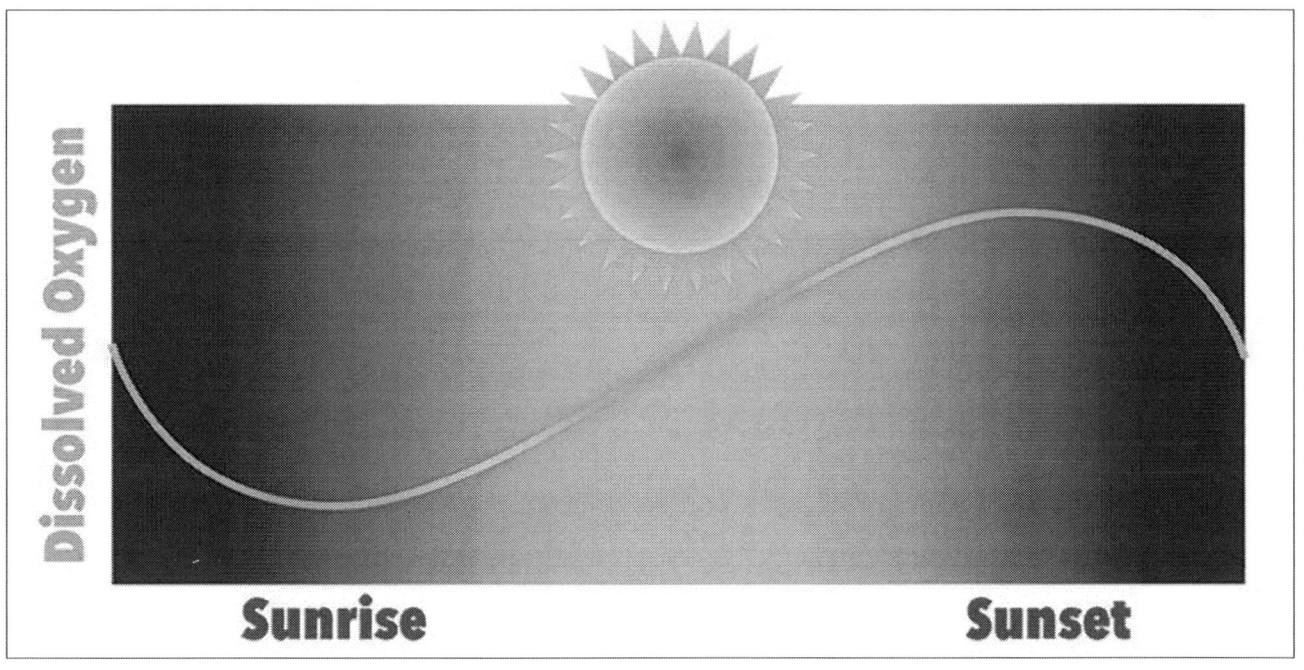

FIGURE 246. *The normal daily cycle of dissolved oxygen production.*

Nitrogen, Potassium, and Phosphorous (NPK)

NPK are the base nutrients your plants will need to thrive. If your system is low in one of these nutrients, the whole system will suffer. To monitor NPK the best thing you can do is watch your plants — they will tell you what they need. A plant chart, such as the figure 247 below, is most useful for diagnosing NPK problems.

Pump Size

The pump(s) should cycle the total volume of fish tank water once every two hours, but ideally once every hour. If the pump is on a timer, then the 'on' and 'off' phases need to be considered, so that the tank volume can still be pumped within an hour.

Clean Water

Don't be fooled by looks. Clean water is not always necessarily clear. Harmful substances such as nitrite, carbon dioxide, as well as other pollutants, do not discolor water. Low DO levels also do not necessarily alter water color. On the other hand, green water, while not clear can actually be healthy for fish, particularly Tilapia. Therefore, monitoring water quality on a regular basis is critical.

Chlorine

Chlorine kills fish, and is also harmful to humans. If water must be added directly to the tank it, is best to use non-chlorinated well water or run water through a filter that removes chlorine prior to adding it to tank. Otherwise set up a temporary holding tank, such as a 55-gallon drum, where the chlorinated water can be stored for a few hours in advance of adding it to the aquaponic system. This will provide an opportunity for the chlorine to dissipate from the water before becoming part of the system (this process is also referred to as 'off gas').

It is prudent to keep spare dechlorinated water readily on hand. It is recommended that you store at least 10 percent of your system water volume in a separate container. If using municipal water, be sure to let it sit for a minimum of 24 hours before adding it to your system so it can properly dechlorinate. Water will need to be added to the system on a regular basis to offset evaporation. Adding chlorinated water directly to your system will kill off much, if not all of, the beneficial bacteria.

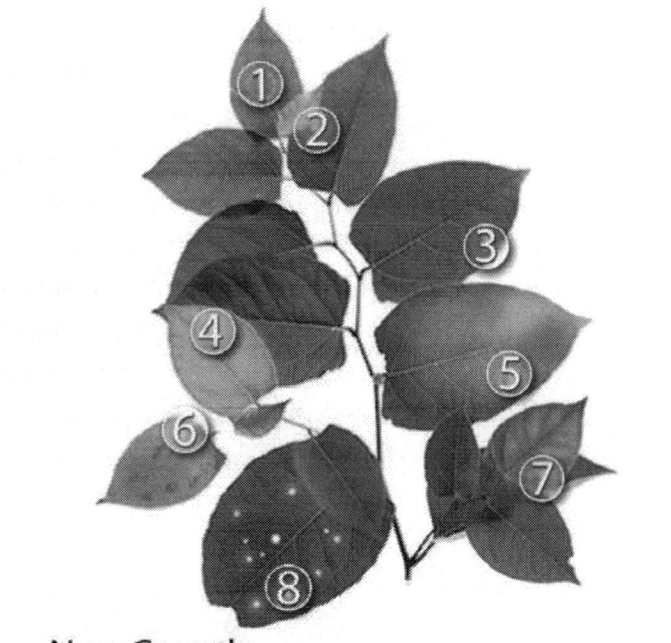

New Growth
1. Calcium/Traces Deficiency (or might be K, Mg overdose) twisted pale new growth
2. Severe Nitrogen Deficiency (white/yellow tiny leaves)

Old Growth
3. Normal leaf growth
4. Iron Deficiency (yellowing of entire plant)
5. Early Signs of Nitrogen Deficiency (old leaves are reabsorbed from tip to stem)
6. Phosphate Deficiency (older leaves yellow, and parts of the leaf are reabsorbed leading to dead patches, the leaf falls off rather quickly, looks similar to early nitrogen deficiency; green spot algae on older leafs)
7. Magnesium Deficiency (dark veins lighter leaf tissue)
8. Potassium Deficiency (pin-holes form in leaf that enlarge with a yellowing edge, leaf is otherwise normal looking)

FIGURE 247. Plant Chart

Light

Note that fish are sensitive to light, and some much more than others (i.e. largemouth bass). Avoid any direct lighting to the fish tank. Even a fish tank positioned under a typical room light can stress the fish. Therefore, it is prudent to ensure that the only lighting of the fish tank, or fish tank area, be that of the ambient lighting type. Ambient lighting is a general illumination that comes from all directions in a room that has no visible source. This type of lighting is in contrast to directional lighting. Even so, the ambient light should not be too bright.

Water Hardness

Aquarium water hardness is a part of the aquarium water chemistry that is often not fully understood. However, don't let the subject intimidate you, it is not that complicated. Fish tank water hardness is measured in degrees of hardness. Many home aquarium water test kits will give you measurements in either degrees of hardness (dh) or in parts per million (ppm).

When we discuss aquarium water hardness, we are simply looking at the amount of dissolved minerals in our aquarium water. There are two distinct measurements of aquarium water hardness. When you test your water, you will test for general hardness (GH) and carbonate hardness (KH).

General Hardness (GH)

The general hardness (GH) of your aquaponics water is a measurement of dissolved magnesium and calcium. The (GH) of aquaponics water can have great effects on your fish, so it is very important to ensure that the fish you choose to keep will thrive in your water or adjust the water accordingly.

There are species of fish that live in soft water and others that live in hard water. Like with many other aspects of raising fish, it is important to do your research in advance. You need to know your water quality, pretences of the fish you desire to raise, and what would need to be done to the water to accommodate the fish. This will minimize losses. You can also keep water treatment cost down by going with a species that is most closely compatible to your water supply.

Most fish will survive in the average water from our tap (chlorine, chloramines, and other contaminates removed) with no problems. Unless tests reveal unusually hard or extremely soft water, there should not be a problem. Most fish will adapt just fine. However, successful breeding and the overall coloration of the fish can be directly linked to the water hardness.

Carbonate Hardness (KH)

The second component of water hardness is the carbonate hardness (KH). (KH) is the measurement of carbonate and bicarbonate ions in your water. In simple terms, the (KH) of water is a measurement of the buffering capacity of your water. The (KH) of water will determine how much your pH will fluctuate. The higher the (KH) is in your water, the more stable your pH will remain. When (KH) drops, so too will your pH. There is a distinct relationship between (KH) and pH. If you have ongoing problems with the pH dropping, check the (KH) levels in the tank water.

TABLE 41. **General Hardness Chart**

DEGREE OF HARDNESS (DH)	ppm	HARDNESS
0–3	0–50	Very Soft
6–Mar	51–100	Soft
12–Jun	101–200	Slightly Hard
18–Dec	201–300	Moderately Hard
18–30	301–450	Hard
30+	450+	Very Hard

How to Soften Aquaponics Water

There are several ways to soften your aquaponics water if your your tap is too hard for your fish. However, if you do need to soften your aquaponics water, do so slowly. Any drastic changes can shock the fish, cause injury, or even result in death.

Reverse Osmosis

The most economical and popular method to soften your aquaponics water is to use a Reverse Osmosis (RO) system. These units remove heavy metals, minerals, and contaminants from your water source.

Water Softening Pillows

Water softening pillows can be used on very small systems. There are several different manufactures of water softening pillows. These pillows work well for aquaponics systems that are less than 50 gallons. You simply channel the water through the pillow material. It softens the water through ion-exchange. Many pillow softeners can be recharged and reused by soaking the pillow in a salt water solution. It is then placed in the tank where the sodium ions are released into the water and replaced by calcium and magnesium ions. After a few hours or days, the pillow (along with the calcium and magnesium) is removed, and the pillow recharged again.

How to Increase Aquaponics Water Hardness

If you find that your water source is too soft, you may need to harden your aquaponics water. There are several things you can do to harden your aquaponics water.

Crushed Coral

Adding crushed coral to your tank can help increase the water hardness. Some operators burdened with a hard water sources, choose to use crushed coral as their substrate media or integrate it into other media types creating a media mix that resolves the problem.

Limestone

Using limestone allows for calcium and other minerals to leach out into the water column. The most popular limestone used in cichlid tanks is Texas Holey Rock Limestone.

Buffer Additives

There are several buffers on the market that will help raise GH and KH while maintaining the pH of the water. Most use a combination of different salts including carbonate salts to increase water hardness. Be sure to closely follow the directions on the manufacture's label.

Aquaponics Water Hardness Tips

- Maintaining a regular schedule of water changes to ensure high water quality will typically prevent your pH, GH, and KH from swinging back and forth so much.
- If you decide to dose additives, be sure they are safe to enter your food chain, and follow the directions on the manufacturer's label.
- When you use limestone or crushed coral, ensure that you clean it thoroughly.
- When adjusting any water parameters such as aquaponics water hardness, adjust slowly. You can kill your fish with extreme changes.
- Purchase a high quality aquaponics water test kit.

Algae

Photosynthetic growth and activity by algae in aquaponic systems affect the water quality parameters of pH, DO, and nitrogen levels. Algae are a class of photosynthetic organisms that will readily grow in any body of water that is rich in nutrients and exposed to sunlight. Some algae are microscopic, single-celled organisms called phytoplankton, which can colour the water green. Macroalgae are much larger, commonly forming filamentous mats attached to the bottoms and sides of tanks.

It is important to prevent algae growing in an aquaponic system because they are problematic for several reasons. First, they will consume the nutrients in the water and compete with the target vegetables. In addition, algae act as both a source and sink of DO, producing oxygen during the day through photosynthesis and consuming oxygen at night during respiration. They can dramatically reduce the DO levels in water at night to the point of even being fatal to fish. This production and consumption of oxygen is related to the production and consumption of carbon dioxide, which causes daily shifts in pH as carbonic acid is either removed (daytime — higher water pH) from or returned (night time — lower water pH) to the system. Finally, filamentous algae can clog drains and block filters within the unit, leading to problems with water circulation. Brown filamentous algae can also grow on the roots of the hydroponic plants, especially in deep water culture, and negatively affects plant growth. It should be noted that some aquaculture operations benefit greatly from culturing algae for feed, referred to as green-water culture, including tilapia breeding, shrimp culture, and biodiesel production; these activities should never be conducted within an aquaponic system.

Algae Control

Algae are unsightly and, on drying, it can emit a foul odor. As mentioned above they will also rob the water of minute elements, including oxygen. Algae are not harmful to plants, except on rare occasions where there is stagnant water when it can harbor insects and diseases. Some people will even argue that algae gives off certain enzymes which are beneficial to your plants, and they openly encourage its growth. The main problem with algae is that it pulls oxygen out of the water and, as a result, can be harmful and even deadly to fish.

Again, algae require two things to flourish: light and oxygen. If one or the other is not present, algae will not grow. In aquaponics, algae growth can be limited or excluded by minimizing light. Do not let any direct light shine on the dish tank (only ambient lighting).

For open aquaponic systems that use a growing substrate (i.e. grow bed surface of the flood-and-drain system), black/white plastic film can be cut to size and placed over the grow bed so that light cannot reach the medium (have holes where the plants can grow through). Fish tanks should be shaded. Outdoor fish ponds can have a shade cover or D.W.C. raft(s) floating on the surface of the water to provide shade.

Nitrite

Nitrite enters a fish culture system after fish digest feed and the excess nitrogen is converted into ammonia, which is then excreted as waste into the water. Total ammonia nitrogen (TAN; NH3 and NH4+) is then converted to nitrite (NO2) which, under normal conditions, is quickly converted to non-toxic nitrate (NO3) by naturally occurring bacteria. Uneaten (wasted) feed and other organic material also break down into ammonia, nitrite, and nitrate in a similar manner.

Brown blood disease occurs in fish when water contains high nitrite concentrations. Nitrite enters the bloodstream through the gills and turns the blood to a chocolate-brown color. Hemoglobin, which transports oxygen in the blood, combines with nitrite to form methemoglobin, which is incapable of oxygen transport. Brown blood cannot carry sufficient amounts of oxygen, and affected fish can suffocate despite adequate oxygen concentration in the water. This accounts for the gasping behavior often observed in fish with brown blood disease, even when oxygen levels are relatively high.

Nitrite problems are typically more likely in closed, intensive culture systems due to insufficient, inefficient, or malfunctioning filtration systems. High nitrite concentrations in ponds occur more frequently in the fall and spring when temperatures are fluctuating, resulting in the breakdown of the nitrogen cycle due to decreased plankton and/or bacterial activity.

A reduction in plankton activity in ponds (due to lower temperatures, nutrient depletion, cloudy weather, herbicide treatments, etc.) can result in less ammonia assimilated by the algae, thus increasing the load on the nitrifying bacteria. If nitrite levels exceed that which resident bacteria can rapidly convert to nitrate, a buildup of nitrite occurs, and brown blood disease is a risk. Although nitrite is seldom a problem in systems with high water exchange rates or good filtration, systems should be monitored year-round and managed when necessary, to prevent severe economic loss from brown blood in any fish culture facility.

Nitrite — Susceptibility of Fish Species to Brown Blood Disease

Largemouth and smallmouth bass, as well as bluegill and green sunfish, are resistant to high nitrite concentrations. Catfish and tilapia, are fairly sensitive to nitrite, while trout and other cool water fish are sensitive to extremely small amounts of nitrite. Goldfish and fathead minnows fall in between catfish and bass in their susceptibility to brown blood disease resulting from high nitrite levels. Striped bass and its hybrids appear sensitive to nitrite, but little is known about the relative sensitivity compared to other species.

Fluoride

Water fluoridation is the practice of adding industrial-grade fluoride chemicals to water for the purpose of preventing tooth decay. The fluoride being added to public drinking water is actually an industrial waste product. One of the little known facts about this practice is that the United States, which fluoridates over 70 percent of its water supplies, has more people drinking fluoridated water than the rest of the world combined. Most developed nations, including all of Japan and 97 percent of western Europe, do not fluoridate their water.

Contrary to popular belief and despite US government propaganda, comprehensive data from many unbiased scientific studies (as well as the World Health Organization) have proven that there is no discernible difference in tooth decay between the minority of western nations that fluoridate water, and the majority of nations. In fact, the tooth decay rates in many non-fluoridated countries are lower than the tooth decay rates in fluoridated ones.

Fluoride is extremely toxic to fish, plants, and humans. Check out the warning label on the back of your toothpaste, then immediately throw it in the trash, and start using non-fluoridated toothpaste from now on.

Harmful Effects of Fluoride

Most fluoride that is added to municipal water is an unnatural form of fluoride that contains sodium. It is over 80 times more toxic than naturally-occurring calcium fluoride.

The fluoride ion (F-) is extremely reactive and strongly attracted to calcium. Its preference for calcium overrides its attraction to other ions. In nature, fluoride is most often bound to calcium. When sodium fluoride is ingested, it rapidly robs the body of calcium. In fact, sodium fluoride poisoning results when calcium is stolen from the blood.

Fluoride has the ability to affect other chemicals and heavy metals, in some cases making them even more harmful than they would be on their own. For example, when you combine chloramines with the hydrofluorosilicic acid added to the water supply, they become very effective at extracting lead from old plumbing systems, promoting the accumulation of lead in the water supply.

Studies have shown that hydrofluorosilicic acid increases lead accumulation in bone, teeth, and other calcium-rich tissues. This is because the free fluoride ion acts as a transport of heavy metals, allowing them to enter into areas of your body they normally would not be able to go, such as into your brain.

Fluoride has long been known to be a very toxic substance. This is why, like arsenic, fluoride has been used in pesticides and rodenticides (to kill rats, insects, etc). It is also why the Food and Drug Administration (FDA) now requires that all fluoride toothpaste sold in the U.S. carry a poison warning that instructs users to contact the poison control center if they swallow more than used for brushing.

Excessive fluoride exposure is well known to cause a painful bone disease (skeletal fluorosis), as well as a discoloration of the teeth known as dental fluorosis. Excessive fluoride exposure has also been linked to a range of other chronic ailments including arthritis, bone fragility, dental fluorosis, glucose intolerance, gastrointestinal distress, thyroid disease, possibly cardiovascular disease, and certain types of cancer.

Certain subsets of the population are particularly vulnerable to fluoride's toxicity. Populations that have heightened susceptibility to fluoride include infants, individuals with kidney disease, individuals with nutrient deficiencies (particularly calcium and iodine), and individuals with medical conditions that cause excessive thirst.

For a complete breakdown of all the harmful effects of fluoride, please refer to the Fluoride Action Network (FAN). FAN's work has been cited by national and international media outlets including the *New York Times*, *Wall Street Journal*, *TIME Magazine*, *National Public Radio*, *Scientific American*, and *Prevention Magazine* among others. *FAN is an official project of the American Environmental Health Studies Project (AEHSP) — a registered non-profit (501c3) organization. FAN's website address is: http://fluoridealert.org/faq*

Fluoride negatively impacts health and it is prudent to avoid it. The following is a brief list of some of the detrimental health consequences of ingesting fluoride:

- Gastrointestinal Effects
- Bone Fractures
- Brain Effects
- Cancer
- Cardiovascular Disease
- Diabetes
- Endocrine Disruption
- Acute Toxicity
- Hypersensitivity
- Kidney Disease
- Male infertility
- Pineal Gland
- Skeletal Fluorosis
- Thyroid Disease

Fluoride Side Note

We all need to beware of other ways we are constantly exposed to fluoride. Not only is one source bad enough (i.e. drinking water, toothpaste, etc.), but the accumulation from multiple sources is especially harmful.

One of the primary sources of fluoride exposure is non-organic foods, due to the high amounts of fluoride-based pesticide residues on these foods. Non-organic foods may account for as much as one-third of the average person's fluoride exposure. Foods particularly high in fluoride include non-organic fresh produce, breakfast cereals, juices (particularly grape juice), deboned meats such as lunch meats, other meats through food chain accumulation, and black or green tea (even if organic).

How to Remove Fluoride from Water

Unfortunately, water filtration does not remove fluoride. Many water filters (e.g., Brita & Pur) use an "activated carbon" filter that does not remove fluoride. The fluoride molecule is smaller than the water molecule, therefore it cannot be removed by filtration.

If you live in a community that fluoridates its water supply, there are several options to avoid drinking the fluoride that is added. Unfortunately, each of these options will cost money (unless you happen to have access to a free source of spring water). These options include:

- **Spring water:** Most spring water contains very low levels of fluoride (generally less than 0.1 ppm).
- **Reverse Osmosis:** Reverse Osmosis can removes between 90 and 95 percent of fluoride. Contaminants are trapped by the RO membrane and flushed away in the waste water. The process requires between two and four gallons of water to produce one gallon of RO water (depending on the quality of the water and the efficiency of the RO unit).
- **Water Distillation**: Distillation is capable of removing just about anything (except volatile compounds) from water. Distilling water is an effective way of removing fluoride from water. The drawback to distillation is that the process is time and energy consumptive. Distillation also leaves the resulting water empty and lifeless. If you use distilled water, you need to do research on how to add minerals back to the water following the distillation process, often referred to as the 'full-spectrum living water'.

Sources of Water for Aquaponics

On average, an aquaponic system uses 1–3 percent of its total water volume per day, depending on the type of plants being grown and the location. Water is used by the plants through natural evapotranspiration as well as being retained within the plant tissues. Additional water is lost from direct evaporation and splashing. As such, this water loss will need to be replenished periodically. The water source used will have an impact on the water chemistry of the unit. This section will review some common water sources and the common chemical composition of that water. New water sources should always be tested for pH, hardness, salinity, chlorine, and for any pollutants in order to ensure the water is safe to use.

Salinity

Now is a good place to discuss a water parameter not yet addressed in detail—salinity. Salinity indicates the concentration of salts in water, which include table salt (sodium chloride—NaCl), as well as plant nutrients, which are in fact salts. Salinity levels will have a large bearing when deciding which water to

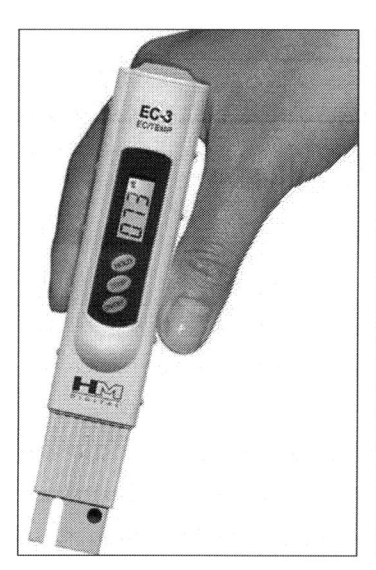

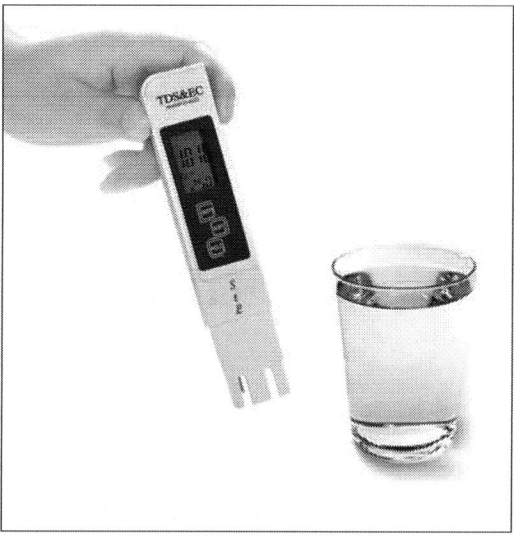

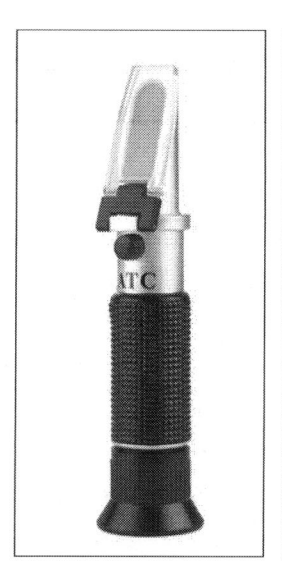

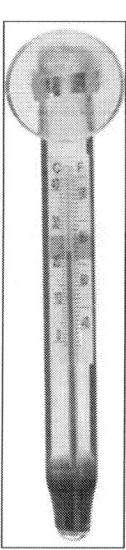

FIGURE 248. (Left to Right): EC meter, TDS meter, Refractometer, Hydrometer

NOTE: All above meters, or an equal, can be acquired for less than $25 each. Several of these meters measure multiple water constituents.

use because high salinity can negatively affect vegetable production, especially if it is of sodium chloride origin, as sodium is toxic for plants. Water salinity can be measured with an electrical conductivity (EC) meter, a total dissolved solids (TDS) meter, a refractometer, or a hydrometer or operators can refer to local government reports on water quality.

Salinity is measured either as conductivity, or how much electricity will pass through the water, in units of microsiemens per centimetre (µS/cm), or in TDS as parts per thousand (ppt) or parts per million (ppm or mg/liter). For reference, seawater has a conductivity of 50,000 µS/cm and TDS of 35 ppt (35 000 ppm). Although the impact of salinity on plant growth varies greatly between plants, it is recommended that low salinity water sources be used. Salinity, generally, is too high if sourcing water has a conductivity more than 1 500 µS/cm or a TDS concentration of more than 800 ppm. Although EC and TDS meters are commonly used for hydroponics to measure the total amount of nutrient salts in the water, these meters do not provide a precise reading of the nitrate levels, which can be better monitored with nitrogen test kits.

Rainwater

Collected rainwater is an excellent source of water for aquaponics. The water will usually have a neutral pH and very low concentrations of both types of hardness (KH and GH) and almost zero salinity, which is optimal to replenish the system and avoid long-term salinity buildups. However, in some areas affected by acid rain, as recorded in a number of localities in eastern Europe, eastern United States of America, and areas of southeast Asia, rainwater will have an acidic pH. If this is the case in your location, it is good practice to buffer rainwater and increase the KH.

If you desire to catch rain water for aquaponics use, make sure that the gutters and roof do not contain any chemicals that may leach into your water. Also, depending upon your location, you made need to beware of the potential for harmful concentrations of air pollution residue on the roof, which would also negatively impact your system. Lastly, in some jurisdictions it is illegal to catch rainwater. These government jurisdictions have decided that homeowners don't own the rain that falls on their property.

A well-designed system includes an overflow pipe to protect against damage caused by the tank overflowing during periods of heavy rains or low usage. All overflow water should be discharged away from foundations and other structures.

- One inch of rain falling on a square foot of surface yields approximately 0.6 gallons of water.

Using the following equation, it is easy to determine how much rainwater can be collected.

- Rain caught (gallons) = (inches of rain) x 0.6 x (portion of building footprint)

For example, if your home's footprint is 1,400 square feet and you want to know the amount of water that comes from a ¼ inch (.25") rain event:

- Rain caught (gallons) = (.25) x (.6) x (1,400) = 210 gallons (or less if you're only gathering from one part of the roof).

Rainwater collection can be easily achieved by connecting a large clean container to water drainage pipes surrounding a building or house (see figure 249 on the following page).

Collecting rainwater is relatively easy, but storing rainwater can be a bit more challenging. The water has to be retained until needed, and the water has to be kept clean. The storage container(s) should be covered with a screen to prevent mosquitoes and plant debris from entering. Depending on the intended uses

PART VI: OPERATION AND MAINTENANCE

FIGURE 249. *Rainwater collection examples*

of your collected rainwater, some form of filtration and/or disinfection of the water that comes from the storage tank may be necessary. Do your research and take the necessary protection measures.

Cistern or Aquifer Water

The quality of water taken from wells or cisterns will largely depend on the material of the cistern and bedrock of the aquifer. If the bedrock is limestone, then the water will probably have quite high concentrations of hardness, which may have an impact on the pH of the water. Water hardness is not a major problem in aquaponics, because the alkalinity is naturally consumed by the nitric acid produced by the nitrifying bacteria. However, if the hardness levels are very high and the nitrification is minimal because of small fish biomass, then the water may remain slightly basic (pH seven to eight) and resist the natural tendency of aquaponic systems to become acidic through the nitrification cycle and fish respiration. In this case, it may be necessary to use very small amounts of acid to reduce the alkalinity before adding the water to the system in order to prevent pH swings within the system. Aquifers on coral islands often have saltwater intrusion into the freshwater lens, and can have salinity levels too high for aquaponics, so monitoring is necessary, and rainwater collection or reverse osmosis filtration may be better options.

Tap or Municipal Water

Water from the municipal supplies is often treated with different chemicals to remove pathogens. The most common chemicals used for water treatment are chlorine and chloramines. These chemicals are toxic to fish, plants and bacteria; they are often used to kill bacteria in water and as, such are detrimental to the health of the overall aquaponic ecosystem. Chlorine test kits are available, and if high levels of chlorine are detected, the water needs to be treated before being used.

The simplest method is to store the water before use, thereby allowing all the chlorine to dissipate into the atmosphere. This can take upwards of 48 hours, but can occur faster if the water is heavily aerated with air stones. Chloramines are more stable and do not off-gas as readily. If the municipality uses chloramines, it may be necessary to use chemical treatment techniques such as charcoal filtration or other dechlorinating chemicals. Even so, off-gassing is usually enough in small-scale units using municipal water. A good guideline is to never replace more than 10 percent of the water without testing and removing the chlorine first. Moreover, the quality of the water will depend on the bedrock were the initial water is sourced. Always check new sources of water for hardness levels and pH, and use acid if appropriate and necessary to maintain the pH within the optimum levels indicated above.

Filtered Water

Depending on the type of filtration (i.e. reverse osmosis or carbon filtering), filtered water will have most of the metals and ions removed, making the water very safe to use and relatively easy to manipulate. However, like rainwater, deionized water from reverse osmosis will have low hardness levels and may need to be buffered.

Water Testing

In order to maintain good water quality in aquaponic units, it is prudent to perform water tests once per week to make sure all the parameters are within the optimum levels. However, mature and seasoned aquaponic systems will have consistent water chemistry and do not need to be tested as often. In these cases, water testing is only needed if a problem is suspected. In addition, daily health monitoring of the fish and the plants growing in the unit will indicate if something is wrong, although this method should not be a substitution for testing the water.

Access to simple water tests are strongly recommended for every aquaponic unit. Color-coded freshwater test kits are readily available and easy to use (see photo above). These kits include tests for pH, ammonia, nitrite, nitrate, GH, and KH. Each test involves adding five to ten drops of a reagent into 5 milliliters of aquaponic water; each test takes no more than five minutes to complete. Other methods include some of the meters and water testing tools referenced in the sections above or water test strips, which are inexpensive and moderately accurate (see figure 251).

The most important tests to perform weekly are pH, nitrate, carbonate hardness and water temperature, because these results will indicate whether the system is in balance. The results should be recorded each week in a dedicated logbook so trends and

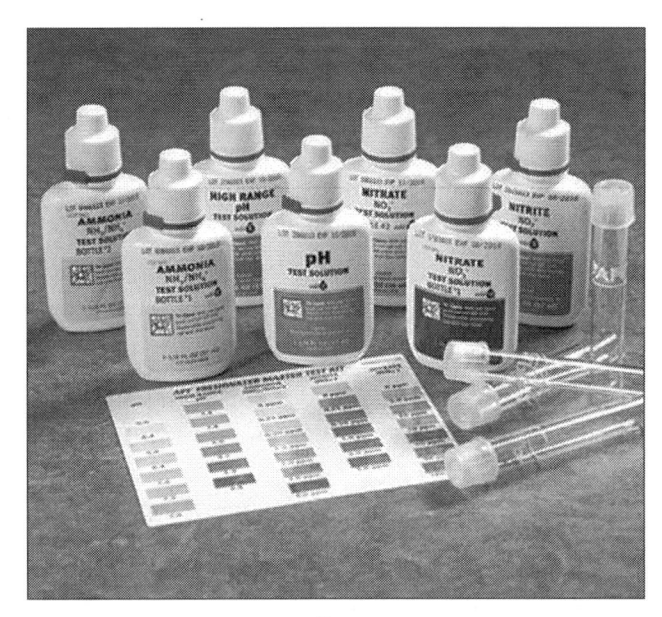

FIGURE 250. *Rainwater Collection*

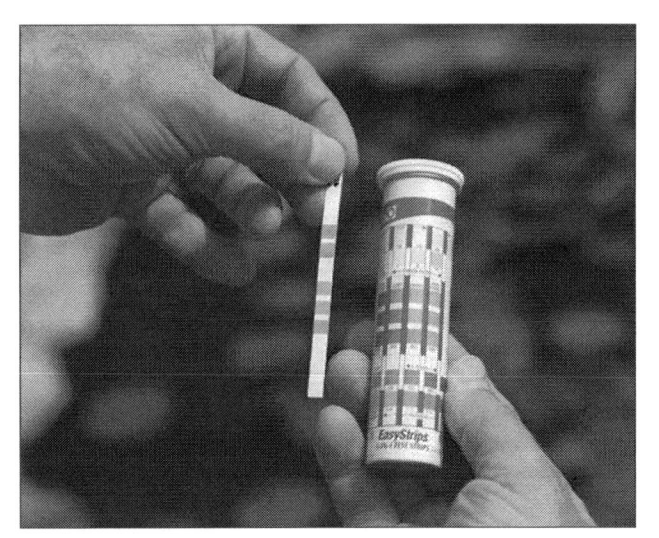

FIGURE 251. *Water test strip kit.*

changes can be monitored throughout growing seasons. Testing for ammonia and nitrite is also extremely helpful in order to diagnose problems in the system, especially in new systems or when significant changes occur, such as with a major fish harvest, or if there is increase in fish mortality raises toxicity concerns in an ongoing system. Although weekly monitoring may not be necessary in established systems, the testing can provide very strong indicators of how well the bacteria are converting the fish waste and provide feedback as the overall health of the system. Testing for ammonia and nitrate should be the first priority if any problems are observed with the fish or plants.

Summary

- Water is the life-blood of an aquaponic system. It is the medium through which plants receive their nutrients and the fish receive their oxygen. It is very important to understand water quality and basic water chemistry in order to properly manage aquaponics.
- There are five key water quality parameters for aquaponics: dissolved oxygen (DO), pH, water temperature, total nitrogen concentrations and hardness (KH). Knowing the effects of each parameter on fish, plants and bacteria is essential.
- Compromises are made for some water quality parameters to meet the needs of each organism in aquaponics.
- The target ranges for each parameter are shown in table 42.
- There are simple ways to adjust pH. Bases, and less often acids, can be added in small amounts to the water in order to increase or lower the pH, respectively. Acids and bases should always be added slowly, deliberately, and carefully. Rainwater can be alternatively used to let the system naturally lower the pH through nitrifying bacteria consuming the system's alkalinity. Calcium carbonate from limestone, seashells, or egg shells increases KH and buffers pH against the natural acidification.
- Some aspects of the water quality and water chemistry knowledge needed for aquaponics can be complicated, in particular the relationship between pH and hardness, but basic water tests are used to simplify water quality management.
- Water testing is essential to maintaining good water quality in the system. Test and record the following water quality parameters each week: pH, water temperature, nitrate and carbonate hardness. Ammonia and nitrite tests should be used especially at system start-up and if abnormal fish mortality raises toxicity concerns.
- It is important that the beginner not be intimidated or overwhelmed by the water testing and water quality management process. Everything discussed in this chapter can be easily learned. Furthermore, experience is an excellent teacher.

TABLE 42. **Compromise of Ideal Water Quality Parameters for Fish, Plants, and Bacteria**

	TEMPERATURE	pH	AMMONIA	NITRITE	NITRATE	DO	KH
Tolerance Range	18–30 °C	6–7	< 1 mg/L	< 1 mg/L	< 5-150 mg/L	> 5 mg/L	60–140 mg/L
	64–86 °F		< 1 ppm	< 1 ppm	< 5-150 ppm	> 5 ppm	60–140 ppm

CHAPTER 32

Fish Breeding, Fish Reproduction, and Raising Your Own Crop of Fish

Breeding and Raising Fish Overview

Breeding fish successfully requires knowledge, effort, and attention to details, but it also provides many rewards. Besides the personal gratification one acquires through the process, a great deal of money can be saved, and gained, as a result. The increasing demand for fish and fish protein has resulted in widespread overfishing, increased prices, and diminishing supply. Throw in the fact that the general public is becoming more knowledgeable about our fish food supply being heavily contaminated with heavy metals (including mercury) and radioactive isotopes, it makes raising your own fish a very lucrative endeavor. Also, replenishing your aquaponic fish tank with your own bred fish, rather incurring the cost of another supply of fingerings (as well as taxes, shipping, and handling fees) can greatly reduce your operational expenses.

It is difficult to raise fish from young in an aquaponic fish tank. Using separate tanks for breeding and raising the fry is the best approach. The more conditions that are right within the tank, the greater your fish breeding success rate, although some species, such as Tilapia, are quite forgiving and make fish breeding a relatively easy process.

Many fish will spawn if you place a male and female in their own tank and give them a green "spawning mop" or create a dark "pot cave". To make a pot cave, simply place a terracotta pot on its side, and then fill it approximately ¼ full with sand. This setup will work for many different species. Most will use the sand, some will use the hard, under-surface, while others will use the crevice created on the outside of the pot. Some fish require a certain speed of water current. Temperature, pH, and overall water quality are also important. For some species, the male and/or mother will need to be removed at some point in the process. Generally, fry can be fed small pieces of flakes, brine shrimp, small worms, and soaked oatmeal. Do your research and know your fish species to understand their natural habitats for determining which methods and food options best suit them.

There are already many resources readily available which provide detailed, systematic instructions on breeding and raising all aquaponic friendly fish. This book will not attempt to address all the specifics of breeding for each species. Rather, this chapter will describe what it takes to successfully breed Tilapia, the most common aquaponic species. Although most of what is being described in this chapter will also apply to other species, it would be prudent for you to investigate the reproduction needs of your fish species if you are not raising Tilapia.

Breeding Tilapia

Tilapia are classified as either mouth brooders or substrate spawners. All Tilapia are prolific breeders. With the proper environmental conditions, Tilapia can easily reproduce and provide an ample fish supply for consumption and commercial success.

Mouth Brooders

Members of the Oreochromis genus are maternal mouth brooders and are a common choice for aquaponics or aquaculture. In terms of popularity, the Nile Tilapia (*O. niloticus*) is the most widely cultured tilapia, followed by Blue Tilapia (*O. aureus*), and Mozambique Tilapia (*O. mossambicus*).

The *Oreochromis* display an elaborate courtship behavior. After building a nest, the male aggressively repels other males that enter into proximity of the nest. When ready to spawn, the male displays a darkened color and leads a female to the nesting area. The fish then swim around the nest and the male will butt against the female genital area to induce egg laying. The courtship is often brief, lasting only a few minutes in many cases and seldom more than a few hours.

The female Tilapia lay their eggs in pits (nests) and after fertilization by males, the female collects the eggs in her mouth (buccal cavity) to maintain them until hatching.

Other Tilapia display different mouth brooding behavior. *Sarotherodon galilaeus* are bi-parental, with both parents brooding the eggs and defending the newly hatched fish.

The male *Sarotherodon melanotheron* is the parent that performs the mouth brooding, while the female leaves the nest.

Substrate Spawners

Tilapia rendalli and *Tilapia zillii* are two popular, commercially-raised species that are substrate spawners. The male and females will build a nest and defend it together. A male and female will typically form a bonded-mating-pair and courtship can last up to a week, but usually takes place over several days.

Females will first lay their eggs in pits (nesting area) dug in the bottom of a lake or pond. You can simulate this condition in a tank by adding some

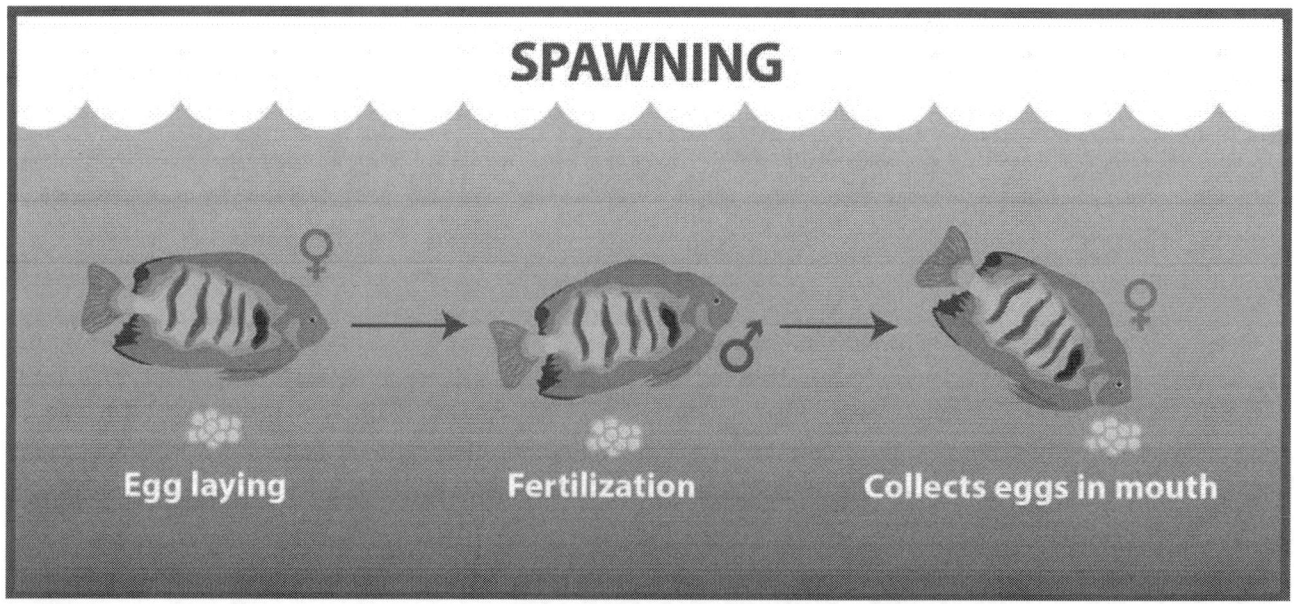

FIGURE 252. Tilapia spawning process.

substrate (e.g. gravel) which allows the Tilapia to evacuate a nesting area. The male will then spawn and fertilize the eggs. After fertilization, the parents guard the eggs, chasing away predators and making sure proper aeration is maintained for hatching.

Frequent Breeding and Mouth Brooding

With the proper set up, at temperatures of 85°F (29°C), they can produce baby Tilapia (fry) almost every week, year round. The mouth brooding and maternal protection of the fry helps to create a high survival rate. This combination of continuous production and high survival rate allows the Tilapia aquaponics operator to have a constant supply of fingerlings to replace those that get big enough to eat.

Tilapia Fish Farming Stages

The process for farming Tilapia includes the following stages:
Breeding → Fry sizing → Fingerling production → Grow-out to plate/market size → Purging → Harvesting → Processing → Packaging → Marketing → Cooking → Eating

Tilapia Breeding Fundamentals

Tilapia will reproduce profusely if adults are well fed and the young can find refuge in the tank. If hungry, the adults will cannibalize their young to some degree, but rarely will they control their own population. They prefer many other food items instead of their own young. However, non-spawning adults do have a seemingly insatiable appetite for eggs. Juveniles from previous spawns will actually be the most cannibalistic fish in the tank. They will eat any sibling they can fit in their mouth.

One female will typically produce about 200-1,000 eggs per spawn, and she will spawn every four to five weeks or so if conditions are right. Even with low survival, that is still a lot of Tilapia recruitment. In an average small system, a single female could have a tank filled to the brim with young Tilapia in a relatively short period.

In large-scale Tilapia farming operations, only the male population is raised, to avoid wild-spawning and eliminate the small size of females. These commercial growers use hormones to convert female fry to male fry. This practice is rarely done in most aquaponic operations, especially those that are on the small to mid-size scale.

Tilapia Breeding Made Easy

Breeding tilapias is relatively easy once you have a pair. The natural way to find a breeding pair is to raise several young tilapias together and observe them as they pair up. Feed them well, maintain water quality, provide the right conditions, and they will spawn readily.

Most tilapias are open substrate spawners who dig large pits in which the male and female will clean and prepare before laying their eggs. The female will select her spot to lay her eggs and the male will follow behind to fertilize them. The female will use her ovipositor, which is a short, wide tube that dispenses her eggs. The male uses his own ovipositor, which is longer and thinner than the female's. He uses this ovipositor to fertilize the eggs with his sperm.

Tilapia parents have the tendency to be extremely aggressive toward all other fish when spawning. If spawning several pairs in the same tank, be sure to add plenty of hiding places.

Large quantities of fry are produced in each brood. Much like many other cichlids, Tilapias use their great parenting skills to protect and provide for their young. Tilapia fry grow quickly and can begin eating as soon as their egg sacs are absorbed. In just a few months, these fry will be able to produce their own young.

Tilapia Breeding Methods

A male and female Tilapia pair can be separated out to have their own tank, or several pairs can be bred in the same tank, so long as there is an abundance of space for each pair to have their own territorial area.

Place a flowerpot (laying on its side) into the tank with the open end of the flowerpot facing the tank wall, about 8-inches away from the tank wall. The male will make his territory between the open end of the pot and the tank wall. This allows females who are on the other side of the flowerpot to be out of site when he is in his territory. The male Tilapia claims a territory usually 2.5 times his body length in all directions that he can see from the center of his arena.

Place some pea size gravel in and around the flowerpot before adding the fish to the tank. Be sure to rinse the gravel before adding it to the tank. Make sure the tank also has the preferred water quality before adding the fish. The temperature of the water needs to be within the desired range and similar to the tank temperature where the fish are being removed from.

Tips from Commercial Tilapia Breeding Professionals

If breeding several pairs of Tilapia within the same tank, it is wise to take a page out of the professional breeders playbook. Professional Tilapia breeders cut the upper lip off the male with a sharp pair of scissors before adding them to the breeding tank. Pulling the upper lip out will reveal the articulation between the edge of the lip and the front area of the Tilapia.

The reason for trimming his lip is that the male is very aggressive with any fish that is within the area he claims as his territory. Breeding multiple pairs of Tilapia in a tank often leaves the females with nowhere to get away from the breeding efforts of the males. The constant harassment from the males can be fatal to other fish.

The flowerpot and anything else that you can add to help define his area, such as a small piece of plywood standing vertically near the base of the pot, is helpful, but it typically does not completely prevent the male Tilapia from continuously chasing the females, even if they are not ready to breed. The male bites them and after so much of this, he ends up scraping a lot of skin and scales off of them. Removing the male tilapia's upper lip prevents him from causing injury to other fish, but does not interfere with his ability to breed with the females that are ready.

To perform this procedure, take the scissors while holding the male firmly (with a towel or wash cloth), then cut a line across the hinge of the upper lip through the thin membrane and the center cartilage over to the opposite hinge, making a clean cut. This cut heals quickly and the male is capable of breeding within a few minutes of being placed with the females. Once the males are trimmed, they can be placed in the tank with the females.

Breeding Tank Conditions

Automatic controls need to be set up for the breeding tank that maintains the optimal conditions for successful breeding.

1. **Lighting:** If you are in a closed room rather than an open greenhouse, the lights should come on at six a.m. and go off at midnight. If you place the tank in a greenhouse or near a brightly lit window, then the lights should come on at six p.m. and go off at midnight.
2. **Heating:** Plug in an aquarium heater with a thermostat. Set the thermostat 85 to 88 degrees F (29 to 31 degrees Celsius).
3. **Thermometer:** A thermometer needs to be installed so that the temperature of the tank water can be checked at least once a day.
4. **Oxygenation:** Plug in an air pump and attach a line to the filter and air stones, which are located in the tank as far away as possible from where the flowerpot(s) is placed. Once a day, the air pump should be checked to ensure the bubbles are being delivered out of the top of each filter

and that the air stone is properly connected to it. If the oxygen level is low, the Tilapia will usually warn you in advance by rising to the surface and skittering. The filter should be taken out of the tank about every three days and washed out under running water and then put back into the tank and hooked up to the air.

Feed for Breeders

Professional Tilapia breeders claim that they have better success using a type of feed for breeding that is different from that used to grow or maintain Tilapia. The feed used by commercial Tilapia breeders has a higher concentration of protein and vitamins, which better meets the needs of the breeders for producing healthy eggs and sperm. The feed also needs to have a higher utilization rate to reduce the amount of waste so that cleaning up after the breeders is not a constant chore.

Commercial Tilapia breeders usually use a good quality salmon or trout breeder chow booster. Other commercial breeders use Tuna Tender Vittles cat food from Purina.

Feeding Process for Breeders

Feeding should be done two or three times a day. Start off using a teaspoon of feed for every ten Tilapia each feed session. If all the feed is being consumed, then add a little more each feed session until you observe that there is left-over feed. This will be the threshold and the point in which you will know the correct amount of feed to use each feed session.

Breeder Transfer

As the male dances around a female and she joins in the dance, it is a good indication that they will breed soon. Once they have bred, the female will have a mouthful of eggs and will not show much interest in eating. When a female is observed in this state, a note should be made and then in three or four days, she should be gently removed from the breed tank and placed in a separate tank until her eggs have developed into free-swimming fry. The best way to do this is to take two 6 by 6 inch nets and very slowly herd her into one. Gently hold the net against the side of the tank wall and bring it up over the side, transfering her immediately into a bucket with water that was taken directly from the breeder tank. Quickly and gently, transport the female into the tank exclusively for her. This is referred to as the nursery tank. The nursery tank needs to have the water at the same temperature as the breeding tank.

Leaving a female in the tank after she has bred will result in fewer eggs reaching the hatching stage. Other female adults will eat many free-swimming fry that do develop from the remaining eggs. Therefore, to obtain maximum breeding results, having a separate nursery tank very important.

If the female spills her eggs during the transfer, don't fret. Although this is not ideal, all is not lost. Catch the female, preferably with two nets, and transfer her and her eggs to the nursing tank. If she has eggs in her mouth and spits them out, simply pick them up with the net and transfer those eggs as well. Once the female and her eggs are moved, she will usually pick them up relatively soon and continue the incubation process.

Sometimes, especially with young females, the male does not properly fertilize the first brood, and the female will not swallow them because she instinctively knows that they are not developing properly. Each female incubates her eggs by slowly and continuously rolling them gently in her mouth. She can tell whether each egg is sick or dead, and then separate those eggs from healthy live eggs while swallowing any that are not right. The eggs begin development almost immediately after the female picks them up in her mouth and within 48 hours, the beginnings of eyes and tails can be seen on the eggs. By the fourth day the fry begin to resemble small fish attached to little yellow balls, which are the egg sacks.

Post-Breeding Behavior

Tilapia males in general are aggressive and territorial. To a certain extent, the female Tilapia also becomes territorial for a period after breeding. Besides the obvious full mouth, the female develops a dark marking on her forehead and the vertical stripes on her body become darker when brooding eggs. She also becomes more sensitive to many things and can be upset rather easily.

Nursery Tank Management

When a female is stressed in the nursery tank, she may get overly excited and try to escape. This may cause her to spit all or part of her eggs out. The best thing, of course, is to move very slowly and gently when near the tank. The more you can do to minimize unnecessary disturbances, the better. For instance, don't play music, slam doors, clang pipes and avoid turning lights off and on unnecessarily. This is being a responsible, animal friendly operator.

Tilapia Motherly Behavior

The mother Tilapia behavior changes as the young eggs develop into fry (baby fish). During the first two or three days, she simply swims around in a group with the rest of the females and young males. By the fourth or fifth day, she begins to look for a place that she can claim as her home area. During this time, she turns a darker shade on the front of her head and sometimes develops darker bands running vertically (kind of like a zebra). Scientist believe that this darker forehead is a warning signal to other Tilapia and fish to stay away, because if they make the mistake of coming too close to her, she will dart at them and even bite them if she can.

By the fifth to ninth day, the fry can navigate well enough on their own to be released by the mother for brief excursions out of her territorial area. At first she lets them out for a brief swim and sucks them back in within a few minutes. By the sixth day, she allows them to feed on zooplankton, bacteria, algae, and fungal growths on the surface of plants, rocks, or tank walls. While the fry are free swimming, the mother Tilapia keeps a sharp eye out for any intruders such as other fish and will aggressively chase them away if they approach the area where the fry are feeding.

If the mother perceives any danger to the fry that she cannot chase away, she will signal the fry by with a sideways wriggle of her body and an open mouth. The fry will immediately swim towards and into her mouth.

The older the fry get, the more time the mother allows them to spend outside. By the tenth day, she will no longer tend them or allow them oral sanctuary. It is also true that the older the fry are, when the mother is disturbed, the more likely she is to spit them out to fend for themselves if she feels her life is in danger.

Tilapia Mother Post Nursing Phases

Once the mother Tilapia begins to allow the fry out to browse on microscopic plankton, the mother can be caught and put back into the breeder tank to breed again. However, for ethical reasons, I believe it is more appropriate to wait unto the mother and fry are more comfortable about being on their own, generally two to three weeks after birth. It is best to catch her with a net with at least 1/4-inch holes in it so as to allow any fry to escape. When picking up the mother Tilapia, hold her gently and firmly with a wet towel or a cloth glove. It is also advisable to place your fingertip on her lower lip and pull downward to open her mouth to ensure she is not harboring any fry in her mouth.

If she is holding fry, continue holding her mouth open and place her back into the tank or into a large shallow pan with water from the tank and swish her forward and backward until all of the fry are washed out. You may then place her back into the breeding aquarium and return to the fry.

Tilapia Fry to Fingerlings

Immediately after the mother is removed from the nursing tank, fry frantically search for a hole or nook that will take the place of their mother's mouth. This behavior typically last a few hours. To be kind to your animals, and reduce their stress, place something in the nursing tank (i.e. rinsed concrete blocks with holes, thick leafless branches, etc.) a couple of days in advance of taking the mother out so that the fry can take refuge and feel secure.

Feeding can begin on their first day, but keep in mind that they will not consume large quantities of food, or some may not eat at all. Some Tilapia fry may not eat the first day or two because the yolk sac egg is still showing in their own stomachs. Any feed that is not eaten by the second feeding should be taken out of the tank and disposed of. After day four, for best results, feed the fry three times a day. Continue to keep the tank as clean as possible. An aquarium cleaning suction device works great for cleaning the tank.

By the second or third day, the fry will swarm around the feed and gobble it up with astonishing tenacity. Feed them as much as they will eat within a fifteen-minute period. Once you know about how much they eat on an individual feeding, try to give them that much for the next two feedings that day. Each day they will eat a little more as long as the oxygen levels, temperature, pH, and water quality parameters remain within the Tilapia's desirable range.

Tilapia Fry Food

The food for the fry needs to be either live food like zooplankton, brine shrimp or high protein powder of flake food such as is used for trout or salmon fry. Obviously, the closer one can stay to a healthy and organic food source, the better.

Many operators claim that the best food they have found is a diet consisting of a mixture of dried spirulina and artificial zooplankton. This type of feed can be purchased online or from most local tropical fish stores. If your local store does not carry it, you can ask them to order it for you. This approach will often enable you to acquire the feed you are after without having to incur the shipping cost.

There are also a number of other ways to feed Tilapia fry. One method is to take a quart of the same food that you use for the larger tilapia and soften it in water. Next, add two eggs and blend it until it becomes soupy. Mix it with two cups of boiled water and stir-in one ounce of knox gelatin. Place the mixture into a bucket, pan, or bowl and refrigerate it until needed. When feeding it to the fry, drop a half-teaspoon into the tank for each feeding.

Moving the Fry out of the Nursing Tank

When the fry reach about 1-inch (2.54 centimeters) in length or more they can be moved into a larger growing area, such as a pond, a larger tank, or into the aquaponics fish tank.

Tilapia Production Goal

How many pounds of fish do you want to have each week? The amount of tank space you can provide for growing fish can then be designed to produce the amount of fish that you wish to harvest or eat each week. The production per week goal is usually figured by the cubic foot of growing space—defined by the size of the tank—and the type of system being used. These factors determine how many fish can be stocked.

For instance, in a small tank or a tank with negotiable water exchange, the weight of fish that can be maintained is very small (around one to two ounces of fish per cubic foot). This type of set up is precarious because the amount of delicate balance of life supporting parameters. A minor disruption of dissolved oxygen levels, a little left over feed, and/or waste generated can easily upset the water quality conditions. Therefore, fish density must be kept to a minimum to ensure life support conditions are not compromised.

If an aeration device is added, or the water can be aerated via an aquaponics media bed, a much higher density of fish can be achieved. This will enable production to be increased substantially, with up to a half pound of fish per cubic of foot water being possible.

For example, considering a tank that is 10 feet by 10 feet with 3 feet of depth is 300 cubic feet. A tank with limited aeration can only hold one ounce per cubic foot of water and would only achieve a production amount of 300 total ounces (18.75 lbs) of fish in the tank at any one time. Having this same tank integrated into an aquaponics system, producing a half-pound of fish per cubic of foot water, would result in a total tank production of 150 pounds at any one time. That is a significant difference of 131.25 lbs (59.5 kg). In a well-designed aquaponic system, the media bed serves as an effective means of eliminating waste and increasing dissolved oxygen.

Tilapia Weight Gain and Measurement

Per industry nomenclature, a standing crop is the total weight of all of the fish in a tank or pond at any chosen moment of time. This number is generally expressed in pounds per cubic foot.

It is important to note that most literature discussing the rearing, breeding, and harvesting of Tilapia is based upon large commercial scale operations. These operations typically discuss Tilapia in regards to larger scale measurements pertaining to ponds referenced in acre-feet or hectors. Fret not, for the oxygen required per cubic foot, the pH, dissolved oxygen parameters, temperature, feed required per cubic foot, feed conversion rates, as well as other life sustaining parameters referenced will be the same for a tank as it is for a pond. The numbers simply have to be adjusted for your tank size via of some basic arithmetic.

The growth rate of Tilapia is determined by several factors. It is affected by water quality, temperature, oxygen levels, and the general health of your fish. The type of food, along with the quantities provided, is also of imperative importance. Ensuring stocking density does not exceed the optimal range is a critical factor as well.

Furthermore, it is important to choose a species, hybrid or strain, that is a good 'fit' for your particular operation and goals. Many Tilapia vendors advertise strains with a super-fast growth rate. However, the purported growth rate will not be attained unless the living environment is ideal and suitable for their particular needs. The tank environment *must* be taken into account.

Mixed-Sex versus Mono-Sex Culture

When male and female tilapias are kept together, they will readily breed and produce a lot of offspring. This will hamper the growth rate of the adult fish, as they will be forced to compete for food with fry and fingerlings. The three following methods are commonly utilized to prevent this from happening:

1. Harvesting the mix-sexed culture before they reach sexual maturity or soon afterwards.
2. Raising the mix-sexed culture in cages or tanks that disrupts preproduction.
3. Raising a mono-sex culture consisting of males only.

Growth Rate in Mixed-Sex Culture

In a mixed-sex Tilapia culture, fish are typically harvested before they reach sexual maturity or soon afterwards. This restricted culture period makes it even more important than normal to facilitate the fish growing as quickly as possible since they have to reach their proper size within a limited time frame. It is therefore common to avoid dense stocking of mixed-sex Tilapia cultures. It is also important to avoid using stunted fish since such fish will reach sexual maturity while they are still too small for the food market.

Blue Tilapia (*Oreochromis aureus*), Nile Tilapia (*Oreochromis* **niloticus**), and their hybrids are common in mixed-sex cultures since they will attain a marketable size before they commence to spawning. Species such as Mozambique Tilapia (*Oreochromis mossambicus*) and Wami Tilapia (*Oreochromis urolepis hornorum*) are avoided by most operators since they will be too small when they reach sexual maturity.

By choosing the right strain of Tilapia, and providing the fish with a suitable environment and proper nutrition, it is possible to achieve a growth rate fast enough to allow fry produced in the spring to reach a marketable size by autumn in temperate regions. For a four to five month long culture period, it is common to stock one month-old fry in grow out tanks. The average weight at harvest can then be expected to be around 0.5 pounds (220 grams), when supplemental feedings with protein rich food is carried out. If you will recall from previous chapters, the recommended stocking density is one pound of fish per five to seven gallons of tank water (.5 kg per 20-26 liters). Therefore, for the most part, the quantity that can be raised is largely dependent upon the tank size.

Tilapia Growth Rate for All-Male Fingerlings

All male fish are grown by operators using a mono-sex cultures method, since the male Tilapia grows faster and reaches a larger size than the female. All male batches can be obtained through hybridization, hormonal treatment, or manual sexing and separation. It is important to note that none of these methods can guarantee 100 percent males in every batch. If you desire large Tilapia, the amount of females in the growing unit should not exceed 4 percent. Many operators use more than one method to ensure a low degree of females in the growing unit. Predator fish of a suitable size can also be added to the growing unit to devour any offspring.

All-male Tilapia cultures are often densely stocked. This decreases the individual growth rate of each fish, but it normally results in a higher yield-per-unit area. Densely stocked cultures are more susceptible to ill-health, so careful water management is critical since poor health can have a devastating effect on growth rate and lead to massive losses.

In a suitable environment with an adequate supply of nutrition, it is possible for 50-gram fingerlings to become 500-gram fishes within six months. This means an average growth rate of 2.5 grams per day. You can expect the average weight gain to be 1.5-2.0 grams/day. The culture period needs to be at least 200 days, often more, if you want to produce fish that weighs almost 500 grams. Keep in mind, that maintaining water quality at higher densities becomes more challenging.

Table 43 Below: Average production values for male mono-sex Nile and Red Tilapia in an aquaponic system. Nile Tilapia are stocked at 0.29 fish/gallon (77 fish/m3) and Red Tilapia are stocked at 0.58 fish/gallon (154 fish/m3).

TABLE 43. ***Average production values for male mono-sex Nile and Red Tilapia in an aquaponic system***

HARVEST WEIGHT PER TANK (lbs)	HARVEST WEIGHT PER UNIT VOLUME ((lb/gal)	INITIAL WEIGHT (g/fish)	FINAL WEIGHT (g/fish)	SURVIVAL (%)
1,056 (480 kg)	0.51 (61.5 kg/m3)	79.2	813.8	98.3
1,212 (551 kg)	0.59 (70.7 kg/m3)	58.8	512.5	89.9

CHAPTER 33

Greenhouse Energy Management

Best Greenhouse Management Practices

There's no cheaper energy than the energy you don't have to use, so if designing a new greenhouse, build it so that it does not require much heating and cooling in the first place. This means using building a air-tight, insulated structure, using proper roofing materials, and orienting the greenhouse with the glazing facing South—where all our light in the Northern hemisphere comes from. If growing in an existing greenhouse, you can insulate your greenhouse and weather-strip air leaks among other things. Reducing your energy requirements to a minimum is always the first step.

For large commercial greenhouse operations, energy is typically the largest cost in the production of greenhouse crops in temperate climates. Of the total energy consumed, roughly 65 to 85 percent is for heating. In an industry with an increased desire for sustainable production, greater emphasis is being placed on producing greenhouse crops in an energy-efficient and environmentally friendly manner. The following sections discuss several strategies and technologies that large greenhouse growers can use to reduce energy consumption and improve greenhouse production efficiency. These concepts can apply to virtually any greenhouse-grown crop located in temperate climates.

Keep in mind, though, that not all strategies may be applicable and some may not deliver a favorable economic return on investment in your specific geographical climate. For optimal results, you may wish to consult with an energy auditor with commercial greenhouse experience. Furthermore, many local utility companies will provide a complimentary energy audit for their customers.

Some energy-saving strategies may reduce energy costs, but they may also create a less favorable environment for plant growth and development. For example, adding a third layer of plastic to a double poly greenhouse will reduce heat loss, but it will also reduce light transmission. During production in the winter and early spring, that reduction in light transmission can delay rooting of cuttings, reduce plant growth, and disrupt production.

Energy Savings through Strategic Lighting Methods

You can save a substantial amount of energy by replacing incandescent lamps with more efficient photoperiodic lighting strategies. Here are some options:

1. Replace incandescent lamps with compact fluorescent lamps (CFLs). CFLs consume about one-fourth of the energy compared to incandescent lamps. There are a few crops, such as petunia and pansy, in which flowering of some varieties is delayed under light from CFLs, especially when used as night-interruption lighting. Therefore, you may wish to only replace every other incandescent lamp with a CFL, at least for these two crops.
2. Provide cyclic lighting, where light is delivered to plants on an intermittent basis during the night. There are three common techniques to deliver cyclic lighting:
   a. Turn incandescent lamps on for six to Ten minutes every half hour during the desired lighting period. This can reduce energy costs by two-thirds or more, and is effective on most crops.
   b. Install high-intensity discharge lamps on a moving boom that runs back and forth above crops for at least four hours during the night. Some growers have used this method with success, although little scientific information is available to support this specific recommendation.
   c. Install high-intensity discharge lamps that have a rotating reflector (such as the Beamflicker from PARsource1) above crops. Operate these lamps for at least four hours during the night. Based on Michigan State University (MSU) research, one 600-watt lamp is recommended for every 1,500 square feet (140 square meters) of growing area.

High-Intensity Lighting for Growing Young Plants

Use high-intensity lighting as a supplement to increase photosynthesis. The objective is to increase the photosynthetic daily light integral (DLI) to increase plant growth. The DLI is the cumulative number of photons of photosynthetic light delivered to a plant canopy each day. The common DLI unit is moles per square meter per day (mol·m-2·d-1), and since it is a value that accumulates, it cannot be determined instantaneously. The DLI is particularly important in greenhouse crop production because plant biomass (e.g., roots, stems, flowers and fruit) is generally a function of the amount of light available to plants; the higher the DLI, the greater the plant growth. In temperate climates, the DLI inside greenhouses can be a limiting factor from late autumn to early spring.

During the year, the average DLI in a greenhouse can range from low values (5 mol·m-2·d-1) to high values (25 mol·m-2·d-1), due to factors such as the seasonal angle of the sun, cloud cover, day length, use of shade curtains, and light transmission of the greenhouse structure. When producing crops during winter in the northern United States, the DLI in the greenhouse is often below ten mol·m-2·d-1, and supplemental lighting is beneficial to maintain plant quality and crop schedules. Remember that the greenhouse structure, glazing material, overhead equipment, etc. can also reduce the DLI inside the greenhouse. Hanging baskets placed overhead will reduce the amount of light reaching the crop below considerably, possibly causing poor plant quality, especially in early spring.

In contrast to photoperiodic lighting, lighting to enhance photosynthesis requires a much higher intensity and lamps usually operate for a longer period (up to 20 hours) each day. A typical lighting installation delivers 60 to 80 µmol·m-2·s-1 (450 to 600 footcandles) at plant height, although even higher intensities (up to 200 µmol·m-2·s-1) are provided for some vegetable crops. The high-pressure sodium (HPS) lamp is presently the most widely used light source for supplemental lighting in greenhouses (Figure 2). On a relative basis it is moderately efficient, has a long bulb life, and emits light that is rich in orange light. Two added benefits to using HPS lamps are: (1) a significant amount of heat is emitted from these lamps, which can save on heating fuel, and (2) the radiation emitted by the lamps can increase plant temperature, thereby accelerating crop growth and development. Much less commonly used is the metal halide lamp, which is slightly less efficient at converting electricity into photosynthetic light, but has a shorter bulb life than HPS lamps. Although numerous light-emitting diode (LED) products are marketed for plant applications, in most cases, they are not yet cost effective. However, as the efficiency of LEDs increases and costs decrease, they will become more economical for greenhouse operations.

Adding supplemental lighting to a greenhouse is a relatively expensive investment and operational costs can be high, approaching or surpassing the cost of heating in some applications. One of the most economical uses of supplemental lighting is on plugs and liners. Adding supplemental light during this stage is especially important in the northern United States because a majority of plugs and liners are produced late in the winter and in early spring, when the natural DLI is low. Providing supplemental light to young plants has many advantages including faster growth, shorter internodes, thicker stems, increased root development, and improved quality (Figure 3).

The cost of lighting during the young plants stage is low on a per-plant basis since they are grown at a high density. The reduced production time from using supplemental lighting during the plug stage provides the opportunity for increased revenue because more crop turns are possible. In most situations, supplemental lighting of crops during the finish stage is not economical because of their lower density and per-container cost.

To minimize operational costs and save on electrical energy, only provide supplemental lighting when irradiance is low, such as at night and on cloudy days. Computers that control the greenhouse environment can be configured so lights only turn on when ambient light levels outside or inside are below a minimum value and for a specific time period (to avoid equipment cycling), e.g., for at least 15 minutes below 300 µmol·m-2·s-1 (1,500 footcandles).

Temperature Management based on the Crop and Finish Date

Most crops are able to be grown under a wide variety of environmental conditions. One principle that applies to all crops is that their developmental rate (such as the rate of leaf unfolding or rate of flowering) decreases with the temperature. As temperature decreases, plants develop progressively slower and, at some point, they stop growing. The cool temperature at which plant development stops is referred to as the base temperature. The base temperature varies among species, although for most crops the estimated values range from 32 to 50 °F (0 to 10 °C).

The rate of development will continue to increase above the base temperature to a certain point. The temperature at which development is maximal can be called the optimum temperature. Again, the optimum temperature varies from one plant species to the next. The optimum temperature does not necessarily mean that a crop will grow as fast as possible. For instance, under light-limiting conditions, plants may only be of moderate or even poor quality if grown at a crop's optimum temperature.

As discussed in 'Chapter 21: Greenhouses' Virtual Grower, a free computer program developed by the U.S.D.A. (www.virtualgrower.net), has enabled growers throughout the United States to predict heating costs for their greenhouses. You can use the program to help make decisions on growing temperature set points, use of alternative fuels, and energy-saving investments. One of the significant uses of the program is the ability to predict the amount of energy needed to maintain a desired temperature at different times of the year. When combined with information on the temperature's effects on crop timing, you can identify the most energy-efficient growing temperatures.

With this information, growers can determine transplant dates so plants are in optimal production for predetermined market dates when grown at different temperatures. Because of the substantial delay in flowering when grown cool, more energy may be consumed by growing a crop at cooler temperatures compared to when grown in warmer temperatures with a shorter finish time.

Growing warmer allows you to turn crops more quickly, which opens space for another crop. This can be particularly beneficial for operations that have space constraints. So why doesn't every grower turn up the heat and produce crops more quickly? It depends on location, greenhouse characteristics, and the crop(s). That's why using Virtual Grower for your own greenhouse is so important; results vary from one greenhouse to the next, so the program's utility hinges on spending a little more time to generate the most meaningful results. Second, under light-limiting conditions in the early spring, growing some species warm can produce low-quality plants. Therefore, plants that are typically grown cool should generally not be grown warmer than the low 70s, until light conditions are higher (beginning in March in the northern U.S.).

Temperature Integration Strategies

As mentioned above, research has shown that plants develop in response to the average daily temperature. Therefore, you can adjust greenhouse temperatures on an hourly or daily basis in response to outdoor conditions without influencing the time to flower as long as the target average temperature is provided. One method of managing greenhouse heating costs is dynamic temperature control in which heating set points are lowered when the greenhouse energy loss factor is high (i.e. outside temperature and incoming solar radiation are low) and increased when the energy-loss factor is low. This strategy maintains a target average temperature over a period of several days. A research greenhouse in Denmark using this strategy consumed substantially less fuel for heating compared to using typical day/night temperature set points. To utilize dynamic temperature control, a greenhouse environmental control computer with sophisticated software is required.

However, not all greenhouses use environmental control computers, and of those that do, very few (in the U.S.) use dynamic temperature control. An alternative and simple energy-saving approach is to use a warmer day than night temperature regimen in which the difference between the day and night temperature is positive (+DIF). With a +DIF, the heating set point is lowered during the night, when energy consumption for heating is highest. A low night temperature is compensated by increasing the day temperature so that the target average daily temperature is achieved. A negative consequence of using a +DIF is that stem elongation is promoted in many crops. Many growers actually use a cooler day than night temperature (–DIF), which inhibits stem extension but also consumes more energy for heating.

Regardless of the temperature set points, energy can be saved by increasing the dead band of the environmental control computer settings. The dead band is a range of temperatures in which neither heating nor venting occur. By setting a fairly wide dead band (5 to 7 °F or 3 to 4 °C), overheating or excessive ventilation is less likely to occur and thus, avoids cycling of temperatures and unnecessary energy inputs. However, increasing the size of the dead band reduces the system's ability to maintain the greenhouse temperature at or near the desired set point.

Reducing Air Leaks

It is quite common to find areas where the warm greenhouse air is escaping to the cold outdoors. Air can escape in many different places. Some energy auditors use an infrared (IR) sensor to identify where air is escaping. However, you can identify many air leaks simply by closely inspecting the greenhouse glazing, walls, doors, fans, vents, and other areas. Pay special attention to where the covering material attaches to the foundation, side and walls, and around fans and vents. Here are some of the primary areas on which to focus:

- Patch holes in the plastic covering and side walls, or replace cracked or missing glass panes.
- Keep the doors closed and make sure they shut completely.
- Weather-strip doors, vents, and fan openings.
- Ensure louvers are sufficiently lubricated so that they close tightly.
- Shut off some of your exhaust fans from late fall through early spring, then cover openings with insulation or plastic to reduce air infiltration.

If you have a double poly greenhouse, it is also important that the space between the two layers is properly inflated (Figure 5). Always use outside air to inflate the two layers of plastic film. The value of having two layers of plastic becomes almost nonexistent if the fan that maintains the air gap fails to operate. One USDA-ARS energy auditor was recently contacted by a grower who had raccoons tear into his outer layer and nest repeatedly between the layers. Frustrated, the grower finally gave up and allowed the double layer to deflate, effectively giving him a single layer plastic greenhouse. An energy audit was performed for the grower in that section, and his energy losses were so high that the grower could pay someone round the clock to prevent the raccoons from deflating the double-layer plastic and the grower would still make money! Thus, especially during the winter, routinely check to make sure the inflation fan is blowing air into the gap between the two layers of plastic.

Insulation of Side and End Walls

A greenhouse structure loses heat (energy) at night and during the winter months through physical processes called conduction, convection, and radiation.

- **Conduction** is the process of heat transfer through a material. When you put a metal pot over a fire, eventually the handle will get hot too. Thus, as long as there is a temperature difference between the inside and the outside of the greenhouse, energy will flow through the structural materials from the warm side to the cold side.
- **Convection** is the process of heat transfer by the movement of a fluid (e.g., air or water). A heating pipe heats the air surrounding it and this warmer air rises and moves away from the pipe. The warmed air is replaced by colder air that is not yet warmer by the pipe, starting a continuous process.
- **Radiation** is the process of heat transfer resulting from the temperature difference between surfaces that are in the line of sight from each

other. Standing next to a hot fire (or standing outside on a hot sunny day), you can feel radiation heating the surface of your skin. Similarly, in a greenhouse hot water pipes radiate energy to their surroundings, including the plants.

Greenhouses are designed for maximum light transmission and, as a result, often have limited insulating properties. Some portions of the side walls and end walls can be modified to improve their insulating capacity with very little impact on light transmission. These portions are called the knee wall or curtain wall, and are the sections that are often constructed of brick or concrete block. They typically rise from the foundation to the first two or three feet. Additional insulation can be attached or incorporated into these wall sections to reduce the overall heat loss through the above-ground greenhouse structure. These wall sections are often used to attach perimeter heat pipes, and the higher the insulating value of these wall sections, the more of that heat stays in the greenhouse.

In the past, some growers opted to add temporary insulating boards to the entire north-facing side wall in an attempt to reduce greenhouse heat loss during the winter when the sun entered the greenhouse at low angles (northern hemisphere). However, the winter period is often relatively dark with many overcast days. During cloudy days, the remaining sunlight is highly diffused and, as a result, also enters the greenhouse through the north-facing side wall. Therefore, a better option might be to increase the insulating value of the north-facing side wall with little impact on light transmission. One solution is to cover the inside of a glass-clad north-facing side wall with plastic bubble wrap.

In addition to insulating opaque sections of greenhouse walls as best as possible, it is also important to consider the potential heat loss that occurs at ground level along the perimeter of the greenhouse. Wet soil is a good conductor of heat. When the soil immediately outside the greenhouse perimeter is wet and in direct contact with the soil or concrete floor inside the greenhouse, a conduit is created for heat to flow from the inside to the outside of the greenhouse. To reduce heat from escaping through this route, an insulating barrier can be installed. For example, a two inch thick polystyrene board extending vertically down for two feet can significantly reduce heat loss at ground level along the perimeter of the greenhouse.

Retractable Curtains: Installation and Maintenance

With the many greenhouse designs in use, there are many ways to install and operate retractable curtains. In gutter-connected greenhouses, growers typically install retractable curtains inside and horizontally above the trusses. Retractable curtains serve two purposes: they create some amount of shading for the crops underneath, and they block some amount of heat radiation to the outside environment. Shading is useful during sunny days from spring through autumn when the plants may become stressed from excessive solar radiation. Blocking heat radiation is useful at night, particularly during the winter, when heat is needed to maintain the temperature set point. Growers have reported significant seasonal energy savings of up to 30 percent, often resulting in a quick return on investment. In most installations, the curtain serves these two purposes and a compromise is made between the effectiveness of either control strategy. In some cases, growers install multiple curtains to increase the level of light and energy control (e.g., multiple curtains allow access to two different levels of shading).

Depending on greenhouse orientation, design, and grower preferences, you can install curtain systems to operate from gutter to gutter or from truss to truss. The mechanisms used to deploy the curtains include various push-rod systems or systems that use cables that are rolled up on a metal post. These mechanisms are operated by motors attached to gear boxes that ensure slow but steady deployment or retraction of the curtain. A computer system that evaluates light conditions at plant level (for shading purposes) and outside weather conditions (for energy conservation purposes) often controls the motors. Curtain motors generally operate several curtains (through mechanical arrangements to distribute power take-off) at the same time and generate significant amounts of torque to do so. Therefore, it is important to set the physical limit switches on these motors correctly to prevent damage to the curtain system or even the greenhouse structure. Install stationary skirts around the outer edges of the curtains to prevent unwanted openings.

In shading mode, curtains are sometimes operated in stages (i.e., deployed or retracted at a certain percentage of full deployment). However, this practice can result in distinct shadow bands on the crop below, depending on whether the plants were shaded or not. For some crops, these shadow bands, if relatively stationary during the day, can result in uneven plant growth. In energy savings mode, curtains are either retracted or fully deployed, except for early in the morning, when curtains are often cracked open a little to allow the cold air that accumulated above the curtain to mix with the warmer greenhouse air before reaching the plants. Once installed, curtain systems should operate satisfactorily for many years, provided regular checking and maintenance is performed. Curtain materials are available in many different configurations, including fire-retardant fabrics that significantly reduce the risk of a rapidly spreading fire. It is highly recommended to have a retractable curtain installed by a professional.

Fish Tank for Heating and Cooling the Greenhouse

The easiest and most common way to even out the temperature of your greenhouse is utilize thermal mass. Thermal mass is any material that stores thermal energy. Most materials do this to some extent, but some do it much better than others. Water for instance, holds about twice as much heat as concrete, and about four times as much as soil.

Incorporating mass does two things: First, it absorbs excess energy during the day, creating a cooling effect. When the temperature drops at night, it starts releasing that energy, thereby 'heating' the greenhouse. The thermal mass is not actually providing the energy; it is simply storing it and releasing it later, like a battery. The fish tank serves as an excellent thermal mass energy storage unit, greatly helping to regulate greenhouse temperatures.

The fish tank also helps cool the greenhouse through evaporative cooling. Much like those restaurant patios equipped with a mist spray unit to help keep patrons cool, the fish tank provides similar cooling benefits. A fish tank located within the greenhouse reduces the demand for external temperature control methods and lowers utility cost.

Infrared Anti-Condensate Polyethylene Film

Plastic film has been used to cover greenhouses for almost 50 years. Different plastics are used, each with its own physical properties and cost. The plastic most often used is clear polyethylene with a film thickness of six mils (0.15 mm). To reduce heat loss, install two layers of polyethylene and keep them apart by a layer of air provided by a small squirrel cage fan that uses outside air for the inflation. Using this method, growers have reported significant energy savings from the relatively high insulating value of the stagnant inflation air. The cost of operating the small inflation fan continuously is very minor compared to the potential energy savings that can be realized. You can accomplish similar or even higher energy savings by covering the greenhouse with rigid twin or triple walled plastic panels made from polycarbonate or acrylic. These rigid panels do not require an inflation fan.

Using polyethylene film to cover greenhouses has several advantages: it is a relatively inexpensive cladding material that's fairly easy to install, and it diffuses incoming solar radiation, resulting in enhanced penetration of light into the plant canopy. Obviously, using a double-layer cladding material reduces the amount of light reaching the crop compared to a single layer design. Polyethylene film as a greenhouse cladding material has a life expectancy of three to four years because of the degrading of the plastic due to the ultraviolet component of sunlight, the inevitable rubbing of the film against the greenhouse structural elements, and the high temperatures of contact surfaces between the film and the greenhouse structure on sunny days.

Over the years, several improvements have been incorporated into polyethylene films used as greenhouse cladding material.

1. Films are now manufactured in a tube shape allowing for easier installation in the double layer configuration. These tubes come in various widths that match most common greenhouse roof dimensions, whether freestanding or gutter connected.
2. Films were developed that block some of the IR (heat) radiation. As a result, during the day, less of the heat component of solar radiation enters the greenhouse and thus, reduces the need for ventilation. During the night, less of the greenhouse heat provided by the heating system escapes to the outdoor environment. This IR blocking feature dramatically reduces greenhouse energy consumption. For example, a Minnesota study estimated that the return on investment for installing an IR double poly plastic was less than two months, even when production was seasonal. When films are manufactured in tubes, both the top and bottom film will have the IR blocking feature. When films are installed as two separate layers, some growers opt to install a film with the IR blocking feature only as the bottom layer, which is done to reduce installation cost.
3. To prevent beading of water droplets on the surface of double layer polyethylene film installations (particularly on the inside of the greenhouse), the film surface was given a treatment so that water droplets run off (the so-called anti-condensate treatment). Water droplets on the inside film surface reflect some of the incoming sunlight, which is undesirable in many situations. Therefore, when plastic covering needs to be replaced, energy can be saved if at least the inside layer of a double poly glazing blocks IR and has an anti-condensate treatment.

Heating Equipment Maintenance

Greenhouses can be heated with a variety of heating systems and fuel sources. Price and availability are the two most important factors that determine the choice of fuel source. As a result of high energy prices, there has been a lot of interest recently in alternative fuel sources, biomass in particular.

While biomass may sound like an attractive fuel source for greenhouse heating, you will need to consider several issues before switching to an alternative fuel source. These issues include, but are not limited to: reliability and quality of the fuel supply, handling and storage, possible modifications to or replacement of current heating equipment, air quality permits, fuel cost, impact upon global food prices, maintenance demands, and waste disposal.

Common greenhouse heating systems include hot-air, hot-water, and IR systems. Hot-air heating systems produce hot air (typically as point sources) that has to be evenly distributed throughout a greenhouse. Non-uniform heat distribution results in non-uniform plant growth and development. You can distribute heat using a fan-jet (poly tube) system, horizontal airflow fans or both. Hot air-heating systems are often less expensive compared to hot-water systems, and offer a degree of redundancy since multiple heating units are typically installed to provide heat to the greenhouse.

Hot-water heating systems produce hot water that is distributed through a plumbing system that is installed throughout the greenhouse environment. As a result, hot-water systems often provide better uniformity of the heat distribution and can be placed closer to the plants through methods such as bench or floor heating systems. However, as heat gets delivered closer to the plants, the maximum allowable water temperature needs to be reduced so that plants are not damaged. In that case, additional heating capacity such as overhead heating, perimeter heating or both are needed to maintain acceptable greenhouse temperatures. Aquaponics helps alleviate this problem, as the fish tank, holding warm water fish, will radiate thermal heat energy in the greenhouse.

By providing heat closer to the plants, the air temperature of the greenhouse could be maintained at a lower set point, resulting in potential energy savings. Hot-water heating systems often have a certain buffering capacity, extending the time between a malfunction of the heating system and catastrophic crop failure.

IR heating systems distribute heat by radiation (and some convection from heated surfaces). Radiators are placed above the crop and any part of the canopy with a direct line of sight to the radiator will receive heat depending on the radiator surface temperature and the temperature of the plant surface. Plant surfaces that are shaded from the radiator, including lower layers of leaves, will not receive the radiation, which can result in uneven heating. IR heating systems can deliver heat very quickly once units are turned on. In older greenhouses, it is not always possible to install the radiators at sufficient distances above the crop to sufficiently cover large enough growing areas.

Heating greenhouses is relatively expensive because they are designed for maximum light transmission and not for maximum insulating properties. As a result, it is important to regularly inspect and maintain heating equipment, including the operation calibration of the temperature sensor(s), and the environment control system. For growers that do not have a service contract with a licensed heating contractor, we recommend training a dedicated employee and maintaining an adequate supply of spare parts. Growers can effectively use computer control systems to monitor and troubleshoot heating systems, and to send out warnings by phone, e-mail, or both to alert growers of potential problems.

If the aquaponics fish tank is housed in the greenhouse, it will radiate heat as well. As a matter of fact, a large fish tank located in the greenhouse serves as an excellent low cost radiator.

Heating System Efficiency

It is almost inevitable that some of the heat contained in the fuel is expelled from the greenhouse with the exhaust gasses. However, the goal is to keep these losses to a minimum. Several reasons why a heating system becomes less efficient over time include:
- The unit is not properly maintained or adjusted.
- Deposits have formed on components of the combustion chamber.
- A heat exchanger is dirty.
- A fan distributing hot air or a pump distributing hot water is not working properly.
- The combustion process is not receiving enough oxygen.

To obtain the highest efficiency from a heating system, a regular check and maintenance schedule is essential. In addition, new developments in the design and construction of heating systems may have resulted in models that have a higher efficiency compared to the ones that were available many years ago. Therefore, compare new model specifications with the specs of the unit(s) that are already installed in the greenhouse. Finally, you can implement control strategies that are better adapted at minimizing energy consumption. Many environmental control companies regularly update their programs, and implementing these updates in a timely fashion can result in significant savings.

The following feature some newer heating system designs are worth considering when retrofitting or installing a new greenhouse:

1. **Condensing Boiler Technology.** When fuels combust, water vapor is a by-product of the combustion process. Because it took energy to convert (liquid) water to water vapor, the conversion energy can be recaptured by allowing the water vapor to condense back to water. These boilers are made with stainless steel components to allow for corrosion-resistant condensation. Condensing boilers have a higher efficiency compared with conventional boilers and can be operated on demand (i.e., they have no stand-by losses, like the much larger conventional boilers that constantly keep some water volume at a predetermined set point temperature).

2. **Direct-Fire Unit Heaters.** These heaters do not have a heat exchanger and thus, have very high efficiencies. However, they need to burn very clean to prevent any contamination of the greenhouse environment from unwanted by-products of the combustion process. An added advantage of these units is that the carbon dioxide produced during the combustion process can be released into the greenhouse environment, which can increase plant growth.

3. **Combined Heat and Power Systems.** These systems generate electricity to run greenhouse equipment, export to the local grid when there is excess, and capture (often by using heat exchangers) the heat contained in the combustion gasses for heating purposes. By using the original fuel-for-two purposes (the simultaneous production of electricity and heat), the overall system efficiency is much improved. Growers in northern Europe routinely use these systems, but their use in the Unites States has been limited for various reasons, including the inability to sell excess electricity back to the power grid at a reasonable price and challenges obtaining permits. However, recent changes in the electrical utility industry is making this a more favorable option (i.e. utility companies purchasing power from homeowners and small businesses).

4. **Heat Pumps.** A heat pump is a refrigerator that can also be operated as a heater. By reversing the flow of the refrigerant in a refrigeration cycle, the same unit can operate as a heater. Thus, one system can be operated as a heater or cooler (air conditioner). Ground-source heat pumps, which are systems that extract heat from the ground during the winter and dump excess heat into the ground during the summer, are especially attractive for greenhouse operations. Their overall system efficiency can be further enhanced by incorporating energy storage (e.g., insulated water tanks).

Horizontal Air Flow Fans: Installation and Maintenance

Horizontal airflow fans (HAF) are installed in greenhouses to help mix the air. Air mixing is sometimes necessary to improve the uniformity of temperature, humidity, and carbon dioxide. Loss of uniformity of these parameters often results in non-uniform plant growth and development. The risk of non-uniform conditions is particularly high when little or no ventilation is needed to maintain temperature, humidity set points, or both (such as during the winter).

HAFs are typically installed in so-called raceways and direct the air in horizontal jets. In each greenhouse section or bay, fans point in a longitudinal direction on one side and in the opposite direction on the other. The length of the section or bay determines the number of fans on each side. The recommended fan capacity is approximately 3 ft^3 per minute per ft^2 (3 m^3 per minute per m^2) of growing area. The airflow capacity of HAFs is very small compared to the capacity of ventilation fans. HAFs are typically

FIGURE 253. Horizontal Air Flow Fan

mounted just below the trusses, but high enough to keep them out of the way for people and equipment moving through the greenhouse. HAFs should be shielded with a screen for safety reasons, and some designs include a shroud for improved efficiency. Fan motors should be rated for continuous operation and have thermal overload protection, especially when the HAFs are mounted in close proximity to retractable curtains. Use a sturdy mount to prevent rocking during operation. Do not use chains.

HAFs provide benefits when the air would otherwise be still. When there is air movement in the greenhouse from other causes such as ventilation fans or open roof vents, HAFs have little or no utility. Therefore, you can save energy if the HAFs are automatically turned off when the greenhouse is being vented.

Ventilation System Operation for Maximum Efficiency

To maintain optimum growing conditions, warm and humid greenhouse air needs to be replaced with cooler, and typically drier, outside air. To accomplish this, greenhouses use either mechanical or natural ventilation. While air conditioning of greenhouses is certainly technically feasible, the installation and operating costs are prohibitively high. Mechanical ventilation requires inlet openings, exhaust fans, and electricity to operate the fans. When designed properly, mechanical ventilation is able to provide adequate cooling under a wide variety of weather conditions. Typical designs specify a maximum mechanical ventilation capacity of 10 ft^3 per minute per ft^2 (10 m^3 per minute per m^2) of floor area for greenhouses with a shade curtain, and 12 cfm per ft^2 for those without a shade curtain. Below are some highlights of energy-efficient ventilation systems:

- Multiple and staged fans can provide different ventilation rates based on environmental conditions. Variable speed fan motors allow for a more precise ventilation rate control and can reduce overall electricity consumption.
- Natural ventilation works based on two physical phenomena: thermal buoyancy (warm air is less dense and rises) and the so-called wind effect (wind blowing outside the greenhouse creates small pressure differences between the windward and leeward side of the greenhouse, causing air to move towards the leeward side). All you need are strategically placed inlet and outlet openings, vent window motors, and electricity to operate the vent motors. Compared to mechanical ventilation systems, electrically operated natural ventilation systems use a lot less electricity.
- Ultimate natural ventilation systems include the open-roof greenhouse design where the very large ventilation opening allows for the indoor temperature to almost never exceed the outdoor temperature.

> ### Utility Rebates
>
> Many gas and electric utility companies offer rebates on new and retrofit equipment, systems and controls that save energy. Contact your utility companies for more information. Examples include:
>
> - T5, T8 and compact fluorescent lamps
> - IR polyethylene film
> - High-efficiency heating systems such as power-vented unit heaters and condensing boilers
> - High efficiency unit heaters
> - Automated systems that turn off equipment when not needed
> - Thermal screens
> - Perimeter and wall insulation
>
> In addition, state and federal grant and loan programs can provide additional support for energy-efficient investments.

FIGURE 254. Utility Rebates

Resources for More Information

For expanded and more detailed information on many of these topics, check out the following resources:

Books

- *Energy Conservation for Commercial Greenhouses* by John W. Bartok Jr., Extension professor emeritus and agricultural engineer, University of Connecticut. For ordering information, visit **www.nraes.org**.
- *Greenhouse Engineering* by Robert A. Aldrich and John W. Bartok Jr. For ordering information, visit **www.nraes.org**.
- *Lighting Up Profits, Understanding Greenhouse Lighting*, edited by Paul Fisher, University of Florida, and Erik Runkle, Michigan State University. For ordering information, visit **http://meistermedia.com/store/ books.html**.

Websites

- *Energy Conservation* resources at the University of Wisconsin Extension Learning Store: **http:// learningstore.uwex.edu/Energy-Conservation-C29. aspx**
- *Greenhouse Energy Cost Reduction Strategies* developed by Matthew Blanchard and Erik Runkle, Michigan State University: **http://hrt.msu.edu/Energy/Notebook.htm**
- *High Tunnels and Greenhouses*, part of the Vegetable Crops Online Resource Center of Rutgers University: **http://njveg.rutgers.edu/html/gc-4high-tunnels.html**
- *Horticultural Engineering* at Rutgers University: http://aesop.rutgers.edu/~horteng

CHAPTER 34

Canning and Saving Produce from Your Harvest

Since the beginning of time, preserving food was a necessity. The process could take considerable time and energy, but could mean survival during harsh seasons if food was scarce. Early methods of preservation included drying, smoking, fermenting, or cooling/freezing foods. Later methods were pickling in an acid (such as vinegar), curing with salts, and using honey and sugar to make jams and jellies. However, besides the time invested, these approaches can be unhealthy. Either the process itself causes harm to the food (i.e. the smoking process generates carcinogens), or the process involves adding enormous amounts of unhealthy compounds, such as sugar and salt. So the search continued for methods with more reliability, ease of storage and transport, and increased health benefits.

Napoleon Bonaparte himself catalyzed the search for a better food preservation method in the late 1700's as a way to better feed his armies, offering a fortune to anyone who developed a method of preserving food on a large scale. In 1810 Nicholas Appert succeeded, but it wasn't until 1858 when John Mason invented the iconic, reusable "Mason Jar" that Appert's "canning" method trickled down to the average family. (Glatz)

Canning is the process of applying heat to food that's sealed in a glass jar in order to destroy any microorganisms that can cause food spoilage. Proper canning techniques stop this natural spoilage by heating the food for a specific period of time and killing these unwanted microorganisms. During this process, air is removed from the jar and a vacuum is formed as the jar cools and seals. The seal then protects the food from new microorganisms entering and from oxidization from the air. After this, the food can be conveniently stored and enjoyed at a later date.

Canning became a way of life and common in nearly every household. That is until the arrival of our modern grocery stores, pre-packaged foods, additives and preservatives. Now the art of food preservation has been lost to the vast majority of people, as has its numerous benefits. Fortunately, interest in food preservation, especially canning, is growing and is seeing

FIGURE 255. Canned vegetables.

a resurgence of popularity. The following benefits are worth considering, despite the initial investment of time and money (due to buying jars, a canner, etc.).

Health

When you grow and can your own food, you know exactly what you are eating. You can be assured that the food was fresh and high quality, and are able to harvest at the peak of ripeness to help preserve the vitamins and minerals. You will also be enjoying food that is free from harmful additives, preservatives and BPA, which is found in most of today's factory-produced foods.

Preserve harvest

One way to make the best of a bountiful garden is to preserve the food by canning. It's quite common that everything in a garden becomes harvest ready at once. Canning what you won't immediately consume is a sensible way to avoid waste and enjoy your produce year-round, even in the off-season.

If you're not a gardener, canning makes it worth the trip to pick organic fruit in an orchard or local farm to get a bushel of organic vegetables. Canning the produce means you can take advantage of these in-season fruits and vegetables and extend their season beyond the growing season, all while supporting local farmers.

Quality taste

Homemade food simply tastes better. You can't beat a quality home-canned product made from fresh, locally grown ingredients. Another benefit is that you'll be able to tweak recipes to your exact tastes and even experiment with new flavor combinations, leaving you with a tastier product stocked in your pantry.

Save money

One very good reason to preserve food by canning is to save money. When you grow your own food, buy in bulk, and take advantage of the plentiful seasons you will easily save money without compromising the quality of the food on your table. Store bought food is expensive. In the store, you could easily pay double over the cost for canning the same product, and you'll have healthier, tastier, quality foods in your pantry.

Prepare for bad economic times

With recent economically trying times, many people are worried about the future. If something drastic happens to our economy or our ability to affordably purchase food, people want and need to be prepared. Learning to can is just one of the steps people can choose. While freezing is also a healthy option, you have to use energy to keep it stored. And if there should be a power outage you run the risk of losing what you have frozen. In the event of a natural disaster you will be self-reliant if you have a pantry full of food that you have canned.

Eco friendly

Canning your own food reduces the environmental impact to a minimum.

If the food is home grown, you remove the emissions pollution caused by the transportation miles from the farm, to the factory, and then to the distributor and local stores. Canning also reduces the waste associated with pre-packed foods since canning jars are reusable and will last for years.

Sentimental connection / gifts

Many people enjoy canning because of the powerful connection to the past — to family or culture. Additionally, canned foods make great gifts. The work and care that went into homemade food products is worth much more than the food itself. Although hard to quantify, the satisfaction that you gain from canning your own food can be one of the most significant benefits.

Regardless of your motivation, the benefits of canning are too numerous to ignore. Now you just need to find the resolve to learn and begin canning. Your

CHAPTER 34: CANNING AND SAVING PRODUCE FROM YOUR HARVEST

first step is to find a few good recipes. It is always best to use those from reliable, tested resources to ensure safety and quality in the outcome. Then prepare the appropriate supplies, depending on your method.

Based on the acidity level in the food you are planning to can, you decide which canning method to use. Although you may hear of various canning methods, only *two* are approved by the United States Department of Agriculture (USDA) — water-bath canning and pressure canning. Remember, older canning methods are unreliable and, for that reason, aren't used or recommended today for home-canning. Occasionally, these methods are said to be faster and easier, but using other methods is questionable to your food safety. Although there are many things you can preserve, understanding the two approved canning techniques and when to use them will help you to get started.

Preserving of High-Acid-Foods

Water-bath canning, also referred to as *hot water canning,* uses a large kettle of boiling water. Filled jars are submerged in the water and heated to an internal temperature of 212 degrees for a specific period of time. Use this method for processing high-acid foods, such as fruit, jams, jellies and other fruit spreads, pickles, pickled food, salsas and tomatoes.

To begin, you need the following items:

- Boiling water bath canner or a large, deep sauce-pot with a lid, and a rack
- Glass preserving jars, lids and bands (always start with new lids)
- Common kitchen utensils, such as wooden spoon, ladle and wide-mouth funnel
- Fresh produce and other quality ingredients
- Jar lifter (optional, but helpful)

Water-Bath Canning Instructions

1. Read through recipe and instructions. Assemble equipment and ingredients. Follow guidelines for recipe preparation, jar size, preserving method and processing time.
2. Check jars, lids and bands for proper functioning. Jars with nicks, cracks, uneven rims or sharp edges may prevent sealing or cause jar breakage. The underside of lids should not have scratches

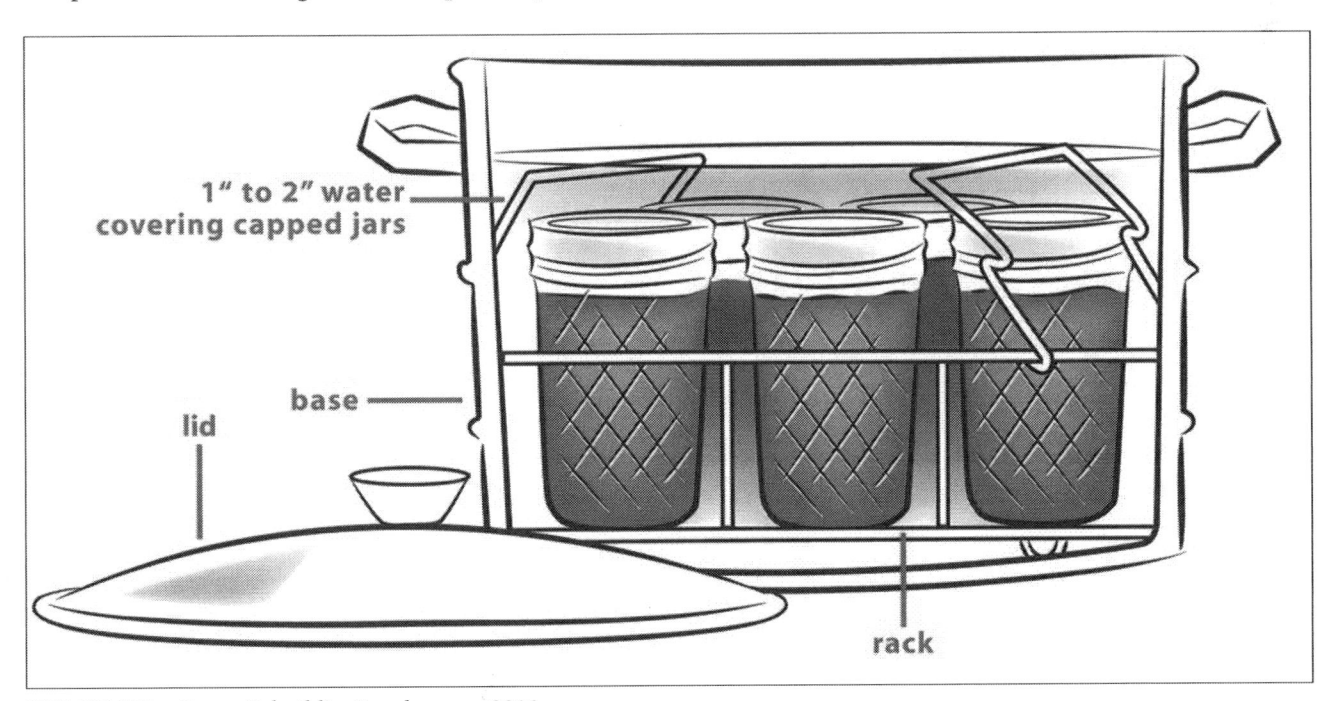

FIGURE 256. Boyer. RebuildingFreedom.org, 2016.

or uneven or incomplete sealing compound as this may prevent sealing. Bands should fit on jars. Wash jars, lids and bands in hot, soapy water. Rinse well. Dry bands.

3. Heat home canning jars in hot water, not boiling, until ready for use. * Fill a large saucepan or stockpot half-way with water. Place jars in water (filling jars with water from the saucepan will prevent flotation). Bring to a simmer over medium heat. Keep jars hot until ready for use. Keeping jars hot prevents them from breaking when hot food is added. *Leave lids and bands at room temperature for easy handling.*

4. Prepare boiling water bath canner by filling half-full with water and keep water at a simmer while covered with lid until jars are filled and placed in canner. Be sure your rack in resting on the rim of the canner or on the bottom, depending on the type of rack you are using. Most kitchens have pots that can double as boiling water bath canners, which is simply a large, deep saucepot equipped with a lid and a rack. The pot must be large enough to fully surround and immerse the jars in water by 1 to 2 inches and allow for the water to boil rapidly with the lid on. If you don't have a rack designed for home preserving, use a cake cooling rack or extra bands tied together to cover the bottom of the pot.

5. Remove hot jar from hot water, using a Jar Lifter or tongs, emptying water inside jar. Fill jar one at a time with prepared food using a wide-mouthed funnel leaving headspace recommended in recipe (1/4 inch for soft spreads such as jams and jellies and fruit juices; 1/2 inch for fruits, pickles, salsa, sauces, and tomatoes). Remove air bubbles, if stated in recipe, by sliding rubber spatula between the jar and food to release trapped air and ensure proper headspace during processing. Repeat around jar 2 to 3 times.

6. Clean jar rim and threads of jar using a clean, damp cloth to remove any food residue. Center lid on jar allowing sealing compound to come in contact with the jar rim. Apply band and adjust until fit is fingertip tight. Place filled jars in canner until recipe is used or canner is full. Lower rack with jars into water. Make sure water covers jars by 1 to 2 inches.

7. Place lid on water bath canner. Bring water to a full rolling boil. Begin processing time.

8. Process jars in the boiling water for the processing time indicated in tested preserving recipe. When processing time is complete, turn off the heat and remove the canner lid. Allow jars to stand in canner for 5 minutes to get acclimated to the outside temperature.

9. Remove jars from canner and set upright on a towel to prevent jar breakage that can occur from temperature differences. Leave jars undisturbed for 12 to 24 hours. Bands should *not* be retightened as this may interfere with the sealing process.

10. Check jar lids for seals, ensuring lids do not flex up and down when center is pressed. Remove bands. Try to lift lids off with your fingertips. If the lid cannot be lifted off, the lid has a good seal. If a lid does not seal within 24 hours, the product can be immediately reprocessed or refrigerated. Clean jars and lids. Label then store in a cool, dry, dark place up to 1 year.

Preserving of Low-Acid-Foods

Low-acid foods are easy to preserve, yet require special handling to eliminate the risk of spoilage caused by the bacteria *Clostridium botulium* and its toxin-producing spores. Pressure canning uses a large kettle that produces steam in a locked compartment. The filled jars in the kettle reach an internal temperature of 240 degrees (eliminating the risk of foodborne bacteria) under a specific pressure (stated in pounds) that's measured with a dial gauge or weighted gauge on the pressure-canner cover. Use a pressure canner for processing vegetables and other low-acid foods, such as meat, poultry, and seafood, vegetables, soups, stews, stocks, or when you're mixing high acid foods with low-acid foods.

To begin, you need the following items:
- Pressure canner
- Glass preserving jars, lids and bands (always start with new lids)
- Common kitchen utensils, such as wooden spoon, ladle and wide-mouth funnel
- Fresh vegetables, meat, poultry or seafood and other quality ingredients
- Jar lifter (optional, but helpful)

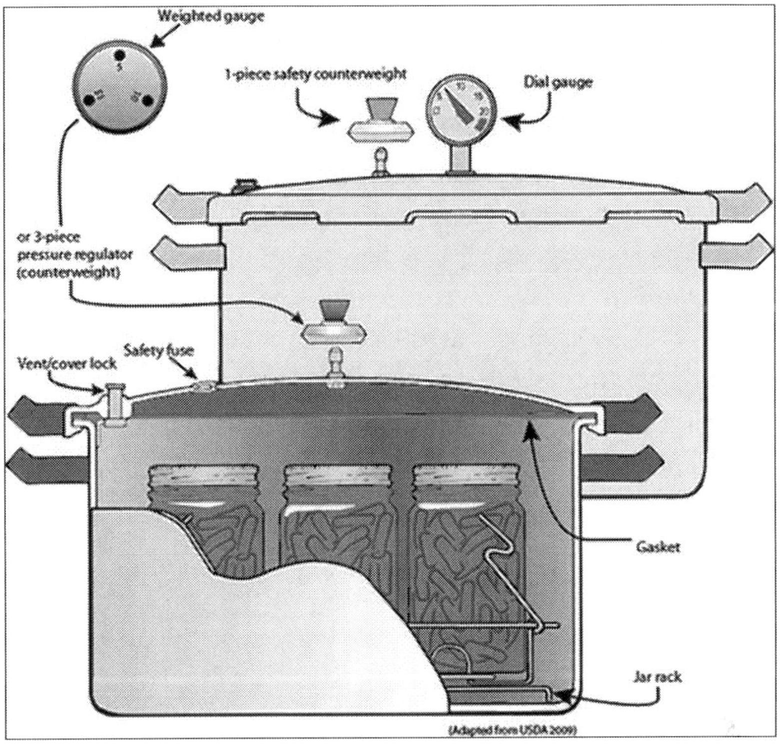

FIGURE 257. Boyer & Chase. Virginia Cooperative Extension, Virginia Tech, and Virginia State University, 2016.

Pressure Canning Instructions

1. Read through recipe and instructions. Assemble equipment and ingredients. Follow guidelines for recipe preparation, jar size, preserving method and processing time.
2. Check jars, lids and bands for proper functioning. Jars with nicks, cracks, uneven rims or sharp edges may prevent sealing or cause jar breakage. The underside of lids should not have scratches or uneven or incomplete sealing compound as this may prevent sealing. Bands should fit on jars. Wash jars, lids and bands in hot, soapy water. Rinse well. Dry bands.
3. Heat home canning jars in hot water, not boiling, until ready for use. * Fill a large saucepan or stockpot half-way with water. Place jars in water (filling jars with water from the saucepan will prevent floatation). Bring to a simmer over medium heat. Keep jars hot until ready for use. Keeping jars hot prevents them from breaking when hot food is added. *Leave lids and bands at room temperature for easy handling.*
4. Prepare for pressure canning. Fill the pressure canner with 2 to 3 inches of water. Place over medium-high heat. Bring to a simmer. Keep water at a simmer until jars are filled and placed in canner. Follow manufacturer's instructions for usage instructions.
5. Remove hot jar from hot water, using a Jar Lifter, emptying water inside jar. Fill jar one at a time with prepared food using a wide-mouthed funnel leaving headspace recommended in recipe. Remove air bubbles, if stated in recipe, by sliding a rubber spatula between the jar and food to release trapped air and ensure proper headspace during processing. Repeat around jar 2 to 3 times.
6. Clean rim and threads of the jar using a clean, damp cloth to remove any food residue. Center lid on jar allowing sealing compound to come in contact with the jar rim. Apply band and adjust until fit is fingertip tight. Place filled jars in canner until recipe is used or canner is full. Check that

PART VI: OPERATION AND MAINTENANCE

water level is about 2 to 3 inches high or that recommended in manufacturer's manual.

7. Lock the pressure canner lid in place, leaving vent pipe open. Adjust heat to medium-high. Allow steam to escape through vent pipe. Once there is a steady stream of steam escaping, vent for 10 minutes to ensure there is no air (only steam) left in canner. Close vent using weight or method described for your canner. Gradually adjust heat to achieve and maintain recommended pounds of pressure.

8. Process canning jars at the recommended pounds pressure for the processing time indicated in tested preserving recipe. Cool pressure canner by removing from heat. Do not remove the weighted gauge. Let canner stand undisturbed until pressure returns to zero naturally. Follow manufacturer's instructions. Wait 10 minutes. Remove weight and unlock lid, tilting away from yourself. Wait 10 more minutes to allow jars to begin to cool.

9. Remove jars from pressure canner and set upright on a towel to prevent jar breakage that can occur from temperature differences. Leave jars undisturbed for 12 to 24 hours. Bands should not be retightened as this may interfere with the sealing process.

10. Check lids for seals. Lids should not flex up and down when center is pressed. Remove bands. Try to lift lids off with your fingertips. If the lid cannot be lifted off, the lid has a good seal. If a lid does not seal within 24 hours, the product can be immediately refrigerated. Clean canning jars and lids. Label and store in a cool, dry, dark place up to 1 year.

After many years of research, it was determined that preheating lids is no longer necessary. The sealing compound used for our home canning lids performs equally well at room temperature as it does pre-heated in simmering water (180 degrees Fahrenheit). Simply wash lids in hot, soapy water, dry, and set aside until needed.

Other food preservation methods

While canning reaps the most benefits overall of the food preservation methods, it's important to note two other healthy methods of preserving food.

Freezing Food

Freezing is likely the healthiest method of food preservation because it preserves more nutrients than heating methods. Freezing foods is the art of preparing, packaging, and freezing foods at their peak of freshness. You can freeze most fresh fruits and vegetables, meats and fish, along with baked items and clear soups and casseroles. The keys to freezing food are to make sure it's absolutely fresh, that you freeze it as quickly as possible, and that you keep it at a proper frozen temperature (0 degrees).

Properly packaging food in freezer paper or freezer containers prevents any deterioration in its quality. Damage occurs when your food comes in contact with the dry air of a freezer. Although freezer-damaged food won't hurt you, it does make the food taste bad.

To avoid freezer burn:
- Reduce exposure to air by wrapping your food tightly
- Avoid fluctuating temperatures by keeping the freezer closed if possible
- Don't overfill your freezer, which could reduce air circulation and speed freezer damage

Drying food

Drying is the oldest method known for preserving food. When you dry food, you expose the food to a temperature that's high enough to remove the moisture but low enough that it doesn't cook. Good air circulation assists in evenly drying the food.

An electric dehydrator is the best and most efficient unit for drying, or dehydrating, food. Today's units include a thermostat and fan to help regulate temperatures much better. You can also dry food in your oven or by using the heat of the sun, but the process will take longer and produce inferior results to food dried in a dehydrator.

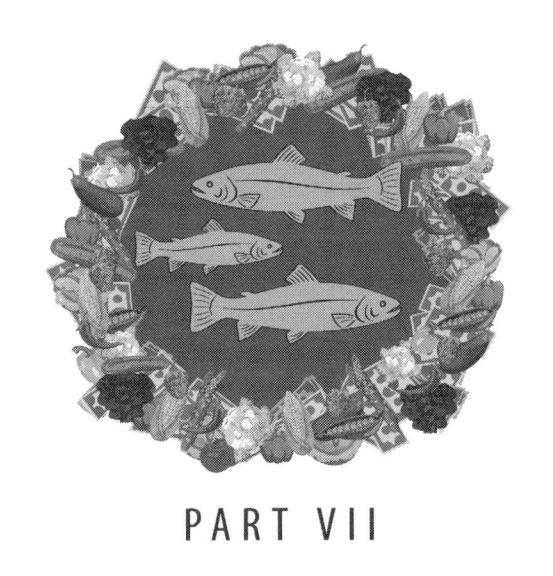

Making Money and Earning a Profit from Aquaponics

CHAPTER 35

Cost-Benefit Analysis (Aquaponic Economics)

This chapter is intended to describe the costs and benefits of a small-scale aquaponic system. The information in the tables is meant to provide the reader with an understanding of the expenses necessary to build and run an aquaponic system, as well as the expected production and returns in the first year. The aquaponic system referenced is based upon the Intermediate Bulk Container (IBC) unit shown below. Design plans, instructions, materials-needed list, and construction details for this aquaponic system are provided in Chapter 29.

Table 44 summarizes the total cost of materials for the initial installation (capital investment) for a small-scale media bed aquaponic system (refer to the design chapters of this book for construction details). Table 45 details all the yearly running costs involved. The details of the running cost calculations can be found in the notes section of the table. Table 46 details the expected production of vegetables and fish in one year. The forth table summarizes the costs, returns, and shows the total profit on the initial investment as well as the payback period.

FIGURE 258. Media-Bed (flood-and-drain) system with sump using IBC containers.

PART VII: MAKING MONEY AND EARNING A PROFIT FROM AQUAPONICS

TABLE 44. **Total capital cost for a small-scale media bed aquaponic system**
- Grow Bed: 10 square feet (3 square meters)
- Fish Tank: 264 gallon (1,000 liters)

ITEM DESCRIPTION	COST (USD, YEAR 2016)
IBC Tank (fish tank) — Used 275 gallon	$145
Electrical Equipment: Water Pump, Air Pump, and Misc. Accessories	$115
Media Beds	$150
Media (mix of pea gravel and expanded clay pebbles)	$100
Miscellaneous Items: Fish Net, Water Test Testing Equipment, etc.	$150
Plumbing: Pipe, Pipe Fittings, Plumber's Tape (Teflon Tape)	$70
Total Capital Cost	**$730**

TABLE 45. **Total Monthly Operating Cost (Monthly Average)**

MONTHLY SYSTEM INPUTS	COST (USD, YEAR 2016)
Plants	$3
Fish	$6
Electricity	$3
Water	$1
Fish Feed	$12
Miscellaneous	$4
Total Monthly Operating Cost	**$29**
Annual Operating Cost	**$348**

TABLE 46. **Expected Yearly Production of Vegetables and Fish from A small-scale media bed aquaponic system**
- Grow Bed: 10 square feet (3 square meters)
- Fish Tank: 264 gallon (1,000 liters)

OUTPUT	PRODUCTION (QUANTITY)	AVERAGE UNIT MARKET VALUE (USD, YEAR 2016)	TOTAL MARKET VALUE (USD, YEAR 2016)
Lettuce	360 heads	$1.48 each	$532
Tomatoes	119 lbs. (54 kg)	$1.60 lb.	$190
Fish (Tilapia)	66 lbs. (30 kg)	$7.00 lb	$462
		Total Annual Production Value	**$1,184**

TABLE 47. **Annual cost–benefit results**

ITEM	TOTAL
Initial Capital Cost (total from Table 44 above)	$730
Annual Operating Cost (total from Table 45 above)	$348
Annual Production Value (total from Table 46 above)	$1,184
Total Annual Profit or Value (annual proceeds minus annual cost)	$836
Length of Time to Breakeven (months to breakeven after capital investment)	11

It should be noted that the figures given in the tables are only intended as guidelines. It is difficult to provide accurate figures, particularly regarding production yields and their values as many production and financial factors may influence them; temperatures, seasons, fish type, fish feed quality and percentage protein, markets prices, etc. must all be considered. Lastly, the following chapters will examine and provide more business type specifics, in regards to profiting from aquaponics.

Calculation Assumptions

- All calculations are based on a small-scale media bed aquaponic system with ten square feet (three square meters) of growing space and a 264 gallon (1,000 litters) fish tank.
- The objective of this aquaponic system is to provide domestic food consumption only and not for small-scale income-generating production. The financial benefits can vary and might be larger than the figures shown below, if the operator selects more profitable crops to grow. As the focus is on small-scale aquaponics for domestic food consumption, two crops have been considered in the calculations as they better reflect the production patterns of users growing food for consumption only; one leafy green (lettuce) and one fruiting vegetable (tomato).
- Yield data is obtained from a continuous production of 12 months, feeding the fish with good quality 32 percent protein feed daily in system water temperatures of 73–79 °F (23–26 °C) throughout the year.
- The system has a constant standing fish biomass of 22–44 pounds (10–20 kilograms).
- The fish cultured are Tilapias. They are fed 5.3 ounces (150 g) of feed per day. The stocking weight of juvenile fish is 1.8 ounces (50 g); the expected harvest weight is 17.6 ounces (500 g) per fish in six to eight months.
- The average yields for amateur growers have been considered in the calculations: 20 heads of lettuce per square metre per month, and seven pounds (3 kg) of tomatoes per square yard (square metre) per month.

Notes:

- This analysis is based on a staggered production of fish in an established aquaponic system. The expected production is lower from a newly established system stocked only with juvenile fish of the same age. For new systems, it is suggested that fingerlings be stocked in greater numbers in order to supply enough nutrients to plants. In this case, harvesting of the first fish (Tilapia) can start from the third or fourth month onward, with fish at 5.3–8.8 ounces (150–250 g), in order to maintain a steady biomass.

This is a very conservative cost-benefit analysis. Unit costs in Table 46 are averages based on 2016 USA prices. Using organic produce values and raising a different species of fish would significantly increase total production value (profit). Furthermore, this analysis is based upon a 'small' aquaponic system. A larger system would have larger start-up and operational cost, but the production value (profit) would increase exponentially. Lastly, it is difficult to put a value on having your own readily available food security supply, as well as the assurance that you and your family are truly eating the very best organic healthy food available.

CHAPTER 36

Creating a Profitable Aquaponics Business

Profiting from Aquaponics

It is a relatively easy transition an aquaponics operation—initially set-up only to provide you with healthy food—into a very profitable venture. Not only can you provide your family with tasty, healthy, safe fish as well as vegetables, but you can also sell the excess crops at a respectable profit margin. Fresh organic products are in great demand, and will be even more so in the future as people become better educated about all of the unhealthy ingredients that are being added into our food. Furthermore, as will be shared below, there are additional ways you can earn revenue from your aquaponics system.

In comparison to the vast majority of businesses, an aquaponics start-up (and maintenance) is low cost, space requirement is minimal, and the manual labor requirements are marginal. Furthermore, productivity is far greater than a traditional organic farm in comparison of space being used for each.

This chapter will cover some general guidelines and other fundamental principles which will help you be more successful. The Appendix section of this book provides additional resources that will also facilitate your progression into operating a profitable aquaponics business.

Know Your Market

The first step is to do some research on the types of herbs or vegetables that are in demand at your local market. Preferably, you would also want those that face less competition and sell at the highest prices.

Growing crops that are out of season can generate an even higher revenue. In some cases, there can be greater costs in terms of heating, additional lighting, and insulation for growing some off-season crops... but not always. In some areas, growing ornamental fish that can be sold as pets, such as goldfish or Koi, can generate revenue that exceeds food fish.

Refer to 'Chapter 10: Plants/Crops — Keys to Success' and 'Chapter 11: Fish — Everything You Need to Know' for product line ideas. Marketing and selling your product will be covered extensively in Chapters 34-37.

Know Your Customer

In addition to knowing what is in demand in your local market, it is important to determine ways in which to market and sell your product. You can sell your produce directly to the customers through a variety of means, such as online, through your network of people, door-to-door, a roadside stand, organic food groups, vegan and vegetarian clubs, at a farmers market, or even co-ops. Selling your produce direct to the consumer has the highest profit potential.

However, you can also sell your produce to other businesses such as local grocers, fish mongers, a farmer's market vendor, or a local pet store (if you are raising Koi or goldfish). Although, selling your product via these methods often means selling them at a lower price than what you could charge if selling directly to the consumer. On the positive side, it can be a more convenient approach, as it generally takes less time. Going wholesale usually provides the opportunity to sell your entire product in bulk, which equates to a more efficient means of getting paid following harvest.

Bartering

Selling your product is only one method to come out ahead. Whenever one has a product that is in great demand, many options become available that can be quite advantages; some even more than cash. Bartering is one option that can be done to your benefit. A matter of fact, bartering your product for other products or services has the potential to increase your return even further than what may be obtained via a cash sale. For instance, carpet cleaning, auto repair, beauty salon, landscaping, yard work, painting, and other services providers charge a certain amount for their expertise. Often these same services can be acquired for much less in a barter trade for produce than what it would cost in cash. In addition, bartering with other gardeners can help you acquire many other items that you may not be growing in your system (i.e. fruit tree produce, radishes, onions, potatoes, nuts, etc.). You can also barter with bakeries for healthy grain products. Another wonderful benefit of bartering is that taxation can sometimes be avoided. Bartering eliminates the middleman, banker, and other side businesses, providing an overall better return for your product.

Additional Revenue Sources

Aquaponics is a rapidly growing industry. Even though aquaponics is still relatively uncommon to the public, more and more people are becoming aware of it all of the time. There is an ever increasing segment of the population that is interested in getting into aquaponics, but have little knowledge, don't know how to get started, or are just too intimidated to try it on their own. As a result, they are seeking ways to learn about it, and even get some hands-on experience before moving forward. By offering training, you can acquire additional revenue and obtain some very enthusiastic free labor. You can even help others develop their own aquaponic system at a reasonable consultant fee.

Another revenue-generating stream that is worth considering is that of giving paid tours. You will find that many people will want to see your operation. You might as well earn some income from it, as it does take time. You can even market this option to various groups and organizations such as garden clubs, Boy and Girl Scouts, schools, vegan and vegetarian clubs, environmental groups, foodie groups, and the community at-large. You can charge a set fee for the tours, or give them for free and write the cost off on your taxes as an operational expense or charitable donation.

However, before you have outsiders coming into your operation, it is prudent to review your current insurance coverage, or upgrade it, to ensure you are covered in case of an accident. Even though you are properly covered, it is still your responsibility to make sure that the area is safe, especially so when having inexperienced people around your system. Take extra precautions that there are no sharp or protruding devices, no slipping hazards, eliminate all tripping hazards, ensure electrical devices are properly addressed, and make sure that there is no possibility of an electric current coming into contact with water.

Most utility companies will provide you credit or a reimbursement check if you have an environmentally friendly alternative energy system, such as solar or wind, that creates more energy back into the grid than what is used. Getting a farm credit on your income tax and reduced rates through your utility provider is another means in which you can increase your profit margin.

Business Organization Types

You may have heard people talk about sole proprietorship, partnerships, 'C—Corporation, "Sub-S" corporations, LLCs and PCs and their relative merits. You will need to decide which form of business is right for you based on criteria such as:

- Ease and cost of formation
- Administration and record keeping
- Taxes
- Liability

The following is a brief description of each form:

Sole Proprietorship

A sole proprietorship is a business run by an individual. The owner is the business who controls all of the profits and losses of the business. The owner also has full authority and all the liability from the business operations. Business taxes are paid by the owner through his or her personal income tax return.

Partnership

A partnership is a business which operates like a sole proprietorship, but with several individuals running it. The partners share the profits/losses, have control and liability for business operations. Partnership taxes are paid by the partners on their personal tax returns, in proportion to their share of ownership.

Corporation (or C-Corporation)

A corporation is a business which is set up as a separate legal entity from its owners. The Board of Directors makes operational decisions. Owners are protected (shielded) from liabilities of the corporation, and the corporation pays corporate income taxes.

S-Corporation (or Subchapter-S Corporation)

A small business corporation may elect to be classified as a S-Corporation to have the liability protection of corporate status, but taxed at individual rates.

Limited Liability Company (LLC)

A limited liability company is formed by "members" whose liability is limited to their investment. An LLC is often used in place of partnerships to limit liability, while having the option of being taxed through the personal tax returns of the members. Here is more information on LLCs vs. Corporations.

Determining the Best Business Structure for You

Understanding the different business types is interesting, but it doesn't help you decide which type to select. Here are some factors to consider:

- **Are you just starting your business?** If so, you may want to select a simple setup, like a sole proprietorship. You can always change later.
- **Are you working alone, with no employees?** In this case, a sole proprietorship or single-member LLC might be the best option for you.
- **Do you sell products which could be misused or subject you to liability?** You probably want a corporate structure to shield your personal assets from the liabilities of the corporation.
- **Are you working with other professionals in a group?** A partnership or a multiple-member LLC might give you protection against liabilities and allow all members of the group to have a say in the day-to-day running of the business.
- **Do you have silent partners or investors who don't participate in running the business?** A

limited partnership or a corporation allows shareholders or limited partners.
- **Do you already have a corporation but you want lower taxes and simpler operations?** You may be able to elect an S corporation status.

These are just some of the considerations in selecting a business type. Before you make a decision, talk to several advisors, including an attorney and a tax expert. The Small Business Association (SBA) has a wealth of information, and advisors, that can be tapped into at no cost.

Non-Profit and Not-For-Profit Organizations

Another option you may want to consider is to create a non-profit or not-for-profit aquaponics operation. Established properly, you can still earn income (salary), grow your own food, and revenue can be put back into your system. This type of arrangement can also provide wonderful opportunities to make a positive contribution to the community and those in need.

Laws and Government Regulations

Before making a large investment towards a commercial enterprise, it is wise to make sure that your desired endeavor will be in compliance with local zoning ordinances, home owner's association rules (if applicable), and various government regulations. Your jurisdictional public agency planner and/or local small business assistance organization can serve as very helpful resources in regards to navigating the system. Most libraries have a resources called 'Ask a Librarian' which serve as wonderful no cost method for gathering information. You can find some very helpful resources online which, at no cost, are tremendously helpful, as well. In addition, YouTube has many instructional videos and seminars.

Becoming a 'Certified Organic' Business

It is important to think about becoming a 'Certified Organic' business. However, it can take up to three years to get certified, so keep this in mind when preparing your business plan. Becoming organic has some advantages and disadvantages.

"Organic" is a label regulated by the US Department of Agriculture. Farmers who want to call their products "organic" have to get certified by a USDA-accredited certification agency. The only exception is the very small grower, with less than $5,000 in farm sales per year. However, even those small organic growers who are exempt from certification must follow rules established by the National Organic Program. A business cannot be certified until it can be proven that the land or other growing media has been free of prohibited substances for three years.

The rules for being organic are comprehensive and cover everything from the seed you buy to the boxes you use to take your crops to market. The organic standards require you to use certified organic seeds and plants, unless you can prove they are not available for the varieties you grow. They control all production inputs, including growing media, fertilizers, pest control products, and postharvest products.

FIGURE 259.

To obtain and keep certification, extensive records must be maintained of everything you do to in your operation, every detail pertainig to your crops, and of every product you purchase.

Although some farmers chafe under these regulations, many consider it something that needs to be done anyway (i.e. good record keeping, and being congizant of what is being introduced into the system). For the most part, organic standards are consistent with best management practices recommended for operators who are concerned about the environment and human health. Nevertheless, a few of the requirements are somewhat cumbersome in regards to the type of beauruecracratic time consuming recordkeeping that is necessary to mainatin the label.

The cost of certification is another obstacle. Being certified organic can cost anywhere from a few hundred dollars to several thousand dollars per year. There are dozens of accredited certification agencies. Each agency or program has its own fee structure. The largest cost is for the inspection, which is required annually.

There have also been some federal funds made available in recent years to help growers pay for certification. The program will pay up to 70 percent of the cost of certification, to a maximum of $500 per year. Check with certification agencies and/or you conduct an internet search for "National Organic Cost-Share Program" to learn more. The best place to familiarize yourself with the process of organic certification is on the USDA's National Organic Program website: www.ams.usda.gov/nop. There you will find the standards themselves, all 554 pages of them. You will also find what is called "the National List" of products you can use in organic production. State contacts for cost-share information is also listed.

The USDA's National Organic Program website will also provide you with a list of certifying agencies. At the time of this writing, there are 50 such agnecies in the United States. The USDA lists them by the states where they are headquartered, not by the states where they conduct certifications. If you don't see your state on the list, check with organic farmers in your area to find out what agencies are serving your state. Some certifying agents say they are national, but be sure to comparison-shop on fees.

One of the effects of the federalization of organics is that regional or state-based groups that used to certify farms and provide educational programs are no longer allowed to do both. Therefore, some have divided into two separate legal entities, whereas others have chosen to be one or the other to their members. Another excellent resource for information on becoming certified will be these smaller membership groups.

Other Labeling Options

Many farmer and aquaponic operators don't see the need for organic certification, don't care to deal with the government bureaucratic hassle, and/or don't want to incur the cost for such, even if they use organic methods of production. Others have philosophical objections to participating in a federal program. Whatever their reasons, noncertified farmers have come up with alternative ways to describe and label their products, such as "Beyond Organic" or "Ecologically Grown." These labels may require specific legal requirements or growing conditions, so it is important to know all of the detials in advance of moving forward.

One program is referred to as Certified Naturally Grown (CNG). It is a web-based, low-cost certification system in which the grower agrees to follow the National Organic Program rules. Inspections are done locally, by other farmers, extension agents, and/or customers. Although it is basically an honor system, each year a certain percentage of CNG farms are randomly selected for pesticide residue testing of produce, at no cost to the farmer. Currently there are about 750 farms and apiaries that are certified under the program. An annual financial contribution is required for certification. Cost ranges from $75 to $200 per year. For more information, visit www.naturallygrown.org.

Another sustainability certification program that does not require compliance with national organic standards is the Food Alliance. Standards for crop production allow limited pesticide use in the context of an integrated pest management program. See foodalliance.org.

In addition, many local groups have their own labeling programs, such as the "Buy Fresh, Buy Local" campaign, and some environmental organizations will certify farms that comply with certain standards (for example, "Salmon Safe"). These "eco-label" programs are proliferating despite concerns that they are confusing to consumers.

Consider whether a label will be helpful in explaining your practices, and whether or not the cost and hassle are worth the anticipated benefits. If so, choose the program that fits your needs.

Operating a Successful Commercial Aquaponics Operation

In any business, profit is a primary consideration. If you cannot make a profit, then you cannot stay in business. Some declare that aquaponics cannot be a profitable business. You will also hear the same thing about the carpet cleaning and restaurants, due to the numbers that go out of business within five years. There are always naysayers. However, there are many profitable carpet cleaning companies and successful restaurants. Contrary to those ill-informed opinions and numbers, the facts show that aquaponics can be a quite successful and profitable. There are many profitable aquaponic businesses throughout the world, and in just about every state of the union.

A University of Hawaii study conducted in 2013 found that commercial aquaponic businesses averaged a gross annual profit of $43,065 per acre, or $86,130 per aquaponic farm. The study also revealed the following:

- Commercial scale aquaponics is economically feasible.
- Aquaponics can be more profitable than terrestrial agriculture and aquaculture.
- Vegetable production is the driving force of economic success.
- Price premium is possible for locally produced aquaponic vegetables.

Virginia Tech created a small aquaponics operation in 2011 for the sole purpose of determining whether or not aquaponics could be a profitable business. There summary revealed the following:

- Their model proved economically viable.
- If scaled up to a larger commercial facility, it would be a lucrative investment.
- Fresh and local fish, fruits, and vegetables can be shipped worldwide.
- Aquaponic systems are resource efficient and will help in water and fuel conservation efforts. Local systems and reduced resources will help lead to decreased pollution and a sustainable future for the food industry.

The University of the Virgin Islands created and evaluated three commercial-scale aquaponic operation for the production of Tilapia and leaf lettuce on a sustainable basis at commercial levels. Each commercial operation was sized differently (small, medium, and larger commercial operations). Their conclusion was that an aquaponic farm can be profitable. Each farm size had positive returns but at different rate of return for the investment. The smallest farm has the lowest return, but was still profitable. Higher and more acceptable returns were achieved with the medium and larger size aquaponics farms; the larger the facility the higher the revenue.

Like any other business venture, to be successful at aquaponics, it is prudent to invest your time researching and planning before putting your money into it. You need to conduct an extensive study into all aspects of the aquaponics business to determine which methods would best suit your circumstances and evaluate them in regards to profitability. I commend you for reading this book, as it shows that you are indeed doing your research and due diligence.

In addition to planning, to ensure the best opportunity for success in commercial aquaponics, you need to know what you are doing. It is wise to start small, and build from your knowledge base. Later phases can be planned from the beginning (with adjustments to the plan as you gain experience), but starting small will enable you to absorb errors without going broke, and provide you with opportunities to maximize effectiveness and efficiency.

Starting small will also enable you the opportunity to get a real feel of everything and determine how large you really want to grow. A modest part-time *aquaponics system* can produce up to 110 pounds of fresh fish like Tilapia every six months, and 220 pounds of vegetable yield. This would amount to a satisfactory profit. Or you could use this as a launching pad to grow into a large scale corporation with employees. Until you gain some viable experience, it will be hard to know exactly which business model will best suit your talents, resources, and preferences.

To help you move in the right direction, the following are some questions that will help you get additional insight, and a foundation for moving forward. Having good, solid answers to these questions will be essential if you are seeking funding for your endeavor. Even if you are not seeking funding, in order to achieve optimal success, it is important that you have reliable answers to these questions.

1. Who are the people who will buy what you are producing?
2. How much will they purchase?
3. What are they currently paying for the food they purchase?
4. What will you be able to charge them when your products become available?
5. Work with a market research company if necessary to establish the 'who, where, and how much' that is out there for your food products.
6. Identify your competition.
7. What product features do you have that they don't?
8. How will you take advantage of your unique offering for locally grown 'organic' food?
9. What marketing plan will you follow?
10. What are your expectations to grow your business?

That is the marketing component; now for addressing the actual planning, building, and property you intend to use for your operations.

11. What improvements will be necessary to make it meet codes and be efficient for you?
12. What structure will you use, what will it cost, and what utilities will be needed.
13. What will be your financial resources to accomplish your objectives? This is usually the most critical step in establishing any business, so don't underestimate the time for this.
14. Analyze your expected operations plan and establish a budget for what it will take to staff, maintain, harvest, and distribute your products.
15. What management time and costs will it take to sustain and account for these activities?
16. What taxes, insurance, utility costs, marketing costs, and facility's maintenance costs will you expect?
17. What financial performance will you require to satisfy your investors?
18. How long will it take to achieve this performance?
19. What will be your cash flow projections that you will need to cover?

An Entrepreneur's Guide to the Big Issues

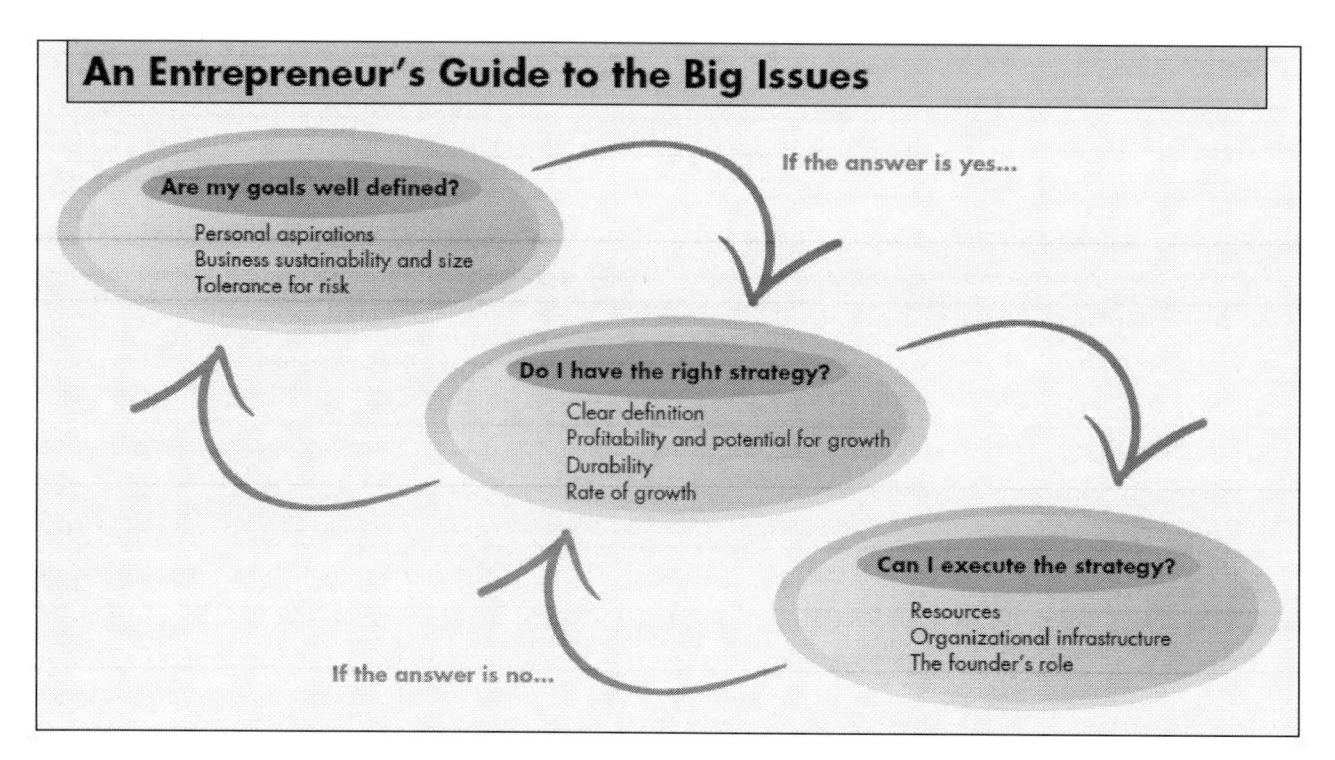

Are my goals well defined?
- Personal aspirations
- Business sustainability and size
- Tolerance for risk

If the answer is yes...

Do I have the right strategy?
- Clear definition
- Profitability and potential for growth
- Durability
- Rate of growth

Can I execute the strategy?
- Resources
- Organizational infrastructure
- The founder's role

If the answer is no...

FIGURE 260. *An entrepreneur's guide.*

Once all of this research completed, it needs to be included in a set of documents that begins with an "Executive Summary" which makes the case for continuing with the project. It should also include:

- A schedule of activities with duration, costs, and expectations of each step.
- Your projected financial performance summarized in a form that can be easily understood by potential investors. These are questions that any investor will ask before providing even a penny of funding. Much of the data generated will be estimated based on experience and research. Hiring the right people, or paying for answers these questions, is also a viable option.
- For most people (namely investors) many of the questions above are about things that they have no experience with, and that is where an expert can help. Be over prepared to answer any questions that are posed. By presenting them with well researched and well documented information, it makes it much easier for them to approve the use of their money.

The Aquaponics Business

Aquaponics comes in a wide variety of designs, sizes, and complexities. The type of aquaponic system (i.e. vertical garden system, nutrient film system, deep water raft system, flood and drain system) must also fit your goals. In addition to desired size and type of operation, other issues must be considered, such as location/climate, temperature control, labor needs, and operational cost. Another factor that comes into play is whether the system is a Do-It-Yourself build, factory kit, or a combination of the two. As you move forward, experience will be one of your best teachers, but never stop doing your research. Try to learn as much as possible from others, and continue to work your plan.

Five-Phases of Building a Successful Commercial Aquaponics Operation

Phase 1: Conceptual
- **Market Study and Analysis** — who, where, what, how much?
- **Needs Analysis** — who is providing these products now?
- **Opportunities for Phased Growth** — start small and grow.
- **Preliminary Business Plan Model** — the basis for your enterprise.

Phase 2: Planning
- **State and Local Requirements** — zoning, business, agricultural requirements.
- **Utilities** — electric, water, gas.
- **Property Investigation** — undeveloped rural, urban, existing facilities, rentals.
- **Engineering Assistance** — site work and zoning assistance.
- **Concept** — analysis of how big you want to grow, what to grow, how to operate.
- **Finalize Business Plan** — the most important thing you can do for suc?cess.
- **Establish Financial Resources** — how much and when do you need it.

Phase 3: Site Preparation
- **Acquisition** — buy, lease, modify.
- **Permitting** — city, county, state.
- **Engineering** — prepare plans and specifications as needed for site.
- **Construction** — prepare property for your facility.
- **Greenhouse** — finalization of design selection and erection of greenhouse.

Phase 4: Initial Operations
- **Training** — onsite staff training onsite and online for operating system and growing food.
- **Initial Planting** — designed for your specific markets.
- **Fish** — species and size selected.
- **Tuning and Insuring Systems Functions** — initial shake down to insure success.

Phase 5: Ongoing operations
- **Feeding and Monitoring** — feed the fish and monitoring the plants; pretty straightforward.
- **Harvest and Distribution** — harvest, package and deliver.
- **Replace Products** — start new seedlings, replacing harvested fish.
- **Business Activities** — accounting, sales, human resources, marketing.
- **Monitoring** — temperature, water quality and levels, light quality, equipment.
- **Facilities Maintenance** — greenhouse, grounds, lights, fans, heaters, shade cloth, media bed, plumbing, and tanks.
- **Celebrate Your Success!**

Production Planning

One of the most challenging aspects in the beginning as an aquaponics business operator is to plan and schedule your production. You have to consider a multitude of factors, such as which species of plants to grow, the best varieties of each plant species, days to maturity of each variety, how you are going to manage your fish harvest (one tank at a time or continuously), harvest time, marketing and selling periods, and the amount of time you will be able to store your product.

The key to managing these activities is to maintain a calendar to choreograph this intricate dance. The first things to consider are the specifics as to what you will grow and where you will sell them. After you have a list of plants and fish suitable to your climate and your specific situation, the next step is to determine realistic production expectations.

The most common mistake new aquaponic business operators make is not growing enough product. There is a tendency to scale up production from the family's current aquaponic system without any concrete idea of how much a planned expansion will yield or how much money it will generate. The result of this approach can be income that is far below expectations. A more logical approach is to set a revenue goal and then work backward to determine how much you need to produce in order to generate enough sales to meet your goal. Then the system can be properly expanded in size to meet that objective.

Predicting Production Yields

Yield can vary considerably by system set-up parameters (greenhouse vs. not greenhouse, grow lights vs. no grow lights, etc.), by plant species variety, geographic location, and climatic conditions. Still, you need to find a ballpark or average figure for planning purposes based upon the best information that you have available.

Since you will be compiling a lot of information, it is best to create a spreadsheet to hold all of the critical information in one location. For those that may not be computer savvy, a notebook will suffice. Your county extension office, the internet, and farming organizations/associations can provide you with some traditional numbers. Although a properly managed aquaponics operation will produce considerably more product than traditional farming, using production quantities based upon traditional methods will be a reliable and conservative approach for planning purposes until you gain the experience and a good understanding of how much your system is capable of producing.

Production Enterprise Budget

Another tool that is useful in planning your production process is a budget. In agriculture, this is referred to as an enterprise budget. An enterprise budget is a projection or estimate of the costs and returns of producing a product (enterprise).

Enterprise budgets are usually prepared on a "per unit" basis. For example, a crop enterprise budget is usually computed for an acre of production. Costs and returns for the entire crop can be calculated by multiplying the budget coefficients by the number of acres that will be produced.

Enterprise budgets are very useful tools. Some of the uses of an enterprise budget for your business are to project the:
- Most viable enterprises.
- Most profitable combination of enterprises.
- Expected production.
- Expected profits.
- Expected capital and cash needs.
- Expected labor needs.
- Expected levels of purchased inputs.
- Expected feed needs.
- Expected machinery or facility needs.
- Other expenditures.

Aquaponic operators may develop budgets for many different crops, in addition to their fish budget, depending on the number of crops within their rotation. An internet search will result in many 'go-by' enterprise budget examples as well as readymade templates that can be downloaded for your use. The author will post an aquaponic enterprise budget on the forthcoming website at **www.FarmYourSpace.com**. The enterprise budgets that can be acquired at present come in both Microsoft Word and Microsoft Excel formats, depending upon your preference. For an aquaponic operation, the best approach is to acquire an enterprise budget from both the aquaculture industry and small-scale farming sector; combine them into one budget for easier access. Tables 48 and 49 are two examples of enterprise budgets.

Be sure to look carefully at available enterprise budgets before drawing any conclusions about a crop. Besides being calculated per acre, they may assume mechanical cultivation, in which plants are spaced farther apart than they would be in a hand-cultivated aquaponic operation. They also may include routine pesticide applications, which you won't be doing if you are using bio-rational pest controls (environmentally friendly and organic pesticide).

Many other variables will likely be different for your operation as well. A good crop enterprise budget should explain the production parameters. Also, check the prices on which the return-per-acre is based. Many of the currently available enterprise budgets use wholesale prices rather than the retail prices in which you may be selling your product.

Crop enterprise budgets are not a guideline to growing a specific crop in a aquaponic setting. However, they can provide you clues in regards to what can be expected. Furthermore, they give you a format that you can use for your own record-keeping. Get acquainted now with how enterprise budgets are generated, because you will

Fish Enterprise Budget 'Go-By' Example

TABLE 48. **Bass Fish Production Cost/Profit**

INVENTORY & INPUT USE	
Beginning number of fish	250
Ending number of fish	212
Beginning biomass (grams of fish, 250 grams at 1 gram each)	250
Beginning biomass (lbs. of fish)	0.6
Ending biomass (lbs. of fish)	265
Max. standing biomass (lbs./gal.)	0.12
Feed used, lbs.	718
Kwh used	1,102

COSTS	
Fingerlings (each)	$1.56
Fingerlings (total quantity)	250
Fingerlings (total cost)	$390.00
Feed	$312.63
Electricity	$208.42
Total of above costs for this unit	$911.05
Cumulative cost per lb.	$3.44

NOVEMBER 2016 MARKET VALUE	
Market value for bass at current retail prices (price/lb)	$13.54

EST. NET PROFIT/2,500 GALLON FISH TANK	
Ending biomass (lbs. of fish)	265
Cumulative cost per lb.	$3.44
Market value for bass at current retail prices (price/lb)	$13.54
Net profit per lb.	$10.10
Total net profit/tank	**$2,676.50**

Multiple Vegtable Product Enterprise Budget 'Go-By' Example

TABLE 49. **Tunnel Greenhouse Operation**
- Tunnel Greenhouse Area 30'x72' = 2,160 sq. ft
- Utilization: 84%

RECEIPTS	YIELD	SQ. FT.	$/LB.	TOTAL	YIELD PER SQ. FT.	GROSS PER SQ. FT.
Cucumbers	567.0	454	$2.00	$1,134.00	1.25	$2.50
Eggplant	204.1	45	$2.00	$408.24	4.50	$9.00
Greens	166.9	363	$7.00	$1,168.47	0.46	$3.22
Herbs	14.5	9	$16.00	$232.24	1.60	$25.60
Lettuce	208.7	181	$7.00	$1,460.59	1.15	$8.05
Peppers—Bell	290.3	181	$2.00	$580.61	1.60	$3.20
Tomatoes—Slicers	1,315.4	454	$2.50	$3,288.60	2.90	$7.25
Tomatoes—Grape	342.9	127	$4.00	$1,371.69	2.70	$10.80
Total Receipts		**1,814**		**$9,644.44**		**$5.32**

ANNUAL EXPENSES	TOTAL
Seeds/Transplants	$135.00
Fertilizers	$108.00
Miscellaneous Supplies	$125.00
Water	$86.40
Water Test	$17.00
Irrigation Supplies	$122.00
Total Annual Expenses	**$593.40**

MARCH — SEPTEMBER

LABOR COSTS	HOURS	$/UNIT	TOTAL
Bed Preparation	17.00	$12.00	$204.00
General Maintenance	23.00	$12.00	$276.00
Planting	10.50	$12.00	$126.00
Pest Management	0.00	$12.00	$0.00
Harvest	48.80	$12.00	$585.60
Total Hours	**99.30**		**$1,191.60**

OWNERSHIP COSTS	ANNUAL
Depreciation—Tunnel	$875.00
Depreciation—Plastic Cover	$113.40
Total Ownership	**$988.40**

TOTAL COSTS	
Tunnel	$2,773.40
Per Square Foot	$1.28
Annual Return Over Total Costs	
Tunnel	$6,871.04
Per Square Foot	$3.18

need to create your own if you desire to have an aquaponics business generating an optimal profit margin

Production Planning for Success

Your production plan should also take into consideration the seasons, daylight hours, supplemental lighting needs, temperature control cost, varieties that work best for you, crop yields, plant cost, fish food cost, etc. Every year you will tweak your plans, based upon previous experience, to meet demand, to maximize efficiency, and to increase revenue. The only way to benefit from your experience is to keep records of everything you do. You may think that you will remember when you planted which variety and when you started to harvest it. It is impossible to remember all the details, so you have to keep and maintain good records as you go.

Many aquaponic operators prefer to use crop sheets in a three-ring binder. Crop production information is alphabetized, with a separate sheet for each variety and each succession planting. Data to be captured includes the source of the seed, the amount purchased, seeding date, planting out date, number of plants set out, spacing, and any special treatment such as supplemental lighting.

The crop sheet also is used to record harvest data: date of first and last harvest, amount picked, amount not saleable, amount sold, and price. To figure your true overall cost, a record of the time spent on each task must be maintained. That way, you have all the data you need to compare crops on an equal footing, which will help you determine which crops make them the most money, considering all aspects of the operation, including labor requirements.

Although many operators still prefer using pencil and paper for this information, which can later be transferred to a computer program, many now use smartphones or tablets to record the information in the field and electronically transfer it to the office computer. Many operators create their own databases and spreadsheets for keeping records. Others prefer to purchase software programs designed specifically for market farms, or download a template from the internet.

Wherever you keep your data, whether it's on paper, your phone, your desktop computer, or the internet, getting these records into a database is valuable, because you can then sort and select records in numerous useful ways.

Sales Records

Although production records are very important, they are only a part of the records that need to be maintained. You also need to keep sales records. Again, you can do it manually, by recording marketable yield when you pick a crop, the price you got for it, and any unsold amounts. Or you can keep these sales records in a software program such as QuickBooks or Quicken. With QuickBooks, the easiest method is to create invoices for every sale, with specific crops in your Accounts list. You also can do the same on a deposit record if you handwrite your invoices and can correlate them with the payment later. In Quicken, a less expensive and easier program, you can do much the same. The key is to create categories for each item you grow, such as Tomatoes, Bell Peppers, Spinach, Kale, Tilapia, and so on. Then when you make a deposit, you can specify how much each crop contributed to the total deposit. You also can create a "New File", which sets up a file that is not part of your regular check book account. Either way, QuickBooks and Quicken can quickly create reports that will help you look at your sales from all angles.

The combination of production records and sales records is essential to improving your profitability in future years. These two sets of records will help you determine whether a crop is really worth growing, as opposed to whether you simply like growing it or think it grows well for you.

How to Obtain a Grant for your Aquaponics Operation

1. National Institute of Food and Agriculture Grant

The NIFA or National Institute of Food and Agriculture, under direct supervision of U.S. Department of Agriculture, are giving grants for aquaponics to projects about research on developments that will benefit the aquaponics industry, as well as use the technology into operations that are large scale. In2010, NIFA awarded seven grant programs, summing up to financial aid of $229 million.

2. Micro Grant Program

The goal of the Micro Grant Program is to increase public awareness, understanding, and knowledge of aquaponics as an educational tool, a hobby, or a business in providing a source of fresh, local, healthy food. Preference is given to educational/outreach based activities that focus on developing aquaponics systems or a plan to move existing growing systems toward aquaponics. Members of the Aquaponics Association can apply for grants of up to $1,000. Deadline for proposals occur quarterly.

3. Smilemundo

Smilemundo is dedicated to connecting fundraising with education.

4. Other Grant Options

Check your state, county, and local agencies to see what type of grants they offer. Garden clubs, technical schools, environmental organizations, and health related organizations sometimes award grants to innovative and positive related projects that are in conformance with their mission. Europe, England, Australia, Canada, as well as many other countries also have grants funds available for aquaponic projects that coincide with their particular criteria. The Small Business Association (SBA) provides grants to non-profit organizations. The SBA also provides assistance in obtaining a small business loan.

Aside from these specifics grantors who are providing financial aids, there are private foundations, organizations, and even individuals that are willing to help and provide grants for aquaponics. For example, an online search will provide more information about the latest international networking opportunities for this emerging industry. These networks can assist you in finding private groups that will provide financial aid to fund new projects regarding aquaponics. These groups are aiming to introduce aquaponics in the one of the major sources of producing foods. You can use the money granted to you to sustain and extend aquaponics technology in your location. In order to get the grants from the organizations listed, you have to carefully read the guidelines, satisfy all the necessary conditions, and provide the required supporting documents.

Insurance Coverage

One of the business matters that business owners must think about, like it or not, is insurance. There is not much enjoyment to be had in visualizing the possible disasters that could impact one's business, and nobody gets excited about spending money on insurance. However, insurance is a necessary evil for most business owners or individuals needing protection of an valuable asset.

Determining your insurance needs is not always an easy task. You need insurance, but what kind? How much coverage do you need? What is the cost? Although insurance companies can be helpful, one has to be cautious about relying solely on them for information, as their advice can be self-serving.

There is no simple answer that is applicable for all situations because every business is different in the amount of risk it faces and the amount of assets it has to protect. However, it helps to know what is available before deciding whether you have the right stuff, and enough of it. If you are just starting out, your business has grown, or your marketing has changed since you bought your insurance, it is a good time to consider your insurance needs.

There are four types of insurance that aquaponic business owners need to think about (besides the personal issues of health, disability, and life insurance): farm owners' insurance (which includes property and liability), product liability, employee coverage, and vehicle insurance. The internet and a trust worthy insurance agent are your best resources for figuring all this out.

The most important thing to know about talking to an insurance agent is that you have to be completely honest about every aspect of your business and make the agent understand exactly what it is you do. Never understate any aspect of your operation in the hopes of saving money on the premium. If you do not disclose the full nature of your business, there is a greater likelihood that the insurance you buy will be inadequate. If something happens you may find out your policy does not cover the situation. Then you are in a worse situation as the money you spent on insurance was not what you needed, and now you have a problem for which you are not properly covered.

Before you go to see an agent and explain your business, though, it helps to know some of the basics about direct-marketing risks and policies.

Some aquaponic business owners do not buy insurance because they don't think they have any problems. They reason that it isn't necessary since they don't have employees or don't have people visit their facility. Some feel that they know their customers and don't worry about them suing.

Before making such a decision, make sure you have all of the facts. Did you know that someone who is injured on your property or by your products may be forced to sue you by his or her own insurance company? They may like you, even be a relative, but they have previously signed an insurance contract that allows their company to seek repayment from you if they get injured. This is known as "subrogation." Under the "subrogation" clause in the policy, the company has a right to seek recovery from someone else if they are responsible for what to their policyholder. If their insurance company believes such recovery is possible, they will likely sue you. Under the subrogation clause, the company can ask their insured to be a "use plaintiff" so the suit will be in their name. Insurance companies usually don't sue in their own names because it might prejudice the jury. The insured is obligated to cooperate with the subrogation and to help with the case, such as by testifying. If the insured party refuses to sign or cooperate, because the third party being sued is a friend, the company can refuse to pay the coverage or seek repayment from the insured. For this reason, it is important to understand that you cannot depend on the fact just because it is close friend, that they won't sue you if something goes wrong. In most cases they will not be making this decision, the insurance company will, and the insurance company is not interested in friendships. They could bring a suit against you on behalf of your friend, despite your friend's opposition.

Although the statistical risk of an accident is probably small, accidents do happen and operators can get sued. With an annual premium of a few hundred dollars, an aquaponics operator can be well protected from accidents at the place of business, on the road, and even when selling products at other locations, such as a farmer's market.

Liability Insurance

The first policy to consider is liability insurance. Many operators, when they first sell produce, assume that their homeowner's and automobile insurance will provide adequate coverage. That may or may not be true. Your homeowner's policy will only cover certain situations. Once your operation turns into a business or an employee is hired many aspects of your policy are no longer applicable, or fail to address the situation in the first place. Some policies will cover small commercial transactions, but if you're making more than a few hundred dollars in sales, have many visitors, or hire an employee, it is best to upgrade your insurance policy. In some cases, you can just add excess liability coverage for your business activities. If you are currently buying only a homeowner's policy, read it carefully for mention of commercial activity, particularly the exclusions, and have a talk with your agent. Once you have a viable aquaponic business, you need a liability policy which covers all activities related to your operation.

Be sure to read the exclusions to find out if roadside markets, off-farm farmers markets, visitors, and employees are covered. The greater the exposure, or potential for someone to be injured, the greater the need for additional coverage. Typical liability policies contain coverage for personal liability and medical payments to others. Property can also be included.

Amount of Insurance Needed

How much insurance coverage should you purchase? The old insurance maxim is "cover your assets." In theory, if someone is injured seriously because of your negligence (in the eyes of the court), the damage award could take everything you own, as well as your future earnings. In the northeast and on the west coast where the price of real estate is high, and on farms with a lot of buildings and equipment, many direct-marketers insure for $2 million or more. Farmers of more modest means might decide to go with a $1 million or $500,000 coverage.

The difference between premiums for $500,000 and $1 million coverage is relatively small. It would cost more to increase the medical payment for non-negligence accidents from $1,000 to $5,000 than it would to increase liability coverage to $1 million because the risk of a small injury is greater than the risk of a big, lawsuit-producing one.

It is unlikely that a court would force you to sell your property, but cash assets would be an easy target for the opposing attorney. Furthermore, there have been cases where defendants' homes, while not taken away from them, have been put in a trust that reverts to the injured person upon the death of the owner.

Insurance Coverage for Employees

Employees to be covered for injuries. This is either achieved through your state workers' compensation program or through your liability insurance.

State laws vary on whether you must purchase workers' comp insurance, so you should call your state labor department to find out. With workers' comp insurance, the employee benefits because in the case of an injury there is a standard recourse to compensation. You, as the employer, benefit as well because with workers' comp, an injured employee is limited to workers' comp as the sole source of recovery. That means the employee can't sue you seeking greater damages or huge punitive damage claims. Unfortunately, workers' comp insurance is relatively expensive compared to the cost of adding employees to your farm or business liability policy.

The negative side to covering employees through the liability policy is that if the employee is injured, he or she would have to sue, and your insurance company might decide to defend the suit. In most cases, you would then be obligated to help with the defense, which can be costly.

Product Liability Insurance

Your general liability policy may or may not cover an incident in which your products make someone sick. If you are doing any value-added products, you may need to purchase separate products liability coverage.

Some stores will not buy from you unless you have a products liability policy. Some insurance companies will not insure for farm-made products like jams, salad dressings, baked goods, and so forth so you may need to shop around to find coverage.

Vehicle Insurance

Make sure that your insurance covers employees driving your vehicles, if that will be the case. In any vehicle accident, the vehicle insurance is considered the primary insurance policy that handles claims first and foremost. If someone other than the people named on your policy is driving at the time of the accident, you are at a substantial risk.

Farmers Market Insurance

Many insurance companies just do not understand the nature of farmers markets, so coverage for such can be difficult to obtain through ordinary means. Fortunately, a new insurance product is available nationwide to provide both general liability and product liability coverage for farmers market vendors. The National Farmers Market Vendor Liability Insurance Program was developed by Campbell Risk Management, a company in Indianapolis, in collaboration with the Farmers Market Coalition, and several state farmers market associations.

The policy provides coverage of $1 million per occurrence with no deductible, with a $2 million annual aggregate limit. It costs $250 to $425 per year, depending on the state of residence and gross estimated annual sales. A vendor can sell at an unlimited number of farmers markets under the policy. It will also cover sales to supermarkets, restaurants, schools, and other markets. The policy does not replace existing homeowner's or farm policies.

Crop Insurance

Until recently, crop insurance programs were limited to just large scale commodities such as corn and wheat. Now there is a specialty-crop insurance program that has already expanded to 35 states so far called AGR-Lite. It is an insurance program that protects a farm's total revenue rather than specific crops. The amount of income that can be covered is based on a farm's previous five years' worth of income as reported on the IRS Schedule F. AGR-Lite provides protection against low revenue due to unavoidable natural disasters and market fluctuations. Several levels of coverage are available. Aquaponic operators can choose to insure 65, 75, or 80 percent of their annual income. If their income drops below that chosen percentage, they can choose to be compensated at a rate of either 75 or 90 percent. Premiums range from about 1 to 3 percent of gross revenue.

Following is an example of how AGR-Lite works. An aquaponics operator has an average annual gross revenue of $100,000. The operator signs up an AGR-Lite annual insurance policy at the 80 percent coverage level and 75 percent payment rate, and pays a premium of about $2,500. A catastrophe reduces gross income by half to $50,000. Because the operator chose the 80 percent coverage level (the highest available), $80,000 of revenue is covered. Actual revenue of $50,000 is $30,000 short of the covered level. The insurance policy will pay 75 percent of the shortfall, or $22,500. The operator still has a bad year, but the blow is lessened by the insurance payment.

Best Way to Obtain Insurance

In the past few years, farmers have been complaining about being dropped by insurance companies and being unable to find others that will cover their business activities. Again, your best approach is to sit down with an independent insurance agent and explain your business thoroughly. The problem with finding insurance may be a computer glitch, as insurance companies often list outdoor markets in the same category as amusement parks. Your insurance agent can clear up that misconception and find the national insurance companies that will cover your operation, and activities at a farmers market.

Many small farmers use Farm Bureau Insurance. Working with an association can often provide a you better rate than obtaining insurance on your own. Farm Bureau bills the association, which then bills its members.

Alternative Labor Supply Idea for Aquaponics

The following article, published on October 20, 2012 in "The News & Observer", provides a good example on how to satisfy labor needs and benefit the community at the same time.

Chapel Hill nonprofit Extraordinary Ventures provides jobs for adults with Autism

Jackson Bono is good at math and is an excellent typist, but his mother thought his chances of finding employment were limited because Jackson also is autistic. Then she heard about Extraordinary Ventures, a Chapel Hill non-profit that starts businesses aimed at employing those with autism and developmental disorders. In 2009, while still in high school, Jackson began watering plants at the business's offices. After he graduated in May 2011, he added washing, folding, and delivering clothes for Extraordinary Ventures' laundry business to his job description.

"There's a start, and a finish, and a 'job-well-done,'" Laura Bono said. "For Jackson, like all of us, having something to do is important. It really provides that critical social component." The non-profit builds its businesses around in-demand services that require skills many with autism spectrum disorders already excel at — like a laundry service ideal for those who like organization and repetition, and a candle-making gifts business for those who enjoy cooking and crafts, said Tom Kuell, director of operations at Extraordinary Ventures. Its goal is to provide jobs to a growing cohort of local autistic adults while creating a model that can be used elsewhere.

"It's not trying to minimize their weaknesses, it's really trying to maximize their strengths," he said of employees, explaining that he thinks the businesses will become sustainable because it produces quality products. "We make candles, and they are better than Yankee Candles."

A Need for Employment

Autism spectrum disorder is the second most common developmental disability following mental retardation. As of 2016 the prevalence of autism spectrum disorder (ASD) in the United States is one in 68 children. In the state of North Carolina alone, there are more than 50,000 individuals with the disorder. Autism spectrum disorders are about five times more prevalent in boys than in girls. As children diagnosed today age, they move out of the school support system and find themselves in need of meaningful employment.

"There are many more families in the Triangle area that have children with autism than in any other part of the state," said David Laxton, director of communications for the Autism Society of North Carolina, explaining that services available to serve those with autism draw families to the area. But as they get older and begin to search for jobs, resources often don't keep up and many with the disorder find themselves unemployed. Kids with autism become adults with autism," Laxton said.

Extraordinary Ventures was established in 2007 when a group of parents concerned about the lack of jobs for autistic children who have aged out of high school joined together to find a solution. Ed Bedford, who has been involved with the Autism Society and has an adult daughter with autism, was one of the five founders who helped form the business.

"Real jobs, with real pay, that's our goal," he said. "We wanted to have different ventures in different businesses."

The non-profit started out with a banquet hall and meeting room rental business and has grown to include a gifts venture that markets bath salts and candles online and at local retailers, a gravesite care business that places silk flowers on graves through purchased packages, Chapel Hill Transit bus maintenance, football parking and the laundry service. Its jobs cater to those who would have trouble finding employment elsewhere, but the group employs people with different work ability levels, Kuell said.

Extraordinary Ventures' roughly 40 developmentally disabled employees — about two-thirds of whom have autism — not only gain skills and job training, but are great workers, said Van Hatchell, a 2011 Kenan-Flagler graduate who is director of marketing and head of the gifts business. "This population is a big workforce that can be tapped if you just give them a chance," he said. The group uses donations from its founders and profits from its businesses to fund its ventures and doesn't take government funding, he said. Job coaches from the University of North Carolina's TEACCH Autism Program provide Extraordinary Ventures' workers with one-on-one, job-related guidance in a Medicaid-funded coaching program, said Laura Klinger, Director of TEACCH and a member of Extraordinary Ventures' board.

Community Engagement

"I think it is a fabulous program. One of the reasons they're so successful is their willingness to collaborate with others in the community," Klinger said. She said the business is unique in that it builds its program around the needs of those with autism while allowing its employees to work as part of the broader public through delivery services and other activities. Other employment programs for autism exist locally, but they often occur in a workshop setting separate from the community and pay below minimum wage.

People with autism spectrum disorders have diverse needs and require individualized solutions, so while socially integrated jobs like Extraordinary Ventures' are great for some, more isolated options may be better for those who prefer seclusion, said Laxton of the state Autism Society.

Life Experiences in Cary is one local outlet for those with developmental disabilities who aren't ready or able to work in society. The non-profit was incorporated in 1978 to create jobs for developmentally disabled adults and today, [it] employs 48 people, said executive director Mary Madenspacher. It has grown from a bakery to include a commercial laundry business. It is approved to pay less than minimum wage and workers are separate from the outside community, but Madenspacher said it does give employees — many of whom have multiple disorders and are nonverbal — the satisfaction of real work. "Everybody needs a reason to get out of bed in the morning," she said. "We give that to our folks."

Helping those with autism spectrum disorders find jobs can be a challenge, but it's an important one to solve, said Kara Hume, a research scientist from UNC's Frank Porter Graham Child Development Institute who has been studying Extraordinary Ventures' model.

"We do know that there is a high rate of depression and anxiety among those with autism," Hume said. Work could alleviate those feelings, and creating jobs catered to autism is a positive step, she said.

Making It Work

Wafts of mulled cider mix with the scent of laundry soap in the brightly lit basement that serves as Extraordinary Venture's workshop. Employees fold laundry and assemble candles — currently autumn-themed — in the room, which is down a flight of carpeted stairs from the business' offices and rental banquet hall. The plain walls are decorated with posters that provide step-by-step instructions for highly specific jobs over full-color photographs.

"No," one reads, depicting poorly folded shorts slashed over in red. The picture to the right of the poor folding shows the same clothing, neatly stacked, under a bold "Yes."

People with autism sometimes struggle with certain tasks, as Extraordinary Ventures' leaders have found — a newspaper delivery service, for instance, proved too stressful and time-sensitive and had to be cut. The team has used methods like color-coding to simplify or clarify other jobs that could be confusing.

"The goal is to get everybody to work as independently as possible," Kuell said. The system evolved over time, he said, with input from experts at TEACHH and people like Hume. Kuell said it seems to be helping employees to master skills.

Even as the workers flourish, word is spreading and the businesses are expanding, Kuell said. Extraordinary Ventures added 22 jobs across its entire portfolio this summer, more than doubling in size. Laundry is turning a profit, and he said other divisions are headed in that direction.

As it becomes successful, Kuell said the goal is that the businesses will become a teachable model for others. "All of our hard work, we're starting to see the fruit," he said. "Things are going from being this idea to businesses that work."

Aquaponic Business Success

Nearly all business experts agree on the importance of drafting a business plan. Yet, there are companies that move forward without a formal business plan. They get carried away and figure their passion and optimism are enough to build a successful company. Others say they were just too busy to develop a formal business plan. However, operating without a plan can prove even more time-consuming in the long run.

The time you invest in a business plan will pay off many times over. Some of the most obvious benefits you can gain from business planning include:

- An opportunity to test out a new idea to see if it holds real promise of success.
- A clear statement of your business mission and vision.
- A set of values that can help you steer your business through times of trouble.
- A blueprint you can use to focus your energy and keep your company on track.
- Benchmarks you can use to track your performance and make midcourse corrections.
- A clear-eyed analysis of your industry, including opportunities and threats.
- A portrait of your potential customers and their buying behaviors.
- A rundown of your major competitors and your strategies for facing them.
- An honest assessment of your company's strengths and weaknesses.
- A roadmap and timetable for achieving your goals and objectives.
- A description of the products and services you offer.
- An explanation of your marketing strategies.
- An analysis of your revenues, costs, and projected profits.
- A description of your *business model*, or how you plan to make money and stay in business.
- An action plan that anticipates potential detours or hurdles you may encounter.
- A handbook for new employees describing who you are and what your company is all about.
- A résumé you can use to introduce your business to suppliers, vendors, lenders, and others.

The Risk of Moving Forward without a Business Plan

The many benefits of having a business plan should be enough to convince you to write one (or hire someone to write one for you). But in case you're still wavering, consider what can go wrong if you don't take time to plan. You risk:

- Running out of cash before you open your doors because you haven't anticipated your start-up costs.
- Missing sales projections because you don't really know who your customers are and what they want.
- Losing customers because your quality or service falls short.
- Becoming overwhelmed by too many options because you never took the time to focus on a mission and vision for your company.
- Going bankrupt because you don't have a rational business model or a plan for how to make money.

Time spent putting together a solid business plan is a worthwhile, if not essential investment. As a matter of fact, the more time you spend, the better prepared you'll be, and your probability of success increases exponentially.

Don't be overwhelmed at the prospect. The basic components of a business plan are fairly simple. Five basic elements make up a business plan are:

- A mission statement that explains the objectives of the proposed business in one sentence, or at most a short paragraph.
- A statement of goals that names specific, usually financial-or sales-based targets and sets a definite time period to reach those targets.
- A financial plan that incorporates the basic financial documents used by all businesses.
- A market analysis, listing the markets you have identified for the products you plan to sell and the percentage of your total sales you expect each market to provide.
- A **competition analysis** that identifies as many of your competitors as possible, together with what you perceive to be their strengths and weaknesses.

On a Mission with Goals

While each of the above noted elements is important, the mission and goals statements should receive your attention first, as they will empower you to execute your plan and keep you focus on your objective. The mission statement presents the big picture, while the statement of goals tells you how to get there. A mission statement might read "We will earn $10,000 in additional annual income by growing tomatoes in a greenhouse." An example of a goal might be, "We will sell organic heirloom tomatoes, out of season, at $25 per box to local retailers between March 1 and July 31."

Several such goals will likely be required to encompass your business objectives, because you will probably have to diversify to succeed. Whatever your particular situation might be, it is difficult to overestimate the value of having precisely articulated goals. If, instead of designating a clearly defined objective, you merely said, "We are going to sell tomatoes when they are not in season," you would leave a lot of questions unanswered. How much space is needed? Do you have enough? When should the plants be planted and harvested? It is impossible to answer any of these by referring to a vaguely worded statement. However, a specific goal statement addresses all.

Compiling all the information about your business into a list of briefly stated goals helps keep you focused and efficiently moving forward. As busy as life is in general, it is all too easy to get distracted and pulled in other directions. You may already have a job, family obligations, and a dozen other things vying for your time. If you are going to embark upon any business venture with any hope of financial reward, it is imperative that you stay focused on your goals. Many people have discovered that the most effective way of managing the big task of "running a business" is to break the large goals down into groups of smaller tasks.

Analyzing the Market and Your Competitors

You will need to do some research to analyze the available markets and the nature of the competition in your area. For starters, ask questions like:

- Are you willing to drive 20 miles to a farmers' market? How about 100 miles?
- What kind of reception will the market have to your products and prices?
- Who are the predominate suppliers in your community?

You can obtain needed information in many ways.

The internet, telephone, and face meetings are great approaches for gathering information. Also, browse your Yellow Pages and visit your nearest farmers' markets and flea markets.

Take three basic steps:
- Define the geographic location and the size of your market area.
- Bring together the available data on all the possible outlets for your products.
- Formulate a good understanding of your competition.

Study this information and come up with three lists. Each list should have answers to the following questions:
- Who will buy my product?
- What venues exist to reach those customers?
- What are your competition's most obvious weaknesses?

The answers to these questions, you will need to identify the "holes" in the current market that will provide wonderful opportunities for you. For example, "None of the other vendors at my nearest farmers' market were selling organic tomatoes." If you are seeking funding with your business plan, be sure to cite evidence from your research to support your contentions in the market and competition analysis sections of your business plan.

Making Money

The financial plan constitutes the heart of your business plan, because the overall objective is, of course, to turn a profit. The financial plan will answer the following basic questions:
- Can I make a profit on this venture?
- How much money will I need to invest?
- How long will it take to recover that investment?
- How much net income will this venture generate per year?

The financial plan consists of four documents: the **balance sheet,** the **profit and loss statement,** the **budget,** and the **cash flow analysis.** Each can be set up in a simple spreadsheet program, or you can purchase software with these documents already set up. As long as the financial data you include is accurate (or consist of your best estimates where accuracy is not possible), it makes no difference by what method you keep track of your assets, liabilities, expenses, and income.

However, you will need records, and it is worthwhile to follow long-established customs when opening a set of books. Furthermore, a well-kept set of business records can be mined for all sorts of valuable data. An examination of the function of each document in detail shows how useful they are, even for a small business.

The Balance Sheet

A balance sheet is a "snapshot" of the value of a business at a specific point in time. Comparing two balance sheets prepared at different times allows you to determine if the business's value, or "net worth," is increasing or decreasing. Despite the size of your venture, you should prepare a balance sheet at least once a year. Otherwise, you will not know if your time and investment is being rewarded. Prepare your balance sheet by first listing all the business assets in a single column. Assets are things of value that the business will use in its quest for profits, including cash, money the business is owed, equipment, buildings, and inventory. You might, for example, open a business checking account with an initial deposit of $100. This amount goes in the asset column of the balance sheet.

A small business may involve few assets other than cash, especially at the outset. At the beginning, you will have made no sales and thus no money will be owed from customers.

As for equipment, you probably already own most of what you need. Keeping track of small items such as used hand tools, garden hoses, sprayers, etc., can be lumped together as a single asset account named "Small Equipment." Estimate the value of each item at the time you place it in service for your business. A good rule of thumb here is to set the value at whatever price you would ask for the item if you were selling it in a yard sale.

More expensive equipment items, such as a power washer, should be listed as individual assets, especially if they are new, or relatively so, because you get a tax deduction for the decrease in their value with use. This decrease is known as "depreciation," and it, too, appears on the balance sheet. Will you be using a vehicle for making deliveries? Do you plan to build a small greenhouse for business use? How about purchasing a portable building for storage? For any of these items, you should keep a record of the following information:

- Name of the item.
- Date of purchase.
- Purchase price.
- Date placed in service for the business.
- The item's fair market value on that date.
- Extent of business use.

Creating and dedicating a spreadsheet with this information (each item purchased for the business or placed into business service) works best. The "extent of business use" entry above refers to things that have both business and personal use, such as a vehicle. Having this information for each piece of equipment will allow you to calculate the annual depreciation for that item when you prepare your tax returns. The IRS provides depreciation tables in the instructions for Form 4562, which is used to report depreciation on your tax return. You can estimate depreciation costs without reference to the tables. All of the equipment you will use in your aquaponic operation is considered to have a life span for depreciation purposes of five years. Dividing the purchase price (for a new item) or fair market value (for used items) by five gives the annual depreciation. If an asset is placed in service later in the year than January, the depreciation for the first year is reduced proportionately. For example, if back-up generator is placed in service on July 1, the depreciation would be reduced by half for that year, because it was not used for business purposes for the first half of the year.

Similarly, if you have something such as a computer, that is used for both business and non-business purposes, you may deduct only the portion of the depreciation corresponding to the percent of business use.

A vehicle is a special case. For a small aquaponic business operation, you likely will use your personal car for making deliveries to customers, rather than having a vehicle dedicated solely to business use. In this situation, most people will find it more advantageous to deduct business mileage rather than compute depreciation. The IRS sets an allowable per-mile deduction that takes into account depreciation, maintenance, and fuel costs, everything except parking fees and tolls. To take advantage of this deduction, keep a mileage log in the vehicle and record the beginning and ending odometer reading each time you take a business-related trip. You will be surprised at the end of the year how much all of your driving, with some element of your business included in the trip, adds up.

Consider with care the total amount of noncash assets you assign to your business. In some jurisdictions, you may be assessed a tax on the value of property held for business use. Sometimes the tax authority will grant an exemption for small, home-based operations. Therefore, claiming new vehicle as a business asset, for example, could add to your tax burden. Not listing your car as a business asset normally would not prevent you from taking the mileage deduction for business use of a personal vehicle. It is also prudent to check with a professional familiar with your local tax laws before you decide how to handle such assets.

After listing all your assets, create a second list of liabilities. Liabilities are amounts of money owed or committed for a specific business purpose. The simplest kind of liability involves purchases on credit. For example, using a credit card to buy fish feed creates a liability. You will be paying off the credit card with the proceeds of your sales. Until that time, you carry the amount as a liability on your balance sheet.

Totaling up the amounts in the assets and liabilities columns and finding the difference determines your business's "net worth." If the total value of your business assets exceeds the total of your liabilities, the net worth is a positive number. As your business grows, you want to see that net worth increase over time. Changes in net worth occur as the money inflows to, and outflows from, the business. This money flow is referred to as, "income" and "expenses", respectively. It is a good business practice, if not essential, to keep track of this money flow. For that, you need another financial document, the profit and loss statement.

A Profit and Loss Statement

The profit and loss statement is a summary of the business's income and expenses for a specific period of time. Businesses typically prepare this statement on a monthly, quarterly, and annual basis. To prepare the statement, you add up your income, subtract your expenses, and the difference is your profit for the time period under consideration. A profit and loss statement summarizes all such income and costs for the business operation.

Costs of production are known as "direct" costs. Direct costs refer to materials, labor, and/or any other expenses related to the production of a product.

Costs that are spread over the entire operation of the business and not attributable to any one product, are known as "indirect" costs. Examples of common indirect costs are advertising, utilities, and office expenses such as postage, paper products, and printer cartridges.

A smart business owner keeps thorough records for the profit and loss statement. Not only will this information tell you how much money the business is actually making, but every expense can also be a potential tax deduction.

Many agricultural business owners find it helpful to arrange the profit and loss statement to correspond with the categories on Schedule F of IRS Form 1040 that must be filed when reporting farm income. You should, therefore, keep a separate tally of both classes of products.

It is also helpful to keep track of the amount of income generated by each specific product in your inventory (i.e. fish, tomatoes, bell peppers, etc.). This practice consumes more time than merely lumping all your sales together, but will enable you to quickly learn which ones generate the highest profit margins and which products are "losers." Because the resources of most small businesses are usually very limited, it is detrimental to waste them on the losers.

Schedule F categorizes farm expenses as follows:

- Car and truck
- Custom hire
- Employee benefits
- Fertilizers
- Insurance
- Labor
- Rent
- Seeds and plants
- Supplies
- Utilities
- Other
- Media
- Depreciation
- Feed
- Gasoline
- Interest
- Pension plans
- Repairs
- Storage
- Taxes
- Tanks

As with all things tax-related, exceptions and rules apply to these categories. Some of the categories simply won't apply to your operation. If you have no employees other than yourself, you can obviously dispense with categories for employee benefits, labor costs, and pension plans.

Tax Considerations

Owning a home-based or small business can provide some very nice tax advantages. Items not formerly deductible become so when they can be shown to have a business purpose. For instance, the interest on financing a greenhouse, along with the cost of water and other utilities to operate it, plus its depreciation, is usually partially deductible if you can show that it was used for business purposes.

Losses sustained from a farming operation can in some instances be deducted from your other income. If the overall operation has a loss, despite the profitable sales on one specific product, you can deduct your loss from your total taxable income.

Because each situation is unique, it is best to consult a professional tax expert about these matters. For more information regarding the tax implications of your micro farm business, go to *www.irs.gov* and download IRS Publication 225, Farmer's Tax Guide.

Labor Costs

Even though a small aquaponics operation is unlikely to require full-time employees, you might occasionally need to hire help for harvesting, sales, or while you go on vacation. However, unless you need a lot of help, don't fret that having a part-time employee who may result in payroll taxes and tons of paperwork. According to the IRS, you can pay a casual worker or workers up to $600 per year without documentation. That includes a relative, or the neighbor's kid. However, once your expenses for this kind of labor exceed $600, you are required to file more forms.

Having actual employees on a regular basis will require you not only to withhold and deposit income and social security tax payments for each person but also to match their contributions to social security and Medicare. In order to justify not only the costs in payroll taxes, but also the time spent keeping records and filing with the government, your business must have significant revenue. If your business idea consists of a labor force, be sure to do your research and know your numbers in advance. If you are thinking about using family members for your labor needs make sure they are on board before moving forward. Otherwise, you may find yourself in a bind and end up doing much more yourself than you anticipated or care to do.

Budgeting

A budget is nothing more than a projected profit and loss statement. In other words, you will be mostly guessing when you try to assign numbers to the expense and income categories set up for your record keeping. In doing so, you will create a budget. Use it, together with the records you amass as your operation proceeds to measure the accuracy of your predictions, with the goal of producing a more refined budget for next year.

No matter how small, businesses have to use budgeting and have a working business budget for good financial management. A budget not only helps you forecast the implications of a business move, it also helps you keep your spending under control.

If you are an established business owner, you can use your own historical data. If you are a new business owner or an entrepreneur with just an idea, you will need to do some research. For example, you will need to research your sources of income. Who will buy your product or service? How much business do you anticipate your advertising plan will pull in? Have you set the price for your product? You may want to look at other businesses selling a similar product or service before you set your price or answer these questions. These become your income figures on your budget. To be conservative, reduce them by one-fourth.

Estimating expenses is probably easier. You have two types of expenses. Fixed expenses are those like rent. They don't change from month to month. Write down all your fixed expenses. Variable expenses do change from month to month. They would be your utility bills, your advertising budget, and other items. Write down all your variable expenses.

What can happen if you don't have a budget? You could easily go out of business. You have to be sure that you can generate enough cash flow and profit from your venture to cover the expenses associated with it. The only way to do that is to establish a budget. The budget gives you some control.

A simple internet search will provide you with many 'go-by' business budgets, as well as budget templates that can be downloaded and edited for your specific operational parameters. The sample business plan in Chapter 40 of this book also has a lot of examples of financial statements.

Cash Flow Analysis

Cash flow analysis takes the budgeting process to the next level by imposing upon it the dimension of time. Cash flow analysis refers to the study of the cycle of one's business cash inflows and outflows. The purpose of a cash flow analysis is to maintain an adequate cash flow for one's business, and to provide the basis for cash flow management. Cash flow analysis involves examining the components of business affecting cash flow, such as accounts receivable, inventory, accounts payable, and credit terms

For example, the first year of your aquaponics operation may consist of mostly expenses with little income. Nevertheless, your product suppliers will expect to be paid. This means you will be laying out cash for seeds, starter plants, fish, fish feed, utilities, etc. in the life of the operation, and only when you have respectable incoming revenue will you replace that outgoing cash with the proceeds generated by sales. Meanwhile, it means you have a cash flow problem. Anticipating such situations is the purpose of a cash flow analysis.

A cash flow analysis for at least twelve months of operation tends to work best. Using a spreadsheet, assign a column to each of the next twelve months. On the first row for a month, enter the amount of cash on hand in the banking account. The subsequent rows will correspond to individual income or expense items for each month. Try to anticipate all such items you may encounter during the forthcoming year. Estimate amounts as accurately as possible. With spreadsheet software, of course, you can always make changes where necessary.

When filling in the data for expenses, enter the amount as a negative number. Done this way, each column can be summed to find the cash balance at the end of the month. That number becomes the beginning balance for the first of the following month.

As you complete the spreadsheet for a full year of operation, you will note that the "Net Cash" balance fluctuates. It may even dip into negative for one or more months. Estimating your cash flow shows you when those low and high points can be expected. This gives you a chance to figure out in advance how to survive those hard times. A clear understanding of the effect of cash flow can make or break your business. For instance, if you are projected to run out of money for fish feed before the fish are big enough to harvest, you could be in big trouble. However, knowing this formation in advance provides you with an opportunity to be proactive on solutions so as to avoid or minimize a situation that could otherwise be a catastrophic event to your business operation.

Getting a Valuable Return on Your Investment

The return on investment shows how much a company made on a particular investment as a percentage. Return on investment (ROI) is a common business term used to identify past and potential financial returns. Any expense a company has can be calculated into terms of ROI.

To maximize your ROI follow these basic principles:

1. Live below your means.
2. Buy low and sell high. Although this may come off as a cliché, there is much truth in it. Comparison shop for the lowest price of things you need for your business. Focus on the activities that generate the greatest revenue with the highest profit margins.
3. Strive to do things as efficiently as possible.
4. Breed your own fish. Use your own plants to generate seeds or propagate plants for future crops.
5. Get sound tax advice to maximize deductions.
6. Implement an alternative energy system, such as solar or wind to reduce energy cost. Take advantage of rebate programs provided by your utility providers.
7. Thoroughly investigate the most profitable ways to sell your product in your geographical area.
8. Have multiple streams of income, such as a variety of vegetable crops along with your fish, as well as offering paid tours of your operation, and running an aquaponics training program.

Aquaponic Business Plan Essentials

Creating a business plan for your new aquaponic operation need not be a daunting task. The following are the three fundamental components of an effective business plan:

- **Create** a mission statement, statement of goals, market analysis, competition analysis, and financial plan.
- **Develop** a balance sheet, profit and loss statement, and budget and cash flow analysis, covering a period of at least one year.
- **Follow** your business plan and budget to help keep you on track when multiple obligations place demands on your time.

A good business plan should be subject to periodic revision. However, the effort to create a plan is wasted if you don't follow it. Review it regularly to ensure you are meeting your goals, and make any necessary adjustments if you are off track. A solid business plan will empower you to achieve optimal success and maximum profitability.

CHAPTER 37

Selling Your Aquaponic Products

Your Market

In the narrowest sense, the term "market" refers to a specific location where sales occur, as in "farmers' market." However, in a broader sense, the "market" consists of all the people who might be customers for your products. Put another way, "marketing," is the act of offering your goods to potential buyers through all available means, not just across the counter. A more appropriate term is "selling." Semantics aside, the important point to keep in mind is that *marketing* involves a lot more than just trucking your goods to *the market*. At a minimum, your marketing should include:

- Business cards
- Advertising
- Website
- Flyers to distribute
- Word-of-mouth (telling others about your product)

Marketing is an important component of an agricultural enterprise. Regardless of size, it is prudent to identify and research your market before you decide on a particular product to grow. Knowing and understanding your local market is an important element of the "being profitable" formula. The better you are at marketing, the better your sales.

Do thorough research on the types of products that can be produced through aquaponics, which are in demand in your local market. Preferably, you would also want those that face less competition and sell at a good price.

Although this chapter talks directly about marketing and selling of plants, much of what is said can also be applied to fish as well. The specific marketing and selling of fish is addressed in the following chapter.

You can either sell your harvest directly to the customers or to your local grocer. A more convenient approach is to sell in bulk, although this approach typically brings in substantially less revenue.

Growing off-season crops is also a great idea. However, keep in mind that you may have to incur greater costs in terms of supplemental lighting, heating, and insulation for growing some off-season crops.

Also, check out your local farmers market, organic food groups, as well as vegan and vegetarian clubs in your area. Craigslist trading can also be a valuable resource for finding customers, and bartering with others for various things you need (i.e.

trade your vegetables for fruit, haircuts, auto repairs, carpet cleaning, etc.).

Your Customer

Knowing and understanding **customer needs** is at the center of every successful business, whether selling directly to individuals or other businesses. Once you have this knowledge, you can use it to persuade potential and existing customers that buying from you is in their best interests.

The more you know about your customers, the more effective your sales and marketing efforts will be. It is well worth making the effort to find out:
- WHO they are
- WHAT they buy
- WHY they buy it

If you're selling to other businesses, you'll need to know which individuals are responsible for the decision to buy your product or service. It's also prudent to keep an eye on **future developments** in your customers' markets and lives. Knowing the **trends** that are going to influence your customers helps you to **anticipate** what they are going to need, and offer it to them as soon as they need it.

You can learn a great deal about your customers by talking to them. Asking them why they're buying or not buying, what they may want to buy in the future, and what other needs they have can give a valuable picture of what's important to them. Visiting other areas in your state or the United States can help you better understand consumer behaviour and produce-marketing practices. Cooperative extension personnel are a valuable source of information about the local marketing situation. In addition, cooperative extension offices often provide training workshops and seminars specifically for direct agricultural related marketers. Data on customer demographics is collected by the U.S. Census Bureau and can be accessed online at census.gov.

Strong sales are driven by emphasizing the benefits that your product or service brings to your customers. You must find out as much as possible about consumers who may buy your products. If you know the challenges that face them, it is much easier to offer solutions that address their needs and preferences.

Ten things you need to know about your customers

1. **Who they are**
   If you sell directly to individuals, find out your customers' gender, age, marital status, and occupation. If you sell to other businesses, find out their size and kind of business they are. For example, are they a small private company or a big multinational corporation?

2. **What they do**
   If you sell directly to individuals, it's worth knowing their occupations and interests. If you sell to other businesses, it helps to have an understanding of what their business is trying to achieve.

3. **Why they buy**
   If you know why customers buy a product or service, it's easier to match their needs to the benefits your business can offer.

4. **When they buy**
   If you approach a customer just at the time they want to buy, you will massively increase your chances of success.

5. **How they buy**
   For example, some people prefer to buy from a website while others prefer a face-to-face meeting.

6. **How much money they have**
   You'll be more successful if you can match what you're offering to what you know your customer can afford.

7. **What makes them feel good about buying**
   If you know what makes them tick, you can serve them in the way they prefer.

8. **What they expect of you**

   For example, if your customers expect reliable delivery and you don't disappoint them, you stand to gain repeat business.

9. **What they think about you**

   If your customers enjoy dealing with you, they're likely to buy more. And you can only tackle problems that customers have if you know what they are.

10. **What they think about your competitors**

    If you know how your customers view your competition, you stand a much better chance of staying ahead of your rivals.

Beating the Competition

Chances are your potential customer is already buying something similar to your product or service from **someone else**. Before you can sell to a potential customer, you need to know:

- who the customer's current supplier is
- if the customer is happy with their current supplier
- if buying from you would offer the customer any benefits and, if so, what those benefits would be

The easiest way to identify a potential customer's current supplier is to simply to ask them. Generally, people are very happy to offer this information, as well as an indication of whether they're happy with their present arrangements.

If you can find out what benefits they're looking for, you stand a better chance of being able to sell to them. The benefits may be related to price or levels of service, for example. Are there any benefits your business can offer that are better than those the potential customer already receives? If there are, these should form the basis of any sales approach you make.

Marketing: Why Locally Grown Food is Best

Evaluating Market Demand

Larger growers, particularly those located in major production areas, can pursue either of the two traditional marketing alternatives: wholesale fresh marketing or processing. Small-scale growers do better using a more direct-to-consumer approach. Both modes of operation require thorough research of the market, and customer behavior before planning crop production.

Some operators generate profits by planting first and then looking for a market, but this is a risky approach to doing business. There is a greater chance for failure than success in this situation. If you are a new grower, or an established one planning to produce a new item, you should first attempt to evaluate the market demand for the product and then decide which direct marketing channel(s) will best meet the needs of your consumers. Your estimates of profitability should include the marketing channel costs as well as production costs.

Aquaponic operators should collect the following information before deciding which crops to produce and market.

- Determine and define the geographic area where you will market your products. Identify potential customers before you investigate consumer demand.
- Assess the level of unfulfilled demand among consumers within the defined marketing area. It is advisable to estimate the amount that consumers (buyers) within that market buy at present. In the process, you will gain insight into how they might be better served.
- Consider the competitive structure of your market. Knowing who your potential competitors are, where they are located, and what services they provide are important pieces of information for you as a new grower-marketer. Some potential

competitors may have marketing advantages (lower costs, better locations, and higher-quality produce) or may provide potential consumers with similar products. Rather than being discouraged at such, look also for their weaknesses and areas where you have an advantage or can capitalize on their weaknesses (i.e. marketing your products as organically and grown through environmentally sustainable methods).

- Determine the likely impact of increased production on future selling prices. If you place more produce on the market, and the products are not of different quality or do not meet some other "unmet need" for which consumers are willing to pay a higher price, then it is likely that prices will drop from current levels. An expected price is a vital piece of information for planning purposes. There is no simple, reliable way of predicting local market prices, but such information is very important to growers. Estimate prices by considering all the available information, as well as good judgment. When using these estimated prices for planning, remember to include marketing costs and the cost of unsold product.

The following are some important questions that are prudent to answer before moving forward with a particular product:

- Who are the likely consumers of your produce and where do these consumers live?
- How many people live within your marketing area?
- Are consumers currently buying the products you have in mind?
- How much of the product(s) do your potential customers currently use? Is this use seasonal?
- What prices are consumers paying for high-quality products?
- Are consumers adequately served at present?

If consumers in the area are being adequately served, here are some additional questions:

- Can you do the job better and draw part of the market away from competitors?
- Can the quantity that consumers purchase be increased by providing better quality than is now available?
- Will your anticipated production come at a time when little else is offered for sale?
- What level of quality must you produce to meet the need unfulfilled demands of consumers?
- How must you prepare and package the produce? What type of arrangement would be most appealing to your customers? What marketing costs will be incurred?

Industry Analysis

The market for organically grown vegetables, herbs, flowers, and fish is vast. The explosion of organic and eco-friendly products on retail shelves is more than just a fad — it is BIG business. If targeted properly, this market can present an environmentally minded entrepreneur endless opportunities. The eco-friendly lifestyle continues to catch on with consumers, which presents enormous growth possibilities for an aquaponics business.

Local Demand vs. Importation

Most produce in the U.S. is picked four to seven days before being placed on supermarket shelves, and is shipped for an average of 1500 miles before being sold. Those distances and times are substantially longer when we take into consideration produce imported from Mexico, Asia, Canada, South America, and other places.

We can only afford to do this now because of the artificially low energy prices that we currently enjoy, and by externalizing the environmental costs of such a wasteful food system. Cheap oil will not last forever though. World oil production has already peaked according to some estimates, and while demand for energy continues to grow, supply will soon start dwindling, sending the price of energy through the roof. In regards to pure economics alone, the world will soon be forced to re-evaluate food systems and place more emphasis on energy efficient agricultural methods, like more energy efficient organic agriculture, and on local production wherever possible.

Cheap energy and agricultural subsidies facilitate a type of agriculture that is destroying and polluting our soils and water, weakening our communities, and concentrating wealth and power into a few hands. It is also threatening the security of our food systems, as demonstrated by the e-Coli, toxic GMO-contamination, and other health scares frequently reported in the news.

When properly educated about these issues people desire to purchase products that produced locally. The vast majority of people also prefer to support local businesses, as a means of supporting local jobs and benefiting their community through the 'ripple' effect.

Reasons Locally Grown Food is Better

1. **Locally grown food tastes better.**
   Food grown in your own community is usually picked within the past day or two. It's crisp, sweet, and loaded with flavor. Produce flown or trucked in is much older. Several studies have shown that the average distance food travels from farm to plate is 1,500 miles.
2. **Local produce is better for you.**
   Fresh produce loses nutrients quickly. Locally grown food, purchased soon after harvest, retains its nutrients.
3. **3. Local food preserves genetic diversity.**
   In the modern industrial agricultural system, varieties are chosen for their ability to ripen simultaneously and withstand harvesting equipment. Only a handful of varieties of fruits and vegetables meet those rigorous demands, so there is little genetic diversity in the plants grown. Local farms, in contrast, grow a huge number of varieties to provide a long season of harvest, an array of eye-catching colors, the best flavours, and their crops have a much higher nutrient density content.
4. **Local food is GMO-free.**
   GE crops are promoted on the basis of ambitious claims—that they are safe to eat, environmentally beneficial, increase yields, reduce reliance on pesticides, and can help solve world hunger. However, solid research studies show that genetically modified crops have harmful effects on laboratory animals in feeding trials and on the environment during cultivation. They have increased the use of pesticides and herbicides, increase the toxicity of soils, and have failed to increase yields. This is the very reasons most of Europe and many other developed countries have banned GMOs. GMO crops have shown to have clear signs of toxicity—most notably disturbances in liver and kidney function and immune responses; as well as many other health issues.
5. **Local food supports local farm families.**
   With fewer than one million Americans now listing farming as their primary occupation, farmers are a vanishing breed. When buying at a large supermarket, only 18 cents of every dollar goes to the grower. The middle man gets 82 cents out

of every dollar. Local farmers who sell direct to consumers cut out the middle man and get full retail price for their crops.

6. **Local food builds a stronger community.**
   When you buy direct from the farmer, you are re-establishing a time-honored connection between the eater and the grower.

7. **Local food preserves open space.**
   As the value of direct-marketed fruits and vegetables increases, selling farmland for development becomes less likely. The rural landscape will survive only as long as farms are financially viable.

8. **Local food helps to keep your taxes in check.**
   Farms contribute more in taxes than they require in services, whereas suburban development costs more than it generates in taxes.

9. **Local food supports a clean environment and benefits wildlife.**
   A well-managed family farm is a place where the resources of fertile soil and clean water are valued. Aquaponics takes these benefits to a whole new level.

10. **Local food is about the future.**
    By supporting local farmers today, you can help ensure that there will be farms in your community tomorrow, so that future generations will have access to nourishing, flavorful, and abundant food.

Economic Benefits to Buying Locally

1. **Buy Local—Support Yourself:** Several studies have shown that when you buy from an independent, locally owned business rather than a nationally owned one, significantly more of your money is used to make purchases from other local businesses, service providers and farms, continuing to strengthen the economic base of the community. A matter of fact, there are many case studies showing that locally-owned businesses generate a premium in enhanced economic impact to the community and our tax base. A long list of these studies can be seen at: http://www.ilsr.org/key-studies-walmart-and-bigbox-retail/

2. **Support Community Groups**: Non-profit organizations receive an average 250 percent more support from smaller business owners than they do from large businesses.

3. **Keep Our Community Unique:** Where we shop, where we eat and have fun, all of it makes our community home. Our one-of-a-kind businesses are an integral part of the distinctive character of this place. Our tourism businesses also benefit. "When people go on vacation they generally seek out destinations that offer them the sense of being someplace, not just anyplace." ~ Richard Moe, President, National Historic Preservation Trust

4. **Reduce Environmental Impact**: Locally owned businesses make more local purchases requiring less transportation and generally set up shop in town or city centers as opposed to developing on the fringe. This generally means contributing less to sprawl, congestion, habitat loss and pollution.

5. **Create More Good Jobs:** Small local businesses are the largest employer nationally, and provide 55 percent of all jobs in America and 53 percent of all jobs in Europe.

6. **Get Better Service**: Local businesses often hire people with a better understanding of the products they are selling and take more time to get to know customers.

7. **Invest in Community**: Local businesses are owned by people who live in this community, are less likely to leave, and are more invested in the community's future.

8. **Put Your Taxes to Good Use**: Local businesses in town centers require comparatively little infrastructure investment and make more efficient use of public services as compared to nationally owned stores entering the community.

9. **Buy What You Want, Not What Someone Wants You to Buy**: A marketplace of tens of thousands of small businesses is the best way to ensure innovation and low prices over the long-term. A multitude of small businesses, each selecting products based not on a national sales plan but on their own interests and the needs of their local customers, guarantees a much broader range of product choices.
10. **Encourage Local Prosperity**: A growing body of economic research shows that in an increasingly homogenized world, entrepreneurs and skilled workers are more likely to invest and settle in communities that preserve their one-of-a-kind businesses and distinctive character.

Why Your Aquaponics Organic Produce is Better than Other Organic Produce

The most common recommendation made by health experts in the area of produce selection is to choose "organic" varieties for maximum health benefits. We trust organic label certifications because they are supposed to comply with standards set by national governments and international organizations. Synthetic pesticides and chemical fertilizers are not allowed in organic practices, but that's not always the case. Research by a Food Inspection Agency in Canada has found that nearly half the organic fresh fruits and vegetables tested in the past two years contained pesticide residue, violating maximum allowable limits for the presence of pesticides, the data shows. As a matter of fact, 45.8 percent of samples tested showed at least trace amounts of pesticides. The data released to CBC News under the Federal Access to Information Act includes testing of organic fruits and vegetables sampled between September 2011 and September 2013. Most of the fresh produce sampled was imported, with only one-fifth of it grown in Canada.

A U.S. Department of Agriculture found nearly 20 percent of organic lettuce tested positive for pesticide residues piqued our interest. Lots of the lettuce contained substantial amounts of spinosad, a pesticide marketed by Dow Chemical under the brand name.

Pesticides can get onto organic produce through contamination of water or soil through pesticide spray-drift from neighboring farms, and through contact with non-organic produce after harvest. However, industry experts have suspicions that some of the larger residue measurements suggest an organic producer deliberately used a pesticide that is not allowed.

The more you can educate your potential customers about these facts, the more they will desire your produce. Therefore, incorporate this information into your marketing campaign to win customers and increase sales.

Sustainability for More Effective Marketing

The aquaponics industry is focused on operating sustainably. In effort to achieve eco-friendly status, the aquaponics operator should research the options to reduce its impact on the environment. Some options include buying energy efficient bulbs for the facility, using alternative energy sources such as solar or wind, and recycling to minimize waste. All of these initiatives will enable a business to be considered a truly sustainable enterprise. The efficient use of energy becomes more important as the concerns for the environment continue to grow. Besides being responsible, these initiatives are not only healthy, but also are environmentally friendly, even if it increases costs somewhat.

Small-Scale Operations vs. Large Commercial Producers

Small-scale operations are much more accessible and easier to start up. This type of set-up requires less money, facility space, equipment, energy costs, and system maintenance. Although the production output may be greater for commercial aquaponics, small-scale aquaponics operations can still be produce a significant amount of food, enough to feed a family, plus extra.

Harmful Viruses Sprayed on Meats, Cheeses and Organic Foods

Clever labeling laws have made it possible for companies to spray a bacteria-eating virus, the food additive called "Listex," on organic food and get away with it. This cocktail of six bacteria-eating viruses (bacteriophages) is found on everything from meat and cheese to fruits and vegetables. This additive is a combination of "trained killers" that have been concocted in a lab to target *Listeria monocytogenes*, a bacterium responsible for human sickness. It's just another Band-Aid approach to dealing with Concentrated Animal Feeding Operation (CAFO) pathogens, with no requirements being mandated that would tell us this unhealthy toxin is being put on our food.

As consumers are becoming savvier about reading labels, food companies figure out new ways to get around being noticed. Changing names, using extremely small-fine print, and jumping to use FDA-approved toxins are several strategies being used against us. Even USDA-certified organic food does not have to pass any heavy metal toxicity tests or have any warnings about mercury, lead, or aluminum content. Such protocol is not safe for the environment or human health.

The American conventional meat industry uses ammonia and bleach to kill viruses, virus combinations to kill bacteria, DNA from insects to kill other insects in corn and soy, insecticide inside seeds to kill beetles and antibiotics to kill pathogens in bred-for-slaughter animals. Nevertheless, these procedures and methods are not always affective. Cases of E. coli and Salmonella go widespread and lead to recalls of all kinds of products. Cases of super-bugs and super-weeds are spreading in the U.S.

You may be familiar with the acronym "GRAS," meaning "generally recognized as safe." This is the FDA's favorite way to approve anything that they don't want to spend money on researching for safety. Another phrase you may recognize, "This statement has not yet been evaluated by the FDA." For natural and organic vitamins, minerals, herbs and tinctures, that usually means that it works and the establishment doesn't want you to believe it, use it, or be healed from your Listex poisoning, GMO food, fluoridated water, or lab-made pharmaceutical medicine addiction.

The virus "safety precaution" statement used on labels addresses substances known to provoke allergies, asthma, autoimmune disorders, inflammation and elevated cholesterol, and even work as a catalyst for colon cancer. Viruses are highly adaptable, so the samples presented to the FDA simply can't be reliable, even if endotoxin levels were undetectable at the time they were approved. This is a complex matter with very scary implications. Intralytix, Listex's manufacturer, claims to have "purified" the viruses, but this is simply not the case.

Bacteriophages battle with friendly bacteria in the digestive system, making it harder to digest food. This can severely damage your immune system and your first line of defense, which is your gut flora.

Viral fragments in genetically modified foods contain viral "promoter genes" that help foreign DNA infect the host. This means that live viruses inside Biotech's GMO corn, soy, cottonseed, canola and sugar beets can recombine to create more aggressive viruses (hence the term *recombinant*). Big Ag, with the government's assistance, is basically inserting live viruses into our food, and spraying our food with bacteria.

Educate yourself, your family members, friends, kids, and neighbors about these dangers. Producing truly organic food will greatly increase the value of your operation.

Best Places to Sell Aquaponic Products

Several factors must be considered when choosing how to sell your products. Location can have a major impact on an enterprise's profitability because it affects the direct marketing channel used as well as the ability to attract customers.

Bringing fruits and vegetables to market requires special handling. To preserve quality and maintain marketability, each crop must be harvested, prepared for market, packaged, and shipped under favorable conditions. Therefore, your exposure to risk is increased.

Each method varies in the amount of labor and capital the operator must put forth to bring the product to market. Following are several viable options:

Local Customer Base

The trend toward locally grown food and organic food bodes well for the would-be aquaponics operator. Selling directly to the consumer is the ideal way to maximize profits and remove costly intermediaries that distribute our products. This is also the simplest way to market the product and establish a loyal customer base. Establishing strategic alliances with local vendors such as newspapers, colleges, farmers markets, and associations can be used as advertising and promotion channels. The downside of this strategy is the time it takes to establish a client base, but the gains are unlimited as "Word of Mouth" is an excellent way to build clientele, as it has a multiplying effect.

For some operators, the best place to market is on location. If zoning code, location, and space allow, selling products at your location is a convenient and fun way to sell your products. The best on location selling operators are typically fine with some mild interruption of their family's privacy, and they don't mind having strangers watching them at work. They are tolerant of obnoxious customers, such as the ones who come up your driveway even when the sign says you are closed or who lay on the horn if your gate is locked. The best markets also are in a location that is convenient for the public, either on a paved road or not far off one, close enough to a population center or tourist traffic to draw sufficient customers.

Selling on location can run the gamut from a seasonal affair, just during a harvest, or be a year-round endeavor with an extensive array of products for sale.

Whatever the scale, on-location selling creates an entirely new layer of work for the aquaponics operator. With such a set-up, you are a storekeeper, with all the additional responsibilities of buying products, keeping inventory, ensuring freshness, complying with labor and tax laws, making your building handicapped-accessible, maintaining displays, creating signs, and so on. Growing the product becomes secondary to marketing it.

Even though an inviting atmosphere and creative displays of excellent produce are important, experienced on-location selling operators will tell you that equally high on the list of attractions are parking and available clean restrooms. The larger your operation, the more important it is to have these amenities. For mid-size to large operations, all of the other business's positive attributes may go unnoticed if these components don't measure up to your customers' expectations.

An onsite selling operation can also be a good job for a family member who is more interested in working with the public than in the production area. On a small scale, you might consider an honor-pay stand. Farmers who have sold that way say they rarely have problems with people stealing from the cash box or taking food without paying. Put up clear signs with prices, bolt a payment box to the wall, and empty it every few hours. An honor-pay stand may not provide enough revenue to support you, but it can be a nice addition to your other sales, without a lot of extra work.

If you are considering creating a retail market, a great resource will be the North American Farmers Direct Marketing Association (NAFDMA). Many NAFDMA members are also involved in entertainment farming ventures, so you'll find plenty of ideas about this type of business. Contact information is www. nafdma.com or 413-529-0386.

Two business matters you should address before moving forward with an on location sales operation are:
1. Talk to your insurance agent about your plans to be sure you have adequate liability coverage.
2. Talk to the zoning department in your local government to be sure you will comply with all the regulations.

Sales to Local Restaurants

Most metropolitan areas have an unlimited number of restaurants that specialize in preparing meals from local producers. Even if you only have one or more small gourmet restaurants in your area, you may be able to sell your products to them. The price for these meals is usually high. Customers are willing to pay the premium for fresh, locally produced organic products. Selling products to local restaurants and building business relationships with them will help you established a solid customer base. Providing organic quality is the key.

Your most suitable targets will be ethnic, vegetarian, and/or chef-owned establishments. Most ethnic-type restaurants are seeking high-quality produce of certain preferred varieties. Restaurants catering to vegetarians are usually very interested in local produce, heirloom varieties, organic vegetables, products grown through sustainable farming methods, and artisanal products. Typically, the more upscale the restaurant, the more likely the chef will appreciate the special qualities of your finest produce offerings (and be willing to pay an appropriate price).

First, make a list of all the possibilities you discover in the Yellow Pages or via the internet. Next, call and ask if the chef is interested in local produce. Try to meet with the key person in charge. Setup a time that is convenient for the chef. Take along samples of your offerings. Your samples should be a fair representation of what you have to offer, not just the best you have to offer. Reputable chefs are known to be picky and obsessive about their work, but a good chef can be a great long-term ' customer.

Farmer's Markets

A farmers' market is a physical retail market featuring foods sold directly by farmers to consumers. Farmers markets typically consist of booths, tables or stands, outdoors or indoors, where farmers sell fruits, vegetables, meats, and sometimes prepared foods and beverages directly to the customer. Farmers markets add value to communities. They exist worldwide and reflect their local culture and economy. Size ranges from a few stalls to several city blocks. In some areas live animals, imported delicacies unavailable locally, and personal goods and crafts are sold. Selling at farmers markets is an excellent option for an aquaponics operator to generate revenue.

When the USDA started collecting information about farmers markets in 1994, there were 1,755 markets in the United States. By 2012, the number had jumped to more than 7,800. Those figures are subject to some debate because collecting the data is difficult given the independent nature and widespread distribution of markets throughout the country. The USDA has not tallied the number of farmers who sell at these markets, nor the amount of revenue generated, so it is difficult to assess the total economic value of farmers markets nationwide. Nevertheless, there is no question that the United States has experienced an explosion of farmers markets in recent years.

Farmers markets come in many themes and variations. You should visit numerous farmer markets to fairly assess the available opportunities. In most cases, you can also sell "value added" products, such as jams and pickles. Some only allow growers to sell their produce. Others have tight restrictions on what can be sold (i.e. no crafts, merchandise, etc.).

Farmers markets can operate weekly, bi-monthly, or seasonally depending upon the location and the way it is operated. Also, rules and terms for participating in farmers markets vary widely. For instance, some include one-time inspection fee, annual dues, rental charges, and/or a percentage of each day's gross sales. In most cases, some of the funds the farmers market earns will benefit you, such as going towards regional advertising of the market. At the other end of the spectrum are public farmers markets, where anyone with something to sell can set up a table.

Benefits of Farmers Markets

Farmers markets can offer farmers increased profit over selling to wholesalers, food processors, or large grocery firms. By selling directly to consumers, produce often needs less transport, handling, refrigeration, and time in storage.

By selling in an outdoor market, the cost of land, buildings, lighting and air-conditioning is also reduced or even eliminated. Farmers may also retain profit on produce not sold to consumers, by selling the excess to canneries and other food-processing firms. At the market, farmers can retain the full premium for part of their produce, instead of only a processor's wholesale price for the entire lot. You can also trade some of your product to other vendors (i.e. trade a bucket of your vegetables for another vendor's fruit).

Where consumers perceive the farmers' produce as of equal or better quality than produce available through grocers, farmers may retain most of the cost savings to themselves. Some farmers also prefer the simplicity, immediacy, transparency and independence of selling direct to consumers. By contrast, relations with agricultural conglomerates can be burdened with quite complex contractual details.

One advantage is the generation of high levels of customer traffic. Staffing needs are simple to plan because operations generally occur only during specified hours. One big disadvantage, however, is the need to predict sales so that enough produce can be harvested and prepared each market day. At a farmers market, you generally cannot replenish stock by quickly harvesting more, which can sometimes be done at roadside stands. Also, inclement weather can impact sales. In addition, farmers markets typically have several sellers on the site with the same products. This can be good or bad for you. The other vendors can help attract customers, but they may also be direct competition. You should look at the market to see how your products will compete. Don't assume that you can't succeed at a particular market just because there will be a lot of competition.

An aquaponics operator can be selling at a farmers market and make several thousand dollars every weekend. The most successful markets are located in urban areas and have dozens of vendors and thousands of customers. Many growers at those markets can virtually sell out every week, so sales depend on their production capacity. At smaller markets, the usual sales figures are less than $1,000 per market, sometimes only a few hundred dollars.

Selling at farmers markets takes skill and perseverance. You need to grow significant amounts of produce, and preferably offer a wide variety of choices for your customers. The better you display your produce, the better the results. You must price it appropriately so that you make a profit. Don't make the mistake of thinking that you can get ahead by selling at lower prices than everyone else.

You must have a pleasant personality that attracts customers. You need to be enthusiastic and knowledgeable about your products. There is definitely a personality type that is well suited to selling at farmers markets. A seller who is gregarious, cheerful, makes a strong effort to genuinely love people, takes pride in his/her work, and is proud of the represented products will have the best results. If that doesn't describe you or you have no desire to be that way, then you should limit the number of markets you attend or hire staff to handle that aspect of your operation.

Selling Produce and Fruit Products at Farmer's Markets

Farmers' market produce is normally grown within a geographical region that is deemed local by the market's management. The term "local" is defined by the farmers' market and usually represents products grown within a given radius measured in miles. Many farmers' markets state that they are "Producer Only" markets and that their vendors grow all products sold. Some farmers markets do not use the term "Producer Only" and may allow resellers of produce and other food products.

Some allow vendors to resell vegetables and fruits if they are not available locally due to the time of the year. Vegetables, fruit, meat, and other products resold at farmers' markets are available to vendors through food distributors. This is a common practice and provides consumers with produce and fruit that are unavailable at certain times of the year. In many markets, resell items are a permanent part of the vendor's inventory.

FIGURE 261. Selling from a Pickup Truck

Roadside Sales

Taking your products directly to the public is viable means of generating revenue, especially if your area lacks existing market venues or if you cannot meet the requirements for participation in sponsored markets. Roadside selling means you load up your wares and haul them to a location where you'll be seen by passersby's. This approach to marketing fresh produce is typically during harvest season.

Roadside marketing is common and effective. Location and your rapport with customers are extremely important. Roadside selling can be as simple as selling right off the tailgate of your pickup truck or involve selling from a structure. Overhead can be negligible or substantial, depending upon the type of roadside arrangement being considered. For instance, selling out of the back of a pickup truck parked in an empty lot is less expensive than selling from a roadside stand, and significantly less expensive than selling from a building. Be leery of paying rent to sell from someone else's property, as this will drastically reduce profits. Instead, point out to the owner or manager that your presence can increase his or her store traffic. Some sellers simply choose a vacant lot, hoping the owner either won't mind the trespass or is absent. In a busy retail zone, this may be the easiest way to find a spot.

However, keep in mind that there are many regulations related to operating any small business, and a roadside market is no exception. Most states have small business development centers that offer publications and workshops outlining these regulations. Potential regulations you may have to deal with include: worker's compensation, unemployment compensation, building inspections, weights and measures, plant pest laws, sales tax, and perhaps others. State departments of agriculture will also have publications that will help you understand and satisfy these regulations. A list of some of these publications and web sites may be found at the end of this publication. Depending on the zoning regulations in your area, permits may be required for driveways or entrances to a business; check with local officials or your local state highway office before beginning construction. Also, check with your local municipality to determine if building/occupancy permits are required for a roadside market, and how any existing zoning regulations may impact your plans. In many areas, permits are required before erecting any roadside signs.

No matter how nice your market looks, people must be able to find it. Signs directing the customer to the market should be attractive, eye-catching, and easy to read from a distance. It is less important to list all available products on the sign than to direct the customer to the market. Signs should be visible far enough in advance to give travelers adequate time to decide to stop at the market and to safely enter the parking area. Be sure to check with your local municipal authorities for any regulations regarding roadside signs.

To draw customers to your market, the parking area and exterior of the market must be clean, attractive, and inviting. There should be ample parking for customers and the parking lot should be easy to navigate. Handicapped parking and accessibility points need to be considered. Ideally, the parking lot should be paved and marked with lined spaces. Gravel lots are also acceptable, but bear in mind that loose gravel can be difficult to walk on (particularly for the young and old), and is often quite dusty during extended dry periods.

Give a lot of consideration to the design of displays in your market. You can enhance your displays by using contrasting colors, shapes, and sizes. Appeal to your customers' sense of smell by using aromatic herbs and fruits. Consider combining soft, firm, smooth, and fuzzy items, thus stimulating your customers' senses and creating a positive purchase environment and helping increase sales. Place high demand items in strategic, high-traffic locations throughout the market. Make displays that look like they came from the farm. Wooden crates, boxes, and baskets work well. Where possible, slant displays towards the customer. Even an attractive tablecloth can add to sales. Stair-stepped displays create an array of depth, color, and texture. However, they may not be easy for the customer to reach or easy to restock. Utilize vertical space by hanging products from slings or hangers.

Remember, roadside markets are often selling an image and experience as well as physical products. When shopping in such an environment, you want your customers to think about traditional agriculture and the many "warm" thoughts that this implies. Old tools, crates, or scales hung on the walls can reinforce this impression. Similarly, pictures showing earlier days on the farm can help "sell the story" of your operation. If it is a family market, consider putting up a picture of the family.

Prices should be clearly marked on or near displays. Many modern shoppers are in a hurry or are bashful, and therefore will be reluctant to ask for help to find out how much something costs.

Agritourism

Agritourism (also called to as 'agritainment') refers to agricultural related activities that bring visitors to your operation. Agritourism can refer to something as simple as a u-pick farm or as complicated as a farm bed-and-breakfast. They can be designed to appeal to local folks or to tourists from afar. There is growing interest in visiting farms; people want to get closer to the source of their food, to learn how it is grown and to share that with their children. This is especially the case when it comes to aquaponics. In addition, theme outings are increasingly popular — hence the growth of ecotourism and culinary tourism. These groups are continually looking for places to visit.

If you think agritourism would work well for your operation, and it is something that would enjoy, be prepared to invest some resources to properly bring it to fruition. Areas that may require additional funds include spending money on staff to supervise visitors, liability insurance, facility upkeep to allow your operation to remain attractive and safe, and signage. You may also need to invest in infrastructure such as restrooms, buildings, additional lighting, and parking improvements. With this type of commitment, it only makes good economic sense to have lots of products to sell to visitors.

Flea Markets

A flea market typically consist of weekend community gathering of buyers and sellers. In some areas, flea markets are operated as a commercial venture, meaning you pay a fee to gain access to selling space. Other flea markets are set up where there is no selling fee or even a formal organizational structure. The sale of anything and (almost) everything is allowed, and participation is open to any vendor with something to sell.

Flea markets may or may not work for you. They tend to be dominated by cheap merchandise and to be patronized by people looking for bargains rather than healthy food. Many flea markets are basically just a large "yard sale". Others are quite organized and elaborate. Some flea markets can be a viable option for selling a large harvest, introducing people to your operation, and earning some extra money. The best strategy is to visit several flea markets to get an impression of the clientele it attracts as well as what's being sold.

Special Events

Event markets are sales venues open only in connection with a special event. Craft fairs, holiday celebrations, and events sponsored by public institutions may all offer lucrative, one-time-only opportunities. While a food or garden themed event might be an obvious choice, events such as an art show or music festival has the potential of generating surprising revenue.

Check local news listings and community calendars for upcoming events. Be sure to inquire early as some events have limited space and the high traffic selling areas always go first.

If you find an event market in your area that will accept you as a vendor, has reasonable rates, and that offers a high likelihood of a profitable day(s) of selling, by all means give it a try. Dedicated patrons of the event's sponsor can become loyal customers for you in the future. Even if sales are slow, the event can advertise your presence in the community. Take advantage of every opportunity to do essential marketing with a good banner, flyers, handouts, and business cards.

Grocery Stores

The sheer volume of food that moves through a local grocery store is amazing. In 2010, the typical grocery store in the United States generated an average of $300,000 in sales weekly. About 20 percent of all produce offered in grocery stores is thrown away before it is sold due to spoilage, resulting in a 50 to 75 percent markup so stores can recover their loss. In essence, grocery stores are in need of produce.

Fresh produce retailing in the United States has seen significant changes and shifts in recent years. For instance, towards the end of the 1990s, many industry insiders incorrectly predicted future trends. At that time, the fresh produce distribution system seemed to be moving toward fewer, and larger, centralized packaging and distribution centers. At the beginning of the 2010s, however, increased transportation costs and changing consumer preferences had grocers large and small considering the purchase of produce from growers closer to their individual stores. Combined with the popularity of "local" produce among many American consumers, opportunities have risen for farm growers to sell produce to local grocery stores. Therefore, aquaponics operators may find this as a possible market for their product.

There are generally three avenues for selling produce to local groceries, based mainly on the size and scale of the store:

- Niche or Specialty Stores may often carry smaller product selections, with management and ownership entirely in the local area. The number of specialty food retailers has increased in recent years due to consumer interest in smaller store formats and specialty food items (like certified organic and local foods).
- Independent Grocers function as a full-scale grocery, making purchase decisions at the store or regional level. These may be stand-alone stores or regional chains focused on a single city or metro area.
- National Chain Grocery Stores typically make purchase and distribution decisions beyond the local level, often employing regional distribution centers. However, many national chains are now shifting to allow direct store purchasing and acceptance of local produce.

Developing the Producer-Retailer Grocery Store Relationship

Selling wholesale produce is a highly relational business. Retailers are accustomed to placing orders with companies and growers, and a relationship of mutual trust is cultivated between clients. Many of the larger produce wholesalers and retailers trace their roots to small, family businesses that have grown across the generations. Personal relationships between buyer and seller in the produce industry is very important.

However, that doesn't mean that opportunities are not available. Aquaponics operators will be rewarded by cultivating a relationship with the produce manager or store manager responsible for making purchases of fresh produce. Aquaponic operators should present themselves professionally and be willing to tell the story of how their produce is grown. This will allow the person purchasing your produce to get to know your operation, and it can create an environment for moving more quantities of your crops. Retailers who understand and appreciate where a food crop is coming from tend to be more likely to promote those products to their consumers. Some stores will include "point of purchase" materials, such as signs and recipes, to promote local crops. Others even invite producers into the store to hand out samples. Producers more willing to contribute to such displays and efforts will help cement the relationship with the retailer.

Pricing for Grocery Stores

Unfortunately, grocers only pay wholesale prices, even for locally grown produce. This makes negotiating price a key consideration for selling to a local grocer. Producers should understand two things about pricing produce for sale to local grocers:

1. Prices received for produce will be wholesale, not retail.
2. Know how much your produce costs to grow and deliver.

Bottom Line: Revenue received from local groceries will be at wholesale price levels. This can be a surprise for aquaponic operators accustomed to selling produce at retail prices. Retailers work on margin, the difference between the wholesale price paid and the retail price. Some retailers may be willing to incur smaller margins to feature local produce. However, prices paid by grocers will still be at wholesale levels.

Since producers will be receiving wholesale prices, it is crucial that they understand how much a particular produce crop costs them to grow, harvest, and deliver to the grocery store. Knowing the costs of production will allow a grower to discuss potential prices with the local grocer. Be sure to factor in costs of packaging, delivery, and extra handling time that may be incurred when delivering produce. Some aquaponic operators find it worth their extra time to sell produce through local grocers, despite smaller profits, because it provides a form of advertising for their farm to local consumers.

Packaging for Grocery Stores

Another difference between selling to grocery stores compared to other markets involves product packaging. Grocers are accustomed to handling produce such as tomatoes, cucumbers, and squash in corrugated cardboard produce boxes. Other produce crops, like lettuce and greens, may be delivered in reusable plastic containers. Some crops are expected to be delivered in wooden crates. Still other produce, such as cherry tomatoes may be packaged and labeled in plastic or paper pint or quart containers. Furthermore, grocers may be accustomed to handling other crops, like melons, that carry a sticker on each fruit with a numerical code. More produce is now labeled with individual Universal Product Code (UPC) or Product Lookup (PLU) symbols for scanning at checkout. For tips and resources on UPC and PLU symbols, producers may refer to Tips for Selling to: Grocery Stores published by the National Sustainable Agriculture Information Service (ATTRA).

While some local grocers may be willing to work with local growers, producers need to bear the responsibility of determining how to package the product in a specific manner that will allow the grocer to handle local product as safely and easily as possible. Whatever you desire to produce and sell to grocers, be sure to research the way the product is typically graded and packed. Furthermore, produce growers need to budget the cost of packaging for delivery to the wholesale client and make production-marketing decisions depending on their ability to bear the cost.

Aquaponic operators also need to realize that packaging can add value to their product. Attractive product labels carrying the farm's name and address can raise awareness among local consumers. Labels and signage indicating that products are locally grown can help generate interest in local products at the retail store. Even bulk boxes or bins may be imprinted with a farm's name so that consumers can see where the produce has been grown as it is stocked in the produce section. Aquaponic operators need to manage costs of production, but money carefully spent on quality packaging may generate positive additional returns.

Quality and Quantity Grocery Store Expectations

Produce is not only graded and packed in standard-sized containers, but it is also sold according to standard quality grades. While some local retailers may afford some leeway for local growers, aquaponic operators desiring to sell through grocers should always adhere to commonly accepted grades and standards for produce crops. The operator and employees may need to be trained on proper picking, grading/sorting, and packing techniques.

The USDA website page titled "U.S. Grade Standards" has a great deal of helpful information on grading, certification, and verification of vegetables and many other crops. Single copies are free from:

Fresh Products Branch
Agricultural Marketing Service
Fruits and Vegetable Division
U.S. Department of Agriculture
P.O. Box 96456, Room 2056-South Bldg.
Washington, D.C. 20090-6456

Many produce crops require cooling and/or cool transport to maintain product quality and safety. Some local groceries may allow local producers to deliver crops directly after harvest for refrigerated storage. Others, especially larger chains, may require growers to maintain certain cool chain standards for different crops.

Aquaponic operators willing to keep their crops at ideal postharvest temperature levels will benefit their grocer customers with improved product quality. Finally, no discussion of selling produce to local grocers is complete without touching on product quantity. Failure of aquaponic operators to deliver adequate and consistent quantities of produce is one of the main barriers for purchasing local produce frequently cited by grocers, foodservice, and other wholesale produce buyers. Producers will need to clearly communicate how much and how often they can deliver to the grocer before the season starts. Grocers understand that unexpected events, such as weather, can impact harvest, but they expect clear communication from the producer in any event of a delivery delay. Remember, the retailer is already accustomed to receiving wholesale produce at regular times and uniform quality. Local producers need to match (or exceed) the quality of existing wholesale options, especially if producers are trying to obtain any premiums in wholesale price.

There is an old adage that, "If you package it right, you can sell just about anything." It's no different for packaging fruits and vegetables; they must be packaged so customers will buy them. Proper packaging is especially important when a grower is selling to a wholesale buyer. Individual crops have specific industry packaging standards. If the crops are not packaged accordingly, wholesale customers probably will not buy them. Reasons for specific industry standards are twofold:

- Everyone in the industry speaks the same "packaging" language. In other words, everyone knows what they are buying and selling.
- Crops have different requirements for handling, so different types of containers are made to fit them

The main functions of a produce container are:
- To prevent and reduce injury to the crop during transit and handling.
- To provide ventilation to hasten cooling and escape of heat caused by respiration.
- To reduce water loss from the crop.

Containers may be made of wood, styrofoam, and plastic, but corrugated fiberboard is the most popular rigid container. Each material has advantages and disadvantages. Stacking strength, length of storage, storage treatment, precooling method, and cost influence the choice of material. If the container has contact with water or ice, it should be made of water-resistant material.

Final Word about Grocery Stores

More local grocers are open to the idea of sourcing local produce. Aquaponic operators willing to develop a face-to-face relationship with a retail buyer, budget all cost, negotiate fair prices, and deliver produce of consistent quality and quantity can develop profitable wholesale market niches.

Selling from a Cart

With a relatively small up-front investment, a lot of money can be made selling from a cart. Some towns and cities even encourage street vendor operations as they recognize them as an additional source of sales tax revenue. Most municipalities require that you purchase a permit and have a tax identification number. At the right location, you can generate a nice revenue stream selling items produced through aquaponics, such as flowers, organic produce, and fish. You can also add other items to your inventory for sale to further entice customers to your cart. Selling from a cart takes a small investment and a time commitment, but it can be a very successful means of selling your products, and further marketing your business as a whole.

Internet and Mail-Order Selling

Some farmers do very well through mail-order sales. The internet has revolutionized this market for the savvy aquaponics operator. With minimal technical expertise and expense, it is now possible to set up a web-based business, collect payments, and communicate with your customers online. The challenge is obtaining some visibility among the millions of other web pages out there. Social media is an excellent marketing tool that can be implemented to overcome this challenge. Also, a good web-designer will integrate features into your website that will enable it to come up early through a number of different searches; otherwise known as Search Engine Optimization (SEO).

One particular benefit of mail-order is the ability to deal with customers on your schedule. Your internet-based store is open 24-7. Even if you post a telephone number, you can always let the call go to voice mail, and most, if not all, of your correspondence will be via e-mail. Eliminating the need for customers to come to you automatically avoids one of the principle drawbacks to a home-based business, where the added traffic generated is seldom welcome. The only traffic engendered by mail order is the UPS truck picking up your packages. You can even avoid this interruption by taking packages to the nearest drop-off point yourself.

While paper mail-order catalogs are becoming more of a thing of the past, you can promote conservation and save thousands of dollars by posting an online catalog or by sending contacts a catalog via a PDF format. This digital technology even allows you to include photos without additional expense.

Unlike many most other markets, the internet and mail-order selling places no limits on your potential market. Even if selling resource options are limited in your area, you can reach thousands of potential customers online.

Admittedly, mail-order marketing has its drawbacks. There are extra expenses for packaging materials, and time is required to pack products so they will not be damaged in transit. With this, as well as with other aspects of your business, it is important to identify and have a good understanding of all of your costs.

As with all other selling markets, providing good customer service is important when doing business online as well. For example, dealing with the inquires and troubleshooting problems requires care and patience. In addition, creating and maintaining a web-site requires a commitment of time. After you create the site you need to keep it updated, add content, and continue improving it to keep up with changes in the industry. Keeping your site's content and appearance fresh is essential to keep customers coming back.

Weigh these pros and cons carefully when considering selling online. Despite the drawbacks, many growers have found it a profitable way to earn income.

Community Supported Agriculture (CSA) and Subscription Farming

Subscription farming, also called Community Supported Agriculture (CSA), is essentially grower contracting, and is an excellent option for making money by growing aqauponic products. You "contract" directly with the consumer to produce and deliver a specific quantity of vegetables and/or fish. Customers typically pay an up-front fee in exchange for products being delivered periodically or at different times of the year (i.e. harvest).

The customer base provides the capital you need for operational cost. The consumer benefits by getting the fresh, organic, locally grown food without the hassle of going shopping. This type of farming has been around for years, but it is becoming very popular as more people are becoming "locavores" (someone who eats foods grown locally whenever possible). Therefore, CSA programs are on the rise.

This method requires integrity to deliver what you promise, a reliable delivery vehicle, and good consumer relations. Start by asking neighbors to become customers. Getting money up front for future product delivery is hard to beat. Furthermore, knowing your customers and how much they will purchase greatly simplifies your planning.

Avoid the mistake of overselling your production capacity. Keep in mind that you can always team up with other growers to provide your CSA customers with more quantity and a wider variety of products than you can grow by yourself. This will help strengthen your business, reduce your risk, and support your community.

Getting and Keeping Customers

It is extremely important to not only listen to your customers, but also earnestly seek their input on how you can improve your business. If customers suggest that you could do something better, want something you don't have, or want information about something you sell, do your best to accommodate them, even if it means buying product from elsewhere to resell. You want the customers coming back and recommending your business to others.

Customers may not know how to use a product or what product will be best for what purpose. A knowledgeable suggestion will be appreciated and go a long way in winning over a customer. Suggest something they may not have tried. Having recipes available can be very helpful; a successful meal will win a long-term customer.

Anyone working with customers should be friendly and courteous. Everyone likes premium products and bargain prices, but great customer service and a friendly disposition goes a long way. As a matter of fact, people will shop with you more often and spend more money at your business if they like you and feel appreciated. Get to know frequent customers by name. The famous Dale Carnegie states that the most pleasant sound people can hear is that of their own name. Always greet customers will a smile and a "hello." Customers are not a distraction; they are the reason you are in business. Make your marketplace a friendly, helpful, and enjoyable experience for your customers.

No matter what you do, you will always have some complaints. Many times when a customer makes a criticism, it's because they want to keep shopping at your market, but they want something to be improved. So listen, remain calm, address the customer's concerns, and take their comments to heart. Perhaps their complaint is evidence of something that could be done better.

These days, customer service is frequently poor or even non-existent. Often times the person at the counter neglects to say 'thank you', much less look you in the eye, smile, and ask you need any additional help. If you will do these things, you will far outshine your competitors and do much to win over a customer's loyalty towards your business.

Keys to Providing Excellent Customer Service

1. Keep in mind that customers are never an interruption to your work. The customer is your real reason for being in business. Chores can wait.
2. Greet every customer with a friendly smile. Customers are people and they like friendly contact. They usually return it. No matter their attitude, be sure to always treat your customers with a smile, kind hospitality, good customer service, and you will not only win them over, but will turn them into your best marketing resources (and at no cost).
3. Call customers by name. Make a game of learning every customer's name. See how many you can remember. A gym owner was asked by one of his employees, after observing his boss greet everyone by name as they came in for their workout, how he could remember all the names of his 750 plus members. His boss simply replied, because they are all important to me. Our customers need to be important to us, as well. Remember your customer's name and say it when speaking to them.

4. Remember, you and your employees are the face of the company. The way you represent yourself to your customer is the way your business will be perceived.
5. Never argue with customers. Be a good listener; agree where you can, and do what you can to make the customer happy.
6. Never say, "I don't know." If you don't know the answer to a question, reply with, "That's a good question. I will try to find an answer for you."
7. Remember, every dollar you earn comes from the customer's pockets. Treat them like the boss.
8. State things in a positive way. It takes practice, but will help you become a better communicator.
9. Make every effort to give your customers a good experience so they will want to come back.
10. Always try to do more than the customer expects.
11. Be sure to say "thank you" and "please visit us again" with a smile.

Attracting Customer Interest in Your Product

The old adage 'image is everything' has a lot of truth when it comes to selling and meeting people. A big smile, quality product, and excellent knowledgeable service is indispensible for optimal business success. However, there are also additional measures that you can pursue which have a tremendous positive influence when it comes to attracting customers.

Simple things, such as displaying your product on a table and at eye level, instead of on the ground where it can be kicked and easily become covered with dust, makes your display more attractive. In addition, allowing a space on the table where ladies can set their purse and bags will entice them to stay and shop longer at your location, whether it be a farmers market, roadside stand, street cart, or at any other location where you will have customers visit you in person.

Adding some farm décor is also helpful in making your display attractive. Setting bales of hay around, hanging up old farm implements, or decorating with other farm related nostalgic items gives people a comfortable feeling, entices them to your display, and encourages them to stay longer.

Sprucing up your display with flowers, bright red strawberries, vibrant yellow squash, nice looking wheatgrass entices customers to visit you longer, and spend money. Background music or theme sounds (i.e. sounds of nature, etc.) are fine as long as it remains in the 'background' (not too loud) and appeals to vast majority of people. If in doubt, do without.

Bringing along a friendly farm animal such as goat, small pig, or chickens will work wonders, as well. A small fish tank with one or several of your fish attracts people too. Children will literally drag their parents to your display just so they can look (and hopefully pet) your animal. Have a display of pictures and a write-up alongside of it about your operation.

Accepting Cash for Payment

Cash is the most commonly accepted and reliable form of payment for a business. Many small businesses operate as "cash only" merchants. Years ago this wouldn't have been uncommon, but with advances in technology, business owners must ask themselves if they're hurting their bottom line by limiting payment options. If you're thinking about starting a cash only business or if you're considering expanding your current payment options, be aware of the pros and cons of only accepting cash.

Pros of Accepting Only Cash

- Cash payments ensure that businesses receive funds immediately. With each transaction, your business immediately receives the appropriate payment amount without the worry of waiting periods or not getting paid at all.
- Cash is the simplest form of payment and therefore involves less bookkeeping. For a business, that not only means less stress and hassle, but it also may save money in the time and labor

it would take for a bookkeeper to record other payments methods.
- There is limited risk of fraud when accepting cash only. There are cases of counterfeit cash payments, but compared to other payment methods, fraud is much less common in cash transactions. Be sure to educate yourself on identifying counterfeit money.
- Cash only businesses don't have to worry about third parties or fees associated with other payment options.

Cons of Accepting Only Cash

- Customers who do not have enough cash on them will have to walk away from a purchase they would otherwise make.
- Your business may lose customers by only accepting cash. As card payments become more and more popular, many consumers expect this to be an option when making purchases. If they find that a particular business only accepts cash, they may feel inconvenienced and shop elsewhere.
- Keeping large sums of cash on your business's premises increases the amount of time you will spend managing finances and also creates an added security risk.
- According to a new survey by Vouchercloud.net of more than 2,300 adults in the U.S., 57 percent said they "never" carry cash—compared to the 17 percent who sometimes carry it, and the ten percent who say they "always" have cash on them.
- In 2013, only 27 percent of all point-of-sale purchases were made with cash and that number is expected to drop to 23 percent by 2017, according to a report published by Javelin Strategy & Research, a market research firm.

For some small businesses, it may be wise to stay cash only. You will need evaluate which payment options will create the most success for your business.

Credit and Debit Cards

Credit and **Debit Cards** are popular, convenient, flexible, and have become increasingly important in business commerce. If your business is considering what forms of payment to accept, or if you would like to expand the payment options of your cash-only business, be sure to go over the pros and cons of accepting card payments.

Pros of Accepting Card Payments

- Card payments are evolving into the most common method of customer payment.
- The convenience of using credit cards generally increases the likelihood of consumer "impulse purchases," which ultimately contributes to an increase in a business's average sale. Customers are more likely to make these purchases if they have access to credit or their available bank account funds.
- Businesses can easily accept card payments. Overall, there are three main methods of accepting cards payments. They are:

Over the Internet — To accept cards over the internet, look into an internet payment gateway system. It is a service that allows merchants to perform real-time credit card authorizations over the Internet.

With a Credit Card Terminal — A credit card terminal is a type of a point of sale (POS) terminal that can do transactions with a credit or debit card. There are several types of credit card terminals are available. Most have the same basic purpose and functions. They allow a merchant to insert, swipe, or manually enter the required credit card information, to transmit this data to your selected service provider for authorization and, finally, to transfer funds to the merchant. Most newer models not only process credit and debit cards but can also handle gift cards, checks, and so on. The majority of card terminals transmit data over a standard telephone line or an Internet connection (either wired or wireless).

Mobile Application on your Smartphone – This method is becoming more and more popular as a way for businesses to accept credit cards, especially for businesses that are on the go. There are several mobile apps available on smart phones, such as the iPhone, Android, and Blackberry that allow you to key in or even swipe (using a card reader device) credit card transactions and get instant authorizations using the phone's wireless Internet signal. The apps are only about $0.99 to $4.99 to download, and usually require you to sign up for a merchant account and payment gateway. This is an especially good solution for sellers that have a mobile business (roadside or street cart selling) or are selling in at a remote location (i.e. farmers markets, flea markets, etc).

Cons of Accepting Card Payments

- Card payments come with an increased risk of fraud. Although there are laws and security measures that help protect and secure customer information, card payments are inherently more susceptible to foul play than cash. Always be sure to read the merchant service provider contract fully before signing it. Understand your responsibilities and how to protect your customers' privacy, as well as how to secure their personal information.
- Businesses that accept card payments encounter small processing fees for purchase transactions. These fees may seem insignificant but they can certainly add up, especially if your business accepts a lot of small purchases on credit cards.

Setting up the necessary equipment to accept cards also carries additional costs.
- Card transactions add another layer of detail to your business's bookkeeping practices. Your business will have to take into account the additional time and resources it takes to maintain these records.

Accepting card payments will, at least initially, cost your business money and add extra processes in your daily operations. Many small business owners view this as a necessary operating expense. Since card payments have become so popular, customers typically expect a plastic option as a rule, rather than a courtesy.

Accepting Checks

Accepting checks is another option to consider. However, to protect the financial health of your business, it is important that you understand the laws that regulate check payment policies, and then implement your own operating policies.

Policies for Accepting Checks

- If your business accepts personal checks, establish a detailed check acceptance policy to help identify and avoid bad checks. Don't just make a document and file it away; be sure you are prepared to consistently operate according to your policies without deviation and to train your employees accordingly. It is also helpful to post reminders in visible and prominent locations where checks are most likely to be presented for payment.

Common check policies include variations of the following guidelines:
- Checks must be from a local or in-state bank.
- Checks should not be written and accepted for more than the purchase amount.
- Checks should not be accepted that are starter checks, unnumbered checks, or non-personalized checks.

- Accepted checks should be deposited as quickly as possible. Some people, with poor management skills, only write checks while they have funds—so depositing the check early has advantages. Also, banks may refuse to honor checks dated back six months or more.
- Regardless of how nice and/or wealthy a person may come across, treat every check as a business transaction and never assume anything. Many criminals are skilled con artist.

Be sure to carefully examine every personal check for information that is essential for chasing the check:

- **Personalization** — The customer's complete name and address must appear on the check.
- **Date** — The check date must be current. Do not accept post or future-dated checks.
- **Bank I.D. numbers** — The check must have a bank identification number, or routing transit number, that runs across the bottom, along with the customer's account number and check number. This information is used by a bank to identify the transaction and resolve payment issues.
- **Payee** — The "Pay to the Order of" section must indicate your business's name.
- **Dollar amounts** — Both the written and numeric amounts must match.
- **Customer Signature** — The check should be signed in your presence and verified with photo identification.

Verifying Checks

Verifying identification can help your business safeguard against fraud. However, some state laws regulate which forms of identification businesses can view. Depending on your business location, it may be illegal to require customers to show a credit card as a condition for accepting their check. Commonly accepted forms of identification often include a state-issued driver's license, I.D. card, or military I.D.

Follow these tips when verifying customer identification:

- Make sure the signature on the customer's identification matches the signature on the customer's check.
- Use discretion when recording personal information like phone numbers, identification numbers and expiration dates.
- Trust your instincts and be on the lookout for suspicious behavior or fraud "red flags." For questionable transactions, call the customer's bank to verify legitimacy of a check.
- If in doubt, don't accept the check.

Bounced Checks

What should you do if a check is returned because a customer's account is closed or has insufficient funds to pay for the transaction? In addition to instituting a check policy, some businesses are employing the help of electronic check verification companies to identify flagged individuals. For a monthly fee, businesses can compare a customer's name against a databases of individuals that are known to have written bad, stolen, or forged checks.

Even with precautionary measures in place, businesses that accept checks may still receive a bad check occasionally. If a check fails to clear on your first attempt, your bank will generally attempt a second deposit. In some cases, the customer can quickly resolve the problem by transferring or depositing funds to cover a bounced check. If the issue is not resolved by the customer, you can consult your local law enforcement agency to understand your rights and options. Some states require businesses to mail a registered letter and allow a designated waiting period to lapse before further action is taken.

If the issue remains unresolved, consider filing a suit with a small claims court or employing a collection agency to resolve the payment. Many businesses choose to employ a collection agency to avoid a lengthy and expensive court settlement.

Dissatisfied Customers

Occasionally, a business will discover that a customer has stopped payment on a check because they believe the products or services bought did not live up to expectations. In some cases, it may be prudent to provide the customer a refund or a reduction of the amount owed. In other cases, it may be necessary to take legal action. Either way, it is always wise to gather facts and seek advice from an expert.

Online Payment Services

Online payment services allow business and consumers to exchange money electronically over the internet. With an online payment service, your business can receive payment from virtually any customer with an email account. Online payment services have recently become very popular with businesses and consumers.

Advantages of Online Payment Services

Online payment services can either replace or supplement your decision to accept credit and debit cards. Opening an online payment account is often faster and easier than setting up a Merchant Account (which is required to accept credit and debit card payments). Online payment accounts typically incur smaller fees than a traditional Merchant Account, which can have a big impact on businesses with many small transactions. From a customer-service perspective, it is beneficial to have multiple payment options available.

Online payment services are also user-friendly and can simplify the payment process by storing customer card information or billing customers at a later date.

Disadvantages of Online Payment Services

As with all payment methods, online payment services have their drawbacks as well. Most of these services redirect customers to a payment service website to complete a transaction. Being forced to leave your business's website can be confusing for customers, especially those new to online shopping, and could make them abandon a purchase they may have otherwise made.

Your business may not get enough value out of offering both an online payment service and accepting card payments. On the other hand, limited payment options may turn some customers away. Finding the right balance of payment options is something that is unique for every business.

Online Payment Security Concerns

Major providers of online payment services have developed features like two-factor authentication to help businesses enhance e-commerce security. Two-factor authentication requires businesses to enter a six-digit code in addition to their password, making third-party scams rare. As e-commerce becomes more popular, security features will continue to evolve. Be sure to research service provider plans for the most current security technology.

Shopping Cart Services

Online payment services require a virtual shopping cart. Virtual shopping carts allow businesses to accept orders on multiple products from their website. A shopping cart can calculate the total, tax, and shipping costs of an order in addition to collecting customer account and shipping information.

Some online payment service providers offer free shopping cart services to businesses. If your online payment service does not provide a free, secure shopping cart option, third-party shopping cart services can be used.

Extending Credit to Your Customers

When your business accepts credit card payments and personal checks or invoices customers, it is essentially extending credit on the assumption that customers have the funds to pay for the transaction. By extending credit to your customers, you give them the option to purchase products or services today and pay for them at a later date.

When you extend credit to customers through card payments, the credit card company manages the risk. When you extend credit through invoices or personal checks, you are responsible for verifying and accepting payments and managing the risks that come with them.

Extending credit through invoices is common in some industries such as construction or manufacturing, but may not be practical for every business. To decide if extending credit is right for your business, weigh the associated rewards and risks.

- The option of credit enables customers to focus less on prices, enhances customer relations, and has the potential to generate more sales.
- Extending credit costs money. When you sell something on credit, you will not have payment on hand and will need to temporarily recoup the cost from other areas of your operating capital.
- If customers don't pay as they should, it can become a significant drain upon your time, energy, and other resources, and can even result in a long settlement process that may not end in your favor.
- Ask yourself if you have a significant business need to extend credit. Extending credit could be the factor that keeps your business afloat if it makes it easier for your customers to buy from you. Nevertheless, if it isn't necessary it may not be worth the extra time and paperwork.

Establishing Credit Practices

Before you extend credit to customers, be sure to establish detailed policies and understand consumer protection laws.

- Determine to whom you will extend credit such as individual customers or other businesses. Run credit checks on all customers before you agree to extend credit.
- Develop clear, consistent payment guidelines. Your bills should indicate when payment is due, when it will be considered delinquent, and who to contact with questions.
- Determine how you will bill or invoice customers. Will you or your employees mail requests for payment yourselves, or will you hire another company to handle invoicing?
- Create a plan for collecting late or defaulted payments. Regardless of the type of application or documents you use for credit transactions, be sure to get all of your customers' information in writing. In return, provide them with a copy of your payment policy, which spells out how penalties will be applied to late payments and how you will handle unpaid bills. It's important to have this documentation in case a fraudulent or delinquent credit transaction occurs.

Complying with Consumer Credit Laws

If your business extends credit to customers, you should become aware of consumer credit laws. The Federal Trade Commission (FTC) enforces the nation's consumer protection laws. These laws regulate how you advertise interest rates, how much time you have to respond to billing-mistake claims, how aggressive you can be when attempting to collect a debt, and other aspects of extending credit and debt collections.

Dealing With Bankrupt Customers and Collecting Debt

What happens when a customer refuses to pay a bill? When you've gone beyond adding late penalties and you still haven't seen any payment, check with your local consumer protection agency to understand your options and state laws. This information will help you decide if you should report these actions to the police, employ a collection agency, or attempt to settle the payment by other means. Depending on your local laws and the severity of the delinquent transactions, it may be cheaper to simply swallow the debt.

You may find yourself in a situation where a customer to whom you have extended credit declares bankruptcy. In this instance, the debtor then has the benefit of an automatic stay immediately upon filing a bankruptcy petition. This stay stops you from taking any further action of trying to collect the debt unless or until the bankruptcy court decides otherwise.

If a money judgment is awarded to you in court, further action may still be needed to receive payment. Such action may include contacting the defendant, or in some cases, providing information about the defendant to a law enforcement officer so that they can assist you in collecting the debt.

Mechanic's and materialmen's liens have specific regulations that apply to their industries in cases where credited customers fail to make their payments. Liens exist in most states to provide special collection rights to those who provide services or building materials used to improve real property. If the debt is not paid, the lien can be foreclosed and the property sold to pay the obligation. For more information on the specific laws that govern these debts, visit your state's Department of Consumer Affairs or Protection.

The best way to solve these situations is by preventing them from happening through strict credit policies and by conducting appropriate evaluations of credit risks before extending any credit.

CHAPTER 38

Bartering Your Aquaponic Products

Bartering

Bartering is the process of obtaining goods or services by direct exchange without the use of currency. Bartering is an excellent way to ensure the flow of necessary items and services into your household without using precious funds. Bartering is especially effective in times of economic instability or currency devaluation.

Historically, bartering was conducted through face-to-face exchanges. This method is still common in developing countries and is still conducted, to some extent, in developed countries. The internet has opened up a new medium for bartering opportunities for both person-to-person exchanges and third-party facilitated transactions.

Why Barter?

There are many reasons to participate in a bartering process. Numerous people have found themselves unemployed or with a limited cash flow, and bartering is a great way to attain products and services when times are tough.

Bartering can be done to cut costs of a small business or to reduce personal expenses. For example, aquaponic products can be bartered for carpet cleaning, hairstylist, mechanical work, or other products.

Bartering works especially well between farmers. For instance, an aquaponic operator can swap vegetables and fish for another farmer's fruit or root vegetables (i.e. potatoes, carrots, radishes, beets, etc.). Each person is still obtaining something of value, and it opens up another means for which you can be compensated for your product.

What Can Be Bartered?

Any goods or services that is desired by another person can be bartered. Bartering is limited only by one's imagination.

In 2010, 17-year-old Steven Ortiz made national headlines by bartering a cell phone to start a series of trades that ultimately put him in the driver's seat of a Porsche. Kyle MacDonald bartered his way from a single red paperclip to a house in a series of fourteen online trades over the course of a year. Even healthcare isn't out of reach. In 2011, Matthew Wagner of Connecticut was able to exchange his photography services for Lasik eye surgery through a barter exchange.

Although these stories, and many like them, are amusing as well as inspiring, one does not have to "trade-up" to be successful. A successful barter can be a trade for anything, in less or more value, that satisfies you. For instance, trading a sack of your aquaponic vegetables for a sack of fruit can easily be a win-win for both parties. Including cash with a barter for goods or services is also an option. The following are some of the most popular items that can be bartered:

- **Services**—Haircuts, massages, mechanical work, plumbing repairs, landscaping, and a variety of personal care services can be acquired through a trade. Utilizing Craigslist can lead to regular trade arrangements and help build bartering relationships and skills, which can lead to future cash sales.
- **Technology**—Electronic products or repairs.
- **Clothing**—Clothing and accessories. We all need clothes and just about everybody has clothing items that we can do without.
- **Toys and Hobbies**
- **Gifts & Crafts**
- **Food**—Food which is not being grown in the aquaponics system.
- **Materials & Supplies**—This could encompass anything from building materials, cleaning supplies, toiletries, auto parts, etc.

Bartering Methods

Face-to-Face

Occasionally, you may find bartering opportunities arise while you are with a group of friends, co-workers, or acquaintances. Other times it may be more intentional, such as approaching someone that has something you desire in order to inquire (persuade) about trading for what you have available.

Online via the Internet

The internet has facilitated a unique system of bartering with strangers. There is no need for introductions. In most cases it is as simple as posting an ad describing what you have to offer, and what you seek in exchange.

It is important to always exercise caution when utilizing the internet for bartering purposes. Not every bartering or swapping website, nor every person using these sites, are reputable. Several cities in the U.S. have created a "safe spot" for the exchange of items between people. These areas are monitored by cameras and/or police officers.

Craigslist

Craigslist is well-known website that has a section dedicated for bartering purposes. To utilize this feature, you simply go through the same process you would to post any other item for sale. It is completely free. However, there is no one monitoring the barter ads, so you must be aware of potential Craigslist scams, and realize that you are always at risk when it comes to meetups and exchanges. It is best to meet in a public place, and/or to have someone with you.

U-Exchange.com

U-Exchange is a bartering website that allows people to trade goods and services in a specific geographical area. The site requires that you register, but it is a free service that is supplemented by banner advertisements.

Just like Craigslist, U-Exchange is a general posting site, and you assume all risk and responsibility for contacts and exchanges you make. When you see a trade that you are interested in making, click on the member's name and you will be provided with their contact information. From there it proceeds very similar to a face-to-face exchange.

Cautions of Bartering

There are reasons why bartering is not the dominant system anymore, as it does have some disadvantages, such as a the availability of the product or service you're seeking. Simply put, the bartering market is not nearly as liquid as other markets. You may find yourself with something to trade, but no one to trade with at your desired time. This leaves you waiting to make your trade.

Bartering Rules

There are a number of rules for bartering, for reasons of safety and courtesy:

- **Remember, "Safety First."** Meet in a public place or have support with you.
- **Be Inquisitive.** Explore trade options. Remember, it never hurts to ask; the worst they can say is "no."
- **Consider All the Goods and Services at Your Disposal.** A great opportunity can be found by keeping your options open as to what and how much you will trade, and what you will accept.
- **Be Skeptical when Necessary.** Beware of services that do not appear to be legitimate. Also, if you wouldn't pay in monetary value for a good or service, then don't barter either.
- **Don't Barter for Something You Don't Want, Need, or Can't Profit From Later.** You should never trade for something you will later regret. It just simply isn't prudent to trade something of yours that you consider valuable for another's goods or services you deem unnecessary or unwanted.
- **Test Items to Be Sure They Work.** Remember, there are no guarantees. Be sure to thoroughly check out all items that you may be receiving in your return.
- **Don't Blame the Other Party for a Bad Trade.** You can always decline a trade, so the responsibility is always yours. If you make a bad decision, learn from it and move on.

How to Begin the Bartering Process

Although bartering is a fairly easy and straightforward process, there are some simple principles you can implement to maximize your return:

- Be proactive in identifying what you need.
- Identify suitable trading partners and/or networks.
- Make contact with a person to begin your trade. Be very clear and detailed regarding what you have and what you are looking for in return.
- Negotiate the details of the trade including where you will meet, and what you are trading.
- Don't let emotions cloud good judgment. Take a 'time out' to think about it and/or get input from a trusted source.

Final Word about Bartering

With a good plan, following the fundamental principles described above, and little effort, you can use bartering to obtain the goods and services you want or need without impeding your cash flow. The possibilities are endless!

CHAPTER 39

Marketing and Selling Fish

Industry Overview

Innovative approaches to marketing are usually the key to financial success for smaller scale fish producers. In essence, the crop must be sold for more than it cost to grow. Regardless of the size or type of venture, marketing is an essential component and requires a plan. The information contained within this chapter will help the smaller scale fish producer to formulate a marketing plan.

Most producers would like to sell to one or two high-volume buyers, such as a processing plant or distributor. This is a good marketing strategy if you are producing large quantities of fish. However, small-scale producers are not on the same economic level as larger producers are and, therefore, must usually sell for a higher price to remain profitable. Their best option is to establish niche markets for their products.

Niche markets have advantages and disadvantages. The main advantage is that producers become wholesalers and, in some cases, retailers. Consequently, producers have more control over the prices they set for their products, and they retain some portion of the profit that otherwise would have gone to middlemen. The main disadvantage is that considerable time must be spent analyzing and developing these markets. A number of critical factors should be analyzed before marketing begins.

- Best Species to Farm
- Competition
- Product Forms
- Price
- Type of Promotion
- Unique Ideas for Market Share
- Where to Market the Product
- Regulations

Best Fish Species for Success

When deciding on which species to produce, keep three things in mind.

1. **Choose a marketable species.** A good example of a species that is easy to produce, but can be difficult to market, is common carp. It is advisable to consider other, more widely accepted species. Of the sixty or so potential fish species used for food, channel catfish, Tilapia, crawfish, rainbow trout, and salmon have a large and well-established industry in the United States. Other species such as hybrid striped bass and various sunfishes also offer considerable potential.

2. **Know the complete production cycle.** Without complete production information, trying to

raise some species can be a very risky venture. Although species such as walleye, shrimp, and lobsters have an extensive public appeal and are widely consumed, each has production peculiarities and problems.

3. **Raise a variety of species,** if possible. Many market outlets for the smaller-scale fish producer prefer buying smaller quantities of various species. Production of more than one species may offer a competitive edge over single species operations. If production of a variety of species is not feasible, pooling resources with other producers may enhance species availability.

4. **Raise a species that is in high demand.** When choosing what fish to rear, you will need to consider what types of fish are in demand in your local market and whether it suits your climate. You preferably want to identify a viable species that is in high demand. In addition, you may also consider growing fish, such as goldfish or Koi, which can be sold as pets.

Dominating the Competition

The seafood industry has been well established and can be competitive in some areas. Aquaponic operators compete with wild-caught and farm-raised products of both domestic and foreign origin. Understanding your competition helps you develop production and marketing programs around specific species and markets that will provide the highest profits. Also, remember that other seafood products are not the only competition you will have to contemplate. You must consider competition from all protein products such as poultry, beef, and pork.

However, the market for fish is vast and the demand for mercury-free, healthy fish increasing at an exponential pace. This particular market, if targeted properly, can present an environmentally-minded, healthy fish farming entrepreneur with endless opportunities. Furthermore, the price of fish and meat is increasing at an astonishing rate.

The U.S. consumption of fish continues rise. As more consumers become concerned with the preservation of the environment, initiatives are launched to improve the way we use resources. This is integral to food production in America. Monterey Bay Aquarium's Seafood Watch Program has published Chinese-raised Tilapia as an "avoid" in a watch list for seafood. Interestingly, "in 2009 the U.S. imported 404 million pounds of Tilapia, up from 298 million in 2005. Wal-Mart imports nearly 8.8 million pounds every month, although they will not say how much comes from China," (Bloomberg, para. 3, 2012). Currently, 92 percent of all fish consumed in the United States is imported. However, as consumers are becoming more aware of the dangers of consuming foreign and ocean caught fish, the demand for organically raised American fish becomes even greater. Since domestic fish farmers do not have the capacity to meet the demand, excellent opportunities exist for aquaponic operators to gain market share and develop a profitable industry.

Use all of your resources—industry experts, the telephone book, and your own energy—to help evaluate the competition. Talk with potential customers to determine their level of interest. Realize that the development of a new market requires substantial effort.

Product Forms

A unique product form can make your business stand out. The size of the product can also be important to the selected market. One of the best ways to select a product form is to find out what the customers want and give it to them. For instance, channel catfish are usually sold after reaching a live weight of one to two pounds. At this size, a 1.5 pound fish will yield two 4.5-ounce fillets.

The following is a list of the more common fish product forms, with descriptions and specific information related to each one.

- **Live** fish are sold to live-haulers who stock fee-fishing lakes or farm ponds, or sell to consumers who dress them at home for consumption.
- **Fish in the round** are put on ice and sold just as they come out of the water.
- **Drawn** fish have their entrails removed and are usually sold on ice.
- **Dressed** fish are sold completely cleaned with the entrails removed. Heads may be left intact, as trout are often sold, but generally the head is removed. Fins and tails may be removed or left intact.

    Species such as channel catfish have the skin removed. On trout and other scaled fish, the skin is usually left intact.
- **Steaks** are cross sections of dressed fish around 1-inch thick. Larger catfish (more than three pounds) are sometimes sold as steaks.
- **Nuggets** come from the belly flap after it is cut free from the fillet. Channel catfish nuggets are common in supermarkets. Their popularity may be a result of the lower price. In general, these nuggets have a stronger flavor than fillets.
- **Fillets** are boneless pieces of fish.
- **Flank fillets** are the two sides of the fish cut away from the backbone. Rib bones and skin are usually removed.
- **Butterfly fillets** are the two, skin-on flank fillets held together by the belly flap or across the back (with the backbone removed). Trout are sometimes sold as butterfly fillets.
- **Strips** are smaller pieces of fish cut from fillets. Strips are usually breaded, marinated, or used for other value added treatments.
- **Deboned** fish have the rib and back bones removed, with the rest of the body intact.
- **Smoked** fish is a value-added product. Two smoking methods (hot and cold smoking) are employed. Hot smoking never produces enough drying to ensure safe keeping without refrigeration.
- **Hot Smoking** involves temperatures of 250° to 300°F for a period of four to five hours. Cold smoking, on the other hand, preserves fish by drying. Cold smoking requires as little as 24 hours or as long as three weeks at temperatures never exceeding 80°F. If you decide to get involved in smoking, there are a number of potential regulations that must be addressed. Proposed FDA regulations are described in subpart A of 21 CFR part 123 of the Federal Register.
- **Breading** fish also adds value (and weight) to a fish product. The fish are generally dipped in liquid batter (usually milk or egg mixtures) and rolled in seasoned bread crumbs or corn meal.

The most common processed product forms are dressed, fillets, nuggets, and steaks. The preferred product size will depend on an individual customer's preferences. Fillets, for example, are generally cut into prescribed proportions that yield a single serving (four to eight ounces) from one or two fillets. As a rule, the whole fish needs to be at least 1.25 to 2.5 pounds to obtain the appropriate size fillets. The dress-out percentage, or yield, on fish such as channel catfish, hybrid striped bass, Tilapia, or trout range from 33 percent on some fillets to more than 60 percent for whole dressed fish. Frozen or refrigerated are also forms that need to be considered.

Regardless of the product form you choose to offer, it is very important to establish and maintain a reputation for quality and reliability. Be sure to gain an accurate understanding of each customer's needs before delivering the first fish.

Pricing for Success

Putting a price on your product is not as simple as you might think. Often, pricing a product is an agonizing, lengthy decision that will likely require periodic adjustments to reflect new market environments.

The lowest price to charge would be equal to your cost per pound, including both fixed and variable costs. The highest price would be what one or two customers could be talked into paying. The following are a number of factors to consider when establishing a product's price:

- How will the product be positioned in the food fish market? Is it more like caviar or carp (i.e. local sustainably raised premium organic fish vs. fish caught off Japan/China in heavily polluted waters laced with mercury, Fukushima radiation, and other contaminants?
- Who are the customers? What are they accustomed to paying? Are they individual consumers, up-scale restaurants, or food wholesalers.
- What species and prices are competitors offering?
- What quality perceptions and uniqueness, if any, are associated with the chosen species or culture method?

Many systems have been developed to aid in pricing products. The following are descriptions of the systems most relevant to small-scale marketing.

- **Cost-Plus pricing** simply adds a constant percentage of profit above the cost of producing a product. The problem with cost-plus pricing is that it is difficult to accurately assess fixed and variable costs. This pricing system works fine in the absence of severe competition.
- **Competitive pricing** is probably the easiest and, in retail marketing, the most common form of pricing. In this system, producers gather market information on prices and quantities of competing products and then price their products accordingly.
- **Skimming** involves introducing a product at a relatively high price for more affluent, quality-conscious customers. Then, as the market becomes saturated, the price is gradually lowered.
- **Discount pricing** offers customers a reduction from advertised prices for specific reasons. For example, a fish farm advertises in the local newspaper that prices will be 25 percent less if they bring the advertisement from the paper. Or, a producer who advertises on local radio offers customers a discounted price when they mention the advertisement. Discount pricing can often apply to purchases of larger quantities.
- **Loss-Leader pricing** is offering a limited selection of the products, at a reduced price, for a limited time. The goal is to attract more customers to the producer's place of business so that they might also buy non-discounted products as well. This pricing method is seen at farmers markets and supermarkets to introduce a new product or to create consumer interest.
- **Psychological pricing** involves establishing prices that look better or convey a certain message to the buyer. For example, instead of charging $7.00 per pound, the producer charges $6.99 per pound. This will make the product appear to be more of a bargain. Or, instead of charging a price close to production costs, the producer charges a higher price that buyers associate with a higher quality or a more desirable fish species.
- **Perceived-Value pricing** is positioning and promoting a product on non-price factors such as quality, organic, farmed sustainably, or grown locally. Then, the producer must decide on a price that reflects this perceived value. An example of this strategy would be promoting local, organic raised versus imported fish or any species you could portray as having a high probability of being contaminated with mercury.

Effective Promotion and Advertising Methods

After a product and price have been decided upon, a promotional strategy needs to be developed.

Promotion is a way to attract customers. Ideally, a high-quality product in demand will sell itself. However, if no one is aware that the product is for sale (when and where), no sales will be made. Time allocated to promoting a product typically results in a worthwhile payoff. The two general methods of promoting fish products are generic and personal promotions.

- **Generic promotion** is commonly performed by large commodity groups such as The Catfish Institute, the National Aquaculture Association, etc. This type of advertising promotes a certain type of product, but does not endorse any particular brand or company.
- **Personal promotion** is used to distinguish your product from other products. A number of methods of personal promotion are available to small-scale fish marketers. Word-of-mouth advertising is one of the best types of personal promotion. One customer who is satisfied will tell their friends about your product. The multiplying effect of word-of-mouth promotion can be tremendous, but often slow and further promotion will be required. It is also important to remember that a *dissatisfied* customer will also tell friends.

Other common channels for advertising include radio, newspaper, TV, magazines, handbills, flyers, and posters. The promotional message must be clear, to the point, and focused.

Point-of-purchase materials, such as recipes and information about your aquaponic operation, will help maximize sales. Before creating your own point-of-purchase materials, decide if available materials can be adapted for your use. For example, a variety of recipes and general information are available about farm-raised fish through the library, internet, public agencies, and various associations.

Social media is the interaction among people in which they create, share, or exchange information and ideas in virtual communities and networks. Social media marketing can help spread the word about your business, quality products, and services. Internet users continue to spend more time with social media sites than any other type of site. At the same time, the total time spent on social media in the U.S. across PC and mobile devices increased by 37 percent to 121 billion minutes in July 2012, compared to 88 billion minutes in July 2011.

According to 2011 Pew Research data, nearly 80 percent of American adults are online, and nearly 60 percent of them use social networking sites. Social media marketing is useful for every business that has customers who are using the internet. It's wise to bring in outside help to establish a social media marketing strategy that can be carried out and implemented. Many firms and qualified individuals will do your social media marketing for you at a wide range of prices and levels, reliant upon your needs and budget. Effective social media marketing places your product in front of your target audience.

The form(s) of promotion you choose will depend on the scale of your operation, available resources, availability of the product, and geographic location of the operation. In addition to public advertising, it is important to consider on-site product promotion,

both visual and verbal. Remember to include the non-price attributes of the product that will help develop repeat customers. The following is an itemized list of potential marketing strategies.
- Point-of-Purchase Materials
- Flyers
- Posters
- Word-of-Mouth
- Craigslist
- Clubs and Organizations
- Auto Wrap
- Business Alliances
- Farmer's Markets
- College Campuses
- Facebook and Twitter Account
- Company Website
- LinkedIn
- Social Media
- Newspaper and local shopping guide advertisements
- Local, State, Federal Association Memberships

Unique Ideas for a Large Market Share

The small aquaponics operator often finds it necessary to provide some unique product or service to carve out a piece of the market. This uniqueness can be obtained by providing a custom order service, offering special delivery schedules, providing products not readily available in the area, etc. Be careful not to commit to any schedules or make promises that can't be fulfilled. Where the product is marketed can also provide some interesting possibilities.

Best Places to Market Your Product

There are many different marketing and product outlets for the small-scale aquaponic operator. Your choices will be affected by costs, such as processing, delivery, advertising, overhead, materials, equipment, and personal time. Species selection, product form, target market, and company location will also have a profound effect on this issue. Selling the fish to a large processor is often not desirable or possible for the small operator. This does not mean, however, that there are not available markets.

Direct Retail Sales

Direct retail sales, where the producer sells directly to the customer, is generally where the greatest per-unit profit is realized. Direct retail sales to consumers is a good place to start if supplies are small or availability of the product is uncertain. The following is a list and description of several direct retail sales options.

- **Local Customer Base:** This is the simplest of all direct marketing options. Individual sales are made to customers on a repeat basis. Clients pickup from the farm or you deliver. A customer base takes time to develop, but using advertising materials such as the local newspaper, Craigslist, or a direct mailer containing news on availability, new products, nutrition information, and recipes can speed-up the process.
- **Roadside Market:** This option has many variations. The product can be live, fresh iced, or in some cases, dressed and iced. A small market may be operated at the farm site, or a live tank can be set up at a more populated location with heavier traffic. The fish may be kept in a live tank on the truck or in a tank set up at the remote location, with the permission of the property owners. Off-farm locations may include busy intersections, convenience stores, gas stations, farmers markets, flea markets, or liquor stores. The mobile marketing technique brings the product to the people and increases the potential market area. Check with local officials to determine if permits or other restrictions apply.
- **Fish Fry Fund-Raiser:** Many groups use a fish fry to raise money. They include churches, schools, hospitals, civic groups (scouts, YMCA, Lions Club, Trail Life USA, etc.), political groups, and

other non-profit organizations. Marketing to these groups may require larger quantities of fish of similar size. You may provide just the fish products or cater the entire event for a percentage of the ticket sales. Other aquaponic product catering opportunities include events such as birthday parties, weddings, and other private parties.

- **Office Building Markets:** Tall buildings hold lots of people who go home from work hungry, but often don't want to stop at the store or fish market. Contacts are made in offices through bulletin boards, flyers, word-of-mouth, and direct sales. A sales force can even be recruited from clerical workers who can make sales during the course of the workday. Sales can even be made during the early part of the week with deliveries later in the week. Ice, coolers, and individual packing are required for this type of marketing.
- **Fairs and Festivals:** This is a proven marketing option. County and state fairs are excellent target markets. A list of these events can generally be obtained from the local or state chamber of commerce. These events draw hungry crowds. Much of the food is overpriced and not very good. Good, healthy fish plates provide an opportunity to capitalize on this opportunity, as well as promote and educate the public on the benefits of aquaponics. On the downside, often a commission or fee is paid to the fair organization.
- **Value-Added Market:** Each of these marketing techniques could be considered value-added if the fish are processed to customer specifications. Other value-added products include smoked, breaded, or marinated fish. Customers in this market understand that they will need to pay premium prices for quality products and services.
- **Tank Harvest Sale:** This is a popular marketing technique that works very well for both small and medium size aquaponic operations. By planning ahead and advertising in local papers and radio, a farmer may be able to sell an entire crop in one day. Prepare holding facilities for sale of any left-over fish.
- **Bartering:** Trading fish for other products and/or services is a great way for all parties to get what they need without incurring a tax burden. Carpet cleaning, landscape work, painting, beauty care, automotive service, trading for fruits, etc. are viable alternative markets for obtaining true value from your product.
- **Direct Wholesale Sales:** Wholesaling to other businesses that sell directly to the consumer is another option. Although, the direct-wholesale option usually reduces the per-unit profit it can increase the units sold.

Set up appointments with managers of every restaurant, grocery store, and food wholesaler within a 50-mile radius of the production site. Find out beforehand, if possible, individual preferences for species, product form, size, volume, availability, and prices. Have a strong sales pitch prepared and a fresh sample of your products. Pricing in the wholesale market is usually based on individual negotiation, so determine your price range and have a negotiation strategy.

Some managers will be immediately interested while others will not. For those who are interested, customize the product to fit individual needs. Keep your customers satisfied by supplying the size, form, quantity, and quality of product that the customer expects. Especially important are good human relations skills, showing common courtesy, and building rapport. A list and description of several direct wholesale options follow.

- **Live Hauling:** Live haulers generally purchase fish at the site and transport them to other outlets, including processing plants, pay lakes, recreational lakes, or retail outlets. Small-scale producers often have difficulty working with live haulers because the producers lack the large quantities of fish the haulers need to make it a profitable activity. There are, however, companies that charge a fee for custom harvesting. These companies are generally in large production areas, and it may be difficult to get them to a small production facility. Live haulers prefer not to handle small quantities of fish (less than 1,000 to 2,000 pounds, and in some areas not less than 5,000 to 10,000 pounds).

One advantage of selling to a live hauler is that there is no additional personal investment of time or equipment to process, transport, or sell your fish. It is prudent to deal with live haulers on a cash basis, especially if you have not worked with them before.

- **Sales to Local Restaurants:** Restaurants can be an excellent market for fish farmers. Growing fish to match the desired plate portion as well as the weekly volume can be a good source of revenue. The typical restaurant will take ten to 80 pounds of fish per week. Restaurants like unique and new items for their "catch of the day" menu. Learning to produce a popular fish species and marketing it out of season can bring big dividends. Work with a chef to develop a new dish using your product. It is good advertising for both you and the restaurant.

When deciding which businesses to contact, remember that many businesses serving food are not necessarily identified as "restaurants." Do not overlook the country club, the VFW, caterers, or the corner pub.

Once a restaurant becomes a customer, make a point of helping to educate the staff about your product. Educating the head of the serving staff and providing a short brochure, or other printed information, may be a key to continued success.

- **Supermarkets:** Many seafood markets and supermarkets buy locally produced fish. Retail chain supermarkets offer a good market for larger quantities of fish. Unless a supermarket is locally owned and operated, it might be necessary to supply part or all of the chain stores. This may be more volume than the small aquaponics operator can handle. A number of the large superstore markets now have live fish tanks that need a consistent supply of quality live fish (20 to 50 pounds per week). Smaller supermarkets and seafood stores are generally easier to work with, and more likely to sell local products. Educating the staff about your products in these settings is also extremely important. It is a good idea to offer point-of-sale information for use at the seafood counter.

- **Specialty Stores:** These stores include ethnic grocery stores, gourmet shops, and health food stores. Fish is an important part of people's diets in some cultures. Health food stores may be willing to try your product because the perceived quality and healthy aspects of farm-raised products is usually higher than that of wild-caught fish. Ethnic markets are usually more willing to purchase whole fish. Each of these markets has special demands for equipment, capital, time, and effort.

Licenses, Permits, and Regulations for Raising and Selling Fish

Fish production and marketing activities are regulated at the local, state, and even federal levels. Depending on your operation, health inspections as well as business and sales tax permits may be required. Following are health permit sources for several kinds of operations.

- Retail outlets and restaurants: County Health Department
- Processing facilities: Governed by most state public health agencies and regulated by local jurisdictions (i.e. county health departments, etc).
- In addition to a state's Fish and Game agency, many states also have Food and Drug Branch that oversee fish farming.

- Interstate commerce: U.S. Food and Drug Administration and the USDA (United States Dept. of Agriculture).

Checking with the chamber of commerce (local regulations), city hall (business licenses), and your state's Department of Conservation and Natural Resources ("privilege" license for operating a fish wholesaling establishment) may be a good idea during the formulation of your business and marketing plan. For retail sales, a sales tax permit may be required. For mobile operations, the Department of Transportation should be consulted. You can also check with your county's Extension Office or Fisheries Specialist regarding this inspection policy.

Unfortunately, "Big Brother" can get so involved with oversight, requirements, policies, and periodic inspections along all points of the operation that running a business can be challenging to the small aquaponics commercial operator, due to the additional time drain caused by these regulatory requirements. Operators must keep records of their testing, meet various operating requirements, and record every detail about their business. With all this regulatory oversight and associated cost in accommodating so many different agency requirements, many aquaponic operators choose to run their operations under the radar.

Marketing Plan Synopsis

Before any production begins, it is prudent to establish a solid marketing plan. One of the primary considerations in developing this plan is the time and effort that you can devote to marketing your product. Evaluate the market for the best species options, keeping in mind your personal situation (including finances, experience, and time availability).

In addition to planning, critical issues such product, price, promotion, and location must be evaluated prior to the onset of production. One of the best ways to address these and other critical questions discussed is to develop a written marketing plan.

This plan should include many of the same items as a business plan. The plan should detail goals, financial data including capital required, budgets, and cash flow analysis, how regulatory requirements will be met, a detailed list of necessary equipment, and a feedback system to monitor the progress of the venture. Haphazard business planning will lead to an inefficient and possibly even a failed enterprise. A detailed plan provides direction and helps avoid some of the pitfalls associated with any new venture.

Regardless of the market avenues chosen, it is best to target specific markets. Determine what size market you can service well and limit your initial marketing program to those areas. Develop more than one market outlet. The key to niche marketing success is to develop and maintain a reputation for quality, dependability, and excellent customer service. Rewarding aquaponics business opportunities are always open to those creative individuals who are willing to plan, work hard, and persist.

Note to Reader

Please visit the website "FarmYourSpace.com" to obtain other helpful aquaponic information, join the aquaponics community, see a gallery to aquaponic operations, and post your own aquaponic project.

CHAPTER 40

Sales & Selling: Being Successful

The Importance of Sales in an Organization

In any business organization, sales is the department that generates revenue. No matter how good the product or service, how cutting-edge the technology, or how progressive and forward-thinking the company, without sales everything else is useless.

Sources of Revenue

A business organization can generate revenue from a variety of sources including operating income from sales, royalties, dividends and interest; income from financial assets it owns; payouts from insurance policies, rental income; and capital gains from the sale of owned properties. However, even when organizations derive little or no taxable income directly from operations stemming from sales—such as an investment company—they must still generally have some sort of sales effort to generate investment revenue to fund their investments in income-producing assets. For example, a limited partnership must often engage in a concentrated sales effort to recruit more limited partners.

Sales versus Marketing

While it is sometimes difficult to draw the line where the marketing process ends and the sales effort begins, the sales effort is the effort that actually collects money—or the obligation to buy, in the case of a purchase order or finance arrangement. The marketing effort creates favorable conditions for the sale to take place. In a nutshell, the marketer leads the horse to water; the sales team makes it drink.

Partnership between the Sales and Marketing Teams

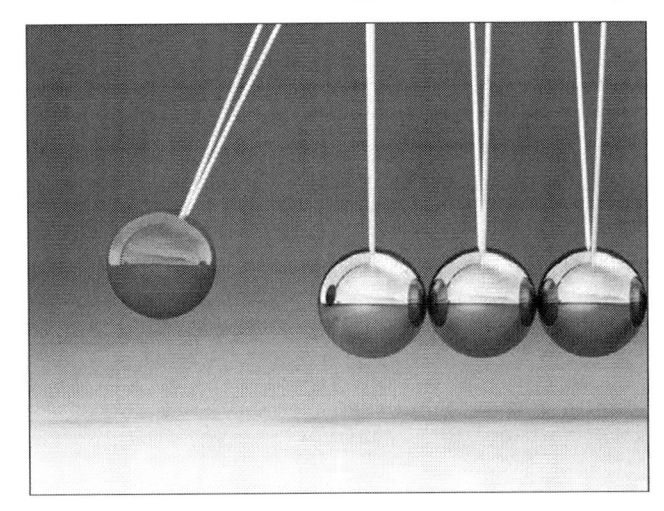

Few products are truly bought on impulse. Even a can of brand-named soda on a shelf is there, because its wholesaler built a relationship with the store manager over time and secured good shelf placement, denying it to a competitor. To get the most out of the sales effort, the sales team needs support from the marketing team to facilitate follow-up contacts, mailings and account service. If the support is not there, the account may not last long, and turnover will increase. If the sales staff is too directly involved in that effort, it may eventually become overwhelmed with account services and find it difficult to grow the business. Sales is so important, then, that it typically behooves management to free its sales staff from some or all of the account services process to generate future revenue.

Investing In Sales
Many companies under-invest in their sales effort, treating sales like an afterthought to be handled after the managers solve all the manufacturing, distributing and financing issues. The best sales forces are built with professional, well-compensated employees. They aresupported by a strong marketing effort and empowered to act, serving key client interests with marketing support, money and time. They have strong personal relationships with key customers, or they learn how to build them.

Big-Ticket Necessity
Sales is more common or more intensive when companies market big-ticket items, such as cars, appliances and furniture. Large purchases usually require more persuasive efforts and communication with prospects. Companies often use advertising and other promotions to build awareness and attract potential customers. Sales reps then either go out to meet with prospects or greet them when they come in looking to buy. Because of the risks of bigger purchases, buyers usually need a more thorough explanation of benefits and the value proposition.

Sales Conversions
A main role of sales is to improve the efficiency of converting prospects into customers. Salespeople can directly interact with prospects, ask questions to help, address any possible buyer concerns and ultimately recommend products or services. Without these important sales steps, a company has to rely on passive marketing messages to carry weight, when buyers are looking to make a purchase decision. Advertising, for instance, can't answer follow-up questions customers have after being exposed to the original ad messages. Salespeople have the benefit of back-and-forth discussion.

Growth
The sales function is a key mechanism for companies to grow through referrals. In fact, a number of sales organizations indicate in collateral materials or thank-you letters that referrals are essential to their businesses. Leveraging the fact that a new or established customer sees value in the solution, a sales rep can ask if the customer knows other people with similar needs or interests. Referrals are an efficient prospecting tool, because they create a personal connection to the new prospect.

Customer Retention
The personal, interactive nature of selling makes it a key ingredient in company efforts to build long-term relationships with customers. Salespeople can follow up after purchases to ensure customers have a good experience. Without this contact, upset customers often don't complain; they just go away to other providers. This ongoing interaction also allows for opportunities to make additional sales that address future or ongoing needs of the customer.

10 Reasons Why Top Salespeople are Successful
Brian Tracy is Chairman and CEO of Brian Tracy International, a company specializing in the training and development of individuals and organizations. He has consulted for more than 1,000 companies and addressed more than 5,000,000 people in 5,000

talks and seminars throughout the US, Canada and 70 other countries worldwide. As a keynote speaker and seminar leader, he addresses more than 250,000 people each year.

Brian Tracy's research has shown that the top 20 percent of salespeople earn 80 percent of the money. His studies have shown that the top salespeople have integrated the following vital keys for success in sales.

Key to Success #1: Top Salespeople Do What They Love to Do

All truly successful, highly paid salespeople, love their sales careers.

You must learn to love your work and then commit yourself to becoming excellent in your field.

Invest whatever amount of time is necessary to improve your sales career — pay any price; go any distance; make any sacrifice to become the very best at what you do. Join the top 10 percent.

Key to Success #2: They Decide Exactly What They Want

Don't be wishy-washy. Decide exactly what it is you want in life. Set it as a goal for your sales career and determine what price you are willing to pay to get it.

According to the research, only about 3 percent of adults have written goals, and these are the most successful, highest-paid people in every field. They are the mover and shakers, the creators and innovators, the top salespeople and entrepreneurs.

Key to Success #3: They Back Their Sales Career Goals with Perseverance

A **key to success** in sales is to back your goal with perseverance and indomitable willpower. Decide to throw your whole heart and soul into your success and into achieving your **sales career** goal. Make a complete commitment to improve your sales career and become one of the most highly paid salespeople. Resolve that nothing will stop you or discourage you.

Key to Success #4: They Commit to Lifelong Learning

Your mind is your most precious asset, and the quality of your thinking determines the quality of your sales career.

Commit yourself to lifelong learning.

I cannot emphasize this too often. Read, listen to audio programs, attend seminars, and never forget that the most valuable asset you will ever have is your mind. As you continue to learn, you will eventually become one of the most valuable salespeople in your company.

The more knowledge you acquire that applies to practical purposes, the greater will be your rewards, and the more you will be paid.

Key to Success #5: Top Salespeople Use Their Time Well

Your time is all you have to sell. It is your primary asset. How you use your time determines your standard of living.

Resolve, therefore, to use your time well. Begin every day with a list. The best time to make up your work list is the night before, prior to wrapping up for the day. Write down everything you have to do the next day starting with your fixed appointments and moving on to everything else you can think of.

Key to Success #6: They Follow the Leaders

Do what successful people do. Follow the leaders, not the followers. Do what the top salespeople in your company do. Imitate the ones who are going somewhere with their lives. Identify the very best salespeople in your field and pattern yourself after them.

If you want to become one of the best salespeople in your company, go to the top earners and ask them for advice. Ask them what you should do to improve your sales career. Inquire about their attitudes, philosophies, and approaches to work and customers.

Key to Success #7: They Know That Character is Everything

Guard your integrity as a sacred thing. Nothing is more important to the quality of your life in our society. In business and sales success, you must have credibility.

You can only be successful if people trust you and believe in you.

In study after study, the element of trust has been identified as the most important distinguishing factor between one salesperson and another and one company and another.

Key to Success #8: They Use Their Inborn Creativity

Think of yourself as a highly intelligent person, even a genius. Recognize that you have great reserves of creativity you have never used.

Say aloud, over and over, *"I'm a genius! I'm a genius! I'm a genius!"*

This may sound like an exaggeration, but it isn't. The fact is that every person has the ability to perform at genius levels in one or more areas. You have within you, right now, the ability to do more and be more than you ever have before.

Key to Success #9: They Practice the Golden Rule

Practice the Golden Rule in all your interactions with others: Do unto others as you would have them do unto you.

Think about yourself as a customer. How would you like to be treated?

Obviously you would want **salespeople** to be straightforward with you. You would want them to take the time to thoroughly understand your problem or need and then show you, step by step, how their solution could help you improve your life or work in a cost-effective way.

If this is what you would want from a salesperson selling to you, then be sure to give this to every customer to whom you talk.

Key to Success #10: They Pay the Price of Success

Finally, and perhaps more important than anything else, resolve to work hard. This is a great key to success in life.

The key to success in selling is for you to start a little earlier, work a little harder, and stay a little later. Do the little things that average people always try to avoid doing. When you begin your workday, resolve to *"work all the time you work."*

The 10 Laws of Sales Success

Entrepreneur is a North American magazine and website that carries news stories about entrepreneurship, small business management, and business. The magazine was first published in 1977, and provides practical advice on entrepreneurship and small business.

Entrepreneur emphasizes that selling can be one of the most rewarding tasks you'll undertake as a business owner, if you follow these 10 tactics:

Law #1: Keep your mouth shut and your ears open.

This is crucial in the first few minutes of any sales interaction. Remember:

Don't talk about yourself.

Don't talk about your products or services.

And above all, don't recite your sales pitch!

Obviously, you want to introduce yourself. You want to tell your prospect your name and the purpose of your visit (or phone call), but what you don't want to do is ramble on about your product or service. After all, at this point, what could you possibly talk about? You have no idea if what you're offering is of any use to your prospect.

Law #2: Sell with questions, not answers.

Remember this: Nobody cares how great you are until they understand how great you think they are. Forget about trying to "sell" your product or service and focus instead on why your prospect wants to buy. To do this, you need to get fascinated with your prospect; you need to ask questions (lots of them).

Law #3: Pretend you're on a first date with your prospect.

Get curious about them. Ask about the products and services they're already using. Are they happy? Is what they're using now too expensive, not reliable enough, too slow? Find out what they really want. Remember, you're not conducting an impersonal survey here, so don't ask questions just for the sake of asking them. Instead, ask questions that will provide you with information about what your customers really need.

When you learn what your customers need, and you stop trying to convince or persuade them to do something they may not want to do, you'll find them trusting you as a valued advisor and wanting to do more business with you as a result.

Law #4: Speak to your prospect just as you speak to a well-respected friend.

There's never any time you should switch into "sales mode" with persuasion clichés and tag lines. Affected speech patterns, exaggerated tones, and slow, hypnotic sounding "sales inductions" are never acceptable in today's professional selling environments. Speak with friendly enthusiasm, but keep it real.

Also, leave out the profanity. Profanity is offensive and makes one come across as being disrespectful. Many also hold the opinion that those using profanity lack the intelligence and discipline to use a better choice of words.

Law #5: Pay close attention to what your prospect isn't saying.

Is your prospect rushed? Does he or she seem agitated or upset? If so, ask if there is anything you can do to be of assistance, and if you can schedule a better time to meet. Some salespeople are so concerned with what they're going to say next that they forget there's another human being involved in the conversation.

Law #6: If you're asked a question, answer it briefly and then move on.

Be respectful of their time and keep the conversation focused on meeting their needs.

Law #7: Only after you've correctly assessed the needs of your prospect do you mention anything about what you're offering.

Know with whom you're speaking and focus on identifying their needs.

Law #8: Refrain from delivering a three-hour product seminar.

Don't ramble on and on about things that have no bearing on anything your prospect has said. Pick a handful of things you think could help with your prospect's particular situation and tell him or her about them. (If possible, reiterate the benefits in his own words, not yours.)

Law #9: Ask the prospect if there are any barriers to them taking the next logical step.

After having gone through the first eight steps, you should have a good understanding of your prospect's needs in relation to your product or service. Knowing this, and having established a mutual feeling of trust and rapport, you're now ready to bridge the gap between your prospect's needs and what it is you're offering.

Law #10: Invite your prospect to take some kind of action.

This principle obliterates the need for any "closing techniques" because the ball is placed in the prospect's court. A sales 'close' keeps the ball in your court and all the focus on you, the salesperson. You don't want the focus on you. You don't want the prospect to be reminded that he or she is dealing with a "salesperson." You're not a salesperson; you're a human being offering a particular product or service to fulfill their need. If you can get your prospect to understand this fact, you're well on your way to becoming an outstanding salesperson.

Improve Your Sales Conversations

If you want to improve your sales conversations, pay attention to these 7 keys:

1. **Build rapport**: Before you ask questions to get the buyer to open up or talk about how you can help, you have to build rapport. All else being equal, people buy from people they like. Be likable and focus on relationship building, and you'll find your sales conversations will go much more smoothly.

2. **Uncover aspirations and afflictions.** If you've ever read any piece of sales advice, you know you need to ask questions to uncover the prospect's pain. That's a given. What most advice doesn't include is how to harness the power of aspirations. Your job is not only to uncover the prospect's needs and pains but to also uncover their aspirations and goals. Get your prospect to open up and share his/her hopes, dreams, and desires and then demonstrate how you can help achieve his/her goals.

3. **Make the impact clear.** If you don't make the business case, you won't make the sale. You can do everything else right, but if the prospect doesn't see the value of your solution (and you need to be very clear with what that value is), they will not buy it.

4. **Paint a picture of the new reality.** This goes hand in hand with points 2 and 3. Once you know the prospect's needs and goals and the tangible impact of alleviating these pains or attaining his/her goals, you must paint a picture of what his/her new world will look like. How will it be better? In your sales conversations, help visualize the other side and build excitement around it.

5. **Balance advocacy and inquiry**. Sales conversations require give and take. You have to get the prospect talking so you can fully understand his/her situation. You also need to take what the prospect says and communicate recommendations based on your expertise to help him/her see how you can help. In each and every sales conversation (yes, this includes capabilities presentations and demos) you have to balance how much you talk and how much you listen.

6. **Build on the foundation of trust**. Trust is the foundation of sales success. A buyer will not open up and share his/her needs if he/she doesn't trust you. A buyer will not believe in your solution and that you can do what you say you can do, if he/she doesn't trust you. If there is no trust, a buyer will never see the full value of what you propose, and you will not win the sale.

7. **Plan to succeed**. Set the table for success by going into each sales conversation with a plan. Do your homework and know what you want to get out of the conversation. If you go into each conversation well-prepared and planning to succeed, you will be much more likely to make the sale.

If you follow these seven keys in your sales conversations, will you still make mistakes? Absolutely. Will you win every sale? Absolutely not. After all, we're only human, and no human is perfect.

But, these keys will help you avoid the common mistakes many sellers make and help you lead more successful and productive sales conversations.

Seven Ideas for Building Trust in Sales

Many things have changed in the world of sales, but some have not. Building trust was important 50 years ago, and it's just as important today. When buyers trust sellers, they depend on them, listen to them, give them access, and spend time with them.

Trust is critical for sales success, but today's buyers are busier than ever, and, at the same time, have access to more information and choices. This makes their time harder to get, and their trust harder to build.

Building trust is one of six key drivers of client loyalty and one of the top 10 things sales winners do. Trust in sales is built around three factors—competence, integrity, and intimacy.

Competence

Trust in a seller's competence means buyers believe you can do what you say you can. This is not referring to trust in the product or service, but trust in the seller as a person. Buyers need you to bring ideas to the table, to help them find solutions to problems, and give sound advice. If they don't trust your competence, they won't accept the advice.

Demonstrating your competence in the following three ways will go a long way towards building trust:

1. **Be an expert.** Too many buyers report they don't trust sellers, because the sellers don't know their stuff. As a seller you need to know your buyers' industries and businesses, competition, marketplace, full set of customer needs, and more—inside and out. You must answer buyers' questions about your offerings and the market, as well as about the buying process itself. If you want to guide the way, you need to make it your business to be a source of knowledge in all of these areas.

2. **Know your impact model.** Sales winners craft compelling solutions. The key to making solutions compelling is a concrete return on investment case. Be prepared to discuss, in concrete terms, what results buyers can expect to achieve. If you don't know the impact model—how you can affect their business—buyers will not trust your business sense.

3. **Develop and share a point of view.** If a buyer is confused about what to do to resolve their problem, and you don't have an answer or are unwilling to develop and share a point of view, they won't see you as a trusted advisor. Part and parcel of being an advisor is you know your stuff and the buyer wants and values your opinion.

Integrity

Everyone has been sold a bill of goods and not gotten what they were promised. You may think your integrity is off the charts, but buyers have been burned before and are suspicious before ever meeting you. You must demonstrate and prove your integrity. Buyers won't just assume it is there. To demonstrate and prove inegrity:

4. **Demonstrate moral principles.** Successful sellers always do the right thing, even in morally ambiguous situations. This can mean turning down business, suggesting alternative (and less profitable) solutions, or referring business elsewhere. Buyers trust sellers who have the buyer's best interests in mind.

5. **Honor commitments.** Successful sellers earn buyers' trust by showing up and honoring their commitments consistently. Do what you promise and do it well. Make sure buyers have clear expectations for how you operate.

Intimacy

Buyers are much more likely to buy from sellers with whom they have a relationship. Strive to develop a relationship with buyers using the following methods:

6. **Create shared experiences.** Shared work experiences expose buyers to your thinking, your work style, and your work product. Also, the more time you spend with someone, the more opportunities you have to develop a solid relationship as long as you are likeable.

7. **Be a person.** Lots of sellers are told, "Don't talk about politics. Don't talk about anything personal. You can ask about the weather, but that's it. Anything else might get you in trouble." Yeah, it might, but if you don't connect with people on a personal level, you're neglecting a critical component of building trust. Don't be afraid to connect on a personal level when the opportunity presents itself.

Sellers who don't work on building trust are missing out on a powerful differentiator. Trust is a major part of the sales equation.

Five Keys to Effective Networking for Sales

Successful salespeople tend to be diligent and purposeful in networking activities. At its core, networking is about making a human connection. Utilize these five keys and you will quickly find that your network is not only expanding, but it is actively working for you to help you increase sales.

1. **Be a connector.** Do you know someone who appears to know everyone? How can you do that? It starts with meeting people. The more people you meet and learn how you can help them, the more likely you will be able to connect them to someone who is a good match. You add value for them, and they respect you for that. The other benefit is when people you have connected sit down to talk. What is likely to be the one thing they have in common? You! It is good for business to have two people talking positively about you and your business.

2. **Listen and help.** Women are often better networkers than men, because they are typically better listeners. They know asking questions is a great way to engage with someone and draw them out. The objective of asking questions is to learn how you can help the other person — not to set them up for your sales pitch. Even though networking is not about selling, it can, of course, be a great sales tool. Why? Because networking can help you discover how you might help the other person and help them to discover how they can help you.

3. **Utilize the strength of "weak ties."** If you haven't already, you will find that most of the critical successes in your personal and professional life will come through someone who knows the person who will ultimately be responsible for that success (future client, employer, etc.). It will likely not come as a result of a direct or planned contact with that person. Those successful "friend of a friend" contacts most often happen when your "friend" is aware of who you want to meet. Thus, these seemingly weak ties can lead to significant sales over time.

4. **Follow through and follow up.** If you say you are going to do something in a meeting, that is a commitment. Do it right away. If you don't, all the hard work you put into the meeting is lost. Not only should you follow through when you are on the giving end, you should also do so on the receiving end. You will find that nothing will cause your referral sources to dry up more quickly than when you fail to contact the person to whom they recommended you reach out.

5. **Learn to tell stories.** Nothing is more engaging, or more effective, than a well-told story. That's why some of the best speeches always start with a good story. It is hard for people to remember facts and figures, but they can often recall and retell a story in amazing detail. Give an example of how you helped someone, rather than simply explaining what you do.

While networking opportunities should not be used for hard selling, networking can — if approached professionally — set the stage for sales success.

Making the Perfect 'Follow-Up Call'

Typically, it's the follow-up call that really gets the sales cycl rolling. It's here where value truly begins to manifest itself. It's here where substantive information is gathered, and it's here where the relationship begins to establish itself.

Therefore, it is absolutely vital to have superb follow-up strategies and tactics in order to make the most of the opportunity. Following are the keys to success for making a perfect follow-up call.

Tip #1: Obtain commitment for the follow-up

One of the single, biggest mistakes made by sales representatives is not establishing a specific date and time for the follow-up call at the end of the initial call. Vague commitments from the prospects ("call me next week") or the sales rep ("I'll send the proposal and follow up in a couple of days") result in missed calls, voice mail messages, a longer sales cycle, and even missed sales opportunities. Avoiding this is a simple fix: schedule a follow-up date and time. If your prospect is not available for your initially proposed date, continue recommending other dates/times until arriving at a mutually agreeable date. If that doesn't work, get them to establish a time and date. Creating a deadline is a simple, but extremely powerful, tactic.

Tip #2: Build equity and be remembered

After every call to a first time prospect, send a thank you card. Handwrite a message on a small thank you card that simply says, "John, thank you for taking the time to speak with me today. I look forward to chatting with you further on the 16th! Kind regards…" In today's fast-paced world, a hand-written card tells the client that you took the time and effort to do something a little different. This gesture registers in the prospect's mind, creating a degree of "equity" in you. It differentiates you and gets you remembered. It also gives the prospect a reason to be there, when you make you follow-up call.

If you don't think a card will get there in time, send an e-mail with the same note. Just be aware that an e-mail does not have the same impact as a handwritten note.

Tip #3: E-mail a reminder and an agenda.

The day before your follow-up call, send an e-mail to your prospect reminding them of your appointment. In the subject line enter "Telephone appointment for August 16th and article of interest." Note that the subject line not only acts as a reminder, but is vague enough that the prospect will likely open it as there is a hint that maybe the date and time has changed. Your e-mail should confirm the date and time of the appointment and briefly list your agenda: "Hi, John, the call should only take 10 minutes. We will review the proposal, and I will gladly answer any questions. Following we can discuss the next steps, if any. Respectfully yours, …"

The language is very similar to that used initially setting the follow up. In particular, notice the trigger phrase "…the next step." The "if any" will help reduce some of the 'stress,' pressure, or concern a first-time prospect may have. Often a first-time prospect will skip out on the follow-up call, because they are worried they will be pressured into making a commitment. This is a natural reaction. If the prospect senses an easy, informal, no-pressure type of phone call, he is more likely to follow through with the call.

Tip #4: Add value in a "P.S."

Notice in the subject line there is a reference to an article. At the end of your e-mail add a P.S. that says, "John, in the meantime, here's an article I thought you might enjoy reading…"

The article may be about your industry, the market, a product, or, better yet, something non-business related that you had discussed in your initial call. This creates tremendous value even if the client does not open it. Why? Because you took the time to do something extra, better educated them on something related to the call, and brought additional value to the subject. This helps get you remembered and gives the prospect another reason to take your follow-up call. Of course, this means you have to do some work in advance. Start looking on the web for articles of interest and value relative to your market, industry etc. Keep a file of these articles, because they can be used over and over again.

Compelling content can open doors, because it is VALUE ADDED instead of VALUE ASKING. The more you give, the more effective your later 'ask for the sale' will be.

Tip #5: Call on time

Make absolutely certain you don't start your relationship on the wrong foot. Call on time. Never be late with your follow-up call. The promptness and respect you show on a follow-up call reflects on you, your company and your products.

Tip #6: Avoid opening statement blunders

Many telephone sales representatives start off in ways counterproductive to the objective of the call. Following are a few classic follow-up opening statement blunders:

- "I was calling to follow up on the proposal."
- "I am calling to see if you had any questions.'
- "I just wanted to make sure you got my e-mail."
- "The reason for my follow-up was to see if you had come to a decision."

It is not that these opening statements are poor, but rather that they are routine, typical, and common. They do nothing to position you or differentiate you. What this really means is that you are perceived as yet another run-of-the-mill salesperson trying to make a sale. The opening statement needs to have pizzazz by creating some interest or enthusiasm and/or building some rapport with the prospect.

Tip #7: Crafting a powerful follow-up opening statement that produces results

There are four simple steps in creating a powerful opening statement. First, introduce yourself using your full name. Second, share your company name. Up to this pint in the opening it has been a simple and obvious process, but it is the third step where you differentiate yourself.

Remind the prospect of the reason for your call. This means going back to your initial cold call and informing the prospect of the "pain" or the "gain" that was discussed or hinted at in your previous call. For instance, "Cindy, this is _____ calling from ABC Company. Cindy, when we spoke last week you had two concerns: First, you indicated you were concerned about having your current on-line training program renewed automatically, before you had a chance to review it in detail, and second, you mentioned there were several modules with questionable content."

This reminds Cindy why she agreed to this call. You do this, because people are busy and have a tendency to forget. Furthermore, the urgency of last week may not seem so urgent this week.

Prospects like a clear, concise agenda. They want a reputable sales representative who is organized and doesn't waste their time. They want someone to take control and who will move the call forward. This gives them confidence.

Finally, notice how this approach repeats the theme established in the first call and in the follow-up e-mail. Remember the language used: "...determine the next steps, if applicable." It's a nice touch and reduces client resistance.

Tip # 8: Be persistent, polite, and professional, but not a pest

Be persistent in your follow up. Tenacity pays huge dividends. If someone genuinely wants you to stop contacting them, then do not make any further attempts to contact them. However, you can often avoid getting to that point by establishing good rapport and clear expectations.

If you end each conversation, regardless of how motivated the prospect is, by asking for permission to follow up with them in a certain period of time, you're always going to be operating in the realm of mutual respect and consideration when you follow up.

If they say no, then you know where you stand. If they say yes, then you have an open door and clear expectation that you will be following up. **By asking for permission to follow up you lower your odds of being an annoyance and have solid footing in the future.**

Summary

Having a solid follow-up strategy, with productive tactics, will separate you from the competition. It gives you a distinct advantage. Make the most of your follow-up calls, and your sales will grow.

Implement a Follow-up Schedule that Works

Don't leave follow up to chance. Most sales representatives fail to invest in setting and communicating clear expectations for what good follow-up actually looks like beyond some vague generality. *It is important to be specific and create a schedule for follow-up contacts who are most appropriate to the nature of the potential opportunity.* Create a follow-up schedule that outlines when calls and email follow-ups should be happening. Following is an illustration of how a follow-up schedule may be developed. Of course, the ideal schedule will have specific dates as well as a way to track activities for quality assurance purposes.

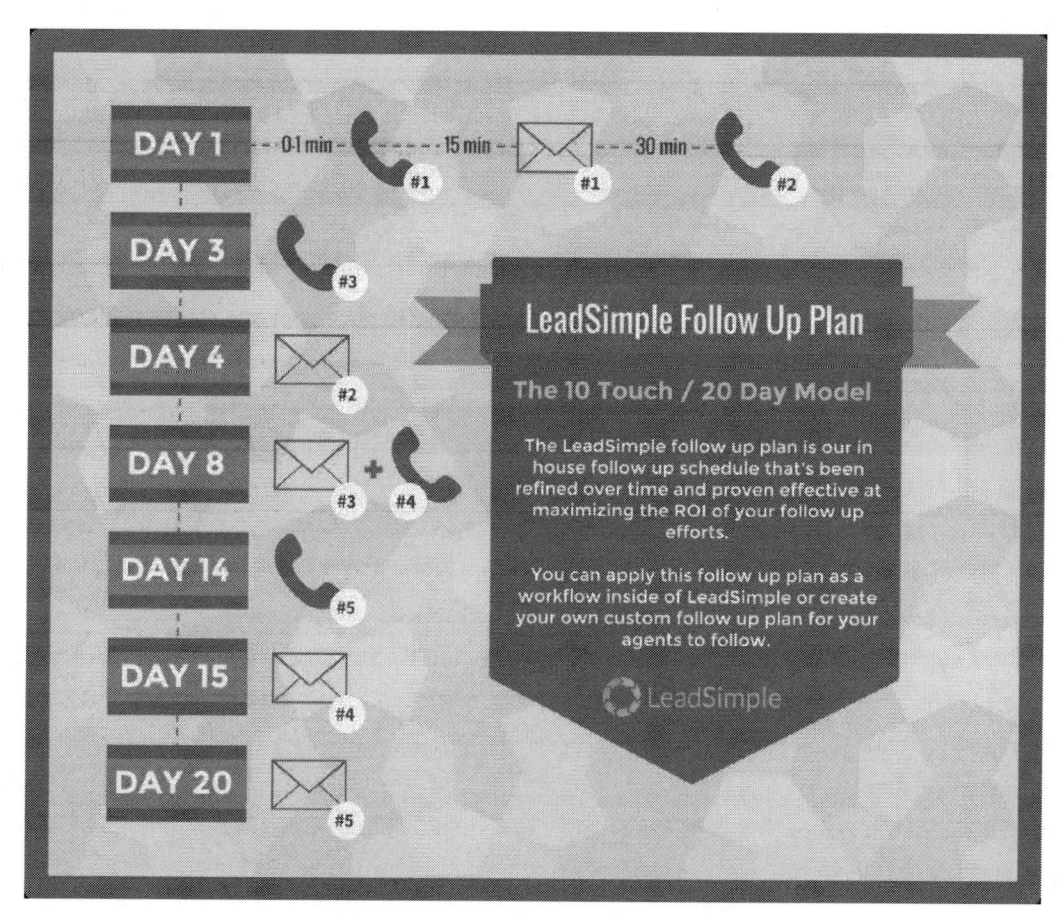

LeadSimple Follow Up Plan

The 10 Touch / 20 Day Model

The LeadSimple follow up plan is our in house follow up schedule that's been refined over time and proven effective at maximizing the ROI of your follow up efforts.

You can apply this follow up plan as a workflow inside of LeadSimple or create your own custom follow up plan for your agents to follow.

- DAY 1 — 0-1 min (call #1) — 15 min (email #1) — 30 min (call #2)
- DAY 3 — call #3
- DAY 4 — email #2
- DAY 8 — email #3 + call #4
- DAY 14 — call #5
- DAY 15 — email #4
- DAY 20 — email #5

Time Your Sales Contact for Best Results

Not all days of the week, or times of the day are created equal. Try to time your follow up to hit the sweet spot — a time where you have the highest odds of getting a response.

Wednesday and Thursday come out on top for both email open rates and phone contact rates. In fact, according to this research:

- Thursday is 26% better for *email open rates* than the worst weekday — Monday.
- Thursday is 49% better for *phone contact rates* than the worst weekday — Tuesday.

Use Multiple Contact Formats

It's as simple as it sounds. Use more than one way to reach out. Email, snail mail, phone, text, social media — it's all on the table. The goal is to touch prospects in different ways in order to stay on their mind and stand out from the competition. Therefore, keep in mind the best methods for reaching each of YOUR specific prospects.

Sales Success: Final Word

Sales success begins with attitude. To optimize success in sales you must understand that the sale is not about you; it's not about commission; it's not about moving an item of off the shelves; and it's not about the hefty paycheck. It is about the customer.

Getting to know your customer's interests and desires is critical to successfully closing a sale. Actively engaging your consumer to not only tap into their wants, but also anticipating their needs, are traits of a great sales person. Don't take it personal if you are unable to sell the product even after seemingly anticipating the needs of your consumer.

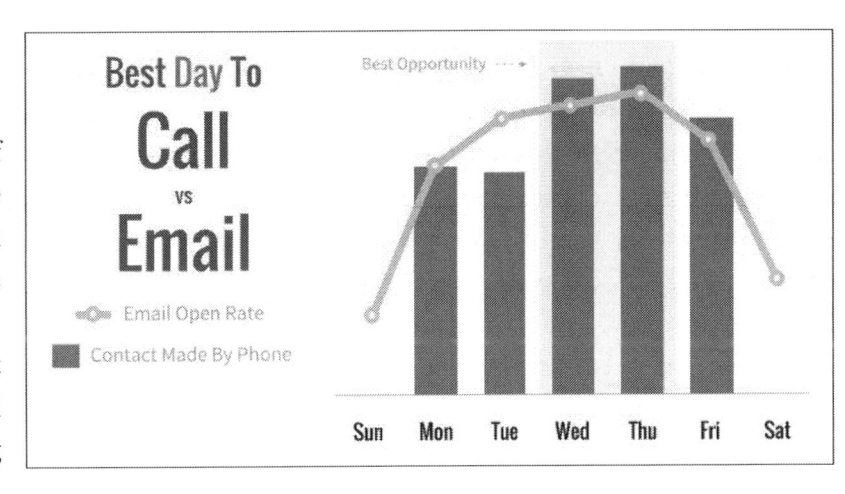

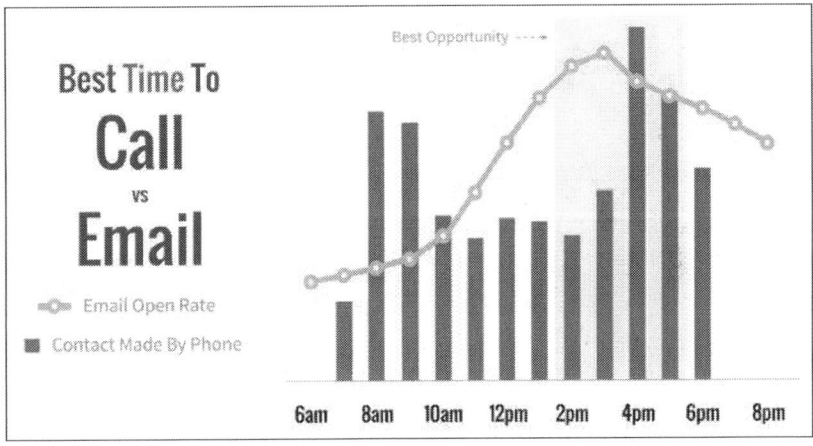

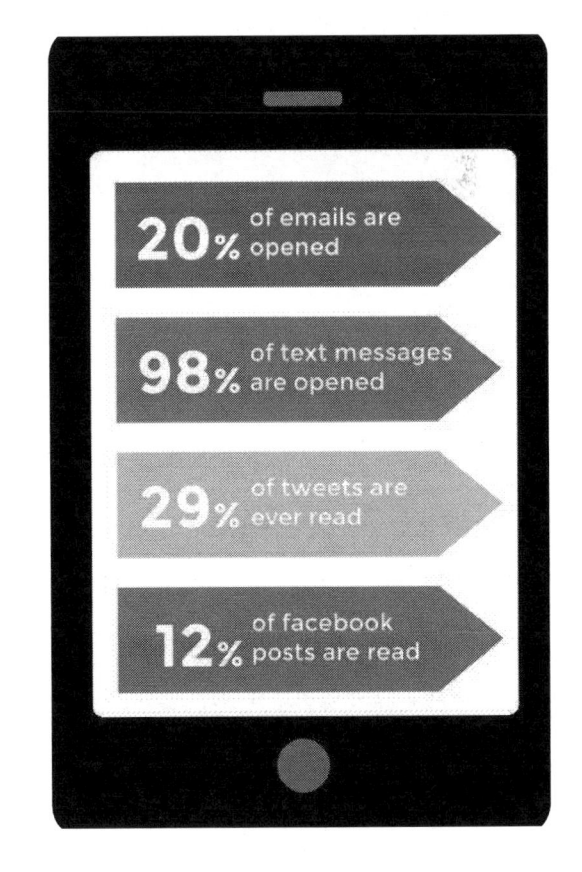

489

Being successful in sales requires having a tough skin. Take rejection as an opportunity to perfect your interpersonal sales tactics and become a better sales person.

If you are engaging your customer over the phone, be upfront regarding the reason for your call, vouch for the integrity of the product or service you are selling, and draw comparisons between your prospect and satisfied customers. When leaving voicemails, emails or follow-up messages, it's important to reiterate the reason for contacting them, as well as summarizing your intent. Finish up the message with action: to make the sale and/or a follow-up engagement of some sort.

CHAPTER 41

Customer Service Recommendations and Rewards

Customer service relates to the service provided to customers before, during and after a purchase.

No matter the size of your business, excellent customer service needs to be at the heart of your business model for you to be successful. It is important to provide good customer service to all types of customers — potential, new and existing customers.

Although it can take extra resources, time and money, excellent customer service can generate positive word-of-mouth promotion of your business, keep your customers happy and encourage them to purchase from your business again. Good customer service can help your business grow and prosper.

Significance

Customer service is important to an organization, because it is often the only contact a customer has with a company. Customers are vital to an organization. Some customers spend hundreds and even thousands of dollars per year with a company; consequently, when they have a question or product issue, they expect a company's customer service department to resolve that issue.

Identification

Customer service is also important to an organization, because it can help differentiate a company from its competitors, according to the article titled "The Importance of Customer Service" on the Drew Stevens Consulting website. For example, it may be difficult for consumers to choose between two small-town drug stores, especially if their prices are similar. Therefore, putting extra efforts into customer service may be thing that gives one drug store a competitive advantage.

Function

A company with excellent customer service is more likely to get repeat business from customers. Consequently, the company will benefit with greater sales and profits. Contrarily, companies with poor customer service may lose customers, which will have a negative impact on business. It costs a lot more money to acquire customers than to retain them, due to advertising costs and the expense of sales calls. Therefore, the efforts that go into maintaining quality customer service can really pay dividends over time.

Publicity

People that have a positive experience with a company's customer service department will likely tell two or three others about their experience, according to the Consumer Affairs website. Therefore, quality customer service can be a source of promotion for organizations. Contrarily, a person who has a bad customer service experience will likely tell between nine and 20 people.

Prevention/Solution

Customer service is important to an organization because of potential complaints. Consumers can file a complaint with the Better Business Bureau, Consumer Affairs or even a class action attorney if they are dissatisfied with a company's customer service. Legally, consumers are protected by the Federal Trade Commission and can sue a company in the courts. Equally as damaging, if not more so, are the negative reviews e and/or rants through social media a customer can post online.

Specifics as to Why Customer Service is so Important

It can help you
- increase customer loyalty
- increase the amount of money each customer spends with your business
- increase how often a customer buys from you
- generate positive word-of-mouth and reputation
- decrease barriers to buying (for example, if your business has an excellent reputation of customer service for refunds, you're more likely to entice a hesitant buyer to purchase from you).

Excellent Customer Service Defined

Excellent customer service is about
- treating your customers respectfully
- following up on feedback
- handling complaints and returns gracefully
- understanding your customers' needs and wants
- exceeding customer expectations
- going out of your way to help your customers.

Types of Customer Service

Issue-Centric Customer Service

Customer service models that focus on addressing issues rather than serving people can be classified as 'issue-centric customer service.' From the company's perspective, an issue is (1) a subject line, (2) a brief description of the problem reported by the customer, (3) the customer's email address, and (4) the issue's status and priority.

CHAPTER 41: CUSTOMER SERVICE RECOMMENDATIONS AND REWARDS

Product-Driven Customer Service

With a product-driven company all functions are focused on the product — its design, features, capabilities, and its subsequent design and manufacture. Customer service operates under the assumption that putting a strong emphasis on great products will result in higher profits and revenue.

Customer-driven customer service

A Customer-driven customer service model revolves around people. Every component of the solution is humanized, acknowledging that customers are real people with daily life struggles like the rest of us. It is a model that recognizes customers as a valuable resource, key to a company's success; therefore, this model is built to respect customers.

> Every contact we have with a customer influences **whether or not they'll come back.** We have to be great **every time or we'll lose them.**
> (Kevin Stirtz)

Developing and Maintaining Excellent Customer Service

There are certain customer service skills every business (management and employees) must master, if they are to achieve optimum success. Without these skills business growth, revenue, and profits will be limited, and customer-related problems will be amplified.

Fortunately, there are a few universal skills that all businesses (management and employees) can master to dramatically improve their relations with customers, which in turn produces positive results. Below are the 17 most-needed customer service skills to master.

The Customer Service Skills that Matter

When most business publications talk about customer service skills, things like "being a people person" tends to take the spotlight. Although such a statement is not exactly wrong, it is so vague and generic that it is of little help. The following are specific practical skills that management and employees can realistically implement to "WOW" the customers, gain loyalty, and grow their business.

1. Patience

Not only is patience important to customers, who often reach out for support when they are confused and frustrated, but it's also important to the business at large. Great service beats fast service every single time. Yet patience shouldn't be used as an excuse for slow service.

Quality time spent with the customer to better understand his/her problems and needs is an essential component of providing excellent customer service. It is imperative to remain patient with customers and take the time to truly figure out what they want, when they come to you stumped and frustrated.

2. Attentiveness

The ability to *really* listen to customers is crucial for providing great service. Not only is it important to pay attention to individual customer interactions (watching the language/terms they use to describe their problems), but it's also important to be mindful and attentive to the feedback you receive *at large*.

For instance, customers may not be saying it outright, but perhaps there is a pervasive feeling that your product or service isn't as great as it could be. Customers aren't likely to say, "Please improve your website;" however, they may say things like, "I can never find the search feature," or, "I am having trouble pacing and order."

493

Listening to your customers can help you focus on things that truly make a difference in improving your business, products, and/or service.

3. Clear Communication Skills

It is very important to remain professional and cautious in your communications. Customers don't need your life story, your opinion on issues, or to hear how your day is going. It is always best to err on the side of caution whenever you find yourself questioning a situation. Positive results from excellent customer service are achieved through simple, concise communications, leaving nothing to doubt.

4. Knowledge of the Product and/or Service

Businesses that have the best customer service ensure all employees interacting with customers have a clear understanding of the product and/or services being offered. It's not that every team member should know everything about the product or service, but rather they need to know the functions of the product or service being offered, just like a customer who uses it every day would. Without such knowledge it is impossible for these employees to help customers efficiently, when they run into problems.

5. Ability to Use "Positive Language"

Sounds like fluffy nonsense but having the ability to make minor changes in conversational patterns can truly go a long way in creating happy customers.

Language is a very important part of persuasion, and people (especially customers) create perceptions about a person and a company based on the language used by company representatives.

Small changes that utilize "positive language" can greatly affect a customer's perception of a company. Following is an example of positive language versus not-so-positive language.

- Without **positive language:** "I can't get you that product until next month; it is back-ordered and unavailable at this time."
- With **positive language:** "That product will be available next month. I can place the order for you right now and make sure it is sent to you as soon as it reaches our warehouse."

The first example isn't necessarily *negative*, but the tone it conveys feels abrupt and impersonal, and the representative could be taken the wrong way by customers. Conversely, the second example is stating the same thing (the item is unavailable), but this example positively focuses on when/how the customer will get his/her resolution.

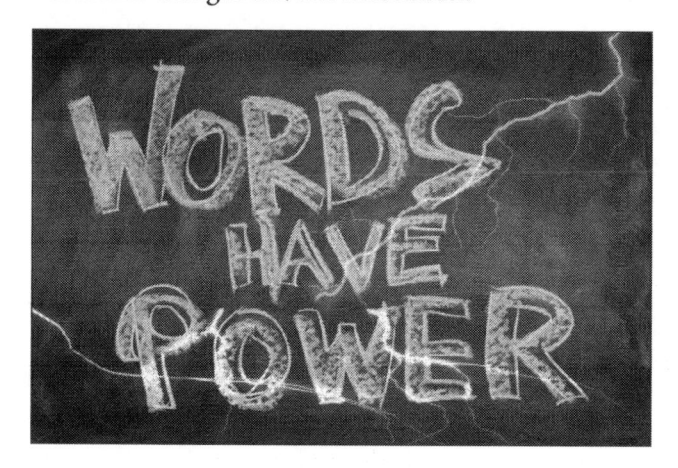

6. Acting Skills

It is impossible to please everyone all of the time. Some situations are outside our control. The other person may be having a terrible day or may just be a natural-born complainer. Some people seem to want nothing else but to pull others down.

A polished customer service representative has the *basic acting skills* necessary to maintain a cheery persona in spite of dealing with people who may be just plain grumpy. Regardless of another's words, tone of voice, or complaints, responding in a polite cheery way exuberates excellent customer service and does much to separate outstanding companies from average companies.

7. Time Management Skills

Despite many publications stating the importance of spending more time with customers, there *is* a limit to how much time you can afford to give a customer. It is also important to be respectful of the other person's time. Therefore, the key is to provide customers what they want in the most efficient way possible.

When a company representative can no longer help or is not making any progress with a customer, it is time for the customer to be connected to someone within the company that can assist him/her further. It is critical to not waste time trying to go above and beyond for a customer when it is not a prudent use of time and resources (where the cost outweighs the benefit).

8. Ability to "Read" Customers

Many customer interactions are not face-to-face. Additionally, because so much business is conducted through the Internet there is often not even a verbal interaction. However, there are some basic principles of behavioral psychology (being able to "read" the customer's current emotional state) that still apply.

This is an important part of the personalization process as well, because it still means understanding customers in order to create a personal experience for them. This skill is *essential*, because misreading a customer and losing that customer due to confusion and miscommunication is costly. To best achieve positive customer interactions, search for subtle clues about the customer's mood, patience level, and attitude in order to respond in appropriate and inviting ways.

9. A Calming Presence

There are several metaphors for this type of personality: "keeps their cool," "stays cool under pressure," etc., but they all represent the same thing. These people have the ability to stay calm and even influence others when things get hectic. The best customer service representatives know they *cannot* let a heated customer force them to lose their cool; in fact it is their *job* to try to be the "rock" for a customer who thinks the world is crumbling due to his/her current problem. Responding to a difficult customer with a negative comment or wrong attitude will, at a minimum, result in a lost customer and, at worst, agitate a customer to the point he or she will leave negative reviews and bad mouth your company to others.

10. Goal Oriented Focus

This may seem a strange thing to list as a customer service skill, but it is vitally important. Several business studies have shown how giving employees' unfettered power to "WOW" customers isn't perceived by customers as good service. Furthermore, the more relaxed or liberal customer service approach doesn't always generate the returns companies initially projected it would.

Rather, relying on customer service frameworks and goals are what help a company achieve the desired results. Instilling structured customer service guidelines — but allowing some freedom to handle customers on a case-to-case basis — efficiently helps customers resolve problems and produces much better results.

11. Ability to Handle Surprises

Whenever interacting with customers the rule of thumb should always be to expect the unexpected. Maybe the problem encountered isn't specifically covered in the company's typical protocol, or maybe the customer isn't reacting as anticipated. Whatever the case, it's best to create guidelines that will best address these situations.

For these instances it is best to create a 'fallback' plan. Following are some examples.

- **Who?** Define who the "go-to" person will be for issues not covered in the company's policies and general practices. Define a logical chain of authority to address problems and answer unusual questions.
- **What?** Define *what* issues will be sent to the people above.
- **How?** Determine the methods as to *how* employees are going to contact those on the chain.

12. Persuasion Skills

The power of persuasion is of extraordinary and critical importance in today's world. Nearly every human encounter includes an attempt to gain influence or persuade others to our way of thinking. In business, it can mean the difference between success and failure — sustainable revenue and extraordinary profits.

For many, the notion of becoming a polished persuader means being forceful, manipulative, and/or pushy. Such an assumption is absolutely wrong. Tactics like these may get short-term results, but Maximum Influence is about getting long-term results. Maximum influence isn't derived from calculated maneuvers, deliberate tactics, or intimidation. Rather, proper implementation of the latest persuasion strategies will allow one to influence with the utmost integrity.

Maximum Influence supplies a complete toolbox of effective persuasion techniques. Most people use the same limited persuasion tools over and over, achieving only temporary, limited, or even undesired results. These tools and techniques are addressed in full detail within the Persuasion chapter of this book.

13. Tenacity

Tenacity in business — more specifically in 'customer service' — involves having a great work ethic and a willingness to do what needs to be done. It is a key skill needed to provide the kind of service people talk about — in a positive way.

The many memorable customer service stories out there (many of which had a huge impact on the business) were often created by a single employee who refused to just do the 'status quo' when it came to helping someone out. This translates into positive 'word of mouth' referrals (free marketing) and higher profits. Putting in the extra effort generates exponential returns and therefore should be a driving motivator to never be lax or too casual in providing excellent customer service.

14. Closing

The 'closing' being referred to here has nothing to do with 'closing a sale.' Being able to close with a customer means being able to end the conversation with confirmed satisfaction (or as close to it as you can achieve) and with the customer feeling that everything has been taken care of (or will be).

It is important to take the time to confirm with customers that each and every issue he/she had on deck has been entirely resolved.

Properly closing shows the customer four very important things:
- That you care about getting it right.
- That you're willing to keep going until you get it right.
- That the customer is the one who determines what "right" is.
- That you care about the customer.

A happy customer is one who can say, *"Yes, I'm all set!"* following the interaction. Such is a win-win for both parties.

15. Adaptability

Every customer is different, and some may even seem to change week to week. Being able to handle surprises, sense the customer's mood and adapt accordingly is critical.

16. Thick Skin

The ability to swallow one's pride and accept blame or negative feedback is crucial. Whether working directly with customers or looking for feedback on social media, great service is doing everything reasonably possible to assist the customer, regardless of the customer's disposition.

17. Willingness to Learn

This is probably the most general skill on the list, but it is necessary. Those who don't seek to continually improve what they do, whether it's building products, marketing businesses, or helping customers, will get left behind by the people willing to invest their time making improvements.

CHAPTER 42

Time Management

Time Management refers to managing time effectively so that the right time is allocated to the right activity. Effective time management allows individuals to assign specific time slots to activities per their importance. **Time Management means making the best use of time** as time is always limited.

What is the importance of time management in your life and work? How much does being able to manage your time well actually matter?

Before we can even begin to manage time, we must learn what time is. A dictionary defines time as "the point or period at which things occur." Put simply, time is when stuff happens.

There are two types of time: clock time and real time. In clock time, there are 60 seconds in a minute, 60 minutes in an hour, 24 hours in a day and 365 days in a year. All time passes equally. When someone turns 50, they are exactly 50 years old, no more or no less.

In real time, all time is relative. Time flies or drags depending on what you're doing. Two hours at the department of motor vehicles can feel like 12 years. And yet our 12-year-old children seem to have grown up in only two hours.

Which time describes the world in which you really live — real time or clock time?

Time management gadgets and systems are designed to manage clock time. However, we live in real time, a world in which all time flies when you are having fun or drags when you are doing your taxes.

The good news is that real time is mental. It exists between your ears. You create it. Anything you create, you can manage. It's time to remove any self-sabotage or self-limitations you have around "not having enough time," or today not being "the right time" to start a business or manage your current business properly.

There are only three ways to spend our waking time: thoughts, conversations and actions. Regardless of the business, your work will be composed of those three items.

Any of the three may be frequently interrupted or pulled in different directions. While you cannot eliminate interruptions, you can manage how much time you spend on them and how much time you spend on the thoughts, conversations and actions that will lead you to success.

Why Properly Managing Our Time Is Important

Whether we assign a dollar value to it or not, time is valuable to us. Think about it: How much of your typical work week do you spend stressed about not having enough time to complete a task or reach a goal?

There are only so many hours in a day, none of which can be reclaimed. How many hours a day do you have left today? Whatever your definition of time management, it can't be stored, saved or borrowed. Once it's gone, it's gone. Time management is about making the most of your time — and the more you value it the better you'll use it.

There are various ways to tackle the issue of time management — you can download apps, adjust your sleep time, create lists, etc. However, if you don't fully understand why it's important for you to better manage your time, those apps and lists aren't going to help you. If you don't have the motivation to use them, you won't.

To appreciate the importance of managing time effectively and efficiently you first must get a good understanding as to what you stand to gain from it. To obtain the motivation and discipline needed to develop better time management skills, review the following 20 reasons why time management is so important.

1. **Time is limited** — No matter how you slice it, there are only 24 hours in a day. If you want to accomplish more and achieve greater success you have to recognize that time is limited and acknowledge the importance of effectively managing this limited resource.

2. **You can accomplish more with less effort** — When you learn to take control of your time, you improve your ability to focus. And with increased focus comes enhanced efficiency, because you don't lose momentum. You will breeze through tasks more quickly.

3. **Improved decision-making ability** — When you feel pressed for time and have to make a decision, you're more likely to jump to conclusions without fully considering every option. That leads to poor decision making.

   Through effective time management, you can eliminate the pressure that comes from feeling like you don't have enough time. You'll start to feel calmer, more confident, and in control. When it is necessary to examine options and make a decision, you are able to do so with clearer thinking, which diminishes the chances of making a bad decision.

4. **Become more successful in your career** — Time management is the key to success. It allows you to take control of your life rather than following the flow of others. As you accomplish more each day, make more sound decisions, and feel more in control, people notice. You gain more respect and admiration from others as someone on whom they can count to get things done, which increases success and the opportunity for advancement.

5. **You learn more** — The more you learn, the more likely you are to succeed. Great learning opportunities are all around you (others, Internet, library,

education system, organizations/clubs, business owners, etc.).

When you work more efficiently, you have that time to better educate yourself and pursue professional development opportunities. You can volunteer or even pick up a part-time job where you can hone your skills and/or identify your passion. There are more opportunities to have lunch with people who can assist or mentor you.

The more you learn about your company and your industry, the better your opportunities for success. Effective time management will provide you the time to learn and grow.

6. **Reduce stress** — When you don't have control of your time, it's easy to end up feeling rushed and overwhelmed. When that happens, it can be hard to figure out how long it will take to complete a task. Learning to manage time efficiently will reduce the amount of unhealthy stress you feel.

Once you learn how to manage your time, you no longer subject yourself to that level of stress. Aside from it being better for your health, it will also give you a clearer picture of the demands on your time, which will allow you to better estimate how long a given task will take you to complete, and know whether or not you can meet a deadline.

7. **Free time is necessary** — Everyone needs time to relax and unwind. Unfortunately, though, many of us don't get enough of it. Between jobs, family responsibilities, errands, and upkeep on the house and yard, most of us are hard-pressed to find even 10 minutes to sit and do nothing.

Having good time management skills helps you find that time. When you're busy, you're getting more done. You accumulate extra time throughout your day that can later be used to relax, unwind, and prepare for a good night's sleep.

8. **Self-discipline is valuable** — When you practice good time management, you leave no room for procrastination. The better you get at it, the more self-discipline you learn. This is a valuable skill that will begin to impact other areas of your life where a lack of discipline may have kept you from achieving a goal.

9. **You're more fulfilled** — People often think getting organized means time management software, lists, planners and diaries, but it goes beyond that. It starts with the choices and decisions you make based on the values you hold. When you know what matters and handle those tasks efficiently, it's time well spent. How you function affects how you feel about the whole of your life.

10. **You have more energy** — Strange but true — the act of finishing tasks often brings a level of satisfaction and energy that makes you feel good. The importance of time management here? It will help you do more of those endorphin-releasing activities. Your ability to manage time has a direct affect on your energy levels.

11. **You develop more qualities** — Once you apply skills, techniques and strategies, you'll find they work in conjunction with qualities we all have, but don't all use: Patience, persistence, self-discipline and assertiveness will all develop. As you expand your awareness of time, your ability to manage it improves, too.

12. **You achieve what you want to and need to faster** — Better time management means you accomplish more.

13. **You enjoy your life more** — The more value you put on your time, the greater your ability to learn how to do what matters, allowing you to enjoy life more.

14. **More Free Time** — While you can't create more time, but you can make better use of it by efficiently managing it. Even simple actions like shifting your commute or getting your work done early can produce more leisure time in your life.

15. **Less Wasted Time** — When you know what you need to do, you waste less time in idle activities. Instead of wondering what you should be doing next, you can already be a step ahead in your work.

16. **More Opportunities** — Being on top of your time and work produces more opportunities. The early bird always has more options, and luck favors the prepared.

17. **Improves Your Reputation** — Your time management reputation will precede you. At work and in life you will be known as reliable. No one is going to question whether you are going to show up, do what you say you are going to do, or meet that deadline.

18. **Less Effort** — A common misconception is that time management takes *extra* effort. To the contrary, proper time management makes your life easier. For example, things such as packing for a trip or completing a project take less effort.

19. **Get More Done** — Of course, being productive is one of the main goals of time management. When you are aware of what you need to do, you are better able to manage your workload. You will be able to get more (of the right tasks) done in less time.

20. **Less Rework** — Being organized results in less rework and mistakes. Forgotten items, details, and instructions lead to extra work. How often do you have to do a task more than once? Or make an extra trip because you forget something? Being organized lessens the frequency of these.

The Positive Cycle of Good Time Management

Looking through the list above, it's easy to see the multiplicative effect of time management. Good time management allows you to accomplish more in a shorter period of time, which leads to more free time, which lets you take advantage of learning opportunities, lowers your stress, and helps you focus, which leads to greater professional success. Each benefit of time management improves another aspect of your life. Effective overall time management benefits you in *all* areas of your life.

The importance of time management depends on the value you place on your time. It can also be argued that the importance of managing time effectively and efficiently is dependent upon the value you place on your life and what you desire to accomplish with your life. Developing self awareness of the value of time in your life is the first step. Learning time management skills is the next.

Time Management Improvements

"Time management" is the process of organizing and planning how to divide your time between specific activities. Good time management enables you to work smarter — not harder — so you get more done in less time, even when time is tight and pressure is high.

The highest achievers manage their time exceptionally well. By using the time-management techniques in this section, you can improve your ability to function more effectively under even the most adverse conditions.

Good time management requires an important shift in focus from activities to results. It is important to realize that **being busy isn't the same as being effective.**

Spending your day in a frenzy of activity often achieves less, because you're dividing your attention between so many different tasks. Good time management allows you to focus on the project, so you get more done in less time.

Ways to Improve Your Time Management Skills

Do you often feel stressed out due to too much work? Do you feel you have more tasks on hand than time to do them or do you feel with effective use of your time you could complete all the given tasks?

The trick is to organize your tasks and use your time effectively to get more things done each day. This can help you reduce stress and increase production. Time management is a skill that takes time to develop and is different for each person. You have to find what works best for you. Below are 17 proven strategies that may help you.

1. **Delegate Tasks:** It is common for all of us to take on more tasks than we can comfortably handle. This can often result in stress and burnout. Delegation is not running away from your responsibilities but an important function of management. Learn the art of delegating work to your subordinates as per their skills and abilities.

2. **Prioritize Work:** Before the start of the day, make a list of tasks that need your immediate attention — too often unimportant tasks consume an inordinate and unnecessary amount of your precious time. Some tasks need to be completed immediately on a specified day, while other less pressing tasks could be carried forward to another day. In summary, prioritize your tasks to focus on those that are most important.

3. **Avoid Procrastination:** Procrastination negatively affects productivity. It results in wasting essential time and energy and should be avoided at all costs. It will become a major problem in one's career and personal life if not addressed. Even small amounts of time lost here or there, due to procrastination, can add up to a significant loss of time in a short period.

4. **Schedule Tasks:** Carry a planner or notebook with you and list all tasks that come to your mind. Make a 'To Do' list before the start of each day, prioritize the tasks, and make sure each is attainable. High achievers typically make their 'to do' list out for the next day at the end of each day. To better manage your time management skills, you may find it beneficial to make three 'to do' lists: work, home/family, and personal.

5. **Avoid Stress:** Stress often occurs when we accept more work than is within our ability to reasonably achieve. The result is that our body starts feeling tired which can affect our productivity. Instead, delegate tasks to your juniors and make sure to allow time for relaxation.

6. **Set Deadlines:** When you have a task at hand, set a realistic deadline and stick to it. Try to set

the deadline a few days before the task is actually due, so you have time to deal with all those tasks that may interrupt you along the way. Challenge yourself and meet the deadline. Reward yourself for meeting difficult challenges.

7. **Avoid Multitasking:** Most of us feel that multitasking is an efficient way of getting things done, but the truth is we do better when we focus and concentrate on one thing. Multitasking hampers productivity and should be avoided to improve time management skills.

    You are not doing yourself, your company, or your friends and family any favors by multitasking. Research shows it's not nearly as efficient as we like to believe and can even be harmful to our health.

    When it comes to attention and productivity, our brains can only handle a finite amount. It's like a pie chart, and whatever we're working on is going to take up the majority of that pie. There's not a lot left over for other things, with the exception of automatic behaviors like walking or chewing gum.

    Moving back and forth between several tasks actually wastes productivity, because our attention is expended on the act of switching gears. Furthermore, it is almost impossible to get fully "in the zone" of any one activity when we're multitasking.

    Contrary to popular belief, multitasking doesn't save time. In fact, research shows that it typically takes longer to finish two projects when jumping back and forth than it does to finish each one separately. According to a study published by the American Psychological Association, switching between tasks can cause a 40% loss in productivity.

8. **Start Early:** Most successful men and women have one thing in common: they begin their tasks for the day early. As the day progresses, energy levels start going down which negatively affects productivity and performance.

9. **Take Some Breaks:** Take a break occasionally. Too much stress can take a toll on your body and affect your productivity. Take a walk, listen to music or do some quick stretches. Or, you could take time off from work to spend time with your friends and family and/or do something you love. Breaks refresh and reenergize us.

10. **Learn to say No**: Politely refuse to accept additional tasks, if you think you are already overloaded with work. Take a look at your 'To Do' list before agreeing to take on extra work.

11. **Allocate time allowed.** Activities, and even conversations, should have an assigned time period. Appointment books work. Schedule appointments with time frames for yourself and your 'to-do' list items. Create time blocks for high-priority thoughts, conversations, and actions. Schedule when activities will begin and end and have the discipline to keep these appointments.

12. **Live the 80/20 Rule.** The 80/20 Rule is one of the most helpful of all concepts of time and life management. It is also called the Pareto Principle after its founder, the Italian economist Vilfredo Pareto, who first wrote about it in 1895.

    This rule says that 20% of your activities will account for 80% of your results. 20% of your customers will account for 80% of your sales. 20% of your products or services will account for 80% of your profits. 20% of your tasks will account for 80% of the value of what you do and so on. This means that if you have a list of ten items to do, two of those items will turn out to be worth as much or more than the other eight items put together.

Make a list of all the key goals, activities, projects and responsibilities in your life today. Which of them are, or could be, in the top 10% or 20% of tasks that represent, or could represent, 80% or 90% of your results?

Resolve today that you are going to spend more and more of your time working in those few areas that can really make a difference in your life and career and less and less time on lower value activities.

13. **Schedule time for interruptions.** Plan time to be pulled away from what you're doing. Build extra time into each of your appointments (meetings, thoughts, tasks as listed above) to allow for the inevitable interruptions that will come your way.

14. **Think about it.** Immediately before every call and task decide what result you want to attain. This will help you know what success looks like before you start. Following each call and activity consider whether or not your desired result was achieved. If not, what was missing? How do you put what's missing in your next call or activity?

15. **Do not disturb.** Put up a "DO NOT DISTURB" sign when you absolutely have to get work done.

16. **Take control of your response to incoming.** Just because the telephone rings doesn't mean you have to answer it. Same goes with text messages. Disconnect all instant messaging, including social media, news feeds, and financial notifications. Don't instantly give incoming your attention, unless it's absolutely crucial in your business to offer an immediate human response. Instead, schedule a time to answer or return phone calls, return messages, and reply to emails.

17. **Block out Social Media distractions.** Distractions like Facebook and other forms of social media, unless you use these mediums as tools to generate business, are a major time drain and should be avoided. Even when using these mediums for your business, be mindful you don't allow them to distract you from the task at hand.

Time Management Tips for Dealing with Email

1. **Add "No Reply Needed."** If appropriate, insert "No Reply Needed" in the "Subject" line or the opening of your email. This can reduce the number of return emails you receive.

2. **Save time by using pre-written responses.** Use pre-written responses to frequently asked questions or requests for information, such as directions, fee schedules, or "how-to" guidelines. Then you can cut-and-paste your reply emails, saving time.

3. **Save time by answering under the question.** When someone send an email asking several questions, reply with, "See my response in bold text under your questions below." Then simply insert your responses after their original questions in bold text.

4. **Attach first, write next.** If you're sending an attached file, attach it when you first start writing the email, so you don't forget to do it.

5. **Use a clear subject line.** Put enough details in the subject line so recipients know the gist of your email right away. This will help reduce replies and questions, while also reminding you immediately of the subject, when there is a reply.

6. **Keep email focused on one topic.** Have one email per subject. People often respond to your first and last questions, but overlook or forget the others, so keep things simple. Use one email to address the meeting reminder, for example, another the department social event, and another for the status of a particular report. Recipients can respond accordingly, as they have the time and necessary information. This also saves you time in organizing your email, prioritizing your responses, and reading responses.

7. **Use the telephone when appropriate.** A phone call is sometimes faster and easier than sending numerous back and forth emails; a phone call is also considered a more personal touch in some situations.

Checking Email

Checking your email regularly during the day can be an effective way to keep your inbox at manageable levels. However, the constant interruption and distraction that comes from multitasking in this way can dramatically lower your productivity and disrupt your ability to enter a state of flow when working on high value projects.

One strategy you can use is to check email only at set points during the day. For instance, you may decide you'll only check your email first thing in the morning, before lunch, and at the end of the day. It helps to set your email software to "receive" messages only at certain times, so you aren't distracted by incoming messages. If you can't do this, at least make sure you turn off audible and visual alerts.

You can also reserve time to read and respond to email after a long period of focused work, or at the time of day when your energy and creativity are at their lowest (this means you can do higher value work at other times).

If you're concerned that your colleagues, boss, or clients will be annoyed or confused that you're not responding to their emails quickly, honestly explain that you only check email at certain times as a means of better managing your time and being more productive. Be sure to let them know they are welcome to call you for urgent matters.

Time Management Tips When You're Drowning in Email

Keep your email box as lean as possible. Having too many emails in your in-box is like having a pile of papers on your desk. It can be stressful just thinking about all that "stuff" with which you have to deal. Instead, organize your email with folders for each project, client, and/or subject. Taking to time organize your email this way will save time in the future. Don't use your 'In-box' or 'Sent' folder as a huge miscellaneous file. If you keep things organized in your email program, you'll also keep them better organized in your mind

Following are a few more ideas to better manage email to save time:

1. Decide what action or response is necessary immediately after reading an email. By deciding right away, you save time not having to reread and rethink it. Make it a priority to handle each email only once when possible.

2. Once you've read an email, make one of these four decisions.

    Option #1: Dump it. Delete the email, unless you have a good reason not to do so.

    Option #2. Delegate it. Forward the email to someone else and have them deal with it.

    Option #3. Delay it. Postpone your response, if you're waiting for more information, if it is a time-consuming issue that is not urgent, or if it is a trivial request. If the sender resends the request, apologize and do what is prudent to promptly end the matter.

    Option #4. Do it. Immediately do what the email requests or requires.

3. Use the "Tools/Organize" or "Tools/Rules" function in your e-mail program to color-code incoming email from key people, so those emails stand out from others.

4. Create an "Action Items" folder for important email that needs attention. As stated above, you want to deal with emails immediately after reading; however, some may require more information or action that must be dealt with at a later time. Review the items in this file daily and save them to your hard drive or delete them when you've finished them.

5. Glance at all new email "Subject" lines and delete the junk mail as you go. As you do this, look for the important ones that you'll read.

6. Use the "Block Sender" option to prevent emails from specific individuals or companies that are always nothing more than a waste of time from cluttering your inbox.

7. Unsubscribe from solicitation emails.

8. If your email system can organize messages according to "threads," read the last message first in a thread that deals with a particular subject. Many times you won't need to read the previous ones.

9. Before you set up auto-filing features consider whether urgent mail might wind up being auto-filed before you see it.

10. Avoid getting on lists for jokes, cute stories, etc. If you like to receive this kind of material, set up an auto-filing function to send them into special files you can review at your leisure.

11. Check your email only at designated times. You should attempt to work until you come to some kind of natural break, a stopping point, or a designated time.

12. If you have to keep complete records of email correspondence, save your "Reply" e-mail. When you reply to people's email, a copy of their entire e-mail is automatically included in the reply.

13. Use one address if you register for something on the Internet (which might attract spam), another for business, and another for personal use.

14. Regularly purge your email of outdated and unnecessary messages. Archive e-mail you need to keep for historical reasons.

15. If you don't want everyone in a group to see each other's e-mail addresses and desire to prevent having to sort through a bunch of 'back-and-forth' traffic, send the e-mail as a blind copy (BCC). BCC will keep the email private.

16. When replying to sender only, don't do a "reply all." Don't "reply all" to an entire group, if your message is not relevant to everyone.

17. If you forward a message, put your comments at the top rather than at the end. This will save confusion and reduce reply questions.

Understanding the Difference between "Urgent" and "Important"

- *"Urgent"* tasks demand your immediate attention.
- *"Important"* tasks matter and not doing them may have serious consequences for you or others.

For example:

- **Getting gas when near empty is urgent.** If you don't do it, there will be significant, immediate consequences.
- **Brushing your teeth regularly is important** If you don't, you may get gum disease, cavities, or develop other problems, but it's not urgent. If you leave it too long; however, it may become urgent.
- **Picking your children up from school is both urgent and important.** If you are not there at the right time, they will be waiting in the playground or the classroom.
- **Reading funny emails or checking Facebook is neither urgent nor important.** A significant amount of valuable time can be wasted on these activities. Even doing so for only a few minutes during the day can add up to a lot of wasted time over a week, month and year. These time drains distract you from getting your urgent and important tasks done and get in the way of your goals.

This distinction between urgent and important is the key to prioritizing your time and your workload, whether at work or at home. To better get a handle

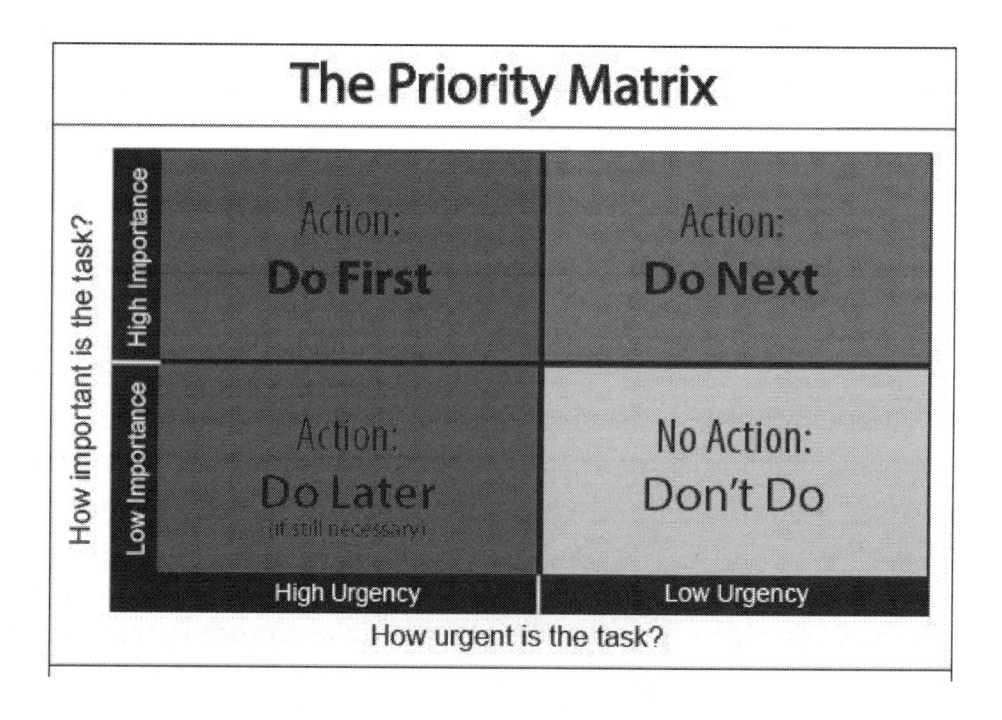

The Priority Matrix

	High Urgency	Low Urgency
High Importance	Action: Do First	Action: Do Next
Low Importance	Action: Do Later (if still necessary)	No Action: Don't Do

How important is the task? / How urgent is the task?

on where items best "fit" in your life try using a grid, like the priority matrix below, to organize your tasks into their appropriate categories:

Remember, also, that your health is important. Just because you have lots of very important things to do doesn't mean you should avoid exercising, preparing a healthy meal, or even taking time for 10-minute walk. You should not ignore your physical or mental health in favor of more "urgent" activities.

Furthermore, urgency and/or importance is not a fixed status. You should review your task list regularly to make sure nothing should be moved up or downgraded, because it has become more or less urgent and/or important.

What can you do if an important task continually gets bumped down the list by more urgent but still important tasks?

First, consider whether it is genuinely important. Does it actually need doing at all, or have you just been telling yourself that you should do it?

If it really is important, then consider delegating it, if possible, or try to make other arrangements (i.e. use automatic bill pay instead of mailing a check each month).

Further Principles of Good Time Management

Clutter can be both a real distraction and genuinely depressing. Tidying up can improve one's sense of both self-worth and motivation. It is also easier to stay on top of things if your workspace is tidy.

Top Tip for Tidying:

Create three piles: Keep, Give Away, and Throw Away.

1. **Keep**, if you need to keep it for your records, or do something with it. If it needs action, add it to your task list.
2. **Give away**, if you don't need it, but someone else might be able to use it. This also includes issues that can and should be delegated.
3. **Throw away** (or recycle) things that have no value to you or anyone else.

Know your Flex, Peak and Weak Times of the Day

Most of us have times of day where we are more alert and productive than other parts of the day. It is best to schedule the highest priority items for the times in which we have the most energy. Other items like meetings, errands, and basic administrative responsibilities can be scheduled during other times of the day.

Another useful option is to have a list of important but non-urgent, small tasks that can be done in that odd 10 minutes we have available here and there throughout the day, such as between meetings, while waiting for someone, etc. These are good times to do things like check the email or revise your "to do" list.

Don't Procrastinate

If a task is genuinely urgent and important, get on with it. If, however, you find yourself making excuses about not doing something, ask yourself why.

You may be doubtful about whether you should be doing the task at all. Perhaps you're concerned about the ethics, or you don't think it's the best option. If so seek counsel, or talk it over with colleagues, family or friends to see if there is an alternative that may be better.

Stay Calm and Keep Things in Perspective

One of the most important things to remember is to stay calm. Feeling overwhelmed by too many tasks can be very stressful. Remember — the world will not end if you fail to finish all the items on your day's "to do" list. Going home or getting an early night so that you are better prepared to tackle things tomorrow may be a much better option than inflicting your body with a lot of unhealthy stress.

Taking a moment to pause, pray, and breathe deeply 10 times can do a lot to put your priorities into perspective. Often, after such a time-out we find our view changes quite substantially, and the things that bothered us aren't such big deals after all.

Plan for Success

What is the point of exercising proper time management without a clear plan to take us where we want to go?

- "By failing to prepare, you are preparing to fail." — Benjamin Franklin

- "He who every morning plans the transaction of the day and follows out that plan, carries a thread that will guide him through the maze of the most busy life. But where no plan is laid, where the disposal of time is surrendered merely to the chance of incidence, chaos will soon reign." — Victor Hugo

- "Lack of direction, not lack of time, is the problem. We all have twenty-four hour days." — Zig Ziglar

- "The best time to start was last year. Failing that, today will do." — Chris Guillebeau

- "Remember that time is money." — Benjamin Franklin

- "Yesterday is gone. Tomorrow has not yet come. We have only today. Let us begin." — Mother Teresa

CHAPTER 43

Aquaponics Business Plan (A Real-World 'Go-By' Example)

> **Note**
>
> This is a 'real-world' business plan, prepared in phases over the course of two years, with the assistance of two different MBA groups, for funding acquisition purposes. The business plans herein has only been slightly 'tweaked' from the original version so as to make it more applicable to multiple users and situations.
>
> An editable copy of this business plan, in WORD format (as well as a second aquaponic business plan, plus additional helpful aquaponics business related documents) can be acquired at: www.FarmYourSpace.com

PART VII: MAKING MONEY AND EARNING A PROFIT FROM AQUAPONICS

AQUALIFE

Aqualife Small Business Administration Loan
Putting the Health and Environment of Sacramento First

(((Your name & contact info here)))

CHAPTER 43: AQUAPONICS BUSINESS PLAN (A REAL-WORLD 'GO-BY' EXAMPLE)

Executive Summary

The reports purpose is to provide the necessary information to our client, Mr./Ms. _____, so that he may start his business tentatively named Aqualife. The company will operate an Aquaponics system, which is the process of growing fish and plants in one integrated system. Both the fish and plants work together in providing essential nutrition for an efficient recirculating system. In no way does the system harm the environment.

Aqualife will strive to be a leading provider of fresh fish and organic vegetables farmed and cultured from the very eco-friendly, self-sustaining system. Aqualife will raise the awareness in the community, and ensure our products are readily accessible to buyers. People of all generations that eat fish and healthy veggies will find a benefit added by purchasing and eating organically grown tilapia from our controlled and balanced farming system. Not only will our fresh tilapia fish provide a high level of quality, but also the organically grown tomatoes will be fresh, crisp, and chemical free.

Aquaponics will penetrate the Sacramento region by implementing a very structured, and focused marketing plan to increase the demand of our product line — tilapia fish and a variety of organic tomatoes. The market niche is thought to be an innovative, environmentally friendly healthy consciences consumer that cares deeply about nutrition and value. Growth potential is unlimited, as fish and vegetables are essential to balanced diet.

Primary data conducted on the campus of California State University provides insightful information about the awareness of the integrated Aquaponics system. In relation to secondary data, obtained from the Census survey, it appears that there is great potential in the Sacramento market. Given growth rates in the Sacramento region, it is feasible that Aqualife will expand in to a large-scale producer. Aqualife will also provide services such as educational courses, workshops and more to its customers.

The legal requirements of starting an Aquaponics business are very important to consider. Fish and Game officials monitor licensing, laws, and regulations in the industry. The FDA, OSHA, and Fish and Game also closely observe property inspections, distribution, and processing of fish.

Of utmost importance is the potential financial health of Aqualife. A Small Business Association (SBA) loan is a practical and viable option for such a start-up company. Pro-forma balance sheets, incomes statements, and break-even analysis are provided to give a financial snapshot to local lenders interested in financing the company.

Mr/Ms. _____ is devoted to this business and maintains that Aqualife; and will be deliberate in producing healthy food for the community while operating in a way that greatly helps the environment.

2

Table of Contents

Business Description and Vision **4**
   1.1 Mission 4
   1.2 Company Vision 4
   1.3 Business Goals and Objectives .. 4
   1.4 History of Aquaponics 5
   1.5 Aquaponics Today 5
   1.6 About Tilapia 5
   1.7 Company Principles 6

Definition of the Market .. **7**
   2.1 Greater Sacramento Market Analysis 7
   2.2 Target Market 9
   2.3 Needs and Profile of Target Market 11
   2.4 Potential Market Share Analysis 11
   2.5 Industry Analysis 13

Description of Products and Services **16**
   3.1 Products and Services Offered 16
   3.2 Competitor Analysis........... 16
   3.3 Competitive Strategy 18

Organization and Management **19**
   4.1 Organizational Chart and Personnel Tasks 19
   4.2 Owner Biography 20
   4.3 Legal Structure 20
   4.4 Licensing and Regulation 21

Sales Strategy **23**
   5.1 Product Distribution........... 23
   5.2 Pricing....................... 24
   5.3 Promotion 25

Financial Management .. **26**
   6.1 Initial Investment 26
   6.2 Projected Balance Sheet 28
   6.3 Projected Income Statement ... 30
   6.4 Break-even Analysis 32

Conclusion and Recommendations **34**

References **35**

Business Description and Vision

This section outlines who our business is and what it stands for. It also outlines how we plan to accomplish goals and objectives so that we may grow into a profitable company. The history of the Aquaponics is provided with a snapshot of the industry as it stands today.

1.1 Mission Aqualife

Will offer organically grown fish and vegetables to the Sacramento region. The focus is to offer and provide high quality fish and vegetables derived from an eco-friendly Aquaponics system. Aqualife is devoted to raising the awareness of the self-sustaining, environmentally friendly aquaponics industry in Sacramento by building strong and lasting business relationships with the local community and its valued customers. Our operational and marketing slogan is:

"Offering Sacramento fresh fish and vegetables at no cost to the environment"

1.2 Company Vision

Aqualife will become a leading and well-known aquaponic business in the Sacramento region by providing the highest quality of tilapia fish and vegetables at a reasonable price. Besides our high value products, Aqualife will be known for providing the local community with educational courses and workshops designed to increase the awareness about aquaponics, where our products can be purchased, and our delivery services.

"With the help of our patrons, and customers we will become Sacramento's leading provider of fish and vegetables. We will educate our customers, promote awareness, and always be a friend to our community"

Starting as a moderately sized commercial producer in the city and county of Sacramento is important to gaining the necessary experience, which will allow us to grow into a large, commercial-size producer that captures business from surrounding cities and regions.

1.3 Business Goals and Objective

Aqualife's keys to success include:
- To grow high quality tilapia from fingerling size to a full grown, marketable size fish.
- To harvest organic and marketable vegetables, such as tomatoes and bell peppers
- To create a recirculating tank system designed to conserve water and grow both fish and vegetables year round.
- To raise the awareness of our business in Sacramento by promoting in promising market segments.
- To establish a market niche and penetrate for growth.

1.4 History of the Aquaponics

Aquaponics has roots in both Aztec and China. Aztecs were one of first to use aquaponics. They developed and cultivated plants on chinampas (called floating gardens). These chinampas were shaped like rectangles that were created by finding a shallow lakebed and fencing it by using a wattle. A wattle is "a composite building material used for making walls, in which a woven lattice of wooden strips called wattle is daubed with a sticky material usually made of some combination of wet soil, clay, sand, animal dung and straw." (Wikipedia — wattle, para.1, 2012) Then the fenced off area was layered with mud, lake sediment, or decaying vegetation.

In ancient China, the Chinese used aquaculture to grow rice. The system consisted of caged ducks that were above fin fishponds. The ducks dropping fed the finfish below them. The finfish then processed the ducks waste. The catfish ponds below the finfish would feed off the finfish's waste. The water from the catfish ponds was used to grow rice (Eden & Eden, 2011).

1.5 Aquaponics Today

Aquaponics emerged from the aquaculture industry because fish farmers were exploring methods of raising fish while trying to reduce waste discharge, and decrease the dependence on land, water and natural resources. In the beginning, aquaculture was done in large ponds but due to advancing technology, it has progressed into a recirculating system, which is now known as Aquaponics.

Aquaponics integrates aquaculture (raising aquatic animals such as fish) and hydroponics (cultivating plants in water). The fish waste feeds the plants and the plants act as filters and cleanse the water that fish live in. Such as system can produce fresh, organic fish and plants at much higher yields than what is produced through traditional agricultural methods.

1.6 About Tilapia

Tilapia is rapidly becoming one of the most popular seafood in the United States, with the National Marine Fisheries Service ranking it, the fifth most consumed seafood. In fact, American's annual consumption of tilapia has quadrupled over the last 4 years, from a quarter pound per person in 2003 to more than a pound in 2007. Researchers predict tilapia is destined to be one of the most important farmed seafood products of the century. Majority of the fish imported to the United States are frozen fish fillet from China. There is, however, an on-going discussion regarding the water quality in Chinese Tilapia Farms. (About Tilapia, 2011)

A most healthy and adaptable fish to cultivate in the Aquaponics technology is Tilapia. Tilapia is a nutritious and healthy fish, which is high in protein, low in fat, and high in Omega 3 and Omega 6 fatty acids. Because of its herbivore eating habits, it thrives in the Aquaponics system. Tilapia thrives on plankton, aquatic macrophytes, algae, and other vegetable matter.

5

Because of tilapias have a vegetarian like diet, tilapia does not have a surplus of toxins or other dangerous pollutants in their bodies that often accumulates through typical commercial fish feeds. Therefore, the plant diet that can be fed to tilapia is very conducive to an Aquaponics system.

1.7 Company Principles
- Provide nutrient rich organic vegetables
- Grow fish and vegetables without any added chemicals, pesticides or growth hormones
- Provide health-promoting Omega nutrient fish crucial to every diet.
- Operate an environmentally friendly business, honoring the central idea to aquaponics—growing healthy food while conserving land and water resources.
- Operate according to applicable regulatory requirements.
- As the public awareness improves, and profit margins reach an appropriate benchmark, offer career opportunities, internships, and workshops for people interested in learning about aquaponics—as a means of helping others, obtaining affordable labor, and/or generating another revenue stream.

Definition of the Market

Aqualife's initial focus is to meet the demand of the Sacramento area. Sacramento County encompasses approximately 994-square miles in the middle of the 400-mile long Central Valley, which is California's prime agricultural region. The County is bordered by Contra Costa and San Joaquin Counties on the south, Amador and El Dorado Counties on the east, Placer and Sutter Counties on the north, and Yolo and Solano Counties on the west.

The table and chart below shows the population estimates for the entire Sacramento region, along with the associated growth rate.

2.1 Greater Sacramento Market Analysis

CITY/COUNTY	2012 ESTIMATE	GROWTH RATE %
Citrus Heights	83,299.00	-0.1
Elk Grove	153,018.00	0.24
Folsom	72,504.00	0.09
Galt	23,760.00	-0.03
Isleton	809.00	-0.12
Rancho Cordova	66,628.00	1.17
City of Sacramento	469,331.00	-0.05
County of Sacramento	558,730.00	0.12
Total Population	**1,428,079.00**	**1.32**

The table above shows that the city and county of Sacramento are growing at a steady rate. Neighboring cities to Sacramento, such as Rancho Cordova and Elk Grove are growing more rapidly. Because of the close proximity to the Sacramento region, the outer cities provide many additional opportunities. The targeted market from the start of business will be the entire county of Sacramento. As sales, profits, awareness and accessibility increase, the business can be expanded to market additional regions.

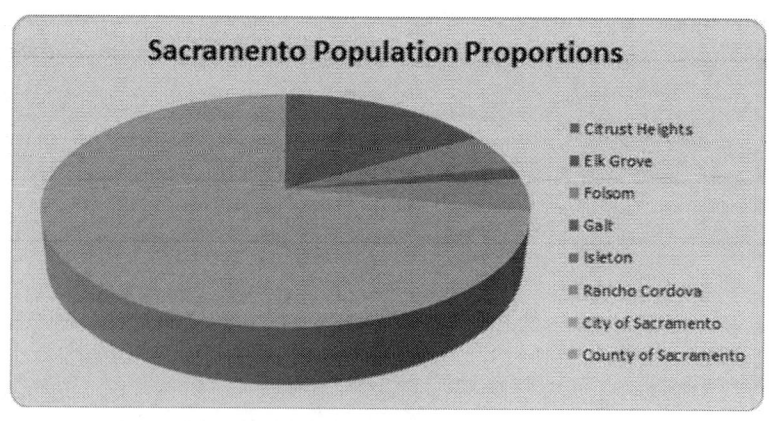

Data retrieved from Sacramento County Website
http://www.saccounty.net/FactsMaps/DemographicsandFacts/default.htm

The table above shows that statistically, there is unlimited opportunity in the Sacramento area market. A closer look at age groups will provide more insight into what percentage of the consumers in the Sacramento Area is environmentally friendly, willing to try our products, and have a taste for fish.

With regard to location, Sacramento spans several major highways. Highway 50, 80, 99 and 1-5 are at the center of Sacramento. Our products will be convenient and easily accessible. The location of our business reduces transportation costs, and these savings will be passed to our customers. The following map of Sacramento County that provides a visual of the Sacramento area and surrounding cities.

The Sacramento area is comprised of several age groups. A closer look at the actual age distributions in Sacramento will be useful in determining what target has the greatest potential for sales and growth.

In general, the population is concentrated with individuals whose age ranges from five to 54. The younger population, aged 19 years or less, is likely the children and grandchildren of the older individuals.

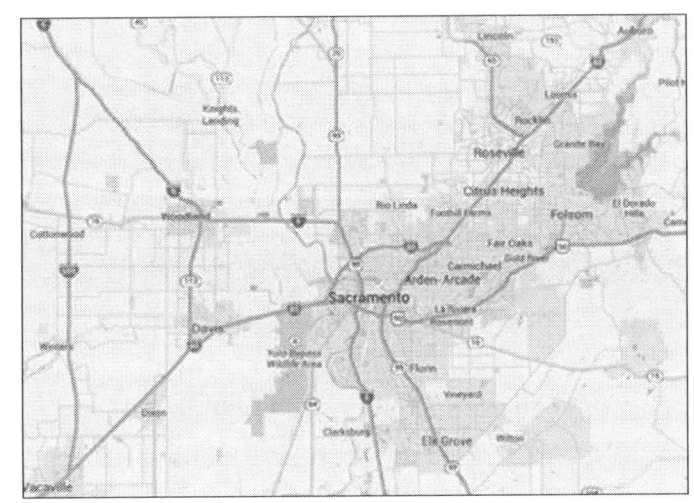

The statistics show that if Aqualife targets consumers 20 and older, there is great potential for growth. The reasons are obvious:

1. Younger populations in the 20+ years age groups are easier to target because they are normally concentrated in one area — school (i.e. colleges, universities, etc.)
2. If we target those in their in their mid-20s, we will become an integral part of their daily diet and grocery shopping behaviors. The population ranging from ages 25 to 39 have a desire to grow healthy families and will prefer healthier food choices for their children.

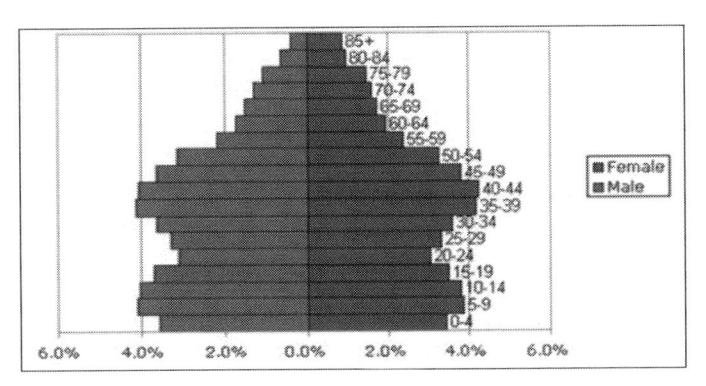

3. The younger population is aware of healthy eating because of the growing concern of obesity and high fat diets that plaguing America.
4. The older age groups in the target range are better educated, more concerned about consuming a healthier diet, and are at a higher income levels.

2.2 Target Market

The table and chart below is an estimation of the segment of consumers in the Sacramento region we want to target. The sample was taken from a population of people on the campus of California State University, Sacramento. The survey was conducted on campus because we expect the survey participants to enter the workforce, earn income, and increase spending habits as discretionary income increases. The distribution of ages relates well to the Sacramento age demographics examined previously.

Students and teachers were asked a series of questions conducive to the mission and vision of Aqualife. For instance, the primary concern is to verify whether consumers are accepting to the

Aquaponics system. It is also crucial to determine if purchasing behaviors will prompt customers to purchase our products available at Aqualife. Questions were specific to the recirculating farming system and eating habits. The objective was to verify how willing and open-minded consumers are to such a unique method of raising fish and growing vegetables.

	20 to 24	25 to 28	29 to 33	34 to 55
Environmentally Friendly	93.0%	72.7%	93.3%	90%
Eats Fish	83.7%	86.4%	86.7%	90%
Eats farm raised fish	65.1%	77.3%	66.7%	100%
Would eat fish/veggies from Aquaponics System	65.1%	63.6%	73.3%	70%
Does Shop at Farmer's Market	72.1%	27.3%	66.7%	80%
Heard of Aquaponics	4.7%	4.5%	26.7%	30%

CHAPTER 43: AQUAPONICS BUSINESS PLAN (A REAL-WORLD 'GO-BY' EXAMPLE)

Survey conducted April 2, 2012

Data extracted from Survey. Participants include students and teachers from CSUS. Sample taken 4/02/2012.

Analysis of Survey

The survey analysis provides that consumers ranging between ages 28 and 55 are most willing to purchase our products and implement them into their daily diet. This same group of people is familiar to the Aquaponics system.

On the other hand, a high majority of the younger group of individuals 27 and younger has never heard of aquaponics. They also seem reluctant to eat the fish and vegetables that come from the system. Because this age group poses such a great opportunity for growth, it is crucial that Aqualife somehow increase their awareness and familiarize them with the benefits of aquaponics as well as Aqualife's products.

Overall, survey results provide that fish eaters are accepting to freshly farmed vegetables. However, taking into consideration that vegetables including tomatoes are a more consolidated industry, the percentage of potential market will be reduced to only half of a percentage. It is an extremely low market share, but the vegetable industry is highly competitive and conservative numbers are used to align with capacity as a newly founded commercial-sized Aquaponics business.

In addition, vegetables are highly perishable and unlike fish typically don';t taste as fresh when frozen and subsequently thawed.

2.3 Needs and Profile of Targeted Customers

The growing awareness of eco-friendly ways to live poses a great potential for Aqualife. Consumers are not only becoming more about the environmental impacts of modern agricultural methods, but are also bombarded with messages from a variety of media sources about the importance of good health. Because of the growing concerns people have for overall wellness, there is an increasing demand for suppliers of organically grown food. Aqualife will position itself to capitalize on this need for quality produce and protein-based products that come from a reliable source.

- Age: 20 to 54
- Environmentally Friendly, also known as "Green"
- Concerned with healthy lifestyle
- Vegetarian, vegan, and/or healthy eater
- Wage earner
- Educated and eager to learn (those in college, as well as thought that are college educated)
- Shops at Farmer's Markets, grocery stores, health-food stores

2.4 Potential Market Share Analysis

The market is unique, but there is potential to grow into larger markets and become profitable. Furthermore, this niche of the market is a strategic sweet spot. According to Hunger and Wheelen, "the strategic sweet spot of a company is where it meets customer's needs in a way that rivals can't, given the context in which it competes" (2010, p. 180). A challenge with niche marketing is that needs are subject to change. A successful business is one that is able adapt to and alter its operations to meet demand.

According to National Marine Fisheries Institute, a member of the International Coalition of Fisheries Institute, Tilapia is rated in the top 10 most consumed fish in the United States. In fact, the consumption per capita is approximately 1.45 pounds per person on an annual basis. Financial projections will be calculated under the assumption that Sacramento consumers demand over 38,924 pounds of fish annually.

POTENTIAL TILAPIA MARKET SHARE Sacramento County and City	PROPORTIONS
Eats Fish	72%
Eats farm-raised fish/vegetables	67%
Would eat fish/veggies from Aquaponics System	61%
Does shop at Farmer's Market/Organic Eaters	61%
Average of Total Proportions	65%
TOTALS	
Total # of Population in Sacramento City/County Combined	1,028,061.00
Estimated Number of Fish Eaters in Sacramento	671,095.38
Estimated Number of Aqualife Consumers (assuming an 4 percent total Market Share)	26,843.82
Estimated Quantity of Fish Demanded (assuming 1.45 pounds of fish consumed per capita)	38,923.53

It is equally important to consider the consumption of fresh tomatoes. Agricultural Resource Center, partially funded by United States Department of Agriculture, conducted a study pertaining to the consumption of tomatoes in United States. They found that "mild, sweet tomatoes established themselves as staples in salads and as an integral component of almost all sectors of American national and regional cuisine." Spicy foods are also a popular cuisine in America. Tomatoes pose an opportunity to gain market share. For purposes of this business plan, tomatoes are the primary vegetable to harvest and sell in the Aqualife business model. Although it is important to note that Aqualife will also focus on bell peppers, which are just as much in demand as tomatoes, and have the potential of generating a greater profit margin.

POTENTIAL TOMATOES MARKET SHARE Sacramento County and City	PROPORTIONS
Eats farm-raised fish/vegetables	67%
Would eat fish/veggies from Aquaponics System	61%
Does shop at Farmer's Market/Organic Eaters	61%
Average of Total Proportions	63%
TOTALS	
Total # of Population in Sacramento City/County Combined	1,028,061.00
Estimated Number of Vegetable Eaters in Sacramento	647,297.67
Estimated Number of Aqualife Consumers (assuming an .05 percent total Market Share)	3,236.49
Estimated Quantity of Tomatoes Demanded (assuming 9.8 pounds of tomatoes consumed per capita)	31,717.59

Overall, it appears that based upon per capita consumption, tomatoes and tilapia fishes are highly demanded product amongst Americans. The extensive number of consumers in the market provides opportunity for a small and mid-size commercial producers, such as Aqualife, to tap in and gain ground as a local producer.

2.5 Industry Analysis

The market for fish and organically grown vegetables is vast. The explosion of organic and eco-friendly products on retail shelves is more than just a fad — it is a BIG BUSINESS. This particular market, if targeted properly can present an environmentally minded entrepreneur endless opportunities. Finding a niche is the key. The eco-friendly lifestyle continues to catch on with consumers, which presents growth possibilities for businesses.

Importation to meet demands of U.S. consumption

The U.S. Consumption of tilapia and tomatoes continue to rise. As more consumers become concerned with the preservation of the environment, initiatives are launched to improve the way we use resources. This is integral to food production in America. Bloomberg *'Business Week'* published an interesting article that discussed the concern for Chinese-raised tilapia. Monterey Bay Aquarium's Seafood Watch has published Chinese-raised tilapia as an "avoid" in a watch list for seafood. Interestingly, "in 2009 the U.S. imported 404 million pounds of tilapia, up from 298 million in 2005. Wal-Mart imports nearly 8.8 million pounds, every month, although they will not say how much comes from China" (Bloomberg, para. 3, 2012).

Unfortunately, domestic fish farmers do not have the capacity to meet the demand, there is opportunity to gain market share and develop a profitable industry. This is where Aqualife must step in.

Sustainability and the environment

The Aquaponics industry is also focused on operating sustainably. In effort to achieve eco-friendly status, a business should research the options to reduce its impact on the environment. Some options include buying energy efficient bulbs for the facility or using alternative energy sources such as solar panels, and recycling to minimize waste. All of these initiatives will enable a business to be considered a truly sustainable. The efficient use of energy becomes more important as the concerns for the environment continue to grow.

There are commercial aquaponic farms in US, Canada, Mexico, and Australia. According to Nelson and Pade, "commercial aquaponics is growing exponentially as innovative entrepreneurs realize that local food production is a profitable venture that is critically important to food-safety and availability "(Nelson, Pade). Large- scale operations may produce more fish and plants but this type of operation is very costly. Some of the costs include equipment, facility space, permits, system maintenance, energy, and packaging, distributing, and marketing.

Small-scale operations vs. commercial producers

Small-scale operations are much more accessible and easier to start up. This type of set-up requires less money, facility space, equipment, energy costs, and system maintenance. Although the profits may be larger for commercial aquaponics, small -scale aquaponics is a good way to test the market and see if customers are willing to buy the products.

Aquaponics is a new industry so it becomes difficult to estimate the demand of this product. According to an article in BBC, " fish continues to be the most traded commodity" Furthermore, "aquaculture production is dominated by the Asia and pacific region accounting for over 89 percent of global production" (Kinver,para. 2 ,2011). Aquaculture is related to Aquaponics and this information may provide some insight into the demand of this industry. Unfortunately, because aquaponics, the integration of Aquaculture (fish farming) and Hydroponics (cultivating plants in water), is still an emerging industry in the United States, there is minimal information pertaining to industry specific trends.

Aquaponics industry leaders

Blue Green Farms is an aquaponics producer located in Brookhaven, New Jersey. It produces vegetables through hydroponics and fish through aquaculture and the combination of growing fish and vegetables through aquaponics. In addition, it offers a wide variety of products, including various types of tomatoes, tilapia, hybrid striped bass, and sturgeon (About Blue Green Farms, 2011)

Green Acre Organics located in Brooksville, Florida. Gina Cavaliero and Tonya Penick founded this company. Both teamed up to start Green Acre Organics. *Green Acre Organics* offers organic food, system sales, farm tours, recipes, and aquaponics training classes that teach the public about aquaponics (Cavaliero, Penick, 2012).

Nelson and Pade Inc. is a supplier of aquaponics equipment/systems (commercial or home use), books, consulting services, and videos. This company has over 20 years of experience with aquaponics. *Nelson and Pade Inc.* have offered its expertise in countries such as Spain, Italy, Greece, Aruba, Jamaica, and Mexico. According to *Nelson and Pade's* website, "Our mission is to continue to lead the industry by providing quality systems, supplies, training and technical support. Quality, integrity, service, and relationship building are key components in our business model. Our goal is food security for all nations, through aquaponics and a controlled agricultural environment" (Nelson, Pade, 2010, para. 1).

Three companies mentioned above are industry leaders in America. Each company sells organic vegetables and fish using the new technology called aquaponics. Only Nelson and Pade and Green Acre Organics offer consulting services and training classes. Both companies offer more than simply selling food but both offers a service. Diversification is "corporate growth strategy that expands product lines by moving into another industry" (Hunger, Wheelen, 2010, p.214). Specifically *Nelson and Pade* and *Green Acre Organics* are using the concentric (related) diversification to grow. The two companies are offering services to relate to its current industry to expand and become more profitable and successful.

Registered aqua culturists in Sacramento

According to the State of California Resources Department of Fish and Game Agency, there is only one registered aqua culturist in Sacramento County. The business is private and noncommercial as of 2011. In addition, this business raises fish for other purposes that do not include farming fish for consumption.

COUNTY: SACRAMENTO
FRESHWATER FISH COMPANY
11520 BRUCEVILLE RD
ELK GROVE, CA 95758

Farmed Species Include: black crappie, black or white crappie, bluegill, brown bullhead bullfrog, channel catfish, common carp, fathead minnow, golden shiner minnow, goldfish, goldfish x carp hybrid, green sunfish, koi, carp koi, carp largemouth bass, red swamp crayfish, redear sunfish, smallmouth bass, striped bass, western mosquito fish, white sturgeon.

15

Description of Products

This section provides information pertaining to the business model of Aqualife. The products offered by Aqualife will be examined along with an analysis of competitors. The Aquaponics system, which develops our products, is very scientific.

3.1 Products and Services Offered

Aqualife will offer products to consumers.

Tangible products include:
- Fresh, filleted, farm-raised tilapia fish
- Fish variety may expand to additional fish types given demand
- Organically grown tomatoes
- Vegetable variety may expand given demand

3.2 Competitor Analysis

The Aquaponics industry is not prevalent in the Sacramento Region. However, the industry for fish and vegetables is fragmented. There are many providers of fish and vegetables in the area. Also, there are foreign competitors that need to be considered.

The price for similar fish and vegetable products were gathered from local competitors. Competitors include farmer's markets, grocery stores, and health food stores in the area.

All items are priced regularly, as 'on sale' prices are not accurate reflections of everyday prices of fish and vegetables.

Included below are four grocery stores selected to represent the general population. Each store targets a different type of consumer.

Competitor #1

RALEY'S, BELAIR, NOB HILL FOODS					
Filleted Fresh Tilapia, imported from China (farm raised)	Filleted Frozen Tilapia, imported from China (farm raised)	Tomatoes	Tomatoes (organic)	Baby Tomatoes (organic)
$6.99/lb.	$4.99 per pound	$1.99/lb.	$2.49/lb.	$1.39/lb.

Raley's pricing was gathered from the Sacramento store located on Marconi Avenue. Also important to our analysis of Raley's is that they promote tilapia on their website. The tilapia is promoted by a summary of its health benefits, followed by recipes. Raley's are normally located in high-cost areas, and therefore sell products at premium.

Competitor #2

WELCO SUPERMARKET			
Whole Fresh Tilapia, imported from China (farm raised)	Tomatoes	Tomatoes (organic)	Baby Tomatoes (organic)
$1.99/lb.	$1.09/lb.	$1.39/lb.	$0.89/lb.

Welco Supermarket is located on Fruitridge Road. It primarily targets the local neighborhood. The Asian population is a primary customer served in this store, they do carry fresh fish and vegetables. At Welco, the fish is served whole (head, bones, fins included) are not removed from the fish. For consumers that prefer to clean and debone their fish, their price is very competitive. This competitor targets a different demographic, but should be considered when pricing policies are determined by Aqualife.

Competitor #3

CORTI BROTHERS		
Wild Halibut (United States)	Baby Tomatoes (non-organic)	Tomatoes (non-organic)
$18.99/lb.	$1.89/lb.	$3.99/lb.

Corti Brothers is a neighborhood grocery store located on Folsom and 58th Street. They specialize in fine meats and cheeses. They have a vast variety of fresh seafood and vegetables. The butcher block only carries halibut fish, and it is priced at a premium. Corti Brothers takes pride in providing high quality products, and spares no expense in the stocking of their inventory. They happily pass the price to consumers because of the high quality, domestic products offered.

Competitor #4

TRADER JOE'S	
Previously Frozen Tilapia (Ecuador)	Baby Tomatoes (organic)
$7.49/lb.	$2.99/lb.

Trader Joes is also located on Folsom Boulevard and is only two blocks from Corti-Brothers. Traders Joes is a well-known neighborhood grocery store that also targets a very specific market. They are concerned with offering high quality, organic produce. Vegans, Vegetarians and meat eaters alike buy meats, vegetables, and more from Trader Joes. They do offer Tilapia and fresh vegetables, but at a premium. Their tilapia fish is previously frozen and imported from China.

Substitutions

Substitutions are threats to our products. Beef, lamb, chicken, and pork are proteins that can replace tilapia in any meal. Prices for these substitutions are across the board. Generally, ground beef is the cheapest substitute, but its health benefits are far polarized from that of our fresh tilapia.

3.3 *Competitive Strategy*

Differentiation Focus

Simply put, our product is better! The vast majority of tilapia stocked in Sacramento are grocery stores is derived from foreign sources such as China; and more often than not, contain heavy metals. There are many reasons our product will differentiated and offer great benefits to our market niche. According the U.S. Food and Drug Administrations, the water quality of fish farmed in Chinese farming systems is highly questionable, thereby offsetting the health benefits of the tilapia. Also important is that research has shown that China is using cheap feed such as fatty soy oil and corn oil.

- Unlike foreign competitors, such as China, our fish will not be delivered as a frozen, processed product.
- Aqualife's water supply with be constantly monitored and free of any harsh chemicals, heavy metals, antibiotics, or illegal treatments. Customers can come and see for themselves.
- Aqualife will stay abreast of regulations and collaborate with important associations to ensure that our fish is meeting the FDA standards.
- A licensed contractor will do fish cleaning, cleansing and packaging.
- No artificial coloring to preserve fresh look of product

Aqualife will set itself apart from the competition. Fish imported from foreign countries regularly stored in a freezer for several weeks at a time; in such conditions the fish loses flavor, nutrients, and texture.

Organization and Management

Aqualife will be owned and operated by Mr. _____. Because of the degree of knowledge required to operate an efficient and profitable Aquaponics business, it will be necessary that Aqualife employ a specialized Operations Manager. Also, to meet the demand of scheduled workshops, seminars and facility tours, additional staff may be necessary.

4.1 Organizational Chart and Personnel Tasks

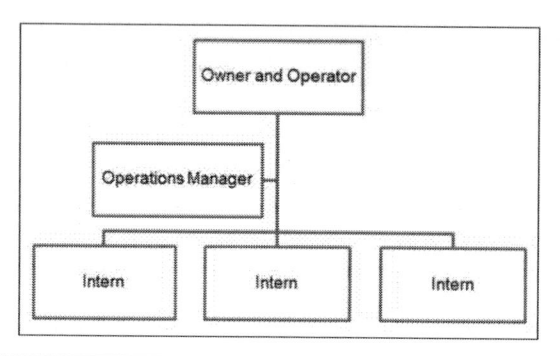

```
              Owner and Operator
                      |
              Operations Manager
              /       |        \
         Intern     Intern     Intern
```

Mr. _____ will oversee and manage the following tasks:

- Human Resources
    ◊ Payroll
    ◊ Candidate Selection
    ◊ Scheduling
- Financial/Account Management
    ◊ All Financial Statements (Balance Sheets, Income Statements, Financial Structure, etc.)
- Operations
  Mr. _____ will oversee the Operations Manager, who in turn will be responsible for:
- Marketing
- Promotion
- Pricing
- Placement
- Product
- Advertising
- Sales
- Forecasting

Operations Manager will be paid approximately $30,000 annually and tasks include:
- Manage daily production/operations
    ◊ Monitor Fish (tanks, feeding, growth, water levels, temperatures, etc.)
    ◊ Monitor Plants (vertical grow beds, fertilizing, harvesting, etc.)
    ◊ Tanks
- Manage Workshops, Educational Courses, and Facility Tours
    ◊ Educate Interns
    ◊ Select Interns
- Facility Safety
    ◊ OSHA
    ◊ Fish and Game
    ◊ Stay abreast of State Regulations
- Interns
    ◊ Tasks will be to assist with workshops, educational sources
    ◊ Administrative work

19

4.2 Owner Biography

Mr. _____ is a resident of _____ looking to start a self-sustaining Aquaponics business geared to target the _____ market. He is excited and hopeful for this business, as he believes that the eco-friendly way of growing organic fish and vegetables is a niche that has great potential for growth in the Sacramento area.

Mr. _____ is currently employed by _____ as a _____, but is committed to healthy living. Mr. _____ is health conscious and has excellent business savvy.

Mr. _____ is also highly educated, as he holds college degrees in _____. Please find Mr. _____'s cumulative vitae attached in the Appendix ((Create an Appendix and include your resume)).

4.3 Legal Structure

Aqualife is a start-up business managed by Mr. _____. Starting as a smaller-scale, commercial producer and will undertake expansion into a large commercial sized provider once profit margins are acceptable and resources are plentiful.

The Aqualife will be a privately owned company, operated by Mr. _____. The legal structure will be a sole-proprietorship. IRS will require the filing of personal 1040 tax returns, coupled with a Schedule C/Ownership, Control, Liability and Taxation is outlined in the table below.

BUSINESS ENTITY COMPARISON TABLE

This table provides an at-a-glance reference to how the most common business entity types—sole proprietorship, general partnership, C Corporation, S Corporation, and LLC—compare a number of key characteristics.

Entity Type	Sole Proprietorship	General Partnership	Limited Partnership (LP)	Limited Liability Partnership (LLP)	C Corporation	S Corporation	Limited Liability Company (LLC)
Formation	No state filing required	Agreement between two or more parties. No state filing required.	State filing required	State filing required. In California the use of LLP is limited to accountants & lawyers	State filing required	State filing required	State filing required
Duration of existence	Dissolved if entity ceases doing business or upon death of the sole proprietor	Dissolves upon death or withdrawal of a partner, unless safeguards are specified in a partnership agreement	Perpetual	Dependent on the requirements imposed by the state of formation	Perpetual	Perpetual	Dependent on the requirements imposed by the state of formation
Liability	Sole proprietor has unlimited liability	Partners have unlimited liability	At least one general partner has unlimited liability	Partners are not typically responsible for the debts of the LLP	Shareholders are typically not responsible for the debts of the corporation	Shareholders are typically not responsible for the debts of the corporation	Members are not typically responsible for the debts of the LLC
Operational Requirement	Relatively few legal requirements	Relatively few legal requirements	Some formal requirements, but less formal than corporations	Delaware, Georgia, Pennsylvania, Texas, and Virginia require an LLP to carry insurance or an escrow account to cover liabilities	Board of Directors, annual meetings, and annual reporting	Board of Directors, annual meetings, and annual reporting	Some formal requirements, but less formal than corporations
Management	Sole proprietor has full control of management and operations	Typically each partner has an equal voice, unless otherwise arranged	Limited partners are excluded from management unless they serve on the Board of Directors	All partners have the right to manage the business directly	Managed by directors who are elected by shareholders	Managed by directors who are elected by shareholders	Members have an operating agreement that outlines management

As Aqualife gains more market share, it is a high probability the company will be converted into a corporation. There are several forms of business structures. Legal advice and a business consultant will be included in formation of specific strategy details prior to expansion.

4.4 Licensing and Regulation

Required licenses, permits, and regulations

According to the State of California's Resources Agency Department of Fish and Game, aquaculture in California is governed by the Fish and Game Commission, which oversee all aquatic fisheries and operations pertaining to usage of water and disposal of water. All aquatic facilities, private or commercial, used for the controlled growing and harvesting of aquatic plants and animals in California must be registered annually. Specifically in order to open a commercial aquaponics operation in California, the business must be a registered aqua culturist. A fishing permit is also required. (www.dfg.ca.gov).

Inspection of marine aquaculture facilities

However, before a marine aquaculture facility can be registered, the state's Marine Aquaculture coordinator may require an inspection of where the fishes are farmed. It is required in Sacramento County to have the Marine Region's Aquaculture Coordinator visit the proposed operating facility for inspection. A section is provided in the Aquaculture registration application to draw a sketch of nearby water sources, location for water disposal, and pipes, which transport water to be indicated. However due to the recirculation of water of an aquaponics system, a detailed sketch may be required to justify water use specific to an aquaponics set up:

Contact:	California Department of Fish and Game
Kirsten Ramey	619 Second Street
Marine Aquaculture	Eureka, CA 95501
	(707) 445-5365

In general, there are specified regulations as to where marine fishes can be obtained. For the purpose of farm raising and selling fish, in this case Tilapia, registered aqua culturists can only obtain fingerling from the following:

- Licensed commercial fishermen, the Department of Fish and Game
- Registered aqua culturists or by a registered aqua culturist under the provisions of Section 15301(b) of the Fish and Game Code and Section 243, Title 14, California Code of Regulations.

The laws and regulations governing the sale of aquaculture products are cited in Fish and Game Code Section 15005 and in Section 238, Title 14, and California Code of Regulations.

Contact: Karen Mitchell Freshwater Agriculture (916) 445-0826 kmitchell@dfg.ca.gov	California Department of Fish and Game 9th and S Street Sacramento, CA 95814

Processing and sale of fish

Farmed fish sold as produce in the State of California is Food and Drug Branch is responsible for approving facilities, equipment, and procedures used for handling, shucking, storing, packaging and shipping of fish and shellfish after harvest. That branch also enforces meat quality standards, and sets requirements for proper packaging and labeling of all fish and shellfish moved in commerce. For this reason, Aqualife will rely on a specialized contractor for the processing and packaging of meat.

Contact: (916) 650-6500	California Department of Public Health Food and Drug Branch 1500 Capitol Avenue MS 7602 PO Box 997435 Sacramento, CA 95899-7435

Sales Strategy

Costs for promotion and distribution have to be contained and resources used wisely. These resources are described in the following sections.

5.1 Product Distribution

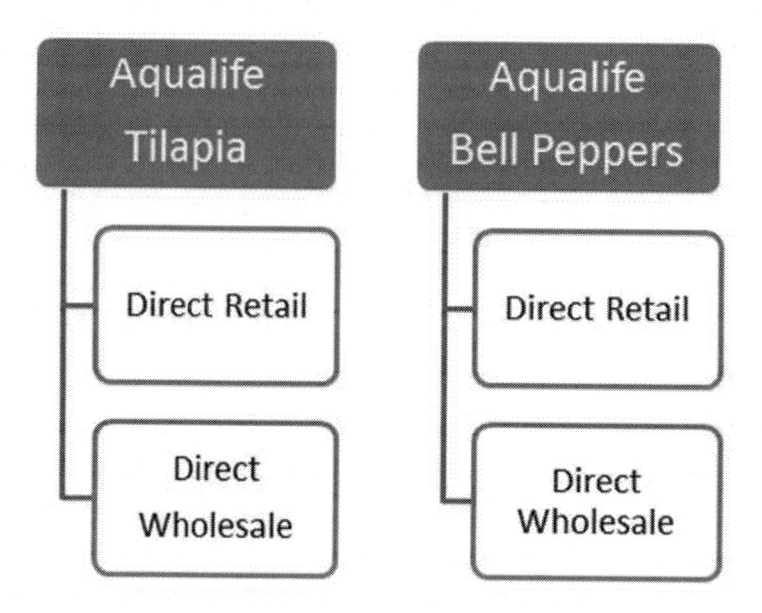

```
Aqualife Tilapia
├── Direct Retail
└── Direct Wholesale

Aqualife Bell Peppers
├── Direct Retail
└── Direct Wholesale
```

Direct Retail Sales

Selling directly to the consumer is the ideal way to maximize profits and remove any costly intermediaries that distribute our products. Since Aqualife is concentrated on targeting the Sacramento region, it is imperative that products and services be available exclusively to our market. Below are recommended retails sales options:

- **Local customer base:** This is the simplest way to market the product and establish a loyal customer base. Establishing strategic alliances with local vendors such as newspapers, colleges, farmers markets, and associations can be used as advertising and promotion channels. The downside of this strategy is the time it takes to establish a client base, but the gains are unlimited as "Word of Mouth" is an excellent way to build clientele, as it has a multiplying effect.
- **Sales to local restaurants:** Sacramento has an unlimited number of restaurants that specialize in preparing meals from local producers. Up-scale restaurants typically have a "Catch of the Day," where fresh fish is a prized item. The price for these meals is usually high. Customers are willing to pay the premium for fresh, locally produced ingredients. By selling our products to local restaurants, business relationships will be established and

the restaurant will market the product in hopes of sales. Cultivating great relationships and providing consistent fresh quality products are essential for continued successful sales to local restaurants.
- **Farmer's Markets:** California State University, Sacramento recently held an ASI Farmer's Market on campus. According to the Hornet Newspaper, it was a success and the only downfall was the "underestimating how many people would show up and how many supplies were needed to accommodate the large crowd.". There were only three off-campus vendors. ASI hopes to collaborate with more vendors, and expand to surrounding community business. The market analysis shows that young people are accepting to Aqualife's products and services — campus collaboration could be a great way to establish clientele and gain product exposure.

5.2 Pricing

Cost Focus Strategy

Aqualife is focusing on a particular market niche in the Sacramento area. Our products—unlike competitors such as Raley's, Trader Joes, Welco, and other local grocery stores—will not come from China or Ecuador, it will come from United States.
- Customers, such as individual buyers, will absorb the cost reduction due to minimal transportation costs, storage, and added shelf life because no time has been wasted shipping our products across the country.
- We will compete with lower prices by providing fresh organic quality products, personalized service, and keep advertising and promotion costs low by focusing on strategic alliances that allow for access to specific markets.

AQUA LIFE	
Tilapia Fillets (Domestic)	Tomatoes
4.50/lb.	$2.09/lb.

- Pricing can be adjusted for inflation, or increased based on sales revenues. Prices range from 2.99/lb. upwards of $7.00/lb.
- Aqualife will appeal to customers via selling healthy organic products grown locally through a unique environmentally friendly farming method.

5.3 Promotion

Aqualife's products are high quality compared to the competitors that are importing potentially harmful fish from foreign producers. The high-quality customer focused services will generate demand, coupled with promotional tools such as:

- Newspaper advertisements
- Flyers
- Posters
- Word-of-Mouth
- Business Alliances
- Farmer's Markets
- Sacramento College Campuses
- Facebook Account
- Company Website
- Linkedin
- Local, State, Federal Association Memberships

Financial Management

Mr. _____ will invest personal funds into the business and will obtain a Small Business Administration (SBA) 7(a) loan. This section provides estimated start-up costs, in addition to a pro-forma balance sheet, income statement, and break-even analysis.

6.1 Initial Investment

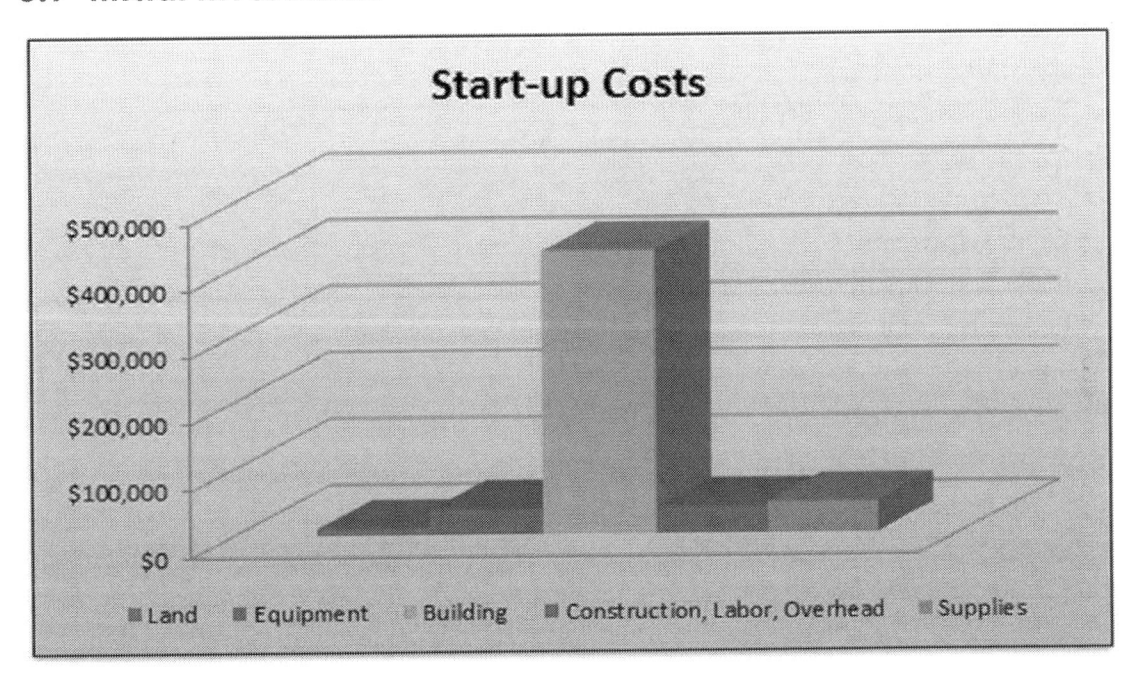

Start-up Costs

Bar chart showing start-up cost categories with y-axis from $0 to $500,000. Categories: Land, Equipment, Building, Construction/Labor/Overhead, Supplies. Building is the largest cost (approximately $450,000), with smaller amounts for the other categories.

Startup costs include the purchase of capital. Building, equipment, construction, labor, and overhead are investments that can be secured by the SBA 7(a) loan. The building purchase is the most expensive cash outlay, followed by equipment and construction costs. Aqualife will construct an aquaponic system that is customized to fit maximize space of the facility and which can be expanded into larger-sized systems as the capacity needs increase. Building our own aquaponic system will also save a substantial amount of money when compared to purchasing a turnkey aquaponic system or a package unit.

INITIAL OUTLAY	
Land	$10,000
Equipment	$37,500
Building	$425,500
Construction & Labor	$40,000
Supplies	$46,599
Total Initial Investment	**$559,599**

Description of Start-up Costs

- Land: Improvements to exterior may be necessary to meet regulations, and to make landscaping aesthetically pleasing.
- Equipment: Approximately 6,144sq foot system will be acquired to produce the demanded amount of fish and vegetables.
- A 4,096 sq. ft. system will be constructed at a cost of approximately $25,000. Another system be custom designed to complete capacity. For another $12,500 for an additional 2,000 sq. ft.
- Building: Aqualife will acquire a building that is approximately 6,500 square feet in order to produce the amount of fish and vegetables to address market demand. There are a variety of viable commercial properties listed in MetrolistMLS, a real estate and property database serving the Greater Sacramento region. Prices range from $200,000 upwards of $1,000,000.
- Construction, Labor and Overhead:
    - ◊ The systems will need to be constructed and built to speculation. The selected building may need remodeling also.
- Supplies:
    - ◊ Supplies include fingerlings, hydroton (an organic material used in place of soil), seeds, water treatment and testing supplies, etc.

6.2 Projected Balance Sheet

AQUALIFE
Projected 12 Month Balance Sheet
January 01, 2013 — December 31, 2013

Assets

Current Assets:

Cash			$85,000
Accounts Receivable		$0	
Less:	Reserve for Bad Debts	0	0
Merchandise Inventory			10,060
Prepaid Expenses			1,250
Notes Receivable			0
	Total Current Assets		$96,310

Fixed Assets:

Office Furniture		10,000	
Less:	Accumulated Depreciation	2,000	8,000
Equipment		37,500	
Less:	Accumulated Depreciation	5,357	32,143
Buildings		425,500	
Less:	Accumulated Depreciation	17,020	408,480
Land			
	Total Fixed Assets		448,623

Total Assets $544,933

Liabilities and Capital

Current Liabilities:

Interest Payable	29,974	
Unearned Revenues	0	
Short-Term Notes Payable	0	
Short-Term Bank Loan Payable		
Total Current Liabilities		$29,974

Long-Term Liabilities:

Long-Term Notes Payable	506,639	
Line of Credit Payable		
Total Long-Term Liabilities		506,639

Total Liabilities		536,613

Capital:

Owner's Equity	0	
Net Profit	8,320	
Total Capital		8,320

Total Liabilities and Capital		**$544,933**

6.3 Projected Income Statement

AQUALIFE
Income Statement
For the Year Ended [December 31, 2013]

Revenue:			
Gross Sales		$241,441.00	
Less: Sales Returns and Allowances		$0.00	
Net Sales		$241,441.00	
Cost of Goods Sold:			
Beginning Inventory		$0.00	
Add: Inventory Cost		$8,000.00	
Distribution Costs		$8,000.00	
Direct Labor		$30,000.00	
Packaging		$5,000.00	
		$51,000.00	
Less: Ending Inventory		$10,060.00	
Cost of Goods Sold			$40,940.00
Gross Profit (Loss)			$200,501.00
Expenses:			
Fingerlings		$22,500.00	
Feed		$6,000.00	
Seeds		$647.00	
Utilities		$15,000.00	
Supplies		$23,452.00	
Maintenance		$2,500.00	
Loan Payments		$38,964.40	
Marketing		$12,000.00	
Depreciation		$22,377.00	
Insurance		$7,000.00	
Property Taxes		$5,319.00	
Miscellaneous		$1,000.00	

Professional Fees	$1,000.00	
Permits and Licenses	$1,000.00	
Total Expenses		$158,759.40
Net Operating Income		$41,741.60
Other Income:		
Gain (Loss) on Sale of Assets	$0.00	
Interest Income	$0.00	
Total Other Income		$0.00
Net Income (Loss)		$41,741.60

6.4 Break-even Analysis

Year 1 Projection

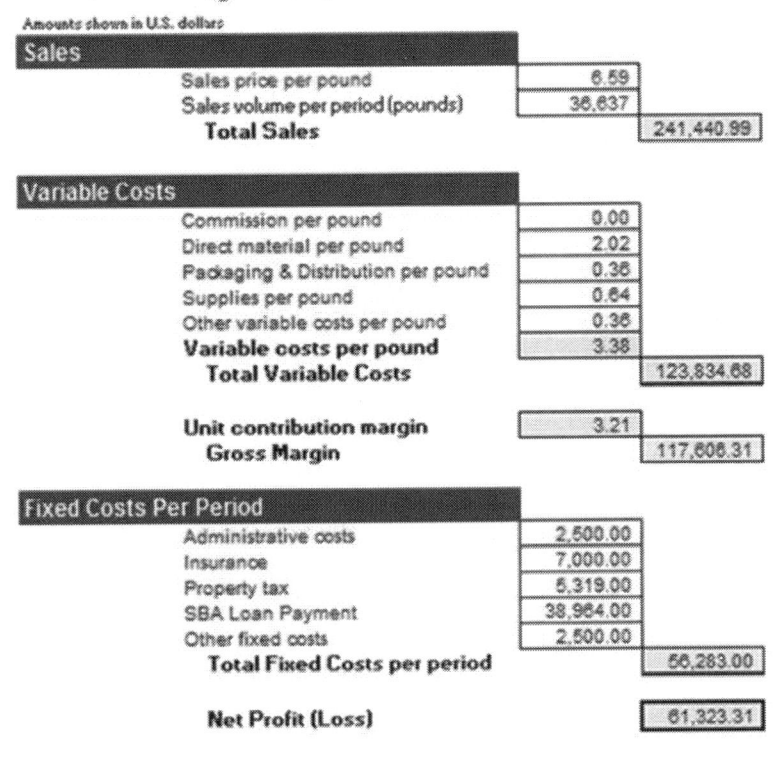

Amounts shown in U.S. dollars

Sales

Sales price per pound	6.59
Sales volume per period (pounds)	36,637
Total Sales	**241,440.99**

Variable Costs

Commission per pound	0.00
Direct material per pound	2.02
Packaging & Distribution per pound	0.36
Supplies per pound	0.64
Other variable costs per pound	0.36
Variable costs per pound	**3.38**
Total Variable Costs	**123,834.68**

Unit contribution margin	**3.21**
Gross Margin	**117,606.31**

Fixed Costs Per Period

Administrative costs	2,500.00
Insurance	7,000.00
Property tax	5,319.00
SBA Loan Payment	38,964.00
Other fixed costs	2,500.00
Total Fixed Costs per period	**56,283.00**

Net Profit (Loss)	**61,323.31**

Unit Contribution Margin

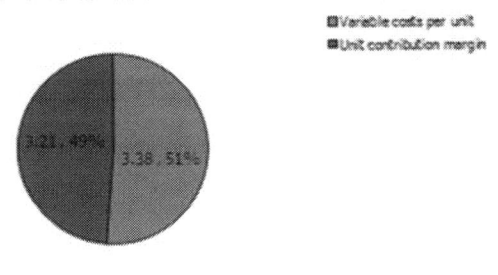

- Variable costs per unit: 3.38, 51%
- Unit contribution margin: 3.21, 49%

Variable Costs Per Unit

- Commission per unit: 0.00, 0%
- Direct material per unit: 2.02, 60%
- Packaging & Distribution per unit: 0.36, 10%
- Supplies per unit: 0.64, 19%
- Other variable costs per unit: 0.36, 11%

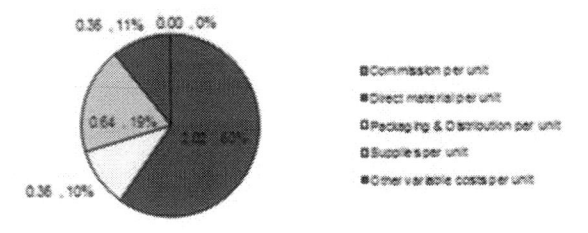

Breakeven Analysis Chart

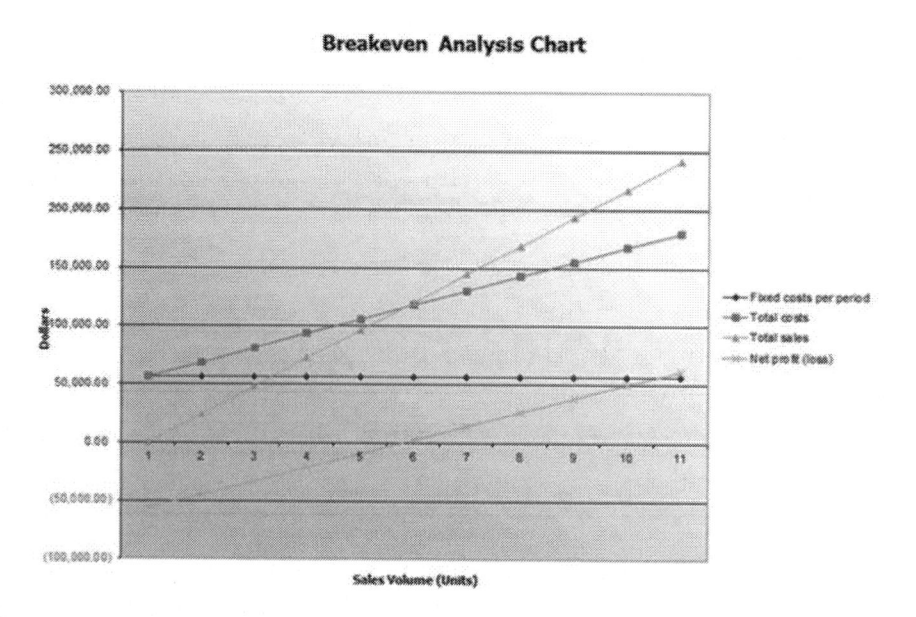

Results:

Breakeven Point (units): 17,534

Sales volume analysis:

Sales volume per period (units)	0	3,664	7,327	10,991	14,655	18,319	21,982	25,646	29,310	32,974	36,637
Sales price per unit	6.59	6.59	6.59	6.59	6.59	6.59	6.59	6.59	6.59	6.59	6.59
Fixed costs per period	56,283.00	56,283.00	56,283.00	56,283.00	56,283.00	56,283.00	56,283.00	56,283.00	56,283.00	56,283.00	56,283.00
Variable costs	0.00	12,383.47	24,766.94	37,150.40	49,533.87	61,917.34	74,300.81	86,684.28	99,067.75	111,451.21	123,834.68
Total costs	56,283.00	68,666.47	81,049.94	93,433.40	105,816.87	118,200.34	130,583.81	142,967.28	155,350.75	167,734.21	180,117.68
Total sales	0.00	24,144.10	48,288.20	72,432.30	96,576.40	120,720.50	144,864.60	169,008.70	193,152.79	217,296.89	241,440.99
Net profit (loss)	(56,283.00)	(44,522.37)	(32,761.74)	(21,001.11)	(9,240.48)	2,520.16	14,280.79	26,041.42	37,802.05	49,562.68	61,323.31

Break even analysis calculated using the sales price of one pound of fish, and one pound of tomatoes. The total sales price is $6.59. The weighted figures are listed below to indicate how many pounds of each item need to be sold based on the break-even point per pounds.

Fish = $4.50/$6.59 = 68.29% or **11,973 pounds of tilapia**

Tomatoes = $2.59/$6.59 = 31.71% or **5,561 pounds of tomatoes**

CHAPTER 43: AQUAPONICS BUSINESS PLAN (A REAL-WORLD 'GO-BY' EXAMPLE)

Conclusion and Recommendations

The possibility that Aqualife can be an industry leader is feasible. Surveys, statistics, and demographics suggest that consumers are accepting to the Aquaponics system. A large majority of the population eats fish — will consume farm-raised fish and vegetables from the same controlled environment.

One concern is that only a small percentage of the population is familiar with or heard of Aquaponics. Adults aged 34 and over seem to be abreast of the new technology offered by the recirculating aquaponics system, while adults in younger age groups have no previous exposure to the industry. This poses great opportunity for Aqualife. As a proper marketing strategy will capture a very large market just by educating our target population about all of benefits associated with aquaponic grown food (i.e. local, organic, healthy, environmentally friendly agriculture).

If promoted and advertised appropriately, Aqualife can be very competitive. It offers exceptional products that are high in quality, generally lower in price, and local. Industry analysis provides that an extremely large percentage of tilapia is imported from foreign sources such as China. Several studies indicate the health of tilapia imported from foreign countries is not recommended for human consumption. Furthermore, most produce is GMO based and therefore contaminated with heavy metals as well as other toxic pesticide, herbicide, and/or insecticide residue constitutes.

Aqualife offers a competitive and unique product to Sacramento, because of the scientific, technical, and regulated nature of the industry, the following steps are being accomplished prior to implementation of this business plan.

1. Complete extensive training courses with well-known companies in the Aquaponics industry.
2. Informed of current and changing policies in the Aquaculture hydroponics industry. Resources include OSHA, FDA, and Fish and Game.
3. Because of changing trends and dietary preferences, additional varieties of fish and vegetables to meet demand will continually be evaluated.
4. Join local associations in the area to stay informed of ongoing developments and for networking purposes.

Business Plan References

About Seafood.(2012). Retrieved from http://www.aboutseafood.com/about/about-seafood/top-10-consumed-seafoods

About Tilapia. Retrieved from http://www.abouttilapia.com/

AG MRC. (2011). Bell and Chili Ptomatoes Profile. Retrieved from: http://www.agmrc.org/commodities__products/vegetables/bell_and_chili_ptomatoes_profile.cfm

Aquaponics. (2012). Retrieved from http://en.wikipedia.org/wiki/Aquaponics

Cavaliero, G., & Penick, T. (2012). Retrieved from http://www.greenacreorganics.biz/what-we-offer/organics-for-you/

Censuscope. (2000). Retrieved from http://www.censusscope.org/us/m6920/print_chart_age.html

Conte, S. F., Permit, Licenses, Laws and Regulations: A guide for aquaculture in California, retrieved from http://aqua.ucdavis.edu/DatabaseRoot/pdf/ASAQ-B01.PDF

Deleo, S. AVP, Business Banker, Capital One Bank. New York Gottlieb, A. (2011) What's the Real Cost to Hire a New Employee? Some Considerations Before Hiring, retrieved from http://frugalentrepreneur.com/2011/05/whats-the-real-cost-to-hire-a-new-employee-considerations-before-hiring/

Demand.(2012) Retrieved from http://www.businessdictionary.com/definition/demand.html

Eden, T., & Eden, S. (2011). Retrieved from http://edenaquaponics.com/2010/04/a-brief-history-of-aquaponics/

Efole, Ewoukem, T.T., Aubin, J. J., Mikolasek, O.O., Corsonx, M.S., Tomedi Eyango, M.M., Tchoumboue, J. J., &...Ombredane, D. D. (2012). Environmental impacts of farms integrating aquaculture and agriculture in Cameroon. *Journal of Cleaner Production, 28* 208-214. doi:10.1016/j.jclepro.2011.11.039

Einhorn,B. (2010). Bloomberg Businessweek. Retrieved from http://www.businessweek.com/magazine/content/10_44/b4201088229228.htm?chan=magazine+channel_top+stories

Fitzsimmons,K.(2011). Retrieved from https://www.was.org/documents/MeetingPresentations/AA2011/AA2011_0246.pdf

Friendly aquaponics.(2011). Building a commercial Aquaponics system. Retrieved from http://www.friendlyaquaponics.com/do-it-myself-systems/commercial-system

Gene Smart (2012). Retrieved from http://www.genesmart.com/pages/tilapia_omega_3_nutrition/84.php

Growing for Market. (2011). Retrieved from http://www.growingformarket.com/

Hunger, D. J., & Wheelen, L. T. (2010). Concepts in Strategic Management and Business Policy. Edition: 12. New Jersey: Prentice Hall.

igrowhydro. (2009).Hydroton. Retrieved from http://www.igrowhydro.com/Hydroton-10-liter-bag.aspx

Interview, March 2, 2012 conducted with: Patterson, J. (Interviewer) & DeLeo S. (Interviewee). (2012).

Kinver, Mark. (2011)Global Fish Consumption hits a record high. BBC news. Retrieved from http://www.bbc.co.uk/news/science-environment-12334859

Madali, M.(2012). Famers Market Draws Campus Community. The Hornet

MorningStar Fishermen. (2011). Retrieved from http://morningstarfishermen.org/

Nelson, R., & Pade, J. (2010). What is Aquaponics. Retrieved from https://aquaponics.com/

Also, "Phone Interview" with *CFO of Nelson and Pade.* 16 November, 2012.

Professional Aquaculture Services. (date was not given) Retrieved from http://www.proaqua.com/aquaculture-projects

Sacramento County. (2012). Retrieved from http://www.saccounty.net/FactsMaps/DemographicsandFacts/default.htm

Tilapia. (2008) Retrieved from http://www.aquaticcommunity.com/tilapia/market.php

Trend. (2012) Retrieved from http://www.businessdictionary.com/definition/trend.html

Wikipedia- wattle, (2012). Retrieved from http://en.wikipedia.org/wiki/Wattle_and_daub

Resource Agency Department of Fish and Game (2011) Retrieved from http://nrm.dfg.ca.gov/FileHandler.ashx?DocumentID=3265&inline=true

PART VIII

Appendix

APPENDIX 1

Aquaponic Resources

Farm Your Space

The author is creating a website **www.FarmYourSpace.com** that will have a great deal of helpful information on it, a Q&A feature where you can get your questions answered, many articles, pics and videos related to aquaponics, vertical gardening, and other helpful ideas to maximize your space; as well as improving the effectiveness and efficiency of your farming operation. I encourage you to check it out. Table 50 lists some of the topics FarmYourSpace.com will cover:

NOTE: Some popular aquaponic resources commonly found on the Internet are not included within the following list of resources, as the author has found that they either do not provide sound advice, lack integrity, or are not reliable enough to recommend.

The Aquaculture Network Information Center

The Aquaculture Network Information Center (AquaNIC) is gateway to the world's electronic resources in aquaculture AquaNIC is maintained at Purdue University in West Lafayette Indiana, and is supported by the Illinois-Indiana Sea Grant Program and Purdue University's Department of Animal Sciences. The AquaNIC site contains links to aquaculture sit at other state Land Grant universities,

TABLE 50.

	"FARMYOURSPACE" WEBSITE CATEGORIES
1	Aquaponics
2	Animals, Poetry, Livestock
3	Community Forum & Expert Advice
4	Do-It-Yourself
5	Fruit & Nut Trees
6	Healthy Soil, Natural Fertilizers, Natural Pesticides
7	Hydroponics
8	Maximizing Your Space
9	Money Saving Tips & Money Making Ideas
10	Nutrition, Health, Organics
11	Raised Beds
12	Surviving Regulations & Big Brother (We will keep you informed and petition for your rights on related government policy and legislation issues.)
13	Sustainable Farming
14	Traditional Gardening & New Ideas!!!
15	Vertical Gardening
16	Water
17	Living off the grid.
18	Cooking
19	Survival
20	WHAT IS WRONG WITH THE WORLD (issues related to Pesticides / Herbicides / Fungicides, GMOs, EMF, Radiation, Pollution, Water contamination, Destruction of natural habitat, Food Contaminants, Drinking Water Contaminants / Fluoride, Health Problems Increasing, Political Corruption, Dumbing Down of the Public, Etc.

USDA sites devoted aquaculture, professional organizations, and other sites with aquaculture information. **AquaNIC can be accessed via: WWW: http://ag.ansc.purdue.edu/aquanic/**

CropKing, Inc.

CropKing, Inc., has been specializing in the business of controlled environment agriculture and hydroponics since 1982 and manufactures greenhouse structures at their facility in Lodi, OH. The company sells to both hobby and commercial growers throughout the United States as well as internationally, emphasizing quality, competitive pricing, and a full range of services to its customers. For more information visit: www.cropking.com; email: cropking@cropking.com; or call 330-302-4203; 134 West Drive, Lodi, OH 44254

Aquaculture in the Classroom

The University of Arizona has extensive experience in hydroponics. They are an excellent source of hydroponic information and are expanding their interest in aquaponics.
http://ag.arizona.edu/azaqua/extension/Classroom/home.htm
http://ag.arizona.edu/azaqua/extension/Classroom/Aquaponics.htm
Aquaponics Library

Greenhouses

How to Build Your Own Greenhouse, Roger Marshall, ISBN: 13-978-1-58017-587-6

Professional Aquaculture Services

559 Cimarron Drive
Chico, CA 95926
PAC is operated by Tony Vaught, with over 30 years experience in production aquaculture. PAC offers a **source of fish** in their Aquaponics System Starter Package for those establishing a new aquaponics system, consulting services and trouble-shooting consultation for system design, feeding and fish health.

Phone: (530) 343-0405
Cell: (530)-519-1051
Fax: (530) 343-0405
tvaught@proaqua.com
http://www.proaqua.com/
http://www.aquaculturedirect.com/

Useful Internet Web Links for More Information

http://www.iasproducts.com/Main.html — Has good prices for some aquaculture components, including auto feeders
http://www.caaquaculture.org/ — The California Aquaculture Association website
http://www.dfg.ca.gov/Aquaculture/ — California Dept. of Fish & Game site for aquaculture (permit included on disk)
http://www.fish.washington.edu/wrac/ — Western Regional Aquaculture Center website
http://aqua.ucdavis.edu/index.htm — The UC Davis Aquaculture website (a lot of publications and links)

Aquaponics Training Programs, Workshops, Seminars & Courses

http://morningstarfishermen.org/
http://aquaponics.com/page/classes-and-seminars
http://www.livingmandala.com/Living_Mandala/Aquaponics_Course.html

SARE Learning Center

The USDA's Sustainable Agriculture Research and Education (SARE) program has funded the publication of many fine books over the years and now offers some of them as free downloadable PDFs. For example, *Building a Sustainable Business* is $17 in print but free as a download. SARE also publishes bulletins, grant project reports, and much other useful information. www.sare.org/publications/.

Reliable Alternative News Sources

A total of 90 percent of all news media outlets in the USA are owned by a total of only six companies. The owners and/or head of all these companies are members of the Bilderberg Group, and attend the annual Bilderberg Club meeting every year. Bilderberg is a highly secretive, elitist, international think tank and policy forming group. This globalist establishment of government leaders and media company heads work together to execute a covert globalist liberal agenda that only benefits them, and which is in opposition to the best interest of the common person.

Also, advertisers support media companies with billions of dollars in ads. These business transactions come with strings attached. Companies spending millions of dollars on advertisements have no qualms about requiring the receiving media outlet to implement policy measures to sway public opinion one way or another, as a stipulation for those advertising dollars.

In addition, the Obama administration dished out tremendous amounts of funding, grants, and special favors to mainstream media companies. These government funds were dispersed with the understanding that the receiving media outlet help promote or discourage various issues.

Lastly, foreign government companies have been purchasing media companies and Hollywood studies at an alarming rate. These are billion dollar transactions. China, widely known for strict censorship and exclusive propaganda media messaging, is one of the largest foreign owners of American media entertainment companies.

As a result of all the above forces, it is next to impossible to obtain unbiased truth from the mainstream media. Mainstream media is pressured to the point where the truth gets buried under a scripted narrative resulting in propaganda, fake news, and twisted truth.

Bottom line: Mainstream news is corrupt beyond measure.

Therefore, to get accurate reporting of the news one must depend upon the other of media sources — the remaining 10 percent of media outlets that are independent. These independent alternative media sources provide more reliable information and news which can be trusted. Several of these valuable media outlets are listed below.

- www.Drudge.com
- www.InfoWars.com
- www.Brietbart.com
- www.DailyCaller.com
- www.NatrualNews.com
- www.fluoridealert.org
- www.fluoridealert.org/articles/50-reasons
- (50 reasons why fluoride in H2O is bad)
- www.organicconsumers.org
- www.NaturalSociety.com
- www.GeoEngineeringWatch.org

The following 'YouTube' channels also provide accurate reporting:
- Paul Joseph Watson, www.youtube.com/user/PrisonPlanetLive
- Stefan Molyneux, www.youtube.com/user/stefbot
- Mike Dice, www.youtube.com/user/MarkDice

APPENDIX 2

References

About Seafood (2014). http://www.aboutseafood.com/about/about-seafood/top-10-consumed-seafoods

About Tilapia. http://www.abouttilapia.com/

AG MRC. (2011). Bell and Chili Peppers Profile. Retrieved from: http://www.agmrc.org/commodities__products/vegetables/bell_and_chili_peppers_profile.cfm

Albright, Louis. "Controlled Environment Agriculture." Cornell University; *Http://www.cornellcea.com/*. N.p., 2011. Web. 18 Nov. 2014.

ATTRA is the National Sustainable Agriculture Information Service. (It was originally called Applied Technology Transfer for Rural Areas and became so well known for its sustainable agriculture information that it kept the original acronym.) It is managed by the National Center for Appropriate Technology (NCAT), a private nonprofit organization founded in 1976, and is funded under a grant from the United States Department of Agriculture (USDA). That's a long way of saying that it's a federally funded information service for sustainable farmers.

ATTRA has compiled a large number of extensive, free publications about topics of interest to market farmers. They are all written by staff members who have expertise in agriculture, and they usually include excerpts from magazines, newspapers, and Extension publications. Under the heading of Horticultural Crops, for example, there are 78 separate publications. These resources offer detailed information on production of specific horticultural crops, focusing on sustainable and organic production methods for traditional produce, and also introducing a range of alternative crops and enterprises. In these publications you can find information on strategies for more sustainable greenhouse and field production of everything from lettuce to trees.

ATTRA publications are free at www.attra.org. If you don't have Internet access, you can also call toll-free to 800-346-9140 (English) or 800-411-3222 (Espanol) to request printed copies of publications for a small fee.

ATTRA also offers the only nationwide internship listing service. If you are a farmer who wants to take on interns, you can list your farm for free with ATTRA.

Censuscope. (2000) http://www.censusscope.org/us/m6920/print_chart_age.html

Conte, S. F., Permit, Licenses, Laws and Regulations: A guide for aquaculture in California, retrieved from http://aqua.ucdavis.edu/DatabaseRoot/pdf/ASAQ-B01.PDF

"Costco Organic Price List." *The Thrifty Mama RSS*. N.p., 6 Jan. 2011. Web. 15 Nov. 2012.

What's the Real Cost to Hire a New Employee? Some Considerations Before Hiring, retrieved from http://frugalentrepreneur.com/2011/05/whats-the-real-cost-to-hire-a-new-employee-considerations-before-hiring/

Demand (2014) http://www.businessdictionary.com/definition/demand.html

Eden, T., & Eden, S. (2011). http://edenaquaponics.com/2010/04/a-brief-history-of-aquaponics/

Efole, Ewoukem, T.T., Aubin, J. J., Mikolasek, O.O., Corsonx, M.S., Tomedi Eyango, M.M., Tchoumboue, J. J., &...Ombredane, D. D. (2012). Environmental impacts of farms integrating aquaculture and agriculture in Cameroon. *Journal of Cleaner Production, 28*208-214. doi:10.1016/j.jclepro.2011.11.039

Einhorn, B. (2010). Bloomberg Businessweek. Retrieved from http://www.businessweek.com/magazine/content/10_44/b4201088229228.htm?chan=magazine+channel_top+stories

Fitzsimmons, K.(2011). Retreived from https://www.was.org/documents/MeetingPresentations/AA2011/AA2011_0246.pdf

Food and the City: Urban Agriculture and the New Food Revolution; by Jennifer Cockrall-King (February 21, 2012); 372 pages; ISBN-10: 1616144580; ISBN-13: 978-1616144586; A global movement to take back our food is growing. This book examines alternative food systems in cities around the globe where people are growing their own food, and taking their "food security" into their own hands.

Gene Smart (2014). Retrieved from http://www.genesmart.com/pages/tilapia_omega_3_nutrition/84.php

Gentle World RSS. "Kale: An Easy Beginner's Guide to Growing." N.p., n.d. Web. 29 Nov. 2012.

GardenGuides "How to Grow Lettuce in a Greenhouse.". N.p., 2010. Web. 15 Nov. 2014.

"Greenhouses Online:: How to Grow a Greenhouse Vegetable Garden." *Greenhouses Online :: How to Grow a Greenhouse Vegetable Garden*. N.p., 2009. Web. 06 Dec. 2012.

Growing for Market, visit **www.growingformarket.com,1-(800)**-307-8949.

Health in Pond Fish: PH and Koi Health. "Koi Health & PH FAQs." N.p., 1997. Web. April 2014.

How to Grow Spinach: Organic Gardening. "Spinach Growing Guide." N.p., 2011. Web. 18 March 2014.

Hunger, D. J., & Wheelen, L. T. (2010). Concepts in Strategic Management and Business Policy. Edition: 12. New Jersey: Prentice Hall.

Igrowhydro. (2009).Hydroton. Retrieved from http://www.igrowhydro.com/Hydroton-10-liter-bag.aspx

APPENDIX 2: REFERENCES

Institute for Local Self-Reliance, Rebuilding Independent Businesses (2014); http://www.ilsr.org/why-support-locally-owned-businesses/

Interview, March 2, 2012 conducted with: Patterson, J. (Interviewer) & DeLeo S. (Interviewee). (2012).

Kinver, Mark. (2011) Global Fish Consumption hits a record high. BBC news. Retrieved from http://www.bbc.co.uk/news/science-environment-12334859

Local Harvest: A Multi farm CSA Handbook is available free from the SARE Learning Center: www.sare.org/Learning-Center/Books.

Madali, M.(2014). Famers Market Draws Campus Community. The Hornet

MorningStar Fishermen. (2014). http://morningstar-fishermen.org/

Nugget Market: "Live Interview" with Produce Manager, Roseville, Ca. Late November 2012.

Organic Gardening. "How To Grow Organic Strawberries." *About.com* N.p., n.d. Web. April. 2014.

Resource Agency Department of Fish and Game (2014); Retrieved from http://nrm.dfg.ca.gov/FileHandler.ashx?DocumentID=3265&inline=true

Tank Culture of Tilapia. "Tank Culture of Tilapia." N.p., 2008. Web. 10 Nov. 2012.

Sustainable Vegetable Production from Start-Up to Market by Vern Grubinger. An excellent overview of vegetable farming, with an emphasis on large-scale production. It covers all aspects of market farming, from planning to marketing, and does a great job of helping the small grower understand what is required to expand.

Tilapia. (2008) Retrieved from http://www.aquaticcommunity.com/tilapia/market.php

The Essential Urban Farmer by Novella Carpenter and Willow Rosenthal shares the experiences of two successful urban growers. It covers all the issues that are particular to urban farming, such as soil contamination, limited space, zoning and neighbors, security, and much more.

The Flower Farmer: An Organic Grower's Guide to Raising and Selling Cut Flowers 2nd Edition by Lynn Byczynski (Author), Robin Wimbiscus (Illustrator); Chelsea Green Publishing; Revised and updated second edition (February 22, 2008); ISBN-10: 1933392657; ISBN-13: 978-1933392653

The Hoophouse Handbook: Growing Produce and Flowers in Hoophouses and High Tunnels by Lynn Byczynski (Author, Editor); December 19, 2006; A collection of articles about all aspects of high tunnels, including how to buy and build one (with photos and illustrations), how to grow various crops, and which crops are the most profitable; 60 pages.

The Landowner's Guide to Sustainable Farm Leases from Drake University Agricultural Law Center can be downloaded at http://sustainablefarmlease.org/the-landowners-guide-to-sustainable-farm-leases/

The New Organic Grower by Eliot Coleman. This guide to small-scale, intensive vegetable production is the foundation of today's local food movement. Its most recent revision was in 1995, but it is just as relevant and important today as then.

The Winter Harvest Handbook by Eliot Coleman. This book is about growing year-round in unheated or minimally heated greenhouses. It is detailed about materials, construction techniques, varieties, planting dates, and more. It is a great resource on season extension. A companion DVD, *Year-Round Vegetable Production,* is a video of a workshop taught by Coleman.

Trend. (2014) Retrieved from http://www.businessdictionary.com/definition/trend.html
"Understanding PH." *Bass Fishing Resource Guide.* N.p., 1997. Web. 15 Nov. 2012.

Walking to Spring by Paul and Alison Wiediger; Describes how to use high tunnels year-round for a succession of profitable crops.

US Department of Agriculture has a huge amount of information on farmers markets, food hubs, and agricultural cooperatives. Start exploring at www.ams.usda.gov

"Whole Foods Market." *Whole Foods Market*. N.p., n.d. Web. April 2014.

Wholesale Success: A Farmer's Guide to Food Safety, Selling, Postharvest Handling, and Packing Produce is an excellent manual for all producer growers. It emphasizes practices that will ensure food safety and a long shelf life—essential for wholesale growers but helpful for farmers market and CSA growers as well. It is a large, spiral-bound book and is available from www.familyfarmed.org

Canning References

Boyer, Renee. "Canning — Boiling Water Bath." *Rebuilding Freedom.* May 30, 2016. http://rebuildingfreedom.org/water-bath-canning/

Boyer, Renee R., and Chase, Melissa. "Pressure Canning." *Virginia Cooperative Extension, Virginia Tech, and Virginia State University.* April 20, 2016. http://pubs.ext.vt.edu/348/348-585/348-585.html

Ewald, Jonathan. "What is Canning and What are the Benefits?" *Life + Health — Your Health Simplified,* August 7, 2014. http://lifeandhealth.org/lifestyle/what-is-canning-and-what-are-the-benefits/172324.html

Glatz, Julianne. "Canning Food, From Napoleon to Now." *Illinois Times*, June 3, 2010. http://www.illinoistimes.com/Springfield/article-7361-canning-food-from-napoleon-to-now.html

Jason. "Benefits of Home Canning." *Brew Plus.* July 23, 2011. http://www.brewplus.com/preserving/benefits-of-home-canning/#ixzz4A1L7GyDb

Jeanroy, Amelia, and Ward, Karen. "Food Preservation Methods — Canning, Freezing and Drying." *For Dummies.* May 28, 2016. http://www.dummies.com/how-to/content/food-preservation-methods-canning-freezing-and-dry.html

"Pressure Canning Low Acid Foods." *Ball*. May 28, 2016. http://www.freshpreserving.com/tools/pressure-canning

"Water-Bath Canning High Acid Foods." *Ball*. May 28, 2016. http://www.freshpreserving.com/tools/waterbath-canning

"What is Canning?" *Ball*. May 28, 2016. http://www.freshpreserving.com/getting-started

Wimbush-Bourque, Aimée. "9 Good Reasons to Can Your Own Food." *Simple Bites*. September 10, 2010. http://www.simplebites.net/9-good-reasons-to-can-your-own-food/

APPENDIX 3

Vendors

Organic Seeds:
- *Seeds of Change*: www.SeedsofChange.com
- *Grow Organic:* www.groworganic.com
- *Seed Savers Exchange* is a non-profit, 501(c)(3), member supported organization that saves and shares the heirloom seeds. Their mission is to conserve and promote America's culturally diverse but endangered garden and food crop heritage for future generations by collecting, growing, and sharing heirloom seeds and plants. At the heart of Seed Savers Exchange are the dedicated members who have distributed hundreds of thousands of heirloom and open pollinated garden seeds since our founding over 37 years ago. Those seeds now are widely used by seed companies, small farmers supplying local and regional markets, chefs and home gardeners and cooks, alike. http://www.seedsavers.org

Plug Suppliers:
Several seed companies serve as brokers for the numerous plug producers that are located around the country. Here are three that will send you a plant catalog:
- Germania Seed Company. 800-380-4721; www.germaniaseed.com
- Gloeckner & Co. 800-345-3787; www.fredgloeckner.com
- Harris Seed Company. 800-544-7938; www.harrisseeds.com

Greenhouse and Hoophouse Manufacturers:
- Agra Tech. 925-432-3399; www.agra-tech.com
- Atlas Greenhouse Systems. 800-346-9902; www.atlasgreenhouse.com
- BWI Companies. 903-838-8561; www.bwi-companies.com
- Conley's Greenhouse Manufacturing & Sales. 800-377-8441; www.conleys.com
- Farm Tek's Growers Supply. 800-476-9715; www.GrowersSupply.com
- G&M Ag Supply. 928-468-1380 or 800-901-0096; www.gmagsupply.com
- Harnois C.P. 450-756-1041; www.harnois.com
- Hummert International. 800-325-3055; www.hummert.com
- Jaderloon. 800-258-7171; www.jaderloon.com
- Keeler-Glasgow. 800-526-7327; www.keeler-glasgow.com
- Ledgewood Farm. 603-476-8829; www.ledgewoodfarm.com
- *Ludy* Greenhouse Manufacturing Corp.; 800-255-5839; www.ludy.com

Net Containers
- B.E. Sustainable; 928-239-9888; www.BlueEarthSustainable.com
- Hydo Galaxy; 800-818-6128; www.hydrogalaxy.com
- HydroBuilder; 888-815-9763; www. http://hydrobuilder.com

Community Supported Agriculture (CSA) Resources

The following software platforms to help CSA farmers manage sign-up and payment, take online orders, create picking lists, and organize deliveries. All have their own features, so explore each in turn to see which works best for your business.

- Member Assembler from Small Farm Central, a business that creates websites for farmers: www.smallfarmcentral.com/memberassembler
- CSAware from LocalHarvest.org: www.csaware.com
- Farmigo, www.farmigo.com

Packaging
- An excellent discussion of produce packaging is available from North Carolina State University at www.bae.ncsu.edu/programs/extension/publicat/postharv/ag-414-8/. The publication describes the various types of packaging, and it lists commonly used packages, weights, and quantities for all the major types of produce.
- Formtex sells clamshells and corrugated boxes. www.formtex.com
- Glacier Valley Enterprises has a wide selection of packaging supplies and containers for fruit and vegetable farmers. And it offers smaller quantities than many produce package suppliers. 800-236-6670; www.glacierv.com.
- Boxes, bushel baskets, clamshells, and so on can be purchased from Monte Package Company. 800-653-2807; montepkg.com
- Hubert Company sells fixtures, displays, and supplies to supermarkets. You will find produce packaging, clamshells, and much more here. Hubert is also a source for the natural plastic containers made from corn, known as PLA, or polylactic acid. 866-482-4357; www.hubert.com
- Produce boxes, bags, mesh bags, bushel baskets, and more are available from Southern Container Corp. 800-261-2295; www.socontainers.com/ProducePackaging.htm.

The Markets

LocalHarvest.org, is a website dedicated to linking producers and consumers. You can create a free listing with LocalHarvest.org if you are a direct-marketing family farm, a producers' farmers market, a business that sells products made from things grown locally by family farms, or an organization dedicated to promoting small farms and the "Buy Local" movement. www.localharvest.org.

Marketing Supplies
- A. Steele Co. sells supplies for farmers market vendors, including portable scales, cash registers, wireless credit card terminals, and E-Z UP tents. 800-693-3353; www.asteele.com
- Eat Local Food produces fine-art graphics of vegetables and fruits, including banners, postcards, tote bags, and other marketing materials. 734-341-7028; www.eatlocalfood.com
- Grower's Discount Labels is a farm-based business that designs and prints custom labels for farm products. 800-693-1572; www.growersdiscountlabels.com
- Produce Promotions sells banners, flags, bags, baskets, and other mar- keting products. 888-575-4090; www.producepromotions.com.

Managing Your Business
- The IRS has remarkably clear publications on farming tax issues. They can be found at www.

- irs.gov, or request the "Farmer's Tax Guide" from 800-829-3676.
- For information about farmland property tax assessments, the American Farmland Trust and the USDA have compiled numerous publications on the website www.farmlandinfo.org.
- How to track electricity usage: The Kill A Watt power measurement tool is available from Real Goods. 888-567-6527; www.realgoods.com.
- Deciding on a legal structure for your business requires the advice of an attorney or accountant. Here is a link to a publication from Kansas State University that describes the various options: www.ksre.ksu.edu/bookstore/pubs/MF2696.pdf.
- **C12 Group**: It's an objective advisory board for brainstorming and decision making, learning what you don't know and focusing on areas you need to sharpen. In a confidential, non-competing trusted C12 peer board, you'll learn from your group's wisdom and insight — and encourage and hold each other accountable to the principles and core values that guide you. http://www.c12group.com

Insurance

- The company that provides our farmers cooperative's products liability coverage (as well as our personal farm policy) is called Goodville Mutual Casualty Company, www.goodville.com. It's based in New Holland, Pennsylvania, and covers a lot of direct-market farmers in these nine states: Pennsylvania, Delaware, Maryland, Virginia, Ohio, Indiana, Illinois, Kansas, and Oklahoma. You can call the company at 717-354-4921 to find an agent near you who sells their policies.
- InterWest Insurance Services, in Sacramento, California. 800444-4134; www.iwins.com.

Labor Supply Sources

- Interns: Most regional and state organic farming associations have newsletters, some with online listings, where you can place a classified ad for apprentices. ATTRA, the sustainable farming information source, allows farmers to list internships on their website: www.attrainternships.ncat.org
- *Growing for Market* runs classified ads for apprentices. www.growingformarket.com
- Willing Workers on Organic Farms is an international organization that allows farmers and interns to advertise for each other. www.wwoof.org.

Payroll services

- QuickBooks Payroll, www.intuit.com.
- ADP. 800-225-5237; www.adp.com.
- Local accounting offices also do payroll for small businesses; be sure to compare prices and services.
- Your state's labor department.

APPENDIX 4

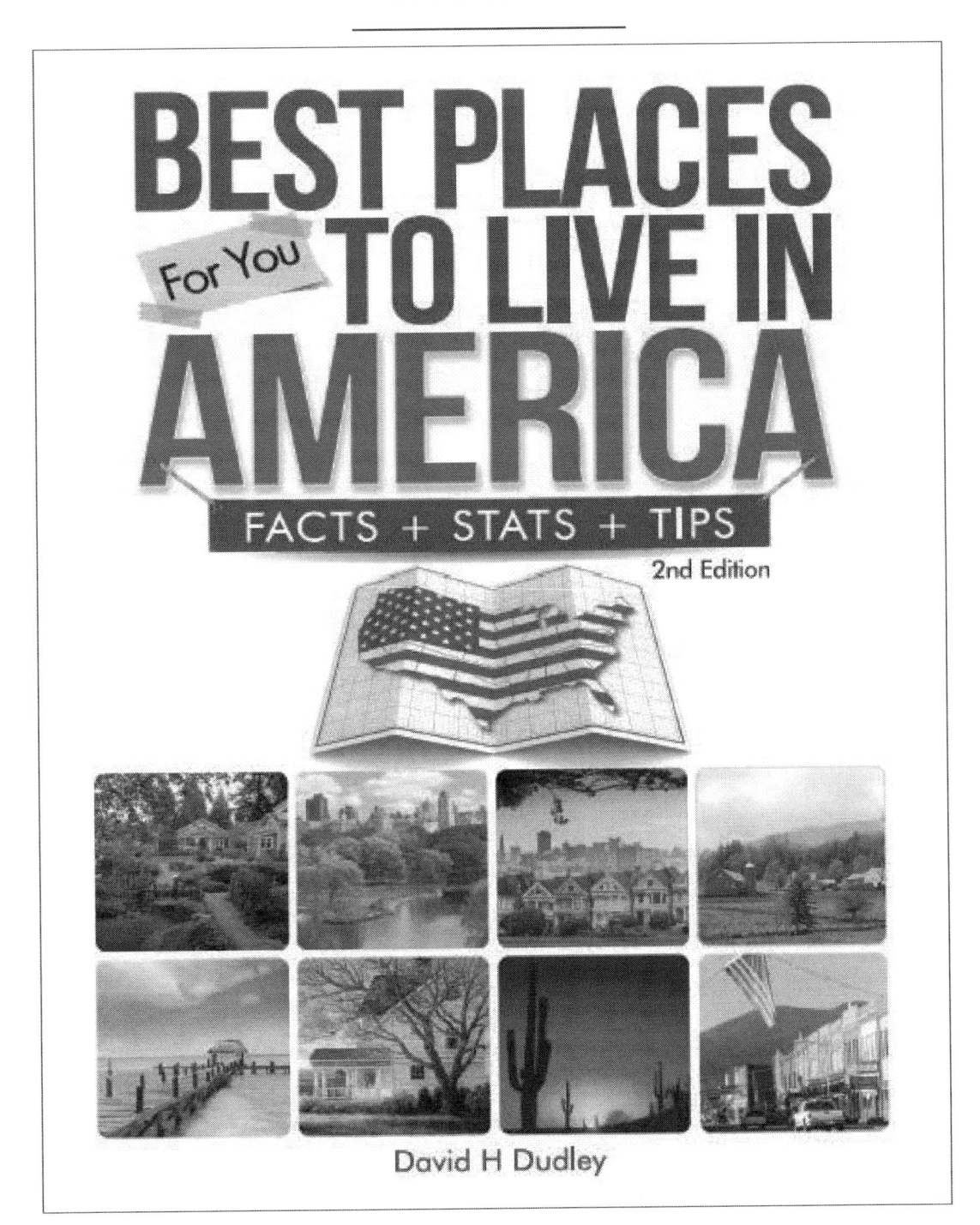

BEST PLACES For You TO LIVE IN AMERICA
FACTS + STATS + TIPS
2nd Edition

David H Dudley

www.bestplacestoliveinamerica.info

APPENDIX 5

BEST PLACES TO LIVE
for
AUTISM
COGNITIVE & PHYSICAL DISABILITIES

David H. Dudley

BEST PLACES TO LIVE for AUTISM COGNITIVE & PHYSICAL DISABILITIES

David H. Dudley

This 300+ page valuable book provides you with the critical information needed to make a smart relocation decision. Hundreds upon hundreds of research results examining data from well over a 1,000 different categories of various criteria are condensed into this user-friendly resource so that you and your family can find the very best place in which to live—which best 'fits' your needs and the needs of your loved ones. Furthermore, this book is an investment that will provide you helpful information on how to best go about relocating, as well as steps for efficiently moving to your best 'fit' location -saving you lots of time and money. Most importantly, you will be able to efficiently identify the ideal special needs care option and location that is best for you and your family.

Part I – How to Find the Best Place & Relocating with Ease

This section shows you how to intelligently go about finding the best place that is truly ideal for you and your family. Also, included are checklist, tips, and professional advice from those in the industry on how to efficiently and successfully relocate to your selected location. Additionally, this section addresses the best approach for transitioning from one employer to another, as well as relocating a business.

Part II - Cities & States Ranked According to YOUR Wants & Needs

Part II examines cities, counties, and states from a multitude of perspectives; everything from cost of living, best places to retire, employment opportunities, tax structures, affordability, recreation, cultural opportunities, business friendliness, growth rates, bicycle & transportation resources, schools, home school regulations, vaccine laws, proximity to coal & nuclear power plants, natural disaster risks, survival/prepper concerns, allergy & asthma problems, air quality & air pollution, crime, commute issues, agricultural issues, happiness levels, plant hardness zones, climate, population demographics, and MUCH more.

Part III - Best Places to Live For Physical Disabilities (Including Help for Veterans & Caretakers)

This section not only shows the best place to live for those with disabilities, but also provides a wealth of very helpful information about those locations so those with disabilities can achieve the best life possible. In addition, history, current statistics, trends, and an abundance of facts are provided about the various kinds of disabilities. Topics such as the best cities for wheelchairs, best locations for disability transportation assistance, and public transportation options are also covered. Furthermore, the best places for veterans and much information to help caretakers is provided. Many other topics to help those with disabilities are also included within this valuable section.

Part IV - Best Places to Live for Autism, Intellectual Developmental Disabilities, and Mental/Cognitive Disabilities

Where are the best places to live for those with autism or other cognitive disabilities? This section will answer that question by showing you which states provide the most assistance, opportunities, and resources. Factual reliable data is provided on each state, along with tables showing how states compare with each other in regards to funding and services. Included are important facts, statistics, resources, and much other useful information to best help those caring for those with intellectual disabilities. The author's daughter has autism and as a result he has invested nearly two decades investigating this topic.

This valuable book is an excellent investment that will save you lots of time and money. It provides the essential information needed to find the best place for YOU and YOUR SPECIAL NEEDS LOVED ONE to live in America.

www.FarmYourSpace.com

APPENDIX 6

Aquaponic Design Plans
Everything You Need to Know *from* Backyard to Profitable Business

David H. Dudley, PE

Aquaponic Design Plans, Instructions & All You Need to Know

Fresh Organic Produce and Plentiful Healthy Fish

Feed Your Family Healthy Food + Barter and/or Sell Surplus

Everything from Beginner Basics to Operating a Profitable Aquaponic Business

Expensive university courses and lengthy on-site training workshops which cost thousands of dollars do not provide as much valuable, comprehensive material as presented in this comprehensive user-friendly 'how-to' book.

Aquaponic Design Plans

Everything You Need to Know *from* **Backyard to Profitable Business**

This how-to resource consists of three important sections:

- Design Plans, Instructions & Everything You Need to Know About Aquaponics
- How to Set up & Operate different types of Aquaponic Systems of any Size
- How to Turn Aquaponics Into a Profitable Venture

This book provides detailed directions to create and maintain different types of aquaponic systems of all sizes so you can consistently feed your family environmentally friendly sustainable healthy organic food and even earn extra income.

The author, David Dudley, is a professional aquaponics consultant who has helped many individuals and companies develop aquaponics systems. His accomplished career in aquaponics, hydroponics, and aquaculture includes serving as the Construction Manager of the Oklahoma Aquarium, Engineering Manager of the nation's largest caviar producing company, overseeing life support systems of four large aquaculture facilities, designing a $5M aquaculture operation for white sturgeon, and Project Manager of a large fishing clinic facility for the U.S. Department of Wildlife. David holds advanced degrees in civil engineering and nutrition/dietetics, owns a commercial nursery, and has several decades of experience in vegetable gardening. David understands every facet of aquaponics and clearly communicates aquaponics in a way that truly helps others.

www.FarmYourSpace.com

ISBN 978-0-9969090-0-6

APPENDIX 7

AQUAPONICS
PLANS AND INSTRUCTIONS
MEDIA-BED (FLOOD-AND-DRAIN) SYSTEMS

David H. Dudley, P.E.

PART VIII: APPENDIX

This 400+ page user-friendly book (with over 350 photos and illustrations) shows you how to easily produce an abundance of Fresh Organic Produce and Fish through a Media-Bed (Flood-and-Drain) Aquaponic System.

This VALUABLE resource has everything from beginner basics to showing you how scale-up as you desire to grow your system. Easy to follow step-by-step Instructions and SO much more.

Expensive university courses and lengthy on-site training workshops which cost thousands of dollars do not provide as much valuable, comprehensive material as presented in this comprehensive user-friendly 'how-to' book.

Feed Your Family Healthy Organic Food, Barter and/or Sell Surplus, Substantially Lower Your Food Cost.

Included are Media-Bed Design Plans, Instructions & Everything You Need to Know about Aquaponics. Also included are chapters on survival, living off the grid, canning, so much more.

This book will teach you how to set-up and operate a productive and successful media-bed aquaponic system of any size; and show you how to scale-up in size to produce even more organic vegetables and fish as you desire expand your operation.

This book empowers you with the knowledge needed to consistently feed your family environmentally friendly sustainable healthy organic food and greatly lower your food bill. Aquaponics is fun, financially rewarding, enjoyable, healthy, and an interesting conversation topic at social functions.

APPENDIX 8

Earning Money from Aquaponics

To learn all the ways in which to earn revenue, or how to create a successful aquaponic business with maximum profits, be sure to order a copy of the author's book "Aquaponics for Profit, Earning Extra Money to a Successful Commerical Business."

Included within this valuable resource are two real-world aquaponic business plans and many other helpful resources, including:

- Cost-benefit analysis for aquaponics
- Aquaponic business questions that must be answered
- Five critical phases to creating a commercial aquaponics operation
- Alternative labor supply options for a commercial aquaponics operation
- The economics of a recirculating tank system
- Where and how to sell fish
- Where and how to sell produce from aquaponics
- Selling aquaponic fish and produce for maximum profit
- How to barter your aquaponic harvest products and gain maximum return
- Marketing aquaponics products
- How to successfully develop and manage a profitable aquaponics business
- Navigating the legal and permitting requirements
- Successfully addressing taxes, insurance, and overhead issues
- Keys to success in business
- Time management methods
- And so much more

If desiring to earn extra money from your aquaponics operation, or successfully barter your harvest for other goods or services, and/or create a successful commercial aquaponics operation that earns maximum profits, then this book is a must have resource and is a wise investment that will save you time and allow you to optimize your revenue from aquaponics.

AQUAPONICS FOR PROFIT

EARN **EXTRA MONEY** OR CREATE A **SUCCESSFUL COMMERCIAL BUSINESS**

David H. Dudley, PE

APPENDIX 8: EARNING MONEY FROM AQUAPONICS

AQUAPONICS FOR PROFIT

EARN **EXTRA MONEY** OR CREATE A **SUCCESSFUL COMMERCIAL BUSINESS**

What are the best methods for making money with aquaponics? Which edible aquaponic fish species will generate the most revenue per pound? Which aquaponic vegetables provide the highest profit margin? Which fish and vegetables are in highest demand by consumers? Do you know about all of the non-vegetable plants that can provide you with greater revenue than vegetable plants? What costs are involved in setting-up and operating an aquaponic system? What is the cost-benefit analysis of an aquaponic system? Where and how can I sell my aquaponic harvest? What regulations and legalities are involved in setting-up and operating an aquaponic business? Did you know that there are other aquatic species that can be grown in aquaponics which can generate more revenue than the commonly grown edible aquaponic fish species? What is the best way to barter my aquaponic harvest? How can I get my products officially labeled as 'organic'? What is the best approach to having a successful aquaponics business that will produce the largest profit margin? How can I earn extra money with my small backyard aquaponic system? Which type of aquaponic system – Media-Bed/Flood-and-Drain, Nutrient Film Technique (NFT), Raft System / Deep Water Culture (DWC) – is the most profitable? What are the pros and cons to each of these types of aquaponic systems? In addition to selling your harvest, are you aware of all the other ways in which you can earn revenue from your aquaponic system? How much time would I need to invest to have an aquaponic system that will feed my family and provide us with some extra income?

This user-friendly easy read book will answer the above a questions. This valuable resources is also packed with the necessary information that will not only show you how to make extra money with your aquaponic system, but to grow it into a successful commercial business; if that is your desire. Also included are two real-world aquaponic business plans. This book is an excellent investment that will reward you greatly with the knowledge needed to earn extra money through aquaponics or optimize revenue from a commercial aquaponic operation.

David H. Dudley is a professional aquaponics consultant who has helped many individuals and companies develop aquaponics systems. His accomplished career in aquaponics, hydroponics, and aquaculture includes serving as the Construction Manager of the Oklahoma Aquarium, Engineering Manager of the nation's largest caviar producing company, overseeing life support systems of four large aquaculture facilities, designing a $5M aquaculture operation for white sturgeon, and Project Manager of a large fishing clinic facility for the U.S. Department of Wildlife. David also holds advanced degrees in civil engineering and nutrition/dietetics, owns a commercial nursery, and has several decades of experience in vegetable gardening. David understands every facet of aquaponics and clearly communicates aquaponics in a way that truly helps others.

www.FarmYourSpace.com

PART VIII: APPENDIX

Aquaponics for Profit

- What are the best methods for making money with aquaponics?
- Which edible aquaponic fish species will generate the most revenue per pound?
- Which aquaponic vegetables provide the highest profit margin?
- Which fish and vegetables are in highest demand by consumers?
- Do you know about all of the non-vegetable plants that can provide you with greater revenue than vegetable plants?
- What costs are involved in setting-up and operating an aquaponic system?
- What is the cost-benefit analysis of an aquaponic system?
- Where and how can I sell my aquaponic harvest?
- What regulations and legalities are involved in setting-up and operating an aquaponic business?
- Did you know that there are other aquatic species that can be grown in aquaponics which can generate more revenue than the commonly grown edible aquaponic fish species?
- What is the best way to barter my aquaponic harvest?
- How can I get my products officially labeled as 'organic'?
- What is the best approach to having a successful aquaponics business that will produce the largest profit margin?
- How can I earn extra money with my small backyard aquaponic system?
- Which type of aquaponic system – Media-Bed/Flood-and-Drain, Nutrient Film Technique (NFT), Raft System / Deep Water Culture (DWC) – is the most profitable?
- What are the pros and cons to each of these types of aquaponic systems?
- In addition to selling your harvest, are you aware of all the other ways in which you can earn revenue from your aquaponic system?
- How much time would I need to invest to have an aquaponic system that will feed my family and provide us with some extra income?

This user-friendly easy read book will answer the above a questions. This valuable resource is also packed with the necessary information that will not only show you how to make extra money with your aquaponic system, but to grow it into a successful commercial business; if that is your desire. **Also included are two real-world aquaponic business plans.** This book is an excellent investment that will reward you greatly with the knowledge needed to earn extra money through aquaponics or optimize revenue from a commercial aquaponic operation.

APPENDIX 9

Recommended Resources

Now that I am entering into my senior years of life I am able to look back and recognize certain things which really made a positive impact on me. Beyond education, life experiences, and people who played an instrumental role in my life, I have been blessed beyond measure by certain documentaries and books. These resources have either greatly inspired me or better educated me to a point of positive change. I wanted to dedicate this portion of this book to pay it forward in the hope that you may be able to benefit from them as well. Following you will find resources that have truly improved my quality of life. I very much recommend the below resources and hope that you will find value in them, too.

Documentaries

I absolutely love documentaries. I learn so much from them. Following are my favorite documentaries that I highly recommend. Most of these recommended documentaries received a review rating of at least 4.5 stars out of 5 stars.

Food / Nutrition / Health Documentaries (Highly Recommended)

- Cowspiracy
- Eating - 3rd Edition (by Mike Anderson)
- Fat, Sick & Nearly Dead
- Fat, Sick & Nearly Dead 2
- Fed Up
- Food Chains
- Food Choices
- Food Matters
- Food, Inc.
- Forks Over Knives
- Fresh
- GMO OMG
- Hungry for Change
- Killer at Large
- King Corn
- Plant Pure Nation
- Processed People
- Scientists Under Attack- Genetic Engineering in the Magnetic Field of Money
- Sugar Coated
- Supersize Me
- The Beautiful Truth (nutrition for cancer patients)
- The Future of Food
- The Gerson Miracle
- The Kids Menu
- The Weight of a Nation
- Vegucated

Social, Environmental, or Nature Documentaries (Highly Recommended)

- (Dis)Honesty: The Truth About Lies
- A Crude Awakening
- Bag It
- Blue Gold: World Water Wars
- Cowspiracy
- Earthlings (by Shaun Monson)
- End of the Line (by filmmaker Rupert Murray)
- Flow: For Love of Water
- Gasland
- God of Wonders
- Happy
- In the Womb (National Geographic)
- Inside Job
- Inside Planet Earth
- Life
- Living on One Dollar
- Minimalism: A Documentary About the Important Things
- Nature's Most Amazing Events
- No Place on Earth
- Planet Earth
- Plastic Paradise: The Great Pacific Garbage Patch
- Poverty, Inc.
- SlingShot
- Tapped
- The College Conspiracy (a documentary on YouTube)
- The Great Rift: Africa's Greatest Story
- The Lee Strobel Film Collection
- The World According to Monsanto
- Vanishing of the bees
- Waste Land directed by Lucy Walker (Arthouse Studio)
- Winter on Fire: Ukraine's Fight for Freedom

Historical Documentaries (Highly Recommended)

- Above and Beyond
- Auschwitz: The Nazis and the 'Final Solution'
- Brothers in War
- Desperate Crossing: Mayflower
- Diaries of the Great War
- Escape from a Nazi Death Camp
- Reader's Digest WWII in the Pacific
- The Civil War
- The First World War (the complete series)
- The Long Way Home
- The Longest Day by 20th Century Fox
- The War
- The World at War
- Treblinka
- World War II - War in the Pacific

Inspirational Films (Highly Recommended)

Below is a list of my favorite inspirational movies.

- Courageous
- Facing the Giants
- Fire Proof
- Flywheel

Classics (Highly Recommended)

Following is my list of favorite classical films. These films received excellent reviews.

- An American in Paris
- Fiddler on the Roof
- It's a Wonderful Life
- Oklahoma!
- Seven Brides for Seven Brothers
- Singin' in the Rain
- The General (*this silent movie filmed in 1926 is the funniest movie I have ever seen*)
- The Great Locomotive
- The Music Man
- The Sound of Music
- White Christmas

Books & Audio Books (Highly Recommended)

The below books are the best books I have ever read in my life. Coincidently, most all received a customer review rating of 5 stars out of 5 stars.

- 1776 by David McCullough
- 50/50: Secrets I Learned Running 50 Marathons in 50 Days
- 7 Habits of Highly Successful People
- A Thousand-Mile Walk to the Gulf by John Muir
- As A Man Thinketh by James Allen
- Awaken the Giant Within
- Band of Brothers by Stephen E. Ambrose
- Born to Run: A Hidden Tribe, Superathletes, and the Greatest Race the World Has Never Seen
- Brian Tracy (all books by Brian Tracy are excellent)
- Bringing Up Boys by James Dobson
- Bringing Up Girls by James Dobson
- Caffeine Blues
- China Study
- Coming Back Stronger: Unleashing the Hidden Power of Adversity
- Desiring God by John Piper
- Disciplines of a Godly Man by R. Kent Hughes
- Don't Waste Your Life by John Piper

APPENDIX 9: RECOMMENDED RESOURCES

- Driven: How To Succeed In Business And In Life by Robert Herjavec
- Eat and Run: My Unlikely Journey to Ultramarathon Greatness
- Eat to Live: The Amazing Nutrient-Rich Program for Fast and Sustained Weight Loss by Joel Fuhrman
- Evangelism and the Sovereignty of God, J.I. Packer
- Extreme Pursuit by John E. Davis
- From Pride to Humility by Stuart Scott
- God's Wisdom in Proverbs by Phillips
- Good to Great: Why Some Companies Make the Leap and Others Don't
- Grace to You by John MacArthur
- Handwriting of the Famous and Iinfamous by Sheila Lowe
- Happy is the man by Robert V. Ozment
- Happy, Happy, Happy
- Have a New Kid by Friday by Leman
- Healthy Eating, Healthy World: Unleashing the Power of Plant-Based Nutrition, by J. Morris Hicks
- How Successful People Think by John Maxwell
- How to Win Friends & Influence People, Dale Carnegie
- Lincoln the Unknown, Dale Carnegie
- Love Dare
- Love for a Lifetime: Building a Marriage That Will Go the Distance
- Making Men by Chuck Holton
- Men Are from Mars, Women Are from Venus
- No Happy Cows by John Robbins
- Nothing to Envy
- One Minute Manager
- Parenting Collection by James Dobson
- Pursuit of Holiness by Jerry Bridges
- Quiet Strength by Tony Dungy
- Raising a Modern-Day Knight by Robert Lewis (book for Dad's with sons)
- Remember Names by Dale Carnegie
- Respectable Sins by Jerry Bridges (*best book I have ever read in my life*)
- Rich Dad, Poor Dad
- Running Man: A Memoir
- Seeking Allah, finding Jesus
- Shaken: Discovering Your True Identity in the Midst of Life's Storms
- Shepherding a Child's Heart by Tedd Tripp
- Strong Willed Child by James Dobson
- The 10 natural laws of successful time and life management
- The 5 Love Languages: The Secret to Love that Lasts
- The Attributes of God, Arthur W. Pink
- The Autobiography of Benjamin Franklin
- The Backyard Homestead
- The Endurance: Shackleton's Legendary Antarctic Expedition
- The Exemplary Husband by Stuart Scott
- The Gluten Connection: How Gluten Sensitivity May Be Sabotaging Your Health - And What You Can Do to Take Control Now
- The Greatest Miracle in the World by O.G. Mandino
- The Greatest Salesmen in the World
- The Guide to Confident Living, by Norman Vincent Peale
- The Human Body Book (Book & DVD)
- The Marriage You've Always Wanted by Gary Chapman
- The Path Between the Seas: The Creation of the Panama Canal, 1870-1914
- The Personal MBA by Josh Kaufman
- The Power of Positive Thinking by Norman Vincent Peale
- The Power of Positive Thinking, Norman Vincent Peale
- The Psychology of Winning by Dr. Dennis Waitley
- The Quest for Character, John MacArthur
- The Success Principles: How to get from where you are to where you want to be, by Jack Canfield with Janet Switzer.
- The Truth War: Fighting for Certainty in an Age of Deception by John F. MacArthur
- Think and Grow Rich
- Through My Eyes By Tim Tebow
- Ultramarathon Man: Confessions of an All-Night Runner
- Undaunted Courage: Meriwether Lewis, Thomas Jefferson, and the Opening of the American West
- Way of the Master by Ray Comfort
- Whitewash: The Disturbing Truth About Cow's Milk and Your Health

Best of the Best (Highly Recommended)

Although I highly recommend all of the above resources, selecting the best of the best out of each category, following is my 'must see/read' list. It you were only going to try a few on my list I would put the following as the highest priority.

- **Environmental** — the documentaries 'Cowspiracy' and 'Plastic Paradise'.
- **Nutrition/Health/Weight Loss** — 'Food Choices' (documentary) and 'China Study' (book or audiobook).
- **Social** — the documentaries 'Tapped', 'Poverty, Inc.' and ''Flow: for the Love of Water'.
- **Business/Entrepreneurship/Success** — all books by Brain Tracy, and the book 'As A Man Thinketh' by James Allen
- **Spiritual** — the books 'Respectable Sins' and 'Pursuit of Holiness' by Jerry Bridges
- **Nature** — the documentaries 'Planet Earth' and 'The Blue Planet'.
- **Inspirational** — the films noted above in the Inspirational category.
- **Relationships** — 'Fire Proof' (film), 'The 5 Love Languages: the Secret to Love that Last (book), Men are from Mars, Women are from Venus' (book), and 'Love for a Lifetime' book by Dobson).
- **Child Rearing** — 'Parenting Collection' (book by Dobson).

Comment and Feedback

As mentioned previously, I am an avid fan of documentaries and non-fiction books. The above resources have greatly helped me grow as a person through the decades and I hope my sharing of these recommendations will help you grow as well. I am still learning and will continue to update this list on the "Farm Your Space" website.

I welcome your comments, recommendations, and feedback on these and other resources. Thank you.

www.FarmYourSpace.com

APPENDIX 10

Conversion Units

Conversions for Units of Volume

To From	cm3	liter	m3	in3	ft3	fl oz	fl pt	fl qt	gal
cm3	1	0.001	1×10^{-6}	0.0610	3.53×10^{-5}	0.0338	0.00211	0.00106	2.64×10^{-4}
liter	1000	1	0.001	60.98	0.0353	33.81	2.113	1.057	0.2642
m3	1×10^6	1000	1	6.1×10^4	35.31	3.38×10^4	2113	1057	264.2
in3	16.39	0.0164	1.64×10^{-5}	1	5.79×10^{-4}	0.5541	0.0346	0.0173	0.0043
ft3	2.83×10^4	28.32	0.0283	1728	1	957.5	59.84	29.92	7.481
fl oz	29.57	0.0296	2.96×10^{-5}	1.805	0.00104	1	0.0625	0.0313	0.0078
fl pt	473.2	0.4732	4.73×10^{-4}	28.88	0.0167	16	1	0.5000	0.1250
fl qt	946.4	0.9463	9.46×10^{-4}	57.75	0.0334	32	2	1	0.2500
gal	3785	3.785	0.0038	231.0	0.1337	128	8	4	1

Conversions for Units of Length

From To	cm	m	in	ft	yd
cm	1	0.01	0.3937	0.0328	0.0109
m	100	1	39.37	3.281	1.0936
in	2.540	0.0254	1	0.0833	0.00278
ft	30.48	0.3000	12	1	10.3333
yd	91.44	0.9144	36	3	1

Conversions for Units of Weight

To \ From	gm	kg	gr	oz	lb
gm	1	0.001	15.43	0.0353	0.0022
kg	1000	1	1.54×10^4	35.27	2.205
gr	0.0648	6.48×10^{-5}	1	0.0023	1.43×10^{-4}
oz	28.35	0.0284	437.5	1	0.0625
lb	453.6	0.4536	7000	16	1

Miscellaneous Conversion Factor for Water

1 acre-foot	equals	43,560 cubic feet
1 acre-foot	equals	325,850 gallons
1 acre-foot of water	equals	271,814 pounds
1 cubic-foot of water	equals	62.4 pounds
1 gallon of water	equals	3,785 grams
1 liter of water	equals	1,000 grams
1 fluid ounce	equals	29.57 grams
1 fluid ounce	equals	1.043 ounces

Temperature Conversions

Centigrade to Fahrenheit =
(C x 9/5) +32 = (C x 1.8) +32

Fahrenheit to Centigrade =
(F — 32) x (5/9) = (F — 32) x (0.5556)

See figure 262

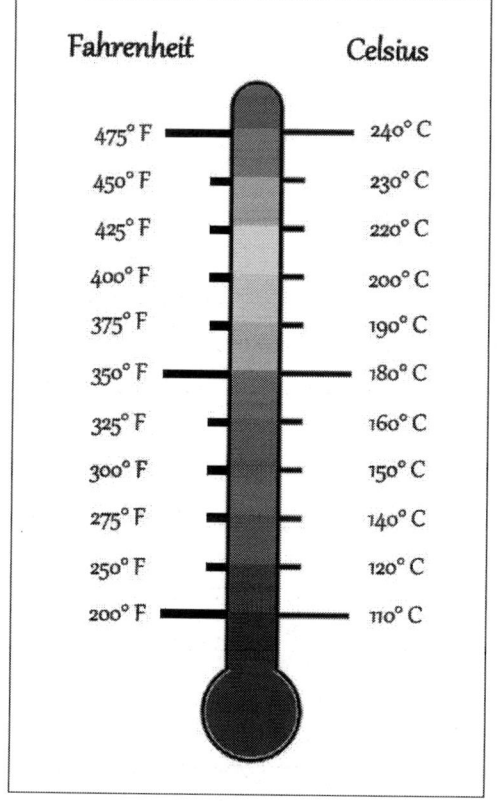

FIGURE 262.

Abbreviations

cm = centimeters	ft3 = cubic foot	kg = kilogram
cm3 = cubic centimeters	gal = gallon	lb = pound
fl oz. = fluid ounce	gm = gram	m = meter
fl pt = fluid pint	gr = grain	m3 = cubic meter
fl qt = fluid quart	in = inch	oz = ounce
ft = foot	in3 = cubic inch	yd = yard

APPENDIX 11

Encouragement & Keys to Success

Do not be afraid of failure. *"Failure is simply the opportunity to begin again, this time more intelligently."* – Henry Ford

Don't let discouragement stop you from pressing on. *"Let no feeling of discouragement prey upon you, and in the end you are sure to succeed."* – Abraham Lincoln

Be brave enough to follow your intuition. *"Have the courage to follow your heart and intuition. They somehow already know what you truly want to become. Everything else is secondary."* – Steve Jobs

Setting goals, whether they are to be achieved in 5 or 50 years, is the first step to success. The next big step is to take actions in life that increase the likelihood of your goal being achieved. Setting goals gives you long-term vision and short-term motivation. It focuses your acquisition of knowledge, and helps you to organize your time and your resources so that you can make the very most of your life.

New Year's resolutions don't work. Get this:

- 25 percent of people abandon their New Year's resolutions after one week.
- 60 percent of people abandon them within six months. (The average person makes the same New Year's resolution ten separate times without success.)
- Only 5 percent of those who lose weight on a diet keep it off; 95% regain it. A significant percentage gain back more than they originally lost.
- Even after a heart attack, only 14 percent of patients makes any lasting changes around eating or exercise.

While New Year's resolutions don't work, writing down goals does work. The research is conclusive. The Dominican University in California did a study on goal-setting with 267 participants. They found that you are 42 percent more likely to achieve your goals just by writing them down.

Harvard's graduate students were asked if they have set clear, written goals for their futures, as well as if they have made specific plans to transform their fantasies into realities.

The result of the study was only 3 percent of the students had written goals and plans to accomplish them, 13 percent had goals in their minds but haven't written them anywhere and 84 percent had no goals at all.

After 10 years, the same group of students were interviewed again and the conclusion of the study was totally astonishing. The 13 percent of the class who had goals, but did not write them down, earned twice the amount of the 84 percent who had no goals. The 3 percent who had written goals were earning, on average, 10 times as much as the other 97 percent of the class combined.

People who don't write down their goals tend to fail easier than the ones who have plans. The Harvard study proves that statement, even if the only criteria was the monetary reward of each graduate in the study. When you don't have a plan, you don't know how you will reach your destination. Sure, you know what your destination is and you have a general idea about how you can reach it, but it's not something that will lead you there for sure.

The secret to accomplishing what matters most to you is committing your goals to writing. This is important for the five reasons.

1. **Putting your goals in writing forces you to clarify what you want.** Writing down your goals forces you to select something specific and decide what you want.

2. **Writing down your goals will motivate you to take action.** Writing down your goals and reviewing them regularly provokes you to take action.

3. **Writing down your goals will provide a filter for other opportunities.** Writing down your goals keeps you on course and keeps you from getting distracted by everything else always popping up in life.

4. **Putting your goals in writing will help you overcome resistance.** Every meaningful intention, dream, or goal encounters resistance. From the moment you set a goal, you will begin to feel it. But if you focus on the resistance, it will only get stronger. The way to overcome it is to focus on the goal.

5. **Writing down your goals will enable you to see—and celebrate—your progress.** Life is hard. It is particularly difficult when you aren't seeing progress. However, written goals are like mile-markers on a highway. They enable you to see how far you have come and how far you need to go. They also provide an opportunity for celebration when you attain them.

S.M.A.R.T. Goals

S.M.A.R.T. goals are designed to provide structure and guidance throughout a project, and better identify what you want to accomplish. That's why setting SMART goals - Specific, Measurable, Achievable, Realistic and Timely - is a critical to bringing your goals to reality.

S - specific, significant, stretching
M - measurable, meaningful, motivational
A - agreed upon, attainable, achievable, acceptable, action-oriented
R - realistic, relevant, reasonable, rewarding, results-oriented
T - time-based, time-bound, timely, tangible, trackable

SPECIFIC – Your goal should be clear and specific, otherwise you won't be able to focus your efforts or feel truly motivated to achieve it. When drafting your goal, try to answer the five "W" questions:

- **What** do I want to accomplish?

- **Why** is this goal important?
- **Who** is involved?
- **Where** is it located?
- **Which** resources or limits are involved?

MEASURABLE – It is important to have measurable goals, so that you can track your progress and stay motivated. Assessing progress helps you to stay focused, meet your deadlines, and feel the excitement of getting closer to achieving your goal.

A measurable goal should address questions such as:

- How much?
- How many?
- How will I know when it is accomplished?

ACHIEVABLE – Your goal also needs to be realistic and attainable to be successful. In other words, it should stretch your abilities but still remain possible. When you set an achievable goal, you may be able to identify previously overlooked opportunities or resources that can bring you closer to it.

An achievable goal will usually answer questions such as:

- How can I accomplish this goal?
- How realistic is the goal, based on other constraints, such as financial factors?

RELEVANT – This step is about ensuring that your goal matters to you, and that it also aligns with other relevant goals. We all need support and assistance in achieving our goals, but it's important to retain control over them. So, make sure that your plans drive everyone forward, but that you're still responsible for achieving your own goal.

A relevant goal can answer "yes" to these questions:

- Does this seem worthwhile?
- Is this the right time?
- Does this match our other efforts/needs?
- Am I the right person to reach this goal?
- Is it applicable in the current socio-economic environment?

TIME-BOUND – Every goal needs a target date, so that you have a deadline to focus on and something to work toward. This part of the SMART goal criteria helps to prevent everyday tasks from taking priority over your longer-term goals.

A time-bound goal will usually answer these questions:

- When?
- What can I do six months from now?
- What can I do six weeks from now?
- What can I do today?

SMART is an effective tool that provides the clarity, focus and motivation you need to achieve your goals. It can also improve your ability to reach them by encouraging you to define your objectives and set a completion date. SMART goals are also easy to use by anyone, anywhere, without the need for specialist tools or training. When you use SMART, you create clear, attainable and meaningful goals, and develop the motivation, action plan, and support needed to achieve them.

Success Advice from Brian Tracy

The following is great advice from Brian Tracy. Brian Tracy is Chairman and CEO of Brian Tracy International, a company specializing in the training and development of individuals and organizations. He is also the author of over 70 books that have been translated into dozens of languages. Prior to founding his company, Brian Tracy International, Brian was the Chief Operating Officer of a $265 million dollar development company. He has had successful careers in sales and marketing, investments, real estate development and syndication, importation, distribution and management consulting. He has conducted high level consulting assignments with several billion-dollar plus corporations in strategic planning and organizational development. He has traveled and worked in over 107 countries on six continents, and speaks four languages. http://www.briantracy.com/

Realize that you have to pay the price. *"The price of success must be paid in full, in advance."*

Nothing you really want in life is free. You have to put in hard work to get it. And usually over a long time period. You have to make hard choices and sacrifices.

Now, doing so can produce a lot of happiness along the way and when you reach your destination. But when you take the step from comfortable dreams about success and happiness to actually start doing things then there is always a price to pay. So be prepared for that.

Keep going. *"Every great success is an accumulation of thousands of ordinary efforts that no one else sees or appreciates."*

How do you put in all that time and effort if no will reward you right now? Well, you find things you love doing, things you do for yourself – rather than to get someone else's attention and appreciation – and when things feel rough you just do what you know is the right thing to do anyway. You keep going with persistence but also simple the joy of doing what you love as two supporting friends.

Take responsibility for your life. *"The happiest people in the world are those who feel absolutely terrific about themselves, and this is the natural outgrowth of accepting total responsibility for every part of their life."*

"The more you like yourself, the better you perform in everything that you do."

"Disciplining yourself to do what you know is right and important, although difficult, is the high road to pride, self-esteem and personal satisfaction."

When you take full responsibility for your own life you will start doing many of these things naturally like making decisions, putting in hard work and really trying to keep your focus in the right place.

When you decide to take responsibility for your life and doing what you know deep down is right – for example, going to gym instead of lying on the couch eating potato chips – you like yourself more and more as your self-respect increases.

When your self-respect goes up you feel more worthy of any success and you are less likely to self sabotage in subtle and not so subtle ways. This is crucial as it impact every area of your life. You tend to behave in alignment with your own self image.

Taking responsibility for your own life and doing the right thing are not the only things you can do to increase your self-respect and success. Another powerful tip is to like/love other people. Why? Because how you view, judge and think about people is usually how you view, judge and think about yourself.

Look at the world through the eyes of a successful person.

"You cannot control what happens to you, but you can control your attitude toward what happens to you, and in that, you will be mastering change rather than allowing it to master you."

Success comes from being uncomfortable so aim to put yourself in a state of discomfort.

"Move out of your comfort zone. You can only grow if you are willing to feel awkward and uncomfortable when you try something new."

APPENDIX 11: ENCOURAGEMENT & KEYS TO SUCCESS

Become one with your thoughts, emotions, and feelings.

"Just as your car runs more smoothly and requires less energy to go faster and farther when the wheels are in perfect alignment, you perform better when your thoughts, feelings, emotions, goals, and values are in balance."

Use the law of attraction to your advantage.

"The key to success is to focus our conscious mind on [positive and edifying] things we desire, not things we fear."

Your actions are the result of habits. If you have successful habits, you will succeed. If you don't, you will fail. Therefore, form good habits and kill your bad ones.

"Successful people are simply those with successful habits."

Be relentless. Life will throw some horrible things your way, but don't lose faith. Don't let life break you. *"Your decision to be, have and do something out of ordinary entails facing difficulties that are out of the ordinary as well. Sometimes your greatest asset is simply your ability to stay with it longer than anyone else."*

Set the bar high. *"We will always tend to fulfill our own expectation of ourselves."*

Request for Input

I invite you to share your input about this book, as it will help me better serve others via improvements to future editions and the accompaniment website. I welcome all contributions, such as your recommended improvements, pics of your system, illustrations, etc. Please send your feedback, suggestions, contributions, etc. via the website: www.FarmYourSpace.com.

Thank you for your interest in this book. My hope is that it has made a positive contribution to your life. In addition to your feedback, please let me know how I can help you with aquaponics and/or pray for you.

Respectfully yours,

David Dudley

www.FarmYourSpace.com

Index

A

Acclimatizing Fish 105
Acid 117, 358, 397-398, 524
Acidity 347
ADA 203
Advertising 437, 470, 496
Aeration 122, 135, 181, 249, 361
Africa 18-20, 24, 27, 124
Agriculture 15, 23-24, 26-27, 31, 37-38, 40-41, 85, 120-121, 206, 215, 397, 412, 422, 443, 452, 454, 475, 489, 499, 518, 521-522, 524, 526
Agritourism 449
Air 82, 187, 204, 242-243, 291, 349, 361, 387, 392-393, 406
Algae 63, 161, 186, 349, 364-365
Alternative News 519
Aluminum 127, 202
Amaranth 78
American Heart Association 29
Ammonia 71, 108, 133-134, 186, 282, 339, 356, 372
Ammonium 339
Ammonium Chloride 339
Annuals 90
Antarctica 24
Aphids 347
Aquaculture 7, 29, 31, 118, 121, 471, 498, 511, 513, 517-518
Aqualife 478-479, 481, 484-491, 493-498, 500-505, 507, 511
Aquifer 25, 370
Arugula 78, 85
Asia 13, 18-19, 27, 30, 38, 97-98, 369, 440, 491
Australia 86, 97, 108, 212, 422, 491
Autism 426-427

B

Backup 136
Bacteria 67, 70-71, 341
Balance Sheet 431, 480, 505, 507
Bankrupt 462
Bark 176
Barramundi 96-97
Barter 463, 465
Bartering 410, 463-465, 473
Basil 78, 84-85
Beans 78, 87
Begonias 78
Bell Peppers 38, 421
Bell Siphon 291-293, 297, 300, 303, 350-351
Biofilter 71, 186, 188-189, 191-192
Biography 497

Black Seeded Simpson 78
Blueberries 37-38, 86
Blue Gill 96
Boron 82
Brim 96
Broccoli 78, 85, 87
Brown Blood Disease 366
Budgeting 434
Bulkhead 148, 293, 317, 322, 327, 330
Business 2, 66, 209, 219, 409, 411-412, 416-417, 422, 429, 436-437, 472, 477-481, 490, 502-503, 512-513, 519, 522, 527
Business Organization 411
Business Plan 417, 429, 436, 477, 512

C

Cabbage 78, 85
Calcium 69, 82, 117, 358-359, 372
Calcium Carbonate 359
Calcium Hydroxide 358
Canada 20, 41, 422, 440, 443, 491
Canning 395-397, 399, 524
Cantaloupe 78, 86
Caribbean 18-19
Carp 96, 98
Carrots 38
Cart 453, 460
Cash 435, 456-457, 505
Cash Flow 435
Catfish 96, 98, 116, 366, 471
Cauliflower 78, 85, 87
C-Corporation 411
Celery 37-38, 78, 85
Central America 18
Certified Organic 412
Chard 78, 85, 87
Checks 458-459
China 18-23, 25, 29, 44, 48, 51, 468, 470, 482, 490, 493-495, 501, 511
Chives 78, 85, 87
Chloramines 370
Chlorine 50-51, 82, 362, 370
Cilantro 78, 85
Climate 27, 29, 31, 221
Climate Change 29
Cod 101
Collard 38, 78, 87
Commercial 85, 102-103, 110, 113, 118, 120, 197, 221, 230, 234, 355, 376-377, 394, 414, 417, 443
Common Chives 78
Community 15, 427, 442, 454, 513, 523, 526
Community Supported Agriculture (CSA)
15, 454, 526
Competition 439, 467-468
Competitor 480, 493-495
Consumer 45, 221, 461-462
Consumption 21, 43, 490, 513, 523
Containers 128, 250, 313, 453, 526
Copper 43, 82, 117
Coriander 78
Corn 49, 78, 85, 87, 117, 127
Corporation 411
Cost-Benefit 405
CPVC 148
Craigslist 181, 194, 299, 437, 464, 472
Crappie 96
CSA 15, 88, 454-455, 523-524, 526
Cucumber 85
Cycle 67, 185, 356
Cycling 281-282, 337-339, 358

D

Daily Maintenance 343
Debt 462
Deep Water Culture 55-56, 59, 162, 227, 245-246, 252
Deforestation 27
Demand 19, 90, 117, 439-440, 512, 522
Design iii, 79, 131, 183-184, 225, 240, 246, 252, 257, 264-266, 271, 278, 299, 303, 313, 405
Developing Nations 21
Dill 78, 85
Direct Retail Sales 472, 500
Discount 470, 526
Disease 45, 49, 52, 109-110, 349, 366-367
Dissolved Oxygen (DO) 70, 133, 359
Distribution 417, 480, 500, 507
DIY 149, 156, 186, 191, 229, 231, 240, 250, 259, 264, 271, 274, 276, 299, 302
DO 70-71, 98, 187, 242, 249, 340, 353, 359-360, 362, 364-365, 372
Do-It-Yourself 2, 186, 231, 240, 259, 264, 416
Duckweed 79, 101, 118, 121-122
DWC 69, 255, 259, 276

E

EC (Electrical Conductivity) 82, 368-369
E. coli 444
Ecological 28, 317
Ecosystem 71
Eggplant 78, 85, 87
Egypt 18
Employment 426
Endive 78, 87

Enterprise 418–420
Enterprise Budget 418–420
EPDM 134, 153, 165
Equipment 131, 390, 406, 432, 504–505
Ergonomic 278
Europe 28, 30, 50, 108, 366, 369, 392, 422, 441–442
Executive Summary 416, 479
Expanded Clay 174, 317

F

Fancy Goldfish 96
Farmers Market 85, 425
FDA 39–41, 43, 46–53, 367, 444, 469, 479, 495, 511
Feed 10, 72, 97–101, 103–104, 107, 109, 113–121, 124, 126, 128, 340, 342–343, 348, 350, 375, 377, 379, 406, 419, 433, 507
Fe (Iron) 82
Fertilizer 236
Fiberglass 170, 199, 302
Filtration 183–184, 192, 239, 274, 276
Financial 417, 480, 488, 496, 503
Fingerlings 100, 119, 338, 379, 381, 419, 507
Fishing 524
Fish Maintenance 109
Fish Meal 117
Fish Tank 119–120, 154, 157, 160–163, 239, 242, 267–268, 273, 277–278, 287, 300, 303, 317–318, 389, 419
Flood and Drain System 162
Florida 35, 86, 213, 394, 491
Flowers 79, 84, 90–91, 523
Flow Rate 140
Food Supply 17, 22, 43–44

G

Germany 15
Ginger 78, 87
Global Warming 29
Glue 303
Go-By 420, 477
Golden Perch 96
Goldfish 96, 99, 120, 366
Grant 221, 422, 517
Gravel 174–175, 203, 317, 449
Greenhouse 65, 193–198, 201–207, 209, 211–213, 383, 389, 394, 417, 420, 518, 522, 525–526
Greenland 24
Grocery Store 451–452
Grommet 303, 317, 320
Grow Bed 131–132, 268, 272–274, 278, 287, 292, 302–303, 406

H

Harvest 83, 231, 244, 343, 395, 417, 473, 523–524
Harvest, Harvesting 83, 231, 244, 343, 395, 417, 473, 523–524
Hauling 473
Hawaii 216, 414
HDPE 134, 153, 156, 160–161, 165, 302
Head 141, 185, 204, 239
Health 35, 41, 43, 45, 49, 51, 72, 344, 348, 366–367, 474, 478, 499, 522, 524
Heat 48, 205, 279, 392, 398–399
Heater 134, 206
Heavy Metals 43
History 13, 61, 480, 482
Homemade 92, 124, 126, 128, 189, 233, 396
Hong Kong 31
HPS 212, 385
Humidity 82
Hybrid Striped Bass 100
Hydrochloric Acid 358
Hydroponics 7, 231, 236, 240, 491
Hydroton 174, 235, 244, 250, 301, 513, 522

I

IBC 156, 242, 249, 254, 274, 302, 313, 317–318, 323–327, 405–406
Image 174
Impatiens 78
Importation 440, 490
Income 480, 496, 507–508, 510
India 18–22
Industrial 30
Industry Analysis 440, 480, 490
Initial Investment 480, 503–504
Insecticide 93
Insects 122–123, 205
Insulation 64, 201, 387
Insurance 422–425, 433, 507, 527
Intermediate Bulk Container 405
Internet 88, 91, 114, 122, 193, 454, 457–458, 464, 471, 517–518, 521
Investment 10, 436, 480, 503–504
Iron 82, 358
Irrigation 145–146, 303

J

Jade Perch 96, 100

K

K 41, 48, 52, 82, 512, 522
Kale 38, 78, 85, 87, 421, 522
Keys 34, 77, 409, 455

Kits 194
Koi 2, 96–97, 100, 105, 190, 280, 409–410, 468, 522

L

Label 39, 398, 400
Labeling 38, 413
Labor 118, 426, 433–434, 504, 507, 527
Ladybugs 347
Largemouth Bass 101
Law 523
Layout 79, 131, 183, 225, 228, 236, 246, 257, 265–266
Lead 43
Leaks 387
LECA 250
LED 118, 134, 385
Legal 480, 497–498
Lettuce 38, 78, 85, 87, 240, 250, 522
Lights 134
Lime 358
Limestone 364
Limited Liability Company (LLC) 411
Liquid Ammonia 339
Local 91, 221, 414, 417, 440–443, 445–446, 453, 472, 474, 500, 503, 523, 526–527
Locally 90, 439, 441–442
Location 63, 162, 196, 219, 445, 448

M

Macronutrients 82
Magnesium (Mg) 82, 358
Mail-Order 454
Maintenance 109, 151, 162, 219, 282, 335, 341, 343, 388, 390, 392, 417, 507
Making Money 85, 119, 403, 431
Manganese (Mn) 82, 117
Marine 170–171, 184, 240, 482, 488, 498
Market 85, 90, 97–101, 103–104, 119, 406, 409, 417, 419, 425, 430, 437, 439, 467, 472–473, 480, 484, 486, 488–490, 501, 513, 522–524, 527
Marketing 97–101, 103–104, 375, 409, 437, 439, 443, 446, 452, 467, 473, 475, 496, 507, 526
Media 45, 55, 134, 173–175, 186, 189–190, 227, 259–260, 263, 274, 278–279, 283, 287, 296, 301, 303, 317, 406, 433, 472
Mercury 43, 46, 50
Metal Halide 212
Methane 27
Mexico 18, 28, 45, 440, 491–492
Mg 82
Micronutrients 82
Mineralization 263, 274

Mint 78
Mission 430, 480–481
Mn 82
Molybdenum (Mo) 82
Municipal Water 370
Murray Cod 101
Mustard 78, 85, 87

N

NaturalNews 43
Net Containers 526
Net House 214
New Tomatoes 78
NFT 59, 69, 259, 276
Nitrates 108, 185, 282
Nitrogen 67–68, 73, 82, 127, 134, 185, 356, 362
Nitrogen Cycle 67, 185, 356
North America 21, 30, 36, 212, 216
Not-For-Profit 412
Nutrient Film Technique 7, 55, 59, 227–228, 240, 263
Nutrients 82
Nutrition 113

O

Okra 78, 84, 87
Operations 234, 417, 443, 496
Organic Pest Control 92
OSHA 347, 479, 496, 511
Overfishing 28
Overflow 268–269, 276
Oxygen 70, 82, 104, 133, 135, 279, 359

P

Packaging 452, 507, 526
Packing 524
Pak Choy 78
Palmetto Bass 100
Parasitic 109
Parsley 78, 85
Partnership 411
Payment 456, 460
Payroll 496, 527
Peas 78, 85, 87
Peat Moss 176
Peppers 38, 78, 87, 421, 521
Perch 96, 100–101, 116
Pest Control 92
Pesticide 35, 93
Petroleum 23–24
pH 63, 69–71, 80, 82–83, 96–97, 101–102, 104–105, 108, 110, 120, 133–134, 161, 173–174, 176–177, 227, 275, 278–282, 338–342, 345, 347–349, 353, 356–359, 363–365, 368–373, 379–380

Phosphorus 82
Photosynthesis 80, 82, 209
pH Range 356
Pipes 141
Plan 270, 417, 429, 436, 475, 477, 512
Planning 277, 417–418, 421
Plant Containers 250
Plants 7, 15, 17, 34, 55, 63, 72, 75, 77–78, 80–83, 85, 90, 92, 134, 173, 190, 194, 209, 212–213, 227, 274, 279, 282, 296, 301, 339, 347, 384, 406, 409, 496
Plug 376, 525
Plumbing 131, 139, 168, 242, 277, 279, 293, 303, 406
Pollution 27, 31
Polyethylene 134, 156, 165, 198–199, 317, 320, 389–390
Ponds 100, 161
Population Growth 19
Potassium 82, 358, 362
Potassium Carbonate 358
Potassium Hydroxide 358
Potassium (K) 82, 358, 362
Potatoes 38, 85, 87
Power 32, 64, 350, 392
Predator 381
Profit 85, 119, 403, 406, 412, 419, 433, 465, 506–507
Profit and Loss Statement 433
Promotion 467, 470, 480, 496, 502
Protein 97–101, 103–104, 113, 117, 124, 126
Psychological 470
Pumps 135, 179, 181, 268, 277, 279, 291, 392
Purina 377
PVC 107, 131, 134, 137, 141–149, 151, 153, 165–167, 186, 190–191, 202, 229–231, 240, 242, 246, 276, 287, 294–295, 301, 303, 317, 320–322, 324, 327–328, 330–332

Q

Quality 25, 70, 133, 279–280, 341, 347, 353–354, 372, 396, 452, 492

R

Radishes 87
Raft System 59, 162, 245–246, 248, 263
Rainforest 27
Rainwater 369, 372
Raley's 493–494, 501
Rapini 78
Recao 78
Redina Lettuce 78
Return on Investment (ROI) 10
Revenue 85, 410, 451, 507

Rice 177
Risk 425, 429
Roadside 448, 472
Rockwool 83, 174, 177
ROI 97, 436
Rooftop 66
Rubbermaid 190, 274

S

Salad 34, 85
Sales 15, 421, 446, 448, 472–474, 480, 496, 500, 507, 525
Salinity 368–369
Salmonella 444
Salt 110
S-Corporation 411
Shopping 302–303, 460
Silver Perch 96, 101
Siphon 285–287, 291–293, 297, 300, 303, 350–351
S.M.A.R.T. Goals 61
Soil 223
Sole Proprietorship 411
South America 18, 86, 440
Spinach 38, 78, 85, 87, 421, 522
Squash 36, 78, 85, 87
Starting 62, 281–282, 333, 337, 351, 415, 481, 497
Startup 340, 503
Start-Up 523
Steps 93, 110, 115, 180, 188, 219, 277
Stocking Density 105, 280
Storage 65, 209, 433
Store 127, 394, 396, 451–452
Strategy 457, 480, 495, 500–501
Striped Bass 100, 118
Subchapter-S Corporation 411
Subscription Farming 454
Sulfur 82
Sump Tank 268, 287, 302
Sunfish 96
Sunlight 209–210, 212
Supermarket 494
Supplements 80
Surplus 122
Survey 44, 46, 487
Sustainability 9, 443, 491
Swiss Chard 78, 87
System Design 131, 183, 240, 246, 257, 266

T

T5 210
Target Market 480, 486
Taro 78
Tatsoi 78
Tax 219, 434, 527
Temperature 70–71, 82, 97–101, 103–104,

108, 133, 213, 279, 341, 354–355, 357, 360, 372–373, 385–386
Temperature Requirements 97–101, 103–104
Testing 73, 133, 142, 371–372, 406
Texas 25, 364
Tilapia 96, 101–105, 107, 116, 119–120, 124, 280, 300, 341–342, 344, 360, 362, 373–381, 407, 414–415, 421, 467–469, 480, 482, 488–489, 493–495, 499, 501, 512–513, 521, 523
Timer 283
Tomato 85
Tomatoes 78, 86, 421, 489, 493–494, 501, 511
Troubleshooting 345, 347–348, 350
Trout 77, 96, 104, 116–117, 469
Tuna 29, 377

U

Uganda 20
Uniseal 148–151, 293, 317, 320
United Nations 24, 121
United States 2, 15–16, 18, 20, 24–25, 29, 36, 40, 45–46, 48, 50, 86, 96–98, 108, 116, 118–120, 145, 206, 217, 219, 221–223, 366, 369, 384–386, 397, 413, 426, 438, 446, 450, 467–468, 475, 482, 488–489, 491, 494, 501, 518, 521
Urine 236
USDA 18, 25, 40–41, 43, 85, 114, 121, 206, 219, 221, 387, 397, 412–413, 444, 446, 452, 475, 518, 521, 527

V

Value 85, 90, 119, 201, 406, 419, 470, 473
Variety 66
Vehicle 425
Vinegar 358
Vinyl 134, 165, 292
Virgin Islands 248, 414
Viruses 444

W

Waste 10, 270
Watercress 78, 87
Water Heater 134
Watermelon 78, 86
Water Use 24, 26
Website 437, 472, 484, 503
Wholesale 473, 524
Wikipedia 482, 513
Wiper 100

Y

Yellow Perch 96
Yield 407, 418

Z

Zinc 82, 117
Zn 82
Zucchini 85, 87